# WHERE IS IT?

## A Guide to the Places Mentioned in this Book

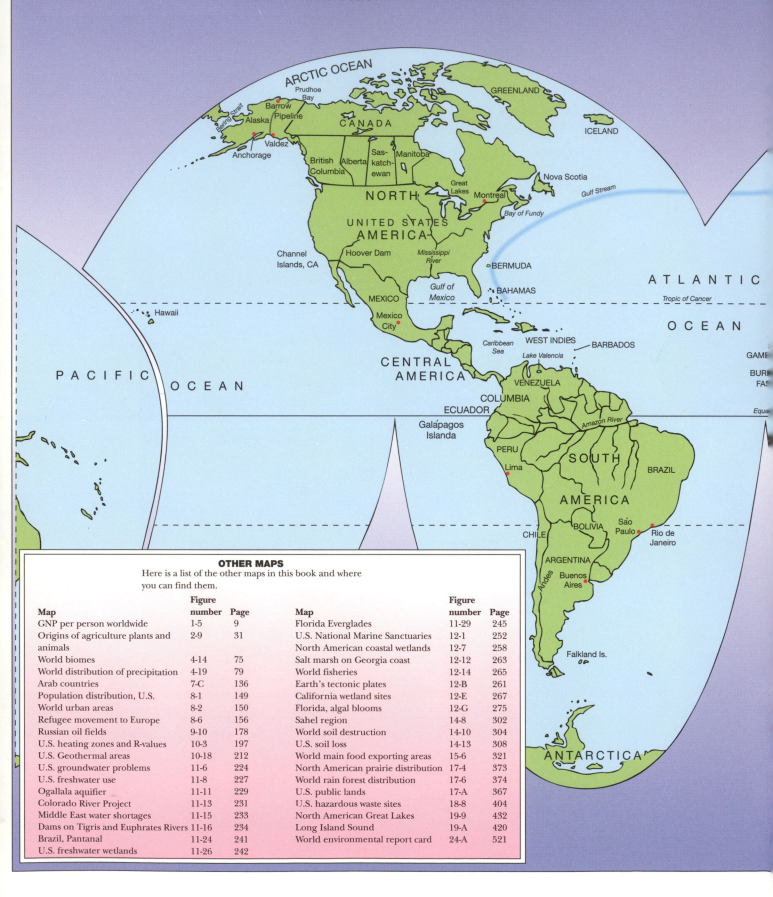

ARCTIC OCEAN

Prudhoe Bay

GREENLAND

Barrow

Pipeline

Alaska

ICELAND

Valdez

CANADA

Anchorage

British Columbia

Alberta

Sas-katch-ewan

Manitoba

Great Lakes

Nova Scotia

Gulf Stream

NORTH

Montreal

Bay of Fundy

UNITED STATES AMERICA

Channel Islands, CA

Hoover Dam

Mississippi River

BERMUDA

ATLANTIC

MEXICO

Gulf of Mexico

BAHAMAS

Hawaii

Tropic of Cancer

OCEAN

Mexico City

PACIFIC

OCEAN

Caribbean Sea

WEST INDIES

BARBADOS

GAME

Lake Valencia

CENTRAL AMERICA

VENEZUELA

BUR
FAS

COLUMBIA

ECUADOR

Equa

Galápagos Islanda

Amazon River

PERU

SOUTH

Lima

BRAZIL

AMERICA

BOLIVIA

Sáo Paulo

CHILE

Rio de Janeiro

ARGENTINA

Andes

Buenos Aires

Falkland Is.

ANTARCTICA

Enlarged maps of parts of North America, Central America, and
Europe can be found on the endpapers at the back of this book.

# ENVIRONMENTAL SCIENCE

*San Joaquin kit foxes. These cat-sized canines are endangered because more than 90 percent of their habitat in California's Central Valley has been converted into farmland and suburbs. (© B. "Moose" Peterson/WRP)*

# ENVIRONMENTAL SCIENCE

## SECOND EDITION

## Karen Arms

SAUNDERS COLLEGE PUBLISHING

Harcourt Brace College Publishers

Fort Worth   Philadelphia   San Diego   New York   Orlando   Austin
San Antonio   Toronto   Montreal   London   Sydney   Tokyo

DEDICATION
To
Richard, Sarah, and Patrick
because they will be part of the solution.

Text Typeface: New Baskerville
Compositor: CRWaldman Graphic Communications
Acquisitions Editor: Julie Levin Alexander
Developmental Editor: Lee M. Marcott
Managing Editor: Carol Field
Project Editor: Laura Maier/Laura Shur
Copy Editor: Ellen Thomas
Manager of Art and Design: Carol Bleistine
Art Director: Robin Milicevic
Cover Designer: Lawrence R. Didona
Text Artwork: Rolin Graphics, Inc.
Layout Artist: Laura Ierardi
Director of EDP: Tim Frelick
Production Manager: Joanne Cassetti
Marketing Manager: Sue Westmoreland

Cover Credit: Sunrise and deer tracks in winter at Crater Lake, Oregon.
          (© Mathias Van Hesemans)

Printed in the United States of America

Environmental Science, 2/e

ISBN: Paper Cover 0-03-098569-2

Library of Congress Catalog Card Number: 93-085659

3456   032     987654321

# PREFACE

Environmental Science is more than just another course, because it deals with problems that confront us every day. It is an introduction to how the world we live in works, how we use and abuse nature, and what we can do to protect our environment for ourselves and for future generations. This edition emphasizes our personal involvement with new boxed features called **Lifestyles**, which illustrate that our everyday actions affect the very future of the world.

We live in challenging times. The human population of Earth has tripled in less than a century, something that has never happened before and will never happen again. The explosion of our population has affected the planet on which we live in ways that we must learn to understand if the world is to support us all in the twenty-first century.

The aim of environmental education is not to fill students with facts and figures, but to show them that, as ordinary citizens, we can analyze environmental problems and weigh the solutions. This book starts by introducing our main environmental prob-

lems. However, problems and solutions change with time as research uncovers new areas of concern and introduces new technology and knowledge. It is more important to know where to look up facts and how to analyze information than it is to learn reams of facts by heart. For this reason, the book emphasizes examples of problems and their solutions because the methods used in one case often suggest ways to solve problems in our own backyards.

Many of the examples we study are depressing. But we are not inevitably doomed by environmental problems. For every area suffering from massive deforestation, we can find another that has replanted its forests; for every country suffering the hardships of rapid population growth, we can find another that has restrained growth and revived its economy. The message of this book is that people can learn and change extraordinarily rapidly.

A solid scientific basis is vital if the content of an environmental science course is to remain useful over the years. The student must be introduced to some of the scientific methodology used by researchers if this is to be more than a "current affairs" course. But science is not enough. When we consider famine relief or a battle over conservation, we have to consider economics, law, ethics, and politics if we are to understand the problem and its possible solutions. All these different approaches are part of environmental science, making it the fascinating and complex subject it is.

## What is New in this Edition?

Preparing the second edition of a book is an exciting task—a chance to bring in new ideas, introduce discoveries of the last few years, and make material easier to understand by reorganization, rewriting, redrawing, and finding new photographs. The first edition of *Environmental Science* was successful because it was scientificaly accurate while avoiding the temptation to deluge the reader with statistics, chemistry, jargon, and technical details. I have tried to maintain this emphasis in the new edition, while adding new examples and altering the emphasis in several places.

Changes in the world since the first edition have led to increased emphasis in this book on ozone depletion, population problems, environmental refugees, and water shortages. We now know that ozone depletion is proceeding much more rapidly than was first realized. New studies show that many recent outbreaks of civil war and huge movements of refugees are indeed the result of environmental degradation, as scientists have suggested since the 1960s. Starvation in Africa and acute water shortages in many countries have brought the problems of explosive population growth home to many people for the first time.

Many of the changes that were made to the second edition were made in response to the comments and suggestions of users of the first edition and reviewers of the manuscript for the second. These are the major changes and improvements in the second edition.

- New **Chapter 8** on Population, Environment, and Disease combines the topics of population and urban land use and includes new sections on pollution and the proliferation of infectious diseases (Section 8-F) and diseases caused by habitat destruction (Section 8-G).

- New **Chapter 21** on Pests and Pesticides increases the coverage of this important topic by detailing the hazards of pesticide use and recommending ways to limit their agricultural and household uses.

- Boxed **Close-Ups**, a popular feature of the first edition, have been retained with nearly 75 percent of them devoted to new topics. These include Scientists and the scientific method (Chapter 1); the Biodiversity Treaty (Chapter 4); Population growth in Arab countries (Chapter 7); Yucca Mountain, nuclear waste facility (Chapter 9); algal blooms (Chapter 12); the Florida Keys and flood insurance (Chapter 16); Costa Rica and the Merck agreement (Chapter 17); methyl bromide and the ozone layer (Chapter 21).

- New **Lifestyle** boxes focus on ways that students can have a positive environmental impact through their everyday activities. Topics include the importance of voting and writing to public officials (both in Chapter 1); energy conservation tips (Chapters 3 and 10); the environmental ramifications of stone-washed jeans (Chapter 4); methods to cut down on water use in the home (Chapter 11); and disposing of used oil from automobiles (Chapter 18), among others.

- New **Equal Time** boxes present several conflicting points of view or a view opposed to that of the chapter. Controversial topics covered include U.S. energy policy (Chapter 10); U.S. and Canadian management of the Alaskan fisheries (Chapter 12); subsidized mining on publicly owned land (Chapter 13); species extinction (Chapter 16); and the activities of the National Agricultural Chemicals Association (Chapter 21).

- New **Facts and Figures** boxes are collections of statistics that should be useful for generating further class discussion or as reference data for the text. About half the chapters include this feature, with additional data for topics ranging from the number of known species to the rate of salinization of agricultural land around the world.

- All the **illustrations** have been redone with particular attention given to finding photographs that are directly related to the text.

## Organization

The main organizational changes in this edition are the integration of the chapter on urban land use into a new chapter on population distribution (Chapter 8), the inclusion of a new chapter on pests and pesticides (Chapter 21), and the removal of the chapter on solving environmental problems to the final part of the book on society and environment (Chapter 22).

Part 1, Overview, provides an overview of environmental science, including the history of human life on Earth and how environmental problems have multiplied as human societies have evolved.

Part 2, Some Basic Ecology, contains the scientific principles that affect environmental problems and their solution. Because we are animals, completely dependent on other organisms for our survival, environmental affairs are directly affected by the principles of ecology. Chapters 4 and 5 are devoted to this subject.

Many will consider Part 3, Populations, to be the most important in the book. Population growth underlies most of today's environmental problems. Here we consider natural populations, their growth, and factors that limit growth (Chapter 6). Then we turn to human populations and the extraordinary gulf that separates countries with fast-growing and slower growing populations (Chapter 7). The final chapter considers some of the consequences of population growth and distribution, including urbanization and the evolution of new diseases (Chapter 8).

Part 4, Natural Resources, turns to the resources with which the earth supplies us, the ways human activities are depleting these resources, and some of the consequences of depletion.

Part 5, Pollution, focuses on the various forms of pollution: what it is, how it threatens life and comfort, and what can be done about it. Included are chapters on toxic and solid waste (Chapter 18) and on pests and pesticides (Chapter 21).

Finally, Part 6, Society and Environment, introduces the social context: the ethical, economic, and political systems that cause environmental problems in the first place and without which we cannot solve them.

### Chapter Features

The book is organized to be flexible. Each of us has a somewhat different syllabus for this course, and the chapters will be taught in various orders. Each chapter can be read independently, and important terms are defined in more than one chapter. Cross references to other chapters point to background material for the subject under discussion. In addition, the book contains a number of features designed to make it easier to study.

**Key Concepts**  The details of environmental situations change from day to day, but the central problems and ideas remain. These, rather than the details, are what stay with you from a study of environmental science. The main concepts developed in each chapter are listed at the beginning of the chapter.

**Boldface Terms**  Each chapter contains key terms in **boldface type**. Technical terms are in boldface where they are first defined. Boldface is also used for key words that should be noted as important vocabulary in the subject under discussion.

**Take-Home Messages**  Psychologists say that after listening to a lecture or reading a chapter, each of us takes away just a few new ideas. These central ideas are highlighted in blue as "take-home messages" within the text. As you review a chapter, a glance at these messages will remind you of the chapter's main points.

**Chapter Summaries**  Each section is summarized at the end of the chapter in which it appears.

**Questions for Discussion**  At the end of each chapter are questions arising from the topics discussed in the chapter. Many of them can be used as essay questions. A number of them ask you to expand on views discussed in the chapter. For instance, there are economists who argue that population growth is a good thing and does not create economic or environmental problems. What do you think of this? Do you agree with those who argue that the free market, rather than governmental regulation, should control activities such as mining, lumbering, and urban development?

### Special Features

Careful attention was given to creating an illustration program that would reflect the beauty of the natural world and help readers to understand the threats to that world and what they can do to help. All new illustrations and several new features have been added to achieve this end.

**Illustrations**  Figures on soil erosion may impress you, but they do not stun you as an aerial view of the avalanche of topsoil washing into the Sea of Bengal stuns you. Full-color photographs and illustrations are used throughout this book to illustrate facts and figures and to explain natural phenomena and technology.

**Maps**  A fuelwood shortage in Sudan, fish for aquaculture from Lake Baikal—where are these places? Inside the covers are maps showing the locations of the countries, towns, regions, lakes, rivers, and oceans mentioned in the text. Other maps throughout the book permit you to find places mentioned in particular chapters.

**Close-Up**    Close-Ups are boxed essays separated from the main body of the text. These cover material that is not central to the theme of the chapter but provides more detail or explanation. Some Close-Ups explain particular technologies. Others describe specific examples, such as California's efforts to save wetlands or a company's introduction to recycling.

**Lifestyle**    New **Lifestyle** boxes emphasize the environmental importance of everyday actions. How much water do we use when we take a shower? Is wearing cotton good for the environment? How we can insulate a house, influence legislation, and save water?

**Equal Time**    New boxes on controversial issues, called **Equal Time**, either detail several points of view on a particular controversy or present a view opposed to the one presented in a chapter—on pesticide use, energy policies, bottled water, and other issues.

**Facts and Figures**    Some chapters contain new boxes full of facts and figures: on soil erosion, pollution in Russia, or the Environmental Protection Agency. These can be used for reference or to generate discussion on particular topics.

## Useful References

Materials provided at the end of the book serve as useful reference for students who may want to read further on a particular subject or simply clarify the meaning of a term.

**Appendix on Units and Measurements**    "1989 dollars, gigajoule equivalents of coal"—what are they? If your interest in environmental science leads you to delve into the massive literature on the subject, you will come across many measurements expressed in unfamiliar units. This appendix will help guide you through the maze.

**References and Further Reading**    Many of the reference sources used in this book contain information that applies to several chapters. These are listed as general sources of information; references that are particularly appropriate to each chapter are listed separately.

**Environmental Organizations and Periodicals**    Those students who would like to become more involved in the effort to improve the state of our environment will find this listing of organizations and periodicals very useful to their efforts. Many of the organizations listed welcome the participation of volunteers in a wide variety of their activities.

**Glossary**    A glossary at the end of the book gives you one list where you can look up many terms. Instead of the vague descriptions found in most glossaries, it gives technically accurate definitions.

## Supplements

The package for *Environmental Science*, Second Edition has been developed to provide both the teacher and the student with a wealth of helpful teaching and learning tools that reflect the key themes of the text and enhance learning beyond this textbook.

1.  **Environmental News: 1993**
    **The Year in Review Videodisc** is designed to promote critical thinking and motivate real-life application of the text material. The disc contains CBS footage of important recent and current environmental events, such as the Environmental Summit in Rio in 1992, and other key events from summer, 1992 through fall, 1993. Photographs, drawings, and videoclips are included with LectureActive™ software for interacting with selected text drawings. The videodisc will be updated quarterly via videocassetes, which will be compiled once a year onto **The Year in Review Videodisc.**

2.  **Infinite Voyage Videos**
    Includes four videotapes from the "Infinite Voyage Series" covering a variety of environmental science topics. This coverage is also available in videodisc format.

3.  **"Meet The People Who Are Conserving America" Videos**
    Includes four videotapes on individuals in environmental science who have made a positive impact on the environment.

4.  **SIMLIFE™ The Genetic Playground**
    **SIMLIFE**™ is creative learning software for Microsoft Windows.™ The program allows you to build your own ecosystem from the ground up. Design plants and animals and test their adaptive abilities by turning their environment into a paradise or a wasteland. An on-screen tutorial assists this simulation program.

5. **SIMEARTH™ The Living Planet**
   **SIMEARTH™** is creative learning software for Microsoft Windows.™ Select from seven different planet scenarios to create and evolve a unique world. SimEarth demonstrates the interrelated nature of the world in a fresh, new manner that intrigues students and engages them deeply in learning. Special features include capabilities to terraform Venus and Mars, and to actuate volcanoes, earthquakes and meteoric impacts.

6. **Regional Environmental Issues Manuals** include articles about environmental issues eight different regions: Northeast, Mid-Atlantic, Southeast, Great Lakes, Midwest, Southwest, Northwest, and Canada. The regional manuals are also available shrinkwrapped with the text. The goal of the manuals is to encourage individual student analysis of local issues in conjunction with concepts they have learned in the text. Thought-provoking questions, commentary, and readings are all included to stimulate further investigate of the issues presented.

7. **Instructor's Resource Manual with Test Bank** was prepared by Virginia Fry of Monterey Peninsula College. The **test bank** includes over 2000 test questions in several formats and levels of difficulty. Question types include fill-in, multiple-choice, essay, and labeling diagrams. For each chapter, the **instructor's manual** includes a chapter outline with key concepts noted in the margin and room for instructors to make his or her own notes; suggestions for projects or reports; and lists of audiovisual resources.

8. **ExaMaster™ Computerized Test Bank** enables instructors to create or modify tests derived from the printed **Test Bank** and has the capability to print out tests with answer keys and answer sheets. Available in 5¼″ IBM, 3½″ IBM, and Macintosh.

9. **Overhead transparencies** include 150 transparencies of full-color figures from *Environmental Science*, Second Edition.

10. The **Laboratory Manual** by Robert Wolff of Trinity Christian College assists in the teaching of laboratory, problem-solving, and thinking skills in environmental science.

\* \* \* \* \* \* \* \* \* \* \* \* \* \* \* \* \* \* \* \* \* \* \* \* \* \* \* \*

The more I learn about environmental science, the more it fascinates me. I hope that this book will launch many newcomers on a voyage of discovery that will prove as rewarding.

KAREN ARMS
*Savannah, September 1993*

# ACKNOWLEDGMENTS

The number of people who have helped to make and improve this book grows with every year that passes. The generous contributions of my teachers, students, friends, and relatives have made my job much easier and cheered me on my way.

I taught at Stanford University when Paul Ehrlich was warning of the consequences of the population explosion. He galvanized all who worked with him into questioning the morality of our own reproduction and our use of the earth's resources. His predictions of environmental disaster from overpopulation have all too sadly been borne out by events. Don Kennedy at Stanford, Dick O'Brien and Will Provine, both at Cornell, taught me the importance of involving students and teachers from traditionally separate disciplines when considering society's problems. They were the guilding lights behind courses that drew biologists, economists, geologists, and sociologists together in a study of how human societies interact, and should interact, with the environment they live in.

I would like to thank instructors who responded to our questionnaire, which was sent out in anticipation of preparing this new edition. Their suggestions for improvement laid the groundwork for many of the changes we have made:

David Armstrong
University of Colorado, Boulder

David J. Cotter
Georgia College

Roger M. Davis
University of Maryland Baltimore County

Lowell L. Getz
University of Illinois at Urbana-Champaign

Kathy Gregg
West Virginia Wesleyan College

Neil A. Harriman
University of Wisconsin-Oshkosh

David A. Lovejoy
Westfield State College

John Peck
St. Cloud State University

Lloyd R. Stark
The Pennsylvania State University

Norman R. Stewart
University of Wisconsin-Milwaukee

John D. Vitek
Oklahoma State University

Ray Williams
Rio Hondo College

James Winsor
The Pennsylvania State University, Altoona Campus

This book's reviewers have taken time from their busy schedules to read the manuscript and to suggest numerous improvements. They have provided me with many fascinating examples from their own experience, corrected mistakes, and helped in dozens of ways. For your many thoughtful contributions, my thanks to:

William G. Ambrose
East Carolina University

Martin B. Berg
University of Notre Dame

Ann Causey
Auburn University

Lu Anne Clark
Lansing Community College

Charles F. Cole
The Ohio State University

Harold N. Cones
Christopher Newport University

John Cunningham
Keene State University

Miriam del Campo
Miami-Dade Community College

Stephen Dina
Saint Louis University

John P. Harley
Eastern Kentucky University

Denny O. Harris
University of Kentucky

David L. Hicks
Whitworth College

David I. Johnson
Michigan State University

James Lein
The Ohio State University

Mark R. Luttenton
Grand Valley State University

Timothy Lyon
Ball State University

Scott McRobert
St. Joseph's University

Elizabeth Nagys
Southwestern Michigan College

Patricia Perfetti
University of Tennessee, Chattanooga

Thomas Reitz
Bradford College

C. Lee Rockett
Bowling Green State University

Barbra Roller
Miami-Dade Community College

James L. Seago, Jr.
State University of New York, Oswego

Randall Stovall
Valencia Community College

Jeffrey R. White
Indiana University

Dennis Woodland
Andrews University

In addition to these reviewers are people who have helped in a variety of ways. They transported, fed, and housed me; argued with me about ecology and grammar; contributed illustrations, information, photographs, and moral support. Without all these people, this book would be but a shadow of itself: Beth and Bruce Bowler; Holden and Ann Bowler; Matthew Gilligan; Paul, Richard, and Sarah Feeny; Virginia Fry; Martin and Eleanora Fry; Tom Kozel; Peter Marks; Joe Richardson; Karen Roeder; Mary Stoller; Edith Schmidt; Thom Smith; Tricia Smith; Steve Webster; and Connie Wingard.

Don Jackson and Ed Murphy at Saunders College Publishing provided the inspiration for this book in the first place; Elizabeth Widdicombe, Julie Alexander, and David Watt organized and launched this new edition. The production, art, and design team of Laura Maier, Laura Shur, Robin Milicevic, Anne Muldrow, Joanne Casetti, and Caroline McGowan brought their bookmaking expertise to bear on the project, with the happy results you see here. All drawings were rendered by the staff at Rolin Graphics. But most of the credit for this goes to Lee Marcott, surely the best editor an author ever had.

# CONTENTS OVERVIEW

# CONTENTS

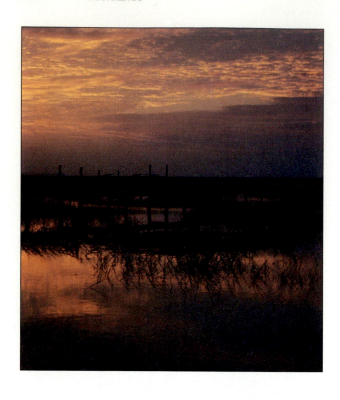

# OVERVIEW

*Bangkok, Thailand.* (Paul Feeny)

# Chapter 1

# SCIENCE AND ENVIRONMENT

**Key Concepts**

- The goal of environmental science is learning how to produce a sustainable world.

- Our basic environmental problem is ever-increasing numbers of people living in a closed system.

- The growing human population is depleting our resources and polluting the environment.

- As people become wealthier, they consume more resources and produce more waste.

- Technology will never be able to solve our most pressing environmental problems.

- Solving our environmental problems requires research, education, and international cooperation.

- Population control and habitat conservation are the most immediate environmental problems.

*"In the middle of the twentieth century, we saw our planet from space for the first time. Historians may record that this vision changed the way we think. From space, we see a small, fragile ball dominated not by human activity and edifice but by a pattern of clouds, greenery, and soils. Humanity's inability to control its numbers and to fit its activities into that pattern is changing our planet and the way we live. Human history has become a race between education and catastrophe."* —(United Nations Report on Environment and Development)

In the 1980s, the "iron curtain" between the Soviet bloc countries and the rest of the world came tumbling down. We had our first close look at industrial countries where the environment has been sacrificed to agriculture and industry. More than 80% of the "fresh" water in the former Soviet Union is dangerously polluted. Smokestacks in manufacturing towns pollute the air and cause so much disease that malformed babies are common, life expectancy is less than 60 years, and more than half the men called up for military service are rejected as total invalids.

In the United States our lives are affected by environmental science every day. We take it for granted. We assume that our food and water are safe to consume, that nuclear power plants are not endangering our health, and that the local landfill is not poisoning the neighborhood. When we petition the local council to stop a factory from spewing foul-smelling clouds over the town, we assume that the technology to prevent the smell is available. We vote for the tax increases needed to upgrade our sewage treatment plants. In all these ways, we show that we believe society has the knowledge and technology to protect our surroundings from environmental damage that would lower our standard of living or endanger our health.

This knowledge and technology are the products of **environmental science**, the study of how humans interact with their environments and of what can be done to improve these interactions. **Environment** is a broad term. It includes all that surrounds us, things produced both by people and by the natural world, which provides our most important resources (Figure 1-1).

The goal of environmental science is to describe a **sustainable world**, a world in which human populations can continue to exist indefinitely with a high standard of living and health. The "pure" side of environmental science is research—discovering the threats to a sustainable world and what can be done about these threats. The problems studied may be scientific: "How does polluted water affect plants in the Everglades National Park?" Or they may be social and economic: "What factors determine how much water people use?" The "applied" side of environmental science develops solutions to the problems that have been identified.

Environmental science cannot produce a sustainable world. It can only tell us how we may reach that goal. The rest of the solution requires political, social, and economic action. It is up to us to tell our governments whether to enact recycling legislation, whether to enforce laws on disposal of toxic waste, and how much money to spend on research and de-

**Figure 1-1**   Some of the resources that sustain life: clean air, trees, and fresh water. These are among the resources we must ensure remain available if we are to build a sustainable world. This view shows the foothills of the Cascade Mountains in Oregon, with a hillside in the background that has been logged and replanted with tree seedlings. *(Richard Feeny)*

velopment for military and nonmilitary purposes. The solutions to many of our environmental problems are known, but they are often not acted on because people and politicians do not understand them or have other priorities. By learning something of environmental problems, we can contribute to creating a sustainable world.

## 1-A  POPULATION GROWTH: THE ROOT OF THE PROBLEM

Our basic environmental problem is a rapidly growing human population living in a closed system, Earth. Earth has been likened to a spaceship, which cannot take on new supplies as it travels. Hardly anything reaches Earth from outside Earth's atmosphere except energy from the sun. (The only other things are radiation, meteorites, and debris from space.) A closed system of this sort has built-in perils. Overcrowding is one of them. If people are born faster than they die, we run out of space. In a closed system too, the dangers are ever-present that we shall use up all our supplies and that we shall pollute our spaceship with waste.

It is not difficult to see why all environmental problems depend ultimately on population size. A household of six people can dump its wastes into the local river without polluting the river, but a town of 100,000 people cannot. The trees in a woodlot grow fast enough to replace all the wood the household cuts each year but not fast enough to supply the town. If only ten people were using it, Earth's supply of oil would last essentially forever, and similar statements can be made for everything that is produced by the natural world and that people use.

The reason most environmental problems are so pressing today is that the human population has grown so rapidly during the twentieth century (Figure 1-2). The human population numbered about 500 million in 1700 and about 2 billion in 1930. Then the population really started to grow. It doubled to 4 billion by 1975. There are now about 6 billion people on Earth, and this number is projected to double to about 12 billion within the next 50 years. Why does this speeding up of population growth in the twentieth century cause such concern?

A nineteenth-century clergyman and economist, Robert Malthus, described the theory behind the problem. In 1803, he wrote that population tends to grow geometrically, by doubling in shorter and shorter periods, whereas food and other resources tend to grow arithmetically (at the same rate over time). The result, he argued, is that every population tends to grow faster than its food supply until reduced by starvation, disease, and war. In practice, food production increased faster than human population for part of the twentieth century. But in principle, Malthus was right. More people are short of

the necessities of life now than at any other time in history. Nobody knows how many people starve to death each year. Estimates range from 10 to 30 million. Even if we accept the lower estimate, the death rate from hunger and hunger-related diseases is equivalent to more than 400 jumbo jet crashes a day with no survivors.

Since the 1960s, scientists have argued that population growth is the most pressing human problem, but the general public and the media have been slow to understand. Now, however, the disaster of unchecked population growth is sinking in. Today, media stories on starvation in Africa or air pollution in the United States usually mention overpopulation as one of the causes. This understanding is at least a step in the right direction.

> The human population is about four times as large as it was 100 years ago and this is the cause of nearly all environmental problems.

## 1-B  OUR MAIN ENVIRONMENTAL PROBLEMS

Producing enough food is the obvious problem for a large population. Perhaps because it is so obvious, the world's governments have thrown vast resources into food production, with considerable success. In the process, however, they have ignored, or even created, environmental problems that now haunt us. Pesticides used to increase food production have poisoned people, other animals, and the soil. Rivers used to irrigate farmland are often polluted by fertilizers and pesticides until the water is no longer fit to drink.

Environmental science studies two main types of interactions between humans and their environment. The first is our use of resources, such as water and plants, that are produced by the world around us. The second is the way in which our actions may pollute the environment.

### Resource Depletion

A **natural resource** is anything produced naturally that is needed by a group of organisms, such as fresh air, fresh water, food, and shelter. The main natural resources are shown in Figure 1-3. A resource is said to be **depleted** when a major fraction of it has been used up.

We shall see that humans depend on other organisms for food, to clean up water, to produce oxygen in the air, and for hundreds of other functions. So

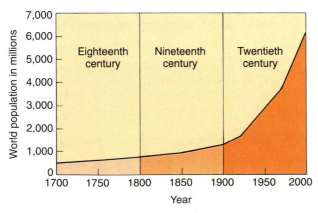

**Figure 1-2**  World population growth from 1700 to 2000. The human population has grown much more rapidly during the twentieth century than ever before.

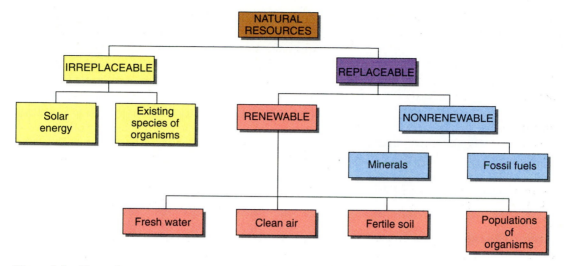

**Figure 1-3** Natural resources may be classified into those that are replaceable and irreplaceable and those that are renewable and nonrenewable.

one natural resource is Earth's **biodiversity**, the enormous variety of organisms that share the world with us. Species of organisms are irreplaceable. Once a species becomes extinct, it is gone forever. Populations of organisms, however, can be replaced. American alligators were once in danger of extinction but are now quite common. The other irreplaceable resource is the sun's energy because life could not exist without it.

Unlike species and sunlight, most natural resources are replaceable, meaning that some of them can substitute for others. These resources are usually classified as renewable and nonrenewable. A **nonrenewable resource** is one that can be used up completely or else depleted to such a degree that it is too expensive to obtain. The main nonrenewable resources are fossil fuels and minerals. Minerals include metals, sand, and clay. Fossil fuels are the partially decayed remains of long-dead organisms. They include peat, oil, and coal. Something is useful as a resource only if we can get hold of it at a price we can afford to pay. This is particularly obvious with minerals. Gold is seldom used to make saucepans because it is rare and therefore expensive. When a mineral is mined, the first miners dig deposits that are concentrated and near the surface because these are the cheapest to extract. If the mineral then becomes expensive because many people want it, it is worthwhile to mine deeper, less concentrated deposits that are more expensive to extract.

**Renewable resources** are those that can theoretically last forever. Either they are produced continuously, or they come from sources that can essentially never be exhausted, such as **solar energy**, the energy

of the sun. Fresh water, air, soil, trees, crops, and other living organisms are all resources that are, or can be, renewed. We need renewable resources more than we need nonrenewable ones. We can survive without copper or oil but not without fresh water to drink or soil to grow food. When human populations get too dense they use renewable resources faster than natural processes produce them. Even worse, we shall see that dense human populations actually reduce the ability of nature to replenish these resources.

One of the simplest ways to reduce the rate at which resources are used is **conservation**, using less of the resource in the first place or reusing it many times. In the 1970s, when the price of oil increased more than fivefold, Americans actually decreased their energy use per person, reversing a trend that had continued for at least 200 years (Figure 1-4). People do not bother to conserve resources that are cheap and easily available, but when people have the incentive to conserve resources, they do it remarkably quickly and easily. Conserving natural resources is much cheaper and easier than trying to replace them, and we shall see that interest in conservation is increasing rapidly.

We need renewable and nonrenewable resources, many of which are depleted.

## Pollution

**Pollution** is an undesirable change in the characteristics of the air, water, or land that can adversely affect the health, survival, or activities of humans or

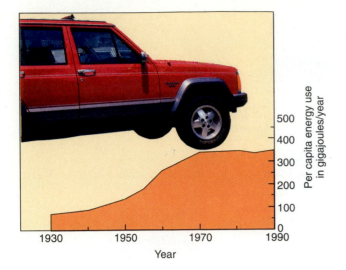

Per capita energy use
in gigajoules/year

**Figure 1-4**  Conservation. This graph shows that American energy use leveled off and started to fall in the 1970s. It had risen for 200 years before this. People started to conserve energy when the price of energy increased enormously. (A note on units: the units used for energy are not standardized. They are discussed in Appendix A. "Capita" is Latin for "head," so "per capita" means "for each person." Per capita figures are produced by dividing figures for the whole country by the number of people in the country.)

other organisms. Although some natural processes cause pollution, nearly all the pollution that troubles us today is **anthropogenic**—produced by people.

We often think of pollution as damage to the environment caused by waste. On the other hand, waste does not always pollute the environment. Sewage does not pollute when it flows into the sewage treatment plant and emerges as "sludge" fertilizer for orchards and as drinking water in the local reservoir. But if the population increases to the point where the sewage plant cannot cope with the volume, raw sewage may flow into the streets or the reservoir, and everybody would agree that we have a case of pollution on our hands. Therefore, pollution is sometimes a matter of quantity: some is all right, more is pollution.

Pollution is partly subjective. New York's Department of Environmental Conservation has poured a yellow chemical into many waterways. The yellow substance is a lampricide, a poison designed to kill lampreys, which are primitive fish that feed on trout and other bony fish. The lampricide is too diluted to harm people who drink the water, but it kills small animals and is thought to be bad for livestock who drink a lot of it. To those who fish and don't like their trout damaged by lamprey scars, this is not

pollution—it is a great way to improve the trout fishing. But to the woman who breeds horses by a New York lake and to the boy with an earthworm farm, it is pollution that may wreck their businesses. Because pollution is a matter of quantity, regulators are always arguing about what level of a particular substance in the environment is safe. Because pollution is at least partly subjective, people often cannot even agree on what is pollution and what is not.

There are two main types of **pollutants**, things that pollute. Pollutants may be degradable or nondegradable, although the two categories are not completely separate. **Degradable pollutants** are those that are broken down completely or reduced to acceptable levels by natural processes. Examples are **biodegradable** things, substances that are broken down rapidly by living organisms. They include biological waste such as human sewage or dead leaves in a lake. Other degradable pollutants are degraded by nonbiological processes such as the decay of radioactive substances. Degradable pollutants become a problem only when we introduce them into the environment faster than natural processes break them down.

**Nondegradable pollutants** are those not broken down by natural processes. Examples are some plastics, mercury, and lead. The trouble with nondegradable pollutants is that the only way to keep them out of the environment is not to release them in the first place, which is expensive, or to remove them once they have been released, which may be very expensive or even impossible.

> Pollutants that make our resources less usable may be either degradable or nondegradable.

## 1-C  A SUSTAINABLE WORLD

A sustainable world is one that can go on indefinitely. This does not mean an unchanging world; it means a world containing enough of the things that people need to support human life forever. In a sustainable world, we would supply fresh water to everyone, while ensuring that we would have just as much water to use next year—by cleaning polluted water before returning it to the water supply or by using no more water than nature purifies each year. Also, we would dispose of our wastes in such a way that the air remains fit to breathe, and no one has to live next to a toxic dump or an unsightly pile of plastic garbage.

At the moment, the world we live in is far from sustainable. Our standard of living is as high as it is because we are depleting the natural resources of our planet. We are using soil, water, air, timber, and minerals faster than they are being replaced, which means that there will be less to use in the future. Unless we can find substitutes for these resources or ways to conserve them, our standard of living will fall in the future.

In 1992, the United Nations sponsored the first worldwide conference on environmental affairs in Rio de Janeiro, Brazil. The participants identified air pollution and extinction of species as two environmental problems that must be addressed by every country in the world if the world is to become sustainable. Population growth, the world's most urgent environmental problem, was not discussed. Powerful groups such as the Roman Catholic Church, which opposes birth control, insisted that it be omitted. Our environment will not become sustainable, however, until two things occur: the human population stops growing and everyone on Earth has a decent standard of living. Both are necessary because prosperous people consume more natural resources than poor ones (Table 1-1). Thus, even if the world population miraculously stopped growing tomorrow, we should continue to use more and more resources for years to come as living standards improved. As people become wealthier, they consume more resources and produce more waste. This is the reason the use of resources in the United States has grown faster than the population: living standards have risen, and each person has used more resources over the years.

We shall see that increasing consumption with increasing wealth can be stopped. People in countries such as Japan and Sweden use much less energy and other resources and produce less waste than North Americans do.

> Wealthy countries consume disproportionately more resources and produce more waste per person than poor countries.

## 1-D THINKING INTERNATIONALLY

One of the most important lessons of the twentieth century is that, as Rene Dubos put it in 1978, "We must act locally, but think globally." No country is

| TABLE 1-1 | Environmental Problems in Rich, Poor, and Developing Countries.* | | | | |
|---|---|---|---|---|---|
| | **Measurement** | **World Average** | **India** | **Brazil** | **United States** |
| **Health** | Life expectancy (number of years a newborn baby can expect to live) | 66 | 59 | 66 | 76 |
| **Population Growth** | Growth rate of population per year | 1.6% | 1.7% | 1.7% | 0.8% |
| **Measures of Wealth** | Gross national product per capita in U.S. dollars (value of all goods and services produced each year divided by the population) | $4,200 | $350 | $2,680 | $21,790 |
| | Portion of land area used as pasture (to produce meat) | 25% | 4% | 20% | 25% |
| **Living Space** | Land area per person (in square meters) | 400 | 37 | 551 | 351 |
| **Consumption** | Oil use (per person per year) | 4.5 barrels | 0.45 barrels | 2.9 barrels | 23.7 barrels |
| | Meat consumed (kilograms per person per year) | (Figure not available) | 2 kg | 47 kg | 111 kg |
| **Pollution** | Carbon dioxide emissions per person per year | 1.5 tons | 0.8 tons | 1.5 tons | 21.6 tons |
| | Garbage produced per person per year | 98 kg | 43 kg | 216 kg | 761 kg |

*The United States is one of the wealthiest countries in the world, India is among the poorest. Brazil is ranked as "upper middle income" by the World Bank. The main environmental problem in poor countries is rapid population growth, which results in overcrowding and poverty. The main problem in wealthy countries is inefficient consumption of resources, which leads to waste and pollution. Developing countries are rapidly joining the consumer nations: rates of trash production and pollution are growing much more rapidly than in the wealthy nations.

wealthy without international trade, and the environment takes no account of national boundaries. Radioactive dust from the 1986 nuclear power plant explosion in Chernobyl, Ukraine, fell on Norway and Sweden. Oil spilled by an Italian ship in the Mediterranean Sea in 1991 killed birds and fish along the shores of France, Egypt, Greece, and Israel. Colorado, Nevada, Arizona, California, and Mexico squabble over who can use how much water from the Colorado River. When the population of one country grows too fast for its economy, many of the unemployed move, legally or illegally, into neighboring countries. Whether we like it or not, the environmental practices of people in many different countries affect the chances of a sustainable world for us and our descendants.

> One country's environmental actions often affect other countries.

This is one reason that the wealthier nations have attempted to assist in the development of poorer nations since before World War I. **Development**, in this context, means the change from a society that is typically agricultural, rural, poor, and illiterate, with a fast-growing population, to one that is largely industrial, urban, wealthy, and educated, with a stationary or slow-growing population. You may not feel wealthy if you are living on $3,000 a year, or even on $30,000 a year, but compared with an Indian farmer with $130 a year, or an urban Kenyan with $40 a year, you are rich. The motives behind aid to developing countries are partly philanthropic and humanitarian and partly selfish. Selfish motives include the search for minerals and crops that only the less developed nations produce and the desire to develop new markets for goods manufactured in the wealthier countries.

Development assistance has been much less successful in equalizing living standards around the globe than people had hoped. In fact, for much of the twentieth century, the rich countries have gotten richer and the poor countries poorer (Figure 1-5). For many of the people who live in the wealthier half of the world, life is better than ever before. These people use most of the food and other resources the human race consumes in a year. Pollution control improves every year, and the wealthy countries also contain large areas where wildlife and natural landscapes are preserved. Of the more than 4 billion people in poorer countries, fewer than half have access to enough food, safe drinking water, and proper sanitation. Here, every year, diseases caused by malnu-

trition or polluted water kill millions and cripple many more.

The poverty of the developing nations is both a cause and a result of rapid population growth. People have large families because poor countries do not provide adequate birth control or old-age pensions to their citizens. Food production and education cannot keep up with population growth, and so each person gets less and less as time goes by. Developing nations have borrowed enormous sums of money from the rest of the world, theoretically to pay for economic progress, industrial development, and increased standards of living. Instead, in many cases, populations have grown so fast that most of the money has gone for necessities such as food and oil, and many of these countries are actually worse off than they were 40 years ago.

## Research and Education

In the twentieth century, we have put large amounts of money and time into research (Close-Up 1-1). We, and the planet we live on, have been the subjects of innumerable studies, experiments, and censuses. When geographers, agriculturists, biologists, and economists look at the facts and figures these studies produce, they realize that overpopulation, depletion, and pollution are reducing our standard of living and imperiling our future. They point out that these problems cannot be solved without greater investment in research and education.

Education is necessary because many problems cannot be solved by governments or other institutions. They can be solved only by individuals, and individuals must be educated to know what needs to be done. There is no point in posting warning notices near a hazardous waste dump if most of the population cannot read. Much of the pesticide pollution of drinking water in the United States is caused by suburban homeowners who spray large amounts of dangerous pesticides on their lawns and gardens without reading the warning labels and guidelines for using these chemicals. Nothing will stop people from misusing pesticides except teaching them how dangerous pesticides can be.

People often clamor for action to prevent a health hazard, preserve an endangered species, or save a beach from erosion. They do not realize that, surprisingly often, we do not know how to achieve the goal. Much of the basic research in environmental science still remains to be done. How large must a national park in China be to preserve a sustainable population of endangered pandas? To answer this,

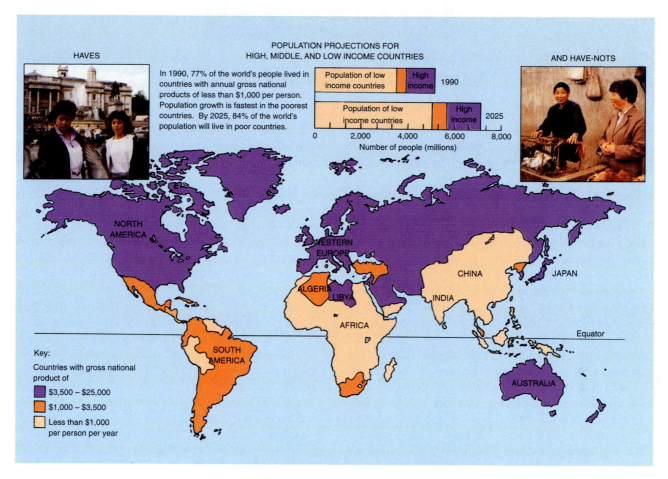

**Figure 1-5** Haves and have-nots. On this map, countries are colored according to their per capita gross national products, the total value of all goods and services produced by the country in a year, divided by the number of people in the country. Populations are growing much faster in the poor than in the wealthy countries. As a result, by 2025, more than four fifths of all the people on Earth will live in countries where the gross national product is less than $1,000 per person per year.

we need to know how much bamboo a panda eats in a year and how fast the bamboo grows; we need to know how large a family group is needed for pandas to reproduce, the birth and death rates of pandas, and hundreds of other facts. Multiply this by the hundreds of species we would like to save from extinction, and you can imagine how much work needs to be done on endangered species alone. Scientists believe that we are increasing the amount of carbon dioxide in the atmosphere and that this will cause Earth to warm up. But we do not even know what happens to much of the carbon dioxide that reaches the atmosphere, and there are scientists who argue that the world will cool down, not warm up, as a result of human activities. We do not even know the extent of soil erosion every year in the United States, the world's largest producer of food.

With our enormous ignorance of vital environmental problems, it is disheartening to realize that most countries still spend much more money on military research and development than on environmental research (Figure 1-6).

## 1-E A LITTLE HISTORY

There is nothing new about environmental problems. Nearly 3,000 years ago, the Greek poet Homer reported that cutting down trees caused flooding and soil erosion, which destroyed ancient cities. Troy was a seaport on a wooded hillside when the Trojan horse was sneaked into the city in an attempt to rescue Helen (she of the ''face that launched a thousand ships''). Today, the ruins of Troy are several

# Scientists and Scientific Method

Science differs from other ways of studying the world in that it deals only with things that can be experienced through the senses. **Scientific method** is a way of answering questions about cause and effect in the natural world. In principle, there are three main steps, although in practice, scientists work in many different ways.

The first step is to collect **observations**, such as measurements of air pollution in two cities. Second, the scientist thinks of **hypotheses**, proposed answers to questions about what has been observed. For instance, a scientist might propose that there was less air pollution in London than in Miami because London is windier and wind blows polluted air away. Another hypothesis might be that Miami is more polluted because Miami contains more cars than London. The third step is **experimentation**, performing tests designed to show that one or more of the hypotheses is more or less likely to be correct. To test our hypotheses about air pollution, we should have to measure pollution in London and Miami on windy and on calm days, find out the density of cars in the two cities, measure how much pollution comes from cars as opposed to other sources, and perform various other measurements.

Experiments must be designed so that their results are as clear-cut as possible. For this reason, they include **controls** as well as experimental situations. The two differ only by the factor(s) being investigated. To test the effect of wind on air pollution, we would need control measurements of air pollution on a day with no wind. We would measure traffic density, industrial activity, and anything else that might affect air pollution on the same day so that we could control for these factors when we measure pollution on windy days. We would then measure experimental situations on days with several different wind speeds but when traffic, industrial activity, and other factors were as similar as possible to those on the control day.

Scientists demand that an experiment can be repeated with the same results. This guards against two kinds of errors. First, we might have accidentally made a mistake in technique. Second, any experiment is subject to **sampling error**, error due to using only a small number of measurements. A hypothesis supported by repeated experiments is generally regarded as a **theory**. Most scientists, however, would argue that it is impossible to be 100% sure that a scientific discovery is "right." A "fact" is really information that we believe in strongly or that seems highly likely to be repeated without change.

Much public support goes to projects on "applied science," investigating problems such as alternative energy sources or food production. However, a great deal of basic, or "pure," research is still needed to discover the principles underlying the behavior of objects and organisms in our world.

**Science and morality**  Scientists often say that science is neither good nor bad; only its use has moral consequences. The discovery that the atom could be split was a scientific discovery with no moral implications. It was the decision to use this knowledge to build an atom bomb that produced the moral dilemma of whether it is ever right to use such a weapon.

It is seldom possible for scientists to foresee all the implications of their work. For instance, in the early 1940s, a graduate student found that a chemical called TIBA could be applied to soybeans to improve their yield. Then the Army Chemical Corps used the student's finding that high levels of TIBA made plants lose their leaves to develop chemical defoliants. TIBA itself was never used in warfare, but it paved the way for production of Agent Orange, which was used to defoliate millions of hectares of forests and farms during the Vietnam war. Who would be brave enough to engage in science with the expectation of being held legally and morally responsible for such unforeseeable consequences?

On the other side of the coin, scientists who discover dangerous situations may be forced to choose between silence and unemployment because the remedy is so expensive. Today, many scientific societies have adopted guidelines of ethical conduct for their members to follow and have pledged legal aid to members who "blow the whistle" on employers that make dangerous products or dispose of hazardous materials unsafely.

There are no simple solutions to these dilemmas. The peaceful coexistence of science with society depends on citizens who understand what science can and cannot do and who do not confuse scientific with moral, economic, or political values.

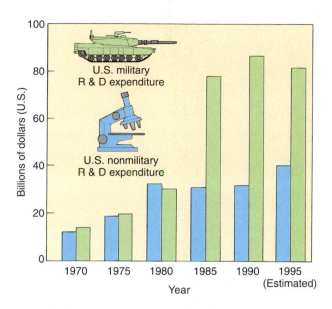

**Figure 1-6** United States spending on military and nonmilitary research and development from 1970 to 1995. Nonmilitary expenditures pay for all research and development in areas including medical care, industrial development, pollution control, forestry, agriculture, and education. At the beginning of the 1980s, military expenditure more than doubled from its previously high level, and expenditure in all other areas actually declined. *(National Science Foundation data)*

miles from the sea. The inhabitants cut down all the trees on surrounding hills for firewood and buildings. Without trees to hold the soil in place, rain washed the soil down into the river, where it "silted up the harbor so that no large ship can come in," and Troy fell into ruin.

The environmental damage of more recent times dwarfs anything that has gone before. Many believe that the cultural background of western civilization is partly responsible for our present overpopulation, resource depletion, and pollution. Seventeenth-century writers saw the human species as above and against nature, entitled to rule it as a tyrant. European settlers viewed North America as supplied with limitless natural resources. When the trees had been felled and the soil exhausted in one area, they could just move further west. The belief that nature exists only to be conquered is still with us. A 1990 book on garden plants from the U.S. National Arboretum starts with this quotation from the Bible's book of Genesis: "And God blessed them, and God said unto them, Be fruitful, and multiply, and replenish the earth, and subdue it: and have dominion. . . ." In fact, most of the problems on Earth today are direct or indirect results of our multiplying much too fruitfully and subduing the earth all too effectively.

## Nineteenth Century

In the nineteenth century, the science of ecology was born. **Ecology** is the study of the interactions between organisms and their environments. People started to realize that natural resources are not unlimited and that human interference can destroy natural systems. In 1858, geographer Mary Somerville clearly appreciated the unexpected results from human interference with natural processes:

*"Man's necessities and enjoyments have been the cause of great changes in the animal creation. A farmer sees the rook pecking a little of his grain, . . . and poisons all his neighborhood. A few years after, he is surprised to find his crop destroyed by grubs. The works of the Creator are nicely balanced, and man cannot infringe His laws with impunity."*

This is a statement of a basic law of ecology: the lives of different species of plants and animals are linked, so that changing one thing in nature changes others.

A book that reached a wider audience and became influential in the crusade for conservation was *Man and Nature* by Vermonter, George Perkins Marsh, published in 1864. Marsh warned that ancient civilizations had fallen as a result of environmental damage. More than a third of the book is concerned with the damage done to freshwater systems by deforestation. Marsh was not totally pessimistic, however, and pointed to what he saw as positive achievements:

*"New forests have been planted; . . . lakes have been drained and their beds brought within the domain of agricultural industry; drifting coast dunes have been checked and made productive by plantation; sea and inland waters have been repeopled with fish, and even the sands of the Sahara have been fertilized by artesian fountains. These achievements are far more glorious than the proudest triumphs of war."*

By 1872, the destruction of natural areas in the United States was so complete that the government realized that some of it must be preserved if it was not to disappear forever. President Ulysses S. Grant signed the act designating more than 8,000 square kilometers of Wyoming as Yellowstone National Park. In the next 15 years, countries such as New Zealand and states such as New York also set aside land to be preserved.

## 1900 to 1920

Theodore Roosevelt, a convinced conservationist, became president in 1901. He founded the U.S. Forest Service to manage publicly owned forest and he tri-

**Figure 1-7**  Grand Canyon National Monument. *(Paul Feeny)*

pled the size of the forest. He protected large areas, including the Grand Canyon, as national monuments (Figure 1-7). He also began the National Wildlife Refuge System by designating Pelican Island off the east coast of Florida as a federal wildlife refuge.

In 1912, Congress set up the U.S. National Park system to conserve areas of natural beauty and historic interest in such a way that they remain unharmed for the enjoyment of future generations. These were not the first land preserves. In 1764, botanist Stephen Hales had performed experiments suggesting that trees caused rainfall. He urged the British rulers of several Caribbean islands to prevent the deforestation that had caused soil erosion and decreased rainfall in Jamaica and Barbados. As a result, forest reserves were established on Tobago, "reserved in wood for rain." These reserves still exist and are the oldest in the world.

By 1910, two conflicting philosophies were apparent among those who believed in saving natural areas from development. On one side were the managers, including Theodore Roosevelt and Gifford Pinchot, first head of the U.S. Forest Service. This group believed in using national land for timber-cutting, mining, recreation, and other activities, regulated so that they did not deplete the natural resources but preserved them for the economic benefit of the country. This is the philosophy that has controlled the management of forests in Europe since medieval times.

The other group, led by John Muir and later by forester Aldo Leopold, rejected the old view of the human species as destined to rule nature. Muir wrote, "Why ought man to value himself as more than an infinitely small composing unit of the one great unit of creation?" These people pointed out that a managed forest does not contain the biodiversity of a natural forest. They believed in preserving wilderness with as little human interference as possible for future generations to study and enjoy. These people were the forerunners of those who fight today to save the remaining old-growth forest of the Pacific Northwest and tropical forests, moves that are necessary if we are to save the area's biodiversity. We now know that millions of species of plants and animals are doomed to extinction if we do not preserve their natural habitats.

> "Preservationists" believed in saving wilderness unspoiled; "conservationists" believed in sustainable use of national land.

## 1920 to 1960

Not until 1920 did a significant number of people search for alternatives to environmental destruction in the name of progress. People began to realize that when we destroy the environment, we destroy ourselves.

During the 1800s, factories fueled by wood and coal had spread across Europe and North America. Without pollution control, they spewed soot and poisonous substances into the air and water, causing disease and death. The problem was most acute in Europe, where people lived close together. Governments began to insist on tall smokestacks that carried smoke away from nearby towns, on sewage works so that human waste did not pollute the waterways, and on the burning of cleaner fuels by more efficient furnaces. Britain's Clean Air Act of 1952 was a landmark. For more than a century, much of Britain had been covered with soot from coal-burning factories. In the north of England, houses looked as if they were built of black bricks. Not until buildings were cleaned after the Clean Air Act did many people discover that Yorkshire towns were built of pale yellow stone. In London, "pea-souper" fogs got their name from their yellow-green color, which looked like pea soup. The fog was largely smoke from millions of coal fires in factories and homes, and it caused the deaths of thousands of people from respiratory disease.

Karl Marx had adopted an environmentalist point of view in *Das Capital*, but to most people in the Soviet Union, nature was still waiting to be conquered. The 1948 "Stalin Plan for the Transformation of Nature" envisioned massive assaults on the environment in the name of economic progress. We are just beginning to learn of the economic and environmental disasters that ensued, which included mass starvation as well as pollution with radioactive and toxic waste to the point where large areas of the former Soviet Union had to be evacuated as unfit for human habitation.

Once the first steps to clean up our air and water had been taken, there could be no going back. People became healthier with the change from industrial pollution to clean air and safe drinking water. They started to demand a safe environment. Because it is cheaper to prevent pollution in the first place than to clean up land and water after it is polluted, Western governments and industries have usually been willing to spend the money needed to control the more obvious forms of pollution. Most Western cities are now much cleaner and pleasanter places to live than they were a century ago (Figure 1-8).

### 1960 to 1990

The 1960s were notable for the publication of *Silent Spring*, by Rachel Carson, and *The Population Bomb*, by Paul Ehrlich. Carson imagined a spring morning that was silent because the birds and frogs and insects that sing in spring were dead, poisoned by pesticides. Ehrlich predicted that tens of millions of people would die of war and starvation before the explosive growth of the human population slowed down and eventually stopped. Scientists had been saying these things in scholarly articles for years, but these two books were written for the general public and were read by many people. They slowly but steadily con-

**Figure 1-8** One of the older parts of Paris. Like many Western cities, this area is much more attractive today than it was at the end of the nineteenth century, when the air would have been filled with smoke from burning coal and the streets heaped with horse dung and human excrement. "Crossing sweepers" would sweep a path across the street for a few pennies so the wealthy might cross without treading in the mess. *(Thom Smith)*

vinced many people that environmental problems threatened human health and even existence. People began to understand that population growth, resource depletion, and pollution are linked.

During the 1980s, people all over the world began to understand and attack environmental problems. The membership of environmental groups grew enormously, and millions of volunteers went to work in environmental projects such as reforestation, recycling, family planning, protecting endangered species, environmental education, restoring degraded habitats, and political struggles of various kinds.

If there is one lesson to be drawn from the history of environmentalism, it is that governments act to protect the environment only when economic interests are directly threatened. Scientists, philosophers, and native peoples warn of environmental threats for many years before governments take any notice.

## 1-F  THE TECHNOLOGICAL FIX

**Technology** is the use of tools and machines. Industrial societies use vast amounts of technology in every aspect of life, from food production to education. The question is whether technology will solve our environmental problems. Some of those engaged in predicting the future consider that we *shall* produce marvelous technologies, which are not yet dreamed of, to overcome our environmental problems. How realistic is this?

The amount of many natural resources is not fixed but can be altered by technology. But surely, you may say, the amount of space, fresh water, and oil on Earth is absolutely fixed. Our planet cannot take on new supplies of these, and they are not being formed by natural processes. No, but they can be created by technology or replaced. Fresh water can be made from sea water, and cars need not use gasoline for energy. They can run on electricity produced in nuclear power plants or alcohol distilled from corn. Similarly, Earth contains a fixed amount of a mineral such as iron, but iron is often replaced by plastic or fiberglass to manufacture computers, cars, or boats. So the amount of most natural resources is not fixed.

If technology can replace or create so many of the resources we need, why is there a problem? The answer is, first, that the economics of the situation often make the technology too expensive to use and, second, that technology cannot replace all natural resources.

We undoubtedly shall see technological innovations in energy supplies, agriculture, and pollution control. Technology has already provided us with vast improvements in all these areas. But most scientists think that there can be no technological solutions to our most fundamental environmental problems. This is because these problems involve physical, chemical, and biological laws that technology cannot change or social and economic conditions to which technology is largely irrelevant.

Our most pressing environmental problems are biological: for instance, too many people and too little fertile soil and clean water. We shall see in Chapter 7 that the things that cause people to have more or fewer children are well understood. They are economic and social factors on which technology has little effect. And it is hard to imagine a technological solution to one of our most acute problems: the loss of fertile agricultural soil. We shall see that soil is

created by living organisms over long periods of time. We can speed up soil production, improve existing soil, and slow down the loss of soil, but soil conservation and restoration require individual effort and time more than they require technological innovation.

If there can be no technological solutions to at least some of our environmental problems, we must work on these problems with the human and financial resources that are available to us now.

> There can be no technological solutions to environmental problems that involve biological and physical laws, which we do not control.

## 1-G  ENVIRONMENTAL INSTITUTIONS

In June 1992, heads of state and governmental delegations met at the United Nations' Environmental Summit in Rio de Janeiro to discuss worldwide environmental problems. At the same time environmental activists and enthusiasts met at a "Global Forum" in the same city. The lunatic fringe was well represented at Global Forum, which featured a session on "Underdevelopment and Sex" by the World Association of Sexology, "Yashalom," who proclaimed herself "Goddess of Peace" and gave back rubs, and a native Amazonian who was caught selling the skins of endangered jaguars to tourists. Both the official summit and Global Forum were **environmental institutions**, groups of people who work for environmental goals of various kinds. There are thousands of these institutions, including volunteers fighting to prevent pollution of a local river, national and international environmental organizations, government agencies, and international bodies such as the United Nations and the World Bank.

At the local level, environmental activity is widespread and effective all over the world. Even in poor countries such as India, people tie themselves to trees to prevent tree-cutting and protest against environmental policies they disapprove of. In those countries where governments suppress free speech, environmental activism is still alive and well. In 1986, scientists and writers in the Soviet Union led a letter-writing campaign. The campaign induced officials to abandon plans to divert several rivers in Siberia so that they would flow south to agricultural land in Kazakhstan instead of north to the Kara Sea. In 1987, 8,000 residents of Irkutsk in Siberia demonstrated successfully against a plan to pipe untreated factory waste into the river that supplies the city's drinking water. (See the front endsheet to this book for the locations of these places.) Sometimes local groups are so popular and successful that they grow into large national, and even international, organizations such as the Nature Conservancy, Planned Parenthood, or the Oxford Committee for Famine Relief (Oxfam). They may even receive government funding to help with their work.

Every country has government agencies in charge of environmental affairs. In the United States, these include the Environmental Protection Agency, the Forest Service, and the Bureau of Land Management. All of these have local branches, and their functions overlap somewhat with similar agencies in every state. Also, there are government bodies that have indirect effects on environmental affairs, such as the National Science Foundation, which distributes government money for research and funds much of the basic and applied environmental research performed in the United States. Many government agencies have environmental as well as other functions. For instance, the National Aeronautics and Space Administration (NASA) photographs Earth from space (Figure 1-9). These photographs

can be interpreted to show changing patterns of rainfall and agriculture and to give early warning of food shortages. In 1990, space agencies notified the United Nations that famine was imminent in Somalia and Sudan. Little was done, however, to prevent the resulting deaths until 1992, when the media called international attention to the situation. The media can rapidly attract public attention to an environmental situation and can almost be considered part-time environmental agencies.

If environmental affairs are well represented at the local and even at the national level, international cooperation has hardly begun. We shall see that international cooperation is difficult. Wealthy countries want to protect their own economies and induce poor countries to curb population growth and preserve their forests. Poor countries want to raise the living standards of their people and believe that if they preserve their forests for the benefit of the rich countries, they should be paid for it. Because rich and poor countries must get together to protect the global environment, history may well find that of the two meetings in Rio de Janeiro in 1992, Global Fo-

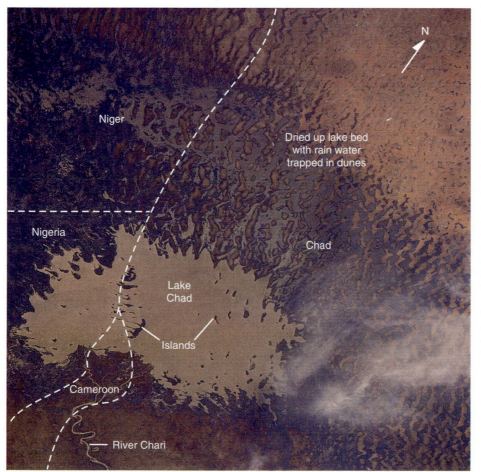

**Figure 1-9** An important research tool: photographs of Earth from space. This 1990 photograph shows Lake Chad in Africa. The lake was once more than twice as large as in this photograph. It provided a route for transporting goods between countries, livelihoods for hundreds of fishing families, and homes for millions of wild animals. Now it is drying up, largely as a result of deforestation, which decreases rainfall. Space surveillance can be used to monitor large-scale changes in the environment like this one as well as changes in food production and the extent of forests. Some areas of the world are still not adequately covered by satellite monitors, and many remote sensing systems are used to produce military, not environmental, information. *(NASA)*

rum was more important than the United Nations summit.

Behind the chaos, Global Forum permitted representatives of international nongovernmental organizations such as Greenpeace, Human-Rights Watch, and the World Wildlife Fund, to write draft treaties outlining steps toward a sustainable world. Most Global Forum delegates were younger than those at the United Nations summit and therefore more likely to determine what happens to the world. They were also free to discuss issues that were considered too controversial for the official summit, such as birth control and human rights. The delegates agreed that a healthy society is necessary for a healthy environment, and therefore "sustainable development" means more than economic growth that does not pollute or deplete natural resources. Without basic human rights, people cannot protest against environmental destruction. Without education, people do not know enough to limit the size of their families. Without economic justice, farmers deprived of their land must continue to cut down forests to grow food. The delegates found themselves in surprising agreement on many of these points and wrote "treaties," parts of which will almost certainly end up as international laws.

### The World Bank

The idea of rethinking economics and politics along environmental lines is not new. One example was a reorganization of the World Bank in the late 1980s. The World Bank is one of the Multilateral Development Banks responsible for channeling money to developing nations. It lends money from developed countries for projects that promote economic development.

Established after World War II, the Bank has long been run by economists with little understanding of environmental science. People have criticized the Bank for its huge investments in unsound projects, such as irrigation dams that have caused floods and soil loss, and in industrial projects, such as factories and airports, in countries where the most pressing need is for agricultural development. Those concerned with environmental affairs have long argued that a country's economic development must be based on population control and sound agriculture if it is to succeed and that the World Bank has wasted enormous amounts of money on projects doomed to failure by environmental considerations.

The Bank's leadership now agrees. It has created new environmental staff positions and a new environmental department. The Bank has announced a campaign to combat soil destruction in Africa, to aid research on tropical forests, and to double its lending for tree-planting projects (Figure 1-10). It is committed to taking environmental considerations into account in all decision making. As a result, the Bank's chances that the money it lends will eventually be repaid will be increased.

## 1-H  PRIORITIES: POPULATION AND HABITATS

There are so many environmental problems to be solved that it is sometimes hard to know where to

**Figure 1-10**   A tree nursery in Ghana, a World Bank project. Trees to replace local forests are bred and grown here. This type of investment in sustainable development is new to the World Bank, which, in the past, funded mostly projects such as infrastructure and industry. *(Biophoto Associates/Holt Studios)*

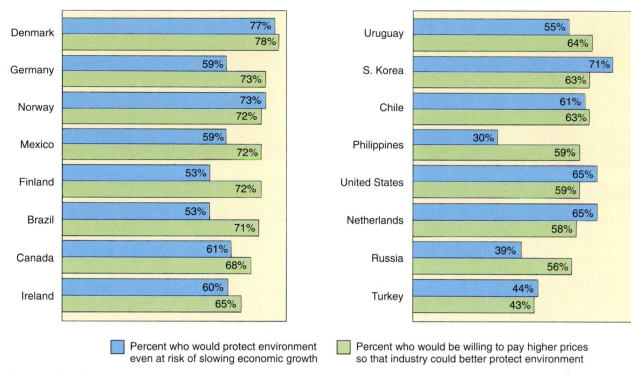

Percent who would protect environment
even at risk of slowing economic growth

Percent who would be willing to pay higher prices
so that industry could better protect environment

**Figure 1-11** World public opinion. This bar graph shows the results of an international public opinion survey on environmental and economic issues conducted between January and March, 1992 by the Gallup International Institute. The margin of error is ±3%.

start. As individuals, we start in our own backyards, with changes in our everyday life, from recycling aluminum cans to turning off lights to save energy. But when we vote for politicians or influence our governments in other ways, it is important to start with the problems that cause irreversible damage if they are not tackled (Figure 1-11).

Environmental damage is often reversible. If sewage is dumped in a river it kills the fish and makes the water unfit to drink. But if you stop dumping the sewage, the river restores itself by natural processes, the fish come back, and the water is once again fit to drink. In contrast, the extinction of a species is irreversible. A species once lost is lost forever. It therefore appears that the problems countries must tackle first are population control and the conservation of habitats that wildlife need to survive.

Why is it essential to control population instead of merely producing more food so that we can feed everyone in the world? One answer is that the amount of food produced in the world is no longer increasing, or is increasing very slowly while the population is growing rapidly. Another is that producing food depletes many natural resources and causes environmental problems such as pollution. Predictions say that if present trends continue the world population will double in 50 years, but this will not happen because there will not be enough food. The effect of further population growth will merely be that more and more people starve to death each year, while the population grows slowly.

Species usually become extinct when we alter or destroy their habitat, the place where they live. Cutting down the old-growth forests in the northwestern United States would lead to the extinction of the spotted owl because the owl nests only in old trees containing rotting wood. When all the old trees are gone, the owl will be gone too. We destroy habitat by filling in marsh or cutting down forest to build houses, roads, and factories, and we also destroy it by polluting rivers, seas, and soil. Much more habitat, however, has been destroyed by being turned into farmland than by other means. In all countries, farmland now covers much of the land that was once forest or prairie, and the plants and animals that once lived in these habitats are extinct or endangered.

# LIFESTYLE 1-1

## Voting

In a democracy, the single most important way in which we influence public policy is by voting. Only 70% of college students were registered to vote for the 1992 elections, and only about half actually voted. In contrast, those over 65 vote in huge numbers. As a result, retired people are actually determining the type of country in which young people will spend most of their lives. So register to vote, and then vote! You can register at either your home or college address. Find out how your representatives voted by sending for the League of Conservation Voters "environmental scorecard." (See Appendix B.)

# LIFESTYLE 1-2

## Writing to Public Officials

One important type of environmental action that young people sometimes neglect is writing to public officials to express an opinion. Officials report that letters from the public are extremely influential and often alter the fate of a proposed piece of legislation. The Sierra Club tells members that legislators take one letter to represent the opinion of ten people, including those who might have written on the same subject but did not get around to it. The official you write to might be an elected representative, a staff member of your state's department of environmental affairs, or an official of the federal Environmental Protection Agency.

Where do you find the names and addresses you need? The names and addresses of many elected representatives are posted in local post offices. Otherwise, go to the local library and ask the librarians. They are helpful people and usually keep this kind of information readily at hand because they are often asked for it.

Writing letters to officials is an acquired skill, and some techniques work better than others. Here are some guidelines for writing letters to influence political action:

1. Cover one subject, in your own words, on a single page.

2. Do not send preprinted cards or form letters.

3. Tell the official what you want him or her to do, and mention anything relevant that you have done yourself.

4. Give your reasons for the action, but do not exaggerate the case.

5. Do not write in anger, and do not be apologetic. Make certain that your letter does not make you sound like a member of the lunatic fringe. Sound reasoning has greater impact than does emotional appeal.

6. If a legislative bill is involved, refer to the name and number of the bill if you can find it.

7. Relate your letter to the legislator's district.

8. Show awareness of the legislator's past actions.

9. Do not mention your membership in a citizens' organization (because the person you are writing to probably already knows the main organizations that favor or oppose the action in question).

10. Ask for the legislator's position on the matter of concern if you don't already know it.

Try drafting a letter to an official expressing your opinion on a particular environmental issue. If you think it is good enough, mail it!

Here is an example of an effective letter:

7 Hilite Road
Anywhere, FL 44444

September 12, 1993

The Honorable Jane Bloggs
United States Senate
Washington, DC 20501

Dear Senator Bloggs:

I urge you to support the Wetlands Conservation Act, Senate bill number 123. This act would prevent destruction of approximately 7 million acres of wetlands in all parts of the country. At the moment, wetlands are being destroyed by development at an alarming rate. For instance, when the Anywhere airport was expanded last year, nearly 5,000 acres of wetlands were destroyed.

We must try to preserve our remaining wetlands. In my county, wetlands provide us with flood control and waste water treatment. They are also the ecosystems where shrimp, flounder, and other marine species breed. About one quarter of the wetlands in this area has already been destroyed, mainly by road-building. As a result, this county has had to spend more than $30 million on flood control projects in recent years, and the value of the shrimp harvest has declined from about $10 million to about $5 million a year.

While wetlands conservation may delay or block some development projects and cause some people financial loss, the financial loss caused by destruction of wetlands is much greater. For economic and ecological reasons, we should preserve our remaining wetlands.

I was pleased to read of your vote to cut federal subsidies for logging roads in national forests. It is good to know that we can count on you for actions that preserve our environment.

Yours sincerely,

Maria J. Lewis

## SUMMARY

Environmental problems are created by human populations too large for their natural environments to sustain. Such populations use up the resources they need faster than these can be replaced. Environmental science studies and invents ways to solve these problems of resource supply and demand.

### Population Growth: The Root of the Problem

The world's human population has more than tripled in the twentieth century. Supplies of resources do not grow as fast as this. Therefore, as time goes by, more and more people become short of the resources needed for life.

### Our Main Environmental Problems

Most environmental problems can be classified as depletion of resources or pollution. Natural resources may be renewable or nonrenewable. Renewable resources, such as plants and fresh water, are those most important to life. Using resources produces pollutants. Pollutants damage health and other resources such as air, soil, and water. Nondegradable pollutants have created new and difficult problems.

### A Sustainable World

The goal of environmental science is a world in which the supply of food, water, building materials, clean air, and other resources needed to sustain a high quality of human life remains constant. At the moment, the world is not sustainable.

### Thinking Internationally

International cooperation is vital if we are to produce a sustainable world. Pollution and resource depletion often affect more than the country in which they occur. The growing gap between rich and poor nations is morally intolerable and threatens the economies of all nations. We are still astonishingly ignorant about environmental matters. More research is urgently needed. Environmental problems cannot be solved without massive education on such subjects as birth control, soil conservation, and pollution prevention.

### A Little History

Westerners have traditionally viewed the world as so large and full of natural resources that they did not have to worry about environmental problems. Instead, they encouraged people to farm the land and clear forests. By the late nineteenth century, few natural areas remained, and governments began to preserve parts of what was left as public lands. In the twentieth century, attention has turned to preventing pollution of air and water, mainly by industry. Governmental support for environmental protection varies, but its public support has increased steadily.

### The Technological Fix

Technology is helping to solve many environmental problems. Technology can be used to replace or create many resources. However, in many cases, replacement costs more than we can afford. But technology cannot solve the most basic environmental problems, which involve physical and biological facts and social and economic conditions.

### Environmental Institutions

Many different social institutions work to solve environmental problems. These include international bodies, such as the United Nations and World Bank, national and local governments, and nongovernmental organizations.

### Priorities: Population and Habitats

The most urgent environmental problems are those that are changing the world in ways that cannot be reversed, particularly population growth and the destruction of habitat for endangered species.

## DISCUSSION QUESTIONS

1. What arguments can be made for the idea that the world is not overpopulated? (Some people believe that it is not.)

2. If present birth rates continue, the human population of Earth (about 2 billion in 1930; now about 6 billion) will be about 24 billion by the year 2100. Or will it? If not, why not?

3. Some economic models assume that populations will continue to grow: there will be more and more people to consume the goods that manufacturers produce, and the economy can continue to grow. Can an economy grow even if the population is stable or declining?

4. Some people believe that the world will never run out of resources because research and technology will find replacements for depleted natural resources. Do you agree? Why?

5. List some environmental advantages and disadvantages of the following inventions: sink garbage disposal units, television sets, drugs that slow human aging, a drug that a woman can use to abort a pregnancy safely, electric cars.

6. Look at Figure 1-11, which shows the results of an international public opinion survey on environmental and economic affairs. What trends can you identify? For instance, are people more likely to be prepared to pay extra for products if they live in wealthier nations? Did the percentage prepared to protect the environment at the risk of economic growth reflect the state of the economy in the country? Does educational level or population growth rate have anything to do with the responses? You may find the information in Figure 1-5 and Table 7-2 useful in your answers.

# How Humans Affect Their Environment

# Chapter 2

**Key Concepts**

- Humans evolved from other primates.
- *Homo sapiens* is a bipedal primate with a large, versatile nervous system, manipulative fingers, a tool-using culture, and language.
- Environmentally, the most important developments in human history were the spread of agriculture and the industrial revolution.
- Modern problems that result from agriculture include overpopulation, habitat destruction, food simplification, and soil erosion.
- Problems that result from the industrial revolution include overpopulation, the energy crisis, and pollution.

*St. Augustine, Florida.*

*"Our numbness, our silence, our lack of outrage, could mean that we end up the only species to have monitored its own extinction. What a measly epitaph that would make: 'They saw it coming but hadn't the wit to stop it happening.'"*

Sara Parkin, U.K. Green Party

Like other organisms, human beings interact with their environment. Because of our numbers and our technology, however, we affect our environment more dramatically and more permanently than do populations of any other organism. In this chapter we consider how humans evolved into the highly successful species that they are today and how this has led to environmental problems.

The story of human evolution is important to understanding how we affect our surroundings. (Evolution is discussed in Close-Up 2-1.) We are descended from intelligent, social animals that lived millions of years ago. Our bodies have not changed much in that time. Cattle have evolved horns and hoofs, bats have developed wings, giant pandas have come to eat nothing but bamboo. Compared with these specialists, humans have unspecialized, conservative bodies but enormously versatile behavior patterns. Our most important advances have been the development of language, the use of tools, and changes in the ways we feed ourselves. We are faced now with one of our most important evolutionary milestones. We must change from a species that destroys its environment to one that does not. Our history, the history of a very adaptable species, offers hope that we shall be able to do this.

## 2-A HUMAN EVOLUTION

Humans have existed for only a tiny fraction of the history of Earth. Yet in this short period of time we have changed the world more than any other species.

### The Time Scale

The solar system where our planet, Earth, lies formed about 4.6 billion years ago. Slowly Earth cooled and water formed on it. Eventually, about 3.5 billion years ago, molecules became organized into the first living things, and evolution began.

Environmentalists sometimes point out that humans are a trivial part of Earth's history. Earth existed without us for billions of years and will doubtless survive for billions of years after we are gone.

This can be illustrated by looking at the time scale of life on Earth up to the present (Figure 2-1).

If we compress the history of our solar system into a single 24-hour day, Earth formed between midnight and 1 a.m. and the earliest forms of life appeared about 7 a.m. Fourteen hours later, at 9 p.m., the only living things were still bacteria, algae, and a few worm-like creatures, all living in the ocean. Not until after 10 p.m. did our ancestors first crawl onto land. The dinosaurs ruled Earth from about 11:15 to 11:40, by which time our mouse-like primate ancestors were hiding from them on the ground and in trees. By about 11:45, most of the dinosaurs were gone, wiped out in some massive disaster that still puzzles scientists. Many think that a large meteorite from space hit Earth, causing a huge cloud of dust that blocked out the sun and lowered the temperature, killing most reptiles, while warm-blooded mammals and birds survived. Some 3 minutes before midnight, human ancestors walked out of the forest onto the African grasslands and started using tools. By 1 minute before midnight, modern humans had evolved and were spreading all over Earth. Less than 1 second before midnight, humans began living on the first farms. The industrial revolution and the explosion of our population started only a few milliseconds before midnight, which represents the present.

## Human Origins

All the people on Earth today belong to the species *Homo sapiens*, a member of the mammalian order Primates, which also includes the monkeys and apes. (Species and classification are described in Close-Up 4-2.) Many primates live on the ground, but some of our distinctive features are inherited from arboreal (tree-living) ancestors. Arboreal animals tend to have well-developed nervous systems and good muscular coordination, adaptations to climbing and leaping among the branches of trees. Birds and monkeys usually have excellent vision that permits them to jump and land on branches. In most primates, both eyes face forward and therefore see the same thing. The superimposed images from the two eyes provide stereoscopic (three-dimensional) vision. Stereoscopic vision is necessary to depth perception. If you do away with stereoscopic vision by closing one eye, it is more difficult to judge how far away something is, which makes it more difficult to catch a ball or to land on a branch.

Primates have five digits (fingers or toes) on each limb. One digit (the thumb) can touch or nearly touch the other four. This enables the feet and hands

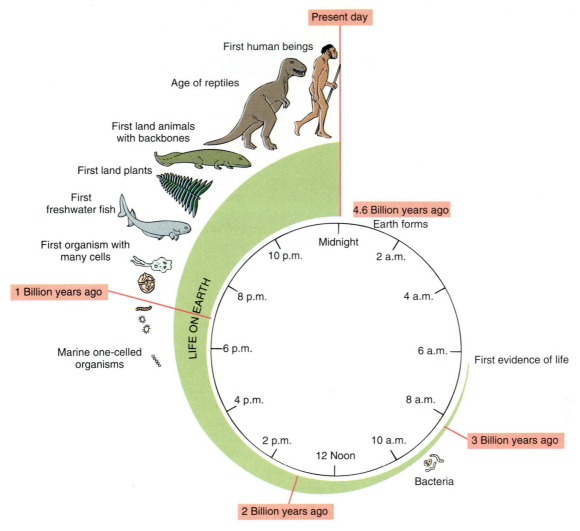

**Figure 2-1** The time scale of life on Earth. The green curve represents the increasing numbers of types of organisms that have evolved.

to grasp tree limbs, food, and other objects, something most animals cannot do. The digits of primates end in sensitive tips, often covered by flattened nails rather than the curved claws of most mammals (Figure 2-2). Claws are used mainly to grip the ground or food, whereas nails act mainly to protect the tips of the digits and are not often used for gripping things.

**Figure 2-2** Hands of primates. This series, from *left* to *right*, shows the evolutionary trend from relatively stiff digits with claws to the flexible human hand, with a thumb that can touch the other four fingers and with fingertips protected by nails.

# How Organisms Evolve

The theory of evolution states that today's organisms have arisen by descent and modification from more ancient forms of life. In modern terms, **evolution** is the process by which the members of a population come to differ genetically from their ancestors. Charles Darwin is usually thought of as the father of evolutionary theory (Figure 2–A). Darwin was not the first person to suggest that organisms evolved, but he was the first person to explain how evolution is brought about.

We can deduce that evolution occurs from three observations:

1. Organisms are variable. Even the most closely related individuals differ in some respects.

2. Some of the differences among organisms are inherited. Inherited variations are caused by differences in genetic material, which organisms inherit from their parents. Parents and offspring tend to resemble each other more closely than they resemble organisms to which they are less closely related genetically.

3. More organisms are produced than live to grow up and reproduce. Fish and birds may produce hundreds of eggs, oak trees thousands of acorns, but only a few of these survive to reproduce in their turn.

Some of the inherited variations among organisms are bound to affect the chances that an individual will reproduce. For instance, if an animal is born with a severe genetic disease, it is not likely to live long enough to pass this gene on to its offspring. The fact that some genes are more likely to be reproduced in the next generation than others is called **natural selection**. Natural selection produces evolution, which can be defined as a change in the proportions of different genes from one generation of a population to the next. The result of natural selection is that populations undergo **adaptation**, or changes appropriate to their environments, over the course of many generations.

Evolution is often discussed in terms of body structure and function. Behavior also evolves, however, and is often more important than body structure in determining whether a population survives. Consider the case of the Canada geese. These big water birds nest in the northern United States and Canada. Until recently, they migrated to wetlands farther south to spend the winter. But in the 1960s, people living near lakes and rivers in the North noticed that more geese than usual were spending the winter. Instead of the few birds who usually stayed behind when the rest migrated, hundreds of birds were feeding in snow-covered cornfields and roosting (sleeping) in groups floating on the lakes.

Nowadays, the number of birds that do not migrate is very large. In 1989, more than 30,000 Canada geese spent the winter on Cayuga Lake in upstate New York. Meanwhile, ducks and geese have been disappearing from the South. One Georgia newspaper called it a "winter hunting crisis." Southern wildlife managers feared they were witnessing the decline and extinction of the geese. In fact, they were watching the evolution of a successful and far from extinct species.

Canada geese have a genetic tendency to migrate. This does not mean that all birds always do it. Humans have a genetic tendency to go to sleep at night. But we don't always do it. Before humans interfered, the geese that traveled south each winter raised more offspring, on average, than those that risked starvation and freezing by staying in the North. In most years, natural selection favored migration.

Then the environment changed. States such as Maryland permitted more geese to be shot. This changed the natural selection acting on the birds. Now, the dangers of the long journey, combined with the chance of being shot, mean that in most years birds that stay in the North produce more offspring than birds that migrate. Offspring inherit their parents' tendency to migrate or not to migrate. The part of the population that does not migrate is growing faster than the part that migrates, and so the number of geese wintering in the North increases.

The decrease in geese in the South happened even faster than this description suggests because geese, like humans, learn much of their behavior. Geese are monogamous (taking one mate for life), teach their goslings (babies), and follow experi-

**Figure 2-A**   Charles Darwin in 1839. *(Royal College of Surgeons)*

enced leaders when they migrate. When leaders appeared among the birds that stayed in the North, other birds with no strong tendency to go or to stay would follow the leaders' example and stay in the North. So an entire group of hundreds of geese could disappear from a southern marsh from one year to the next without a hunter raising a gun.

The parallel with human evolution is interesting. The evolution of nonmigratory behavior in Canada geese is partly a genetic change. Genes that give birds a strong urge to migrate are becoming less common in the population. But it is also an example of change as a result of learning, often called cultural evolution. Cultural evolution sounds like a different type of evolution, but it is not really. It too is based on genetics and caused by natural selection.

An organism's genes determine the types of things it can learn and how likely it is to learn them. It is because humans and geese have different genes that humans do not learn to sleep floating on lakes and geese do not learn to fire guns.

The spread of a behavior pattern through a species is as much an example of evolution as is a change in the average brain size. Such changes can usually be traced to some new environmental situation that exerts selective pressure on the population. Some organisms do not survive environmental changes. They become extinct. Populations survive only if they change their lifestyles as their environments change. This book is largely a study of how the lifestyle of the human population is changing and must change if, like the geese, we are to survive.

**Figure 2-3** A tarsier, a primate from Indonesia. Note the large forward-facing eyes, reduced snout, and mobile fingers. *(Gary Milburn/Tom Stack & Associates)*

> Primate features important to human evolution include an unspecialized skeleton, stereoscopic vision, and mobile fingers.

Early in primate history, a mouse-like primate started living in trees. The most advanced member of this group still living is the spectral tarsier from Indonesia, with huge eyes, stereoscopic vision, and nails instead of claws (Figure 2-3). In addition, the upper lip is free of the gums, which it is not in other mammals such as dogs. This feature is the main reason higher primates have such mobile and expressive faces. During primate evolution, the snout (nose) became progressively reduced. This is probably an adaptation that gives the forward-looking eyes a clear view of the world.

Monkeys, apes, and humans are anthropoid primates (Table 2-1). An obvious feature of anthropoids is their upright posture. Even monkeys that walk on all four legs sit upright for long periods, freeing their hands to manipulate food, handle their young, and perform other tasks (Figure 2-4). There are only four modern apes: gibbon, orangutan, gorilla, and chimpanzee. All live in Africa and Asia, and their structure and behavior bridge the gap between monkeys and humans.

There are more similarities than differences between apes and members of the human family of species. Humans and chimpanzees have body proteins and genes that are about 99% similar. It is clear that

| TABLE 2-1 Classification of the Order Primates |
|---|
| Suborder Prosimii ("before apes"): Tree shrews, lemurs, lorises, bush baby, tarsier |
| Suborder Anthropoidea: Monkeys, apes, humans |
|    Superfamily Ceboidea: American monkeys, including marmosets, capuchin |
|    Superfamily Cercopithecoidea: Eurasian and African monkeys, including macaques, baboons |
|    Superfamily Hominoidea |
|       Family Anthropoidii: The anthropoid apes— gibbon, orangutan, gorilla, chimpanzee |
|       Family Hominidae: *Australopithecus* (extinct prehumans); *Homo habilis, H. erectus, H. sapiens* |

chimpanzees are our nearest living relatives. We are not as closely related to gorillas or orangutans.

## The First Tools

Few areas of research have produced as much argument and confusion as the search for the fossils of our ancestors. The fossil evidence of human ancestry is fragmentary. A find seldom consists of more than part of a jawbone and a few teeth. Much of the muddle, however, can be ascribed only to human vanity.

**Figure 2-4** A white-faced monkey (*Cebus capucinus*) from Central America. Like most primates, this species can comfortably sit upright, freeing its hands to carry things and manipulate objects.

**Figure 2-5**   An artist's impression of the life of a *Homo habilis* family.

Researchers inevitably hope to discover vital clues to human ancestry. As a result, many fossils that are now identified as monkeys (and even modern humans!) were at first hailed as "the missing link" between apes and humans.

Since our nearest living relatives are the African apes, the search for fossils of the presumed common ancestor of apes and humans has centered in Africa. Many forested areas in Africa turned into open grassland some 15 million years ago. Between 10 and 4 million years ago, human ancestors apparently moved out of the forests and roamed the open grassland in bands, like modern baboons. In contrast, the ancestors of the apes remained in the forest. In Tanzania, Mary Leakey found footprints made 3.75 million years ago by primates that were bipedal, meaning that they walk on two feet. Walking upright leaves the hands free to carry food and babies, throw stones, and do all kinds of things. This was the beginning of technology, the use of tools and machines.

The first members of the human primate family are called *Australopithecus* and appeared about 4 million years ago. In 1979, the first almost complete skeleton of a member of this group was discovered in the Ethiopian desert. This was a female, nicknamed "Lucy," a member of the species *Australopithecus afarensis*. There is some debate as to whether australopithecines were completely ground-dwelling or spent some of their time in trees because, compared with humans, they had long arms and short legs. Australopithecines had brains that were little larger than those of modern apes, and they were ape-like in having large, heavy jaws, showing that they ate a lot of tough plant food.

There is no convincing evidence that australopithecines used tools, but we would not expect to find remains of the earliest tools. If they were like the tools that chimpanzees use, the first prehuman tools must have been "found" materials—rocks, bones, sticks, large thorns, or lengths of vine— which either would not have been preserved or, if they were, would not bear any marks of having been used as tools.

At least one species of australopithecine lived at the same time as the first definite species of *Homo* (*Homo habilis*), which had a larger brain than an australopithecine (Figure 2-5). Piles of animal bones found with *H. habilis* skeletons show that meat had been added to plants as a regular part of the diet by this time. These hominids also used crude stone

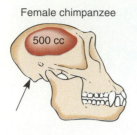
Female chimpanzee

500 cc

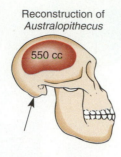
Reconstruction of *Australopithecus*

550 cc

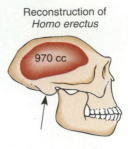
Reconstruction of *Homo erectus*

970 cc

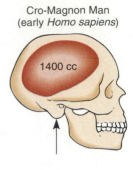
Cro-Magnon Man (early *Homo sapiens*)

1400 cc

**Figure 2-6**  Changes in proportions of the skull from ape to human (*left* to *right*). Note that the brain case and brain (*red*) increased in size. The angle and position of the attachment between the neck and the skull (arrows) changed as primates became more bipedal and upright. The size of the teeth and jaws was relatively reduced during the change from a largely plant-eating diet to a mixed diet of plant and animal food.

tools. We have no idea whether language had evolved by this time. We can only speculate that the advantages of cooperation in hunting and group defense may have encouraged the development of language for communication.

*Homo habilis* lasted only a few hundred thousand years and was then replaced by a similar species,

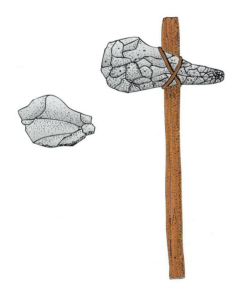

**Figure 2-7**  Stone tools of the type used by early humans. *Left:* A flake tool like those found with *Homo erectus* in Tanzania. Flake tools were made by splitting flakes off larger rocks, either by hitting one rock with another or by heating the rock in a fire until it cracked. *Right:* Stone axe produced in Europe by early *Homo sapiens*. The axe was shaped by chipping at a rock of roughly the right shape and size until it had a sharp blade and a shaft that could be lashed to a wooden handle with leather thongs.

*Homo erectus. H. erectus* was fully bipedal, used tools, and ate both plants and animals. The skull was thick, with heavy jaws and teeth, brow ridges over the eyes, and a low forehead. Some individuals had brains almost as large as those of modern humans (Figure 2-6). Some *H. erectus* bones were found in caves, suggesting this species used at least temporary homes. Besides animal bones and quite advanced stone tools, some of the caves contain heaps of charcoal and charred bones, showing that fire had been domesticated and brought indoors by this time. Presumably this habit started with the use of natural fires (perhaps started by lightning) to keep warm, cook food, or split stones (Figure 2-7).

Human changes from their ape-like ancestors include bipedalism, an enlarged brain, eating plants and meat, tool use, and language.

## 2-B  THE SPREAD OF HUMAN POPULATIONS

Tools and fire, the beginnings of technology and culture, contributed to the success of *Homo erectus*. The species spread widely, and emigrants from Africa colonized other, colder areas in Europe and Asia. New behavior or technology was necessary before human beings could survive winters as cold as those of Central Europe and China. Plainly, the brain of *H. erectus* could produce social and technological solutions to the problems of surviving cold winters, such as fire, clothing, stored food, and communal living in caves.

Human ancestors invented technological solutions to the problems of living in cold climates.

*Homo sapiens* probably evolved from *H. erectus* only a few hundred thousand years ago. The Neanderthals, with human-sized brains but heavier skulls and teeth, appeared about 100,000 years ago, but nothing is known of the origin of this short-lived group. Remains of completely modern *H. sapiens* go back about 50,000 years. Human populations spread from Europe and Asia to the Americas and Australia some 25,000 years ago.

## 2-C HUNTER-GATHERER POPULATIONS

For most of our evolutionary history, human beings were hunter-gatherers, people who obtain their food by collecting plants and killing wild animals. Our knowledge of hunter-gatherers comes from archeological finds and from studies of modern hunter-gatherer groups such as Eskimos, Australian aboriginals, South American Indians, and the !Kung bushmen of the Kalahari desert in southern Africa.

By cooperating with each other, hunters can herd, spear, or trap large game animals. But the main source of food in hunter-gatherer groups is plant parts—seeds, fruits, leaves, nuts, roots, and berries—gathered as they are used or preserved and stored for later. The use of fire opened up a new range of plant foods. Cooking can remove poisonous molecules that occur in many plants and can also soften plants, making them more digestible.

Seafood is another source of food for hunter-gatherers. Indians in the southeastern United States left piles of shells up to five meters high from the oysters that formed a major part of their diet. They also caught fish by spearing and trapping them. Remains of paddles and canoes date from about 5,000 years ago in northern Europe. Danish remains about 3,000 years old contain the bones of deep-sea fish such as cod, showing that these people could exploit ocean resources. Although fish are farmed to some extent today, most of our fish still comes from "hunting" these animals in their wild habitats.

Most early hunter-gatherer groups were undoubtedly nomadic, moving from place to place according to the availability of plant and animal food at different times of the year (Figure 2-8). Later, permanent settlements developed. A settled society involves social organization and communication between individuals that would be a strong selective pressure for the development of language, social rites, laws, and customs. We find these reflected in the decorated tools, pots, and dwellings that began to appear in Europe and Asia about 20,000 years ago and in America at least 12,000 years ago.

Settlements, permanent or temporary, usually lead to environmental problems such as pollution. A village well or a nearby stream is easily polluted by human waste or the runoff from a rubbish pile. Scientists have found deformities in the skeletons of hunter-gatherers in the southeastern United States that they believe were caused by drinking polluted water, which can cause many diseases (discussed in Section 8-F).

The Old Stone Age extends from the first use of tools by human ancestors until the establishment of agriculture about 10,000 years ago. Sometime during this period, *Homo sapiens* evolved, and eventually all other species of *Homo* became extinct. Were our hominid relatives the first species we drove to extinction? It is perfectly possible, but we do not know.

Even the small populations of ancient hunter-gatherers affected their environment in many ways. Human activities such as starting or spreading fires do not need many people to produce major effects. As they traveled to new areas, early humans probably carried plant seeds with them, altering the distribution of plants (and possibly animals). Many animal species, including African game species, the Irish elk, mammoth, steppe bison, wooly rhinoceros, and mastodon, became extinct during the Old Stone Age. Many more species became extinct then than in earlier or later periods. The traditional explanation has been that these extinctions were caused by rapid changes in climate during the major glaciations (Ice Ages) that occurred between about 2 million and 10,000 years ago. Some researchers, however, believe that many of these extinctions were caused by humans.

The environmental impact of early hunter-gatherers included changing the distribution of plants, starting fires, and possibly also causing the extinction of species they hunted.

## 2-D THE AGRICULTURAL REVOLUTION

**Agriculture** is the practice of breeding and caring for animals and plants that are used for food, clothing, housing, transportation, and other purposes. Agri-

**Figure 2-8**   The life of a hunter-gatherer: a Minnesota Chippewa boils down sap from sugar maples to make maple sugar. The most important Chippewa plant foods were wild rice and maple sugar because both could be stored through the long northern winter. In early spring, Chippewa left their winter residences and walked or paddled canoes to sugar maple groves, where they lived in lodges made of birch branches and bark while they made maple sugar. In the summer, the group moved to a camp by a lake to harvest wild rice, which grows in shallow water. The rice seed was harvested from canoes and then dried and stored in bags of deerskin or birch bark to be carried back to the winter camp. Before metal pots and cotton were introduced by settlers, the maple syrup pots were made of clay and the women dressed in deerskins. Women were in charge of processing maple sugar and wild rice, while men made the canoes and hunted, trapped, and fished for meat, which was added to the diet. Wild rice was the staple food and was seasoned with maple sugar or with broth made from fish or meat.

culture originated, probably independently, in many different places about 10,000 years ago (Figure 2-9). Fossils of domesticated dogs dating from 11,000 years ago have been found in Iraq, and cultivated plants date back 10,000 years in America. The changeover from hunting and gathering to agriculture had such a dramatic impact on human societies and on the environment that it is often called the **agricultural revolution.**

At first, agriculture merely provided part of the food for a hunter-gatherer society. For example, when Spanish missionaries first landed in the southeastern United States, they found hunter-gatherer tribes living largely on deer, seafood, and local plants. In addition, many cultivated a few plants around their huts. They also cleared temporary plots in the forest, where they grew a single crop of maize (corn)[1] and beans, using seeds that were originally imported from Mexico.

We are so used to thinking of agriculture as a superior way of life that the relative advantages of hunter-gatherers may come as a surprise. Hunter-gatherers do not face the constant battle with pests, droughts, and famine that beset all agricultural communities. Studies in southern Africa during a drought in the 1970s showed that farmers starved, while the population of hunter-gatherer bushmen in the Kalahari remained stable and the people well fed. This is probably because hunter-gatherers control their population size, while most agricultural societies do not. Hunter-gatherers keep their popula-

[1]Much of the English-speaking world calls wheat "corn" and corn "maize." To avoid confusion, this book calls wheat "wheat" and corn "maize"!

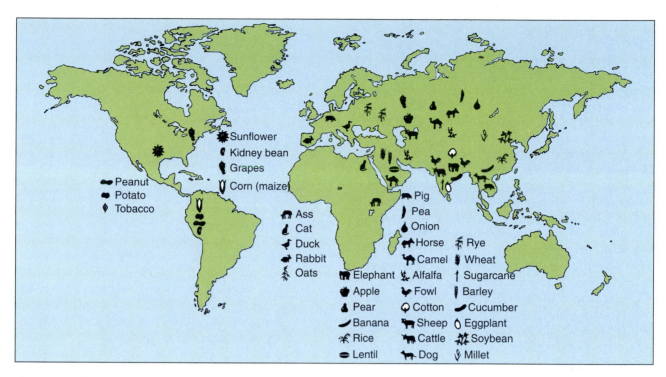

**Figure 2-9**   Parts of the world where plants and animals we use today were probably first domesticated. Some species are believed to have been domesticated independently in two or more areas, and these are shown in both places.

tions well below the size that their territory can support. The people make a conscious effort to keep the population stable by such practices as abstaining from sexual intercourse, abortion, infanticide, late marriage, and late weaning.

People from agricultural societies are shocked when they learn that Inuits and other hunter-gatherers used to practice infanticide, which is usually considered immoral in farming societies. The killing of baby girls, considered less useful than boys, is nevertheless quite common in the history of farming societies. In fact, infanticide, the killing of very old people, and other forms of population control were logical responses if not enough food had been gathered for the whole tribe to survive the winter. The choice then was between the whole tribe becoming half-starved, with perhaps none of them surviving the winter, and saving the lives of those who could reproduce next year. In contrast, when famine strikes a farming society that does not control its population, the entire village becomes malnourished, as we know from the photographs of famines in Africa and elsewhere. Infants are always much more likely to die in famines than are older people, simply because they are smaller and more delicate. Most of the millions of people who starve to death every year are

infants and young children. U.S. troops in Somalia also reported that boys survived starvation much more frequently than girls, probably because they were considered more important and received more of any food that was available.

Studies of hunter-gatherers show that they have a more balanced diet than most farmers, and their incidence of diseases is no higher. Their life expectancies are comparable to those of agricultural peoples in less-developed countries. Even though they live in inhospitable deserts, Kalahari bushmen and Australian aboriginals devote only about 15 hours each week to collecting and preparing food. Children do not have to work until they are married, and the aged are cared for and revered. In contrast, most people in agricultural societies work for at least 60 hours each week and spend around 70% of their income on food. Thus, about 42 hours of work a week are devoted to acquiring a subsistence (survival) diet. Even in the affluent West, we devote at least a third of our incomes (equivalent to 13 hours of work a week) merely to buying food, and this does not include food preparation time. Clearly, then, some hunter-gatherers have an easy life, in some ways, compared with most members of even affluent agricultural societies.

It is not clear what pressures induced early human populations to surrender their hunter-gatherer existence for life on a primitive farm. There may have been no single reason. Agriculture is merely an extension of what people already knew. Studies show that modern hunter-gatherers know all they would need to know to settle down as farmers and that this knowledge is centuries old. Farming was combined with hunting and gathering at first, and complete dependence on agriculture for food was a later development. Population pressure may well have been one reason for this, as for so many other dramatic changes in society. The same area of land can support up to 500 times as many people by farming as by hunting and gathering.

The development of agriculture was undoubtedly the most profound revolution in human history, and its consequences were far-reaching. One of the most important of these is that it permitted the accumulation of material goods. Nomadic people travel with few possessions. Farmers, living in one place, can accumulate as much as they can afford. Even land can be owned and passed on by inheritance or sale.

A striking consequence of agriculture is that a new type of division of labor grows up within the group. In prosperous agricultural societies, a few people can produce food for everyone. The rest are free to become builders, bakers, and merchants. Finally, the population may even be able to afford the luxury of poets, scholars, and students, who contribute little to the group's physical well-being but are the basis of its cultural life.

Once farming had begun anywhere on Earth, it inevitably spread over the face of the globe. Because farming supports larger populations than hunting and gathering, an agricultural community always expands into the land of any nearby hunter-gatherer group, fighting for the territory if necessary and driving the hunter-gatherers to extinction or to become integrated into the farming community. Similarly, settled agriculture overwhelms nomadic livestock herding as a way of life. The battle for land between farmers and ranchers in the West is part of American history. The same forces are destroying the lifestyles of Masai and Bedouin herders in Africa today.

Agriculture has had such profound effects on the environment, direct and indirect, that much of this book is concerned with it. The most important direct effects can be summarized as follows:

1. **Habitat destruction.**   The area of land devoted to farming has increased steadily, converting natural habitats to farmland and causing the extinction of thousands of species of plants and animals. Clearing natural vegetation, particularly forests, to produce farmland has had numerous environmental effects, including altering the climate, producing a widespread shortage of fuel, and causing floods and water shortages (Figure 2-10).

2. **Soil erosion.**   Agriculture does not need to destroy the soil, but it usually does. The amount of soil on the surface of the globe has steadily decreased since the first digging tool was invented thousands of years ago.

3. **Population increase.**   Not only can agriculture support more people on a given land area than hunting and gathering, but also the population control practiced by hunter-gatherer societies is usually abandoned by agricultural communities. This is partly because children, who are not important food collectors in a hunter-gatherer society, are useful as unskilled labor on a farm. In addition, the inheritance of land and goods becomes more important. The desire to have children who will inherit the property and care for their aged parents is a recurrent theme in mythology and literature. Only within the last 50 years has effective population control been practiced by people in agricultural societies.

4. **Food simplification.**   Humans once ate a much greater variety of plant and animal species than they do in agricultural societies today. The Cherokees of North America alone ate more than 400 different species of plants (Figure 2-11). Nowadays, the vast majority of human food comes from just four species: wheat, rice, maize, and potatoes.

## Origins of Agriculture

Once people begin to plant seed deliberately, cultivated strains of plants soon come to differ from their wild counterparts. For instance, grain (annual grass) seed heads that do not burst open are easy to collect, whereas seeds that have been shed from their seed heads are scattered and difficult to gather. So the human planter collects and plants the seeds of grain plants that do not shed their seeds easily, and this trait rapidly increases in the population of cultivated plants.

Similarly, docile animals, with small horns and wooly coats, would have been those most likely to become the domesticated breeding stock. We can imagine that an early agricultural society would rapidly have pushed a vicious ram with big horns into the wild again or, more likely, would have made it the centerpiece of a tribal feast. By a combination of conscious and unconscious selection, early farmers

(a)

**Figure 2-10** Examples of environmental damage caused by agriculture. (a) Habitat destruction. The forest in the background has been replaced by this hay field, reducing the number of different plants and animals that the area supports. (b) Soil erosion. The muddy water in this irrigation ditch contains topsoil, which has washed off the fields. The water is also polluted by pesticides and fertilizer, which it carries into a nearby river.

(b)

rapidly produced animals and plants that differed considerably from their wild ancestors.

Irrigation was an important development because it has had such profound effects, not only in increasing agricultural productivity but also in degrading the environment (see Chapter 14). The earliest evidence of artificial irrigation is an engraving from 5,000 years ago, showing an Egyptian king cutting an irrigation ditch. The plow, used to turn over soil, is another agricultural introduction notable for its environmental impact and also invented about 5,000 years ago.

Agricultural societies soon moved from a technology based on wood and stone to one based on metals. Copper tools date from about 5,700 years ago in Iran

and possibly from as long as 7,000 years ago in Thailand. By the Bronze Age, 3,500 years ago, products made of bronze (an alloy of copper and tin) stretched from Britain in the West to China in the East.

> The environmental impact of agriculture includes population growth, destruction of natural habitats and species, destruction of hunter-gatherer societies, soil erosion, and food simplification.

## 2-E THE INDUSTRIAL REVOLUTION

The **industrial revolution** is the change that spreads through societies as they start to use fossil fuels as energy sources to power their technology (Figure 2-12). The fossil fuels—oil, natural gas, and coal—produce enormous amounts of energy compared with the amount available from an equal weight of traditional fuel such as wood. The development of steam engines powered by fossil fuel in the eighteenth century and of internal combustion engines in the twentieth century vastly increased the amount of energy that humans could use for all purposes.

Large industries began to develop even before the widespread use of fossil fuel. The building of large ships capable of crossing oceans in the sixteenth and seventeenth centuries was part of this development. International travel became common and meant that previously separated areas of the world were now in frequent contact so that the world became a "global village."

Like the agricultural revolution, the industrial revolution has reduced the area of land needed to sup-

Arrowhead   Milkweed   Jerusalem artichoke   Wild rice

| Plant | Type | Uses |
|---|---|---|
| Arrowhead *Sagittaria latifolia* | Tall plant that grows in mud and shallow water | Starchy roots boiled; can also be dried for later use |
| Bearberry *Arctostaphylos uva-ursi* | Trailing plant with bell-like flowers found in sandy and rocky places | Berries cooked with meat for flavoring; leaves were smoked like tobacco |
| Black cherry *Prunus serotina* | Large deciduous tree in eastern forests from Nova Scotia to Florida and Texas | Twigs boiled to make tea; cherries cooked and then formed into cakes, which were dried for winter use |
| Bur oak *Quercus macrocarpa* | Large deciduous tree found in southern Canada and the eastern U.S. | Nuts (acorns) boiled, or roasted, split, and eaten as vegetable or with grease |
| Canadian hemlock *Tsuga canadensis* | Evergreen coniferous tree of northern forests | Leaves boiled to make tea |
| Hawthorn *Crataegus* | Small tree found in most of the U.S. | Raw berries, rich in vitamin C, squeezed into cakes and dried |
| Jerusalem artichoke *Helianthus tuberosus* | Tall sunflower found in open spaces in eastern, midwestern, and southern U.S. | Roots eaten raw like a radish or fried |
| Labrador tea *Ledum groenlandicum* | Woody shrub found in northern bogs | Leaves boiled to make tea, which might be sweetened with maple sugar |
| Milkweed *Asclepias syriaca* | Tall flowering plant found in open areas | Flowers cut up, stewed and eaten like preserves |
| Mountain mint *Pycnantheum virginiana* | Small perennial flowering plant found throughout the eastern U.S. | Flowers and buds used to season meat and broth |
| Virginia creeper *Parthenocissus quinquefolia* | Woody vine found in southern half of eastern North America | Stems boiled then peeled to reveal the sweet layer under the bark which was eaten like corn on the cob |
| Wild ginger *Asarum canadense* | Flowering plant found in northern and eastern woods | Roots used as flavoring; also used for indigestion |
| Wild rice *Zizania* | Cereal (grass) that grows in shallow water of lakes from Canada to Texas | Processed seed boiled or fried; eaten with meat, maple sugar, or blueberries |

Hawthorn   Black cherry   Virginia creeper   Bur oak

**Figure 2-11**   Some of the hundreds of North American plants that Native Americans used to eat but that are seldom part of our diet today.

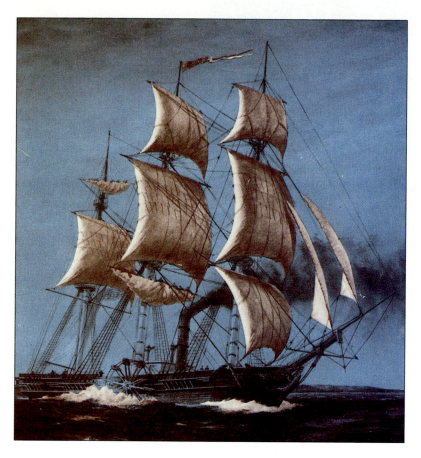

**Figure 2-12**   The *S.S. Savannah,* a steamship launched in 1819, symbolizes the early stages of the industrial revolution. Designed to travel cheaply under sail when there was plenty of wind and under steam power when there was not, she was supposed to replace slower sailing ships in the tea and cotton trade. The *S.S. Savannah* proved a bit of a white elephant. She could not carry enough fuel (wood or coal) to cross the Atlantic under steam, and her owners never recovered the cost of her construction. *(Painting by C. Lundgren; courtesy States Marine Lines)*

ply each individual with food, clothes, and housing and has increased the use of natural resources. Modern science and modern medicine have compounded these effects so that the rate at which the human population grows has increased rapidly since the industrial revolution even in societies that are not industrialized, such as rural areas of Africa, South America, and Asia.

The accelerating use of resources made possible by the industrial revolution has led to many environmental problems. For instance, industrial societies use vast amounts of metals, which may deplete mineral reserves. A vast increase in consumption has led to the snowballing problem of waste disposal. Many industrial processes produce by-products that pollute air and water. The number of ways in which humans affect their environment is increasing. For instance, nearly all the most powerful pesticides date from after World War II, and so do the problems caused by nuclear weapons and nuclear power stations.

Environmental problems that were once local have become regional and even global as a result of industrial capacity. DDT in the water and lead in the air are found thousands of kilometers from the societies that produce them. Finally, the complexity,

magnitude, and frequency of human effects are increasing. A modern copper industry has a very different level of impact from a Bronze Age copper smelter.

> Industrialization has accelerated every kind of human impact on the environment, including population growth, resource depletion, and pollution.

## 2-F  CULTURAL EVOLUTION

Ecologists have pointed out that one way to classify human cultures, and their impact on the world around them, is by their energy sources. The hunter-gatherer relies mainly on human muscle power and fire. This type of energy use cannot support a large human population, and the culture's impact is limited.

Agriculture usually coincides with the domestication of animals, which are used for additional energy, whether to pull plows, turn mill wheels, or transport

# Young People and Cultural Evolution

*"The Agricultural Revolution began some 10,000 years ago and the Industrial Revolution has been under way for two centuries. But if the Environmental Revolution is to succeed, it must be compressed into a few decades."*

Lester R. Brown, President, Worldwatch Institute, 1992

Students come home from college and tell their parents not to water the lawn or flush the toilet every time. Children come home from school and ask their parents to start saving paper and aluminum cans for the school recycling project. Parents are amazed. Where have these activist children come from, and what are they talking about? Environmental science was not taught in school when the parents were educated. They know little about it except for an occasional television program or newspaper article.

Scientists have been warning since the 1960s of the dangers of overpopulation and pollution. Many of them had given up in despair by 1990. Governments continued to give and lend poor countries money for arms but not for family planning; state agencies continued to let people cut down forests and build houses on wetlands. The world was going to hell in a handbasket and nobody seemed to care. Many scientists became convinced that the world a century or two from now would be populated by very few people with a poor and declining standard of living—all because the public did not know enough or care enough to force its leaders to act in the twentieth century.

In the last few years, however, a change in attitude has begun to spread. "No generation has a freehold on this earth. All we have is a life tenancy." That was then–British Prime Minister Margaret Thatcher in 1988 in a speech that *The Economist* magazine said marked her conversion from Iron Lady to Green Goddess. In that same year, George Bush campaigned to be elected the "environmental president." Then–Senator Al Gore wrote a book about environmental problems. Mikhail Gorbachev of the then-Soviet Union used the environment as the theme of an address to the United Nations. Czecho-

**Figure 2-B** Politicians are slowly responding to public pressure by becoming more aware of environmental affairs. Then–Senator Al Gore (*right*), shown at a political rally, has even written a book on the subject.

slovakia was one of dozens of new nations that wrote environmental protection into their constitutions. Are these politicians beginning to understand at last? Possibly. At the least they are doing what politicians always do—responding to public pressure. The children have educated their parents, and both are pressuring politicians (Figure 2-B).

Young people understand that environmental problems are related. They understand that you cannot cure resource depletion and pollution while subsidizing energy costs and why the extinction of one species may mean the extinction of dozens of others. They understand the biology of ecosystems and why Earth cannot produce ever-increasing amounts of food. They see the connection between environmental destruction and the ever-growing stream of illegal immigrants into the wealthier countries. They understand why environmental damage and poverty go hand in hand and that you cannot cure one without curing the other. If the world is saved from environmental disaster, it will be largely because of the political efforts of young people.

**(a)**

**(b)**

**Figure 2-13**    Preindustrial sources of energy that are still used today. (a) An ox cart with wooden wheels is used for transporting lumber in Costa Rica. (b) A water-driven mill. Once used to grind grain, mills today have been converted to generate electricity.

people and goods. The use of boats for fishing and for transporting people is undoubtedly very ancient, but the settled life of agricultural societies led to more extensive exploitation of the power of water. Water power is used to move irrigation water; to float rafts of timber down rivers; and to power mill wheels, spinning wheels, and, eventually, electric generators (Figure 2-13). Windmills, harnessing the energy in wind, also have a long history as power sources. All these extra sources of energy may be added to wood-burning to produce considerable energy use, pollution, and environmental damage even in preindustrial agricultural cultures. Deforestation and soil erosion are problems of this stage of cultural evolution. Their impact has accelerated in modern times mainly because the industrial revolution has permitted our population to grow so fast.

The discovery of energy from fossil fuel gave people hundreds of times more energy than had ever been available before. This energy has fueled agriculture, permitting the population to explode, and has caused life-threatening pollution.

Not until coal smoke started to pour from chimneys in the 1800s did most people grasp how easy it was to make their own communities unfit to live in. It is hard for us even to imagine the time, fewer than 50 years ago, when wash hung outside to dry was black with coal soot in a few minutes, as it still is in many parts of eastern Europe today. Our great-grandparents may well have preferred the layer of soot, however, to the horse manure and human excrement that littered city streets before the mid-nineteenth century.

We have used modern technology to clean up as well as to pollute our environment. Many cities in industrialized areas are more pleasant to live in than they have ever been (Figure 2-14). This is of prime

**Figure 2-14**    St. Augustine, Florida. The problems caused by the industrial revolution should not make us forget how much better life is for many city dwellers than it was in preindustrial times. City streets are no longer awash in sewage and manure, streets and homes are well lit and heated, and transportation is (sometimes) more rapid.

importance because a major effect of the industrial revolution was to hasten urbanization—the movement of people from the countryside, where relatively few people now find jobs, into cities, where industries are established and most jobs are to be found. The most important environmental effects of the industrial revolution can be summed up as urbanization, population growth, and pollution.

We are using up the world's supply of fossil fuel.

Will another energy source fuel the next stage of human cultural evolution? Many people believe that if we are to survive at all, human energy use, impact on the environment, and population size must all peak within the next 100 years. If this model is correct, success will depend on our evolving into a species with multiple energy sources and fewer people, with massive, but carefully controlled, effects on our environment.

## SUMMARY

Like other organisms, humans interact with their environment. Because of our numbers and our technology, we affect our environment more acutely and more permanently than does any other organism.

### Human Evolution

Humans evolved very recently in Earth's history from tree-dwelling primates. Because of this ancestry, we have well-developed nervous systems, excellent muscular coordination, stereoscopic vision, manipulative hands, and related physical features that have contributed to our success as a species. Tool use, language, and complex social behavior developed as human ancestors evolved from a forest-dwelling, plant-eating species into hunter-gatherers living in more open country.

### The Spread of Human Populations

Humans evolved in Africa. Technology, such as the use of fire and the construction of clothes and shelter, permitted us to colonize Asia and Europe and later Australia and America.

### Hunter-Gatherer Populations

For most of their evolutionary history, *Homo sapiens* societies have been hunter-gatherers, living mainly on collected plant food, supplemented by animal food from hunting and fishing. Hunter-gatherer populations generally control their populations to sizes that can find enough food and other resources within their territories. These populations may suffer from environmental problems such as the extinction of animals they feed on and water pollution. Nomadic hunter-gatherers carried plants with them as they traveled, distributing organisms to new habitats.

### The Agricultural Revolution

Agriculture has spread through the world since its origin more than 10,000 years ago. Its spread has wiped out most hunter-gatherer societies and has had a greater impact on our lives and environment than any other development in history. Agriculture led to population increase; to the development of towns with their extreme specialization of labor; and to habitat destruction, soil erosion, and our reliance on a very few plants for food.

### The Industrial Revolution

The speed with which environmental problems have increased recently is a result of the development of heavy industry since about 1700. One of industry's most important effects has been a vast migration of people from the countryside into cities. Another has been the pollution of air, land, and water.

### Cultural Evolution

Human societies can be classified by the types of energy they use, which have changed during the course of human evolution. Hunter-gatherers use mainly the energy from human muscle and from small fires. Early agricultural societies add to these sources the muscle power of domesticated animals and the energy of wind and moving water. Industrialized societies release enormous amounts of energy from fossil fuels. Fossil fuel energy has increased the impact of humans on their environment. It has fueled agriculture to support the population explosion, increased the speed with which we deplete natural resources, and led to modern levels of pollution.

## DISCUSSION QUESTIONS

1. Do you think that *Homo sapiens* will be extinct within the next thousand years? Why or why not?

2. If we imagine the evolution of the human species toward a sustainable society, we encounter a theoretical problem. In general, the people who reproduce most rapidly today are those who are poor, with least education, with the most environmentally destructive agriculture, and with little concern for environmental problems. Are these, therefore, the evolutionarily successful people? If so, how can we imagine that humanity will evolve toward an environmentally responsible society with a low rate of reproduction?

3. The world contains a few remaining hunter-gatherer societies in places such as Australia, the Philippines, New Guinea, Brazil, and Peru. The governments that control the countries where they live generally believe these people should be left alone and perhaps given title to the land where they have lived for centuries. In some places this may actually happen. In others, hunter-gatherers are rapidly disappearing. What do you think are the pressures that determine one fate rather than the other? Is it worth trying to save modern hunter-gatherers?

4. This chapter stresses the environmental problems caused by agriculture and industry. But agriculture and industry have brought inestimable benefits to human societies. What are these benefits?

# SOME BASIC ECOLOGY

*Desert vegetation by the Colorado River,
Canyonlands National Park, Utah.*
(David Muench)

# Chapter 3

# ENERGY, CHEMISTRY, AND CLIMATE

## Key Concepts

- The availability and cost of energy affect almost every aspect of environmental science.

- Energy cannot be created or destroyed, but some useful energy is lost each time energy is transformed from one form to another.

- Nearly all the energy we use comes originally from the sun.

- The composition of Earth's atmosphere is determined partly by the photosynthesis and respiration of organisms, and, in turn, the atmosphere determines the climate and the temperature on Earth.

- Human activities are altering Earth's atmosphere in ways that are causing Earth to heat up and permitting more and more ultraviolet radiation to reach the surface of Earth.

*Fire in a hay rick. Energy reaches living things when green plants use the sun's energy to produce their food. It leaves when plants and other organisms live and die, are eaten, decompose, or burn.*

We often talk as if energy is our most important resource. Directly and indirectly, the supply of energy affects the supply of all our other resources, and it affects our ability to deal with problems such as pollution. This chapter is an introduction to energy and other physical and chemical facts that determine what goes on in the world around us. Chapters 5 and 6 apply these facts to living things and the ways they interact with one another.

## 3-A  THE LAWS OF THERMODYNAMICS

Energy from the sun determines climate and weather and powers human endeavors. As a result, energy shortages may cause environmental disasters. Plentiful energy may prevent them. For instance, where there is a shortage of fuel to cook with, people may cut down trees to use for fuel faster than the trees can regrow. The forests that once covered the slopes of the Himalayan Mountains have been cut down for this reason. The destruction of the forest has led to soil erosion, floods, and loss of life. To prevent such disasters, the Indian government is trying to supply kerosene for cooking fuel. The trees can then be saved for functions only trees can perform, such as holding the soil on steep slopes in place.

**Energy** is usually considered to be the capacity to do work and transfer heat. Energy comes in many forms, but all of them are stored work and heat. Energy can be converted into movement, as when you use energy to lift up this book.

The work you do to lift the book can be considered the product of the force you use times the distance over which you exert the force. If you carry the book upstairs, you do more work than if you just lift it up because the force required is the same, but you are moving the book over a greater distance. Once you have applied energy to lift it, the book has **potential energy**, energy due to its position. If you lift the book higher, it gains more potential energy because you have applied the same force over a greater distance.

Suppose that having lifted it, you now drop the book. The book's potential energy is converted into **kinetic energy**, energy of movement. The amount of kinetic energy depends on the mass (weight) and velocity (speed) of the moving object. Suppose you have accidentally dropped the book into a tub full of water. At the moment the book hits the water, its velocity changes rapidly (it slows down), and much of its kinetic energy is lost. However, energy cannot disappear. The total energy (potential plus kinetic) does not change. Most of the kinetic energy of the falling book is converted into work when the book hits the water, and that work pushes the water out of place. In hydroelectric plants, the kinetic energy of falling water is used to generate electricity.

The **law of conservation of energy** is the first law of thermodynamics. It states that energy cannot be created or destroyed; it can only be transformed from one form into another. This law is sometimes expressed, "you can't win" (because you cannot create energy).

> Energy cannot be created or destroyed; it can only be transformed from one form into another.

Energy becomes important to us only when it is changed from one form to another. When a car burns gasoline, the chemical potential energy of the gasoline is transformed into the kinetic energy of the car's movement (Figure 3-1). Energy transformations obey the laws of thermodynamics. We have already encountered the first law, the law of conservation of energy. The second law of thermodynamics states that, in any change of energy from one form to another, some useful energy (energy available to do work) is lost. In other words, in the search for energy that can be used to do work, "you can't even break even." In any energy transformation, some useful energy is converted into useless forms such as heat that escapes into the environment, and so the energy available to do work decreases (see Figure 3-2).

> Whenever energy is changed from one form to another, some useful energy is lost.

Oil and natural gas do not power machinery by kinetic energy. They contain **chemical potential energy**. When their molecules are burned, they release light and heat (thermal energy) Light, a form of **radiant energy**, is one type of energy the sun supplies to Earth or a light bulb supplies to a room. Most thermal energy comes from the movements of atoms and molecules in solids, liquids, and gases. The faster the molecules move, the higher the thermal energy of the material.

### Heat and Ocean Currents

Heat and temperature are not the same thing. **Heat** is the total energy of all the moving atoms in a substance. **Temperature** is a measure of the average speed of motion of the atoms at any given moment.

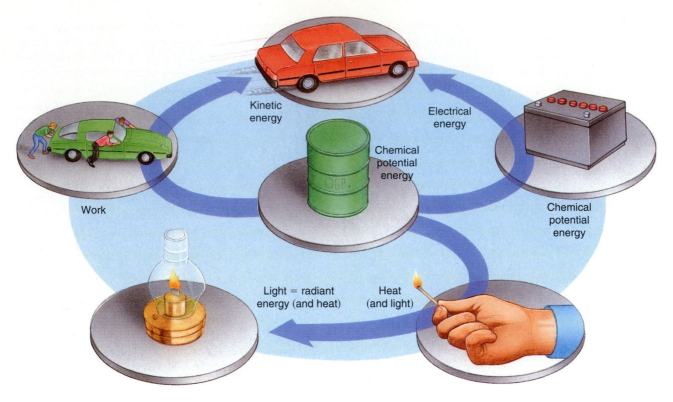

**Figure 3-1**   Transformations of stored chemical energy. The chemical energy in oil can be used to do various kinds of work when it is activated by heat (from the match), electrical energy (from the battery), or muscular energy (from people pushing the car).

Thus the temperature of a cup of hot coffee is higher than that of a cold lake, but the lake contains much more heat because it is so large.

The heat contained in large bodies of water affects the climate. Temperatures near the ocean are higher in winter and lower in summer than in inland areas because the ocean retains heat even after the air has cooled in autumn, and it warms up more slowly than the air in spring. Ocean currents may carry large amounts of heat from one part of the world to another. The most dramatic example is in the climate of western Europe, which is affected by the Gulf Stream. The Gulf Stream is a huge current that carries warm water from the Caribbean Sea across the North Atlantic Ocean to the coasts of Ireland and England. Wind blowing across the Gulf Stream produces warm west winds in western Europe. The Gulf Stream contains so much heat that Ireland seldom suffers from frost, although it is further north than Winnipeg, where the temperature may fall to 20° below zero. We shall see in Chapter 12 that movements of currents that cause El Niño winds in the Pacific also have major effects on the life of marine organisms.

## Human Energy Use

Nearly all the energy we use every day to walk around, heat buildings, run vehicles, power factories, or cook comes originally from the sun. This is because the sun powers photosynthesis by plants (Section 3-C). Figure 3-2 shows how the sun's energy is converted, via photosynthesis, into the muscle energy that humans use to move.

The fuel we burn was made by photosynthesis. Sometimes we burn plants directly, as when we burn biomass. **Biomass** is any biological material. Wood is the most important biomass fuel, but animal dung and the residues left after crops are harvested are also used. Fossil fuel was also produced by photosynthesis. Oil, natural gas, and coal are the fossils of organisms that lived long ago. They either stored energy themselves by photosynthesis or they ate other organisms that did. Even the energy in wind and fall-

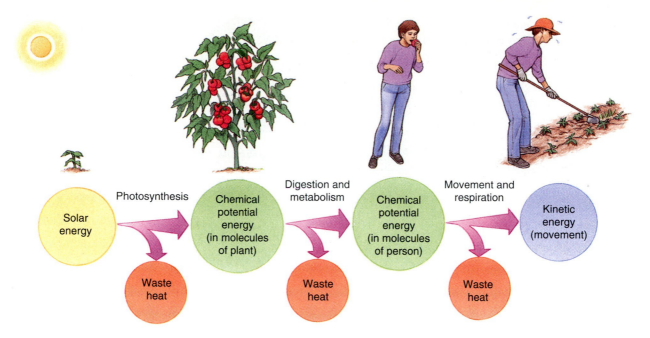

**Figure 3-2**  The energy transformations by which solar energy moves human muscles. Solar energy powers photosynthesis by which plants made food molecules, containing stored chemical energy. A person eats part of a plant and digests it, converting plant chemical potential energy into human potential energy. Muscles break down food by respiration, releasing the stored energy as muscle movement. At each transformation, some energy is wasted as heat.

ing water come from the sun because the sun causes the weather (Section 3-D). Sometimes we capture the sun's energy directly, using it to heat a house or water in a pool or to run a solar-powered calculator. The only important source of energy people use that does not get its energy from the sun is uranium, the fuel used in nuclear power stations.

Uranium and fossil fuels are nonrenewable sources of energy; all the others, such as wind, water, and wood, are renewable. At the moment the world uses much more nonrenewable than renewable energy. However, many more people depend on renewable energy sources because most people live in the world's poorer countries where biomass is the main fuel.

### Energy Use in Developed Nations

One of the main arguments between rich and poor nations today concerns energy use. The developed nations contain relatively few people but use most of the energy. In 1990, Americans used more energy for air conditioning alone than four times as many Chinese did for all purposes, including heating, cooling, cooking, manufacture, agriculture, and transportation. The United States wastes amazing amounts of energy. Other industrialized nations, such as Japan, Germany, and Switzerland, use less than half as much energy as the United States to produce a book, a washing machine, a meal, or any other type of domestic product. And remember that using energy nearly always means burning fuel, which produces air pollution. In Chapter 10, we shall see that the United States could cut energy use and air pollution enormously without lowering our standard of living (see Lifestyle 3-1).

### 3-B  ELEMENTS AND MOLECULES

You, your food, and this book are examples of **matter**, the stuff that things are made of. The amount of matter in something is its **mass**, which is the same as its weight on Earth's surface. Things have weight only when there is gravity. In space, where there is no gravity, objects have no weight so they float around in a spaceship, but they still have mass. Matter is made up of molecules and atoms that are attached to each other by various kinds of bonds.

Much of the energy that we use is stored in the form of chemical potential energy in the bonds between molecules. Most of these are organic molecules, large molecules containing carbon. (They are called "organic" because they are made by living organisms.) All other molecules are inorganic, and they are usually smaller than organic molecules. The molecules in coal and oil are organic because they were made by living things: coal and oil are the remains of plants and marine creatures that lived millions of years ago.

> Fossil fuels contain chemical potential energy stored in organic molecules.

Carbon is the chemical element on which life is based. An **element** is a substance that cannot be broken down into other kinds of substances. If you take a piece of the element silver and divide it into ever-smaller pieces, you will ultimately have a single atom—still of silver. An **atom** is the smallest unit that shows all the properties of the element.

Ninety-two elements occur naturally, but living things contain only about 20 of these (Table 3-1). The chemicals that organisms need are not always the most common. For instance, silicon is about 300

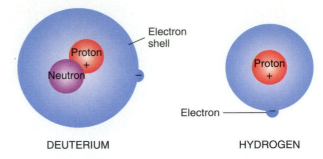

**Figure 3-3** Models of the atoms of two isotopes of hydrogen: deuterium and hydrogen.

times more common than carbon in Earth's crust; yet carbon is vital to every living thing, whereas silicon is found in very few. Other elements are important in the environment because they are toxic to organisms (Close-Up 3-1).

One dramatic achievement of the twentieth century was the splitting of atoms into their component parts. The simplest atoms are those of the element hydrogen. A hydrogen atom is made up of one positively charged particle, a proton, and one negatively charged particle, an electron. Some hydrogen atoms also contain a third atomic particle, a neutron, which has no electrical charge. These atoms are called deuterium. Hydrogen and deuterium are both isotopes of hydrogen. **Isotopes** of an element contain different numbers of neutrons in their atoms (Figure 3-3).

The protons and neutrons in an atom are clumped together in a central nucleus. Electrons whizz around the nucleus in orbits called shells or clouds. Because opposite electrical charges attract one another, the negatively charged electrons are attracted to the positively charged protons in the nucleus, and this attraction holds the atom together. Each atom contains as many electrons as protons: the electrical charges cancel one another out, and the whole atom is electrically neutral.

## Bonds Between Atoms

In nature, very few atoms exist singly because the atoms of most elements are not stable. For an atom to be stable, its outer electron shell must be filled by a certain number of electrons: two for hydrogen and eight for most other atoms.

Atoms with unfilled outer shells may form **bonds** that join them with other atoms to form various chemical compounds. After it has formed bonds, each atom ends up with a filled outer shell and therefore is more stable (Figure 3-4).

---

**TABLE 3-1 Some Chemical Elements Found in Humans, Their Chemical Symbols, and Main Sources**

| Element | Symbol | Source |
|---|---|---|
| Oxygen | O | Air |
| Carbon | C | All food: sugars, fats, carbohydrates |
| Hydrogen | H | All food, water, air |
| Nitrogen | N | Protein foods |
| Calcium | Ca | Milk, cheese, leafy vegetables, whole grains, legumes* |
| Phosphorus | P | Milk, cheese, meat, eggs, legumes, fruits, vegetables |
| Sulfur | S | Protein foods |
| Potassium | K | Whole grains, meat, legumes, fruits, vegetables |
| Chlorine | Cl | Table salt (NaCl), protein foods |
| Sodium | Na | Table salt, milk, meat, baking soda, carrots, spinach, celery |
| Magnesium | Mg | Whole grains, nuts, meat, milk, legumes |
| Iodine | I | Seafood, iodized table salts |
| Iron | Fe | Liver, eggs, meat, whole grains, green vegetables, legumes |

*Legumes are members of the large plant family Leguminosae, which contains the beans, peas, alfalfa, and acacias.

# CLOSE-UP 3-1

## Heavy Metals and Other Elements

Environmental science is full of elements, such as toxic heavy metals, salts in the soil, or phosphorus and nitrogen for plant fertilizer. Figure 3-A shows the elements. From top to bottom, each rectangle contains the element's atomic number (the number of protons it contains), its symbol, name, and atomic weight (which is almost the same as the number of protons plus the number of neutrons). The elements are arranged so that related elements are near each other. For instance, the alkali metals, the elements in Group 1 (left-hand column), give off colored light when they are heated, so they are often used in fireworks. Lithium burns with a red light and potassium with a violet light.

In environmental discussions, we often hear of pollution by heavy metals, such as copper (Cu), zinc (Zn), mercury (Hg), and cadmium (Cd). These are all poisonous (Chapter 18). They are the metals toward the right-hand end of the transition elements in the periodic table. Many of them are common and are found in vitamins and other molecules in living things. They are used in many industrial processes, notably to produce metal products from coins to cars. Because of their many uses in society, heavy metals often find their way into waste water and solid waste, where they are difficult to get rid of and may become dangerously concentrated.

Metals

Nonmetals

| Group I | II | | | | | | | | | | | III | IV | V | VI | VII | VIII |
|---|---|---|---|---|---|---|---|---|---|---|---|---|---|---|---|---|---|
| 1<br>H<br>Hydrogen<br>1.0079 | | | | | | | | | | | | | | | | | 2<br>He<br>4.0026 |
| 3<br>Li<br>Lithium<br>6.941 | 4<br>Be<br>9.0122 | | | | | | | | | | | 5<br>B<br>10.811 | 6<br>C<br>Carbon<br>12.011 | 7<br>N<br>Nitrogen<br>14.007 | 8<br>O<br>Oxygen<br>15.999 | 9<br>F<br>Fluorine<br>18.998 | 10<br>Ne<br>Neon<br>20.180 |
| 11<br>Na<br>Sodium<br>22.990 | 12<br>Mg<br>Magnesium<br>24.305 | | | | | | | | | | | 13<br>Al<br>Aluminum<br>26.982 | 14<br>Si<br>Silicon<br>28.086 | 15<br>P<br>Phosphorus<br>30.974 | 16<br>S<br>Sulfur<br>32.066 | 17<br>Cl<br>Chlorine<br>35.453 | 18<br>Ar<br>Argon<br>39.948 |
| 19<br>K<br>Potassium<br>39.098 | 20<br>Ca<br>Calcium<br>40.078 | 21<br>Sc<br>44.956 | 22<br>Ti<br>47.88 | 23<br>V<br>50.942 | 24<br>Cr<br>Chromium<br>51.996 | 25<br>Mn<br>Manganese<br>54.938 | 26<br>Fe<br>Iron<br>55.847 | 27<br>Co<br>58.933 | 28<br>Ni<br>Nickel<br>58.69 | 29<br>Cu<br>Copper<br>63.546 | 30<br>Zn<br>Zinc<br>65.39 | 31<br>Ga<br>Gallium<br>69.723 | 32<br>Ge<br>72.61 | 33<br>As<br>Arsenic<br>74.922 | 34<br>Se<br>Selenium<br>78.96 | 35<br>Br<br>79.904 | 36<br>Kr<br>83.80 |
| 37<br>Rb<br>85.468 | 38<br>Sr<br>87.62 | 39<br>Y<br>88.906 | 40<br>Zr<br>91.224 | 41<br>Nb<br>92.906 | 42<br>Mo<br>95.94 | 43<br>Tc<br>(98) | 44<br>Ru<br>101.07 | 45<br>Rh<br>102.91 | 46<br>Pd<br>106.42 | 47<br>Ag<br>107.87 | 48<br>Cd<br>Cadmium<br>112.41 | 49<br>In<br>114.82 | 50<br>Sn<br>118.71 | 51<br>Sb<br>121.75 | 52<br>Te<br>127.60 | 53<br>I<br>126.90 | 54<br>Xe<br>Xenon<br>131.29 |
| 55<br>Cs<br>Cesium<br>132.91 | 56<br>Ba<br>137.33 | 71<br>Lu<br>174.97 | 72<br>Hf<br>178.49 | 73<br>Ta<br>180.95 | 74<br>W<br>183.85 | 75<br>Re<br>186.21 | 76<br>Os<br>190.2 | 77<br>Ir<br>192.22 | 78<br>Pt<br>195.08 | 79<br>Au<br>Gold<br>196.97 | 80<br>Hg<br>Mercury<br>200.59 | 81<br>Tl<br>204.38 | 82<br>Pb<br>Lead<br>207.2 | 83<br>Bi<br>208.98 | 84<br>Po<br>(209) | 85<br>At<br>(210) | 86<br>Rn<br>(222) |
| 87<br>Fr<br>(223) | 88<br>Ra<br>Radon<br>(226) | 103<br>Lr<br>(260) | 104<br>Unq<br>(261) | 105<br>Unp<br>(262) | 106<br>Unh<br>(263) | 107<br>Uns<br>(262) | 108<br>Uno<br>(265) | 109<br>Une<br>(266) | | | | | | | | | |

Lanthanide series

| | | | | | | | | | | | | | |
|---|---|---|---|---|---|---|---|---|---|---|---|---|---|
| 57<br>La<br>138.9 | 58<br>Ce<br>140.1 | 59<br>Pr<br>140.9 | 60<br>Nd<br>144.2 | 61<br>Pm<br>(145) | 62<br>Sm<br>150.4 | 63<br>Eu<br>152.0 | 64<br>Gd<br>157.3 | 65<br>Tb<br>158.9 | 66<br>Dy<br>162.5 | 67<br>Ho<br>164.9 | 68<br>Er<br>167.3 | 69<br>Tm<br>168.9 | 70<br>Yb<br>173.0 |

Actinide series

| | | | | | | | | | | | | | |
|---|---|---|---|---|---|---|---|---|---|---|---|---|---|
| 89<br>Ac<br>(227) | 90<br>Th<br>232.04 | 91<br>Pa<br>231.04 | 92<br>U<br>Uranium<br>238.03 | 93<br>Np<br>(237) | 94<br>Pu<br>(244) | 95<br>Am<br>(243) | 96<br>Cm<br>(247) | 97<br>Bk<br>(247) | 98<br>Cf<br>(251) | 99<br>Es<br>(252) | 100<br>Fm<br>(257) | 101<br>Md<br>(258) | 102<br>No<br>(259) |

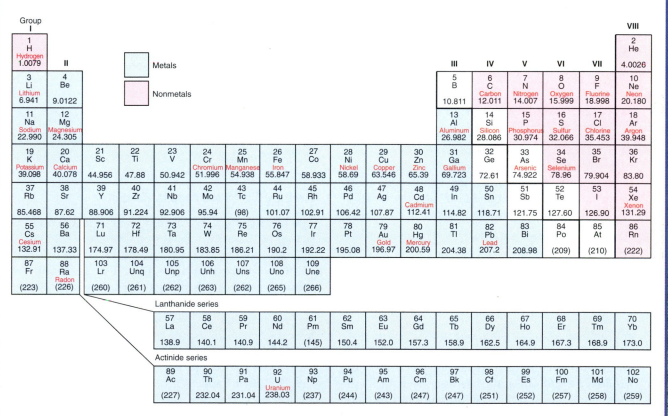

**Figure 3-A** Periodic table of elements. Some of the elements mentioned in this book are labeled in red.

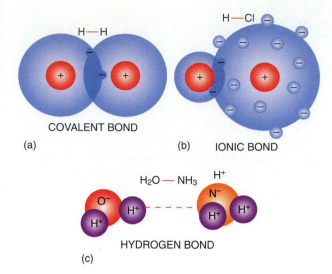

COVALENT BOND

(a)

IONIC BOND

(b)

HYDROGEN BOND

(c)

**Figure 3-4**   Types of bonds between atoms. (a) The covalent bond in an atom of hydrogen gas ($H_2$) is between two atoms of hydrogen that share their electrons. (b) The ionic bond in hydrogen chloride (HCl). Hydrogen and chlorine share two electrons, which are more strongly attracted to the chlorine atom. The chlorine atom ends up with a slight negative charge and the hydrogen atom with a slight positive charge. (c) A hydrogen bond (*red dashes*) is a weak attraction between a hydrogen with a weak positive charge and a nitrogen or an oxygen with a weak negative charge.

Three types of bond between atoms are important in living things:

1. A **covalent bond** forms when two atoms share pairs of electrons. For instance, two hydrogen atoms may form a covalent bond by sharing their electrons to form a molecule of hydrogen ($H_2$). $H_2$ is a molecule of hydrogen gas.

2. A **hydrogen bond** is an electrical attraction between a hydrogen atom bearing a partial positive charge and another atom bearing a partial negative charge. A molecule of the genetic material, DNA, is partly held together by thousands of hydrogen bonds.

3. An **ionic bond** forms when one atom gives up one or more electrons to another atom so that each atom ends up with a stable set of electrons. For instance, a sodium atom has one electron in its outer shell, but it is more stable without this electron. An atom of chlorine, on the other hand, requires one more electron to make up a stable electron shell (Figure 3-5). It can accept an electron from sodium. After an electron has passed from sodium to chlorine, the sodium atom has one more proton than electron and so has a pos-

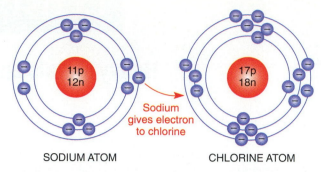

SODIUM ATOM        CHLORINE ATOM

**Figure 3-5**   Sodium chloride (NaCl). Sodium becomes stable by giving up the only electron in its outer shell (*red arrow*). Chlorine adds one electron to its outer shell of seven and ends up with a filled shell of eight. After losing a negatively charged electron, sodium is a positively charged ion. Chlorine, by accepting an electron, becomes negatively charged.

itive electrical charge of $+1$. It is now called a sodium **ion** (an electrically charged particle) instead of an atom. The chlorine atom, now with an extra electron, has a negative electrical charge of $-1$. It is called a chloride ion. The oppositely charged sodium and chloride ions are attracted to each other. Together, one ion of each element makes up a molecule of sodium chloride (NaCl), common table salt.

> Most atoms are attached to other atoms by bonds, forming compounds that are more stable than the isolated atoms.

### Electrical Charge and Pollution Control

The electrical charge on many substances is used in pollution control devices. Opposite electrical charges (positive and negative) attract each other. If a stream of air or water is passed over a positively charged surface, negatively charged particles in the stream will be attracted to the surface.

This principle is used in **electrostatic precipitators**, used to clean the air coming from cement mills and plants that burn coal (Figure 3-6). An electrical wire gives negative electrical charges to molecules in the air. For instance, an oxygen molecule picks up an electron from the wire to become a negatively charged ion. This oxygen ion may then attach to a dust particle in the gas coming from the plant, giving the whole particle a negative charge. The charged particle is attracted to the wall of the precipitator, which is kept positively charged. The particle precipitates (falls out of the air) on the wall and runs down

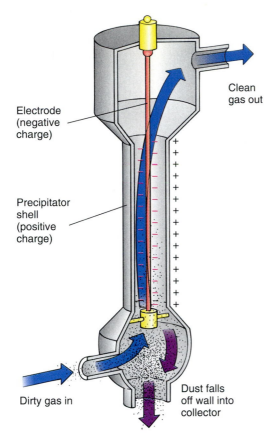

Clean gas out

Electrode (negative charge)

Precipitator shell (positive charge)

Dirty gas in

Dust falls off wall into collector

**Figure 3-6** An electrostatic precipitator, a pollution control device for cleaning gases produced by industrial processes. The central electrode imparts a negative charge to particles in the gas. These are attracted to the positively charged wall of the precipitator, where they slide down into a collector.

the wall into a collector. In the United States alone, electrostatic precipitators remove 20 million tons of "fly ash" from the air that leaves coal-burning power plants each year.

A Colorado corporation has invented a system that works in a similar manner to clean dirty waste water. The waste water is passed through tubes with negatively charged walls to precipitate metal ions such as copper ($Cu^{2+}$) and zinc ($Zn^{2+}$). It is then passed through positively charged tubes to remove negative ions such as cyanide ($CN^-$) and fluoride ($F^-$). The process is being used to clean water seeping from a coal mine in New Mexico and to remove fat and other pollutants from a meat processing plant in Denver. This electrical technique costs much less than the filters and chemical methods formerly used to clean up the waste water. This is an important invention because metal ions are among the most widespread modern pollutants, and they are particularly difficult and expensive to remove from liquids.

> The electrical charge on many substances is used in pollution control devices.

## Chemical Reactions

Sometimes a collision between molecules rearranges the electrons in the two molecules, forming a new set of bonds and making new compounds by a chemical reaction. For example, when methane (marsh gas) burns, it undergoes a chemical reaction that can be written:

$$\underset{\text{methane}}{CH_4} + \underset{\text{oxygen}}{2\,O_2} \rightarrow \underset{\text{carbon dioxide}}{CO_2} + \underset{\text{water}}{2\,H_2O}$$

starting materials          end products

It took energy to form the bonds between the carbon and hydrogen atoms in methane, and some of this energy was stored in the bonds themselves. When the bonds are broken, the stored energy is released, and some of it goes to form the new bonds in carbon dioxide and water molecules; the rest is released as heat. The arrow in a chemical equation points in the direction of lower total energy and therefore shows which way the reaction tends to go. Every reaction goes "downhill" in terms of energy, forming end products with lower energy than the starting materials.

Even though this reaction releases energy, it will not happen unless some energy is put in to start it. The reaction can be started by holding a burning match to the methane to supply heat energy. Once the reaction starts, it releases enough energy to make many more molecules of methane and oxygen react. (If enough methane is present, this may cause an explosion.)

Sometimes the arrows in a chemical reaction point in both directions:

$$\underset{\text{carbon dioxide}}{CO_2} + \underset{\text{water}}{H_2O} \rightleftharpoons \underset{\text{carbonic acid}}{H_2CO_3}$$

The double arrows mean that there is little energy difference between the two sides of the equation, and therefore this reaction is reversible: it proceeds from left to right or from right to left depending on the conditions.

## Solutions, Acids, and Bases

Chemical substances are often of interest when they are in solution: dissolved in the water of oceans, rivers, rain, soil, or the bodies of animals and plants.

Acid rain is caused by air pollutants that dissolve in rain. It does a lot of damage to buildings, forests, lakes, and rivers. When it rains on the Hudson River Valley in New York, fifth and sixth graders at the Poughkeepsie Day School hurry outdoors with buckets and pH test strips to set up their acid rain monitoring station in the playground. Rainwater in the first two storms of the 1992 school year, the students report, was as acidic as Coca Cola (see Figure 3-7). A student reports the results to a center in Manhattan, where results from 300 volunteer groups around the United States are collected for research purposes. In West Lafayette, Indiana, Ann Piechota calls her readings in to the local TV weather reporter, who announces the acidity of each storm the day after it passes. In Poughkeepsie, teacher Pete Salmansohn reports calls from surprised parents whose ten-year-olds have been discussing acid rain at the dinner table—another example of children educating their parents (see Close-Up 2-2).

Many substances come apart, or **dissociate**, into ions when they dissolve in water. An **acid** is a substance that releases hydrogen ions ($H^+$) when it dissociates in water. Hydrogen chloride (HCl) dissociates into hydrogen ions ($H^+$) and chloride ions ($Cl^-$) in water to produce hydrochloric acid. A **base** (also called an **alkali**) is a substance that releases hydroxyl ions ($OH^-$) in water. The base sodium hydroxide (NaOH) dissociates into sodium ions ($Na^+$) and hydroxyl ions ($OH^-$) in solution. (As it dissociates, NaOH also gives off a lot of energy as heat, as anyone who has used a drain cleaner containing it knows.) A **salt** is a substance that gives off neither hydrogen nor hydroxyl ions when it dissociates. Sodium chloride (dissociating to $Na^+$ and $Cl^-$ ions) is an example.

The acidity or alkalinity of a solution is indicated by a number known as **pH**. The pH scale goes from 0 to 14. A pH of 7 is neutral (neither acidic nor basic). The pH values below 7 are acidic, and those above 7 are alkaline (Figure 3-7). Pure water is neutral, with a pH of 7, because it gives off equal numbers of hydrogen ions and hydroxyl ions when it dissociates:

$$H_2O \rightleftharpoons OH^- + H^+$$

The pH scale is logarithmic. As a result, a solution with a pH of 5 is ten times as acidic as a solution of pH 6 and 100 times as acidic as a solution of pH 7.

> In solution, acids release hydrogen ions, bases release hydroxyl ions, and salts release neither.

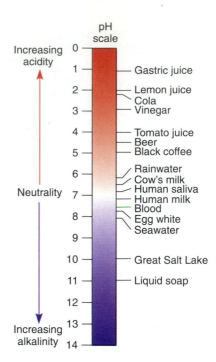

**Figure 3-7**   The pH scale, with the pH values of some substances. (The units of the pH scale are the exponents to the base ten of the hydrogen ion concentration.)

## 3-C  HOW THE EARTH GOT ITS ATMOSPHERE

The main things that make life possible on Earth, while there is no life on nearby planets, are Earth's climate and atmosphere. **Climate** describes the weather conditions in an area, including the temperature, precipitation, and winds. We shall see that energy from the sun is largely responsible for the weather. Much of the weather occurs in the **atmosphere**, the layer of gases that surrounds Earth. We shall see that living organisms determine the composition of the atmosphere.

A few million years after Earth formed, it began to stabilize, surrounded by an atmosphere in which hydrogen ($H_2$) is believed to have been the most common element (Figure 3-8). The hydrogen reacted with other elements to produce gases that contain hydrogen, such as water vapor ($H_2O$), ammonia ($NH_3$), and methane ($CH_4$). Hydrogen itself is so light a gas that it would have escaped Earth's gravity and gone off into space. Sunlight would have decomposed the ammonia into hydrogen (which would have escaped) and nitrogen gas ($N_2$), which makes up about 80% of our modern atmosphere. Methane

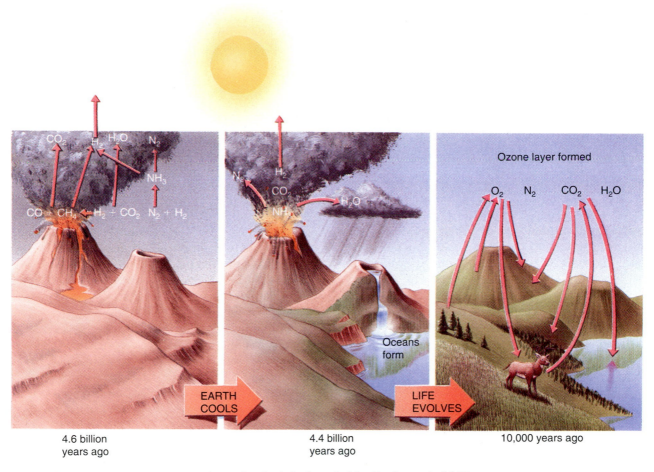

**Figure 3-8** How Earth got its atmosphere. On the left, the primitive Earth nearly 5 billion years ago. In the middle, the formation of water as Earth cooled. On the right, the changes caused by living organisms.

lost its hydrogen and reacted with oxygen to form carbon dioxide ($CO_2$).

As Earth cooled, the water vapor in its atmosphere condensed, forming the oceans about 4.4 billion years ago. Then, apparently, the conditions for life to originate existed. This was a very different form of life from today's, for there was no oxygen in the atmosphere, and few modern organisms can survive without oxygen. The origin of life has been copied in laboratory experiments. When water (to represent the early ocean) together with the gases that composed the early atmosphere is exposed to electrical discharges, representing lightning or other natural sources of energy, organic molecules are formed. (Organic molecules are formed by such processes today, but they do not last long. They are rapidly destroyed by oxygen in the atmosphere or eaten by living things.)

Life on Earth evolved, probably many times, some 4 billion years ago. And once life was there, it began

to change the world, at ever-increasing speed. The first organisms probably fed on organic molecules in the primordial soup of the early ocean. As nutrients in the soup became scarce, competition for them became fiercer, and there was strong selection for any organism that could feed in some other way. These organisms became the first **autotrophs**, organisms able to manufacture their own food from inorganic molecules.

### Origin of Green Plants

Green plants are very advanced autotrophs. They make their food by **photosynthesis**, the production of sugars using light energy, which is trapped by the green plant pigment chlorophyll. Sugars contain carbon (from carbon dioxide in the air), oxygen, and hydrogen. Green plants differ from other kinds of autotrophs in that they get the hydrogen they need for photosynthesis from water. Because there was plenty of water on Earth, autotrophs that used this system were very successful. Photosynthesis takes hydrogen from water and releases oxygen gas as a by-product. When plants started to multiply, they began to release large quantities of oxygen into the atmosphere. Virtually all the oxygen in the atmosphere today comes from plant photosynthesis.

Oxygen from photosynthesis also produced the ozone layer in the atmosphere, which prevents much of the sun's ultraviolet light from reaching Earth's surface (Section 3-E). Ultraviolet light damages living things. The earliest organisms lived in water, which absorbs ultraviolet light and protects its inhabitants, but formation of the ozone layer permitted organisms to live on land as well as in water.

The appearance of an atmosphere containing oxygen must have been a shock to other forms of life because oxygen destroys many biological molecules. But the destructiveness of oxygen was probably the selective pressure that brought about the next important evolutionary advance, respiration. Most modern organisms supply their energy needs by **aerobic respiration**, a process that breaks the chemical bonds in food to release energy and that uses oxygen.

Respiration produces carbon dioxide gas as a waste product. Life would not have lasted long if this carbon dioxide had collected in the atmosphere, but it did not. Plants use up carbon dioxide as one of the raw materials for photosynthesis, and carbon dioxide also leaves the atmosphere because it dissolves in water. The ocean acts as a "sink" for carbon dioxide, absorbing what plants do not use and converting it into various salts. We shall see in the next section that carbon dioxide is once again collecting in the atmosphere because we are burning so much fossil fuel.

> Oxygen and ozone in the atmosphere come from plant photosynthesis.

## 3-D  SOLAR ENERGY AND CLIMATE

The atmosphere is not the only part of Earth that organisms have modified. They also keep the temperature on Earth fairly stable. Life can exist only within a narrow range of temperatures. Most of the planets near Earth are closer to the sun and therefore much too hot for life. But the universe is still cooling down. If there were no life on Earth, Earth would probably be too cold for life by now. The oxygen and carbon dioxide produced by organisms, along with the water vapor in the atmosphere, keep temperatures on most of Earth's surface high enough for life.

Although the center of Earth is hot, nearly all the energy on the surface of Earth comes from the sun, not from inside Earth. The sun is a nuclear fusion reactor similar to one type of nuclear power plant considered in Chapter 9. The sun gives off several forms of energy. But the only form of energy from the sun that crosses space and reaches Earth is **electromagnetic radiation**. Three main kinds of electromagnetic radiation reach Earth: visible light, near infrared radiation, and ultraviolet radiation (Figure 3-9). Solar energy warms Earth, evaporates water that will fall as rain, and drives the winds. Earth's shape, its atmosphere, its orbit around the sun, and its rotation mean the difference between night and day, winter and summer, tropical rain forests and deserts. Let us consider some of the details.

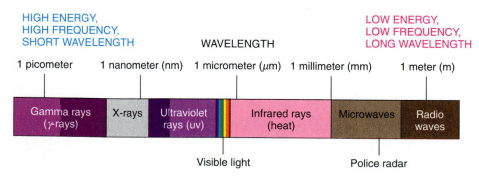

HIGH ENERGY,
HIGH FREQUENCY,
SHORT WAVELENGTH

WAVELENGTH

LOW ENERGY,
LOW FREQUENCY,
LONG WAVELENGTH

1 picometer    1 nanometer (nm)    1 micrometer (μm)    1 millimeter (mm)    1 meter (m)

Gamma rays (γ-rays)  |  X-rays  |  Ultraviolet rays (uv)  |  Infrared rays (heat)  |  Microwaves  |  Radio waves

Visible light    Police radar

**Figure 3-9**  The electromagnetic spectrum. Our eyes can detect light with wavelengths of about 400 to 700 nanometers, and so this range is called visible light. Wavelengths just shorter than visible light are called ultraviolet. Even shorter are x-rays (which overlap with ultraviolet and gamma rays). Wavelengths on the longer side of the spectrum are known as infrared, then microwaves, then radio waves.

> Solar energy warms Earth, evaporates water that will fall as rain, and drives the winds.

### Air and Temperature

The layer of the atmosphere that lies closest to Earth contains the air we breathe and is called the **troposphere** (shown in Figure 3-12). The atmosphere is held in place by the pull of Earth's gravity, which is strongest toward the center of Earth. As a result, air is denser near Earth's surface than it is at greater heights. The sun causes weather mainly by the way it heats up gas molecules in the air. Where there is less air, high in the troposphere, the weather is cold, even though the top of a mountain is closer to the sun than the surface of Earth is.

Two properties of air are particularly important in discussions of weather:

1. Cold air sinks and hot air rises.
2. Warm air can hold more water vapor than cold air. This means that if warm air is cooled, the water vapor gas it contains condenses to form liquid water, which will fall to Earth as rain, dew, or snow.

The sun heats the atmosphere largely by warming Earth, which then heats up the air. Much incoming solar radiation never reaches Earth. It is reflected back into space, largely by Earth's **albedo**, its shininess. Light, shiny surfaces reflect radiation, and Earth has huge light-colored surfaces in the form of clouds, deserts, and ice caps. Another 10% of solar energy never reaches Earth because it is absorbed by gases in the atmosphere. Less than half of incoming solar energy finally hits Earth's surface. Here, it can bounce back to heat the atmosphere in three general ways: conduction, radiation, and heat transfer. In conduction, heat passes from Earth to the air by direct contact. During radiation, heat from Earth increases the kinetic energy of gases in the air. Heat transfer occurs when warming of the surface causes water to evaporate. When this water later condenses, it transfers heat directly to the atmosphere.

The temperature of the ground and of the air varies from time to time and from one place to another. These variations depend largely on Earth's rotation around its axis and its orbit around the sun. These movements mean that different parts of Earth are exposed to different amounts of solar radiation at different times of year. Near the equator, the sun's rays strike almost vertically throughout the year, giving these tropical areas fairly steady, high temperatures with little seasonal difference in day length and

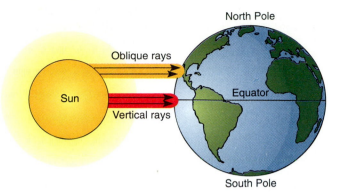

**Figure 3-10**   A beam of sunshine striking Earth away from the equator is spread over a wide area. It is, therefore, less intense at any one point than a similar beam near the equator, which strikes Earth vertically.

temperature (Figure 3-10). Outside the tropics, the sun's rays strike Earth obliquely. Nontropical areas receive different amounts of solar energy at different times of the year, and this produces the temperature differences that mark the seasons.

The atmosphere heats unevenly. Hot air rises and cold air falls, moving to fill the spaces left by the hot air. Uneven heating of the atmosphere produces the winds. The prevailing winds on Earth and their effects on local climate are discussed in more detail in Close-Up 3-2.

### Precipitation

In addition to temperature, the other important component of climate is moisture, and this also depends on energy from the sun. Air heated at the equator evaporates water from the forests and oceans and rises, cooling as it goes. As it cools, it releases some of its moisture. The result is the steamy rains of tropical jungles. The air moves on, high in the atmosphere, both north and south from the equator, until eventually it sinks to Earth again. The descent of this dry air creates the world's great deserts. Still farther north and south, in the **temperate zones** that include most of the United States and Europe, swirling winds pull masses of air, sometimes from warm tropical areas, sometimes from frigid polar regions, giving us varied weather patterns.

Mountain ranges may also affect precipitation, producing areas called **rain shadows**, where there is little precipitation on their leeward sides, the sides away from the prevailing wind. As air rises up the side of a mountain it expands and cools, and much of its moisture condenses as rain or snow (Figure 3-11).

# Wind and Weather

Air is warmest at the equator, where it rises, leaving low pressure areas with little wind, called the "doldrums." As it moves north and south from the equator, this air is cooled until much of it falls to the ground again at about latitudes 30°N and 30°S. As it falls, air contracts, forming regions of high pressure. One of these high pressure zones is usually present in the Pacific off the coast of California and causes the dry, stable climate of that state. Some of the descending air is forced back toward the equator as it hits Earth's surface, creating the steady trade winds and completing the tropical circulation cell (Figure 3-B). At the poles, air falls and rises in similar cells, and the wind blows fairly steadily from the east as a result of Earth's rotation.

Between the tropical and polar cells, the winds are more variable. At Earth's surface, some of the tropical air that descends is deflected toward the

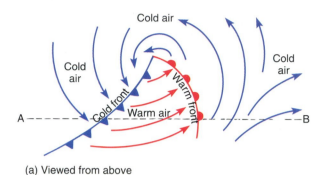

(a) Viewed from above

(b) Cross section along line A–B

**Figure 3-C** Cold and warm fronts. (a) Frontal waves start as irregularities between warm and cold air masses. A counterclockwise circulation develops around the point where the two masses join (because of Earth's rotation). (b) The cold front moves faster than the warm front, pushing the warm air up away from the ground, where its water vapor usually condenses to form precipitation.

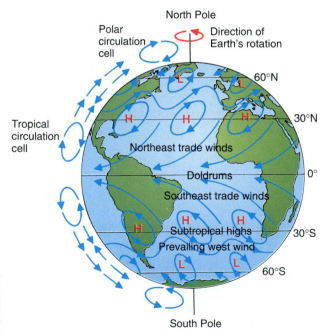

**Figure 3-B** Circulation of air within the lower atmosphere to create the prevailing winds. H (*red*) indicates an area of high pressure and L (*red*) an area of low pressure.

pole. As it travels north, it absorbs water from the ocean. Eventually it meets cold dry air from the pole blowing south. The area where these two bodies of air meet is called the polar front. The behavior of this front dominates the weather over most of Europe and North America. When the cold polar air pushes under the warm moist air, a cold front moves in. When the warm air pushes the polar air back and rises over it, a warm front forms. In both cases, precipitation usually occurs (Figure 3-C).

The position of the polar front is strongly influenced by the position of the jet stream, which blows between the polar cells and the tropical cell high in the atmosphere. Here, the prevailing westerly wind blows at up to 320 kilometers an hour. In the Northern Hemisphere, the jet stream tends to move south in winter and north in summer, affecting the position of the air masses beneath it.

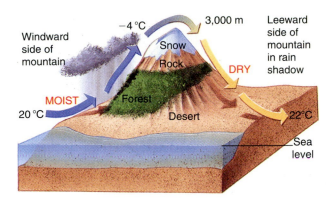

**Figure 3-11**  The way in which a rain shadow forms on the leeward (downwind) side of a mountain.

Coming down the leeward side of the mountain, the air warms up so that it can hold more water vapor. Rain shadows from prevailing west winds are the main reason for the large areas of deserts and grasslands found on the eastern side of the Sierra Nevada and the Cascade and Rocky mountains.

## 3-E  HUMAN IMPACT ON CLIMATE

Since the dawn of human society, people have tried to change the weather. Rain dances and prayers for rain were important ceremonies in ancient civilizations. Here, however, we are concerned with the ways human activities affect the weather unintentionally.

On a local level, cutting down a forest reduces rainfall. Plants hold large quantities of water that they have taken up from the soil. Trees hold more water than crop plants because they are larger. This water evaporates from the plants into the air and falls to Earth as rain. For instance, tropical forest in Brazil supplies the water for rainfall in agricultural areas farther south. As the forest is cut down, leaving few plants, rain that falls on the soil does not stay in the area but drains immediately into the nearest river. Less rain now falls on the area downwind.

Towns are hotter than the surrounding countryside because they lie in **urban heat bubbles** formed by the energy used in the town for heating and lighting as it rises into the air. In windy areas, the heat given off by a town is soon blown away, but when there is no wind, the heat forms an invisible bubble in the air over the town. Clouds tend to bounce sideways off urban heat bubbles so that they travel around the town instead of over it, and the town gets less rain than the surrounding countryside. In Phoenix, Arizona, the temperature in the town has in-

creased steadily and the rainfall has decreased during the last two decades.

International concern today centers on two important changes in the atmosphere that affect not just local areas but the whole world: ozone depletion and the greenhouse effect.

### The Ozone Layer

The ozone layer in the atmosphere reduces the amount of ultraviolet radiation from the sun that reaches Earth. We shall see in Chapter 20 that air pollutants rising into the atmosphere are destroying the ozone layer, and this permits more ultraviolet radiation to reach Earth. Ultraviolet radiation damages genetic material, causing skin cancer in humans and actually killing some smaller organisms, such as tiny plants that float near the surface of the sea and produce much of our oxygen.

Ozone molecules contain three atoms of oxygen ($O_3$) instead of the two found in ordinary oxygen gas ($O_2$). When oxygen gas is irradiated with intense ultraviolet light, small amounts of ozone form. Little ozone forms in the troposphere because there is not enough ultraviolet radiation. Above the troposphere, however, radiation from the sun is sufficiently intense that ozone forms in the stratosphere (Figure 3-12). Once formed, the ozone itself is a good ab-

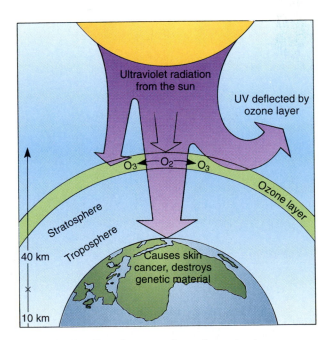

**Figure 3-12**  How the ozone layer forms in the atmosphere.

sorber of high-energy ultraviolet radiation, a process that helps to prevent this radiation from reaching Earth.

> Pollution is destroying the ozone layer in the atmosphere, and this permits increasing amounts of ultraviolet radiation to reach Earth.

## The Greenhouse Effect

Imagine that by 2030 our summers are much hotter than they are today. Imagine that the homes of millions of people living on the East Coast of the United States are under water. Imagine four months without rain every summer in the farm belt. Imagine the Rocky Mountains without snow. These projections come from the finding that human activities are heating up Earth's atmosphere.

Nitrogen and oxygen make up about 99% of clean, dry air in the troposphere. Dust, smoke, pollen, salt, ash, and organic molecules are present in varying amounts. In addition, air contains up to 2% (20,000 parts per million by volume) of water vapor, about 335 parts per million (ppm) of carbon diox-

ide, 1.6 ppm of methane, and several other gases in even smaller amounts. These tiny traces of water and other gases trap heat within the atmosphere, preventing it from escaping into space. This "greenhouse" of gases warms Earth's surface from −18 °C to habitable temperatures (Figure 3-13). Without the greenhouse effect, Earth would be too cold for life. But human activities are increasing the level of greenhouse gases in the atmosphere. As a result, the atmosphere is warming up. We intensify the greenhouse effect by increasing the rates at which gases that cause the greenhouse effect enter the atmosphere or by reducing the rates at which they leave.

## Methane

After remaining stable for thousands of years, the concentration of methane ($CH_4$) in the troposphere has doubled in the past 250 years. Burning biomass produces some methane, but most of it is produced by **methanogens**, bacteria that produce methane as a by-product of the reaction they use to obtain energy. Methanogens can live only in environments that contain no oxygen, such as the bottoms of wetlands, sewage-treatment ponds, rice paddies, garbage dumps, and the digestive tracts of livestock.

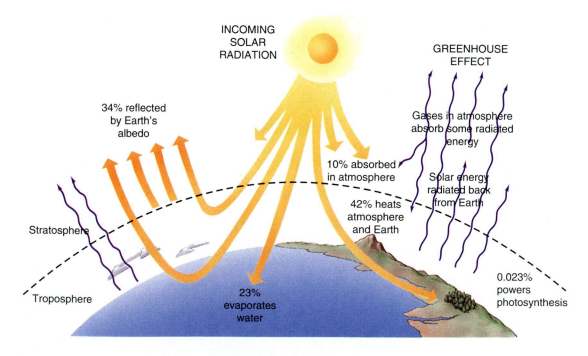

**Figure 3-13**   The fate of solar energy reaching Earth's atmosphere from the sun and the greenhouse effect. The greenhouse effect is caused by dust and gases in the atmosphere that stop energy being radiated back into space from Earth. The more concentrated the greenhouse gases, the greater the greenhouse effect.

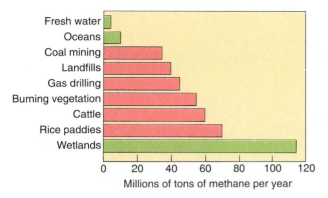

**Figure 3-14**   The sources of methane released into the atmosphere each year. Sources that are largely natural are shown in green, those created mainly by human activity in pink. *(Source: Cicerone and Oremland, Biogeochemical Sources of Atmospheric Methane, Collins 1991.)*

Methane is removed from the atmosphere in several ways. One of these is useful: methane can react with a hydroxyl radical (OH) in the troposphere to produce ozone. Methane also leaves the air when it is broken down by various bacteria in the soil. Although the methane content of the atmosphere is very small, methane makes a relatively large contribution to the greenhouse effect because each molecule of methane absorbs infrared radiation 20 times more effectively than a molecule of carbon dioxide. The main reason that there is more methane in the air nowadays is that methane from agricultural sources is increasing rapidly as we struggle to feed our exploding population (Figure 3-14).

### Carbon Dioxide

The biggest villain of Earth's increasing temperature is carbon dioxide ($CO_2$). The amount of carbon dioxide in the atmosphere at any one time depends on the balance between reactions that release carbon dioxide and those that use it up. Nearly all the carbon dioxide that is removed from the atmosphere dissolves in the oceans or is used up by plants during photosynthesis.

The main things that release carbon dioxide into the air are fires and respiration. (We release carbon dioxide when we breathe.) Forest fires, grass fires, cooking fires, and burning fossil fuels all produce carbon dioxide.

Occasionally, geological disturbances release carbon dioxide. In 1986, a huge cloud of carbon dioxide was released from Lake Nyos in Cameroon. We do not know how the gas originated deep underground, but it seeped to the surface and dissolved in surface water until the lake was a saturated solution of carbon dioxide. Some disturbance to the lake then released a cloud of the gas from its surface. Seventeen hundred people died from asphyxiation. Carbon dioxide is not poisonous. It kills people because it pushes aside air and people suffocate for lack of oxygen. Let us hope that we are not putting so much carbon dioxide into the air that other bodies of water will become saturated with it.

## 3-F  CLIMATE, TOPOGRAPHY, AND POLLUTION

The cities that suffer worst from air pollution nearly all lie in valleys where it is not often windy. The climate is very important in determining how air pollution affects an area. Even places such as Mexico City and Los Angeles do not suffer all the time from dangerously polluted air. The degree of pollution varies with the weather.

During the day, the sun heats the ground and the air near it. This warm air rises through the cooler air above, carrying smoke from factories and other pollutants upward into the atmosphere. If nothing interferes, the air may rise more than 15 kilometers from Earth. Frequently, however, a temperature inversion interferes. A **temperature inversion** exists if there is warm air some distance above the ground. When air rising from the ground meets the warmer air of the inversion layer, it stops rising and is trapped under the inversion (Figure 3-15).

A temperature inversion is often set up at night. When the sun sets, the ground and the air near it cool faster than the air higher up, creating an inversion. This type of inversion usually disappears when the sun warms the ground in the morning, but it often persists for days in cold, cloudy weather.

Another kind of inversion occurs when a high-pressure weather system compresses the upper air, thereby heating it. Inversions caused by stationary high-pressure systems are common in late summer and winter in northern latitudes. On the West Coast of the United States, they are common throughout the year. Any city in an area where inversions last for days on end invariably has a serious problem with air pollution.

The topography, or surface structure, of the land interacts with inversions to increase or decrease pollution. An inversion prevents pollutants from dispersing upward from a city. If the city also lies in a

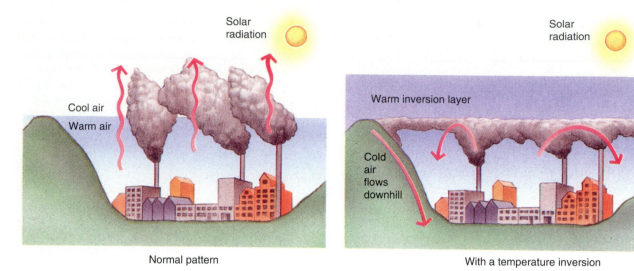

**Figure 3-15**   The effect of a temperature inversion on air quality near the ground. The inversion intensifies pollution by trapping cool, dirty air near the ground.

basin-like valley (as does Missoula, Montana) or in the bowl of an old volcano (as does Mexico City), the pollutants cannot escape sideways either. Surrounding mountains can even help to create inversions. This is because heat radiates from the tops of mountains faster than from valleys. As the top of a mountain cools, the cold air flows down, filling the valley and creating an inversion.

## How Pollutants Travel Around the World

Most pollutants remain in the air within 1 kilometer of the ground. They eventually fall to Earth again in rain or snow. However, some move beyond the local atmosphere into the upper troposphere, where they are preserved by the low temperature and absence of rain. They will be dispersed around the world by the high winds at these altitudes. For instance, part of the radioactive cloud from the 1986 nuclear accident at Chernobyl and the clouds of ash and dust produced when volcanoes erupted in the 1980s were detected thousands of miles from their source. The eruption of Mount Pinatubo in the Philippines produced a dust cloud that turned the sunsets bright red for weeks at my home in the southeastern United States, more than 15,000 kilometers away. Radiation from Chernobyl was detected in North America, 11,000 kilometers away. Some rising pollutants even reach the stratosphere, where they may contribute to destruction of the ozone layer.

> Most pollutants remain in the air within 1 kilometer of the ground and fall to Earth again in rain or snow.

Pollutants generally move upward very slowly. They take months to reach the upper atmosphere so that most of them are removed by rain or snow before they get there. Some people have argued that this meant that changes such as ozone depletion could not be due to pollutants. Scientists were also puzzled that air pollutants known to have originated in one place sometimes end up at the other side of the world within a few days. The Chernobyl cloud was detected in Sweden within 48 hours.

We now know that thunderstorms create a fast route that can carry pollutants to the upper atmosphere in a matter of hours. Scientists have studied how thunderstorms act using radar, satellite data, and weather balloons. Finally, they flew a research airplane in and out of a severe thunderstorm over Oklahoma, sampling gases as they went. These studies confirmed that rapid updrafts through thunderstorm clouds can carry gases and particles from near the ground to the stratosphere at enormous speeds. In the upper atmosphere, high winds, such as the jet stream, whirl pollutants around the world. Eventually, the downdraft that may also accompany a thunderstorm may suck the pollutants back down into the lower atmosphere, depositing them at the other side of the world.

# LIFESTYLE 3-1

## Energy Conservation

One way to reduce the cost of energy is to increase the efficiency with which we use it. The energy efficiency of a device is the percentage of the energy put into the device that does useful work. For instance, the human body converts about 20% of the energy it takes in into the useful work of muscle movement and metabolism. Internal combustion engines and incandescent lights are very inefficient, converting only 5% to 10% of the energy that reaches them into light or movement. We can save ourselves money by buying household appliances and heating systems with high energy efficiencies.

An efficient appliance is often more expensive to purchase than a less efficient device, but it saves money in the long run with lower **lifetime cost**, the total cost of the useful work performed during the lifetime of the device. Having enough money to buy an efficient appliance is one of the many ways that "the rich get richer."

Consider the efficiencies of heating a house by various means. The most efficient method involves building the house in the right position and planting trees to protect it from chilling winds, insulating the house, and heating it by the effect of the sun on south-facing windows. The body heat of the people in the house and waste heat from appliances supply part of the heat. The only energy input involved to operate the system is the human muscle power used to pull blinds over the windows to prevent heat loss or gain, or to open the windows when necessary. Such a system is 90% efficient (90% of the energy that reaches the house heats the living space) and can heat a properly built house even in winters where temperatures drop to −30 °C—as long as it is in an area that gets plenty of winter sunshine. At the other end of the scale, consider heating a house with electricity from a nuclear power station. The efficiency is a mere 15%: only 15% of the energy in the uranium used to fuel the power station ends up as heat in the house. In between cheap passive solar heating and expensive electricity as ways to heat a building come various stoves and furnaces. The most efficient of these are new natural gas furnaces with remarkable 95% efficiencies. The combination of such a furnace with good insula-

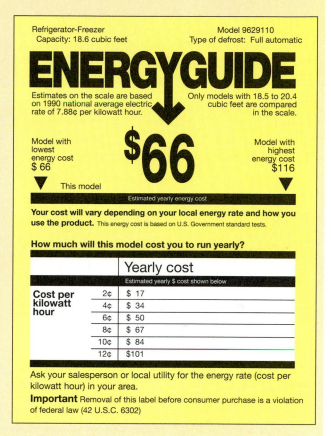

**Figure 3-D** An Energy Guide label.

tion and passive solar heat can be used to heat a house very cheaply.

By law, many appliances sold in the United States must bear labels disclosing their energy efficiencies or annual energy costs on yellow Energy Guide labels attached to new appliances (Figure 3-D). Such a label must be attached to every new boiler, clothes washer, dishwasher, heat pump, refrigerator, refrigerator/freezer, room air conditioner, and water heater. The label compares the energy efficiency of the appliance with that of the most efficient version available. In the figure, this refrigerator-freezer uses about $66-worth of energy a year (depending on the local cost of electricity) compared with $66 for the most efficient available model and $116 for the least efficient. So this appliance is very efficient.

# SUMMARY

Energy is the resource that affects our lives most directly. We use it, for example, for heat, cooking, food production, transport, and industry. Abundant energy can provide substitutes for other resources and solve some environmental problems.

### The Laws of Thermodynamics

Energy can neither be created nor destroyed. It can be transformed from one form to another. Some useful energy is lost each time energy is transformed.

### Elements and Molecules

Much of the energy we use is stored as the bonds between elements in organic molecules. Atoms are made up of nuclei containing protons and neutrons, and of electrons. When atoms bond, they share, lose, or gain electrons. The electrical charges of ions are used in pollution control devices to precipitate pollutants. In solution, acids release hydrogen ions, bases release hydroxyl ions, and salts release neither.

### How the Earth Got Its Atmosphere

Living things on early Earth changed the composition of the atmosphere by removing carbon dioxide and releasing oxygen. An atmosphere containing oxygen made possible the evolution of modern organisms.

### Solar Energy and Climate

Solar energy warms the ground (which heats the atmosphere), evaporates water to cause rainfall, and drives the winds. Variations in the amount of solar energy that reaches Earth at different times of the year and in different places cause the seasons and determine the climate in different places.

### Human Impact on Climate

The ozone layer in the atmosphere, caused by the effect of ultraviolet radiation on oxygen molecules, protects living things from ultraviolet radiation, which destroys the genetic material. We are destroying the ozone layer.

As a result of human activities that intensify the greenhouse effect, the atmosphere is warming up more rapidly than it ever has before. We intensify the greenhouse effect by increasing the rates at which gases that cause the greenhouse effect enter the atmosphere and reducing the rates at which they leave. Anthropogenic global warming will probably disrupt human lives in disastrous ways within the next 100 years.

# DISCUSSION QUESTIONS

1. Earth has been through several cycles of warming and cooling in the last 2 million years—during the Ice Ages. List as many reasons as you can think of why the global warming we are now experiencing will do much more damage to the environment than did the Ice Ages.

2. During the Ice Ages, some populations of animals and plants became extinct. Others moved north and south so that they always lived within areas with climates and vegetation to which they were adapted. Why do biologists fear that these slow migrations, which saved many populations in the past, will not have the same result as Earth warms up this time?

3. Do you think it likely that within our lifetimes we shall significantly slow depletion of the ozone layer or warming of Earth? If not, why not? If not, what long-term consequences do you predict?

4. Are there any possible technological solutions to the problems of global warming and ozone depletion?

# BIODIVERSITY AND HABITATS

## Key Concepts

- The biosphere contains a huge variety of organisms that can all be classified as bacteria, protists, fungi, plants, or animals.

- Microorganisms (bacteria, protists, and fungi) are the food of many other organisms, act as decomposers, and cause many diseases of plants and animals.

- The main groups of land organisms are arthropods, vertebrates, gymnosperms, and flowering plants, which have adaptations to support the body and prevent them drying out so that they can survive out of water.

- All organisms live in only a few different types of community. Similar communities are found in many parts of the world.

- On land, the biome in a region is determined largely by the pattern of precipitation and temperature.

- Habitat destruction has exterminated thousands of species, threatening us with the loss of fertile soil, food plants, and new products.

*A staghorn fern and other epiphytes in a tropical rain forest.*

Population pressure is destroying many things that we can live without if we have to. Beaches, seafood, and maple syrup make life more pleasurable, but we can survive without them. There are thousands of species of organisms, however, without which we should rapidly die. These include the plants that we eat and the organisms that create the soil plants need. Because so many species are vital to human survival, biologists are deeply concerned about the loss of **biodiversity**, the variety of organisms on Earth. Close-Up 4-1 describes an international treaty designed to preserve biodiversity.

Are there 10 million species of organisms on Earth? Or 100 million? We do not know, but biologists estimate that almost 1 million species will become extinct during the twentieth century, mainly because humans have destroyed their habitats (Facts and Figures 4-1). Population experts project that the number of people on Earth will stop growing toward the end of the twenty-first century. When population growth stops, we shall probably be able to save most of the remaining habitats and species on Earth from destruction. The question is whether sufficient biodiversity will remain to prevent widespread human deaths. We shall explore the ways in which overpopulation is destroying species of organisms in Chapter 16. Here we consider what those organisms are, why we need them, and where they live. Close-Up 4-2 describes how scientists name and classify organisms.

Living things need a moderate temperature, water, a source of energy, and various chemical nutrients found in the soil, water, and air. One hundred kilometers beneath our feet, Earth is white hot. Thirty kilometers above our heads, the air is too thin and cold for organisms to survive. Suitable combi-

## FACTS AND FIGURES 4-1

### Species

About 1.4 million species have been named and described. About 750,000 of these are insects, 41,000 are vertebrates, and 270,000 are plants.

Although rain forests cover only 7% of Earth's surface, they house more than half of all species on Earth, and only about 1% of these species have been studied. The Rainforest Action Network estimates that between 4 and 15 species become extinct each day in the world's rain forests.

nations of the things organisms need are found only in a narrow layer that forms a rough sphere around the surface of Earth. This layer is called the **biosphere** because it is, as far as we know, the only place where life can exist. The biosphere extends about 8 kilometers up into the atmosphere (where insects, bacteria, and plant seeds may be found) and as much as 8 kilometers down into the depths of the ocean.

Everywhere we look around us we see life in incredible variety. The pond teems with plants so small that it takes a microscope to see them. We cannot see the forest for trees so large their tops touch the sky, and all around us, animals crawl, fly, burrow, gallop, and wriggle. How can we make sense of this bewildering diversity? Since prehistoric times, people have named and classified the types of animals and plants in many different ways. In this book organisms are divided into five major kingdoms: bacteria, protists, fungi, plants, and animals (Table 4-1). (Viruses

*text continues p. 65*

| TABLE 4-1 | The Kingdoms of Organisms | |
|---|---|---|
| **Kingdom (Biological name)** | **Characteristics** | **Examples** |
| Bacteria (Monera) | One-celled organisms without nuclei | *Streptococcus*, blue-green bacteria |
| Protists (Protista) | Organisms with nuclear membranes; most with only one cell living in water. | Diatoms, dinoflagellates, ameba, trypanosomes, paramecium, *Plasmodium* |
| Fungi (Fungi) | Saprobes, which absorb their food, with cell walls; most living on land | Yeasts, mushrooms, toadstools, bracket fungi, molds, mildews, rusts, *Chlamydia* |
| Plants (Plantae) | Many-celled, photosynthetic organisms with cell walls; most living on land | Ferns, mosses, horsetails, duckweed, trees, flowers |
| Animals (Animalia) | Many-celled organisms that ingest their food; without cell walls; living in water and on land | Corals, sponges, worms, insects, fish, reptiles, birds, cattle, rodents, humans |

# The Biodiversity Treaty

Eli Lilly, a drug company, has made hundreds of millions of dollars from the rosy periwinkle, a plant from Madagascar, an island nation off the east coast of Africa (Figure 4-A). In 1954, an Eli Lilly chemist extracted vinblastine and vincristine from the rosy periwinkle. These proved to be powerful cancer-fighting drugs. They have saved many lives, including those of children who used to die from childhood leukemia, which can now usually be cured by these drugs.

Shall we find hundreds more valuable drugs in tropical forests? Certainly not in Madagascar, which is one of the poorest countries in the world. Most of its natural habitats have been destroyed for farmland and fuel. Anyway, why should Madagascar save its forests so that Western drug companies can get rich? Madagascar never saw one penny from the sale of vincristine and vinblastine.

This is the situation that the Convention on Biological Diversity is designed to prevent. The Convention, commonly known as the biodiversity treaty, was signed at the United Nations Conference on Environment and Development in Rio de Janeiro in 1992. It provides that developing countries shall be rewarded, with money and technology, for preserving habitats in which valuable plants may be discovered.

This sounds like a good idea, so why did the United States bring international scorn on itself by refusing to sign the treaty in 1992. No one denies that countries should be compensated for products developed from their species, perhaps by paying the country a small royalty on each product sold. But U.S. biotechnology firms do not want countries to have automatic rights to new drugs or foods developed from their wealth of plant and animal species.

Most observers, including more than 100 countries that signed the treaty, think such fears are silly. They say that the treaty is vaguely worded so that details can be worked out later. The important thing is to sign the treaty so that developing countries see it as worthwhile to preserve their vanishing habitats. In 1993 the United States declared that it intended to sign the treaty with an accompanying note of interpretation designed to make the treaty more acceptable to biotechnology companies. It probably makes little difference whether the United States signs or not. From now on, developing countries will insist that any company exploring for new plants compensate the country involved.

**Figure 4-A** The rosy periwinkle, or vinca, *Catharanthus roseus*, originally from the island of Madagascar and now often planted in gardens.

# CLOSE-UP 4-2

## Naming and Classifying Organisms

We need internationally recognized names for organisms because the common names of organisms are not precise. Different people use different names for the same animal or plant, and conversely, the same name may be used for different organisms in different places.

Scientific names are Latin (often with Greek roots) because, for hundreds of years, Latin was the language of Western scholars. Many Latin names are full of information and folklore. A gardener finds it useful to know that *Gypsophila* (baby's breath) means "to love lime," a reference to this plant's need for lime in the soil. *Hepatica*, a spring flower whose name means "liver," was believed to cure diseases of the liver because its leaf is liver shaped. You can imagine how *Ilex vomitoria* affects the human body (Figure 4-B).

We need some sort of classification scheme to make sense of the diversity of organisms around us. The basic unit by which organisms are classified is the species, a group of organisms that resemble one another in appearance, behavior, chemistry, and, ultimately, the genes they contain. Each species is then assigned to a genus, which may contain other, similar species. Each genus is placed in a family, each family in an order, and so forth. Each successively higher group generally includes a larger number of more distantly related species. So that you can see how this works, here is the classification for one animal and one plant (Figure 4-C):

| Taxon | French Marigold | American Alligator |
|-------|-----------------|--------------------|
| Kingdom | Plantae | Animalia |
| Phylum | Tracheophyta | Chordata |
| Class | Angiospermae | Reptilia |
| Order | Asterales | Crocodilia |
| Family | Asteraceae | Crocodilidae |
| Genus | *Tagetes* | *Alligator* |
| Species | *patula* | *mississippiensis* |

Species and genus names are conventionally italicized or underlined: the human species is *Homo sapiens* or <u>Homo sapiens</u>. When you name a particular species, you must give both the genus and the species: *Tagetes patula*. This is because the species name is often trivial (*patula* means "spread out"), and many organisms have the same species name.

(a)

(b)

**Figure 4-C**  (a) *Alligator mississipiensis*, an American alligator. (b) *Tagetes patula*, a French marigold. (You might expect a French marigold to come from France, but this is actually a native of Central America.)

**Figure 4-B**  *Ilex vomitoria*, yaupon holly.

are not considered living organisms because they do not contain the structures necessary to reproduce themselves. They can multiply only by using the machinery of a living cell.) The first organisms were probably tiny, relatively simple creatures living in the sea. Larger, more complex organisms evolved later. So we start with a visit to the sea.

## 4-A PROTISTS AND ALGAE

Most life in the sea is found on the **continental shelves**, areas of shallow water that surround continents. Continental shelves constantly receive nutrients that wash off the land into the sea and support large populations of organisms (Figure 4-1). If we tow a net of very fine mesh through the sea, it will trap marine **plankton**, floating organisms so small that we can see them only using a microscope. As we examine plankton, we notice that many are green or yellow. These are **algae**, simple plants and protists that make their own food by photosynthesis, using light from the sun. **Protists** are organisms made up of only one cell. They include diatoms with shells like glass and dinoflagellates, protists with flagella, whip-

like appendages that whirl about and propel the dinoflagellate through the water (Figure 4-2). Diatoms and dinoflagellates contributed to our deposits of petroleum because these protists store excess food as oil. Over millions of years, the bodies of dead plankton drifted to the sea floor and were later pressed down by the weight of mud and silt settling over them. The oil accumulated and underwent chemical changes that turned it into petroleum, or crude oil. Geologists searching for oil examine rock samples for the fossils of protists to determine where petroleum deposits are likely to be found.

A plankton net will also collect a great many protists that cannot make their own food by photosynthesis but live by eating other organisms. Some have flagella. Some are covered with shorter projections (cilia), which beat like oars and move the cell through the water. Many protists make shells of protein, silica, or calcium carbonate (lime). These shells sometimes accumulate in great numbers to form limestone rocks: the White Cliffs of Dover in England are chalky deposits of plankton shells. Many of these marine protists have relatives that live in fresh water or as parasites in animals, where they often cause diseases.

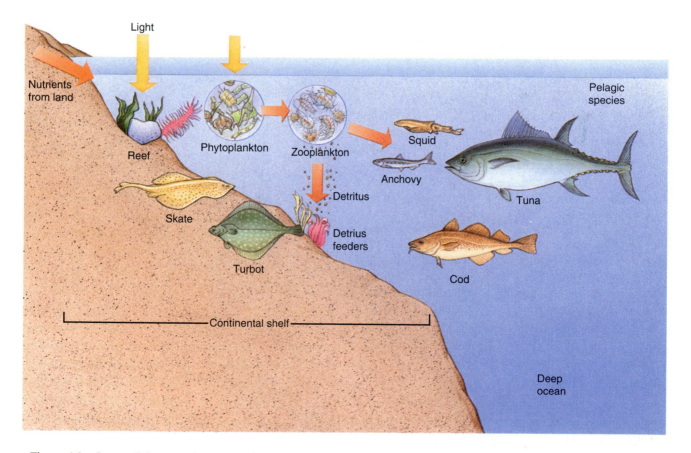

**Figure 4-1**   Some of the organisms found in the ocean on a continental shelf.

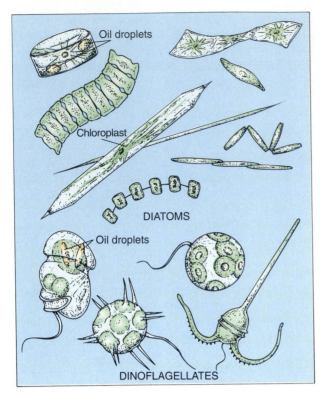

**Figure 4-2** Some protists floating in the plankton. These organisms are tiny and transparent, diatoms at the top and dinoflagellates at the bottom. They all contain green chloroplasts, where they make their own food by photosynthesis. Some of the food is stored as drops of oil.

Some protists with flagella, called **trypanosomes**, are parasites in animals, causing high fever and sometimes death. Trypanosomes cause sleeping sickness (Africa) and Chagas's disease (South and Central America). Sleeping sickness is transmitted by bites of the tsetse fly, where the trypanosome lives for part of its life. Chagas's disease affects about 12 million people. It, too, is transmitted by bloodsucking insects, which inhabit cracks in the mud and adobe walls of houses, crawling out at night to feed on people, dogs, and cats. Leishmaniasis is caused by parasitic flagellates transmitted by sand flies. This disease afflicts millions of people in Africa and Asia. Symptoms range from skin sores to death. Trypanosomes also cause nagana, a disease of cattle that makes cattle farming impossible in parts of Africa. The most destructive human disease caused by a protist is malaria, which kills millions of people every year (Section 8-F).

> Protists cause some of the world's most destructive diseases in humans, such as malaria, trypanosome infections, and leishmaniasis.

Some members of the plankton have more than one cell. These include animals of many kinds, called the **zooplankton** and floating plants. The zooplankton include the young stages of various sea animals, their **larvae** (singular: **larva**). The larvae float in the plankton and spread the species around the world to wherever the water takes them. **Phytoplankton**, the photosynthetic plankton, are the ones most important to human life because they use up carbon dioxide and produce oxygen. Plankton make up the main food for fish, which provide more than 10% of the protein in the human diet.

When the tide goes out, it leaves pools and fingers of water among rocks and sand banks. Here, and in shallow water around the coast, live large algae, the seaweeds, sea grasses, and kelps, all ancient members of the plant kingdom. These plants need light for photosynthesis, but water absorbs light and so the ocean depths are dark. Most large plants in the sea live near shore, especially where they can attach to sunlit rocks.

Algae are used as food for humans and livestock. Alginate, an extract from kelps, is used in about half the ice cream produced in the United States, giving it a smooth texture and preventing the formation of ice crystals. The alga known as Irish moss produces carrageenan, an ingredient in puddings, candies, and ice cream.

## 4-B  SEA CREATURES WITHOUT BACKBONES

As we look at the animals of the ocean, we find that the simplest ones are small, with every cell close to the water that provides food and removes body wastes. As animals evolved, many became larger, and the inner cells were farther and farther from the environment. Animals that evolved means of circulating food and oxygen to their inner cells and of removing wastes to the outside were able to increase in size, whereas those without such systems remained small.

As we examine various animals, we shall see that the members of any group of organisms show adaptations to a particular way of life. Fish, for instance, are adapted to life in the water, birds to life in the air. Members of every group, however, have evolved adaptations that suit them to ways of life different from those of their ancestors. Thus, the ancestors of protists undoubtedly lived in the sea, but other members of the group have evolved adaptations to life in fresh water, as parasites in animals, and in damp places on land.

Animals without backbones are called **invertebrates**, and the many groups of invertebrates cover a wide range of size and complexity. **Sponges** are the simplest of the multicellular animals, and they move so little that the ancient Greeks thought they were plants. But unlike most plants, they have flagella on some of their cells and produce tiny, mobile larvae. The "natural" sponges sometimes used for bathing are actually sponge skeletons.

Swimming in the sea carries with it the hazard of jellyfish stings (Figure 4-3). Jellyfish, corals, and sea anemones are common members of the **Cnidaria**.

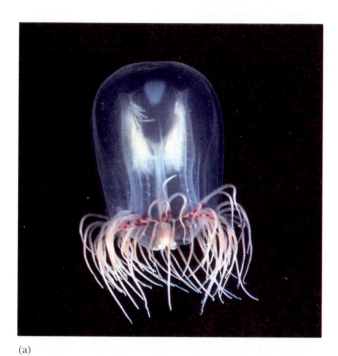

(a)

(b)

**Figure 4-3**  Invertebrates of the ocean. (a) A transparent jellyfish floats in the plankton, its tentacles ready to capture for food any small organism it encounters. (b) A colorful worm, related to the garden earthworm, crawls across a coral reef. (a, *Langdon Quetin*, b, *Steven Webster*)

(Don't pronounce the "C.") Cnidarians are simple carnivores (meat eaters), and the tentacles around the mouth are armed with small organs that sting prey with a poison that paralyzes it or entangles it in sticky threads.

The world is full of worms, in the ocean and on land. We find nematode worms and earthworms in garden soil and worms crawling through tide pools and on coral reefs in the ocean (see Figure 4-3). Worms differ from cnidarians in having definite left and right sides, top and bottom, and head and tail ends. This permits a streamlined shape and efficient movement. In addition, the nerve cells and sense organs (organs that respond to things such as light, chemicals, and sound) are concentrated in the head end, allowing the animal to sample an area before it moves its body there. The segmented worms, including earthworms and leeches, are the first animals we have encountered with bloodstreams, with pumping hearts to move the blood round the body. Many worms are disease-causing parasites in humans and other animals, and many others help to create soil.

The seashore abounds with molluscs—snails, limpets, scallops, clams, mussels, and their relatives. A mollusc has a soft body with a muscular foot that it uses to move. Most also have shells, which are protective outer skeletons, although some molluscs, such as slugs and octopuses, have no shells. In clams, oysters, and mussels, the shell is double and hinged. In snails and limpets, it is single.

Arthropods, which include the insects, are the most successful group of animals today, with the greatest number of individuals and of different species. Crabs, shrimps, and lobsters are marine **arthropods**—animals with jointed external skeletons. An arthropod's armor plating, reminiscent of a knight's armor, is fitted together so that it bends at the joints (Figure 4-4).

Sea urchins, sea stars, sand dollars, and sea cucumbers are all **echinoderms**. Sea urchins look like brittle pincushions and must be turned over carefully, for they are protected by sharp spines that can pierce fingers and break off, leaving splinters. On the underside is the sea urchin's mouth, with large, tough teeth for scraping algae off the rocks. Most echinoderms have curious tube feet with suction-cup tips, used for moving around and, in the case of sea stars, for prying open mussels and clams for food. Despite their peculiarities, study of their embryonic development shows that echinoderms are more closely related to vertebrates—animals with backbones, including humans—than any other group yet discussed.

**Figure 4-4** An arthropod: a ghost crab, so-called because it looks ghostly as it scuttles across a beach at night. Like insects and other arthropods, the crab is protected by an external skeleton that has many joints so that the animal can move. To grow, an arthropod has to shed its old skeleton and grow a new, larger one.

## 4-C  VERTEBRATES OF THE SEA

The **vertebrates**, animals with backbones, are represented in the ocean mainly by fish. The long axis of the vertebrate body is the backbone—a string of vertebrae with projections to which muscles attach. Fish swim by bending the body from side to side in S-shaped undulations, using their many muscles and the flexibility of the vertebral column. A fish can move where it will in the open sea, unlike most invertebrates, which live in the plankton or have to hold onto rocks if they are not to be swept away.

In coastal waters, we might catch a dogfish, a small shark often examined in biology classes. If we dissect a dogfish, we find that the skeleton is composed of cartilage, the same flexible material as in the ends of

our noses, instead of bone, which makes up most of the skeleton of other vertebrates, including most fish. Skates and rays, flattened, bottom-dwelling predators, are the other cartilaginous fish, close relatives of the sharks.

Fish with bones evolved about 340 million years ago (Figure 4-5). We can recognize bony fish without dissecting them. While sharks have a row of gill slits opening separately to the outside behind the head, bony fish have a single gill covering with one opening, where water leaves the body after giving up its oxygen to blood in the gills. Bony fish show an incredible variety of body shapes and sizes. Their bewildering variety is a result of evolutionary modifications of a very successful body plan for underwater living.

## 4-D  LIFE ON LAND

As early organisms evolved in the sea, there came to be more and more competition, among plants for a place in the sun, among animals for food and refuge from predators. The land was barren, but it had possibilities. For plants there was abundant light and rocks made of the minerals that plants need.

However, the land presented problems that kept it uninhabited for a long time. One is the problem of support. Many aquatic animals collapse if placed on land because they do not have skeletons that can prop them up in air. But the greatest problem was

**Figure 4-5** A fossil of a fish with bones from 10 million years ago, looking much like a modern bony fish.

desiccation, or drying out, because water evaporates into the air. Land organisms have evolved various waterproof coverings, but these created fresh problems. Most waterproof coatings are also gas-proof, and all organisms must exchange gases with their environments. Both plants and animals need oxygen for respiration and must rid their bodies of carbon dioxide. In addition, plants need carbon dioxide for photosynthesis. Most land-dwelling organisms evolved gas exchange surfaces inside their bodies, such as our lungs, which allow for gas exchange and also limit water loss.

Most of the vertebrates in the ocean are fish. The only other marine vertebrates are turtles and snakes (reptiles), seals and whales (mammals) and penguins (birds).

## 4-E  LAND ANIMALS

Many invertebrate groups have members that live on land: flatworms under logs near streams, earthworms in soil, snails and slugs in vegetation. However, these animals lead restricted lives because they need a lot of moisture. Only two groups—vertebrates and arthropods—have members able to move about freely on a dry, sunny day. Therefore, these are the animals we see most often.

**Amphibians** (toads, frogs, salamanders, newts) are vertebrates that are at home in water and on land at different stages of their lives. Nearly all lay their eggs in water, and the egg develops into an aquatic larva that turns into a land-dwelling adult.

The first truly terrestrial (land) vertebrates were the **reptiles**—represented today by turtles, lizards, snakes, and the alligator clan (Figure 4-6). The secret of their early success was evolution of an egg with an almost waterproof shell, allowing it to survive on land, well away from egg-eating predators that lived in the water.

**Birds** evolved from the dinosaur branch of the ancient reptiles. Adult birds have feathers, and the front limbs have been modified as wings (or sometimes, as in the case of penguins, as flippers). Having wings as well as feet gives most birds at least two modes of locomotion, such as walking and flying or swimming and walking. Bird eggs have harder shells than reptile eggs, and the parents usually sit on the eggs and young, keeping them warm until they de-

(a)

(b)

**Figure 4-6**   The first land vertebrates. (a) A tree frog, an amphibian. The adult lives on land and feeds on insects, but it must lay its eggs in water. (b) A large iguana from Central America, *Ctenosaurus similis*. Reptiles get most of their heat from the sun, and this species is often seen sunbathing on fence posts and even in trees. The large bluish circles at either side of its jaw are eardrums.
(a, *Mark Rausher*)

**Figure 4-7**   A little blue heron on its nest. A blue egg can be seen near the bird's feet.

**Figure 4-8**   Rodents are one of the most successful groups of mammals: a mouse eating an apple.

velop the insulating layer of fat and feathers needed to hold body heat (Figure 4-7).

**Mammals** are another group of land vertebrates that evolved from reptiles. Like birds, mammals are warm-blooded, but they retain body heat with a layer of fur or hair rather than feathers. Early mammals laid eggs, but most modern mammals gestate the young inside the mother's body. Early mammals were carnivores, but the mammalian body plan has proven versatile, and many herbivorous (plant-eating) mammals have also evolved. Herbivorous mammals, such as deer, antelope, and cattle, have always been a vital part of the human diet. Rodents, such as rats and mice, are a successful group of mammals because of their small size and rapid reproduction (Figure 4-8). The ability to maintain a high body temperature has permitted mammals, like birds, to colonize a wider range of habitats than most organisms, ranging from deserts to oceans, from the steaming tropics to the polar circles.

**Insects**   The only animal group apart from vertebrates with members fully adapted to dry land is the arthropods—notably the insects. Of more than 1 million species of insects, almost all live on land. Al-

though humans have waged war on insects for thousands of years, as far as we know we have never intentionally exterminated even a single insect species. No wonder many people conclude that the insects are destined to inherit the world.

Many factors contribute to insect success. Their external skeleton is waterproofed, reducing water loss. Their small size permits them to live on little food and to hide from enemies in tiny cracks, on the undersides of leaves, in the hair of an animal, or inside a seed. Most adult insects have wings, and so they can fly to new food sources. And finally, insects reproduce rapidly and in vast numbers. Their eggs resist desiccation, and the young develop rapidly.

Many insects and plants have evolved together and cannot survive without each other. Insects may carry pollen from male flowers to fertilize a plant's egg, which then develops into a fruit (Figure 4-9). Without insect pollinators, we should not have tomatoes, cucumbers, apples, and many other crops. We also obtain several useful products from insects, such as honey, silk, and shellac. One of the most important functions of many insects is to eat other insects that we consider pests. However, humans and insects are usually enemies. Insects attack human beings directly with bites and stings. Bloodsucking insects transmit many human diseases, such as malaria, river blindness, and sleeping sickness. They transmit plant diseases, such as Dutch elm disease, and many crop diseases caused by viruses. Insects probably do more damage indirectly, however, by competing with people for food and other crops (Section 21-A). Ironically, after spending billions of dollars on pesticides without making a dent in the diversity of insect pests, we have now, by destroying habitats with bulldozers and chain saws, unintentionally exterminated thousands of species of insects.

**Figure 4-9** Insects and plants: a pipevine swallowtail butterfly on a milkweed flower. The flower produces sugary nectar, which attracts the butterfly. As the butterfly drinks, its hairy body picks up pollen grains from the flower's male organs. When the butterfly moves to another plant, it carries the pollen, which brushes off on the plant's sticky female organ and fertilizes the plant's eggs.

> Insects and vertebrates are the only animal groups with many members fully adapted to life on land.

## 4-F LAND PLANTS

The plants around us are as well adapted to life on land as are insects and vertebrates. In land plants, division of labor among the body cells is more pronounced than in seaweeds. This is because the resources a land plant needs are segregated: sunlight, oxygen, and carbon dioxide are above the surface of the ground, and water and minerals are below. The different parts of the body are connected by **vascular tissue**, a transport system that carries water and food.

Vascular tissue is reinforced by strengthening material, so that a wheat plant or a tree is like a building supported by its plumbing. A waterproof, waxy cuticle covers the parts above ground, slowing down water loss. Many tiny pores in the leaves and stems allow gases to enter and leave the plant with minimal water loss.

A land plant moves water from the soil up through its body by **transpiration**, the evaporation of water through pores in its leaves. As water evaporates from the leaves, it is replaced by water drawn up through the tubes of the vascular system from the soil (Figure 4-10). A large tree may transpire as much as five tons

**Figure 4-10** Water movement through a plant. In sunlight, the plant opens pores in its leaves so that carbon dioxide can enter the leaf and oxygen can leave it during photosynthesis. Whenever the pores are open, water evaporates from inside the leaf, a process called transpiration. The water that is lost is replaced by water in the soil, traveling from roots to leaves by way of the vascular system.

**Figure 4-11**   Lower plants: a tropical tree fern in Puerto Rico. In temperate regions, all the ferns are much smaller than this.

of water on a hot day, moving it from the soil, through the tree, into the air. Transpiration is the reason that cutting down forests may change the climate and cause floods. When rain falls on a forest, much of it is taken in by trees and transpired through their leaves into the air, where it eventually forms rain clouds. If a forest is cut down, or replaced by smaller crop plants, most of the rain that falls is not absorbed by plants, so it does not reach the air to fall again as rain. Instead it runs off the soil, causing flooding and soil erosion after heavy rains. The climate downwind from the felled forest becomes drier, and water and soil wash downstream.

## Lower Plants

The earliest land plants had no vascular tissue and so had to remain small and live in moist places. These early plants are represented today by small plants such as mosses. As with algae, the sperm of these plants must swim to the eggs, and therefore water is necessary for sexual reproduction.

Club mosses and ferns represent early groups of vascular plants. The fossil record shows that ancient relatives of the ferns reached tree-like size, and indeed some small tree ferns survive today in the tropics (Figure 4-11). Coal is the decayed, compressed remains of ancient forests of tree ferns and their relatives.

## Gymnosperms

**Gymnosperms** are the conifers, best known as the firs, pines, and spruce trees of forest and garden. They are relatives of ginkgo trees, often planted on city streets because they survive air pollution better than most plants. Two obvious features adapt conifers to life on land: strong, woody stems and small leaves. Strong stems enable many forms to grow into very tall trees. Indeed, the tallest living plants are redwoods and Douglas firs, conifers of West Coast forests in the United States. A conifer's needle-like or scale-like leaves have a heavy, waxy cuticle that cuts water loss and enables many conifers to thrive in areas of the southern United States where the soil is dry and in the far north where plants cannot obtain moisture from the frozen soil in winter.

In their reproduction, conifers are also adapted to life with little water. Here, for the first time we find plants with sperm carried in airborne pollen instead of swimming sperm. Another reproductive advance is the **seed**, a small plant embryo surrounded by a supply of food and enclosed in a protective covering. The embryo's food supply keeps the new seedling alive until its roots can reach water in the soil and its leaves can reach sunlight and make their own food.

## Flowering Plants

Most trees and many low-growing ground plants are flowering plants, called **angiosperms**. The most obvious feature of a flowering plant is its flower, which is a reproductive structure. Those with large showy petals and sweet odors attract insects or birds that pollinate them. Most flowering plants without large flowers, such as maples and grasses, produce pollen that is carried from male to female reproductive organs by the wind.

Many flowering plants depend on animals, including rodents, birds, and human farmers and gardeners, to disperse their seeds. Some seeds hitchhike by fastening onto an animal's fur, feathers, or clothing. Others grow inside fruits, which an animal eats. Angiosperm fruits and seeds are vital sources of human food (Figure 4-12). Colorful animal-pollinated flowers and the equally colorful animal-dispersed fruits reflect the fact that flowering plants evolved when the land was already well populated with animals. Animal life has played a crucial role in the evolution of the spectacular array of flowering plants, from duckweed to dogwood, onions to oak trees.

We are just as dependent on angiosperms as they are on us. Most important, we cannot survive without eating them. But we also make clothes and houses out of them and use them to cure our illnesses. Twenty-five percent of prescription drugs sold in the United States contain chemicals extracted from angiosperms. The World Health Organization esti-

(a)

(b)

**Figure 4-12**  Angiosperms. (a) A maple tree, source of hardwood valued for building and furniture making. (b) Grasses and other perennial flowering plants on a dune. The most important human food plants are cereals, members of the grass family.

mates that 80% of people in developing countries rely on substances they extract directly from plants for their primary health care. This does not mean merely that the poor use herbal medicine because they cannot afford anything else. The wealthy city of Hong Kong contains nearly 1,500 herb shops.

## 4-G  BACTERIA AND FUNGI

Bacteria and fungi are often grouped with protists and viruses as the "microorganisms" (tiny organisms). The main thing bacteria and fungi have in common is that most live on the dead remains of other organisms. Otherwise bacteria and fungi are not alike. Fungi have cells like those of all the other organisms we have considered, but bacterial cells have a simpler internal structure with no nucleus. The oldest known fossils are those of bacteria.

Microorganisms rank with higher plants and insects as the most obvious group of organisms that humans need to survive. Many bacteria and fungi are among the most important **decomposers**, organisms that break down dead bodies into their chemical components. When we say something is **biodegradable**, we mean that it can be broken down by decomposers. Without the activity of decomposers, Earth would be knee-deep in corpses, our sewage would be with us forever, and the soil would contain too few nutrients for plants to grow. Decomposers break down the remains of other organisms, releasing the nutrients they contain back into the soil and water, where they supply the nutrients plants need.

### Bacteria

Most bacteria are too small to be seen without a microscope. Most people, however, have seen polluted water containing green or bluish scum, which is caused by long strings of blue-green bacteria (also known as blue-green algae).

Despite their simple structure, the chemistry of bacteria is incredibly varied. There are photosynthetic bacteria, bacteria that live as parasites on the bodies of plants and animals, bacteria that live on dead bodies, bacteria that need oxygen, and bacteria that can survive only in the absence of oxygen. Our own bodies are inhabited by bacteria—on the skin and in the nostrils, mouth, large intestine, and vagina—which keep us healthy by crowding out disease-causing bacteria.

All the food we eat contains bacteria. Although milk is sterile when it leaves a healthy cow, it contains several types of bacteria by the time it reaches the table. Bacteria in food are often a nuisance because they cause food to decay, and some of them, such as *Salmonella* (food poisoning) and *Clostridium botulinum* (botulism), cause human disease. The control of many bacterial diseases, by improved hygiene and antibiotics, has created a health revolution in the past 200 years. It is probably the single most important reason for the human population explosion.

## Fungi

Some fungi are tiny, like the yeasts that are used to make bread and ferment wine. Others, such as mushrooms and bracket fungi, have many cells and are often quite large. In 1992, researchers at the University of Minnesota announced that they had discovered the world's largest known organism, a single fungus hundreds of feet in width, living in the soil. Whatever their size, all fungi are **saprobes**, which means that they absorb their food across their body surfaces—unlike plants, which make their food, and animals, which ingest food by way of a mouth.

The feeding part of a fungus usually lies buried in its food, and we normally see only the reproductive structures: green, white, or pink fuzz on moldy foods, brackets on trees, or puffballs and mushrooms on the ground (Figure 4-13). All of these produce spores, reproductive cells that usually reach new food by traveling through the air. When a dead leaf drifts to the forest floor or an animal dies, fungal spores floating in the air have already settled on it. Each spore quickly germinates to form a new fungus, which secretes digestive enzymes. These enzymes break down organic molecules in the leaf or animal into smaller molecules that the fungus can absorb and into minerals that may be absorbed by the fungus or by nearby plants. Although their decomposer action in the countryside or the town dump is convenient, the decomposer fungi are a nuisance when they break down things such as telephone insulators, clothes, books, leather, and wooden structures.

Many fungi are important parasites of plants and cause enormous losses of crops every year. On the other hand, fungi such as the yeasts used to make bread, cheese, sauerkraut, beer, and wine are important in the production of human food.

## 4-H BIOMES, CLIMATE, AND VEGETATION

Now that we have encountered the main groups of organisms on earth, we turn to their habitats. In this chapter we consider only habitats on land. Aquatic habitats are examined in Chapters 11, 12, and 19.

The surface of Earth contains hundreds of types of habitat. For convenience, ecologists divide these into a few different biomes, such as desert and tropical rain forest (Figure 4-14). **Biomes** are types of habitat described by their vegetation. We often need to take the biome into account when considering environmental problems because different biomes pose different problems for their inhabitants. For instance, water pollution caused by agricultural fertilizers is not likely to bother people living in the arctic tundra, although it may be a serious threat to the water supply in a grassland biome farther south.

As travelers explored the world and cataloged its life, they observed two general patterns. First, each newly discovered area contained previously unknown species of organisms. Second, despite the ever-increasing number of known species, there were only a few basic types of natural habitat. Walking through a tropical forest in South America, we would find tall trees with large leaves and fruit, festooned with immense climbing vines, and we would see colorful butterflies and birds flitting through the gloomy shade. A tropical forest far away in Indonesia would look very much the same, although the particular species of trees and vines, of butterflies and birds, would be different.

Biomes are named by their plant life because plants are the ultimate source of all food. Therefore, the plants that grow in an area determine what other organisms can also live there. But what determines what plants grow where? The main answer is climate. If we look at a map of the world's biomes, we see that the biome varies with the climate. Rhododendrons grow in northern forests and mahogany trees in tropical forests because rhododendrons cannot survive in hot, wet conditions and mahogany trees cannot survive cold or dry weather.

Temperature and precipitation are the climatic factors that determine the type of plants that can grow in an area (Figure 4-15). Large trees need quite a lot of moisture, while progressively lighter rainfall supports communities dominated by small trees, shrubs, grasses, and finally scattered cacti or other desert plants. In extreme cases, lack of rainfall results

**Figure 4-13**   A clump of mushrooms, which are the reproductive structures of much larger fungi that live in the soil. *(Thom Smith)*

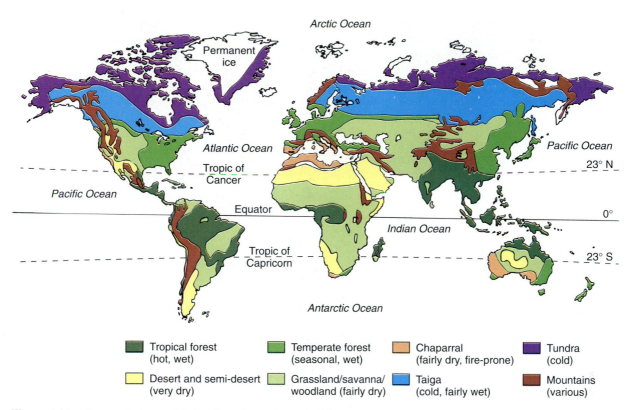

Tropical forest
(hot, wet)

Desert and semi-desert
(very dry)

Temperate forest
(seasonal, wet)

Grassland/savanna/
woodland (fairly dry)

Chaparral
(fairly dry, fire-prone)

Taiga
(cold, fairly wet)

Tundra
(cold)

Mountains
(various)

**Figure 4-14**   A map of the world showing where the main biomes occur.

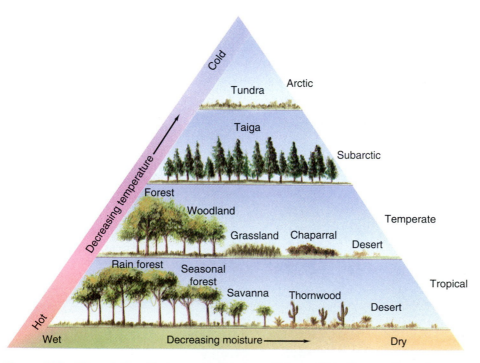

**Figure 4-15**   The relationship between climate and biome. This diagram shows that temperature and precipitation are the main factors determining what biome occurs where.

in a total lack of plants, whatever the temperature. Climate and vegetation both vary with latitude (distance from the equator) and with altitude (height above sea level). This is because the temperature falls both as one moves farther from the equator and as one moves up a mountain.

> The higher the rainfall and temperature, the more plants and the larger the plants an area can support.

## 4-1 TROPICAL BIOMES

The main feature of a tropical climate is the lack of seasons. **Tropical rain forest** occurs in areas where high, fairly constant rainfall and temperatures permit trees to grow throughout the year. In such areas, annual rainfall may exceed 400 centimeters. (For comparison, the eastern United States receives about 125 centimeters of rain or snow each year and the central United States about 75 centimeters.)

Tropical rain forest is both familiar and strange to a visitor from a temperate area. The dominant plants are tall trees with slender trunks that branch only near the top, covering the forest with a dense canopy of leathery evergreen leaves that blocks out much of the light. Whatever the time of year, some of the trees are in flower and some in fruit. Surprisingly, some of the plants are familiar. Plants that can survive in the heat and the dim light of the forest can also survive the low light levels and central heating of our homes and offices (although they would prefer higher humidity). Most of our house plants come from this biome. It is an uncanny feeling to stroll through the forest recognizing old favorites such as species of *Ficus, Callandria,* or *Dieffenbachia.*

Since few plants can survive in the permanent twilight of the forest floor, the lower levels of a tropical forest are fairly open. The kind of impenetrable "jungle" for which you would need a machete occurs only in open areas, along river banks, and in clearings created by fallen trees, where sunlight can reach the ground. Despite its open appearance, however, the forest floor is not easy to walk on, for the ground is spongy and wet. Boots sink in at every step, the tree trunks are wet, and water drips on the traveler (Figure 4-16).

Many plants are not rooted in the soil but on the surfaces of other plants. These are called **epiphytes**. They include a great variety of bromeliads, which look like baby pineapple plants, and exquisite orchids, members of the largest family of flowering plants (Figure 4-17).

The animal life of rain forests is exceedingly rich. Huge, colorful butterflies, beetles, and birds flicker by overhead. From above come the chatter of monkeys and the calls of many species of birds and frogs. Some of the frogs are among the most gaudily colored animals in the world, clad in brilliant pink, yellow, or green.

Because most of the plant food is high up in the forest's canopy, most of the animals, including the reptiles and mammals, also live there, which makes them hard to study. To add to the difficulties, most of the tree trunks are too slim and unbranched to be climbed. Only in recent years have scientists developed lightweight scaffolding and ropes that permit them to work in the canopy.

**Figure 4-16** Tropical rain forest on the slopes of Volcano Poas in Costa Rica. This area is almost always wet. It receives more than 300 centimeters of rain each year and is often shrouded in low clouds, as shown here. Flowers can be seen on some of the trees. Because there are no distinct seasons, some of the trees are in flower at all times of year.

(a)

(b)

(c)

**Figure 4-17**    Species of the tropical rain forest. (a) An orchid, which grows as an epiphyte attached to a tree trunk, absorbing water and nutrients that run down the trunk. The exotic flowers attract one particular species of butterfly for pollination. (b) A macaw, a large member of the parrot family. Macaws were once common in Central and South American forests but are now endangered because so many of them have been shot for their feathers and captured to be sold as pets in North America. (c) Frogs mating, with the smaller male on top. The female sheds her eggs into the damp moss beneath her, and the male releases sperm, which fertilize the eggs outside the female's body. Their bright color suggests that these frogs have poisonous mucus in their skins as do many frogs. Birds and mammals learn to recognize the colors and do not attempt to eat the frogs. (c, *Richard Laval*)

Tropical rain forest is the biome with the highest **species diversity**, the number of different species in a given area. There may be 100 different tree species in each hectare (2.5 acres), as opposed to fewer than 10 species in a hectare of temperate forest. Biologists estimate that most species of trees and animals in this biome have not yet even been described or classified. And most of them never will be because the rain forest is disappearing. Most of it lies in developing countries, where the forests are being cut down at great speed to provide firewood and agricultural land for rapidly growing populations.

> The most diverse biome is tropical rain forest, found where rainfall and temperature are high throughout the year and containing an enormous variety of plants and animals.

Fertile soil is made up largely of litter or humus, the partially decomposed remains of plants and animals. In a tropical rain forest, there is hardly any litter, so there is hardly any soil. Every year, tons of leaves fall to the forest floor, but the high temperature and moisture are ideal for decomposer organisms, so that this organic matter is quickly decomposed. A fallen leaf may disappear completely in less than a month, whereas the same process takes one to seven years in a temperate forest. The minerals released by decomposition are rapidly taken up again by plants, and so almost all the nutrients of the forest are locked within the bodies of living organisms. So efficient are the plants at absorbing nutrients as fast as they are released that essentially no minerals are left in the soil.

As we travel south and north from the equator, we move into areas with distinguishable seasons. Here, rainfall is concentrated during part of the year, and there is an increasingly pronounced dry season. Tropical rain forest merges into tropical seasonal forest (such as the monsoon forests of India). Here, the canopy is lower, and the proportion of deciduous trees (trees that lose their leaves for a season) increases as the length of the dry season increases.

**Tropical savanna** consists of grassland dotted with scattered small trees and shrubs. It extends over large areas, often in the interiors of continents, where there is not enough rainfall for forests to grow or where recurrent fires prevent the development of forests. Some savannas are entirely grassland, whereas others contain many trees.

Savanna grades into **tropical thornwood** (sometimes called shrubland or tropical scrub). The proportion of trees in these biomes depends on the competition between trees and grasses for water. Where rainfall is light, grass roots lying close to the surface absorb all the water during the dry season. Woody plants, which have deeper roots, cannot survive the dry season. As rainfall increases, the grasses are unable to absorb it all, leaving water available for scattered trees. Where rainfall is sufficient to support woodland (or a thick growth of shrubs), shade from the canopy inhibits the growth of grasses, and woody plants outcompete grasses.

Savannas are most extensive in Africa, where they support a rich collection of grazing mammals, such as zebra, wildebeest, and gazelles (Figure 4-18). The spectacular migrations of some of these species are related to shifting patterns of local rainfall. The animals follow the rain, which stimulates the growth of the young, nutritious grasses. Traditionally, human hunters and herders have also moved from place to place, following the rains and the new grasses.

Savanna was probably the habitat of the earliest humans, who lived on a combination of plant roots, fruits, grasses, and mammals. More recent human inhabitants have been herders. As populations have grown in the twentieth century, much of the savanna has been turned over to agriculture, which can feed more people from a given area of land than can herding. The soil of a savanna is more fertile than that of most tropical rain forests, but the rainfall is unreliable. The famines of recent years in Somalia, Ethiopia, and neighboring countries resulted from the failure of agriculture in this biome (discussed in more detail in Section 14-D).

## 4-J DESERT

Deserts generally occur in semitropical and temperate regions having rainfall of less than about 25 centimeters a year. Typical hot deserts are found be-

**Figure 4-18**   Savanna in Kenya during the dry season. *(Dale Corson)*

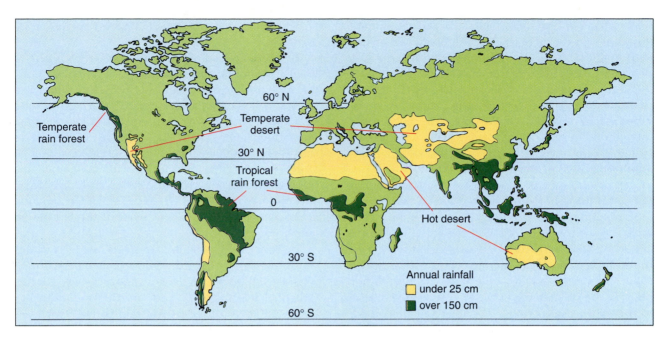

**Figure 4-19** The distribution of major very dry and very wet areas in the world. Note the rain shadow of the mountains on the west coast of South America. Part of North America's temperate desert is also caused by a rain shadow—from the Cascade and Rocky mountains and the Sierra Nevada.

tween latitudes 15° and 30° north and south (Figure 4-19). Because it contains little water vapor, the atmosphere over a desert is a poor insulator, and although the days can be very hot, nights are often cold because the ground radiates heat rapidly.

Desert areas with less than two centimeters of rain a year support little life of any kind, and the terrain is dominated by rocks and sand. Less extreme areas, including parts of the Sahara Desert, have highly specialized plants, many of them annuals (plants that live for less than a year) that grow, bloom, and set seed in the few days when water is available. Most desert perennials (plants that live for more than a year) are small woody shrubs that shed their leaves in the dry season, or else succulents, like the American cacti, that store water in their tissues. Desert animals have adaptations that restrict the loss of water through their skin and lungs and in their urine and feces. Many are active only at night, avoiding the desiccating heat of the day by burrowing into the cooler soil.

Cold desert, or semidesert, occurs in nontropical regions too dry to support grassland, often in the rain shadows of mountain ranges. Cold semidesert occupies much of the Great Basin east of the Cascade and northern Sierra mountain ranges in the western United States. In these regions, large areas are dom-

inated by sagebrush (*Artemisia*) interspersed with perennial grasses.

## 4-K TEMPERATE BIOMES

The world's most valuable real estate lies in the temperate regions. They are called temperate because they typically have moderate temperatures (although we may not think so as we wrestle to start a car on a February morning in Wisconsin). Temperate forest occurs in regions of abundant rainfall and contains both deciduous and evergreen trees. The composition of temperate forests, the proportion of deciduous to evergreen trees, and the spacing and height of the trees depend largely on when it rains and snows, the severity of the winters, and the frequency of fires. Three major categories of temperate forest can be distinguished: deciduous forest, evergreen forest, and rain forest.

**Temperate deciduous forest** occurs in areas where precipitation occurs throughout the year but where winters are cold, restricting plant growth to the warm summers. Since most of the trees lose their leaves in autumn, they lose little water by transpiration in winter, when their roots could not replace it from the frozen soil. Broad-leaved deciduous trees, such as beech, oak, maple, and hickory, dominate this kind of forest. There is an understory of shrubs and small plants on the forest floor, and the soil is rich in minerals and organic matter.

Mammals of North American deciduous forests include white-tailed deer, chipmunks, squirrels, opossums, raccoons, and foxes. Wolves, black bears, bobcats, and mountain lions once roamed widely. In spring, plants such as skunk cabbage, violets, Solomon's seal, and trillium produce their leaves and flower before the tree canopy grows new leaves that prevent most light from reaching the forest floor.

**Temperate evergreen forests** occur where poor soils, forest fires, and frequent droughts favor conifers or broad-leaved evergreens over deciduous trees. In the United States, temperate evergreen forests include impressive stands of ponderosa and other pines in the West as well as the extensive pine forests of the southeastern states. Elsewhere in the world, temperate evergreen forests are to be found in eastern Asia, southern Chile, New Zealand, and Austra-

**Figure 4-20**   A massive redwood tree in Oregon. Old redwoods and Douglas firs of the Pacific Northwest are the tallest trees in the world. *(Paul Feeny)*

lia, where forests are dominated by various species of *Eucalyptus*.

**Temperate woodland** occurs where the climate is too dry to support forest yet provides sufficient moisture to support trees as well as grasses. The dominant trees may be conifers, evergreen flowering trees, or deciduous trees. Pygmy conifer woodlands of piñon pine and juniper cover extensive areas of the American West, between the grassland semidesert biomes at lower elevations and the pine forests higher up. Oak woodlands are common in central California, and evergreen oak and oak-pine-juniper woodlands are extensive in Mexico and parts of the Southwest (see Lifestyle 4-1).

In Figure 4-19, you can see a very wet area at the southwestern tip of South America and on the northwest Pacific coast of North America—in Alaska, Oregon, Washington, and northern California. These are areas of **temperate rain forest**, which occurs in cool climates near the sea with heavy rainfall. The Olympic peninsula in Washington often receives more than 300 centimeters of rain a year. This forest resembles tropical rain forest in that it contains many very tall trees, such as sitka spruces up to 100 meters tall, with giant redwoods in drier areas of California (Figure 4-20). It differs from tropical rain forest in having many fewer species. There are no parrots and monkeys in the Pacific Northwest and relatively few tree species, most of them conifers, such as redwoods, spruce, fir, pine, and hemlocks.

> Temperate forests and woodlands are less productive and diverse than tropical forests because for part of the year the temperature (and sometimes the rainfall) is too low for plants to grow.

The **temperate shrubland biome** is best represented by the **chaparral** communities that occur in all five regions of the world having fairly dry climates with little or no summer rain: coastal California and Chile, the Mediterranean coast, southern Australia, and the southern tip of Africa. Most of the shrubs have leathery leaves. They are often distinctly aromatic, with volatile and flammable compounds in their leaves. Fires are frequent and pose a constant threat to the residents of Santa Barbara and other cities within this biome. After fires, the dominant shrubs regrow from surviving tissues near the ground.

**Temperate grassland**, known variously as **prairie** (North America), **steppe** (Asia), **pampas** (South America), and **veldt** (South Africa), covers extensive areas in the interior of continents where there is not enough moisture to support forest or woodland.

(a)

(b)

**Figure 4-21**   Temperate grassland, one of the most interesting and completely destroyed biomes in North America. (a) Grassland with scattered shrubs in Montana in summer. Trees are growing on the mountain in the distance. Because much of the precipitation falls as snow which takes longer to thaw in the colder air on the mountain, the soil retains just enough moisture for conifers to grow. In the warmer valley, the soil contains less moisture so that only grasses can grow, with occasional trees and shrubs near streams. (b) Bison, or buffalo, are the best-known large mammals of North American grassland. This animal was photographed in Wyoming in early summer when the grass is still green. In the foreground, you can see that the grass and flower plants are widely spaced, showing that this area receives less precipitation than the prairie of the Great Plains, where plants grow close together. Here, each plant has widespread roots to permit it to absorb enough water to survive. (b, *Richard Feeny*)

The world's grasslands are natural ranges for grazing animals, but they must be treated with care or they degrade into scrubland (Figure 4-21). When the grasses are grazed by animals, there are fewer leaves to lose water by transpiration. This may leave just enough water in the soil to support a few tough, woody plants. These include sagebrush and also mesquite, which can outcompete the grasses and which has invaded overgrazed grassland in the western United States, reducing its usefulness as grazing land.

Decomposition is slower than leaf fall in temperate grassland, and so the fertile soil gets deeper and deeper as time goes by, producing the deepest topsoil in the world. Parts of the prairies had as much as 2 meters of soil when they were first farmed. Much of the world's food consists of cereal crops, which are domesticated grasses, naturally suited to life in the grassland biome. Because of this and because of the deep, fertile soil, the grassland biome is often referred to as the world's "breadbasket." The midwestern United States, much of western Europe, and Ukraine have become prime areas for sustained and productive agriculture. As long as the soil and its fertility are conserved, there is no reason why they should not go on producing essentially forever. Because it is such valuable agricultural land, only a few, scattered remnants of unfarmed prairie now remain.

> Temperate grassland produces the world's deepest soils and, therefore, its best agricultural land.

## 4-L TAIGA

"Taiga" (pronounced "tie-guh") comes from a Russian word meaning "primeval forest." The **taiga**, or boreal forest, biome is dominated by conifers that can survive extreme cold in winter (Figure 4-22).

**Figure 4-22**   Taiga, with coniferous trees adapted to soil that is often very cold or very dry.

Trees in the taiga tend to be farther apart than in most forests, and enough light penetrates to the forest floor to support an extensive ground cover of shrubs. The taiga stretches in almost unbroken monotony in a giant circle through Canada and Siberia. The monotony is due to the small number of different tree species and is interrupted occasionally by areas of bog.

Much of the precipitation of the taiga falls as snow, and in winter many of the resident mammals grow fur or plumage that blends in with the white background. Many taiga mammals, such as wolverines, lynx, and wolves, have been hunted to the verge of extinction for their beautiful fur.

## 4-M TUNDRA

**Tundra** is the treeless biome that occurs far north in arctic regions and high up on mountains, where winters are too dry and cold to permit the growth of trees. In many areas, the deeper layers of soil remain frozen throughout the year or the thin topsoil lies on a permanent layer of ice. This permanently frozen layer is **permafrost**. Decomposition is very slow because the temperature is so low for most of the year.

Because the soil is thin and plant growth is slow, tundra takes a long time to recover when it is disturbed. This is why conservationists are so concerned about the effects of oil fields on the tundra. At Prudhoe Bay Oil Field in northern Alaska, buildings, oil wells, pipelines, and roads are raised above the ground so that their heat will not melt the permafrost. Nevertheless, this oil field has had some adverse effects on the landscape. Dust from roads has altered the heat absorption of the surrounding tun-

dra, melting ice and causing floods. The gravel platforms used as foundations to raise buildings off the ground sometimes act as dams, causing floods and interfering with the migration of caribou, musk-ox, grizzly bear, and wolves.

Tundra vegetation is dominated by grasses, mosses, and dwarf woody shrubs. Bogs are common because the permafrost retards drainage. The tundra is beautiful during the short summer when all the plants flower, but the traveler is distracted from the lovely scenery by the hordes of mosquitoes, deerflies, and blackflies. The insects provide food for a horde of migratory birds that nest in the tundra, such as plovers, sandpipers, snow buntings, and horned larks.

> Tundra is a biome of low-growing plants found where water reaches plant roots for only a few months of the year because the soil is frozen for the rest of the time.

Neither taiga nor tundra is found at sea level in the Southern Hemisphere because the continents do not extend far enough south. Antarctica harbors a very scanty population of organisms around its edges. Here, as in the northern parts of the arctic, animals on land (or on ice) feed mainly on plankton in the sea.

A variety of alpine grasslands, alpine shrublands, and alpine semideserts, similar to the related biomes at lower altitudes, are found on mountains. They lie between the timberline (the highest point where trees can grow) and the rocky heights where nothing can live (Figure 4-23). The cattle country of the

**Figure 4-23**   Mountain vegetation near Mount Hood in the Cascade Mountains of Oregon. At lower elevations, the mountains are covered with forest. Higher up on the right you can see areas of grassland merging into tundra, and above this is bare rock, where nothing will grow because there is little soil and the ground is covered by snow for most of the year. *(Richard Feeny)*

South American Andes is partly alpine grassland. Plant communities of similar structure but different evolutionary origins occur in the alpine zones of North American, Asian, and African mountains. African alpine communities also include heaths, and these shrubs dominate the alpine shrublands of the Himalayas, from which come many of our cultivated azaleas and rhododendrons.

## LIFESTYLE 4-1

### Save the Woodland: Give Up Stone-Washed Jeans

Stone-washed jeans get their bleached look when they are tumbled with pumice, a porous volcanic rock. The demand for pumice is destroying large areas of pine-juniper woodland in the Southwest. In New Mexico, the Jemez Mountain range is one of the areas being bulldozed to mine pumice. The local Jemez Action Group calls for a boycott of acid-washed and stone-washed clothing to reduce the demand for pumice.

Levi Strauss & Co., which sells most of this clothing, has responded by starting to replace pumice with chemicals. It plans to give up using pumice entirely as soon as chemicals alone prove satisfactory.

## SUMMARY

In the twentieth century, the biodiversity of Earth has been dangerously reduced by habitat destruction. The organisms on earth can be divided into five kingdoms: bacteria, protists, fungi, plants, and animals. These organisms all live in a small number of biomes.

### Protists and Algae

Protists and tiny plants are most abundant in the plankton that drifts near the surface of the sea and forms the food for marine animals. The dead bodies of plankton accumulate in vast numbers to make geological formations, such as oil deposits and limestone rocks. Protists that live as parasites cause many devastating diseases.

### Sea Creatures Without Backbones

The invertebrates of the sea include sponges, cnidarians, worms, molluscs, arthropods, and echinoderms. Most of these are small and adapted to life in the sea, but a few have relatives capable of living in fresh water or on land.

### Vertebrates of the Sea

Vertebrates have backbones that permit them to move more rapidly than most invertebrates. The most widespread marine vertebrates are sharks, skates, and other cartilaginous fish and the more numerous bony fish. A few other vertebrates live in the sea, notably whales.

### Life on Land

Land was a desirable environment for early organisms in many ways, offering abundant supplies of gases, minerals, and space. To live on land, organisms had to have support for their bodies, ways to avoid drying out, and new methods of exchanging gases with their environment.

### Land Animals

The successful land animals are the reptiles, birds, mammals, and insects. All can reproduce on land, have supporting skeletons, and have ways to reduce the amount of water their bodies lose to the air.

### Land Plants

The first land plants, such as mosses and ferns, need water to reproduce and are restricted to damp places. Gymnosperms and flowering plants are the only plants fully adapted to life on land. Their most important adaptations are vascular tissue, which supports them and transports food and water, and methods of reproducing without swimming sperm.

### Bacteria and Fungi

As decomposers, many bacteria and fungi are vital to human life. Others are parasites that cause many diseases.

### Biomes, Climate, and Vegetation

A worldwide survey of the distribution of organisms reveals two main patterns: (1) different areas of the world are inhabited by different species of plants and animals; (2) plant communities on land in different parts of the world can be divided into a fairly small number of biomes on the basis of their type of vegetation. These biomes are worldwide and are not restricted to single continents. The biome found in an area depends mainly on rainfall and temperature. Similar changes in biomes occur with increasing altitude and increasing latitude.

### Tropical Biomes

The richest biome is tropical rain forest, where high temperatures and rainfall permit plants to grow throughout the year. Most of the plant and animal life is found in the canopy among the evergreen leaves of trees. Decomposition is rapid, and the soil is thin; most of the nutrients are locked in the bodies of living organisms.

### Desert

Deserts have hot days, nights that are often cold, and very little rainfall. Their plant life is mainly annuals with very short growing seasons and succulent perennials adapted to the low rainfall.

### Temperate Biomes

In temperate deciduous forest, the soil is much richer in nutrients than it is in tropical forest because the trees lose their leaves in the fall, creating a litter layer that decomposes only slowly. Deciduous forest is an important biome of North America, Europe, and Asia in areas with warm, moist summers and cold winters. Where the soil is poor or fire is frequent, temperate evergreen forest replaces temperate deciduous forest.

Grasslands receive more rainfall than deserts and less than deciduous forests. Grasslands occur in the drier interiors of continents in the Americas, Asia, and Australia. Shrubs and trees may be scattered among the grasses. Grasslands are replaced by semi-desert scrub or by small woody shrubs in areas where there is too little water for grasses to grow.

### Taiga

North and south of the temperate region, grassland and temperate forest are replaced by taiga, a biome dominated by conifers adapted to growing in sparse soil and to resisting extreme cold and water loss in winter.

### Tundra

North of the taiga and above the timberline lie treeless tundra and alpine grasslands, dominated by cold-resistant woody shrubs or grasses.

## DISCUSSION QUESTIONS

1.  What biome do you live in? What other biomes have you visited?

2.  Why is the biosphere essentially a closed system?

3.  Look at a map of the biomes on Earth, and suggest some changes that may result from temperature increases caused by global warming.

4.  If the temperature in the taiga biome rises far enough to permit wheat to grow there, does that mean that we shall in fact be able to grow wheat in what was once taiga. Why?

5.  The remains of dead plankton formed huge heaps that turned into oil deposits. Yet only the occasional skeleton of a dead dinosaur is found. Does this mean that the mass of dead plankton is much greater than the mass of dead dinosaurs? If so, why?

6.  When an old farm field in the temperate zone is no longer farmed, the field rapidly grows back into forest. The same is not true of a desert or tropical rain forest. If the plants in these biomes are removed to form an agricultural field, it is often impossible to restore the land to its original condition later. Why do you think this is?

7.  Suppose you were given the task of preserving or increasing the biodiversity on a small island off the East Coast of the United States. What steps would you take?

# ECOSYSTEMS AND HOW THEY CHANGE

# Chapter 5

**Key Concepts**

- The organisms in an ecosystem interact with one another and with their physical environment, influencing each other's lives and evolution.

- An ecosystem can be self-sustaining only if it contains nutrients, producers, and decomposers and receives a continuous input of energy.

- Nutrients may cycle indefinitely in an ecosystem, but energy is continuously gained and lost.

- After a disturbance, succession slowly returns an ecosystem to the climax community typical of the biome.

- Suppressing fires, killing carnivores, and cutting timber are among actions that can degrade ecosystems.

*Secondary succession after the eruption of Mount St. Helens.* (Richard Feeny)

**E**cology is the science that examines the interactions of organisms with the environments in which they live. Even a simple landscape, such as a vacant lot, raises questions that ecologists might study. Sumac, ivy, and dandelions grow here. Why only these plants and not the hundreds of other species found elsewhere in the city? A large black beetle trundles across a slab of concrete. What does it find to eat? That butterfly fluttering by will lay hundreds of eggs before she dies. The city is not full of butterflies, so what happens to all the baby butterflies?

To study such questions, ecologists divide biomes into manageable units they call ecosystems. An **ecosystem** is the community of all the different organisms living in the area, along with their physical environment. Ecosystems include vacant lots, lakes, potted plants, woodlots, and fields. For convenience, we usually regard an ecosystem as an isolated unit, but ecosystems usually do not have clear boundaries. Things move from one ecosystem to another. Soil and leaves wash from a forest into a lake, birds migrate from their summer to their winter homes, whiteflies disperse from a neighbor's greenhouse to the plants on your windowsill.

## 5-A  THE COMPONENTS OF AN ECOSYSTEM

The basic components of an ecosystem are mineral nutrients, water, oxygen, and living organisms. Ecosystems also need a continuous supply of energy that comes, ultimately, from the sun (Figure 5-1). If it has all these things, an ecosystem can survive indefinitely.

Everything in an ecosystem interacts with everything else by way of two main processes: energy flow and chemical cycling. Energy moves through an ecosystem, and chemicals are passed around within it.

Let us consider a simple ecosystem, a potted plant in a covered glass jar on a sunny windowsill. The organisms present are decomposers in the soil (fungi and bacteria) and a green plant. The plant's leaves use water, carbon dioxide, and minerals to make carbohydrates. **Chlorophyll**, the pigment that makes plants green, is the extraordinary molecule that traps solar energy for photosynthesis. We have only recently learned to harness solar energy to run solar-powered calculators and heating systems. Plants have been using it to power much more complicated reactions for millions of years. A carbohydrate is stored chemical energy. The plant uses minerals to convert some of the carbohydrate into the other molecules it needs, such as fats and proteins. To do this, the plant needs free energy, and it obtains this by breaking down some of its carbohydrate by respiration,

a

b

**Figure 5-1**   Energy enters ecosystems when plants use sunlight to make their food by photosynthesis. (a) Some of the sunlight shining through this red ginger from Malaya will be used to turn carbon dioxide and water into carbohydrate in the leaves. (b) Banana plants so loaded with fruit that the owners have propped up the branches to prevent them from breaking. Plants use some of their energy to make fruit and seeds for reproduction. When we eat a banana, we are ingesting some of the energy the plant has collected from sunlight by photosynthesis.

using oxygen (and releasing carbon dioxide back into the air). So the plant grows, flowers, sets seeds, and reproduces.

Plants are autotrophs (self-feeders) because they make their own food. Autotrophs trap energy in a usable form. They are the route by which energy gets into the ecosystem. Their role in the ecosystem is that of producers (of food). An ecosystem would be doomed, however, if it contained only producers. Eventually, the plants would use up all the minerals

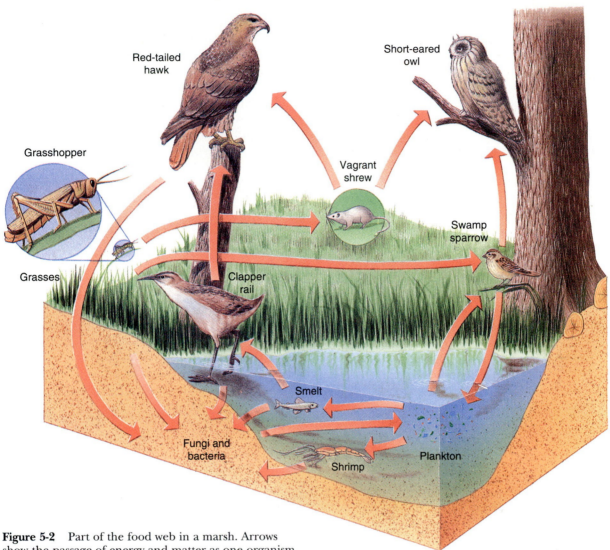

**Figure 5-2**   Part of the food web in a marsh. Arrows show the passage of energy and matter as one organism eats another.

in the soil and die. The decomposers are the solution to this problem. They break down dead leaves and other plant material, extracting nutrients and energy for their own use and releasing the minerals and some of the other nutrients from the dead plant back into the soil, where they can be reused.

Ecosystems usually contain more than just producers, decomposers, nutrients, and energy. In most cases, some of the green plant material is eaten by plant-eating herbivores, such as insects or rabbits, which are the ecosystem's primary consumers. In turn, either these may die and pass directly to the decomposers, or some of them may be eaten by meat-eating carnivores, such as frogs or dogs, the secondary consumers. There may also be carnivores that feed on the secondary consumers.

## 5-B  ENERGY FLOW: FOOD WEBS

Energy enters an ecosystem from the sun when a plant uses sunlight to make food in photosynthesis. It leaves an ecosystem as heat produced when organisms use energy in their food to move, grow, and reproduce. As organisms consume food and use it up, energy travels from one organism to another.

Consider the energy flow in a marsh. If we examined what eats what, we might find, for instance, that shrimp (primary consumers) eat phytoplankton; small birds such as sparrows (secondary consumers) eat the shrimp, and hawks eat some of the sparrows. When a hawk dies, it is decomposed by bacteria. A planktonic protist eats the bacteria, and the shrimp eats the protist (Figure 5-2). This list of organisms is a food chain, a series of organisms, each of which provides food for the next. Food chains in an ecosystem usually interweave in a complicated pattern called a **food web**.

We are often warned that interfering with food webs we do not understand properly may have unforeseen consequences. Ecologist Lamont Cole investigated one such situation in the 1950s. The World Health Organization (WHO) tried to eliminate malaria from Borneo by spraying with the insecticide DDT. The spray did indeed kill the mosquitoes that carry malaria, but there was a snag. The spray also landed on cockroaches. Then insect-eating lizards called geckos ate the cockroaches and suffered nerve damage from the DDT. Their reflexes became slower, and more of them than usual were caught and eaten by cats. Because most of their gecko predators were now gone, caterpillars eating the local thatched roofs multiplied unchecked, and the roofs started to collapse. The cats were soon dying from DDT poisoning, and so rats moved in from the forest. On the rats came fleas carrying the bacteria that cause plague. Most people would rather have malaria than plague, and so the WHO stopped spraying DDT and, in an attempt to remedy the damage, parachuted large numbers of healthy cats into the jungle—an expensive lesson in the importance of understanding a food web before you start pulling out the strands!

> All the organisms in an ecosystem are connected in a complex food web.

## Trophic Levels

The trophic level of an organism describes how far it is removed from plants in the food chain. Autotrophs are the **producers**, which make up the first trophic level. The second trophic level contains **primary consumers**, the herbivores. **Secondary consumers** are carnivores that eat herbivores. Parasites, such as fleas and ticks, that live on herbivores are also secondary consumers. Carnivores in turn may be eaten by other carnivores in still higher trophic levels or by parasites. **Decomposers** are the consumers that recycle organic material by consuming dead organic material (detritus) for energy. Decomposers are sometimes called **detritovores** because they get their energy from detritus.

The main producers in ecosystems on land are plants. In lakes and the ocean, photosynthetic protists and blue-green bacteria are the most important producers. In ponds and streams, much of the organic matter taken in by consumers comes from bits of land plants that wash or blow into the water. Pollutants frequently find their way into water in the same way and may enter the food web, poisoning some of the consumers (Lifestyle 5-1).

An organism cannot always be assigned to one trophic level. Thus, some plants, such as sundews, get some of their food from photosynthesis (first trophic level) and some of it from digesting insects (third or fourth trophic level). Many mammals, such as pigs and humans, are omnivores and also belong to several trophic levels because they eat both plants and animals.

## Why So Few Trophic Levels?

Ecosystems seldom have more than four or five trophic levels. This is because there is not enough food or energy in the top trophic level to feed another level. Why is this? First, not all the food available at one trophic level is actually eaten by animals at the next level. At each level, the **biomass**, the total mass of all organisms present, is only partly consumed. Second, most of the energy an animal eats does not add to the biomass. Most of the food we eat does not make us fatter—it is used in respiration; to

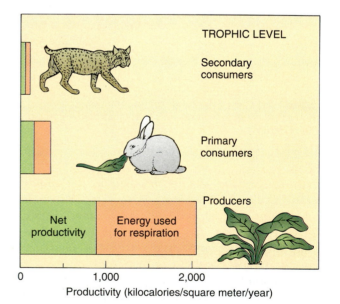

**Figure 5-3**  Energy flow for an actual ecosystem. The graph shows total energy input at each trophic level (gross productivity) and its division into net energy gain (net productivity, in green) and energy lost by way of respiration (orange). This type of graph is sometimes called a ''pyramid of energy.'' The pyramid is so wide at the bottom and so narrow at the top that the bottom layer would be much wider then this book if more than three trophic levels were included.

repair body tissues; and for locomotion, circulation, and feeding. As we would expect from the laws of thermodynamics, none of these processes is very efficient. Some useful energy is lost each time energy is converted from one form to another.

Because of these energy losses from one trophic level to the next, there is not enough energy left to support higher trophic levels (Figure 5-3). A wolf may have to travel 30 kilometers a day to find enough to eat, and a tiger requires a home range of up to 250 square kilometers. An animal that fed on wolves or tigers would have to cover a huge hunting area to find enough of its widely scattered prey. It is not worth expending the energy needed to harvest the small amount of energy available at the highest trophic level. The organisms that do feed on top predators are small, such as parasitic worms and fleas, and they get only a tiny crumb of the ecosystem's energy pie.

> Energy from the sun enters an ecosystem during photosynthesis. Then it passes through from one to five trophic levels in the ecosystem's food web.

## 5-C  MEASURING ENERGY FLOW: PRODUCTIVITY

The flow of energy through an ecosystem can be measured by answering various questions. How much solar energy reaches the system? How much energy is present in the food made during photosynthesis? How much of the energy in plant material can a herbivore use, and so forth?

### Primary Productivity

**Primary productivity** is the rate at which plants make more plant material by photosynthesis. We can think of it as the rate at which solar energy is stored as biomass by photosynthesis. Because plants use about half of the energy they store for their own respiration, only what is left, the **net primary productivity**, is available to other organisms.

Ultimately, net primary productivity limits the number of organisms, including people, that can survive on Earth because net primary productivity is Earth's whole food supply. There is a limit to the rate at which plants can convert solar energy into plant material. Every year, humans use up a portion of Earth's net primary productivity: we harvest crops for food, cut down trees to build houses and burn as fuel, and destroy vegetation when we build roads or buildings. Ecologists have calculated that humans cur-

**TABLE 5-1**  Net Primary Productivity of Some Major Types of Community (In Grams of Dry Plant Material Produced per Square Meter per Year)

| Community | Average Net Primary Productivity |
|---|---|
| Coral reef | 2,500 |
| Tropical rain forest | 2,200 |
| Temperate forest | 1,250 |
| Savanna | 900 |
| Cultivated land | 650 |
| Open ocean | 125 |
| Semidesert | 90 |

rently use up and destroy as much as 40% of Earth's net primary productivity each year. If this figure is correct, Earth cannot possibly support even three times its present human population because this many people would use up more plant material in a year than plants can produce.

Table 5-1 shows that the net primary productivity varies from one biome to another. Productivity is limited by whichever factor needed for photosynthesis is in shortest supply. This becomes the **limiting factor** that prevents productivity from rising. Sunlight is often a limiting factor, and productivity generally increases from polar regions toward the tropics because there is more sunlight in the tropics. Where water is scarce, temperatures low, or soil infertile, these may be the factors limiting productivity, no matter how much sunlight reaches the area. In dry areas, water is usually the limiting factor, and productivity increases with increasing precipitation (Figure 5-4).

Intensive agriculture, using special crop varieties, irrigation, fertilizers, and planting two or more crops on the same land each year, can achieve productivities as high as those of any naturally occurring vegetation on land. The average productivity of agricultural land, however, is not particularly high. Ocean productivity can be higher than any productivity found on land. In most places, however, it suffers from a constant drain of nutrients as plant and animal remains sink below the sunlit upper layer of the sea, where photosynthesis can occur.

Only a tiny fraction of incoming solar radiation is converted into net primary productivity. Much of the solar energy is reflected or absorbed by the atmosphere. In addition, only 45% of solar radiation that reaches Earth's surface is the visible light needed for photosynthesis. A highly productive crop plant such as barley or sugarcane may convert 15% of the visible

**Figure 5-4** Productivity increases with precipitation. This area of central Idaho receives only about 28 centimeters (11 inches) of precipitation a year, most of it as snow. It supports grassland, which has dried up by midsummer. In the background, however, trees are growing, but only on the north-facing slopes of the hill. This is because snow melts more slowly on colder, north-facing slopes, permitting moisture to sink into the soil instead of evaporating or running off the surface as it does in warmer areas. That slight difference in soil moisture permits the north-facing slopes to support trees as well as grass. *(Thom Smith)*

light that reaches it to biomass. But the average net primary productivity of plants on Earth represents the use of less than 1% of the visible light that reaches Earth's surface. You can see why agriculturists and plant breeders think that there is room for improvement in the productivity of most crops.

## Secondary Productivity

**Secondary productivity** is the rate of formation of new biomass by organisms that feed on other organisms. It amounts to only a tiny fraction of primary productivity. Of the net primary production in a temperate forest, herbivores eat only 1% to 3%. In some natural communities, as much as 15% of the vegetation may be eaten. In artificial communities, as when too many cattle or sheep are put on grazing land, up to 75% of the productivity may be eaten (Figure 5-5). Most of the vegetation that is eaten, however, does not go into secondary productivity. In one experiment, grasshoppers in a Tennessee field were found to convert only about 4% of the food they ate into more body mass or offspring. As a very rough estimate, 10% of most primary productivity actually eaten is converted into secondary productivity.

Even if the organisms at each trophic level were able to find, capture, and eat all of the net productivity from the previous trophic level, secondary consumers (such as humans when they eat meat) would

receive only about $1/10 \times 1/10 = 1/100$th of the primary productivity of their food chain. It is obvious, then, why meat is expensive (in terms of money) and that humans feed much more efficiently when they consume plant food. When people live on vegetarian diets, they make it possible to skip the energy losses at one trophic level so that more people can be supported from a given area of land.

> Because energy is lost at each transfer, productivity decreases at each higher trophic level. On average, animals convert only about 10% of the energy they consume into new animal (by growing or reproducing themselves).

**Figure 5-5** Sheep grazing in a field in England. The grass in this field has been eaten short. Herbivores may consume up to 75% of the productivity of a pasture.

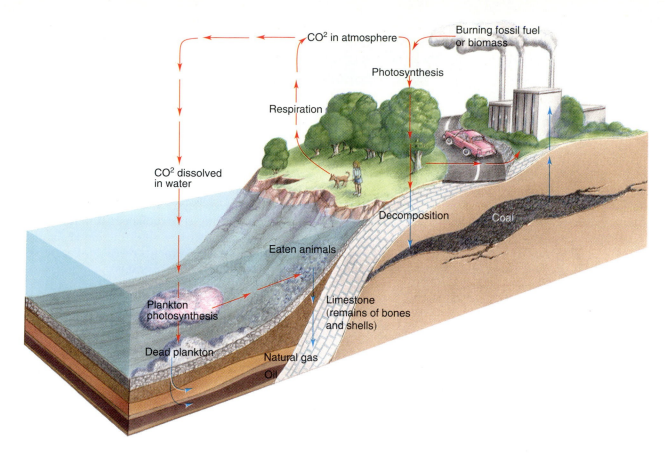

**Figure 5-6**   A simplified carbon cycle, showing how carbon moves through several ecosystems. Red arrows show short-term pathways; blue arrows long-term processes.

## 5-D  CHEMICAL CYCLING

Although the productivity of an ecosystem may be limited by the supply of sunlight or availability of water, in many cases it is limited by the availability of **mineral nutrients**, chemicals other than water that plants need to grow.

Living organisms require six elements in relatively large quantities: carbon, hydrogen, oxygen, nitrogen, phosphorus, and sulfur. Elements needed in smaller quantities include sodium, potassium, manganese, calcium, iron, magnesium, chlorine, iodine, cobalt, and boron. All these elements are present in rocks. They are released, by erosion and weathering, into soil, rivers, lakes, and the oceans. Some elements, such as nitrogen, oxygen, and carbon, are also present in the atmosphere. The movements of nutrient elements through the biosphere are called **biogeochemical cycles**. They are called cycles because minerals are used over and over again by living things unless they are lost from the ecosystem by erosion (Close-Up 5-1). A statistician calculated that on average every breath you breathe contains millions of carbon atoms that were once inhaled by any person in history who lived to middle age, atoms that have been recycled frequently over the years.

Nutrients sometimes cycle rapidly through an ecosystem, entering and leaving plants every few hours or days. Chemicals may also become involved in long-term cycles, in which it is millions of years before an atom in one organism enters the body of another. Every element has a somewhat different fate, depending on its properties and its role in living organisms. We shall illustrate the concept of nutrient cycling with a few examples.

### The Carbon Cycle

Carbon is a particularly important element because organisms are made up largely of huge molecules containing carbon, such as proteins and carbohydrates. Carbon gets into an ecosystem when plants convert carbon dioxide into carbohydrate during photosynthesis. When an organism breaks down carbohydrate by respiration to release energy, carbon dioxide is released back into the air.

In terrestrial ecosystems, most carbon is involved in a short-term cycle between organisms and the atmosphere, taken in during photosynthesis and released during respiration (Figure 5-6). Some carbon, however, enters long-term cycles. For instance, some carbon may be converted into carbonates, found in

## An Experimental Ecosystem

Nutrient cycles, and the effect of human activities on them, are now being studied experimentally in several parts of the world. One such ecosystem is the Hubbard Brook Experimental Forest in the White Mountains of central New Hampshire. The forest consists of a group of valleys, each with its own stream running down the middle. By catching and analyzing the water in a stream, investigators can determine the effects of various ways of treating the hills that drain into the stream.

Data from Hubbard Brook revealed that the forest is extremely efficient at retaining chemical nutrients. Nutrients that reach the forest in rain and snow approximately balance those leaving the forest in the streams, and both quantities are small relative to the total amounts of nutrients present.

The investigators examined what happens when a forest is cut down. One winter they cut down all of the trees and shrubs in one of the six valleys (Figure 5-A). Dramatic effects became obvious almost immediately. The valleys started to lose minerals much faster than they were replaced. Mineral nutrients dissolved in the stream water increased sixfold to eightfold. Other experiments revealed that nutrient losses are reduced if the forest is cut in horizontal strips, leaving strips of standing trees, rather than being clear cut. The remaining trees absorb many of the nutrients that enter the soil water after their neighbors are felled.

Nutrients recycle indefinitely in a natural ecosystem, with small losses into the air and in runoff and sedimentation. Disturbances by humans or by natural events, such as fires and landslides, can increase the ecosystem's loss of nutrients enormously.

**Figure 5-A**    A V-notch weir at Hubbard Brook. This device catches all the water that runs off the hill behind. Scientists analyzed the water for several years while the hill was covered with forest and then cut down all the trees and analyzed the water for several more years. *(Peter Marks)*

hard parts such as bones and shells. When the organism dies, bone and shell take a long time to break down. Carbonates from dead organisms may be washed into rivers and the ocean. Here, they may be incorporated into limestone rocks, which are made up of the shells of long-dead marine organisms. Geological processes may heave the rock back to the surface of the earth millions of years later. Then the carbon may rejoin the short-term cycle in various ways. For instance, acid rain falling on limestone in a rock or building releases carbon dioxide into the air. When we sprinkle lime on the garden to raise the pH, acids in the soil react with the carbonates and also release carbon dioxide back into the air.

Some carbohydrate in organisms is converted into fats, oil, and other storage molecules. These too may survive after the organism dies. Deposits of coal, oil, and natural gas are the remains of dead organisms. They represent part of the net productivity of bygone times. As we burn them, we release their carbon back into the atmosphere as carbon dioxide.

"Missing: 1 billion tons of carbon." That headline dramatizes the fact that there is still a lot we do not know about how carbon cycles through the biosphere. The amount of carbon taken in each year by plants is roughly equal to the amount that leaves organisms in respiration. But human activities release an extra 7 billion tons of carbon each year as carbon dioxide formed by burning tropical forests and fossil fuels (Figure 5-7). About half of this remains in the atmosphere, where the amount of carbon dioxide is steadily increasing. What happens to the other 3 to 4 billion tons? Most scientific estimates agree that about 2 billion tons of it dissolve in the ocean. Carbon dioxide dissolves in water, and the ocean acts as a vast **sink**, a reservoir, for carbon dioxide.

Between 1 and 2 billion tons of carbon a year are still unaccounted for, however. Probably plants are absorbing some of this. In parts of the world, particularly in Europe, forests that were cut down in the last century are now regrowing. As the trees grow larger, these ecosystems take in more carbon dioxide

**Figure 5-7**   Sources of some of the carbon added to the atmosphere in 1991. (a) Carbon emissions in selected countries. Countries emitting more than 3 tons per person per year are colored red; those emitting between 1 and 3 tons are green, and those emitting less than 1 ton are blue. (b) and (c) Sources of carbon emissions in the United States and western Europe. Can you explain why both rich and poor countries (e.g., Canada and Czechoslovakia) appear in red and in green (e.g., Romania and Japan)? Why do you think emissions from Middle Eastern countries such as Kuwait and Saudi Arabia are so high? Why do you think the U.S. emissions from transportation are so high compared with those of European countries?

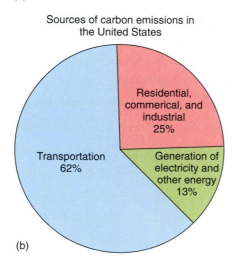

(b)

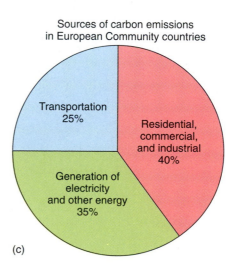

(c)

than they release. Other people have suggested that rising carbon dioxide in the atmosphere may be "fertilizing" forests so that trees are growing faster than they used to and taking in more carbon dioxide.

Researchers want to find out what is happening before one of these mysterious sinks stops working. Then the missing carbon might suddenly appear in the atmosphere, causing massive devastation, including global warming (Section 3-E). There are probably also limits to the amount of carbon dioxide the ocean can absorb. Remember that carbon dioxide dissolved in a fresh water lake was suddenly released in 1986, killing many people (Section 3-E). If the ocean were to release its carbon dioxide in a similar big "burp," we shall all go together when we go, as the song says.

> Carbon is fixed into organic form during photosynthesis, and most of it is soon released by respiration. The carbon balance of the biosphere is moderated by the exchange of carbon dioxide between the atmosphere and the oceans.

## The Nitrogen Cycle

Nitrogen is an essential constituent of many organic molecules, especially proteins and DNA (Figure 5-8). It is often in short supply and becomes the limiting factor for the growth of plants and animals (Close-Up 5-2).

Nitrogen appears to be all around us in vast quantities. Molecular nitrogen gas ($N_2$) makes up 78% of

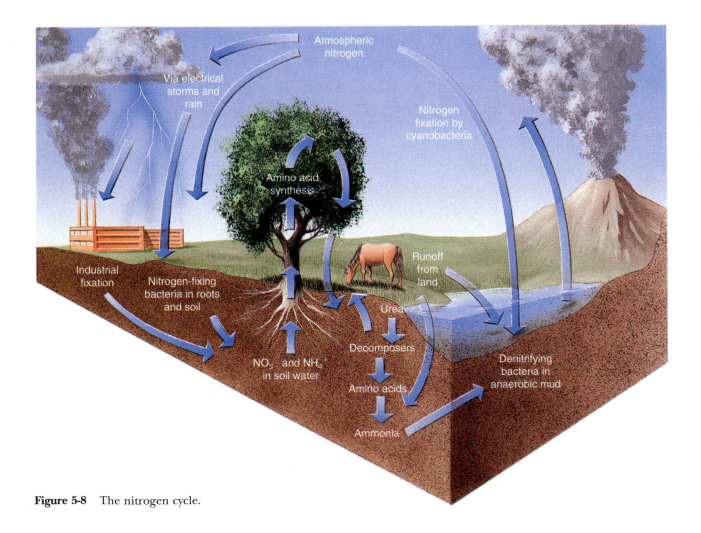

**Figure 5-8** The nitrogen cycle.

# Scientific Method and Limiting Factors in Plant Growth

How can we discover which factor is limiting plant growth in a given area? One way is to supply more of a particular factor and find out if the plants grow faster. Consider Table 1. This table shows what happened when nitrogen and phosphorus fertilizers were applied to trees in a Tennessee forest. The same amount of nitrogen fertilizer, with or without additional phosphorus, was applied to each hectare of forest. The growth rate of various tree species was measured before and after the fertilizer was applied. Glancing at the middle column shows us that nitrogen increased the growth rate of all the species studied. This is not surprising since most soils do not contain as much nitrogen as plants can use.

The right-hand column is more interesting. It tells us that extra phosphorus with the nitrogen increased the growth rate of cucumber trees and dogwoods but not of hickory and northern red oak, which grew faster with nitrogen alone. We can conclude that forest soil contains all of the phosphorus that most tree species can use. What about the figures for hickory and northern red oak? These species grew more slowly with phosphorus and nitrogen than with nitrogen alone. Could phosphorus be preventing the trees from using all the extra nitrogen? Or is it merely a "margin of error," as they say in public opinion polls? Scientists call this a "sampling error." It means that you have not measured enough trees or for a long enough period to get an accurate measure of growth rate. You would have to perform more studies to find the explanation for these figures.

How accurate are these results anyway? If you measure how much a tree grows this year, then fertilize it and measure growth again next year, can you be sure that any change is due to the fertilizer? No you can't. Perhaps there was a drought the first year so that the trees did not have enough water to grow much no matter how much fertilizer you gave them. Perhaps it was hotter in one year than the other, and plants generally grow faster in hot weather. To rule out these possibilities, you need controls for your experiment.

A **control** is a sample that you treat exactly like the experimental sample except for the experimental treatment. In this case, the controls would be nearby trees of the same species that never receive fertilizer but are measured in both years. If the control trees grow at about the same rate in both years, it seems likely that the faster growth in the fertilized trees is indeed due to fertilizer and not to temperature, rainfall, or some other factor. It is much easier to perform a controlled experiment in a laboratory than in a complex natural setting such as this.

**TABLE 1 Growth of Trees in a Tennessee Forest with Fertilizer**

| | Increase in Growth (%) | |
| --- | --- | --- |
| Tree Species | Nitrogen Alone | Nitrogen + Phosphorus |
| Hickory | 261 | 172 |
| Northern red oak | 167 | 100 |
| Cucumber tree | 107 | 207 |
| Chestnut oak | 61 | 70 |
| Black cherry | 52 | 71 |
| Yellow poplar | 48 | 69 |
| Dogwood | 36 | 173 |

the atmosphere. It is much more abundant than carbon dioxide, so why is it in short supply? The answer is that although producers can use carbon dioxide directly, the only organisms that can use molecular nitrogen are a few species of bacteria. The only source of nitrogen except the air is a limited pool of nitrite ($NO_2^-$), nitrate ($NO_3^-$), and ammonium ($NH_4^+$) ions in soil and water. Nitrogen is added to this pool by nitrogen-fixing bacteria, as nitrate formed from nitrogen in the air by lightning, by the decay of organic matter, and by the erosion of rocks. In addition, human activity is increasing the amount of usable nitrogen in the biosphere by manufacturing ammonia-based fertilizers from atmospheric nitrogen. These fertilizers are often washed out of lawns and fields and end up in rivers, lakes, and oceans.

All organisms release carbon dioxide into the air, but most do not release nitrogen gas. Nitrogen first reaches the environment as waste or in dead bodies. When organisms break down protein, they produce nitrogenous waste. The urea in human urine is one example. Urea is decomposed by bacteria and fungi, usually in the local sewage plant. Similarly, the proteins in dead organisms are broken down by various decomposers. Unlike carbon, nitrogen returns to the air by a long series of decomposition reactions performed by various fungi and bacteria. The reactions end with bacteria in anaerobic ("without oxygen") mud at the bottom of oceans, lakes, and bogs (or the sewage works) that release nitrogen gas back into the air.

Since plants need a lot of nitrogen and it is added to an ecosystem only slowly, loss of nitrogen is a serious matter. All terrestrial ecosystems lose some of their nitrogen as it washes away in ground water and streams, but many ecosystems are very efficient at retaining nitrogen. When organic matter reaches the soil, bacteria convert it into nitrogen compounds, which are immediately absorbed by plant roots. This recycling often breaks down when human activity disrupts an ecosystem.

### Nitrogen Fixation

The bacteria that can "fix" nitrogen gas from the air are enormously important in making this element available to plants. **Nitrogen fixation** is the conversion of gaseous nitrogen into dissolved ammonia ($NH_3$). It has been known for centuries that you can improve soil fertility by growing plants of the legume family. The legumes include trees and shrubs, such as mesquite (*Prosopis*) and acacias, as well as peas,

**Figure 5-9**  A legume. This broad bean, also known as a fava bean, is one of the large legume family of plants. Legumes carry nitrogen-fixing bacteria on their roots, adding nitrogen to the soil and making them a relatively high-protein source of human food.

clover, vetch, and beans (Figure 5-9). These plants improve soil fertility because bacteria of the genus *Rhizobium* live in nodules on their roots. The bacteria use sugars produced by the legume and supply the plant with ammonia. Legumes have an almost unlimited supply of nitrogen, which is a vital component of proteins. As a result, legumes make more of their body parts from proteins than do most plants, making legumes very nutritious for animals to eat.

Various nitrogen-fixing bacteria other than *Rhizobium* live in the soil, and there are nitrogen-fixing bacteria in the ocean. Nitrogen fixation, however, requires a lot of energy and proceeds slowly compared with the rate at which plants can use up fixed nitrogen.

Nitrogen is abundant in the air but in a form that few organisms can use. The scarcity of fixed nitrogen compared with plants' ability to use this vital resource makes nitrogen a limiting nutrient in many ecosystems.

## The Phosphorus Cycle

Phosphorus is a constituent of many biological molecules and membranes. In addition, many animals need large amounts of phosphorus to make shells, bones, and teeth. The phosphorus cycle is a sedimentary, or earthbound, cycle because phosphorus almost never occurs as a gas (Figure 5-10). The phosphorus cycle is simple, with many fewer steps than the nitrogen cycle.

When rocks are eroded by weather, small amounts of phosphorus dissolve, usually as phosphate

$(PO_4^{3-})$ in the soil water. Phosphorus leaves the bodies of organisms when excess phosphorus is excreted and when they die and decompose.

As with nitrogen, some phosphorus washes out of any terrestrial ecosystem and eventually ends up in the sea. Because some phosphate salts are not very soluble in water, the ocean's phosphorus compounds continually fall to the bottom. The main reason few plants grow in large parts of the oceans is that there is little phosphorus in the water.

The phosphorus cycle is, in the short run, a one-way flow—from rocks to land ecosystems to the ocean and finally to ocean floor sediments. The main way that phosphorus returns to land is by way of slow geological processes in which sea floor sediments again become terrestrial rocks. However, ecosystems on land can retain much of their phosphorus for short-term cycling since soil particles absorb phosphate and provide a reservoir of this mineral. Soil

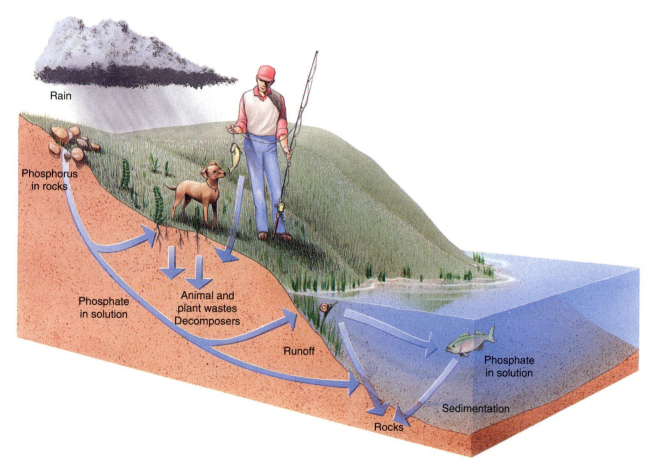

**Figure 5-10** The phosphorus cycle. Plants take in most phosphorus in the form of phosphate. Fishing and the droppings of sea birds bring trivial amounts of phosphorus from water to land ecosystems. Most flow is one-way, however, from terrestrial rocks to the sea floor.

erosion robs an ecosystem of its phosphorus, and it may take thousands of years to recoup this loss through the weathering of rocks.

**Mycorrhizae and the Absorption of Phosphorus**
Most plants cannot absorb phosphorus through their roots without help from fungi. If soil is sterilized by heating so that all the organisms in it die, plants grown in the soil do poorly because they are short of phosphorus, even if the soil contains plenty of phosphorus. In nature, fungi in the soil live on plant roots. The fungi absorb phosphorus from the soil and transport it into the root. In turn, the plant supplies the fungus with sugar for food.

The combination of a root with its particular fungus is a symbiotic association called a **mycorrhiza**. A **symbiotic relationship** is one between two species in which both species benefit. How do we know that each plant makes a mycorrhiza with a particular species of fungus? This was discovered some years ago when pine trees from North America were transplanted to Puerto Rico and Australia. The trees grew poorly because the soil did not contain the right species of fungi for them to form mycorrhizae. When the trees were treated with a handful of soil from North American pine forests, they perked up right away and grew rapidly.

This story suggests that the soil in an area is an ecosystem in itself, containing particular organisms as well as particles of rocks and the roots of trees and other land plants.

> Phosphorus moves through ecosystems in a one-way, sedimentary cycle. It is usually in short supply but tends to be retained efficiently.

### The Sulfur Cycle

Organisms need small amounts of sulfur to make proteins. Plants obtain sulfur by absorbing sulfates from the soil. Sulfur is returned to the soil when organisms die. But sulfur has numerous other effects in ecosystems. Sulfur can alter the pH of the soil, which determines how easily plants can take up other nutrients.

Some of the most important members of the sulfur cycle are photosynthetic bacteria. Green plants use carbon dioxide and water for photosynthesis. Sulfur bacteria also use photosynthesis to make their food, but they use a sulfur compound, hydrogen sulfide ($H_2S$) instead of water ($H_2O$) to supply hydrogen atoms. Instead of giving off oxygen like green plants, these bacteria give off sulfur. In the presence of oxygen, this sulfur is oxidized to sulfites ($SO_3^{2-}$) and sulfates ($SO_4^{2-}$). Sulfates are absorbed by plants. Sulfides may form hydrogen sulfide, which is recycled by sulfur bacteria, or they may react with iron in the soil to form insoluble iron sulfides.

Coal and oil often contain iron sulfides. That is what is meant by ''high-sulfur'' oil or coal. These sulfides cause problems when the oil or coal is burned or exposed to air. Then burning or bacteria convert the sulfides to sulfur dioxide ($SO_2$). Sulfur dioxide dissolves in water to form sulfuric acid ($H_2SO_4$), which makes rain acid if it is in the air. It can also make water draining from a mine very acidic so that it dissolves toxic chemicals and pollutes rivers, lakes, and soil water.

## 5-E  HOW ECOSYSTEMS CHANGE

A superficial look suggests that ecosystems remain unchanged for long periods of time. The forest that covers much of Canada, for instance, has been there for thousands of years. But is it really unchanging? If we examine the forest more closely, we see that it is not. Different parts change all the time. Here, an aged white pine has died and crashed to the ground, creating a gap in the leaf canopy. Sunlight through this gap reaches this part of the forest floor for the first time in 200 years. The sunlight causes seeds that have been lying on the ground for years to germinate, and soon a collection of herbs and shrubs and a grove of spindly pin cherry trees fills the space where the pine once stood. The animals here also change: insects, mice, and birds that feed on pin cherry move into the area, displacing those that lived on the pine tree. The death of this pine tree has caused a dramatic change in the species of organisms in this area.

When the species in an area change, the ecosystem changes. The plants that grow in an undisturbed prairie are different from the species that can live there once the prairie has been grazed by domestic animals or plowed up and turned into a wheat field. It is important to understand the changes that ecosystems undergo so that we can understand how ecosystems react to the changes that human activities tend to bring about.

## 5-F  SECONDARY SUCCESSION

In Chapter 4 we saw that similar climate produces similar vegetation in different parts of the world. Much of Earth's surface, however, is not covered by the plant community that we would expect in that

**Figure 5-11** Secondary succession following the eruption of Mt. St. Helens, Washington. The volcano erupted in 1980, spilling hot lava down the mountain and releasing a thick cloud of ash into the air. A lava flow occupies the middle of the photograph, which was taken in 1991. Most of the lava is covered with plants, including some small shrubs. Note trees in the foreground that were killed by the heat. The fallen trees are now covered with plant growth. In the background are trees that were untouched by the eruption. *(Richard Feeny)*

biome. We know that much of coastal California is covered, not by chaparral, but by farms, roads, and buildings, and we know why: human civilization has disturbed the natural communities, clearing the vegetation to make room for human affairs. So when we say that climate determines the type of plant and animal community in an area, we mean the community that would exist if the area were left alone for long enough, rather than what is actually there. The community that forms if land is left undisturbed, and that perpetuates itself as long as no disturbances occur, is called the **climax community**.

When an ecosystem is disturbed, either by human activities or by natural means, it begins the slow process of returning to its original state by a process known as succession. **Succession** is a progressive series of changes that ultimately produces a climax community, provided that no further disturbances occur (Figure 5-11).

A familiar example of succession is "old field succession," by which abandoned farms return to the climax community. For instance, in New England when a farmer stops cultivating the land, grasses and weeds quickly move in, clothing the earth with a carpet of black mustard, wild carrot, and dandelions. These plants are the "pioneers" of new sunny habitats. They grow rapidly and produce seeds adapted to dispersal by plants or animals over a wide area. Soon, taller plants, such as goldenrod and perennial grasses, move in. These newcomers shade the ground and their long roots take up most of the soil water,

so it becomes difficult for the seedlings of the pioneer species to grow. But even as these tall weeds choke out the sun-loving pioneer species, they are in turn shaded and deprived of water by the seedlings of pioneer trees, such as pin cherries and aspens, which take longer to become established but command most of the resources once they reach a respectable size. Succession is still not complete, for the pioneer trees are not members of the species that make up the mature climax forest. Slower growing oak and hickory or beech and maple trees eventually move in and take over, shading out the saplings of the pioneer tree species. After perhaps a century, the land returns to the approximate species composition of the original climax forest (Figure 5-12).

Repeated disturbances prevent an area from returning to the climax state. Suburbanites spend considerable time, energy, and money creating a continuous series of disturbances that maintain a lawn of a few species of short grasses, continuously interrupting the succession of tall weeds, shrubs, and tree seedlings that would take over if we stopped mowing the lawn.

### Patches and Fugitive Species

In any tract of land, we can always find at least small patches that are undergoing succession—a spot where a tree has fallen, a slope where a landslide has occurred, or a burned area. The existence of patches undergoing succession ensures a steady supply of fu-

| Year 1: annual weeds | Year 2: perennial weeds and grasses | Years 3–10: shrubs | About year 20: young pine forest | After about 150 years: mature oak forest |

**Figure 5-12**   Old field succession on an abandoned Tennessee farm.

gitive plants, the fast-growing, here-today-and-gone-tomorrow weeds. These species have seeds that can spread over appreciable distances, carried by wind or animals. In addition, the seeds of many of these fugitive plants are adapted to live for long periods in a dormant state, germinating only when a disturbance provides the proper conditions, such as increased light.

Animals as well as plants may be fugitive species. Insects that specialize in eating a particular plant species may travel far and use their keen senses to smell out new patches of their food plant some distance away. Some of our agricultural pest problems stem from the fact that many crop plants originated as fugitive species, depending on their sparse distribution and their nomadic habits (never in the same place for many seasons in a row) to protect them from their insect predators. By planting fields exclusively with one crop year after year, farmers create a paradise for such fugitive animals as tomato worms and cucumber beetles, which no longer have to spend energy to find food and have nothing to do but eat and multiply.

### The Importance of Preserving Patches

This discussion of succession explains why we are destroying habitats more rapidly than people without an understanding of ecology could have anticipated.

If we ever want an ecosystem to return to its original condition, we need to preserve patches of species that can recolonize it after a disturbance. An excellent example comes from the destruction of forests.

Traditional agriculture in forested regions consisted of cutting down trees to provide space for crops. The plot was cultivated intensively until the soil was exhausted, and then the plot was abandoned and a new one created. The abandoned plot underwent succession. The fertility of its soil was rebuilt as fugitive plants were killed by competition and broken down by soil bacteria. Within about 50 years, the forest was as good as new, and the soil could be used for agriculture again.

Under the pressure of population explosions, this cycle of cultivation and succession has been abandoned, particularly in tropical regions. Instead of small areas, thousands of hectares of tropical forest are being cut down at one time, particularly in Latin America (Figure 5-13). The area is used for agriculture until the soil is exhausted, which takes only a few years in many places because the soil is so thin. Then the area is abandoned. But this time it does not return to climax forest because no patches of forest are left to supply the seeds of successional plants. It becomes a new biome—we might call it "tropical desert."

One way to preserve patches, at least artificially, is to encourage homeowners to grow native plants in

**Figure 5-13**  Permanent destruction of tropical forest in Costa Rica. In the foreground is a field of coffee plants; in the background, the remains of forest. The soil in the coffee field will last longer than most tropical soil before its fertility is exhausted because it originated as volcanic lava, which is quite rich in plant nutrients.

their gardens. Highway departments also often plant native flowers on roadsides. These areas can act as reservoirs of seeds as well as insects and other small animals that can repopulate local ecosystems.

## 5-G  PRIMARY SUCCESSION

Old field succession is an example of secondary succession, which occurs relatively rapidly because it uses the soil and microorganisms left by the original forest. The soil may already contain the seeds of some of the successional species, and others will blow or be carried into the field from surrounding areas. But there are many situations in which succession is much slower because the soil has been changed or because the habitat is some distance from a source of seeds and animals to colonize the newly available area. The extreme case is primary succession.

**Primary succession** is succession that starts where there is no soil. Many farms in the northeastern United States have been abandoned because essentially all the topsoil has eroded. In places, nothing but naked rock remains. Felling some tropical forests permits the soil to wash away, leaving behind a hard crust. The scoured rock left by a retreating glacier, sand washed up on a beach, and an abandoned parking lot are all areas with no soil.

### Soil Formation

Soil can be defined as weathered material that supports the growth of rooted plants. It contains rock particles, decaying organic matter, water, air, and living organisms. Depending on the climate, it takes several hundreds to several thousands of years to produce fertile soil naturally.

Soil can form in two main ways. Sedimentary soil, including much fine agricultural soil, is made up of particles formed somewhere else. It consists of organic remains and rock particles, picked up by wind or water and deposited by rivers, wind, floods, glaciers, or lakes.

Primary succession is the way in which soil forms on the spot. Five factors influence the type of soil formed during succession. These are climate, rock type, shape of the land, time, and biological activity. The most important of these is climate because climate determines the organisms that occur in the area, and organisms form the soil. The organisms that live in a particular climate produce a soil type that is typical of the climate rather than of the rock. The reverse is also true. A particular type of rock will give rise to one kind of soil in a temperate region with high rainfall and to a completely different kind of soil in a semitropical arid region.

Very few organisms can grow without soil. Those that can are important in the early stages of primary succession. Bacteria and some protists can grow on naked rock, surviving on the nutrients in rain and those that dissolve out of the rock. **Lichens** are also important early pioneers (Figure 5-14). These organisms are actually pairs of species consisting of one fungus and one alga. Lichens are autotrophs, obtaining the nutrients they need from rain and from the rock. Water may freeze and thaw in cracks, breaking up the rock still further. Soil slowly accumulates as dust particles in the air are trapped in cracks in the rock and as the dead bodies of lichens, protists, and bacteria gather. Mosses and ferns may gain a hold in

**Figure 5-14** A yellow lichen growing on the dead branch of a pine tree. Lichens are self-sufficient and get most of their nutrients from rain. *(Thom Smith)*

even a thin layer of lichen remains and rock dust. As the mosses break up the rock even more and add their own dead bodies to the pile, the seeds of small, rooted plants are able to germinate and grow, beginning a process much like that of old field succession.

Primary succession can be seen in any city street. Mosses, lichens, and weeds establish themselves in cracks in the sidewalk, quite large plants may grow in a corner where leaf litter and dirt have been deposited by a rain gutter, and fungi and mosses invade a roof that needs repair. If we stopped cleaning and repairing it, even the center of Manhattan would turn into a rock-filled woodland within our lifetimes.

## 5-H FIRE-MAINTAINED COMMUNITIES

Fires, set by lightning, storms, or human activities, occasionally sweep through large areas. Burned areas undergo secondary succession. In the spruce and fir forests of the Rocky Mountains, for example, burned areas are rapidly colonized by wind-borne seeds of fireweed, which clothe the slopes with its purple flowers in summer, until it is displaced by spruces and other trees.

In many communities, fire is sufficiently frequent to determine the nature of the climax vegetation. In the United States, such communities include chaparral, temperate grassland, and many southern and western pine forests. Seedlings and saplings of decid-

a

b

**Figure 5-15** Plants that live with fire. (a) A fire-adapted pine. This is a young longleaf pine (*Pinus palustris*). For the first six years of its life, the pine stays at about this size, storing food in its roots so that in the seventh year it can grow very rapidly. This is an adaptation to ground fires in southern forests. The tree is resistant to fire in its seedling stage and after it is about 20 feet high. Its peculiar manner of growth permits it to grow very rapidly through the stage when it might be killed by fire.
(b) Fireweed (*Epilobium*), a pioneer species that springs up rapidly on ground scorched by fire, flowering and dispersing its seeds before it is shaded out by the regrowth of trees and shrubs.

uous trees are especially susceptible to fire, whereas many pines are relatively resistant and so survive in regions where fires occur often. Indeed, many species are adapted to exploit fire (Figure 5-15). Some pines have seeds that will not germinate until exposed to temperatures of several hundred degrees during a fire, which will kill their competitors and give them room to grow.

Grasses also regenerate readily after fires that kill trees. In some places, recurrent fires prevent grassland or savanna from turning into woodland. Areas of the Midwest were prevented from reverting to climax forest by Native Americans, who set fires after they found that herds of bison, which live on the prairies, could be hunted more efficiently than the white-tailed deer of the young forest.

## 5-I DISTURBANCES WITH LONG-TERM EFFECTS

If all ecosystems, when left undisturbed, eventually return by succession to the climax community of the biome, maybe we are worrying too much about our impact on the environment. Perhaps all we have to do is leave an area alone and it will recover from human interference. There is an element of truth in this suggestion, but for three main reasons it does not solve all our problems.

First, we cannot afford to leave much of Earth's surface undisturbed for any length of time. We need the land too badly, particularly as farmland to produce food. In many cases, we are not interested in letting devastated land return to climax forest. We want to turn it into fertile agricultural land.

Second, it is possible for human activities to irreversibly alter a particular ecosystem. If we do not leave patches of land containing organisms that can colonize an area, a disturbed ecosystem may never be repopulated. Even more extreme, an animal or plant species, once extinct, can never be recovered. Once a lake or marsh has become irrevocably polluted or filled in, it will never become a lake or marsh again, whatever else may develop in its place. So if we care about a particular local ecosystem, we cannot always rely on succession to restore it.

Third, the length of time it takes to restore the vegetation and the fertility of the soil may be hundreds or thousands of years, so succession is of little practical use to us.

### Ecosystem Degradation

Ecosystems are usually degraded when humans disturb them. The usual effect is that some of the species that once lived in the ecosystem can no longer

survive so the biodiversity of the ecosystem is reduced. This can happen in hundreds of different ways. We consider just a few examples here.

**Putting out Fires**   When people start to manage forests, they usually do two things: they put out any fires that start, and they remove older trees for timber. These actions might seem harmless, but in fact they alter the forest ecosystem drastically. Consider the southeastern coast of the United States. The climax community is a mixture of stands of loblolly and slash pine, interspersed with areas of hardwoods such as oaks, southern magnolias, tulip trees, sweetgums, red maple, and hickory. The distribution of these stands depends on fire, which is not very frequent. When a fire passes through the forest, it kills more of the hardwoods than the pines, which are quite fire-resistant. Pine seedlings germinate in the open spaces left after a fire. This is one reason that the pine trees in an area often appear to be all the same size—because they all germinated after the last fire (Figure 5-16). But a stand of pines cannot survive forever because pine seedlings cannot grow in the shade of their parents. The seedlings that can live under pine trees are those of hardwoods and palmettos. Over the centuries, the stand of pines is replaced by hardwood forest until the next fire starts the cycle all over again. When we suppress fire in a forest of this type, the pines eventually die and do not grow again.

This fate is also befalling the beautiful pine barrens of New Jersey and New York. Because these are near civilization, fires in the barrens are put out and the pines are slowly being choked out by shrubs and hardwood seedlings.

**Timber Cutting**   Cutting the old trees in a forest for timber converts an "old growth" climax forest into a "second growth" forest that contains only young trees, most of them less than 50 years old. This has numerous effects on the forest because old trees are quite different from young trees as habitat for other organisms. The most obvious difference is that old trees have more, and larger, dead and dying branches than young trees. Dead wood, whether it is on the tree or has fallen to the ground, is the habitat for many species that are becoming scarce as dead wood disappears. One study found that old growth forest in Tennessee contained many more bats than young forest. Why? Apparently because bats eat insects that live on dead wood, and they roost and give birth in hollows in trees. Dead wood contains many species of insects that are food for larger animals, and it also decomposes into woodland soil where many species of ferns and flowers thrive.

**Figure 5-16** Forest at the edge of salt marsh in the southeastern United States. The forest consists mainly of pines, hardwood species, and palmettos.

Many animals need large holes in trees for at least part of their lives. Wood ducks, bluebirds, spotted owls, bats, and raccoons nest in them; woodpeckers both nest and feed on insects in them. An old tree may contain holes that are a meter or more deep. We hardly ever see a hole of this size in a tree today, and the species that need these holes, including the western spotted owl, are disappearing (Figure 5-17).

We can help. Encourage your community and family to leave dead branches on trees whenever it is safe to do so. When branches fall off trees, don't toss them out with the garbage. Tuck them behind shrubs

b

**Figure 5-17** Destroying old forest in the Pacific Northwest. (a) A logging road through newly cut forest. Once old forest has been cut, it will never be allowed to turn into old forest by succession, which would take hundreds of years. It will be cut again at least every 75 years. The trees will never develop the rough bark and holes that are home to animals such as the spotted owl (b). You can see that many animals and small plants that live on the ground will have been killed by the passage of logging machines. (a, *Richard Feeny,* b, © *1993 Jack Wilburn/ Animals Animals*)

a

**Figure 5-18**   An alligator in its wallow. *(Steve Bisson)*

in the yard or leave them lying on the ground. Your local woodpecker thanks you.

**Killing Carnivores**   Nearly every ecosystem in the world has lost many of its carnivores, which have been killed by trapping and hunting. For instance, bobcats, wolves, and cougars (also known as mountain lions and panthers) once roamed most of the United States and are now restricted to a few small areas. The American alligator was only just rescued from extinction.

The loss of these carnivores affects different ecosystems differently. Consider merely a few effects on the east coast of Florida. Bobcats and panthers once kept populations of raccoons, rabbits, mice, and squirrels under control. Raccoons are particularly fond of eggs, and they can climb trees so they eat the eggs of woodpeckers, wood storks, herons, and dozens of other species whose numbers are declining. They also dig up the nests of endangered sea turtles that lay their eggs in holes in sand dunes. Floridians spend large sums and lots of time protecting sea turtle nests from raccoons by moving the eggs or covering the nests with wire netting. But raccoons are not easily stopped, as you know if you have ever tried.

It is an odd experience to stroll through a coastal forest in the middle of a long drought and find a pool of fresh water surrounded by a large area of mud. The pond is an alligator wallow or "gator hole." It originated as a small depression filled with rain water. Alligators love fresh water, which is not common on that sandy coast, and the alligators enlarge the pond when they come to drink or wallow in the cool mud so that the pond gets larger and holds more water (Figure 5-18).

Ponds maintained by alligators are the only source of fresh water on many islands and coastal areas in the Southeast. Without them, frogs would be unable to reproduce because frogs lay their eggs in fresh water. The ponds are resting and feeding areas for migratory waterfowl. Many species of herons nest in the surrounding trees. Eggs and babies falling out of the trees provide food for the alligators and for several species of snakes. You will see the mud around the pond pockmarked by the hooves of deer, turtles, wild pigs, raccoons, bobcats, and birds that come to the pond to drink. Without the alligators to keep them open, these freshwater ponds would soon be filled in by the secondary succession of shrubs and trees, and dozens of species would disappear from the coast.

# LIFESTYLE 5-1

## Poisonous Fish and the Food Chain

When you buy a fishing license these days, you are quite likely to receive a notice telling you not to eat more than one fish a week from this river or that lake because the fish contain poisonous mercury. Why are the fish poisonous when the level of mercury in the water is so low that the water is perfectly safe to drink?

Some poisons become concentrated as they move up the food chain. In aquatic food chains, the most important poisons involved are heavy metals, such as mercury, and stable, persistent pesticides. Such poisons have killed humans as well as birds that have eaten fish containing high concentrations of these substances.

The poison is often not distributed equally through the water. Much of it is absorbed onto the surfaces of particles, including algae and bacteria, which are then eaten by fish or shrimp. Most of these poisons are more soluble in fat than in water, so they dissolve in the organism's fat and do not diffuse back out into the water. Some larger animals concentrate the poison until it is as much as 10,000 times as concentrated in their bodies as it is in the surrounding water. In the late 1970s, for instance, fish from Lake Ontario, once an important recreational fishing ground, became unfit for human consumption because they contained so much mercury, even though the concentration of mercury in the lake water was not high enough to endanger human health (Figure 5-B). Similarly, the pesticide Endosulfan was ac-

**Figure 5-B**   A fish from Lake Ontario. This one looks too small to keep, but even if it is not, the fisherman is advised not to eat more than one fish a week from the lake because of the fish's high mercury content.

cidentally spilled into Europe's River Rhine in 1968. The concentration was so low that it would not have harmed a human drinking the water. Nevertheless, millions of fish died from the poison after feeding on small organisms and organic matter in the water.

# SUMMARY

An ecosystem consists of a community of organisms that are dependent on one another in various ways, plus their physical environment.

## The Components of an Ecosystem

The simplest ecosystem that can survive indefinitely consists of mineral nutrients in soil, water, oxygen, producers, and decomposers. Most ecosystems also contain primary and secondary consumers.

## Energy Flow: Food Webs

All the organisms in an ecosystem are connected in a complicated food web that describes what eats what. Autotrophs make up the first trophic level. In addition, most ecosystems contain organisms that can be assigned to up to five trophic levels. An organism may belong to more than one trophic level.

Energy passes from one trophic level to another, with only a small fraction passing on to the next trophic level. In consequence, the ecosystem supports less and less biomass at each trophic level, and energy flow is essentially one way. Because of the energy lost at each trophic level, many more people can be fed from a given area of land when the people eat plant food rather than meat.

### Measuring Energy Flow: Productivity

Energy enters an ecosystem when producers use it to synthesize their food. They use about half of this energy in their own respiration. What is left is net primary productivity, available to consumers in the ecosystem. The productivity of an ecosystem depends on the supply of whatever limits plant growth—usually water or sunlight or a mineral nutrient.

### Chemical Cycling

Nutrients cycle through an ecosystem. They are taken in by organisms as inorganic substances and may remain as minerals or be incorporated into organic molecules. Nutrients may pass through the food web for a time, but eventually they are once again released into the environment as inorganic substances. An ecosystem may be very efficient at conserving and recycling its nutrients. The availability of nutrients often limits the productivity of an ecosystem.

### How Ecosystems Change

In any biome, there are always areas in various stages of ecological succession following disturbances. Organisms adapted to living in the unstable communities of early successional stages have effective dispersal mechanisms and perpetuate themselves by continuously colonizing new habitats as they arise in the surrounding climax community. An area will return to its natural state after it has been disturbed only if there are patches of the species involved in each successional stage living near enough to reach the area after the disturbance stops.

### Secondary Succession

When an ecosystem is disturbed, species adapted to the disturbed area move in to replace the climax species. These fugitive species are eventually crowded out by slower-growing species. An area cannot recover from a disturbance if there are no patches where fugitive species can survive.

### Primary Succession

If a human or geological disturbance destroys the soil of an area, the area can return to the climax vegetation characteristic of that region only after undergoing primary succession, which involves the formation of soil. Primary succession may be relatively rapid or may take thousands of years.

### Fire-Maintained Communities

In some areas, fires are so frequent that they determine the climax community. When humans set or prevent fires, they alter the vegetation of an area.

### Disturbances with Long-Term Effects

Many human disturbances of ecosystems have long-term effects, such as destroying the soil or eliminating all patches of native plants and animals. In these cases, they prevent the ecosystem from returning to its climax state at least for many years.

## DISCUSSION QUESTIONS

1. Why is so much of the food on Earth never eaten? Consider a lawn or a forest. All those green leaves contain nutrients that herbivores could use. Yet the lawn and forest remain green, which means that most of those leaves are not eaten. Why not?

2. Is more energy lost from an ecosystem when a herbivore eats a plant or when a carnivore eats an animal? Why?

3. Is there one or more than one food web in any ecosystem?

4. How is the flow of energy in an ecosystem linked to the flow of nutrients? How do energy flow and nutrient flow differ?

5. The productivity of an ecosystem increases during the course of ecological succession. Why is this so?

6. Robert MacArthur found that the number of bird species in forests is correlated not with plant species diversity but with the amount of layering of foliage at different heights. Can you account for these findings?

7. How can frequent fires increase the species diversity of a region?

8. Suppose you were a member of a congressional committee developing legislation to set aside 60,000 hectares of land within a particular region for a biological preserve or national park. The

committee must decide whether to recommend government purchase of one large area or several smaller areas. What would your advice be if the prime objective was to preserve (a) an endangered large mammal population (such as the Texas red wolf), (b) as many species as possible, (c) as many local habitats as possible, and (d) the best possible compromise of these?

9. Examine Figure 5-7 on carbon emissions and their sources. Why do you think carbon emissions per person are so much higher in the countries with red bars than those with green bars? Note that the tax on gasoline is more than $500 per thousand liters in every European country ($965 in Italy) and less than $200 in the United States. But this does not explain all the differences.

# POPULATIONS

*Jaipur, India.* (Paul Feeny)

# Chapter 6

# NATURAL POPULATIONS

*Populations of shorebirds on an Atlantic beach.*

**Key Concepts**

- Some ecosystems have greater species diversity than others.
- The number of species in an ecosystem tends to remain fairly constant, although the particular species present may change.
- The reproductive potential of a population is determined largely by the age of first reproduction.
- Although a population invading a new area may grow exponentially, a population in nature seldom grows as fast as its reproductive potential would allow.
- In nature, the size of most populations remains fairly constant.
- The size of a natural population is determined by many factors, including the supply of resources, competition for those resources, and predation.

M uch environmental planning involves predicting how large populations will be in the future. How many toilets will empty into the local sewage system in 15 years' time? Will agriculture produce enough food for the population in 2010? Is this national park large enough to support a growing population of an endangered species? How many tuna can fishermen catch each year without reducing the future supply of tuna? These are all questions about the sizes of populations and how they change.

A **population** is all the members of a species in a particular area at any one time. Examples are the brown trout population of a stream, the deer population of Vermont, and the human population of Canada. Each population has its own properties. It has a particular size, density, and age structure. These properties can be used to describe the population and to predict how its size is likely to change in the future.

To manage our natural resources wisely, we need to know not only how large but also how stable populations are. How many deer can we permit hunters to take from a Vermont forest this year without reducing the deer population in the future? How long will it take the local oyster population to increase to its previous level after oysters are killed by an oil spill?

Growth of the human population lies at the root of most environmental problems. To understand human populations and how they will behave in the future, we need to know something about the properties of populations in general. In this chapter, we examine the factors that determine the sizes of all populations and those that determine how many populations can coexist in an area. In Chapter 7, we consider human populations.

## 6-A  SPECIES DIVERSITY

The ecosystems discussed in Chapter 5 all contain populations of species of producers, consumers, and decomposers. What determines how many populations of different organisms exist in an area and how many individuals each of these populations contains? The number of species in any ecosystem is its **species diversity**.

The species diversity of an ecosystem is important to its human managers. If we set up a state park designed to preserve redwood trees and an endangered species of bird, we should like to know that existing populations of redwoods and birds will not disappear from the park. Surprising as it may seem, we shall see from the examples in this section that there are no guarantees. Unless we understand the ecosystem very

well indeed, populations of organisms may come and go in ways that we do not predict. Consider the following example.

### Chobe National Park

Chobe Park in Botswana was set up to preserve an area of Africa's tropical savanna and thornwood. It brings in enormous revenue from tourists who come to admire the scenery and large mammals. Chobe contains patches of acacia woodland and grassland with a large population of herbivores, including giraffe, impala, buffalo, elephant, zebra, and wildebeest, as well as predators such as lions, hyenas, wild dogs, and jackals. In 1986, the park's managers sought advice because the acacia woodland is disappearing. The trees are suffering from heavy browsing (tree-eating) by elephants and other herbivores, and no new acacias are growing because elephants destroy the seedling trees (Figure 6-1).

Studying this problem, ecologists discovered that the acacias in Chobe are there only because in the 1890s the river through the park dried up in a drought. This sent the elephants elsewhere in search of water. Soon afterward, disease wiped out most of the other herbivores in the area. With no animals to eat them, acacia seedlings thrived, and many trees grew to maturity, forming woodland that has supported large populations of animals ever since.

The only way to restore the acacia groves would be to exclude elephants from the area as completely as possible for 10 to 15 years, while reducing the populations of other herbivores as well. This would also drive lions and other carnivores out of the area in search of food, which would make the park much less attractive to tourists. There is no ideal solution to this problem. The coexistence of large populations of herbivores and of thriving acacia woodland is inherently unstable. Populations of the food plants and of the animals that eat them come and go over the centuries.

One partial solution to the Chobe problem is possible. This is **exclosure**, fencing an area to keep animals out. If fences are erected around some of the acacia trees, elephants and other herbivores will be unable to reach the tree seedlings that germinate in the exclosures, and the trees will be able to mature. When the little acacias in an area are large enough, the fence can be removed and placed in another area where seedling trees are being eaten. Of course, fences that keep elephants out are expensive, and rotating exclosures will not permit enough trees to grow to feed very large herbivore populations. The park's managers will probably have to settle for fewer

**Figure 6-1**    Elephants in acacia woodland. *(Pamela and William Camp)*

herbivores. Or can you think of other possible solutions? Perhaps trees that grow faster and have higher nutritive value than acacias could be grown in the exclosures? With this, as with other environmental problems, a little imagination and some money for research might produce a more desirable solution than merely letting nature take its course.

## Species Turnover

Environmental experts sometimes invoke the notion of the stability of an ecosystem, meaning its ability to return to its previous state after a change. But do ecosystems really do this? The answer is sometimes. If a single load of sewage or oil is spilled into a river, the ecosystem will be changed drastically for a time, but it may eventually return to its original condition. Research has shown, however, that ecosystems often do not return to their original state after they have been disturbed. Succession returns an ecosystem to the climax state of its native biome. But this may not be identical to the ecosystem that existed before the

disturbance because it often contains different species.

Experiments on the nine Channel Islands off the coast of southern California showed that the number of species in an ecosystem may be more or less constant, whereas the actual species change with time. In a 1968 survey, biologists found that the total number of bird species breeding on each island had changed little since a similar survey 50 years earlier. However, the species had changed markedly. On San Nicolas Island, for instance, there were eleven species of birds in 1917 and the same number in 1968, but only five of these species were the same. Six species had disappeared from the island, but six other species had colonized it.

> The species diversity of an undisturbed area remains relatively constant, although the species present may change.

Studies such as these have shown that the population of a particular species in an ecosystem is often surprisingly small. Many of the populations of birds studied on these islands contained fewer than 15 individuals. There may be no more than 50 deer on a particular New England mountain. There may be a population of 100 magnolia trees in a South Carolina forest. Biologists used to think that these populations were stable for long periods of time, but studies like those of the Channel Island bird populations and the Chobe acacias show that natural populations change more than we had imagined.

## Are Complex Ecosystems More Stable?

It is often said that populations in simple ecosystems, with fewer species, are less stable than those in ecosystems containing more species. An example that is often quoted is the case of agricultural ecosystems, which contain only a few species and may be severely damaged by one pest. The nineteenth-century potato famine in Ireland was caused by a fungus (*Phytophthora infestans*) that wiped out the potato crop for several years. Tundra is often spoken of as simple and unstable—as exemplified by rapid changes in population sizes. Islands tend to contain fewer species than neighboring mainland ecosystems, and the introduction of competitors such as rats and goats has caused the extinction of many island species.

On the other hand, there are many cases in which disturbances have massive effects in complex ecosystems. For example, in the 1980s, insects ate the leaves of virtually all the Brazil nut trees in Bolivian rain forests, and rain forest monkeys died in large numbers. Conversely, simple ecosystems may be resistant to change: a salt marsh is a very simple ecosystem in that it is dominated by only one species of plant, but it is remarkably resistant to the disturbance caused by pollution (Figure 6-2).

There really seems no reason to believe that complex ecosystems normally recover from disturbances faster or more completely than simple ecosystems.

Some types of ecosystem are more vulnerable to disruption than others because of features that have little to do with their complexity. For example, pollutants are more likely to damage a lake, where they may be trapped indefinitely, than a river, through which water flows continuously, or an ocean, which is larger. We have to search beyond simplistic notions such as "stability" of an ecosystem if we are to predict the species diversity of an ecosystem and the sizes of its populations.

## How Is Diversity Maintained?

Why does an ecosystem contain so many different organisms anyway? Why doesn't one kind of decomposer become so efficient at digesting and absorbing dead matter that it drives competing species of decomposers out of the ecosystem? Similarly, we might expect that one species of "super-predator" would beat all other predators to the available prey.

What prevents one species in a trophic level from eliminating all the others by competition? One answer is specialization. When species might compete for a food supply, they often instead divide the food source up between them. For instance, different insect-eating warblers (forest birds) hunt for food at different heights in a forest (Figure 6-3). Each warbler species is a specialist at feeding on insects at a particular range of heights above the ground. Similarly, flycatchers (birds) and bats (mammals) feed on the same general food (flying insects) but at different times of day. Each specialist species becomes so efficient at using its particular resource that no other species can compete with it. An ecosystem is not uniform, and different species are best suited to different parts of it.

This explains why ecosystems containing many different plant species tend to contain more species of other organisms than those with fewer plant species. In general, the more productive an ecosystem, the

**Figure 6-2** Salt marsh, an ecosystem made up almost entirely of one plant species, cordgrass (*Spartina porpoises*). Two dolphins are fishing in the creek in the foreground.

Blackburnian warbler     Bay-breasted warbler     Myrtle warbler

**Figure 6-3**   Several species of warbler of the genus *Dendroica* hunt for insects in coniferous trees in the same New England forests. Each usually forages in a different part of the tree (dark green), reducing the competition for food.

more likely it is to contain a large number of species. Tropical rain forest has higher productivity and greater species diversity than tundra or desert. Table 6–1, however, shows that productivity may be high and the number of species low. Marsh and floodplain forest are both highly productive, but the forest contains four times as many bird species in a given area than the marsh. This is because the marsh is made up almost entirely of only one species of plant so that all the vegetation is the same height. There is no room for the kind of specialization on different heights of vegetation that we saw in the case of the warblers.

The consumers in an ecosystem also help to maintain species diversity. For instance, if you exclude rabbits or larger mammals from grassland by fencing it, the number of species of grasses in the fenced area decreases with time (Figure 6-4). In the fenced area, the fastest growing plants smother slower growing species. Where rabbits or deer are present, fast-growing grasses are eaten by these herbivores more than slower growing wildflowers. In addition, the consumers create more different habitats. They trample the soil and leave droppings. Species specialized to grow on trampled land or in animal feces flourish, whereas they die in the fenced grassland. This illustrates the fact that the species found in an area may determine what other species are also found there and that removing one species may have unforeseen effects.

Two of the factors that maintain species diversity in an area are the different specializations of species and the actions of consumer organisms.

| TABLE 6-1   Plant Productivity and the Number of Species of Birds in Some Temperate Zone Habitats* | | |
|---|---|---|
| **Habitat** | **Productivity (Grams of Plant Material per Square Meter per Year)** | **Average Number of Bird Species in 5 Hectares** |
| Marsh | 2,000 | 6 |
| Grassland | 500 | 6 |
| Shrubland | 600 | 14 |
| Desert | 70 | 14 |
| Coniferous forest | 800 | 17 |
| Floodplain deciduous forest | 2,000 | 24 |

*Data from Rickleffs, 1990.

## 6-B  HABITAT AND NICHE

In general, species diversity increases with the number of habitats that an area contains. The first requirement for a population is a suitable habitat with the proper range of temperature, humidity, soil, and so on. A population's habitat is not to be confused with its **niche**, which is its functional role in the ecosystem. To use a human analogy, the niche is a population's profession or way of life, whereas the habitat is where this way of life is carried on—the population's address.

**Figure 6-4** An exclosure in a national forest in Montana. The area in the background was fenced to keep elk out to examine the effect of elk browsing on dwarf willows.

To illustrate niche, consider the niche of a cottontail, a common rabbit inhabiting rural and suburban areas. The rabbit is a herbivore. To live, it must eat plants, and it eats some plants more than others. It tramples and digs burrows in the soil, which alters the soil. It drinks water and eats salt and other minerals, removing these from wherever they are found and depositing them in urine or feces. It attracts flies that feed on its feces. It may be eaten by foxes or hawks or shot or trapped by people. All these and every other interaction of the rabbit with its environment are part of its niche and determine where it can live and what other organisms can live in the same habitat with it.

## 6-C POPULATION SIZE AND GROWTH

Explosive growth of pest populations sometimes makes front-page news. Spectacular changes, however, are probably the exception. The size of most populations seems to change very little from year to year. In most habitats, the rare species remain rare and the common species common. Spiders, houseflies, and weeds in the lawn show up in much the same numbers every year.

Organisms produce many offspring, some of which do not live to grow up and reproduce. A herring may lay a million eggs a year; a housefly, several hundred. It is obvious that most of these do not survive. Why not? Various factors kill organisms before they reproduce and act as selective forces, shaping

the evolution of the population in the long run. In the shorter time scale of ecology, these factors control the sizes of populations. Before we examine what normally controls population size, let us first see how rapidly a population could increase if nothing checked its growth.

Each population has a characteristic **reproductive potential** (also known as **biotic potential**), which is the rate at which it could grow if it had unlimited resources. Experiments show that under ideal conditions, any population grows exponentially, or geometrically, meaning that a larger number of individuals is added to the population in each succeeding time period (Figure 6-5). Population explosions of

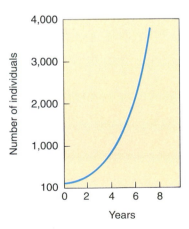

**Figure 6-5** Exponential (geometrical) growth of a hypothetical population. In exponential growth, the population increases at an ever-increasing rate.

# CLOSE-UP 6-1

## Hurricane Andrew and Introduced Organisms

Dandelions, house sparrows, starlings, and yarrow were all introduced into the United States from Europe and underwent population explosions when they first arrived (Figure 6-A). The predators, parasites, and pathogens that control the size of a population in its native land are usually absent in a new area. So if the organism can survive at all, its population may grow rapidly. After a time, the population explosion ceases. Predators and pathogens learn, or evolve, the ability to feed on the new species, and density-dependent control takes over. While this is happening, however, the introduced species can do a lot of damage, sometimes causing the extinction of native species.

Florida has been particularly hard hit by the introduction of foreign plants, usually for landscaping purposes. Florida's native plants are adapted to the unusual conditions of that state, and many of them are not found anywhere else in the world. If they are wiped out by competition with invading plants, they can never be replaced. For instance, the Everglades have been invaded by an Australian tree called melaleuca, which crowds out native pines and palms. Many of the state's waterways are choked with Asian water hyacinth (Figure 6-B).

**Figure 6-B**   Water hyacinth, *Eichhornia crassipes*, a floating plant, native to Asia, that chokes many Florida waterways. It poses no danger in colder parts of the country where it is killed by cold weather in winter, but it is illegal to sell water hyacinth in Florida or even Georgia, where it might occasionally survive the winter.

The Exotic Pest Plant Council warns that Hurricane Andrew in 1992 made the problem worse by blowing seeds into new areas. After hurricanes in the 1960s, Australian pines spread along many of Florida's beaches. Now the woods are full of foreign plants. The Department of Natural Resources estimates that it will cost up to $50 million to remove the foreign plants that threaten native vegetation in the Everglades National Park alone.

Homeowners replanting landscapes ravaged by Hurricane Andrew are warned not to plant exotics. It is illegal to plant Australian pine, melaleuca, and Brazilian pepper in Dade County, but these ordinances are not very effective because these plants are already widespread. What is needed is a program to educate people on the dangers of introduced plants and the value of their native vegetation—which needs no watering and survives storms better. Native wax myrtles, oaks, and palmetto trees in home landscapes suffered much less damage from the hurricane than exotic species, such as African tulip tree, eucalyptus, and tabebuias.

**Figure 6-A**   Yarrow, a European native (*Achillea millefolium*), photographed in Utah. Yarrow is found along roadsides all over the United States and is particularly common in dry areas of the West.

people, Mediterranean fruit flies, or gypsy moths are examples of the exponential growth that occurs when a population has plenty of the resources it needs. Exponential growth is particularly common when a species is imported into a new area where it has no natural enemies or competitors (Close-Up 6-1).

> Although populations can grow rapidly, most populations are fairly small and do not vary in size much from year to year.

Each species has its own reproductive potential, meaning that populations of some species are capable of growing faster than others. A population of houseflies can increase faster than a population of humans because houseflies reproduce when they are younger than humans and produce more offspring each time they reproduce. In general, an individual's contribution to population growth is higher if:

1. It produces many offspring each time it reproduces.

2. It has a long reproductive life so that it reproduces more than once.

3. It reproduces early in life.

The first two of these are obvious and self-explanatory, so it may come as a surprise that the third factor is by far the most important (Figure 6-6). For example, a bacterium reproduces only once, and it produces only two offspring (it reproduces by dividing into two). Yet a population of bacteria can grow faster than a population of oak trees because the bacterium may reproduce when it is less than an hour old, whereas an oak tree cannot reproduce until it is many years old. Populations of insect pests also tend to grow very rapidly because most insects reproduce when they are only a few days old (Close-Up 6-2). As an example of the way early reproduction can offset having fewer offspring, it has been calculated that a human population in which each woman bore 3.5 children (if such a thing were possible), starting at age 13, would grow just as fast as a population in which each woman bore 6 children, starting when she was 25. This is why people concerned with human population growth are so interested in the age at which a woman has her first child.

> Age at first reproduction is the most important factor determining reproductive potential.

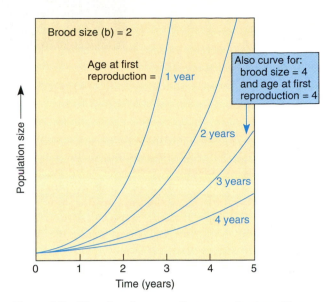

**Figure 6-6** How female age at first reproduction affects population growth. In all these curves, females produce two offspring per year, but the age at which females first reproduce differs in each curve (first reproduction at 1, 2, 3, or 4 years of age). The boxed note indicates that moving the age at first reproduction up from 4 to 3 (one year) has the same effect as doubling the brood size from two to four *(After Cole, 1954).*

The greater the reproductive potential of a species, the faster populations of that species can grow if they have the chance. But populations seldom get the chance to grow as fast as they can, and population explosions are the exception rather than the rule.

## 6-D CARRYING CAPACITY

Exponential growth is of particular interest to us because the human population is growing almost exponentially, and we want to know how and why such growth will end. No population can grow exponentially for very long. Eventually, all the food (or some other resource) is used up, and the death rate increases.

When rabbits were introduced to Australia, their numbers increased rapidly. Eventually, they became so numerous that some could not find food, local predators learned to catch them, and disease spread rapidly through the dense population. The rabbit population declined rapidly, but thanks to the species' high reproductive potential, it was soon growing again. The population went through a series of ups and downs that were progressively less drastic, eventually stabilizing around a size known as the carrying

# California and Medflies

The Mediterranean fruit fly, or medfly, is a small fly that usually lays its eggs on ripening fruit. The larvae, looking like cream-colored worms, burrow into the fruit and eat it, leaving a mushy mess. The medfly was introduced into California accidentally, and the authorities have fought a running battle to prevent it from wiping out California's valuable fruit-growing industry ever since.

The battle is not particularly sophisticated, although it has cost billions of dollars. It involves spraying Malathion on infected trees and over large areas from airplanes and releasing sterile male medflies. Some who object to this program point out that Malathion is not particularly good for children and other living things (see Section 21-B). Biologists also question whether the effort is cost-effective. The state has several times declared the medfly eradicated, and every time it comes back. Was it really eradicated and then reintroduced from abroad? That is possible, of course, but it is even more likely that the insect was never eradicated in the first place.

If you think of the difficulty of spraying every crack and cranny where a few medflies might be lurking, you might conclude that you would have to cover California with a sheet of plastic and then fumigate it to make sure you had gotten all the insects. Medflies have a huge reproductive potential since they reproduce when they are only a few days old and lay many eggs. In addition, California has a mild climate that makes reproduction possible almost year round. Even if only half a dozen medflies are left after an eradication program, simple calculations show that it will not be long before the next population explosion.

The other problem with this method of pest control is that it does not allow time for predators to evolve. Malathion not only wipes out medflies, but it also wipes out other insects that might eat or parasitize medflies and insect-eating birds (who die of starvation). The California authorities would have a hard time explaining to the fruit industry that they proposed to leave the medflies alone for a couple of years to let predators evolve, but this would probably be the most cost-effective solution in the long run.

Large-scale cropdusting programs, such as this one near Ciudad Constitucion on Baja California, can disseminate pesticides over a wide area. (© C. Allan Morgan/Peter Arnold, Inc.)

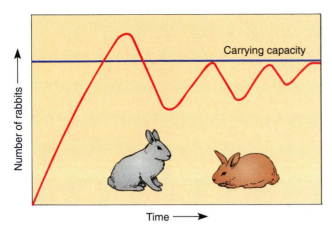

**Figure 6-7** The red line shows how the size of the rabbit population changed after rabbits were introduced into Australia. At first, the number of rabbits grew exponentially. Then came a population crash, then smaller and smaller oscillations until the population stabilized around a size that is known as the carrying capacity of the environment for the rabbit population.

capacity (Figure 6-7). The **carrying capacity** of the environment for a particular population is the number of individuals of that species that the environment can support indefinitely. The population size may increase beyond this number, but it cannot stay at this increased size for very long.

A species reaches its carrying capacity in a particular ecosystem when it is using up some natural resource as fast as the ecosystem can produce it. The particular resource involved is the limiting resource for that species in that area. We noted in Chapter 5 that net primary productivity (plant growth rate) is limited by the supply of water, sunlight, or a particular mineral nutrient. The supply of the limiting factor determines the carrying capacity of the environment for the species.

We have defined carrying capacity as the number of members of a species that the ecosystem can support indefinitely. If we take that literally, we would have to conclude that the carrying capacity of Chobe National Park for elephants is zero because no elephants could survive in the park during the drought of the 1890s. This is a ridiculous conclusion when we know that the park has supported large populations of elephants for hundreds of years. It is more useful to think of carrying capacity as changing when conditions change. Thus, Chobe's carrying capacity for elephants was zero during the drought, but since about 1910 it has been several hundred elephants.

However you define carrying capacity, it is clear that populations can grow larger than the carrying capacities of their environments. Chobe's elephants have been eating the acacias faster than they can regrow for many years now. Eventually, there will be too little food for all the elephants, and many of them will have to leave the park in search of food or die of starvation. A population that is greater than the carrying capacity degrades its environment. An ecosystem in which starving elephants (or deer, or people) are scrambling for food is an ecological disaster area as the animals try to tear every last edible leaf from the trees and blade of grass from the soil. When the last animal has emigrated or died, the habitat's carrying capacity will have been reduced because it contains no food. The ecosystem and its carrying capacity may eventually recover, but this may take many hundreds of years.

Human populations today are consuming Earth's resources faster than they can be replaced. For this reason, many people believe that our population is already greater than Earth's carrying capacity for *Homo sapiens*. The big, unanswered question is whether this means humans are destined to undergo a population crash. (This would occur in the twenty-first century, probably as a result of widespread starvation or war over the remaining resources.) The argument that a crash must come is that, unlike the elephants, we cannot just move elsewhere in search of food. Our ecosystem is the whole Earth. Ecologists are divided as to whether or not populations invariably undergo population crashes before they stabilize at the carrying capacities of their environments (see Figure 6-7). Estimates of the carrying capacity at which the human population of Earth must eventually stabilize range from about 5 to 20 billion people.

> A population that exceeds the carrying capacity of its environment degrades the environment and lowers the environment's future carrying capacity, at least temporarily.

## 6-E HOW DENSITY AFFECTS POPULATION REGULATION

The average size of most populations changes relatively little over the years. This suggests that population size is generally regulated in such a way that small populations grow fast, larger populations grow slowly, and still larger populations get smaller.

### Density-Dependent Regulation

Population size would change in this way if at least some of the causes of death in the population were density-dependent factors, killing a higher proportion of a crowded population than of a sparse one. Predation, parasitism, and disease are factors that can cause higher rates of death in dense populations. A disease-causing organism or parasite is more likely to encounter its host, or a predator its prey, when there are more hosts or prey in the area.

> The effects of many factors that may limit population size depend on the population's density.

## Density-Independent Factors

Some causes of death are density-independent, killing a proportion of the population without regard to its density. Bad weather and natural disasters sometimes act in this way. A severe winter or a drought may kill most of a plant population no matter what its density. Bad weather, however, may also cause death in a density-dependent manner, particularly in vertebrates. If it is possible to survive by finding shelter, and if the number of shelters is limited, more may survive in a sparse than in a dense population. Similarly, earthquakes, floods, and tornadoes may kill a higher proportion of people in an urban than in a rural area because the density of people makes escape more difficult in the urban area.

## 6-F  HOW COMPETITION LIMITS POPULATION SIZE

Competition between individuals of the same or different species is one factor that often prevents populations from growing in size.

## Competition Between Members of the Same Species

Since the members of a population share the same niche, they use much the same resources in the same ways and are bound to compete with one another. A good example is the fate of mealworm maggots in a sack of flour. The first beetles to find the flour lay their eggs, and most of the maggots that hatch will have enough food to grow to maturity. However, a sack of flour provides only so much food, and mealworms from eggs laid later may run out of food and die.

Plants sometimes die from the effect of competition, or they may survive but fail to grow to their normal size and produce few seeds. We see this when weeds choke flowering plants in an unweeded flower bed. In one experiment, seeds of white clover were planted at three densities. Half the plants at each density were watered regularly, but the other half

were watered only for the first 18 days. After seven weeks, most of the seedlings that had been watered regularly were still alive. Among the seedlings kept short of water, however, the proportion of seedlings that survived was three times as great in the low-density as in the high-density area (Figure 6-8). This is dramatic evidence of density-dependent death, presumably caused by competition among the seedlings for the limited water supply.

In another experiment, researchers found that the crop yield from a well-watered clover field remained the same even when five times as many seeds as usual were planted in the same area. Each plant in the dense plot was much smaller and produced fewer seeds. In this case, the plants were supplied with all the water they could use, and the yield was determined by the amount of sunshine, no matter what the density of plants. The lesson learned from this experiment is that crops should be planted at particular densities. Once the plants are dense enough to produce maximum yield from the plot, planting more seeds is merely a waste of money. Much agricultural research is devoted to determining the most cost-effective population density for a particular crop.

Instead of competing directly for a limited resource, members of many species compete indirectly

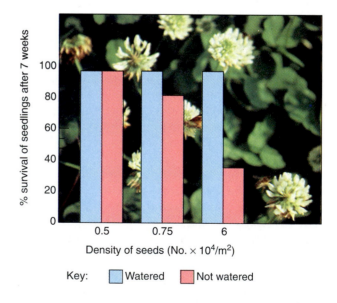

**Figure 6-8**   The effect of seed density on survival of white clover seedlings subjected to the stress of water shortage. When water was available, survival was about the same at all population densities (blue bars). Pink bars represent the survival of seedlings that were not watered after the eighteenth day. Many fewer of these seedlings survived in the dense population. *(After Harper, 1961).*

for social dominance or for a territory. A territory is an area occupied by one or more individuals and defended by its occupants against other members of the same, and sometimes other, species. The territory is of value not for the space itself, but for the shelter, food, or nesting sites it provides. Humans compete indirectly in a similar fashion. In countries with inadequate welfare systems, people compete for jobs and the money they bring. The competition is not really for the money but for the food and shelter the money can buy.

> Members of a species usually compete for food, space, or mates.

### Competition Between Members of Different Species

Members of different species often use the same resources (ecologists would say that their niches overlap), and so competition occurs between them. Sometimes the competition is so acute that one of the species becomes extinct. A species of giant tortoise lived on one of the Galápagos Islands off the west coast of South America. In 1962, an expedition reported that this species was extinct, although they found the remains of tortoises that had been dead for only about two years. The reason for the extinction was clear: the tortoises' food plants had all been eaten by goats, introduced onto the island in 1957 by a party of fishermen who wanted to be able to obtain fresh goat milk and meat when they were in the area.

Competition between species does not usually lead to the extinction of one of them. Humans coexist, if not happily, with insects such as the corn borers that compete with us for each year's corn crop. But competition between two species reduces the population of each species that a given area can support.

Many crop plants have been bred for maximum yield and have lost some of the competitive ability of their wild ancestors. Crops have to be protected from competition with weeds if they are to yield a good harvest. Sometimes, however, this is difficult or impossible. The wild oat (*Avena fatna*) is a weed that competes with other cereal plants for light, nutrients, and water (Figure 6-9). It often reduces the agricultural yield of cereal crops significantly. The weed is hard to get rid of because it is a wild cereal, and so herbicides that will kill the weed will also kill the crop. The wild oat sheds its seeds before the harvest so that its seeds persist in the field and are not removed when the crop is harvested.

**Figure 6-9** Cereal, such as this field of wheat, often contains wild oats, which reduces the yield.

> Competition between members of two species may lead to the elimination of one of them from the area or to their partitioning the resource for which they compete.

In general, we can say that competition for food, space, or any limited resource can determine the maximum size of a population. If something reduces the population to below this level, more juveniles than usual will live to grow up, and the population will return to its original size, when it will again be limited by competition between members of the species or with members of other species.

We have also seen that when a population of a species dies out in an ecosystem, it is often replaced by immigration of individuals from other populations of the same species. This sometimes permits us to restock ecosystems with species that have disappeared locally or with species with similar niches.

## 6-G  HOW PREDATION LIMITS POPULATIONS

When we think of a predator, we usually imagine an animal like a lion or wolf—a carnivore that kills and devours prey animals for food. In fact, insects are the most common carnivores, although they are so small we seldom notice them (Figure 6-10). In ecological interactions, however, it is convenient to define predation as members of one species feeding on members of another. This includes a parasite feeding on

**Figure 6-10** Insect carnivores. Red ladybugs feeding on aphids on yucca flowers in a Colorado field. The yellow-green spots all over the flowers are aphids, insects that suck the sap from inside plant cells.

its host, a herbivore eating a plant, and a carnivore eating another animal.

Most well-known carnivores, such as dogs and lions, are generalized predators feeding on many species. Sometimes these do appear to limit the populations of prey. For example, the survival rate of young sockeye salmon in a lake in British Columbia increased threefold after predatory squawfish and trout were removed. More often, however, such pred-

ators have little effect on the numbers of their prey. This is because a generalized carnivore feeds most efficiently when it feeds on one species at a time, eating the prey species that is most abundant and easy to catch. The predator develops a "search image" for the prey it is feeding on at the moment and ignores perfectly good food species that do not fit the image. In the same way, if we scan a bookshelf in search of a biology textbook that we think has a blue cover, we form a particular search image and may overlook the book, even with "Biology" written in large letters on its spine, if the cover is, in fact, yellow.

Another reason large predators seldom control the sizes of their prey populations is that the reproductive potential of a generalized predator is often much lower than that of its prey. Wolves, for instance, reproduce so much more slowly than mice that wolves could never control a mouse population explosion without assistance from other predators. The most the predator usually does is to "take the top off" a population explosion of one of its prey species.

After mountain lions (cougars) and wolves were eliminated from much of their former range in the United States, deer populations increased. This led many to believe that the deer had been limited by these carnivores. We now know that the rise of the deer populations was correlated with human activities, such as deforestation, which increased the amounts of the deer's favorite food—shrubs and young trees (as opposed to the trees of mature forest). Many people, however, still believe, wrongly, that animals such as tigers, lions, and wolves kill significant numbers of domestic animals and wild herbivores. Partly for this reason, most large carnivores have been hunted to the verge of extinction.

It may occur to you that *Homo sapiens* is an example of a generalized predator that limits the population size of its prey. We obviously do control the sizes of populations of cattle and other domesticated animals, and we have controlled the populations of other species when we have hunted them to extinction. However, we cause most extinctions by destroying habitats and not by predation.

Specialized predators, which feed on only one or a few species, can control the population size of their prey. This discovery was important to the development of biological pest control, pest control that depends on living organisms instead of artificial pesticides (see Section 21-F). For example, a scale insect feeds on citrus trees in California. After other methods of controlling the scale failed, a biologist visited Australia and brought back 129 vedalia beetles, natural predators of the scale insect. These were used to start a breeding colony of the beetles in California. Ten thousand vedalia beetles were released in the orchards, and by the end of that year the cottony cushion scale had been virtually eliminated from southern California. In areas where pesticides have not destroyed the beetle, it continues to control scale in California's citrus orchards. This remains one of the most striking successes of biological pest control.

Millions of dollars worth of fruits were saved for a very low cost each year.

> Generalized predators seldom control the size of a prey population, but specialized predators may hold the size of their prey population to a low level.

## SUMMARY

Populations of various species of organisms are the basic biological units of ecosystems. An understanding of how population size is controlled is essential to the solution of many environmental problems. Many factors influence the size of a population, and the main factor actually controlling the size of a population may vary from time to time and from place to place.

### Species Diversity

The study of islands shows that the species diversity in an area tends to remain relatively constant. However, the actual species present may change with time. New species move in as resident species disappear. There is no easy way for the human managers of an area to guarantee that particular, desirable species will remain in the area.

### Habitat and Niche

A population can survive only in a suitable habitat where it can find a niche.

### Population Size and Growth

Given ideal environmental conditions, the number of individuals in a population increases at its reproductive potential. This reproductive potential is determined mainly by the age of the (female) parent at first reproduction but is also influenced by the number of individuals produced at each reproductive event and by the parent's reproductive life span. A population seldom, if ever, reproduces at its reproductive potential even when it is growing exponentially. Among the fastest growing populations are organisms introduced into new, favorable environments where there are few natural enemies or competitors.

### Carrying Capacity

When growth of a population ceases, the size of the population usually stabilizes at approximately the carrying capacity of the environment for that species. Carrying capacity may change as the environment changes.

### How Density Affects Population Regulation

Most populations remain about the same average size from year to year. Some populations living in areas with extreme climates may be kept in check by density-independent natural events (such as storms or freezes). However, most populations are generally limited by density-dependent factors, such as predation, disease, or competition for resources.

### How Competition Limits Population Size

Competition between members of the same species often plays an important part in regulating the size of a population. Competition between members of different species leads either to the elimination of one species, the weaker competitor, or to selection for adaptations that reduce the competition. For example, each species may become specialized in such a way that it uses only part of a limited resource.

### How Predation Limits Populations

Predation may reduce the size of a population. Many specialized predators and parasites are known to keep their prey species at low density. Generalized predators tend to prey on the most abundant prey available, and so they may check outbreaks of a prey species, but these predators seldom appear to be the main factor limiting the population size of a prey species.

## DISCUSSION QUESTIONS

1. Ecologist Paul Ehrlich has said, ''It is quite possible that the penalty for frantic attempts to feed burgeoning populations in the next decade may be a lowering of the carrying capacity of the entire planet.'' Ecologist Lee Talbot has said, ''We haven't inherited the earth from our parents. We've borrowed it from our children.'' What do each of these statements mean? Do you agree? Why?

2. What are some lines of evidence indicating that the human population has already exceeded Earth's carrying capacity?

3. How does a project such as filling in a marsh for a housing development or building a four-lane highway affect the populations of organisms in an area?

4. Why does the rate of population growth slow as the carrying capacity is approached? Does the current rate of increase in the human population tell us anything about the planet's carrying capacity for humans?

5. People worry about overpopulation of a number of species besides humans. What are some of these species? Why should we worry about their overpopulation?

6. Consider two women born in the same year, each of whom will give birth to twin girls as her only children. However, one woman (A) will have her twins at age 18, the other (B) at age 36. Each daughter will have twin daughters at the same age her mother gave birth, and so on. All mothers will die at age 72.
   a. How many descendants does A have when she dies?
   b. How many descendants does B have when she dies?
   c. Construct a graph to show the growth of the populations descended from A and B.
   d. How do the rates of increase compare in the two populations? (Find a numerical answer if you can.)

# HUMAN POPULATIONS

# Chapter 7

## Key Concepts

- The human population has grown more rapidly during the last century than it ever has before or ever will again.

- There are more than three times as many people on Earth today as there were at the beginning of the twentieth century.

- The future size and growth rate of a population depend largely on its age structure, survivorship, and fertility rates.

- Increasing dependency ratios are a growing economic burden in countries where populations are still increasing.

- All prosperous countries have declining, stationary, or slow-growing populations.

- Countries with fast population growth are suffering from rapid environmental degradation. They face economic decline and rising death rates.

*A fishing village in the Shetland Islands.* (Richard Arnold)

Recently, a group of scientists was listening to a record by Tom Lehrer, a popular comic of the 1960s. Lehrer sang of Earth's population as "nearly 3 billion" people. Those present looked at each other in amazement. Was it possible that when we first heard that song the human population was really less than 3 billion? It is now about 6 billion and still growing rapidly. Even if birth rates keep falling, United Nations projections show that the world's human population will grow past 12 billion within the next century.

**Demography** is the study of human populations. This is a fascinating time to be a demographer because we are witnessing rare events. The human population has grown faster in the twentieth century than it ever has before or ever will again (Figure 7-1). Our species is approaching or exceeding carrying capacity, the maximum number of people Earth can support. There is no more new land to explore. We can no longer "go west" or anywhere else when we run out of land, or wood, or water. We have to stay where we are, figuring out ways to conserve and replenish our natural resources and to make our fresh water, soil, and forests provide a decent living for everyone.

Slow population growth may lead to economic prosperity. Countries such as France, Canada, and Romania with declining or sparse populations may encourage immigration and childbearing in the hopes of increasing their populations. But growth is destructive when a country cannot afford to feed, educate, and employ a growing population without destroying natural resources.

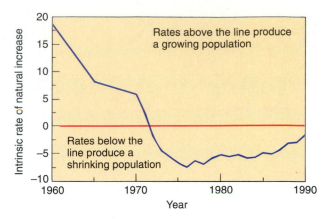

**Figure 7-2**   The difficulty of population forecasts: changing reproductive behavior in the United States from 1960 to 1990. The intrinsic rate of natural increase is a measure of what would happen to the population if the birth rate in that year were continued for a long period of time. If the birth rates from 1960 to 1970 were to continued for many generations, the population would shrink fairly rapidly. If the birth rates from 1975 to 1990 continued into the future, the population would grow slowly.

Part of the difference between countries with severe environmental problems and those that have built more sustainable economies is due to natural factors. Some countries lie in biomes where there is more fertile soil and fresh water than elsewhere. But much more of the difference is due to the impact of a country's population on the environment. The most important aspect of population nowadays is how fast it is growing. Almost without exception, countries with rapidly growing populations face serious economic problems, and those with stable populations are prosperous and engaged in solving their environmental problems.

> Countries with rapidly growing populations face serious economic problems.

## 7-A  POPULATION FORECASTS

A frequent reason for studying a population is to forecast what will happen to it in the future. Such forecasts are notoriously unreliable because human reproductive behavior often changes rapidly (Figure 7-2). People do not reproduce randomly. From educated Americans to illiterate African villagers, many people consciously decide how many children to have and when to have them and often achieve their goals. Large groups of people, like the American "baby boomers," often make similar reproductive decisions at the same time in response to a particular

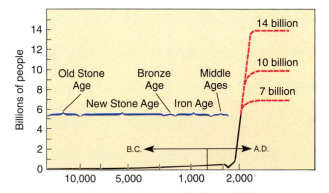

**Figure 7-1**   World population growth, with three projections into the future. The bottom curve assumes that the averge worldwide birth rate will continue to fall, reaching replacement level in 2035. (Replacement level means that each couple is having only the number of children needed to replace them when they die.) The middle curve may well be the most realistic.

economic or cultural situation. Because human beings have so much control over their reproduction, population growth rates can change very rapidly.

> Forecasting population change is difficult because human reproductive behavior is very variable and sometimes changes rapidly.

What factors determine how fast a human population grows? The rates at which people are born and die are two factors, but the migration of people is also important. **Migration** is movement from one place to another. For instance, the migration of people from the countryside into the cities has been an important trend in the twentieth century. Migration affects the size of a country's population only when people enter or leave the country. **Emigration** is the movement of people out of a country or particular population. **Immigration** is the movement of people into a country or population. The numerical size of a population at any one time depends on the balance between birth and immigration, which add individuals, and death and emigration, which remove them:

Change in population size =
  (Births + Immigration) − (Deaths + Emigration)

If (Births + Immigration) exceeds (Deaths + Emigration), the population will grow. If the terms are equal, the population will remain the same size. If the second term exceeds the first, the population will decrease. Because birth rates have been historically high and death rates low in the twentieth century, births and deaths have been much more important than migration rates in determining the population of most countries. One exception is the United States, which has added large numbers of immigrants to its population (Figure 7-3).

Because human behavior is so unpredictable, population forecasts are constantly updated. The characteristics that demographers find particularly useful in making forecasts are the population's age structure, survivorship, fertility, and doubling time.

### Age Structure

Age structure histograms for three human populations are shown in Figure 7-4. The histogram is a bar graph that shows the age and sex composition of a population at the time of a **census**, the procedure by which countries gather facts about their populations. The age structure permits us to draw some tentative conclusions about the population's future development. The parallel sides of the Swedish graph tell us

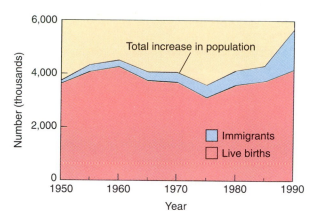

**Figure 7-3** Adding to the U.S. population: legal immigrants and live births from 1950 to 1990. Since 1900, the immigration rate has been about one tenth to one quarter of the birth rate.

that there are roughly the same number of people in each age group. The graph will have this shape if two adults produce two children each, who will replace their parents. If this pattern continues, the population will neither grow nor decrease, and we would predict that it will remain the same size into the foreseeable future.

The narrow base of the U.S. histogram is typical of a population in which growth is slowing down but has not yet stopped. During this transition, the average age of the population is increasing.

Predictions from age structures are not always as easy as they appear. Consider the broad base of the Mexican graph. The population contains more children who have yet to reproduce than it does adults. The obvious conclusion would be that when these children move up to reproductive age, the population will grow rapidly. Without birth control and with low death rates for children, this would be true. In Malawi (Africa), however, which has an age histogram very similar to the Mexican one, 75% of deaths occur in people under 14 years old. Therefore, most of the juvenile population will never grow up to reproduce, and the population is not expanding nearly as fast as one would predict from the age structure. To predict the future of a population, we need to know not only the age structure and the overall rates of birth and death but also whether death occurs before or after people reproduce.

### The Dependency Burden

The age structure of a population gives us a measure of the economic impact of the population's age structure. This is the **dependency ratio**, the ratio of

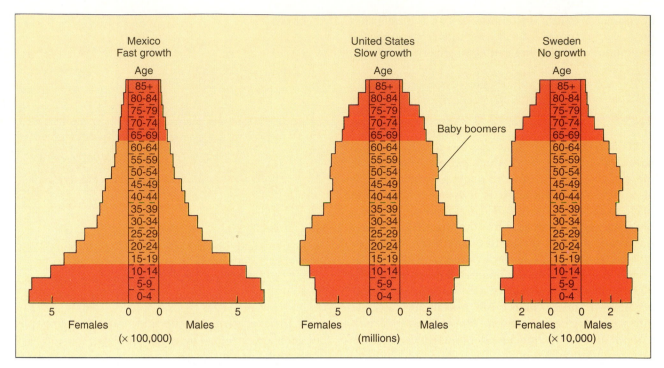

**Figure 7-4**   Age structure histograms for three countries with different demographic profiles. The histograms show the number of males and females in each age group. It is interesting to notice that the "baby boomers" no longer make much of a mark on the U.S. population.

people over 65 and under 15 to the rest of the population. People over 65 and under 15 contribute little to the economy and must be supported by the working population. Japan, for instance, has 47 people over 65 or under 15 years of age for every 100 working people: about two workers to support every dependent. The ratio is only slightly higher in the United States. Although the number of children under 15 is falling in developed countries, the dependency ratio is still rising in many of them. This is because the age at which people retire has not been raised to reflect longer life.

The age at which people are eligible for the old-age pension provided by Social Security was set in the United States at 60 for women and 65 for men in the 1930s, when hardly anyone lived long enough to collect this pension. That situation has now changed completely. Demographers predict that by 2030, the United States will contain fewer than two workers for each dependent, and about half of the dependents will be over 65 rather than under 15. Social Security taxes, already greater than income taxes for most Americans, will eat up almost one quarter of the income of most Americans. In 1990, 24% of the population was under 15 and 13% was

over 65, yet only 3% of the federal budget was spent on those under 15, while 27% was spent on the much smaller number over 65. It is obvious that we are investing in our old people while neglecting young people. Social Security has been called the largest transfer of money from one generation (working people) to another (retired people) in the history of humankind. The steadily increasing Social Security tax alone is enough to account for the fact that working Americans have a lower standard of living now than they did in 1975. A household now has to have two working members to achieve the standard of living a household with only one income enjoyed then.

If the dependency ratio is a fearful burden on the economies of developed nations, it is a much worse problem for less developed countries (where the dependents are nearly all young rather than old). Already many developing countries, such as Ghana in Africa, have 104 dependents for each 100 people of working age, and that dependency ratio is increasing.

Dependency ratios are increasing in most countries, adding to the demands on national income.

## Survivorship

The ages at which death occurs are given by the **survivorship data** for the population. To determine survivorship, we start with a group of individuals born at the same time and follow them to determine the age at which each member of the group dies. We can then plot the number of survivors against age in a survivorship curve (Figure 7-5).

In a Type I curve, most individuals survive for a long time and die as a result of the diseases of old age. Human populations in developed countries show Type I survivorship. In a Type III survivorship curve, most individuals die at an early age. This type of survivorship is found in many invertebrates, fish, plants, and fungi, which produce large numbers of offspring. It is also the pattern of survivorship in very poor human populations where people do not have access to adequate birth control. Here, many babies are born, but most people die before they reach reproductive age. In the intermediate Type II survivorship curve, an individual is just as likely to die at any age. This type of curve is typical of human beings exposed to poor nutrition and hygiene.

Survivorship data can be summarized in **life tables**. Table 7-1 shows a life table for the United States. Note the relatively high infant mortality. **Infant mortality** is death before the age of one year. More deaths occur in the first year of life than in any other five years until after the age of 40. High infant mortality is found in the survivorship of every species of organism. Genetic or developmental defects and birth accidents make getting started in life a risky business.

## Fertility Rates

There are many different ways to describe a population's fertility, each with its own advantages. The **general fertility rate** of a population is the number of babies born each year to 1,000 women of reproductive age. However, the rate of population growth depends heavily on the average age of first reproduction. Therefore, whether a population is increasing, decreasing, or stationary is more accurately discovered from the **age-specific fertility rate**, the number of births per year per 1,000 women of each age group. Figure 7-6 shows the age-specific fertility of women in the United States. From this graph we can see that women between the ages of 20 and 24 have most of the children and that women today are having fewer babies and starting later in life than their mothers did. This confirms our prediction that the U.S. population is becoming both stable (unchanging) and stationary (showing no signs of future change).

One of the most useful measures of fertility is the **total fertility rate**, the number of children that an average woman bears during her lifetime.[1] The total fertility rate for the United States was 3.4 in 1960 and 2.0 in 1990 (1.8 for whites; 2.6 for all other races). In 1972, for the first time, the total fertility rate dropped below 2.1, which is **replacement-level fertility**, the number of children each couple must have if they are to replace themselves. (The number is more than 2.0 because not all children born will survive to reproduce themselves.) If this fertility rate continued, the U.S. population would eventually start to decline (if we ignore immigration). The total fertility rate can, of course, change. It predicts future population size most accurately when it remains constant over a period of time.

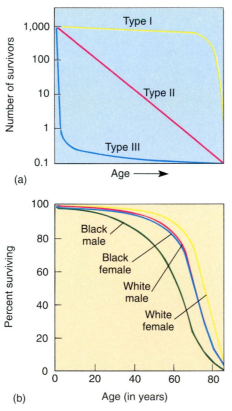

(a)

(b)

**Figure 7-5** Survivorship: (a) Theoretical survivorship curves. (b) Survivorship curves for Americans, based on data from the 1990 census. Survivorship in developed countries approximates a theoretical Type I curve.

[1]Total fertility rate is defined as the number of children that would be born to a woman if, at each year of age, she experienced the birth rate that actually occurred in that year. This produces a hypothetical woman who reflects the age-specific fertility for her lifetime.

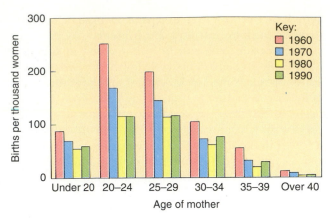

**Figure 7-6** The age-specific fertility of American women in 1960, 1970, 1980, and 1990. Births per 1,000 women in each year are plotted against the age at which the mother gave birth. Note that the number of births in all age groups fell from 1960 to 1980 and then rose slightly in 1990. The average age of the mother has increased during this period (compare women in their twenties in 1960 and 1990).

The age structure, survivorship, and fertility of a population permit fairly accurate predictions of its size and growth rate for about two generations into the future.

## Doubling Time

Wherever we live, it is hard to escape the expensive nuisance of population growth. Houses are going up in the field where I used to walk the dog, a marina now occupies the old fishing hole, and traffic jams get worse every year. Local governments confront taxpayers who balk at increased taxes to pay for a new landfill, expansion of the sewage system, and new roads. Yet most of us live in towns where the population has increased less than 20% in the last 20 years. Suppose the population had doubled in that time. Our existing problems would seem trivial.

In many places, the population *has* doubled in the last 20 years or less. The population of Mexico City doubled between 1960 and 1970 and then doubled again in the next 10 years. The populations of whole

| **TABLE 7-1** Life Table for the Population of the United States, 1990* | | | |
|---|---|---|---|
| **Age Interval** | **Of 100,000 Born Alive** | | **Average Remaining Lifetime** |
| *Period of Life Between Two Exact Ages (in Years)* | *Number Living at Beginning of Age Interval* | *Number Dying During Age Interval* | *Average Number of Years of Life Remaining at Beginning of Age Interval* |
| 0–1 | 100,000 | 981 | 75.3 |
| 1–5 | 99,019 | 196 | 75.0 |
| 5–10 | 98,823 | 88 | 71.1 |
| 10–15 | 98,735 | 132 | 66.2 |
| 15–20 | 98,603 | 491 | 61.3 |
| 20–25 | 98,112 | 550 | 56.6 |
| 25–30 | 97,564 | 605 | 51.9 |
| 30–35 | 96,959 | 736 | 47.2 |
| 35–40 | 96,223 | 935 | 42.5 |
| 40–45 | 95,288 | 1,219 | 37.9 |
| 45–50 | 94,069 | 1,764 | 33.4 |
| 50–55 | 92,305 | 2,724 | 28.9 |
| 55–60 | 89,581 | 4,229 | 24.7 |
| 60–65 | 85,352 | 6,210 | 20.8 |
| 65–70 | 79,142 | 8,706 | 17.2 |
| 70–75 | 70,436 | 11,622 | 13.9 |
| 75–80 | 58,814 | 14,704 | 10.9 |
| 80–85 | 41,110 | 14,879 | 8.3 |
| 85 and over | 26,231 | 26,231 | 9.2 |

*Data courtesy of the National Center for Health Statistics.

countries (such as Syria and Kenya) are doubling in less than 20 years. The contrast between developed and developing countries in doubling time is dramatic. When population increases at a rate of less than 0.5% a year, as it does in many developed countries, the population takes more than 150 years to double. When it increases at more than 3% a year, as in many developing countries, it takes less than 25 years to double.

Doubling time predicts future population size only if the annual rate of population increase remains the same in the future as in the past. This is not very likely. However, doubling time is an excellent way of dramatizing the adverse effects of population growth. People can imagine the difficulty of doubling the supply of water, food, and energy to their own neighborhoods within 20 years and the strain such a population increase would impose on schools, hospitals, jails, police protection, and every other vital service.

Most developed countries have population doubling times of more than 100 years. The populations of many developing countries double in less than 25 years.

## 7-B DECLINING DEATH RATES

People leave populations when they die or emigrate. The death rate is usually expressed as the number of deaths per year for every 100 members of the population. In most countries, death rates and birth rates are much more important than immigration and emigration in determining the size and growth rate of the population. Modern population problems arise because we have reduced the death rate but not the birth rate.

The number of people on Earth has increased steadily since at least 10,000 years ago, but the greatest population growth has taken place in the last 200 years. This is largely because death rates have been dramatically reduced, particularly among infants. The discovery of antibiotics in the twentieth century has made a slight contribution to the decline of the death rate, but much more important were improvements in nutrition and hygiene.

Reducing death rates without reducing birth rates has caused the human population explosion.

## Life Expectancy

Infant mortality is the most important factor determining **life expectancy**, the average number of years that a newborn baby can be expected to survive. In 1900, life expectancy in the developed world was 46 years, largely because the infant mortality rate was about 40 per thousand births. In 1991, the infant mortality rate for developed countries had declined to fewer than 10 per thousand, and life expectancy was more than 70 years. The world's highest life expectancies are those of Japan, Israel, Canada, Australia, and the European Community, where people can expect to live beyond age 77 (with women outliving men by about six years). In 1991, life expectancy in the United States was 75.7 years, lower than that of several much poorer countries, such as Costa Rica.

One reason for low life expectancy is that the United States has one of the higher infant mortality rates in the developed world. In 1991, it was 10.3 per thousand, compared with 4.4 for Japan and 4.7 for Switzerland. This is because the quality of American education and medical care is worse than in other developed countries, all of which have national health services. Studies on a Navajo reservation in Arizona showed that the presence of a modern hospital within easy reach saved only one life in 10 years. In contrast, a district nurse, who instructed parents on infant care, reduced infant mortality to 6 deaths instead of the 15 expected on the basis of the infant mortality rate before the arrival of the nurse.

In 1991, developing countries had infant mortality rates ranging from 17 per thousand in Chile to 163 per thousand in Afghanistan. Most infant deaths in low-income countries are due to respiratory and digestive tract infections, and such deaths can be reduced without expensive medical care. The simple lack of clean water, nutritious food, and elementary education in hygiene is the reason that life expectancy for most of Africa remains about 45 years, whereas the average for Latin America and East Asia has risen to 65 years since 1950.

The biggest reductions in infant mortality are achieved by improving education, nutrition, and hygiene. Infant mortality is the main factor determining life expectancy.

## 7-C THE DEMOGRAPHIC TRANSITION

In 1945, demographer Frank Notestein outlined the theory of the demographic transition. The theory states that the economic and social progress caused by the industrial revolution affects population growth in three stages.

1. In the first stage, in preindustrial societies, birth and death rates are both high, and the population grows slowly, if at all.

2. In the second stage, the population explosion occurs. Death rates fall as education and public health measures are introduced, but birth rates remain high. The population grows very fast. It increases by about 3% per year, which means it doubles every 25 years, and is 20 times its original size by the end of 100 years.

3. In the third stage, birth rates fall until they roughly equal death rates, and population growth slows down and stops.

The theory of the demographic transition was based largely on what had happened to European populations. It is clear that the developed countries of Europe and parts of Asia have indeed passed through these stages as Notestein described them. Populations in these areas have grown as much as fivefold since 1850 but today are growing slowly or not at all. The United States is unusual among developed countries in that its population is still growing rapidly, although the growth rate has slowed since 1960 (Figure 7-7).

The single factor most clearly correlated with a decline in birth rate during the demographic transition is increasing education and economic inde-

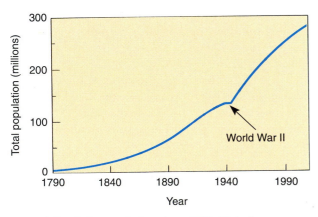

**Figure 7-7**   U.S. population since 1790. This does not give an accurate picture of the excess of births over deaths because of additions to the population from immigration and the addition of new territory to the country. The dip in population growth around 1940 is not due to deaths during World War II (many more Americans were killed during the Civil War, which ended in 1865). It is due to low immigration rates during the Great Depression of the 1930s and to very low birth rates from 1941 to 1945.

pendence for women (Close-Up 7-1). The spread of knowledge that lowers the death rate is usually part of a general program to improve education. Educated women find that they need not bear a large number of children to ensure that a few survive, and they also learn contraceptive techniques. In addition, women find that they can contribute to the family's increasing prosperity by holding a job and by spending less time and energy on raising children. This is attractive to women, even in countries where religious doctrine and tradition dictate large families.

> Low birth rates are found only where women have at least some education and economic control over their own lives.

The demographic transition has taken from one to three generations to spread through the populations in most developed countries. During its progress, death rates are low, but birth rates remain high, and so the population grows enormously. The population of the United States will have almost quadrupled in the twentieth century.

> A country's population usually doubles at least twice as it passes through the demographic transition.

### The Demographically Divided World

The idea of the demographic transition has been widely used to predict and explain national and global population changes. But some countries are not proceeding through the demographic transition as the theory predicts they should. They appear to have stopped in the second stage, unable to make the educational and economic gains that would reduce the birth rate.

If the vast rate of population growth in the second stage is maintained for any length of time, it begins to overwhelm food production, medical care, and education. When the demands of the population exceed the production of local agricultural land, forests, and freshwater supplies, people begin to consume the resource base itself. Vegetation disappears, soils erode, wells run dry, and productivity declines.

Natural resources are degraded when the population is growing rapidly, even in developed countries. Millions of hectares of forest and billions of tons of topsoil have disappeared from the United States during the last 200 years. But in the developed countries, growth has now decreased or stopped, giving these countries the chance to slow the destruction of resources and even to repair some of the damage. In many developing countries, populations continue to grow, while resource degradation has reached the point at which the carrying capacity of the land has been significantly reduced. This in turn reduces food production and incomes in a downward spiral.

> While its population is growing rapidly, a country's natural resources are degraded rapidly.

Population trends appear to be driving about half the world's people toward a better future and half toward economic decline. The world is divided into countries that have completed the demographic transition and countries stuck in the second stage, with only a few countries still in the process of passing through the demographic transition. Table 7-2 illustrates this for Asian countries. The countries are divided into two groups based on their 1990 fertility rates to see whether fertility rates (births per woman during her lifetime) were correlated with wealth and literacy. The countries at the bottom of the table have fertility rates of more than 2.5. The countries at the top of the table have fertility rates of 2.5 or less. (The rate of 2.5 is slightly higher than replacement rate fertility.) How close is the correlation between fertility rate and gross national product per person?

In slowly growing areas, mainly Western Europe, North America, Australia, China, and Japan, populations are growing on average 0.5% per year, and living conditions are fairly stable. Some ecological resources may actually be increasing, as these countries invest in pollution control, reforestation, and restocking wild populations of plants and animals. In complete contrast, populations in high-growth countries, chiefly in Southeast Asia, Latin America, India, the Middle East, and Africa, are growing more than three times as fast—on average by 2.5% per year (Close-Up 7-2). The slow-growth countries add 19 million people a year to the world's population, and the fast-growth countries add 80 million. In many of the fast-growth countries, population growth and falling incomes reinforce each other so that living conditions decline.

Southeast Asia is the only part of the world where several countries are probably still proceeding through the demographic transition toward slow growth and prosperity. Fertility is falling rapidly in Thailand and Indonesia, which have good family planning programs. In the same region, however, Vietnam and the Philippines have high birth rates and falling standards of living.

## 7-D FAST-GROWTH COUNTRIES AND SUSTAINABLE POPULATIONS

One problem in fast-growth countries is that falling standards of living lead to rising death rates, pushing these countries back toward the preindustrial

**TABLE 7-2** Population Growth and Wealth in Asian Countries, with Estimates for the Year 2000

| Country | Population (Millions) | | | Total Fertility Rates* | | | GNP per Person, $† | Literacy Rate, % |
|---|---|---|---|---|---|---|---|---|
| | 1965 | 1990 | 2000 | 1965 | 1990 | 2000 | | |
| China | 729 | 1,134 | 1,299 | 6.4 | 2.5 | 2.1 | 547 | 73 |
| Hong Kong | 4 | 6 | 6 | 4.5 | 1.5 | 1.5 | 11,490 | 90 |
| Japan | 99 | 123 | 128 | 2.0 | 1.6 | 1.6 | 23,558 | 99 |
| Singapore | 2 | 3 | 3 | 4.7 | 1.9 | 1.9 | 11,160 | 88 |
| South Korea | 29 | 43 | 46 | 4.9 | 1.8 | 1.8 | 5,400 | 96 |
| Thailand | 31 | 56 | 64 | 6.3 | 2.5 | 2.1 | 1,420 | 93 |
| Afghanistan | 12 | 18 | 27 | 7.0 | 7.2 | 6.8 | 150 | 29 |
| India | 495 | 850 | 1,042 | 6.2 | 4.0 | 3.0 | 350 | 48 |
| Indonesia | 107 | 178 | 219 | 5.5 | 3.1 | 2.4 | 570 | 77 |
| Malaysia | 10 | 18 | 22 | 6.3 | 3.8 | 3.0 | 2,320 | 78 |
| Nepal | 10 | 19 | 24 | 6.0 | 5.7 | 4.6 | 170 | 26 |
| Pakistan | 57 | 112 | 162 | 7.0 | 5.8 | 4.6 | 380 | 35 |
| Philippines | 32 | 61 | 77 | 6.8 | 3.7 | 2.7 | 380 | 35 |

*Births per woman during her lifetime.

†In 1990. For comparison, Switzerland had the highest gross national product (GNP) per person in the world in 1990 at $35,081. GNP per person for the United States was $21,863.

## The Position of Women

"Educating girls probably yields a higher rate of return than any other investment available to a developing country," said Lawrence Summers, Chief Economist at the World Bank. Summers used figures from Pakistan to illustrate the point. In Pakistan, educating an extra 1,000 girls for an additional year cost about $40,000 in 1990. Studies show that an extra year of education for Pakistani girls produces a return of more than 20% on this $40,000 investment, including:

1. An increase of 15% in each woman's wages.

2. A 10% reduction in fertility, preventing 660 births (which would cost $43,000 by medical means).

3. A 10% reduction in deaths among children under five born to the women, preventing 60 deaths (which would cost an estimated $48,000 by medical means).

4. The prevention of four deaths of women in childbirth (about $10,000 worth of medical care).

Parents in low-income countries do not invest in their daughters because they do not expect them to make a financial contribution to the family: girls grow up to marry into someone else's family, so daughters are kept home to do chores while sons are sent to school. Equalizing just the primary education of boys and girls in developing countries would mean enrolling an extra 25 million girls a year

**Figure 7-A**  A school in Costa Rica, one developing country that believes in educating its girls and provides free primary and secondary education for all.

in school. Equalizing secondary education would mean educating an extra 21 million girls. Summers estimates that eliminating educational discrimination would cost $2.4 billion a year, less than 1% of the gross national product of the low-income countries and less than 10% of their defense spending (Figure 7-A).

The educational level and economic power of women affects not only population growth but also many other aspects of environmental affairs. For instance, African women gather most of the wood used for fuel and carry home the family's water (Figure 7-B). They also perform more than half the agricultural labor, confronting the problems of soil erosion, drought, overgrazing, and the destruction of habitats. The United Nations found that in Tanzania the average woman works for more than 3,000 hours a year, the average man for 1,800 hours. Kathryn Scott Fuller, president of the World Wildlife Fund, reported, "Women perform two thirds of the world's work, but they receive only 10% of the world's income and they own only 1% of the world's land."

Fuller has found that women have so little power that it is difficult to include them in decisions that affect conservation. In Zambia, the World Wildlife Fund enlisted villagers to prevent the killing of elephants for ivory. Profits from tourist visits to local elephant preserves were used to hire local men to patrol for poachers. Elephant poaching stopped, and a survey of village leaders showed that they were happy with the new prosperity from the villagers' added income. Then the survey was extended to include women, and a different story emerged. The women's gardens, which provided food for the

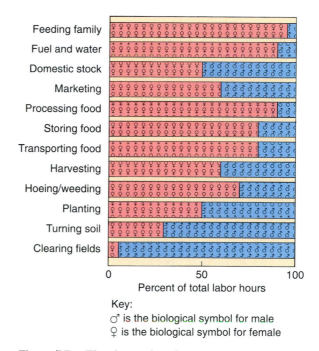

Key:
♂ is the biological symbol for male
♀ is the biological symbol for female

**Figure 7-B**  The share of work performed by women and men in African countries.

whole village, were being trampled by growing numbers of elephants, and the women felt their standard of living had fallen as a result of the program. Zambian conservationists are now attempting to provide elephant-proof fences for the gardens. But Fuller reports that even as the Fund attempts to include women and the resources they manage in conservation projects, the women can have little effect on village decisions until they can attain some political power by improving their socioeconomic status.

# Population Growth in Arab Countries

Walk along Avenue Habib Bouruiba on any working day and you will see cafes brimming with bored young men: 40% of Tunisians are unemployed (Figure 7-C). Food riots in Morocco, millions of homeless in Cairo, and infant mortality rates of up to 100 per 1,000 all attest to one of the world's fastest growing population groups (Figure 7-D). No Arab country has a growth rate of less than 2.5% a year, and Arab fertility rates are twice the world average. The population of the Arab world is now about 200 million; it is estimated that this will double in the next 23 years.

To combat population growth, Arab leaders have to overcome societal beliefs that large families are a blessing and that contraception is not compatible with Islam, the religion of most Arabs. Both beliefs are associated with the powerless position of women. Jordan and Egypt are among the leaders of attempts to control their population explosions. Their efforts are opposed by fundamentalist Muslims, who oppose changes that would give more power or education to women and who have gained political power with the increasing economic problems that have resulted from rapid population growth.

**Figure 7-C**   The countries of the Arab world. Arabs are those who speak Arabic as their native language. Although most Arabs are Muslims (Islamic), most Muslims are not Arabs. For instance, the largest Muslim population in the world is in Pakistan, a non-Arab country.

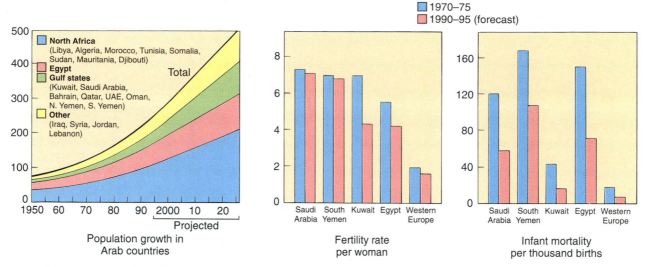

**1970–75**
**1990–95 (forecast)**

North Africa
(Libya, Algeria, Morocco, Tunisia, Somalia, Sudan, Mauritania, Djibouti)
Egypt
Gulf states
(Kuwait, Saudi Arabia, Bahrain, Qatar, UAE, Oman, N. Yemen, S. Yemen)
Other
(Iraq, Syria, Jordan, Lebanon)

Total

Projected

Population growth in
Arab countries

Fertility rate
per woman

Infant mortality
per thousand births

**Figure 7-D** Demographics of the Arab world. Note the decline in infant mortality from the 1970s to the 1990s shown by the right-hand graph. This decline, due to the introduction of modern hygiene and medicine, is a large part of the reason for the population explosion since 1970. The middle graph shows that Egypt's population policy has had some success in lowering fertility rates. The reduction in fertility rates in Kuwait in recent years may be due to the fact that women probably have more economic power in Kuwait than in any other Arab country.

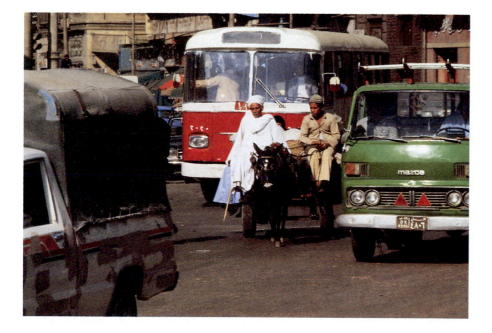

**Figure 7-E** Middle Eastern cities, such as Cairo, Egypt, have seen explosive growth in recent years resulting in massive traffic jams. The blare of automobile and truck horns is routine in downtown Cairo. (Iverson/Gamma Liaison)

conditions of the first stage of the demographic transition.

Consider Ethiopia, Nigeria, and Kenya, African countries with growth rates that would cause their populations to quadruple by the middle of the twenty-first century. In fact, these countries cannot possibly support four times their present populations. They already appear to have populations higher than their carrying capacities, known to demographers as the **sustainable population**, the number of people the land can support indefinitely.

The sustainable population of a particular area depends on the availability of the vital resource that is in shortest supply. For instance, in parts of Arizona and the southeastern United States, the freshwater supply limits the number of people who can live in an area (Chapter 11). United Nations and World Bank officials find that wood (for use as fuel) is the vital resource limiting the populations most African countries can support. In seven African countries, which actually contain 31 million people, it has been calculated that the net productivity of the forests could supply fuelwood for a sustainable population of 20 million people, although local agriculture could support about 36 million.

The populations in these countries are already larger than the sustainable wood supply can support. This means that people are cutting down more trees each year than are being replaced by reforestation or natural growth. When trees and shrubs are cut down and not replaced, roots no longer hold the soil in place and it blows or washes away. When the soil is gone, it may never be possible to grow trees in that place again: the sustainable population of that area will have been reduced.

Why cannot developing countries continue to support populations that are higher than the carrying capacity of the land? Economies do not depend entirely on the local supply of natural resources. Countries can exchange the labor of their people for natural resources they need that are produced elsewhere. Densely populated countries, such as Japan and Taiwan, do just this, importing food, raw materials, and fuel and paying for them by exporting manufactured goods. Such a strategy, however, depends on an educated work force to build industries that produce manufactured goods. In many developing countries, population growth has overwhelmed the educational system. Most of the work force is illiterate, and these countries educate hardly any of the engineers, designers, and industrialists needed to produce competitive manufactured goods. These countries have relied on exporting natural resources, such as timber and minerals. As these are used up, the economy declines. In addition, the scale of the problem is so vast (more than half the world's population lives in fast-growth countries) that the wealthy half of the world probably does not have the human and financial wherewithal to halt the erosion of resources in places such as Africa and India.

When a population grows beyond carrying capacity, it may follow what has been called an "ecological transition" that is almost the opposite of the demographic transition. In the first stage, human demands for natural resources such as wood, water, and agricultural land are growing but are lower than the sustainable yield of the area. In the second stage, when the population exceeds the country's carrying capacity, demand is still growing but can be satisfied only by importing food and fuel and consuming the resource base (forest, soil, or water). As the country's international debt (trade deficit) grows, imports slow down, living standards fall, and natural resources are consumed even faster. Countries such as Malawi, Mozambique, and Somalia may even have reached the theoretical third stage, in which the death rate would rise beyond the birth rate. Africa is not the only place where populations higher than carrying capacity are found. India, Central America, western South America, and Brazil are probably in the same situation and are also doing irreversible damage to their natural resources.

> Countries that do not complete the demographic transition seem to become trapped in an ecological transition that will eventually lead to declining populations and increasing poverty.

It is very difficult to predict how many people a country could support indefinitely because there are so many variables. The number depends on how much food is imported, how much fossil fuel and machinery are used to modernize agriculture, how many trees are planted each year, unpredictable technological advances, and so forth. A World Bank team concluded, however, that in many African countries it is not economically feasible to increase the production of food or wood much above its present level, largely because so much of the soil is already seriously eroded and so much forest destroyed. Seriously eroded soil produces a low yield of crops no matter how much money is spent on irrigation and fertilizer. And we shall see that deforestation is irreversible in some tropical areas because cutting down the trees destroys the soil (Chapter 14).

It seems inevitable, then, that many fast-growth countries face economic decline. They will continue to use up their natural resources faster than they can be replaced and regress to the first stage of the demographic transition. Most demographers believe that, as a result of this, the world population will peak sometime in the next century and must then decline to a lower level than the population on Earth today.

> The ecological transition may well lead to a world population that is smaller in 2100 than it is today.

## 7-E ECONOMICS AND POPULATION GROWTH

In the 30 years after World War II, the global economy grew briskly, raising incomes everywhere. Between 1950 and 1973, the world economy grew at about 5% per year. Incomes rose in every country, whatever its population or stage of economic development.

Since 1975, however, the world economy has been sluggish for many reasons. For instance, during the 1970s, the price of oil rose, and almost every country is economically dependent on oil. To keep their economies expanding, many countries borrowed heavily to pay for oil and, like the United States, are now crippled by debt. Another contributor has been agricultural decline. Despite new, productive crop varieties, poor agricultural policies and soil erosion have slowed the growth of agricultural production.

This worldwide economic slowdown is bad enough, but where it is reinforced by rapid population growth, it is disastrous. Consider grain production in Western Europe and Africa. In 1950, the two were not far apart. Western Europe produced 234 kilograms of grain per person and Africa 157. In 1990, however, Western Europe produced 511 kilograms of grain per person, whereas Africa produced only 146. This dramatic difference is not due to a decline in African productivity. Since 1950, African grain production has grown by 129%, little less than the 164% growth in Europe. The difference is in population, which grew by about one fifth in Europe, while it more than doubled in Africa. The result is economic boom for Europe and decline for Africa. Western Europe produces more and more grain per person, leaving an ever-larger surplus that can be exported. Africa, which could barely feed its population in 1950, now cannot do so and must spend money to import food or see malnutrition spread. Either way, the income of most Africans goes down.

When population grows faster than productivity, incomes fall. The demographic differences between countries today are so great that the rich are getting richer and the poor are getting poorer quite rapidly. During the 1970s, Africa became the first region to experience a decline in income per person since the 1930s. During the 1980s, Africa was joined by Latin America, where incomes are down more than 19% since 1980. Some of the low-growth countries provide a stunning contrast to this decline. The average income of China's 1 billion people has doubled since 1980, and many other countries in this low-growth group have exceeded 10% growth in personal income.

> Between World War II and 1970, incomes rose throughout the world. Since 1980, incomes have declined in many countries with fast-growing populations.

### Political Causes and Effects

Rapid population growth is not the only reason for falling incomes and low productivity in many countries. Incompetent government also contributes. A government that is competent to educate its people and slow population growth is also usually better at managing its natural resources and agricultural programs.

But politics is a two-way street. While government may affect population growth, population growth also affects government. In industrialized countries, most of the population lives in towns and is employed in industry or services. In the countries whose populations are growing most rapidly, most of the population is rural. As the population grows, more and more of those living in the countryside cannot own or rent land or find other work. This situation is extreme in Central America, where 5% of the population often owns 80% of the land. The situation in Brazil and India is almost as bad. Since farming is almost the only work available in the countryside, landless rural people live a precarious existence. Even in the United States, such people have shorter life expectancies and higher rates of malnutrition than other members of society. Rural people who do not own land have the least political power of any group, so they have little to lose. It is not surprising that they sometimes band together in lawless groups that roam the countryside making a living as they can or seeking to overthrow the government, as is the case with the communist bands in the Philippines and the civil war in Somalia.

The solution to this problem is twofold. The problem cannot be solved until population growth is brought under control and agricultural reform permits economic development in the countryside. After World War II, China, Japan, and South Korea instituted land reform programs that have proved very successful. Land reform usually consists of removing land from the hands of a few wealthy landlords, but the problem is sometimes just the opposite. In Bangladesh and Kenya, privately owned land has been divided into ever-smaller plots as it is split up among children, until most plots are so small that they cannot support a family.

Rapid population growth tends to promote social unrest in several ways. If it is accompanied by slow economic growth, it leads to unemployment and to competition for everything from food to space. Even without widespread unemployment, it strains educational and social institutions. One inevitable result is that young people come to dominate society. In many countries, half the population is under 20 years of age—the age group most prone to violence and crime. Disputes between regions and countries are also generated by population growth. For instance, rivers often flow through more than one country. Even within the United States, disputes about how much of a river's water a particular state can use are commonplace. The problem is much worse in cases such as the Nile River, where the amount of water that flows through Egypt, Sudan, and Ethiopia may be a matter of life and death. Riots over food subsidies in Egypt, illegal immigration into the United States and Europe, and tribal conflicts in Somalia are all the direct result of rapid population growth.

## 7-F  BIRTH CONTROL

Since the dawn of history, people have regulated their reproduction so that the number of offspring bears some relation to the parents' ability to raise them. In some traditional societies, the size of the population remains stable because the culture encourages abstention from intercourse, abortion, late marriage, infanticide, and prolonged breast-feeding. Some of these practices are methods of **birth control**, anything that prevents birth. **Contraception** refers to birth control methods that prevent fertilization. The only methods of contraception that are 100% reliable are abstention and sterilization.

### Birth Control Research Needed

The ability to control the number of children is vital to the physical and economic health of a family. A study in Kenya showed that the health of all members

of the family was maximal if a woman had her first child after the age of 20 and had no more than three children, spaced two to three years apart. Studies in developed countries produce the same conclusion. Babies born to teenagers are disadvantaged compared with those born to older mothers. They are usually smaller than the optimal birth weight of 8 pounds, have a greater risk of ill-health and early death, and are more likely to be born into poverty. The mother's health is better and the educational achievement of each child is greater if children are spaced more than two years apart.

It therefore seems astonishing that we have made so little progress toward producing effective, acceptable methods of family planning. The United States is not particularly progressive in this respect, banning or limiting various forms of birth control for political reasons and financial reasons. For years, the French "morning-after" pill, RU486, was illegal in the United States although in other countries it had safely terminated thousands of unwanted pregnancies. Another example is Norplant, a hormone implant that prevents pregnancy for up to five years. Norplant has been used in many countries for more than 10 years. It was not approved for use in the United States until 1990. Another peculiar fact about Norplant is that drug companies charge about $25 for this implant in many countries but sell it for $350 in the United States. Birth control methods are discussed in Close-Up 7-3.

### Importance of Suitable Birth Control

For most people in the world, suitable contraception does not exist or is not available. What is needed is contraceptives, both long-lasting and short-term, that can be purchased cheaply, without prescription, in every neighborhood and then used effectively by people who cannot read instructions. Another necessity is education on contraceptives in school from an early age so that children grow up secure in the knowledge that they have the tools to plan their families. Consider a couple of examples of the inadequacy of birth control methods today.

India, with one of the world's fastest growing populations, has concentrated on the use of sterilization, which is one of the cheapest methods of contraception. But sterilization is not available without health care professionals because it involves a minor operation. It is no use to people who have not yet had as many children as they want, and it does not permit you to space your children. The situation is different, although not much easier, in developed countries. Here, the pill is the most widely used method of birth control, but it cannot be obtained without a prescription and is expensive. One of the reasons for the number of births to unmarried teenagers in developed countries is that effective birth control methods for women require visits to a physician or clinic, even in countries where the contraceptives are cheap or subsidized, which they are not for most people in the United States.

> Methods of birth control suitable for everyone who needs them have not yet been invented.

## 7-G  CONTROLLING POPULATION GROWTH

Health services have played a vital role in the decline of both birth and death rates. The twentieth century has witnessed extraordinary efforts by governments, private agencies, and international organizations to extend health care to almost every person in the world. Education, immunization against diseases, improved hygiene, and antibiotics have all been part of this effort. Mobile clinics and teams of health workers have brought immunization and education for pregnant women and mothers even to widely scattered rural populations.

### Population Policy

Many governments have announced official policies on population (Table 7-3). They promote these policies by publicity campaigns or financial incentives. If they want their populations to grow, governments encourage people to have children and welcome immigration. Many developed countries provide a weekly allowance to families with children as well as tax deductions for children. These are usually not

**Figure 7-8**   Adoption: doing your bit for population control. These are my children, with their 100-year-old great grandmother. One is my biological child; the other we adopted in Costa Rica. We have all enjoyed getting to know that country and learning some Spanish. There are millions of children all over the world whose parents have died or cannot take care of them. If everyone who could afford it adopted one of these children instead of having their own, it would reduce the rate of population growth considerably.

viewed as incentives to having children, however, because they provide much less than it actually costs to raise a child in a developed country.

Governments usually view slow population growth as a good thing. For instance, two thirds of the U.S. gross national product consists of consumer goods. The traditional view is that the more consumers there are, the greater the demand and the stronger the economy. But governments desiring population increases have found that it is difficult to increase the fertility rate of a country where people are reasonably well educated. Romania under communist rule, for instance, outlawed abortion, its main birth control method, and imposed a tax on single adults and childless couples. Women had to submit to monthly pregnancy tests, and persistent nonpregnancy or sudden termination of a pregnancy had to be explained. Despite these measures, the government goal of four children for every Romanian woman was not attained. The population has grown by less than 1% per year since 1970. Despite the low birth rate, Romania was so poor while it was under a communist government that many children were abandoned by parents who could not support them and were later adopted by people from other countries (Figure 7-8).

Effective governments have had much more success with campaigns to slow population growth by

| | Current Total Fertility Rate* | Target Fertility Rate | Target Year |
|---|---|---|---|
| **TABLE 7-3   Population Policy: Fertility Targets for Selected Countries** | | | |
| **Country** | | | |
| Nigeria | 6.5 | 4.0 | 2000 |
| Tunisia | 3.4 | 1.15 | 2001 |
| Jamaica | 2.6 | 2.0 | 2000 |
| Bangladesh | 4.9 | 2.34 | 2000 |
| Nepal | 6.1 | 2.5 | 2000 |
| Turkey | 3.6 | 2.89 | 1995–2000 |
| India | 3.9 | 2.0 | 1995–2000 |

*For comparison, the U.S. total fertility rate in 1991 was 2.1.

# Birth Control Methods

Modern contraceptives can be safe and effective, but each of them has major disadvantages. For instance, the most reliable methods are expensive and require the ongoing services of health care professionals. This means that they are not freely available to those who need them most: those in rural areas far from medical assistance, unwed teenagers, and poor people.

**Contraceptive Pills** ''The pill'' contains synthetic hormones that prevent ovulation, the release of an egg from the ovary. Pregnancy is avoided because there is no egg to be fertilized. The hormones in the pill imitate those of pregnancy, and the medical risks of using the pill are largely the same as those of pregnancy. The pill does not reliably prevent ovulation and pregnancy until it has been taken regularly for at least two weeks. Women who have intercourse during this period or who experience condom failure can protect themselves from pregnancy by taking additional contraceptive pills within 72 hours of unprotected intercourse.

**Implants** Implants of hormones that prevent ovulation are the ideal contraceptive for people in developed countries who wish to postpone pregnancy for any length of time. With Norplant, a set of matchstick-sized implants is placed under the skin of the upper arm. When it is removed, full fertility is restored. Because they last for years, you do not have to bother to remember to take a pill every day, and implants are also generally cheaper than pills.

**Injections** There are several contraceptives that are injected and prevent conception for about three months. They have been used with considerable success in a number of developing countries but are not yet widely available.

**Condoms** The condom is the contraceptive device most commonly used by men. A condom is rolled onto the erect penis shortly before intercourse. It catches the semen so that sperm do not enter the female reproductive tract. Latex condoms are important as the only form of birth control that can reduce or prevent the spread of sexually transmitted diseases, such as acquired immunodeficiency syndrome (AIDS), herpes, gonorrhea, and syphilis, from one partner to the other.

**Diaphragm and Spermicides** Another contraceptive is a rubber diaphragm, which is smeared with a spermicidal (sperm-killing) jelly or cream each time it is used. Inserted into the vagina, it blocks the entrance to the uterus so that sperm cannot reach the egg. A similar device is a foam sponge, impregnated with spermicide, which has the advantage that one size fits all, but it is not very reliable.

**Breast-feeding** The hormones that circulate while a woman is breast-feeding reduce the chance that she will ovulate. In traditional societies, breast-feeding is usually prolonged for at least two years and is an age-old method of birth control. It also provides the only complete nutrition for a baby and supplies the baby with antibodies that protect it from disease. Tragically, this natural, free method of reducing the chance of pregnancy has been undermined in developing countries. Western baby food companies, notably Nestlé, embarked on a campaign to convince mothers that bottle-feeding was the chic Western way to raise a baby. Many of these mothers cannot afford enough formula to feed a baby properly and do not have access to clean water with which to mix the formula. Millions of babies died of malnutrition and disease as a result of this campaign, and millions of unwanted babies were born to mothers who gave up breast-feeding because of it. An international outcry finally caused Nestlé to announce that it would abandon the campaign. But breast-feeding has still not become universal again.

**Male contraceptives** Men are demanding more control over their reproduction. The 1980s witnessed an enormous increase in the number of paternity lawsuits in the United States. In such a suit, the mother seeks to force the father to provide financial support for a child. Men defending paternity suits have frequently claimed they were misled by partners who said they were using birth control but in fact were not. When the woman becomes pregnant and demands that the man support the child, the man feels himself defrauded. (For those interested in the outcome, a man is legally liable for his child's support whether he was defrauded into fatherhood or not.) Research is proceeding into antifertility vaccines for men, reversible sterilization, and antisperm drugs.

**Sterilization**  Sterilization is any more or less permanent change that prevents an animal from reproducing sexually. Sterilization is the fastest growing method of contraception in the world. The most common sterilization operation for men is vasectomy, that is, tying off the tubes that transport sperm into the semen. Sterilization of a woman usually involves tubal ligation, cutting and tying the oviducts that transport an egg to the uterus. After tubal ligation, sperm cannot reach the eggs, and thus fertilization cannot occur. Sterilization that can be reversed without medical assistance is under development and may be the contraceptive technique of the future.

**Abortion**  Women without access to effective contraception, or those for whom contraceptives fail, often turn to abortion, legal or illegal, to limit the size of their families. The number of abortions is an indication of how adequate family planning services are. Thus, in Latin America, where contraception is opposed by the Catholic Church, the number of abortions approaches the number of births. In for-

mer Soviet-bloc countries, where health care is inadequate, the International Planned Parenthood Federation estimates that the abortion rate is up to twice the birth rate. In countries where family planning services are cheap and available, abortion rates are lower than elsewhere.

Abortion was largely accepted, and fairly common, in the United States and Europe until the early nineteenth century. Before this time, a woman had up to a one third chance of dying every time she gave birth to a child, whereas abortions were less risky for the mother, so abortions often saved women's lives. After midwives found that washing their hands and clothes improved their patients' survival rates, childbirth became less dangerous to a woman than a nineteenth-century abortion, and physicians started to oppose abortion. Most religious and ethical opposition did not develop until some time later. The situation has changed since World War II now that the danger of abortion to a woman is again less than the risk of a completed pregnancy. Abortion is, once again, legal in most countries, although in the United States about 83% of women do not have access to legal abortion because clinics that perform the procedure are few and far between. In 1988, about half the world's people lived in countries where legal abortions were freely available.

curbing the birth rate. In 25 developing countries where the government has given high priority to family planning services, birth rates have dropped by 20% to 60%, and contraceptive use approaches that in developed countries. One example is China, where the population growth rate fell from an average of 1.8% annually in the 1970s to 1.4% in the 1980s and is projected to fall to 0.9% by 2000. China's methods involve universal access to family planning (including abortion), economic incentives, and an educational campaign extolling the virtues of smaller families. China achieved the most rapid fall in the birth rate ever recorded. Reports of human rights abuses resulting from this policy filtered out of China, and the methods employed were undoubtedly more coercive than those any democratic government could use. Nevertheless, many noncommunist governments have brought about declines in their birth rates almost as rapid as those achieved by China: Taiwan, Barbados, Singapore, and Egypt are examples. These countries combine access to subsidized family planning services with intensive educational campaigns. Their success is all the more remarkable in that many face religious establishments opposed to birth control.

Singapore's methods of reducing population growth to almost zero can be used as an example. The government supplied intensive sex education and free family planning services, including abortion and sterilization. Having large families was made a burden: women were denied paid maternity leave after the first two children, the delivery fee for childbirth rose with each birth, there were no income tax exemptions for children after the third, and there was no subsidized housing for large families. In 1983, Singapore discovered that educated women were having fewer children than uneducated women. Concerned that this might lower the general intelligence level of the population, the government introduced measures to try to reverse the trend. In a move designed to reduce childbearing among the less educated, a cash payment roughly equal to a family's annual income was offered to any woman who was sterilized after her second child. Educated parents received more tax relief for their children than the uneducated. For a time, top educational opportunities went to the third and later children of women with university degrees. This provision was denounced as discriminatory and was, therefore, rescinded in 1985. The Singapore program shows that financial incentives can be very effective in reducing population growth.

> Effective government policies can rapidly reduce the birth rate in a country.

**Figure 7-9**  Celebrating Children's Day in China, which has the fastest rate of economic growth in the world, partly as a result of its population policies. Most urban families are allowed to have only one child. As a result, children are cherished and a cause for celebration. (If your one pregnancy results in twins, you have really hit the jackpot!) *(Tricia Smith)*

## International Family Planning

Many poor countries with rapid population growth have official policies designed to lower the birth rate, but either because the government is not in effective control of the country or from lack of money and trained personnel, the policy has little effect. Developed nations have poured resources into programs designed to help the developing nations reduce their population growth rates because population growth in other countries affects us all. The destruction of tropical forests contributes to the greenhouse effect, the debt of many countries endangers the international banking system, and migration of environmental refugees is a growing problem (see Section 8-C). Family planning programs are exceedingly cost-effective. International agencies estimate that every dollar spent on population control returns between $3 and $25 in economic savings (Figure 7-9).

Despite international efforts, there is still a huge unmet need for birth control services, even in developed countries. In the United States, programs that make contraceptives freely available have reduced teenage pregnancies by up to 50% but are available in only a handful of cities. A world fertility survey found that half of all women in developing countries said they wanted no more children. But a World Bank survey reported that half of all women who want no more children do not practice contraception—usually because they have poor or no access to family planning services.

> International investment in birth control is still inadequate, although such investment is highly cost-effective.

In some countries, private charities and organizations, such as Family Planning International Assistance and the United Nations Family Planning Agency, supply some birth control services. Both these agencies receive large sums from the U.S. Agency for International Development. It was long-standing agency policy not to fund abortions, but in 1984 the United States adopted a policy prohibiting grants to organizations that engage in abortion-related activities. The policy was designed to reduce the number of abortions, but it actually increased the number. A University of Michigan study showed that this shortsighted policy reduced funds for contraceptives in 35 countries and resulted in thousands of abortions, unwanted births, and deaths from pregnancies that would not otherwise have occurred.

All the countries that supply significant aid to impoverished countries with rapidly growing populations have thrown their support behind population control with varying degrees of enthusiasm. There is widespread agreement that slowing the twentieth-century population boom is vital to any country's future peace and prosperity.

## SUMMARY

Like all other organisms, humans inhabit an environment with limited resources. The number of people in the world has grown exponentially in the last few hundred years. The human population today is at or above the carrying capacity of Earth. We face the challenge of building a sustainable relationship between population and resources.

### Population Forecasts

Population forecasts are unreliable because reproductive behavior is unpredictable. Predictions are usually based on (1) the age structure of the population, (2) the survivorship revealed by life tables, and (3) fertility rates of the population. The age structure of a population also permits us to predict the dependency ratio, which has a major impact on the country's economic future. The doubling time of a population permits predictions of the rate at which the resources people need will have to be supplied.

### Declining Death Rates

The population explosion has been caused mainly by declining death rates throughout the world since about 1800. Infant mortality has fallen, mainly as a result of improved hygiene and nutrition, and has led to a rapid increase in life expectancy.

### The Demographic Transition

The theory of the demographic transition states that population growth increases and then slows as a result of the economic and social changes caused by industrialization. Initially, the population grows rapidly as death rates fall but birth rates remain high. Then, with greater education of women, birth rates fall and the population stabilizes. The nations of the world can be divided into those that have passed through the demographic transition and show little population growth and those that appear stuck in the middle of the demographic transition, with populations that are doubling about every 25 years.

### Fast-Growth Countries and Sustainable Populations

Many fast-growth countries have populations too large to sustain. They are passing through an ecological transition, consuming their natural resources faster than these can be produced, leading to increasing death rates and declining standards of living. Populations that are too high to be sustained degrade the environment and permanently reduce the carrying capacity.

### Economics and Population

After World War II, a strong global economy led to rising incomes in every country. Since about 1970, an economic slowdown has made economic growth difficult to sustain except in developing countries with slow-growing populations. In countries with fast-growing populations, it has led to massive foreign debt and declining incomes. The growing gap between Earth's poor and wealthy inhabitants makes for a dangerous political situation.

## Birth Control

Controlling the number and spacing of children contributes to the health and economic well-being of a family. New cheap, effective methods of birth control are badly needed. Family planning is largely unavailable to those who need it most.

## Controlling Population Growth

Most governments have policies designed to slow population growth. Those governments that have sufficient resources to implement these policies have been quite effective in slowing population growth and increasing the prosperity of their people. Aid from wealthy countries for family planning in poor countries is probably the most important step that can be taken to reduce the world's environmental problems and political tensions.

## DISCUSSION QUESTIONS

1. The education of women is the single most important factor leading to smaller family size. Why does educating men not have the same effect?

2. Do you think the population of your own country is too high? Why or why not?

3. One person says there are too many people in the world, another disagrees. What criteria would you use to decide whether a human population is too large?

4. A law in Georgia prohibits distributing contraceptives on public school property. What is the reason for this law? Do you think it is a wise law? Does your answer have anything to do with population control, or is it based on other arguments?

5. It has been suggested that those who prevent others from gaining access to contraceptives or abortions should have to bear the cost of raising the unwanted children who will be born as a result of their policies. Are you in favor of this idea? Who, in fact, does bear the cost of raising unwanted children?

6. Rapid population growth in developing countries affects people in developed countries in many ways. How many examples can you think of?

7. Look at Table 7–2. Does the correlation between fertility rate and income apply in your experience? Do women with few children live in wealthier households than those with more, or does this apply only to a nation as a whole?

# POPULATION, ENVIRONMENT, AND DISEASE

# Chapter 8

### Key Concepts

- The proportion of the population living in urban areas has increased rapidly during the 20th century. In developed countries, the proportion has now stopped increasing.

- In developing countries, urban areas are still growing rapidly as a result of population growth and rural poverty.

- Cities have grown faster than the ability to provide the services residents need. Inner-city decay and homelessness are worldwide problems. Many city residents in developing countries lack even essential services such as clean water and sewerage.

- Environmental degradation has turned millions of people into refugees.

- Population expansion has displaced aboriginal people from their homelands. Some countries are now trying to improve this situation.

- Air pollution and radiation cause ill health directly.

- Polluted drinking water spreads many infectious diseases.

- Pathogens that live in both human and nonhuman hosts cause a group of diseases that are increasing rapidly as humans alter the environment.

*Almost half the people on Earth live in cities.*

The urban areas where most people live have expanded rapidly with the population growth of the twentieth century—into nearby farmland, forest, and wetlands. Many environmental problems, including disease, can be traced to the expansion of human populations into previously wild areas. This is one reason that in many governments the department of the environment is also the department of human health. For instance, when cities expand faster than essential services can be provided, more and more people find themselves without garbage collection, sewers, and drinking water, leading to all the diseases associated with polluted water and unsanitary living conditions. When people are crowded together, natural disasters such as earthquakes and hurricanes do much more damage than if people lived farther apart. We shall see that it can even be argued that new diseases, such as Lyme disease and AIDS, would never have occurred if we were not occupying more and more of Earth's surface.

In this chapter, we examine how population growth and environmental degradation combine to affect human life. First, we consider the movement of people from the countryside into towns, the movement of aboriginal peoples into areas where they may pursue their traditional ways of life, and the movement of people from one country to another. Then we examine some of the human health problems that are caused or made worse by population pressure and environmental factors.

## 8-A  WORLD POPULATION AND ITS DISTRIBUTION

Table 8-1 gives some idea of how people are distributed over the face of Earth. It turns out that neither population size nor population density is much help in predicting whether a country's environmental problems have damaged its economy. Some of the most crowded nations are the most prosperous. Taiwan, South Korea, Singapore, and Hong Kong all have more than 1,000 people to the square kilometer. Some crowded countries even manage to produce all their own food. The United Kingdom (600 people per square kilometer) exports food, whereas Brazil (44 people per square kilometer) has to import food. Similarly, Europe has all the fresh water and timber it needs, but Africa, with only 95 people to the square kilometer, is desperately short of both.

People are not evenly distributed over the surface of the globe. Most of us live in towns and suburbs, near oceans, lakes, and rivers. The water that covers 70% of Earth is one of the most important means of transporting goods and people from one place to another. This is why most of the world's great cities have

| TABLE 8-1   Where People Live | |
|---|---|
| Area | Number of Inhabitants (very approximate) |
| Central America | 125 million |
| Eastern Europe | 125 million |
| North America | 250 million |
| South America | 250 million |
| Former Soviet Union | 250 million |
| Africa | 500 million |
| Western Europe | 500 million |
| East Asia (including China, Japan, Korea, Philippines) | 1 billion |
| South Asia (including India, Indonesia, Iran) | 2 billion |

grown up on the ocean or on rivers near the ocean (Figure 8-1).

The most common reason for people to live where they do is employment. In medieval times, almost everyone lived in a village that housed the people who worked on the local landowner's estate, together with a few independent farmers and tradespeople. Most of the jobs were agricultural, with people also employed to manage the forests, to work the local coal mine, mineral mine, or water mill, and to manufacture the necessities of life for the village. In Europe, villages were so self-sufficient that in 1900 many people had never visited the neighboring village only a few kilometers away. Genetic defects from inbreeding were common because most people married relatives who lived in the same village. They had little choice because they seldom met anyone else.

As agriculture has become less labor intensive, this social organization has steadily disappeared. Better transportation means that rural necessities can be supplied from large factories or distribution centers anywhere in the world and no longer need to be manufactured locally. The number of jobs in the countryside has fallen, and people have moved to towns in search of work.

## 8-B  URBANIZATION

Most of us now live in an urban area, defined by the U.S. Census Bureau as a place with a population of more than 2,500 people. In 1940, half of the U.S. population lived in urban areas. By 1985, this fraction had risen to more than three quarters.

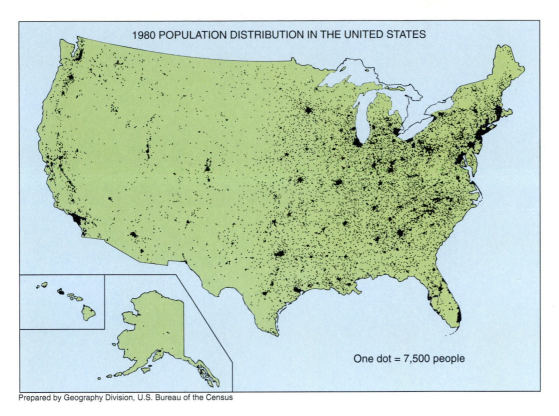

1980 POPULATION DISTRIBUTION IN THE UNITED STATES

One dot = 7,500 people

Prepared by Geography Division, U.S. Bureau of the Census

**Figure 8-1**    The U.S. population clusters around oceans, lakes, and rivers. The same pattern of distribution is found in all other countries.

Although many people cherish a utopian view of country living, life in the town has much to recommend it. For the well-off, cities offer an array of goods, services, and cultural and social activities that no rural community can provide. Even slave-holding plantation owners in the southern United States at the height of their wealth and power lived in primitive squalor compared with their relatives who ran factories in towns such as Philadelphia. Urban life is also attractive to those thrown out of work by the modernization of agriculture. It is sometimes hard to understand why the unemployed choose to move from the rural beauty of a Caribbean island, the Appalachian Mountains, or the Yucatán peninsula to an urban slum in Rome, New York, or Mexico City, but these people are, in material terms, better off in the city. The rural poor, in all parts of the world, suffer from worse malnutrition, receive less medical care, and have less political power than the urban poor. So, rich or poor, the trek to the cities continues. By 2000, about half the world's population will live in cities.

The number and size of cities have increased. An "urban" area with a population of 2,500 seems like a small village to most people who are familiar with "supercities," urban areas containing more than 10 million people. There are 17 of these in the world. By 2000, demographers predict that there will be 19, most of them in developing countries.

The populations of cities have grown much faster than the overall population as the proportion of jobs in the countryside has fallen.

### Urbanization in Developing Countries

In North America and Europe, about 75% of the population is urban, and that percentage is not increasing significantly. **Urbanization**, the movement of people to urban areas, is now proceeding most rapidly in poor countries (Figure 8-2). Here, population growth and the modernization of agriculture

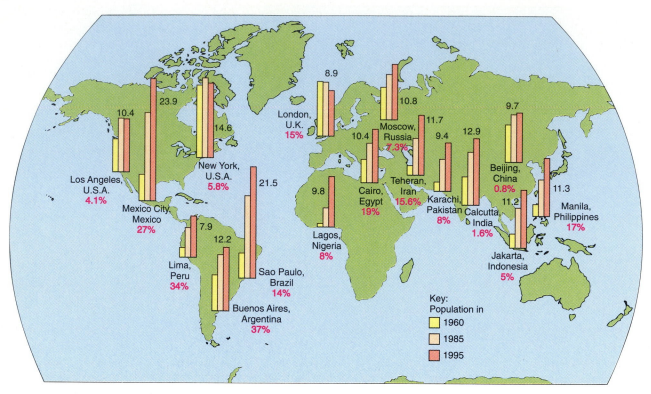

**Figure 8-2** The growth of some of the world's largest cities from 1960 to 1995. The figure by each graph shows the population in 1991 in millions. The figures in red show the percentage of the country's total population living in the city. London and New York are characteristic of developed countries. Developing countries with economic problems include the Philippines, Nigeria, and Pakistan. Compare the cities in these countries with Beijing, China: most of the population lives in small towns and the countryside, and Beijing is growing only slowly.

are still happening, with increasing joblessness in the countryside. The extraordinary rate at which cities such as Mexico City and Calcutta are gaining population is due about equally to migration from the country and to high reproductive rates.

One of the most obvious distinctions between the economically successful developing countries and those that are in economic difficulty lies in their degree of urbanization. For instance, Asia is generally less urbanized than Africa. In China, Malaysia, and Thailand, where economies are growing steadily, less than one third of the population is urban. The countries in deepest economic trouble are more urbanized than developed countries with more than one quarter of their populations wedged into single cities, such as Lima, Santo Domingo, Mexico City, Santiago, and Buenos Aires. These miserably poor cities drain resources. They require services but produce little to pay for them. The urban crisis and the economic problems in many of these places result from the neglect of agriculture and other aspects of rural development.

It is difficult to think of an exception to the rule that countries that develop successfully (including the United States and Japan) pour money and talent into rural areas. Investment in rural areas supports agricultural innovation, soil conservation, family planning and medical care, education, reforestation, and environmental protection. Such investment pays dividends in many ways. It makes the country more self-sufficient, saving money that would otherwise be spent to import food and energy. It may generate income from tourists and research workers who visit national parks, and it lowers population growth and rural unemployment. Above all, it reduces the urban crisis. If people can find worthwhile work in the countryside, they are less likely to migrate to urban areas. Instead, they stay in rural areas, growing food for the cities, producing the fuelwood or charcoal that urban populations need, or building the dams and waterways that supply water to urban areas. Increased investment in rural development is almost certainly the only way developing countries that now face failing economies can solve their problems.

Rapid urbanization in developing nations is a symptom of economic and environmental problems in the countryside. The more successful nations have invested in rural development and are less urbanized.

## The Urban Crisis

Life is not particularly pleasant for the many low-income people who live in or around American cities. Many of them are homeless, largely because the supply of low-income housing actually fell during the 1980s, despite rising demand (Figure 8–3). Many more live in crowded conditions in pest-infested buildings that are expensive to heat because they are not insulated. They receive poor medical care and education, and when young people grow up, about one third of them may be faced with unemployment. Many of them live with violence and drugs, peer pressure reinforcing the notion that crime is the only way to make a living and that someone who does not belong to a prosperous part of society has no responsibilities to it. These people report that they fear crime, are afraid to go out at night, and feel powerless, isolated, and trapped. Despite the enormous wealth of the developed nations, all of them face deteriorating living conditions and increased homelessness in their largest cities, and this is what is meant by the urban crisis.

Conditions in many cities in industrialized countries are bad, but conditions for many city dwellers in developing nations are appalling. Here, the flood of new residents frequently overwhelms the inadequate services. The United Nations projects that by 2020, almost one quarter of city dwellers will be homeless. These people live on the streets or crowd into slums. Some 30% of the urban population in some places live in shantytowns, in houses made from plastic, packing cases, and other garbage. Even if the rightful owners of the land do not show up and demolish the shanties, many shantytowns are located on land prone to flooding and landslides or in the filthiest part of the inner city. Fire from unsafe cooking stoves is a constant hazard, and shanty dwellers seldom have access to safe drinking water, sanitation, education, or health care. Despite the wretched conditions, many shanty dwellers retain the optimism of immigrants, and some remarkable leaders of industry and politics have emerged from backgrounds such as the shantytowns of Hong Kong and New Delhi.

Urban growth has outstripped the ability of cities to provide vital services, resulting in declining standards of living and widespread homelessness.

## Suburban Sprawl

In some places, the local political system has permitted cities to grow haphazardly in the hands of landowners and developers. However, the municipality is usually responsible for providing roads, sanitation systems, garbage collection, and other services to new residences and business premises (Figure 8-4). This can be an onerous burden if the city has no control over how and where the city grows. For instance, in Europe after World War II, many people built new houses along the roads leading out of towns. This "ribbon development" meant that many kilometers of utility lines and other systems were installed to serve just a few houses. Haphazard development of this kind put an undue financial burden on many municipalities, and they reorganized themselves with zoning boards, central planning commissions, and similar mechanisms to ensure that the area grew in a more controlled fashion.

When a town expands, the usual conflicts that arise are over how land that is just outside the existing city should be used. Should the natural ecosystem and farmland that now exist be preserved, or should residential, commercial, or industrial development be permitted? Tokyo's problems provide an interesting example of the kinds of conflicting values that beset planners.

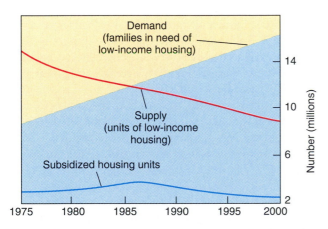

**Figure 8-3** One aspect of the urban crisis in the United States: the growing shortage of housing for those with low incomes. This graph shows the gap between supply and demand that resulted from government neglect of housing during the 1980s. The supply of low-income housing exceeded the demand by several million units in 1975. Now, millions of people are homeless because they cannot afford housing.

**Figure 8-4**    Suburban sprawl: Las Vegas, Nevada. The city's population has almost tripled since 1970. Suburbs, mainly of one-story houses, stretch out into the surrounding desert, straining city services. Leaders say that the city will be unable to grow at all after 2000 unless new sources of water are found.

## Tokyo and Land-Use Planning

The Japanese tend to be incredulous when told they belong to a very wealthy nation. A resident of Tokyo, like anyone else, would feel well-off with a large, well-equipped house surrounded by a large, beautifully landscaped yard, with a car or two in the garage, and a boat, fishing gear, or similar recreational equipment for use on the weekend. For most Tokyoites, such possessions are a hopeless dream. Money accumulates in the bank because it cannot be used to buy the house, the land, or the cars. People in Tokyo cannot register cars unless they can prove that they have off-street parking. Houses and apartments are too small to accommodate much furniture or many appliances. A major reason for the United States's trade deficit with Japan is that the Japanese simply do not have space for the large items Americans have for sale. The nations (like Italy and Germany) that trade most successfully with Japan sell them tiny items such as jewelry, watches, perfume, and leather goods.

Why can wealthy Tokyoites not buy land on the outskirts of the city and build the houses of their dreams? The population density of Japan is little greater than that of Massachusetts. The answer lies in land-use regulations designed to save Japan's remaining farmland from being built over and to protect traditional values against the twentieth century. Farmland and old buildings are protected from development. Small retailers are protected from competition with department stores. The farmland around Tokyo is taxed only on its value as agricultural land, so that a farmer can continue to grow rice on the world's most valuable real estate. The owner of an old house, even if it is on the perfect site for an apartment block, is taxed only on the building's value as a single-family home.

The Japanese could solve their housing problems and alleviate the frustration of Tokyo commuters by repealing some of the laws that restrict change in and around the city. But these laws protect traditions that the Japanese value. They also protect Japan's ability to produce most of its own food. The Japanese are simply not prepared to see farms, small retailers, and old buildings swept away in a tide of ''progress.''

Those who advocate controlling the spread of American suburbs should consider the example of Tokyo carefully. Each year, suburbs spread over another million hectares of land in the United States. Houses and shopping centers are built on prime agricultural land and forest. Municipalities have to supply services over greater and greater distances. Many argue that state or national land-use planning should be used to stop or slow this spread.

Opponents argue that it is not politically feasible to force middle-class Americans to live like Tokyoites, without the spacious houses and yards to which they are accustomed. If you prevent suburbs from spreading, people have to live within the existing boundaries of cities. With greater population density, the cost of housing increases. Some Americans cannot afford housing even at its present price, and Americans have become accustomed to cheap housing. They spend a lower percentage of their income on housing than people in almost any other developed country. It would be a brave politician who would propose land-use regulations that would increase the cost of housing and decrease the size of house and yard that the average American can afford.

If the American population were still growing as fast as it did after World War II, the argument that suburban spread must stop would be compelling. We cannot afford to lose much more of our agricultural land, marsh, and forest to spreading suburbs. But population growth in developed countries has almost ceased. When the population stops growing, suburbs will stop spreading across the countryside. Suburban growth will occur around particular cities whose populations are still growing, but the need for national or state regulation will be less pressing.

## Urban Open Spaces

The world's most attractive cities all include a lot of open spaces, varying in size from the vast green belt that surrounds London to old graveyards and tiny parks (Figure 8-5). Some open spaces are privately owned, perhaps attached to a shopping center or large office building, but most are owned by the city.

Open spaces are pleasant for the city's inhabitants and are often important social gathering places. In addition, they serve several functions that improve the city environment. Open spaces usually contain plants, which absorb carbon dioxide and produce oxygen. Some plants even filter pollutants out of the air. In addition, trees provide shade, which provides

(a)

(b)

**Figure 8-5**   Urban open spaces. (a) Savannnah: this park includes areas for tennis, lawns for Frisbee and other sports, and landscaped areas such as this. (b) Paris: this square, with a fountain and benches, is in the middle of a commercial area. Local office workers often eat lunch in the square. Studies show that small areas, near where people work and live, attract more people than large isolated areas such as Central Park in New York.

some shelter from rain and snow and cools the city in summer. Measurements in New York City in summer showed that while the temperature in the shade in Central Park was 86°F, in the shade on a nearby street without trees it was 120°F—more than 30° hotter.

One of the most important functions of open land and its vegetation is to reduce drainage problems. Land covered with pavement does not absorb water. Heavy rains run down the streets, causing floods and overloading storm drains. In many places, the storm drains run into the sewers. Whenever it rains hard, the sewage treatment plant receives more volume than it can handle, and the plant overflows, discharging raw sewage into the local waterway. One result of expanding a city's open space is to reduce the problem of flooding and the volume that the sewage treatment plant must handle.

An urban area usually contains many types of open space. Some of these are lost if they are not built into the city's design at an early stage, but others can be developed at any time.

A **green belt** is an area around a city that cannot be used for building. The most famous green belt surrounds London. It was set aside in the 1930s as a belt about 15 kilometers wide consisting mainly of farmland. Today, the suburbs have jumped the green belt and developed beyond it, supported by a network of railroads used by commuters. London's green belt saved for the future some planning possibilities that many cities would love to have. For instance, in the 1980s, a ring road that completely encircles London was built, largely in the green belt. The cost of the ring road would have been prohibitive if the land had already been built upon. Ring roads were built into plans for several American cities in the 1950s as part of the federal road-building program that followed World War II, but most of these are farther away from the city center than London's (or Paris's) because the land for them was acquired 20 years later.

Thousands of cities have had the foresight to preserve land for parks. The parks usually contain gardens and recreational facilities, such as collections of trees, tennis courts, ice rinks, and even lakes where the public can rent boats. The Tuileries in Paris, Lake Front Park in Chicago, and Golden Gate Park in San Francisco are notable examples. One of the largest is New York's Central Park, designed in the nineteenth century.

Planners continue to debate whether large areas such as Central Park provide city dwellers with more usable open space than smaller parks. Studies have shown that the most widely used parks are those near busy streets from which they are visible. The city's workers will stroll out to eat lunch in a park only if they do not have to travel far to reach it. There are few places in central London or Paris where one has to walk more than two blocks before encountering an open space, which may be only a paved area with a few plants and a park bench or the graveyard surrounding an old church and provided with a bench for sightseers. Many of the London parks date from World War II, when large areas of the city were bombed into rubble. The city acquired vacant lots cheaply before London was rebuilt and turned some of them into small parks.

## Odd Urban Spaces

Some of the most interesting recent developments of urban spaces involve the odd pieces of land that exist in most cities, sometimes only for a short period of time. Urban gardening projects are one example. These are gardens of vegetables and flowers planted by local residents in vacant lots, often in run-down inner-city neighborhoods, with the aid of seeds, fertilizer, and advisers provided by the city. New York's urban gardening program, for instance, has many goals. It permits residents to grow some of their own food and motivates them to keep drug dealers, muggers, and prostitutes out of the area. It also supplies intellectual and aesthetic food to underprivileged people whose minds as well as their bodies are often starved. It provides job training for the unemployed, who may go on to work for the local parks service.

City governments are developing small open spaces such as disused railway beds and the banks of canals, converting them into areas where people can walk, bicycle, ride horses, or play. The possibilities are almost endless when a city uses its collective imagination.

## Natural Disasters

Urbanization increases the chance that natural disasters will cause wide devastation and death. Disaster-warning systems have improved in recent years, and fewer people are killed by things like hurricanes. But disasters such as tornadoes and earthquakes usually give no warning, and even floods cannot be predicted long in advance. Society accepts disaster losses as unfortunate accidents of fate, but, if we look a little more closely, we can see that human actions cause part of the damage that results from earthquakes, tidal waves, volcanic eruptions, floods, fires, and hurricanes. Population pressure is the reason that so many people now live in areas that are prone to natural disasters, and environmental degradation makes these events more disastrous when they do occur.

For instance, millions of people in Bangladesh live on sandbars in the Bay of Bengal that are washed away every year by high tides and floods. In 1988, 1,200 people died and 25 million were left homeless when the monsoon rains flooded the Ganges River. The floods are annual events, but they are now much worse than they used to be because the hillsides of the Himalayas are deforested, leaving no vegetation to absorb the monsoon rains, which therefore wash straight down the river.

Natural disasters invariably leave people homeless, and many of these people leave the area, never to return. Disasters such as a hurricane in Florida or a flood in India are second only to land degradation in causing environmental refugees.

## 8-C ENVIRONMENTAL REFUGEES

People move around the world for various reasons, but movements of large numbers of people at once have usually been caused by the search for work and by war. Thousands of people moved from Ireland to the United States in the nineteenth century when Ireland's potato-growing economy was devastated by failure of the potato crop. Millions of people moved away from homes bombed into rubble during World War II. Today, however, a new cause of population movement is upon us: environmental destruction that makes an area unfit to live in.

Two years after the 1986 explosion at the nuclear reactor near Kiev, Soviet authorities announced plans to demolish the town of Chernobyl. The city's 10,000 inhabitants will never be able to return to their homes, which are contaminated by radiation. These people are **environmental refugees**, people driven from their homes by severe environmental damage.

The millions who leave their homes in Somalia, Kenya, and Ethiopia in search of food are also called environmental refugees by the relief workers who care for them. The media talks as though these refugees are the victims of drought and civil war, but this is not the whole story. The underlying cause of their plight is destruction of the environment. Intensive farming, caused by overpopulation, has destroyed the fertility of the soil, and the trees on which their animals browsed have been cut down for firewood. The soil contains so little organic matter that it cannot hold the water needed for crops to survive a drought. Relief workers in Africa say that they can tell the extent of environmental damage by the number of refugees who reach them from an area. People leave their homes for environmental reasons only when death from starvation is imminent. Most of these people can never return to their homes because the carrying capacity of the land will take hundreds of years to recover—if it ever does.

> Relief workers can tell the extent of environmental damage by the number of refugees who reach them from an area.

Environmental refugees move from one country to another. They are a growing international problem because immigration laws simply do not cover them. The United Nations defines refugees as people who abandon their homes as a result of war, natural disaster, or political oppression. Refugee laws are not written to take care of people driven from their homes by soil erosion or radiation damage.

This problem has been obvious in the United States for many years. Illegal immigrants from Central America, Caribbean nations, and Mexico are not labeled as environmental refugees, but that is what many of them are. Most are people who can no longer support themselves and their families in overpopulated countries with high unemployment and deteriorating environments. Now the problem is beginning to hit home in Europe. Industrial pollution and joblessness in eastern Europe, starvation and soil destruction in Africa, and war in Yugoslavia and Afghanistan have flooded western Europe with immigrants in recent years (Figure 8-6).

When refugees arrive in a country in large numbers, they may overwhelm the ability of the host country to provide education, health care, and jobs. Since refugees are often willing to work long hours for poor wages, they take over many jobs, making unemployment worse and causing local resentment. In 1992, unemployment in parts of eastern Germany stood at 40%, contributing to the racism and xenophobia (fear and hatred of foreigners) that led to attacks on Vietnamese and other refugees. Anti-Polish riots in Britain after World War II had similar roots, and many Americans have found that racism has been exacerbated by high unemployment in the United States in the 1990s.

What can and should a country do when environmental refugees arrive? The United States has evaded the issue by labeling some of the immigrants as political refugees and by declaring most of the rest illegal immigrants. But such labeling does not solve the problem. As environmental destruction accelerates in countries with fast-growing populations, the number of people driven from their homes by the approach of starvation can only increase. And these people will continue to cross international borders. The only real solution is to improve environmental conditions in the countries from which the refugees come. This would prevent many of them from leaving their homes at all. It would also help to prevent

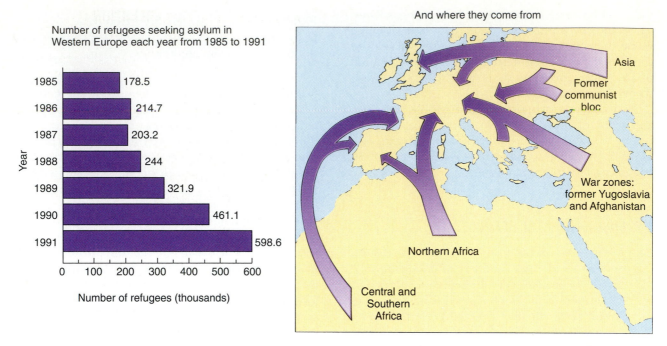

**Figure 8-6**   The growing flood of refugees into Western Europe. The refugees are displaced by civil wars and also by deteriorating environmental conditions in Africa and eastern Europe, where they can no longer make a living.

or stop some of the civil wars that are the other main cause of modern population movements. Wars are usually fought at least partly over the control of natural resources. The Gulf War was fought over who controlled Kuwait's oil fields. The civil war in Yugoslavia is partly over control of agricultural land and fresh water. The flood of environmental refugees in the 1990s is perhaps the most powerful argument that can be made for helping other countries to solve their environmental problems and reduce their populations.

> Refugees from environmental destruction of their homes are becoming more numerous.

## 8-D   ABORIGINAL POPULATIONS

One group that has suffered appallingly from the expansion and movement of populations is the world's remaining aboriginals. **Aboriginals** are the indigenous people of an area, those who have lived there since before the population movements of the last thousand years. They include groups generally known as "Indians" in the Americas, Eskimos in the American Arctic, Aborigines in Australia, Maoris in New Zealand, Bedou (bedouin) tribes in north-

ern Africa and the Arabian peninsula, and various Mongol peoples in Asia.

Most aboriginals are at least partly hunter-gatherers or nomadic herders. Their lifestyles tend to use a lot of land per person and to include population control. As a result, they have been steadily pushed out of their homelands by agricultural societies with exploding populations and an insatiable need for more farmland. Because aboriginals tend to be nomadic, agricultural societies justify their occupation of native lands by claiming that the tribes do not own the land. The destruction of aboriginals by agricultural invaders in many parts of the world amounts to **genocide**: the planned annihilation of a racial or cultural group.

In recent years, many countries have reassessed their treatment of aboriginals. Colombia, for instance, has granted Indian tribes land rights to half the country's Amazon rain forest, roughly one third of the country's area. Martin Hildebrand, an anthropologist responsible for the program, argues that the Indians are the people most likely to protect the biological diversity of the forest, which contains many plants of biological and medical interest. The government has another interest in handing remote, thinly populated areas over to the Indians. The forest borders on the huge neighboring country of Brazil, which is sending people to settle in remote forest areas. This worries smaller neighbors such as Colombia, which have already seen drug laboratories and guerrilla activity spill over into their own countries. "It makes sense to sandbag the border with Indians," says a foreign observer. "The international community would be outraged if a neighboring country invaded the Indians' land."

### The Argument for Aboriginal Conservation

The argument that aboriginals should be granted some kind of legal title to land is partly moral and partly practical. If you cannot convince people that aboriginals have a moral right to land, you can point out the many advantages of giving aboriginals legal authority to protect some of their traditional lands and preserve their own culture. This is one of the most effective ways to protect biodiversity, especially in the tropics. Biologists believe the following arguments are most likely to convince people:

1. **Protected land stores relatives of domesticated plants** that may one day be useful. If plant breeders want a new variety of potato, they look for genetic varieties in the Andes, where potatoes originated. Aboriginals know the plants in their homelands because they use them for food. They are the best guides to plants and animals that might be worthy of research and have the understanding to protect the area from poachers and settlers.

2. **Preserves are vital to improved pest control.** Controlling weeds and pests by using their predators and parasites is becoming more and more important (Sections 6-G and 21-F). The place to search for predators that can control the spread of potato beetles, water hyacinths, corn borers, boll weevils, and fire ants is the original habitats of these organisms, which are often tropical forests and mountains. Research on the interactions of tropical plants and insects in their native biomes is essential to developing biological controls.

3. **Protected areas store plant chemicals.** We have seen that plant chemicals often end up as useful drugs, and protected forests can store the plants until we get around to research on them. Again, aboriginals are the best guides and caretakers because they have used the native plants for medicine for generations.

> Aboriginal peoples are an irreplaceable source of knowledge about habitats and natural resources.

## 8-E DISEASE CAUSED BY POLLUTION

Many diseases are caused, at least partly, by various environmental conditions. These diseases are of many kinds, but for convenience we can divide them into several categories:

1. Diseases caused directly by pollution, such as radiation sickness and the respiratory diseases caused by breathing polluted air.

2. Infectious diseases that are transmitted in polluted environments. These are caused by **pathogens**, organisms and viruses that cause disease. They include the many diseases, such as cholera and hepatitis, caused by organisms found in polluted drinking water.

3. Infectious diseases caused, or made worse, by humans living very close to each other or to natural sources of infection. Here, we include diseases caused by bites from wild animals infected with diseases such as rabies or Lyme disease.

### Direct Effects of Pollution

In 1952, a high-pressure system over southern England trapped stagnant air under an inversion over London for five days. The concentration of air pollutants rose rapidly, and within a week 3,000 more people died in the city than usually die in a week. Hospital admissions were 40% greater than normal. What is it about breathing very polluted air that can cause this kind of illness and death?

Air pollution produces **acute health effects** immediately. For instance, burning fuels, such as wood or gasoline, produces carbon monoxide (CO). Carbon monoxide binds to the oxygen-carrying pigment hemoglobin in the blood, preventing oxygen transport in the body. Because carbon monoxide binds to hemoglobin much more powerfully than does oxygen, a carbon monoxide concentration of one part in a thousand can be lethal. But air pollution, like most types of environmental damage, causes many more health effects that are slow acting.

**Chronic Health Effects   Chronic health effects** are those that last for a long time, lowering the general level of health and resistance to disease. Chronic effects (on trees and fish as well as people) are believed to be responsible for much of the damage caused by air pollution. For instance, one study showed that animals living in polluted air were much more likely to die when they were exposed to pathogens that cause pneumonia than were animals that had previously lived in clean air.

Chronic bronchitis, aggravated asthma, and emphysema are chronic conditions caused or made worse by air pollution. **Chronic bronchitis** is inflammation of the lining of the respiratory tract. **Asthma** is an allergic reaction in which the respiratory tract is irritated and swells so that less air reaches the lungs. **Emphysema** is a reduction in lung area, also caused by blockages of the respiratory tracts. All these chronic conditions, which may be caused by air pollution (including smoking), reduce the air flow to the lungs and make the heart work harder. Many of the people who die at times of high air pollution die of heart attacks that would not have occurred if the chronic respiratory problem had not been present first (Close-Up 20-1).

Chronic diseases caused by pollution are not easy to demonstrate because they come about slowly, and only a few of the people exposed to the pollution will suffer from chronic disease. The fact that pollution causes the disease can usually be shown only by **epidemiological studies**, studies of large numbers of people with a disease and of the factors they have in common (Close-Up 8-1). The same is true of studies showing that pollution contributes to causing cancer. Cancer takes many years to develop and seems always to have more than one cause. Only epidemiological studies can show the connection.

> Pollution directly causes acute and chronic ill health.

## Cancer from Environmental Pollution

Epidemiological evidence is usually difficult to interpret and requires intricate mathematical analysis. The argument over whether environmental pollutants cause cancer is a good example. Since World War II, we have devoted billions of dollars to studying and treating cancer. Some types of cancer can be cured completely, yet premature death from cancer is more common today than it was 50 years ago. Is this because we are exposed to so many carcinogenic (cancer-causing) pollutants?

The difficulty with answering this question is that there are dozens of other possibilities. For instance, we would expect more cancer in developed countries today than 50 years ago because people now live longer, and cancer is largely a disease of old age. People are exposed to more radiation today than they were 50 years ago, largely because they have more medical x-rays. Radiation is also known to increase the cancer rate. Lung cancer deaths are increasing, but so is the group of people most at risk: those who have smoked for more than 25 years. Is all the increase in lung cancer due to smoking? Our diet has changed in the last 50 years. How has this affected the cancer figures?

When all the information has been analyzed and every known variable allowed for, we find that the increase in cancer deaths during the last 50 years cannot be accounted for by any known factor except our increased exposure to known carcinogens in the polluted water we drink, the polluted food we eat, and the polluted air we breathe.

> Exposure to pollutants appears to have increased the rate of cancer in the twentieth century.

A cancer is a malignant tumor. Tumors are clumps of cells that grow and divide abnormally. Some are harmless, like the common wart or fibroid cysts of the uterus. A tumor may become malignant, however, growing uncontrollably, destroying nearby tissues, and eventually causing death.

Cancers arise when a normal cell is transformed into a cancer cell. This cell then divides to form a group of genetically identical cells called a **clone**. The cells may form a solid tumor or may come unstuck from their neighbors and **metastasize**, traveling to other parts of the body, where they may start new tumors.

Much of the mystery involved in the causes of cancer may be explained by emerging evidence that cancer develops by at least two steps, one of which is a genetic change in the tumor-forming cell. Many factors known to increase the risk of cancers are **mutagens**, agents that cause genetic mutations, including various forms of radiation and chemicals such as benzene. However, several cancers are linked to substances that are carcinogenic without being mutagenic, including some viruses and chemicals such as DDT and vinyl chloride.

It is probable that cancerous transformations occur in our bodies quite frequently. But the tumor that results does not usually kill us. The immune system, the body system that fights disease, recognizes and destroys most tumors in their early stages. Part of the reason that cancers are more common in old people and in those with diseases such as AIDS is that the immune system is weakened in both these groups of people.

One reason it is so difficult to track down the causes of most cancers is that a cellular change (such as a mutation or virus infection) may permit a cancer to develop more than 20 years later. This means that evidence on the causes of cancer often takes a long time to collect. For instance, the people exposed to radiation damage when atomic bombs were dropped

# Epidemiology

How do we know that exposure to DDT increases the chance of developing breast cancer? What kinds of evidence convince us that smoking, radon gas, and asbestos cause lung cancer? We can very seldom prove these connections directly because we cannot experiment on human beings. There are three ways to study human diseases: in the laboratory, using diseased cells, pathogens, and animals other than humans; clinically, by studying the disease in sick humans; and epidemiologically. Epidemiologists study the relationships between disease, the population, and the environment as they search for the causes of diseases and ways to prevent them.

We discovered that smoking causes lung cancer when epidemiologists confirmed that smoking was the only factor linking almost everyone who died of lung cancer. In 1974, a physician in Louisville, Kentucky, realized that a surprising number of his patients were dying of a rare type of liver cancer and that all those who died of this cancer worked with vinyl chloride in the manufacture of plastics. Epidemiologists checked death certificates and found that the cancer was indeed a much more common cause of death among people who worked with vinyl chloride than among the population in general. At the same time, Italian workers showed that vinyl chloride causes cancer in laboratory animals. The authorities eventually concluded that vinyl chloride contributed to causing this type of cancer.

Epidemiology is a slow and indirect method of determining causes of illness or death. For instance, a British study showed that nonsmoking letter carriers who lived and worked in northeast London, the most polluted part of the city, had a higher rate of respiratory disease and were more likely to die of lung disease than letter carriers in other parts of the city. Does this prove that the city's air pollution causes lung disease? No. There may be dozens of differences between letter carriers in different parts of the city. Perhaps more of those in the northeast are of a particular race, were raised by parents who smoked, or themselves smoke marijuana, or walk farther each day. The study certainly *suggests* that polluted air contributes to lung disease, but much more evidence would be needed before a cautious observer would be convinced of the correlation.

**Figure 8-A** The first modern epidemiological study, tracing the source of an 1854 outbreak of cholera in London. John Snow mapped the homes of all who died of cholera as shown here. The deaths clustered around a pump in Broad Street, which supplied water for residents of the area. Snow found the well beneath the pump was contaminated with sewage. Authorities ended the cholera epidemic by removing the pump's handle so that it could no longer be used.

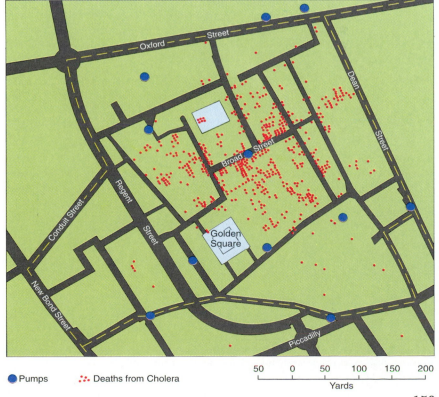

Pumps  Deaths from Cholera

50  0  50  100  150  200

Yards

on Hiroshima and Nagasaki in 1945 can provide valuable information on the frequency and types of cancer caused by radiation. Because radiation is mutagenic, children born later to the Japanese survivors may also provide information on radiation damage. It is obvious that the continuing study of this group of people will not be complete for many decades.

In 1980, the most common form of cancer in American men was lung cancer. By 1989, cancer of the prostate had become more common (although lung cancer rates were about the same). The precise environmental cause of prostate cancer is not known, but epidemiological studies show that this cancer is most common in those areas with the worst local air pollution. The increase in this type of cancer during the last 20 years appears to have been caused by increasing levels of air pollution.

## Teratogenic Effects

A **teratogen** is a substance that causes birth defects. Teratogenesis was one of the most devastating effects

| **TABLE 8-2**   The Infectious Diseases That Caused Most Deaths in 1990 | | | |
|---|---|---|---|
| **Cause of Death** | **Estimated Number of Deaths (in thousands)** | **Symptoms** | **Pathogen and Method of Transmission** |
| Acute respiratory infections | 6,900 | Difficulty breathing, fever, pneumonia | Various bacteria and viruses in the air |
| Diarrheal diseases | 4,200 | Diarrhea, death from dehydration (water loss from body) | Mainly bacteria; from drinking polluted water |
| Amebiasis | 40–60 | Bloody mucous stool; damage to intestine and other organs; death from dehydration | Protistan parasite; from drinking water polluted by human feces |
| Tuberculosis | 3,300 | Coughing, fever; death from lung damage | Bacterial lung infection; from breathing air containing the pathogen |
| Malaria | 1,000–2,000 | Recurrent chills and fever | Protistan parasite; spread by bite of mosquito, which is secondary host* |
| Hepatitis | 1,000–2,000 | Liver damage | Viral infection of liver; from contaminated food or drink |
| Measles | 220,000 | Fever and rash; sometimes brain damage and infertility in men | Viral infection; from breathing air containing the virus |
| Meningitis, bacterial | 200,000 | Fever and brain damage | Bacterial infection of membranes in the brain; often a complication of measles |
| Schistosomiasis | 200 | Damage to intestine, liver, spleen; general debility | Parasitic worm; drinking or bathing in water containing the snail that is the parasite's secondary host* |
| Hookworm | 50–60 | Diarrhea, weight loss, debility; sometimes brain damage | Parasitic worm; usually from stepping in feces of infected pig (or one of several other secondary hosts) with bare feet |
| Rabies | 35 | Fever, nerve damage | Virus; bite from infected animal |
| Yellow fever | 30 | High fever, dehydration | Virus; spread by mosquitoes |
| Sleeping sickness | 20 | Lethargy, coma | Protistan parasite; transmitted by bite of tsetse fly* |

*Humans are the primary hosts for the organisms that cause these diseases, but many of them also pass part of their lives in another organism, such as a snail or mosquito, which is the secondary host.

| TABLE 8-3 | The Increase in Tuberculosis Rates in Europe and the United States | | |
|---|---|---|---|
| Country | Lowest Recorded Rate per 100,000 People with Date It Was Recorded | Most Recent Recorded Rate per 100,000 with Date | % Increase of 1992 Rate Over Lowest Recorded Rate |
| Switzerland | 13.8 (1986) | 18.4 (1990) | 33.3 |
| Denmark | 5.2 (1984) | 6.8 (1990) | 30.7 |
| Italy | 5.7 (1988) | 7.3 (1990) | 28.0 |
| Norway | 7.0 (1988) | 8.5 (1991) | 21.4 |
| Ireland | 15.1 (1988) | 17.9 (1990) | 18.5 |
| Austria | 17.8 (1989) | 20.8 (1990) | 16.8 |
| Finland | 15.5 (1990) | 18.1 (1991) | 16.7 |
| United States | 9.3 (1985) | 10.4 (1991) | 11.8 |
| Netherlands | 8.4 (1987) | 6.7 (1990) | 4.6 |
| United Kingdom | 10.1 (1987) | 10.5 (1991) | 3.9 |

of the herbicide Agent Orange that was used by the United States to defoliate forests in Vietnam. The teratogen involved was dioxin, a very toxic chemical with which the herbicide was contaminated. Most teratogens show their effects more rapidly than do carcinogens, so they are somewhat easier to detect. But the epidemiological problems are similar. Of a given number of men or women exposed to a teratogenic chemical, only a small percentage will have defective children. It is often difficult to distinguish the effect of the chemical from birth defects due to natural causes.

## 8-F INFECTIOUS DISEASES AND POLLUTION

Infectious diseases are caused by pathogens, organisms that cause disease and can be passed from one person to another. Some of them, like tuberculosis and whooping cough, are spread from one person to another through the air. Others come from drinking water contaminated with the pathogen, and others are transmitted by way of secondary hosts. A **secondary host** is another animal in which the parasite must pass part of its life.

Table 8-2 lists the main killers among infectious diseases. Most of these diseases are found mainly in the tropics, in developing nations. As a result, very little money from governments or drug companies is spent for research that might find cures. Most people with these diseases are poor, as are their governments, so very little money is available to pay for drugs or vaccinations that might cure or prevent these diseases. This may change in the future as more and more cases of these diseases occur in developed countries as a result of movements of people around

the world. For instance, in 1991 some U.S. troops returned from Iraq with schistosomiasis. Twelve hundred cases of malaria and 30 cases of cholera were also reported in the United States in 1991, mainly brought in by immigrants and tourists. Consider tuberculosis (TB), a bacterial lung infection. After falling to very low levels in the mid-1980s, rates of tuberculosis in developed countries are rising again, partly because people with AIDS are very susceptible to it (Table 8-3).

### Water That Kills

Each year, at least 25 million people, most of them children, die of diseases contracted from water. About half the world's leading diseases depend on water for their transmission: they either breed in water or are contracted by contact with water containing the infection. In developing nations, where there is not enough water for basic needs, the local water supply is often used for drinking, washing, and sewage disposal. It usually ends up filthy, a perfect breeding ground for pathogens.

Diseases caught by washing in dirty water include trachoma, leprosy, and conjunctivitis, infections that attack the skin and the eyes. Trachoma is a contagious inflammation of the inner lining of the eyelids that often causes blindness, although it does not kill. Some 500 million people suffer from trachoma, many of them in developed countries.

The most deadly group of water-borne diseases are those spread by drinking or washing in water contaminated with human feces. The pathogens excreted by people with the diseases are present in the water. These diseases, including typhoid, cholera, and dysentery, cause the body to lose water by diarrhea and vomiting, killing about 10 million people every year. They are the cause of much of the infant mortality in the world. The main reason is that a baby does not contain as much water as an adult. Feeding the baby water is not the solution because water dilutes the body's salt solution, and this may kill the baby. A simple and lifesaving cure for many cases of dehydration is now known. This is **oral rehydration therapy**, feeding a solution of salt and sugar in water. The sugar and salt help the body absorb the water and add solutes to the body fluids so they are not diluted. Millions of lives have been saved by this simple therapy. But to use the therapy, people have to learn about it, and education is as rare as clean water for millions of people in developing countries.

Dirty water is involved in transmission of the infectious diseases that kill more people than any other cause of death in the world.

**Cholera** In January 1991, the first cholera epidemic in the Western Hemisphere in more than 75 years was reported from Peru. Cholera is a water-borne disease, found in areas with poor water and sewage systems. It is caused by the *Vibrio cholerae* bacterium. Cholera causes diarrhea and vomiting, leading to dehydration that is often fatal, especially in the young and infirm. The bacterium is found in water contaminated with feces from people who contain the bacterium. Laboratory tests showed that drinking water in Lima, Peru's capital, contained the bacterium. Where untreated sewage reaches the ocean or fresh water, the bacterium gets into fish. One of Peru's favorite dishes is ceviche, raw fish marinated in lemon. Some people have contracted cholera after eating ceviche.

The reason for the epidemic is Peru's failure to supply clean water and sanitary sewage disposal to its exploding urban population. This is not surprising since the population of Lima has grown from less than 3 million in 1970 to about 7 million today. More than half the capital's population is housed in shantytowns with no running water, sewage system, or electricity.

How do you tackle a cholera epidemic? The European Community sent emergency medical aid, mainly oral rehydration therapy kits for victims. Surrounding nations banned imports of Peruvian food products and required cholera vaccination for travelers coming from Peru. Ecuador cleaned up a polluted canal that runs along its border with Peru. The Peruvian government closed beaches and unhygienic street food stalls and handed out chlorine tablets for purifying water. Ultimately, however, the only way to stop diseases such as cholera is to make clean water and hygienic sewage disposal available to everyone in the world, a task that is almost impossible for a country such as Peru with a population growing at the rate of 2.7% each year and an urban population growing many times faster.

**Malaria** Insects that breed in water are the secondary hosts that transmit malaria, filariasis, yellow fever, and river blindness. Malaria is the world's biggest killer, causing the deaths of between 1 and 2 million people a year (Table 8-2). An estimated 500 million people become infected each year and suffer from recurrent bouts of fever, anemia, enlarged spleen, abdominal pain, and extreme weakness. Although we seldom remember it, malaria was common in much of the United States and Europe before the days of mosquito control. Now it is largely restricted to the poorer of the tropical countries.

Malaria is caused by *Plasmodium*, parasitic protists, and is transmitted by a bite from the female of about 50 species of *Anopheles* mosquitoes. Figure 8-7 shows the parasite's life history. The mosquito lays her eggs in stagnant fresh water, where the larva develops. There is no vaccination for malaria. The main drug used to treat the disease is chloroquine, made from *Cinchona* trees (Close-Up 8-2). Unhappily, new, chloroquine-resistant strains of malaria have evolved.

Since the 1950s, malaria has been controlled by draining marshes and paddies where the mosquitoes reproduce and by spraying breeding areas with DDT, dieldrin, malathion, and other pesticides. Since the 1970s, however, the mosquitoes have evolved resistance to most of the pesticides used to control them (Section 21-B). In addition, the world contains more and more breeding places for mosquitoes with the spread of irrigated agriculture and hydroelectric power. Irrigation ditches and the reservoirs behind hydroelectric dams are vast new areas for mosquitoes to breed.

Other diseases transmitted by aquatic invertebrates are also spreading with irrigation canals and dams in the tropics. You should never bathe in a tropical freshwater pond, no matter how inviting. It might contain the snail that transmits schistosomiasis, an incurable disease that kills about 200,000 people each year.

## 8-G DISEASES OF THE FUTURE

The world is full of new and evolving diseases, from AIDS to influenza and hemorrhagic fever. Ecologists point out that the way in which we are altering the environment and destroying habitats ensures that these diseases will become more common in the fu-

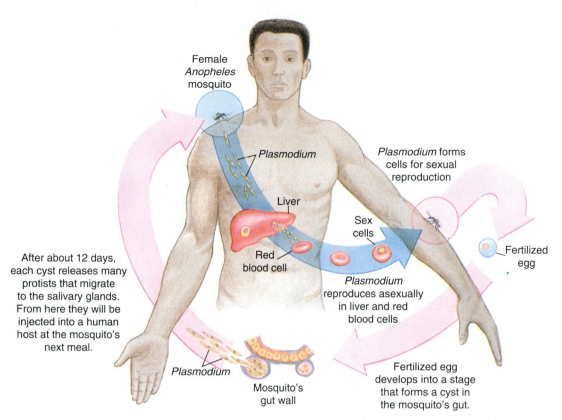

Female
*Anopheles*
mosquito

*Plasmodium*

*Plasmodium* forms
cells for sexual
reproduction

Liver

Sex
cells

Fertilized
egg

After about 12 days,
each cyst releases many
protists that migrate
to the salivary glands.
From here they will be
injected into a human
host at the mosquito's
next meal.

Red
blood cell

*Plasmodium*
reproduces asexually
in liver and red
blood cells

*Plasmodium*

Mosquito's
gut wall

Fertilized egg
develops into a stage
that forms a cyst in
the mosquito's gut.

**Figure 8-7**   Life cycle of *Plasmodium,* the protistan parasite that causes malaria. The parasite has to develop in both a mosquito and a human to complete its life cycle.

ture. To take just one example, in Argentina, herbicides (weed killers) were sprayed to increase maize crops. The herbicides killed the native grasses and allowed other plants to invade, bringing with them a species of rodent that carried the virus for a hemorrhagic fever, which infected many of the agricultural workers. Hemorrhagic fevers, as their name suggests, cause hemorrhages, or internal bleeding, by breaking blood vessels.

While many diseases caused by bacteria have been controlled by antibiotics and vaccines, these treatments are much less effective in controlling diseases caused by viruses and protists because members of both groups may evolve very rapidly. A vaccine protects only against one or a few forms of a disease, and newly evolved forms are not prevented. Influenza (flu) is caused by viruses, and a new vaccine for the most common strain of flu has to be produced every year.

One of the reasons viruses evolve so rapidly is that they sometimes take up genes from the cells they infect, altering their own genetic makeup. Humans encourage this when they invade the habitats of other species. For instance, ducks and other birds are a rich source of genes for flu viruses. As we build over wetlands and other habitat of water birds, we help our

flu viruses to evolve new and different forms. Many other viruses jump from one species to another, sometimes causing disease in humans, although not in the other species. The virus that causes AIDS is believed to have evolved in an African monkey in which it does not cause disease.

> Human alterations of the environment are increasing the number of people who suffer from many diseases and the severity of other diseases.

**Lyme Disease**   Lyme disease is an example of a disease caused when we upset the balance within an ecosystem. The disease causes fever, lethargy, and, sometimes, long-lasting arthritis. It is caused by a spirochete bacterium that is transferred to humans by a bite from a tick. The spirochete has a complicated life history (Figure 8-8). Lyme disease was first identified in the northeastern United States. Here, about one third of deer ticks are infected with the spirochete. But the reservoir, or breeding place, of the spirochete is white-footed mice, which are also bitten by deer ticks. Populations of deer and mice have exploded in New England forests in the last century because most of their natural predators, from

## A Plant that Changed History

Many plants have changed the course of history. The Americas were discovered by Europeans as a by-product of their search for pepper, valued for its ability to mask the flavor of salted meat and stinking fish. A prosperous trade in spices once flourished between Europe and Asia. But by 1480, the Turks had blocked the overland trade route. In reaction, Italian, Portuguese, and Spanish explorers sailed west or south in their attempt to reach the East. They discovered the Americas and, later, ocean routes to Japan and India.

The discovery of a plant that controlled malaria had even more dramatic effects. Malaria was always present in southern Europe, Asia, and northern Africa during the Middle Ages. European travelers introduced it into other countries all over the world. Thousands died of it during the settlement of the United States.

The only effective treatment for malaria was discovered in Peru in the seventeenth century when the Countess of Cinchon recovered from malaria after drinking tea made from bark of the *Cinchona* tree (Figure 8-B). Peruvians had recommended this to Jesuit missionaries as a cure for fever. The bark was imported to Europe, but many Protestants, such as English leader Oliver Cromwell, distrusted the Jesuits, refused their drug, and died of malaria.

West African and some southern European populations had evolved sickle cell hemoglobin, which affords some protection from malaria. Genetic resistance to malaria was a major reason west Africans were prized as slaves in the southern United States and the Caribbean.

The active ingredient in *Cinchona* bark is quinine, usually sold today as chloroquine. Tonic water is quinine dissolved in carbonated, sweetened water. Gin and tonic was invented to make the daily dose of bitter quinine more palatable to northerners living in malarial areas. Quinine permitted northern Europeans, who had no natural defenses against the disease, to establish vast empires in tropical areas more ably defended by malaria than by any human agency. It also permitted Europeans to move about 20 million Indians and Chinese from their homes to tropical areas, where they would have died without quinine. The European empires of the nineteenth century in Africa, Madagascar, Malaysia, and Sri Lanka were based on plantations worked by the

**Figure 8-B**    *Cinchona.* Several species of this tropical tree produce chemicals that are the only partly effective cure for malaria.

cheap labor of these immigrants, many of whom became the ancestors of large modern populations in these areas.

Huge new industries were based on these population movements: sugar in the Indian Ocean and the Caribbean, tin and rubber in Malaysia, and tea in India and Sri Lanka were all made possible by quinine. The drug also probably permitted the Allies to win the Second World War in the Pacific. During that war, 25 million people in the Allied armed forces traveled to areas where malaria was epidemic, areas where many of them would not have survived without quinine.

Extracting quinine from a forest tree was expensive until plantations were established in other countries with seed stolen from the Peruvian Indians, who understood its value. Most quinine is still extracted from *Cinchona* bark. The natural product is cheaper and more pleasant tasting than the synthetic variety and remains one of the world's most important drugs, although many varieties of the malaria parasite are now resistant to it.

If one South American tree affected world history in all these ways, no wonder scientists are sure that in destroying unexplored tropical forest, we are depriving ourselves of products whose value we cannot even imagine.

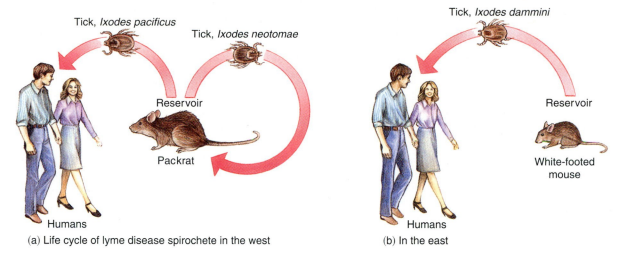

**Figure 8-8**   The different life cycles of the pathogen that causes Lyme disease in the eastern United States and in the West.

owls and hawks to cougars and bobcats, have been destroyed. As the two are forced together, deer ticks pass from deer to mice more frequently, infecting each new generation of mice with the spirochete. As a result, there is a much higher chance that any tick that bites a person will be infected with the spirochete than would have been the case a hundred years ago.

Lyme disease also occurs in the West. Here, however, it is much less common, infecting only a few hundred people a year. In California, fewer than 5% of the ticks that bite humans, are infected with the spirochete, and biologists believed that this was too few to maintain a reservoir of the spirochete. They have now discovered that the animal reservoir is the packrat (dusky-footed woodrat), and the reason

many packrats are infected is that there is another packrat tick, which does not bite humans, that often contains the spirochete. It appears that Lyme disease could exist anywhere that packrats exist along with both species of tick. This includes the mountains, which are popular recreation areas for Los Angelenos.

Packrats get their name from the fact that they collect leaves, aluminum foil, pennies, and other odd items and store them in their nests. This suggests one way of controlling Lyme disease in the West. Give packrats cotton balls soaked in insecticide for them to take home to their tick-infested nests. If even one species of tick were eliminated from the nest, the baby packrats would have no chance of being reservoirs for Lyme disease.

## SUMMARY

During the twentieth century, much of the world's population has moved from rural to urban areas. Making urban areas healthy, efficient, pleasant places to work and live is an acute problem in developing nations and an ongoing struggle in developed nations.

Movement and growth of populations have displaced and destroyed aboriginal populations and increased the incidence of diseases that are at least partly caused by environmental change.

### World Population and Its Distribution

The most populous area on Earth is South Asia. Africa and Western Europe come next, followed by

North America, South America, and the former Soviet Union. About half the world's population is rural and half urban. In all parts of the world, urban populations are clustered near the ocean and nearby rivers.

### Urbanization

Employment is the main factor that determines where people live. In preindustrial societies, nearly all jobs were related to agriculture. The industrial revolution concentrated factories and jobs in urban areas. Urbanization is essentially complete in developed countries, where about 75% of the population is urban. In developing nations, urbanization is proceeding rapidly and many urban areas have quadrupled their populations in 25 years. The urban crisis is a result of a lack of urban jobs and cities having grown more rapidly than services. Rural development and population control are the main solutions to the unmanageable growth of cities in developing countries.

### Environmental Refugees

In the past, war and joblessness have been the main things that created large numbers of refugees. Today, environmental damage also creates refugees. In recent years, people have been driven from their homes by radiation in Russia and by starvation caused by drought and desertification in Africa, South America, and South Asia. Improving environmental conditions, especially in developing nations, is the only solution to the problems that huge numbers of refugees cause.

### Aboriginal Populations

Aboriginal populations have suffered displacement and even genocide as their homelands have been invaded by agricultural societies. Some countries are now trying to improve the situation by giving aboriginals title to land and asking their assistance in protecting endangered ecosystems, a job for which they are well qualified.

### Diseases Caused by Pollution

Environmental conditions such as air pollution and radiation cause ill health directly, often in the form of chronic respiratory problems and an increased incidence of cancer. Some pollutants are also teratogens, which cause birth defects.

### Infectious Diseases and Pollution

The main killers among infectious diseases are connected to water. In diseases such as cholera, hepatitis, and trachoma, the pathogens are transmitted in polluted water. These diseases can be prevented only by clean water supplies. Diseases such as malaria, schistosomiasis, and sleeping sickness are spread by secondary hosts that need water to breed. The spread of irrigation ditches and dams has increased the incidence of these diseases.

### Diseases of the Future

We face an increasing incidence of viral diseases for which there are no cures. Some of these evolve into new, more dangerous forms as they pass between humans and other animals. Human alterations to the environment are also responsible for the increase in diseases like Lyme disease and rabies, which are spread from other land animals to humans.

## DISCUSSION QUESTIONS

1. What are the different advantages and disadvantages of living in the center of a large city, in a suburb, in a small town, and in a rural area?

2. Suppose you were chief city planner for Mexico City. The population is now 13 million. It has doubled in the past 10 years and continues to grow. The city lies in the bed of an extinct volcano, surrounded by steep hills. Millions of squatters live illegally in shantytowns all around the city. The city's inhabitants are less well provided with essential services than they were five years ago. Can you come up with any ideas for improving the situation? Can you think of any ideas that don't cost vast sums of money, which the city does not have?

3. Many people live in the suburbs because they view cities as ugly environments. What makes an environment attractive for a person to live in?

4. Does an attractive environment invariably include large amounts of space, trees and flowers, and cleanliness? Do people differ in the environments they find pleasing?

5. What are the arguments that the only cure for urban problems in most developing countries is rural development? Rural development is expensive because of the distances involved. How can you justify the investment when most of the population is suffering from lack of investment in the cities?

6. It used to be fashionable to argue that cities were poor environments for people because they were crowded. (Studies showed that crowding caused various health problems in rats and other species.) Cities such as Singapore and Hong Kong, where the traffic moves freely, the streets are clean, and the people are educated, have made it clear that people can live at very high densities without suffering physical or emotional problems. Is there a limit to the overcrowding that people can tolerate without stress? Is crowding a factor that should be taken into account in urban planning?

# NATURAL RESOURCES

*Sawtooth palmetto,* **Serenoa repens.** (Thom Smith)

# Chapter 9

# FOSSIL FUELS AND NUCLEAR ENERGY

*Oil wells in California.* Richard Feeny

## Key Concepts

- Large human populations and high standards of living have been made possible by the use of enormous amounts of energy.

- Industrialized societies depend heavily on fossil fuels, which will be exhausted within the next few hundred years.

- Directly and indirectly, burning fossil fuel is the main cause of pollution.

- Conservation is the best way to stretch our nonrenewable fuels and to reduce the adverse environmental effects of using them.

- Nuclear power stations generate electricity from uranium fuel, but they have not proven very cost-effective and they produce large amounts of hazardous waste.

- The prospect of fossil fuel shortages is leading to widespread experiments with other energy sources, both renewable and nonrenewable.

In the next six chapters, we consider the main resources of the biosphere. These are **natural resources**, things produced naturally that we use. In this chapter, we consider the fossil fuels and nuclear power that have fueled economic development since the industrial revolution. In the next chapter, we consider the renewable energy sources that will become more important as fossil fuels are used up. Until other energy sources replace fossil fuels, the technological challenge is to extract energy as efficiently as possible from these fuels while minimizing harm to the environment.

## 9-A RENEWABLE AND NONRENEWABLE ENERGY SOURCES

Our main energy sources have changed during the course of history (Figure 9-1). In agricultural societies, most energy comes from wood, sunlight, streams of water, and the muscle power of humans and draft animals (animals used to pull or carry things). These energy sources are **renewable**, meaning that they are replaced by natural processes or are essentially inexhaustible. Other renewable sources of energy are biomass from plants as well as wind, tide, and solar energy.

As a country develops industries, people make more use of energy sources that are **nonrenewable**, meaning that they can be **depleted**, used up to the point that it is no longer economically possible to use them. The most widely used nonrenewable energy sources are oil, coal, and natural gas (Figure 9-2). Oil, coal, and natural gas are **fossil fuels**—the remains of ancient protists and plants, fossilized and

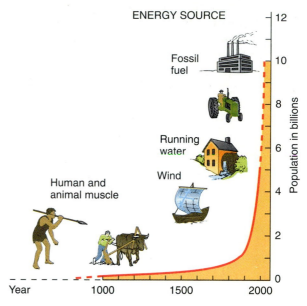

**Figure 9-1** Increase in human population with some of the energy sources used at different times. New sources of energy have probably permitted populations to grow particularly quickly. The human population could not have grown as fast as it has during the last 200 years without the use of nonrenewable energy sources to produce large amounts of food.

compressed by geological processes in Earth's crust. We are so dependent on these fuels, for food, clothing, heating, industry, and transportation, that it is impossible to imagine what our lives would be like without them. Nonrenewable fuels provided not only new types of energy, but also much more of it. People

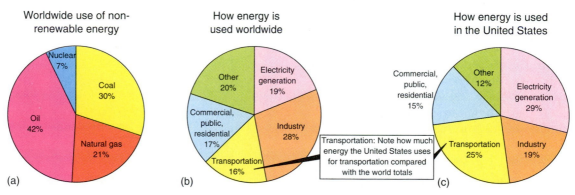

**Figure 9-2** The use of nonrenewable energy. (a) Oil is the main nonrenewable source of energy worldwide. (b) Worldwide, industry uses more energy than any other sector of the economy. (c) In the United States, industry uses less energy than does transportation. The United States also differs from many other countries in converting much of its energy into electricity before using it. (This is partly a result of our relatively high use of nuclear and hydroelectric power.)

**TABLE 9-1**   Energy Consumption per Person in Some Countries That Are Largely Industrial and Some That Are Largely Agricultural (People in industrial societies use much more energy than those in agricultural societies)

| Country | Energy Consumption per Person in 1990 (in Kilograms of Coal Equivalents) |
|---|---|
| **Countries that are largely industrial** | |
| United States | 10,124 |
| Belgium | 5,787 |
| Canada | 10,927 |
| Denmark | 4,413 |
| Germany | 6,511 |
| Japan | 3,995 |
| Norway | 7,181 |
| Switzerland | 3,662 |
| **Countries that are largely agricultural** | |
| Bangladesh | 69 |
| China | 819 |
| Cuba | 1,532 |
| India | 307 |
| Malaysia | 1,278 |
| Mexico | 1,663 |
| Pakistan | 265 |
| Tanzania | 36 |
| Thailand | 637 |
| Tunisia | 761 |
| Vietnam | 132 |
| Zambia | 198 |

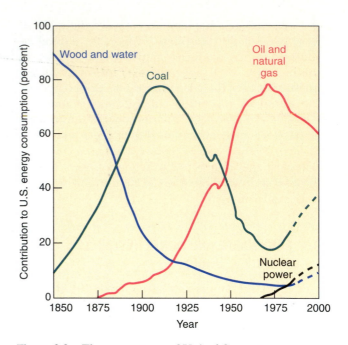

**Figure 9-3**   The percentage of United States energy supplied from various sources since 1850. Note the development lag: it takes many years for a new source of energy to make a significant contribution to the supply. Why do you think this is?

in industrial societies use many times more energy than those in agricultural societies (Table 9-1).

Projections suggest that by the middle of the twenty-first century, worldwide energy demand will be at least twice what it is today, mainly as a result of increased population and the growth of industry in developing countries. Experts believe that there is not enough oil, coal, and natural gas in the world to supply this demand. We shall run out of fossil fuels, and future civilizations will be powered by other means. It is essential to plan now for the energy we shall need later because it takes many years for a new source of energy to make a significant contribution (Figure 9-3).

### Reserves of Nonrenewable Energy Sources

We do not know how much fossil fuel is left on Earth, but estimates suggest that there is enough for at most another couple of hundred years. A **reserve** of a fuel or mineral is a deposit, documented by detailed surveys and judged to be extractable at the present market price.

The reserves of a substance do not remain constant. Figure 9-4 shows how oil reserves increased during the 1980s as new oil fields were explored. When the supply of a substance goes down or the demand for it increases, a number of things happen. First, reserves of the substance become more valuable, so the cost goes up. This makes it worthwhile to search for new deposits, which are often found. During the late 1970s, coal became so expensive in the

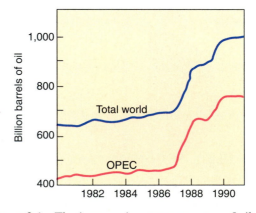

**Figure 9-4**   The increase in proven reserves of oil throughout the world and in OPEC countries during the 1980s. OPEC is the Organization of Petroleum Exporting Countries, founded by Venezuela and Saudi Arabia. *Data from: Cambridge Energy Research Associates/Arthur Anderson*

United States that it became economical to mine for coal in refuse heaps of rock and coal from old coal mines. A high price also promotes research into more efficient technologies for extracting the mineral and finding cheaper substitutes for it. Conservation becomes more cost-effective, and people reduce their wastage of a valuable substance.

As fossil fuels are used up, they will become more expensive, making other energy sources more attractive. Even today, one economic problem with depending on fossil fuels is that some countries have large reserves of these fuels and others have none. Countries without fossil fuels have to buy them from other countries. Many countries, including the United States, have gone deeply into debt as they import fossil fuels that they cannot afford to pay for.

> Fossil fuels supply the enormous quantities of energy that power industrialization and modern societies. As fossil fuels are depleted, the world is turning to conservation and alternative sources of energy.

### Hidden Costs of Using Natural Resources

In 1989, the U.S. Department of Energy spent more than $15 billion to safeguard oil supplies in the Persian Gulf. In addition, the Gulf War cost at least another $30 billion. A Texas oil man said that by paying these hidden costs of using oil, U.S. taxpayers were subsidizing oil from the Middle East so that its cost was artificially low and domestic oil could not compete. If the consumer of foreign oil had to pay directly for the Gulf War, Middle East oil would be very expensive, and oil produced in Texas and other states would be just as cheap. This is what oil producers meant when they accused President Bush of destroying the U.S. oil business.

Hidden costs of products are known to economists as **externalitie**s: costs borne by people who are not buying or selling that particular product at the time. Hidden costs of American energy production and consumption are shown in Figure 9-5. They range from job losses in the Texas oil fields to crop losses caused by air pollution from energy use. If we add just the minimum hidden costs of energy shown in the figure to the price of oil, oil should cost about $75 per barrel, $50 more per barrel than it does now.

Economists agree that if we are to use our natural resources more wisely in the future, we must start pricing them at their true cost instead of at some arbitrary, subsidized cost. If I drive my car 200 fewer miles each week, all I save myself is $10, the price of the gasoline to drive 200 miles. But I save society $20, the externality of the gasoline. If I saved myself $30, reflecting the true cost of the gasoline saved, I should be much more likely to drive less and conserve gasoline. The easiest way to include hidden cost in the price of a resource is to tax it (Close-Up 9-1). A $2

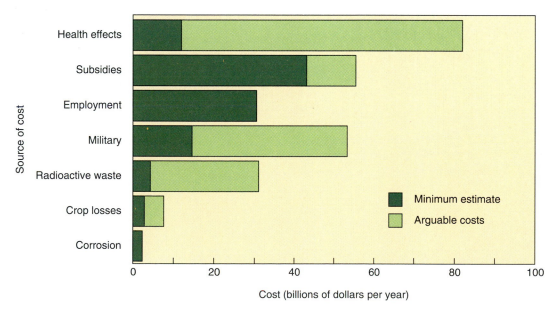

**Figure 9-5**   Hidden costs of energy in the United States. Minimum estimates are those economists agree on. Arguable costs are controversial. For instance, under health effects, what dollar value should you put on the life of a coal miner who dies in a mine accident? *Data from: Resources for the Future*

# CLOSE-UP 9-1

## Gasoline Taxes and the Gulf War

At least one of the United States's reasons for fighting to oust Iraqi troops from Kuwait in the 1991 Gulf War was to prevent Saddam Hussein from controlling Kuwait's oil output as well as Iraq's. The United States feared that this would permit Iraq to raise the price of oil. (This might not have happened. At the moment, oil-producing countries are pumping oil rapidly to increase their incomes. Such high levels of production keep oil cheap.)

Why did the United States, rather than one of the other nations involved, lead the Gulf War? Probably not because we were anxious to restore Kuwait to its dictatorial rulers, who are heartily disliked by their subjects. It is noteworthy that two of the biggest industrial powers, Japan and Germany, did not feel compelled to send troops to the Gulf. Part of the reason is that their constitutions, largely dictated by the United States and its allies after World War II, forbid these countries to fight on foreign soil.

Neither Japan nor Germany has oil reserves. Both import nearly all their oil, while the United States imports less than half of its oil. Are these countries less worried about rising oil prices than the United States? The following figures shed light on this question. According to the U.S. Department of Commerce, the United States consumes 20,000 Btu of energy to produce $1's worth of gross national product. Table 1 shows that Germany uses 93% less and Japan uses less than half as much.

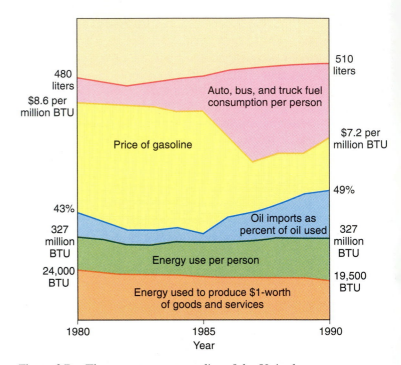

**Figure 9-B**  The strange energy policy of the United States during the 1980s. Business and households became more efficient while transportation became less efficient because the price of gasoline fell. The amount of energy used by each person (green) changed very little. The energy used to produce goods and services (orange) actually fell. But the United States imported more and more of its oil (blue), largely because the price of gasoline (yellow) fell by one fifth. Fuel consumption by vehicles (red) rose as the price of fuel fell, partly because fuel use by trucks rose by one fifth: as fuel gets cheaper, businesses switch their freight from ships and trains to trucks.

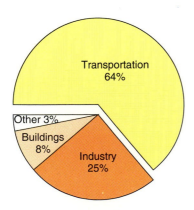

**Figure 9-A**  Where does the oil go? This chart shows the fate of oil in the United States.

172

How about total energy use? Energy use per person in the United States is more than twice that in Japan and 50% higher than in Germany (see Table 9-1). But, interestingly, U.S. energy consumption per dollar of GNP has fallen. It is now 30% less than it was in 1970. Industry is obviously using energy more efficiently (Chapter 10). In contrast, U.S. per capita energy consumption rose between 1970 and 1990. Obviously, the extra energy used for every man, woman, and child in the United States today is not being used to produce more goods and services. Most of it is going for transportation. Partly in airplanes and trains, but largely in motor vehicles, we use more oil per person with every year that passes (Figure 9-A). This is largely because the United States is the only country in the world that has permitted the price of gasoline, adjusted for inflation, to fall (Figure 9-B). Not surprisingly, consumers have used more and more of it. Our low gasoline taxes have allowed us to continue our "love affair with the automobile" and helped to propel us into the Gulf War. Other industrialized countries raised their taxes on gasoline as the price of oil fell on the world market, and their use of oil has not increased as it has in the United States.

You can see why Germany and Japan might feel confident that the United States will do whatever it can to ensure the free flow of cheap oil. It will be hurt worse than any country with an advanced economy (except, perhaps, Canada) should the cheap oil spigot ever be turned off. Note in Table 1 that some countries use energy even less efficiently than the United States. Most of these are countries that produce and export oil (Canada, Mexico, Venezuela, Iran). Others were once members of the Soviet-bloc and received oil from the Soviet Union at subsidized prices (Czechoslovakia, Romania, Bulgaria, Poland). As a result, they used energy inefficiently, which is one reason their industries are unable to compete in the world market now that these countries are independent.

**TABLE 1**  Energy Consumption to Produce $1's Worth of Gross National Product (GNP) in Some Countries in 1990. (Members of the European Community are shown in blue, former Soviet-bloc countries in cream)

| Country | Energy*/$ of GNP | Difference from U.S. ($-100\%$ = half as much; $+100\%$ = twice as much) |
|---|---|---|
| Switzerland | 131 | $-183\%$ |
| Japan | 174 | $-138\%$ |
| Czechoslovakia | 184 | $-130\%$ |
| France | 230 | $-104\%$ |
| Brazil | 252 | $-95\%$ |
| Germany | 258 | $-93\%$ |
| United Kingdom | 346 | $-69\%$ |
| Belgium | 368 | $-65\%$ |
| *United States* | *480* | |
| Canada | 541 | $+56\%$ |
| Mexico | 807 | $+84\%$ |
| Romania | 916 | $+95\%$ |
| Bulgaria | 917 | $+96\%$ |
| Iran | 986 | $+102\%$ |
| Poland | 991 | $+103\%$ |
| Venezuela | 1398 | $+145\%$ |

*Billion tons of coal equivalent.

federal tax on a gallon of gasoline would repay the federal government for money it has spent on the externalities of the gallon of gasoline. Most countries do have high taxes on gasoline that reflect its hidden cost, and states such as New York, Wisconsin, and California are beginning to follow suit in the United States. But people in most countries are still a long way from paying directly for the hidden cost of most energy.

## 9-B   ELECTRICITY

When we run a car on gasoline, we are using an energy source directly to power a machine. In many cases, however, an energy source is not used directly but instead is burned in a power station to generate electricity. Electricity is convenient for the user, and it is also easy to distribute once a distribution grid has been built. An **electricity distribution grid** is a system of wires and substations that carries electricity to people around the country. Once a grid has been set up, electricity can be fed into it from power stations run on many different fuels. The utility company that controls the distribution system usually owns and operates some of these power stations, often burning different fuels as one becomes cheaper than another. In addition, the utility company usually contracts to buy electricity from small producers, which may be individual farmers with small hydro-electric plants or municipalities that generate electricity by burning waste.

### Batteries: Storing Electricity

The main problem with electricity is that it is difficult to store. Despite the invention of tiny batteries for cameras and watches, batteries that can store large amounts of electricity are still bulky and heavy. We cannot yet store kilowatts of electricity as we can store tanks of oil. So a utility company constantly switches electricity producers in and out of the distribution grid as demand rises and falls. Every utility company's nightmare is a spell of extremely hot or cold weather when electricity use in an area may as much as triple over normal use and it is very difficult for the utility company to supply enough electricity. Then "brown-outs" and similar electricity shortages occur. It is so difficult to overcome the problem of storing electricity that many scientists think that the only solution is to use electricity to produce liquid and gaseous fuels that can be stored in tanks.

### How Electrical Energy is Produced

In 1831, Michael Faraday found that he could produce electricity in a coil of copper wire by moving a magnet near the wire. This is the principle behind electrical **generators**, the machines used to produce electricity. A generator converts mechanical energy

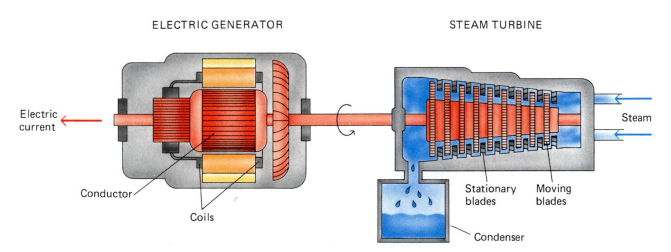

**Figure 9-6**   How a steam turbine works. Steam is forced into the turbine on the right. Stationary blades deflect the steam so that it hits the thousands of moving blades as hard as possible. The steam condenses after it has passed through the turbine, and the water can be reused. The moving blades are attached to a central spindle, which rotates under the steam's pressure. The spindle extends into an electric generator. The generator consists of an electrical conductor that can rotate inside a magnetic field produced by the coils. When the spindle rotates, it forces the conductor through the magnetic field, generating electricity. A steam turbine (and generator) is the most common way of generating electricity and one of the most effective.

(the energy used to move the wire or the magnet) into electrical energy in the wire. If the copper wire is part of a complete loop of wires (a closed circuit), the electricity flows through the circuit of wires as an electric current. A generator can be powered by anything that produces mechanical (kinetic) energy.

**Turbines**   Commercial generators are usually driven by turbines. The simplest turbine is a paper pinwheel that turns when you blow on it. A turbine is a wheel that is turned by the force of a gas or a liquid, changing this force into energy that can do work. Turbines are among the simplest and most useful of machines. They are also ancient. Waterwheels and windmills are turbines, turned by the kinetic energy of moving wind or water and converting that energy into work, such as the movement of one stone over another to grind corn.

The electric generators at hydroelectric plants get their energy from water turbines, which are merely more efficient versions of the old-fashioned waterwheel. But most big electric generators are powered by steam turbines (Figure 9-6). The steam is produced by heating water in a boiler. The heat is produced by burning oil, coal, wood, or the local gar-

bage collection or by the radioactive decay of uranium. It is the fuel used to supply heat for the steam turbine that gives a generating plant its name as a coal-fired, oil-fired, or nuclear plant.

Once electricity has been generated, it must be transmitted to the user. The higher its voltage, the more efficiently an electric current is transmitted over long distances. So the first stage in transmission is large step-up transformers, which increase the voltage of the current produced by the generator. Power is carried from the transformer to a service area by high-voltage lines. When it reaches its destination, its voltage must be reduced again as the current flows through transformers and local substations to the user (Figure 9-7). Energy is lost when electricity is carried in power lines, and the farther it is carried, the greater the loss. As a result, it is most efficient to produce electricity near where it is used.

## 9-C EVALUATING ENERGY SOURCES

When we compare one source of energy with another, we need standards to help us decide which is better or cheaper for particular purposes. Energy is used for three major things:

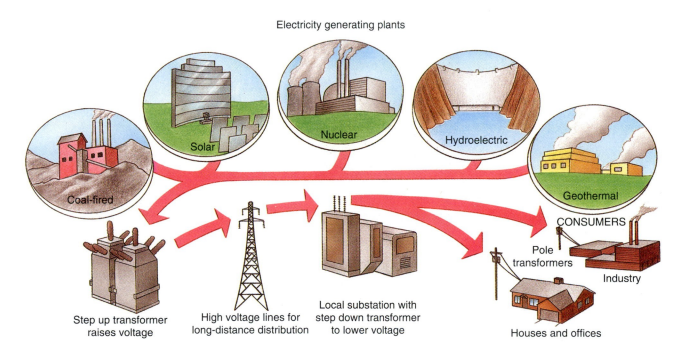

**Figure 9-7**   How electricity reaches us. This diagram shows the distribution grid that carries electricity from the generating plant where it is produced. The plant may be powered by various forms of energy, including nonrenewable fossil fuels such as coal and oil and renewable sources such as solar energy, hydroelectric power, and geothermal power. Transformers raise and lower the voltage for more efficient distribution. Substations distribute the current within a local area.

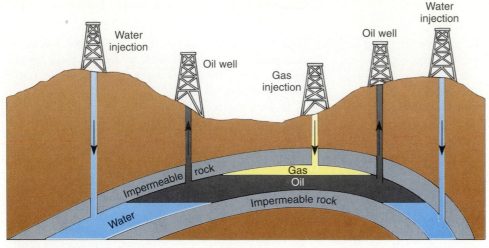

(a)

(b)

**Figure 9-8** Getting oil out of the ground. (a) The easiest oil to extract floats on water that is trapped above a dome of impermeable rock. A quantity of gas, released from the oil, often lies above the oil. A well is drilled into the oil, which can then be forced up the well by pumping water into the water layer or gas into the gas layer. Often the gas is also extracted as an energy source. (b) This oil shale rock contains oil chemically combined with rock. Much of the world's oil occurs in oil shale. It is expensive to extract oil from shale. High temperatures are required, and large quantities of waste rock are produced.

1. As fuel for transportation.
2. To produce high temperatures for industrial processes.
3. For heat, air conditioning, and appliances in buildings.

In all three types of application, electricity is also used.

These three applications require different types of energy. A liquid fuel is the most convenient for a vehicle. (It is difficult to imagine an airplane running on piles of coal.) High-temperature industrial heat is usually powered directly by fossil fuel because sources such as wood or direct solar energy are not concentrated enough to produce the required heat. Building use is less demanding and can be supplied in many ways.

> The three main uses of energy are to supply buildings, to power transportation, and to produce high temperatures for industry.

## Energy Efficiency

One way to reduce the cost of energy is to increase the efficiency with which we use it. The **energy efficiency** of a device is the percentage of the energy put into the device that does useful work. For instance, the human body converts about 20% of the energy it takes in into useful work in the form of muscle movement and metabolism. Incandescent lights and internal combustion engines used to power vehicles are very inefficient, converting only 5 to 10% of the energy that reaches them into light or movement. The most commonly used method of producing electricity, a steam turbine with a generator, is more efficient, converting 45% of its energy input into electrical energy.

Consider the efficiencies of heating a house by various means. If the house is heated with electricity from a nuclear power station, the efficiency is a mere 15%: only 15% of the energy in the uranium used to fuel the power station ends up as heat in the house. Various stoves and furnaces are much more efficient. The most efficient of these are new natural gas furnaces with remarkable 95% efficiencies. In the next few sections, we consider the efficiencies and environmental costs of the various nonrenewable sources of energy.

## 9-D  PETROLEUM (OIL)

Oil supplies more of the world's energy than any other source. It accounts for 45% of commercial energy use. In addition, oil is the raw material from which plastics and many other substances are made.

Geologists search for oil by analyzing rock formations. When a likely site is found, test holes are drilled to see if there is a reservoir of oil beneath the earth's surface (Figure 9-8). Oil is pumped to the surface as **crude oil**, or **petroleum**. The cheapest oil to pump out is that which flows into the hole in the earth by gravity. When this has been extracted, water is injected into the well to force out more of the oil. These processes remove only about one third of the oil in a typical well. The hole still contains thicker, heavy oil. It may be worthwhile to recover some of this by processes such as pumping steam into the hole to soften the heavy oil. But heavy oil takes a lot of energy to recover, so this process is economical only when the price of oil is high.

From the oil well, petroleum is transported in tankers or pipelines to a refinery. Petroleum consists of a mixture of many molecules, which are used for different purposes. At the refinery, groups of molecules are separated by the temperatures at which they

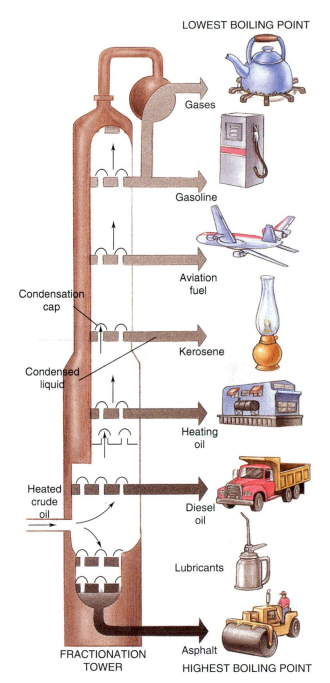

**Figure 9-9**   Refining petroleum to produce fuel. In a modern oil refinery, crude oil is distilled. It is heated and then passed into this tower. The lower the boiling point of the oil fraction, the higher it will rise up the tower before it condenses. This diagram shows only the production of fuel, not of petrochemicals used in the production of things such as plastics.

boil (Figure 9-9). The petrochemical fraction goes to factories as the raw material used to produce many things, such as industrial chemicals, fertilizers, and plastics. Most of the other fractions are used as fuel for furnaces, airplanes, and road vehicles.

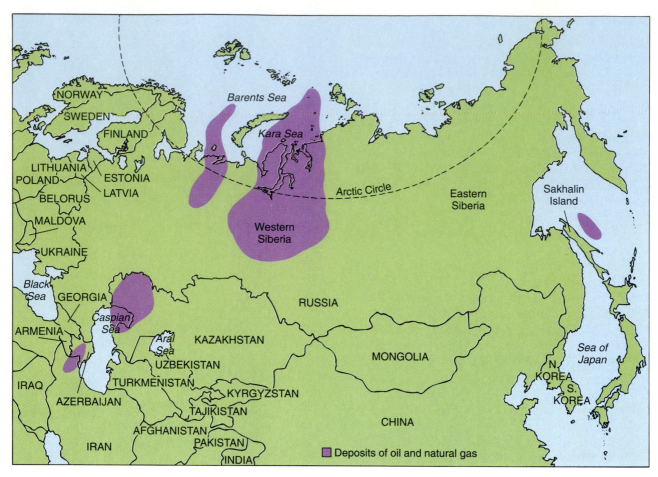

**Figure 9-10**   Oil and gas deposits in Russia and neighboring countries. Since 1975, huge deposits have been discovered and are now being explored by many multinational corporations. Experts believe this area may contain almost as much oil as Saudi Arabia.

Oil is relatively cheap to extract and transport, and it requires little processing. It also has a relatively high useful energy yield. But oil, like all energy sources, has some disadvantages. Its environmental costs are quite high as a result of oil spills and air pollution. In addition, oil burns to release carbon dioxide, nitrogen, and sulfur into the air.

> Oil is relatively cheap to produce and can be used for many purposes. But oil reserves are depleted, and using oil causes much pollution.

### Running Out of Oil

The world will eventually run out of oil, but we do not know when. For instance, huge new reserves of oil were discovered in Russia and Kazakhstan in the

1980s (Figure 9-10). The amount of oil used worldwide is expected to remain about the same for another 20 years or more and then slowly decline as oil is replaced by cleaner fuels. When the price of oil shot up during the 1970s, consumers started shifting to other fuels, and many did not return to oil when the price fell in the 1980s. In the United States, most of the oil that can be pumped out of the ground cheaply has been used up and production is declining. The United States was a net exporter of oil for the last time in 1946. In 1993, the country imported more than one third of its oil.

### Oil Spills

Most of the oil spills that make headlines affect bodies of water, from Alaska to the Persian Gulf. We have seen some devastating oil spills from tankers in recent years, although carelessness and other accidents probably spill more oil into the environment. For instance, in the 1980s, more than 3,500 tankers full of gasoline crashed on U.S. highways, spilling millions of gallons of gasoline into the soil and water. Every time a car or a container leaks oil or gasoline, more oil reaches the environment. Most service stations now have collection tanks so that if you change the oil in your car at home, you can take the oil to the service station and dispose of it safely, but many people still do not know this or are too lazy to bother.

The Gulf War generated the largest oil spills the world has ever seen. Iraqis opened the spigots on pipelines leading to tanker terminals in the Persian Gulf, an act of destruction that was named **ecoterrorism**. As a result, many ecosystems in and around the Gulf have been destroyed, possibly forever. Important fish supplies were destroyed, many people lost their means of earning a living, and several endangered species were brought to the verge of extinction. These include the dugong, a roly-poly relative of the Florida manatee; loggerhead and green turtles; and birds such as a remarkable species of pink flamingo.

After an oil spill, fish die of toxins, such as benzene. Other animals die of cold, asphyxiation, and starvation. Diving birds and seals depend on their feathers and fur to keep them warm. Even small patches of oil on their bodies destroys their insulation and they die of cold. Predators such as falcons and eagles may feast on the oily carcasses. The oil coats their intestines, preventing them from absorbing food, and they die of starvation.

### Cleaning Up Oil Spills

Millions of dollars have been spent to clean up oil spills in bodies of water in the last decade. There is considerable argument about how best to do this and even about whether it should be attempted at all.

Spilled oil floats, at first, and floating booms can be used to stop it spreading. Skimmer machinery and absorbent pads can remove it from the water surface. As the smaller, volatile molecules escape into the air, the oil turns into thick black gunk that sinks. Much of this oil enters the food chain as it is absorbed by plankton, crustaceans, molluscs, and other organisms.

No one disputes that it is worth trying to contain a floating oil spill with booms, absorbent pads, and skimmers. Chemical controls are more controversial. When the *Braer* spilled oil in the Shetland Islands north of Scotland in 1993, fishermen protested against airplanes spraying detergent on the floating oil. The detergent breaks up the oil and makes it sink, but the fishermen complained that the detergent did more damage to young fish than the oil did.

**Bioremediation** is the use of microorganisms to clean up pollution. When the *Exxon Valdez* spilled massive amounts of oil in Prince William Sound, Alaska, workers experimented with bioremediation. They started by spraying one rectangle of oil-covered beach with fertilizer. The idea was to encourage the reproduction of bacteria and fungi that would metabolize the oil. Within days of treatment, the experimental beach was almost clear of oil. However, some scientists think that nature would have cleaned up the oil spill just as rapidly (or slowly) without human assistance. In some places, fertilizer was sprayed on beaches with high-pressure sprays, which alone washed the oil off the rocks to the bottom of the sea. Scientists engaged in the Alaska cleanup agree that bioremediation needs a lot more research and better understanding before it becomes a standard method of cleaning up oil spills in the sea.

Preventing oil spills is much more effective than trying to clean them up after they have happened. As a result, many nations prevent oil tankers from traveling close to the coast. New oil tankers are built with double hulls so that if the outer hull is pierced by a rock, the inner hull may remain intact and contain the oil.

## 9-E NATURAL GAS

About 20% of the world's nonrenewable energy supply comes from natural gas, and this percentage is increasing as the technology for transporting and using gas improves. Natural gas will last longer than oil as an energy supply. In the 1970s, an enormous reserve was discovered in Russia. An even larger reserve was discovered in the 1980s in Qatar, a state on the Persian Gulf. The United States has considerable natural gas reserves, including the country's largest known deposit in Prudhoe Bay, Alaska.

Natural gas sometimes occurs in underground pockets by itself, but it is usually found in the same geological deposits with oil. It consists mainly of methane ($CH_4$), with smaller amounts of propane ($C_3H_8$) and butane ($C_4H_{10}$). When natural gas is processed, the propane and butane can be compressed to turn them into a liquid—**liquefied petroleum gas** (**LPG** or **LP-gas**). The remaining gas is dried, purified, and pumped under pressure through pipes to the consumer. Rural customers, in areas where there are no gas pipes, are often supplied with LP-gas in pressurized tanks. LP-gas tanks are also used as portable energy for cooking in barbecue grills, boats, and campers.

Natural gas produces less air pollution than coal or oil burned by conventional methods. But there are snags. In particular, gas is difficult to transport except in pipelines. It can be compressed into a liquid, which can travel in refrigerated tanker ships, but a liquid gas tanker can transport only about half the energy contained in an oil tanker of the same size. Pipelines are more efficient. Russia has built pipelines to several European countries, and Canada has built a gas pipeline to California.

> Natural gas is cheap to produce and causes relatively little pollution. But it is expensive to transport except in pipelines.

## 9-F  COAL

Coal was widely used during the nineteenth century, but its use declined as oil became cheap and readily available. Now the use of coal is increasing again because there is much more coal in the world than there is oil, it is found in more countries, and less-polluting methods of burning coal have been invented.

Coal and peat are the decomposed, compressed remains of ancient plants. These remains consist mostly of carbon, with smaller amounts of water, nitrogen, and sulfur. Peat is very soft and contains a lot of water. It is burned for fuel in countries such as Canada, Ireland, and Scotland, where it is locally abundant. **Low-sulfur coal** gives off less sulfur and nitrogen into the air when it is burned and can be used with fewer pollution-control devices. The most valuable coal is **anthracite**, hard coal containing little sulfur and water, which produces a lot of heat and little pollution when burned. It is less common than **bituminous coal**, a softer coal with moderate heat content, containing a lot of sulfur.

About two thirds of the coal in the United States is burned to generate electricity. Most of the rest is burned in various industrial processes.

## 9-G  REDUCING THE DAMAGE FROM USING COAL

The main problem with coal is the enormous hidden cost of using it. Since 1900, more than 100,000 Americans have been killed in underground coal mine accidents. Coal dust also destroys the air spaces in the lungs, causing black lung disease, a progressive disease that disables and kills. Despite modern regulations, one third of American underground mines still have conditions that produce black lung, and taxpayers spend more than $1 billion every year in benefits to disabled miners. Scientists are experimenting

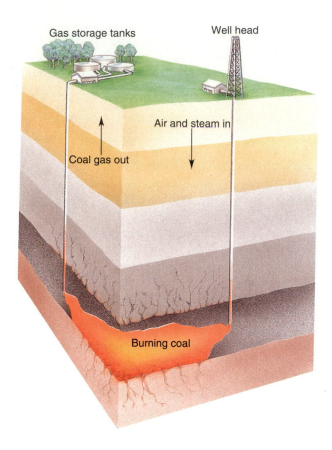

**Figure 9-11**    Coal gasification. This system produces gas from a seam of coal that is too difficult or costly to mine. The coal is set alight, and air and steam are pumped into the fire to keep it alight. The reaction produces coal gas (various compounds of hydrogen, carbon, and oxygen), which is collected for distribution to the consumer. This is one of many small projects that explore methods of extracting energy from deposits of fossil fuel that are too expensive to exploit by conventional methods.

with methods of recovering the energy from underground coal without actually digging it up. Figure 9-11 shows an experimental method of converting underground coal into a gas.

Weight for weight, coal produces more carbon dioxide when it is burned than does oil or natural gas. Burning coal also gives off sulfur and nitrogen oxides that contribute to acid rain. Some of the most dangerous substances in the smoke from burning coal are removed by devices that cause problems of their own. For instance, coal burning produces a fine dust known as fly ash. If this is captured by pollution-control devices, it becomes solid waste that must be disposed of in a landfill. Sulfur dioxide may be removed from coal smoke by pollution-control devices known as **scrubbers**, but scrubbers produce a toxic waste containing fly ash and sulfur that must also be disposed of.

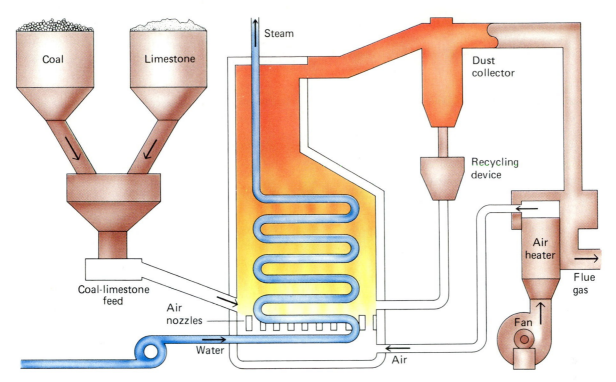

**Figure 9-12** A fluidized bed combustion (FBC) boiler, used to burn coal. (Fuels other than coal can also be burned.) Burning crushed coal and limestone is suspended in air from the air nozzles and behaves like a boiling liquid. Mixing ensures that the coal burns completely. Mixing also permits the limestone particles to reach all the gases formed by burning. The limestone reacts with more than 90% of the sulfur compounds that are major air pollutants formed by burning coal. Pipes run throughout the fluidized bed so that water is converted to steam, which provides energy for industrial processes. The dust and gases that leave the boiler pass through a collector that recycles any remaining dust back into the furnace. Some of the excess heat is used to heat the air on its way to the air nozzles in the furnace.

Research into methods for burning coal is accelerating because coal is already cheaper to use than oil. This means that it is increasingly cost-effective to spend the money required to burn coal in ways that produce only acceptable levels of air pollution. One experimental method of replacing oil with coal is to create a liquid mixture of powdered coal and water that can be stored in oil tanks and burned in furnaces previously powered by oil.

### Fluidized Bed Combustion

Another promising method of burning coal is **fluidized bed combustion (FBC)**, which can also be used for other fuels. Powdered coal and limestone ($CaCO_3$) are blown into a furnace where they are heated with hot air. The coal then burns very efficiently. The limestone is converted by the heat into

calcium oxide (CaO), which reacts with the sulfur released from the burning coal to form harmless solid calcium sulfate. Gases from coal burned in this way are well below federal emission control standards for sulfur and nitrogen released into the air (Figure 9-12). An additional advantage is that existing conventional furnaces can be converted to FBC. FBC furnaces are used, particularly in Europe and China, to burn biomass, such as wood, wood waste, agricultural wastes, heavy oil, trash, sludge from sewage plants, and high-sulfur coal. Small FBC plants are beginning to replace conventional boilers and furnaces in the United States.

### Integrated Gasification Combined Cycle

In Daggett, California, a demonstration plant uses coal to generate electricity, while causing much less pollution than an ordinary coal-fired power station, by the **integrated gasification combined cycle (IGCC)**. (Why do engineers use such tongue-twisting names for their processes?)

The first step is to crush the coal and add water. The resulting **slurry** is combined with oxygen and injected into the gasifier, where it is heated to a very high temperature. This converts much of the coal, water, and oxygen into hydrogen and carbon monoxide. These gases are used to run a gas turbine, which generates electricity. All the heat produced in the plant, from the hot gas, the remaining coal mixture, and the gas turbine, is used to turn water into steam, which drives a steam turbine, creating more electricity. The plant is, therefore, very efficient. It is also very clean. Emissions of gases, toxic metals, and particles are only one tenth that allowed by federal standards and well within California's emission standards, which are the nation's strictest.

## Synfuels and Coal Liquefaction

Coal can replace oil or natural gas for many purposes if it is first converted into synthetic fuels, **synfuels**, that are liquids or gases. Synthetic coal gas can be used exactly like natural gas and transported in pipelines (see Figure 9-11). Coal can also be liquefied to produce fuels such as synthetic gasoline. South Africa will soon get half of its gasoline from coal liquefaction. This has the political and economic advantage of making the country less dependent on imported oil.

Coal is the only fossil fuel that is not already seriously depleted in the United States. It might seem obvious that the United States should invest in synfuel production, particularly for use in vehicles, which consume much of the energy used in this country. Energy from synfuels is still much more expensive than energy from subsidized imported oil, but synfuel production will undoubtedly be renewed as soon as oil prices rise substantially, probably sometime after 2000.

> Coal is the most abundant remaining fossil fuel. But it must be converted into other fuels for many uses. Advanced technology and pollution controls are vital to reduce health and environmental damage from coal use.

## 9-H  NUCLEAR POWER

Electricity generated by nuclear power plants was once hailed as the energy of the future because the fuel it uses was cheap and plentiful. Nuclear power plants actually emit less radioactivity into the air than do conventional coal-fired power stations. They also conveniently generate electricity that can go straight into existing electricity distribution grids. But the dream of powering civilization by nuclear power has faded as costs have escalated and the dangers of nuclear accidents and waste disposal have become more apparent.

## Nuclear Fission Reactors

Most nuclear reactors are fueled by an uncommon isotope of uranium: uranium-235 (U-235). Recall that **isotopes** are atomic forms of an element with different numbers of neutrons in their nuclei. U-235 occurs in small amounts in larger concentrations of uranium-238 (which contains three more neutrons). When the nuclei of U-235 are bombarded by neutrons, they undergo **fission**, which means that they split, giving off large amounts of energy. In a power plant, this energy is used to heat water that drives a steam turbine used to generate electricity. Figure 9-13 shows how a nuclear power plant works.

Uranium-235 gives off neutrons spontaneously. If the U-235 is in the middle of a lump of uranium, this causes nearby uranium nuclei to fission, releasing yet more neutrons so that a **chain reaction** is set off. The same kind of chain reaction in uranium-233 and plutonium-239 produces the explosion in a nuclear bomb. A nuclear power station controls the fission reaction so that the energy is given off as heat over a long period of time. If the fission reaction is not controlled, runaway chain reactions can produce enough heat to melt the reactor in a core meltdown. (An atomic explosion is prevented by keeping the fuel much less concentrated than it is in a nuclear weapon.) The fission reaction in a power station is controlled by **moderators**, such as water or various gases, and by **control rods**, which control the reaction by absorbing neutrons from the **fuel rods** containing the uranium fuel. When control rods are lowered between the fuel rods, they slow down the reaction. Lowered completely, they prevent fission and shut down the reactor. Numerous safety features are built into a nuclear power plant. They consist mainly of vast steel and concrete shields designed to prevent radioactive substances from reaching the environment if the fission reaction should get out of control.

## Disposing of Nuclear Waste

The fuel cycle of uranium produces many environmental hazards. One kilogram of uranium fuel yields as much energy as 2,000 tons of coal so that uranium

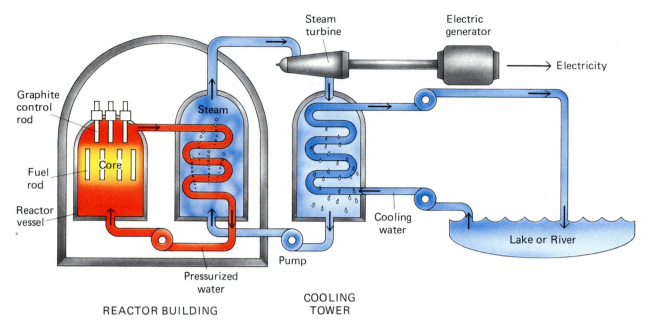

**Figure 9-13**  How a nuclear power station works. The power station produces steam that is used to generate electricity. On the left, the nuclear reaction takes place in uranium-235, contained in fuel rods in the reactor's core. The rate of the reaction is controlled by graphite control rods and by the substance used to remove the reaction's heat (in this case, pressurized water). In the next compartment, heat generated by the reaction is used to turn water into steam. This steam drives a steam turbine and electric generator. Steam that has passed through the turbine is cooled by water from a nearby river or lake. The walls of the reactor vessel and the reactor building are the main protection against a nuclear accident and are actually much thicker than shown here.

mines damage less land than coal mines because the mines are fewer and smaller. However, uranium must be **enriched**, or concentrated, before it can be used as a fuel, meaning that much of the mineral removed from a mine is discarded as **tailings**. Before the 1970s, uranium tailings were just dumped. Some were even used as landfill. In Grand Junction, Colorado, tailings were spread over a site before 4,000 houses were built there. Residents were exposed to radiation equivalent to ten chest x-rays a week. The leukemia rate in Grand Junction was twice that in the rest of Colorado, and the radioactive sand is now being removed at great expense.

Nuclear power stations produce some **low-level waste**, meaning waste that it is not very radioactive. This waste is hazardous for about 300 years. In the United States, more than 3 million cubic meters of low-level radioactive waste (from nuclear power plants, hospitals, and other facilities) have been buried in shallow landfills. These sites should, of course, be cleaned up, but federal and state governments have largely avoided the problem. Much of the ra-

dioactivity has leaked, and will leak, into the soil and ground water.

A more dangerous product of a nuclear power plant is **high-level waste**, which will remain hazardous for tens of thousands of years. Disposing of this waste is a political hot potato because nobody wants to live anywhere near it (Close-Up 9-2: Yucca Mountain). As a result, Congress has put off dealing with the problem, and millions of liters of high-level liquid waste are stored in steel drums near nuclear reactors. Because these drums corrode, the waste must be repackaged periodically. Regulation of nuclear waste has been so inadequate that we do not even know where all the waste is stored. Some waste sites are already known to be leaking. Nuclear waste is part of the general problem of hazardous waste disposal that we shall consider in more detail in Chapter 18.

### Safety of Nuclear Energy

Aside from the hazardous waste it produces, the main danger of a nuclear power plant is that the fission reaction creates radioactive products, many of

# Yucca Mountain, Nuclear Waste Facility

In 1992, an earthquake rustled the sage brush on Yucca Mountain, Nevada. It even rattled casinos 100 miles away in Las Vegas. In 1987, the federal government chose Yucca Mountain as the site for the nation's first storage dump for high-level radioactive waste. Yucca Mountain lies on the edge of the Nevada Nuclear Test Site, and the nuclear industry has already poured $6.7 billion into studies of its suitability for waste disposal. The idea is to seal the nation's 77,000 tons of used fuel rods and other waste in steel canisters and store them in a maze of underground tunnels designed to enclose the radioactivity for 10,000 years (Figure 9-C).

Although the waste repository is designed to withstand earthquakes up to 7.0 on the Richter Scale, the 1992 earthquake has provided ammunition for those who oppose using this site for nuclear waste.

"This earthquake was a wake-up call," said Robert Loux, director of Nevada's Office of Nuclear Waste Management. "It proves this area is active geologically and not the sort of place you want to bury highly dangerous waste."

"No matter what the experts say," said a columnist for the *Las Vegas Review-Journal*, "there is now zero chance of convincing the public that this dump will be safe."

Yucca Mountain is out in the middle of nowhere. If the nuclear power industry cannot convince the public that this site would be safe, there is little chance they could ever persuade anyone to let them install a high-level waste dump anywhere else in the United States—in which case the waste will probably stay where it is, in hazardous dumps scattered unsafely all over the country. Disposing of high-level waste in a manner that the public is convinced is safe has become the biggest barrier to the expansion of nuclear power.

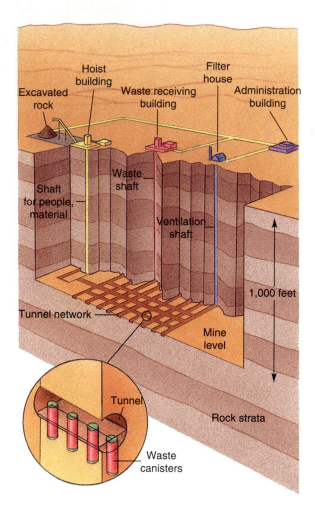

**Figure 9-C** Plan for underground nuclear waste disposal. This is the kind of facility designed for Yucca Mountain. The filter house filters air from the tunnels so that no radioactivity reaches the air. The waste would be moved through the tunnels and handled by trucks and robots.

 This international symbol warns of the presence of radiation levels that endanger human health.

### The Effects of Various Doses of Radiation*

| | |
|---|---|
| 1,000–5,000 rems[†] | Diarrhea, fever. Blood chemical imbalance. Death in 1–14 days. |
| 600–1,000 rems | White blood cells destroyed. Bowel malfunction in 4–6 weeks. 80 to 100% chance of death. |
| 200–600 rems | Low white blood cell count. Blotchy skin in 4–6 weeks. 50% chance of death. |
| 100–200 rems | Not immediately fatal. Long term risk of cancer increases. |
| 0–100 rems | Nausea, vomiting. Not fatal. |
| 0.2 rems | Typical dosage received during a medical x-ray. |

\* Radiation is particles and rays (alpha, beta, and gamma rays) given off by radioactive substances. The rays cause damage by penetrating the body and ionizing biological molecules.

[†] A rem (Roentgen Equivalent Man) is the unit to measure the effect of radiation on the body. It is defined as the amount of ionizing radiation required to produce the same biological effect as one unit of x-rays.

How Radiation Affects Various Organs

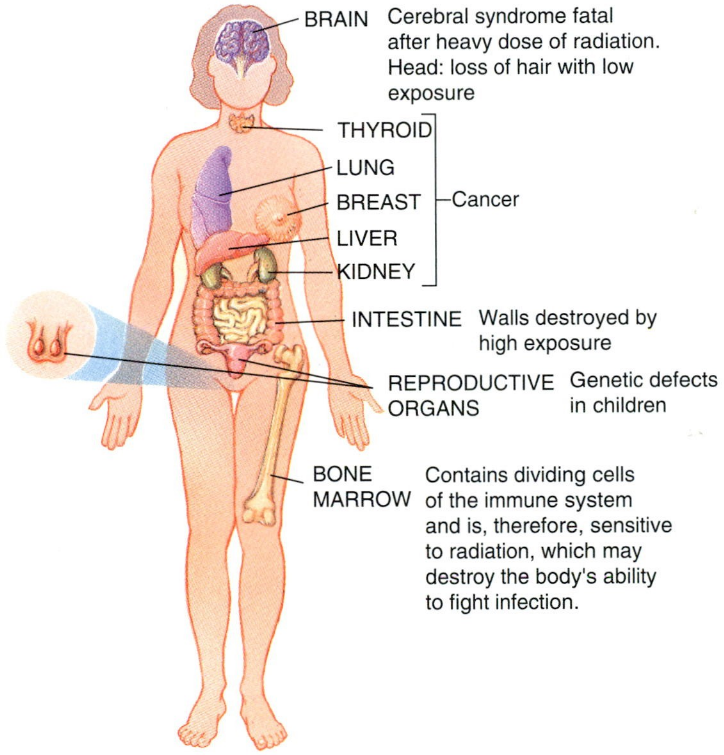

BRAIN   Cerebral syndrome fatal after heavy dose of radiation. Head: loss of hair with low exposure

THYROID
LUNG
BREAST ── Cancer
LIVER
KIDNEY

INTESTINE   Walls destroyed by high exposure

REPRODUCTIVE ORGANS   Genetic defects in children

BONE MARROW   Contains dividing cells of the immune system and is, therefore, sensitive to radiation, which may destroy the body's ability to fight infection.

**Figure 9-14**   Radiation dangers to human health

which are very dangerous (Figure 9-14). If the reaction gets out of control, the enormous heat it generates may destroy the reactor building and spew the radioactive products into the air.

The Nuclear Regulatory Commission (NRC) is responsible for the regulation and safety of nuclear power stations in the United States. Since an accident at Three Mile Island in Pennsylvania in 1979, the NRC has required more than 300 safety improvements. The NRC is doing a much better job than its counterparts in some other countries, but the United States still has a surprisingly long list of nuclear near misses, considering that we generate much less of our electricity by nuclear power than many other countries.

The worst nuclear accident to date occurred at Chernobyl in Ukraine in April, 1986. Two gas explosions in one of the four graphite-controlled, water-cooled reactors at this nuclear power plant north of Kiev blew the roof off the building and set the graphite on fire. The accident occurred when engineers deliberately turned off most of the reactor's automatic safety systems to carry out a safety experiment that they were not authorized to make. The explosions and fire carried radioactive debris into the air and deposited it over thousands of kilometers of land in Ukraine, Russia, and neighboring countries. Immediately after the accident, 31 firefighters and plant workers died of radiation sickness. Medical experts estimate that between 5,000 and 100,000 people will die prematurely over the next 70 years from cancer caused by exposure to radiation from Chernobyl.

Chernobyl was a badly designed plant, and we are told that hardly any radioactivity would have been released in the accident if it had been built to NRC specifications. There are 15 Chernobyl-type reactors still running in Russia, Ukraine, and Lithuania. As *Science* magazine said, "When a Russian reactor sneezes these days, all Europe fears the flu." Western nations are contributing money and technical expertise to increasing the safety of nuclear reactors in the Eastern-bloc countries, many of them designed to produce plutonium for nuclear weapons rather than energy.

What generalizations can we make about the safety of nuclear power plants? First, it is obvious that we now know how to build nuclear power stations that are remarkably safe from a technical point of view. However, recent studies conclude that American power stations are essentially safe in their design and construction: the accidents that have occurred in the United States have been caused by human error, especially by poor management. Although it is possible to redesign a power station, it is essentially impossible to make people more sensible or reliable, so these studies imply that nuclear accidents will always occur, no matter what we do to prevent them.

It is important not to exaggerate the damage that has been done by nuclear accidents. Many more people have been killed during the use of most other forms of energy. More people have undoubtedly been killed while felling trees to burn for energy than by all the world's nuclear accidents, and more people are killed mining coal each year than will die as a result of the accident at Chernobyl. Deforestation and overgrazing make more land unusable each year than radioactive contamination has ever done. The trouble with nuclear power is that even if accidents are rare, they can do such massive damage.

## Breeder Fission Reactors

When uranium ore is mined, less than 1% of it is in the form of fissionable uranium-235. This tiny amount must be separated from the much larger amounts of uranium-238 in the ore. Breeder reactors use the leftover uranium-238 as fuel.

A fast breeder nuclear fission reactor "breeds" its own fissionable fuel, plutonium-239, from uranium-238. The reactor core contains plutonium-239 surrounded by uranium-238. This is an efficient and convenient fuel system because the plutonium-239 in the core is produced by ordinary fission reactors as a by-product of the fission of uranium-235. So the breeder reactor is also disposing of hazardous waste for us. As in a regular fission reactor, the plutonium undergoes fission and the energy is captured to run a steam turbine that generates electricity. The difference here is that the fission reaction is allowed to run "hotter" so that fission produces "fast neutrons." These have so much energy that when they strike the uranium-238 that surrounds them, they convert the uranium into plutonium-239, providing more fuel for the reaction. The fission of each plutonium atom in the core converts uranium-238 into more than one atom of plutonium-239 so that the reactor actually produces more fuel than it uses. It takes a breeder reactor about 30 years to produce as much plutonium as it consumes. Until that time, it must be supplied with plutonium fuel.

Breeder reactors appear to be safer than other fission reactors; they produce little hazardous waste, and they require little mining. Why are they not the energy source of the future? One snag is political. Breeder reactors use and produce plutonium-239, which is one of the ingredients of a nuclear weapon. It would be all too easy for a terrorist or other organization to build a nuclear bomb if it could get hold of the plutonium. Japan transports plutonium for breeder reactors to Japan from a processing plant in France by ship. Many countries protest this transport for fear of a nuclear accident and the possibility that the ship might be highjacked. Another problem is cost. Energy from a breeder reactor is much more expensive than energy from cheap oil or coal. As a result, breeder reactors are attractive mainly to countries that have few energy sources within the country and wish to produce their own energy rather than import it.

Electricity from nuclear fission is expensive. It carries with it the danger of an accident that can contaminate the environment with radiation and dangers from the unsolved problem of disposing of radioactive waste safely.

## Nuclear Fusion Reactors

Nuclear fusion is the atomic reaction that takes place in the sun and the stars and that produces solar energy. In the sun, very high temperatures induce the nuclei of four hydrogen atoms to fuse to form the nucleus of a helium atom. The reaction emits large quantities of high-energy radiation. The technical problem in harnessing fusion reactions for energy on Earth is to create and sustain **plasmas**, states of matter hot enough for fusion to occur. Containing the plasmas is one problem: what would you use to contain a plasma with a temperature of 40 million °C? Plasmas cannot be contained in any substance; they have to be suspended in space. (These high-temperature fires are not dangerous because they go out instantly if they touch the wall of a vessel.)

Despite the many technical difficulties that remain, fusion reactors may become a reality in our lifetimes. Fusion reactors will have many advantages over fission reactors. They will generate very little, and only low-level, nuclear waste. Their fuel (probably water) is essentially inexhaustible and will not require mining. Their main disadvantage is that fusion reactors will probably require huge containers and be large complex structures. This means that they will be of no use in areas such as transportation: you can put a small fission reactor on a nuclear-powered ship or submarine but probably not a fusion reactor.

> Fusion reactors, generating heat by reactions like those that occur in the sun, may one day supply large quantities of energy more safely than fission reactors.

## The Future of Nuclear Energy

Nuclear power is not the energy source of the immediate future in the United States, largely because its price has risen rapidly, while the price of oil remains artificially low.

During the 1960s, utility companies planned and built nuclear power plants at a frantic pace to meet the predicted demand for energy. But population growth was slowing, and a rise in oil prices during the 1970s made conservation attractive. For the first time in 200 years, the demand for energy in industrialized nations slowed down. Utility companies realized that they were not going to need all the power they had predicted and canceled plans to build 100 nuclear power plants.

Canceling plans to build new power stations is not cheap, but abandoning plants that have already been started is devastatingly expensive. For example, the Washington Public Power Supply System (nicknamed WooPPS) canceled plans to build several plants and halted construction on another. It defaulted on $2.25 billion worth of bonds sold to raise money for the project. The CMS Energy Corporation specializes in converting half-built nuclear power stations into conventional power stations. Their slogan is "We turn lemons into lemonade." The first such conversion was of a power station at Midland, Michigan, in 1990, and several more are in the works.

Although several countries are cutting down on their plans for future nuclear power plants, there are political reasons why countries without major fossil fuel reserves will become increasingly dependent on nuclear power. Many countries try to avoid becoming as dependent on foreign oil as the United States, and so nuclear power is becoming increasingly important in places such as Western Europe, Japan, Taiwan, and South Africa.

> Despite the expense and danger of nuclear power, its use will continue to grow in countries with few other sources of energy.

Some environmentalists are rethinking their opposition to nuclear power as the environmental damage from fossil fuel use mounts in the form of air pollution and oil spills. If we decide to reduce the use of fossil fuel, it is hard to see how the world's growing energy needs can be met without the use of at least some nuclear energy.

## 9-1 TRANSPORTATION

Transporting people and goods consumes much of the energy we use, produces much of our air pollution, and destroys farmland and wilderness. "By 2020," wrote U.S. Senator Daniel Patrick Moynihan to his constituents, "it will take 44 lanes to carry the traffic on I-95 from Miami to Fort Lauderdale. Pretty soon there won't be anything left of Florida!" In 1900, the average speed of traffic in Manhattan was 18 kilometers per hour. Today Manhattan's traffic moves at an average of less than 8 kilometers per hour. Despite a frenzy of road building from 1960 to 1980, our transportation systems are in trouble.

Over long distances, air transport is the most efficient means of moving people, who are relatively light, but it is an expensive way to move heavy freight. It is generally cheapest to move freight from where it is produced to a distribution center by large transporters such as ships, trains, or barges. Goods are then carried from distribution centers to their des-

**TABLE 9-2** U.S. Energy Use by the Main Methods of Transporting People on Land: Energy use decreases with more passengers*

| Transport Method | Number of Passengers | Energy Used to Move One Person 1 Kilometer (in thousands of joules) |
|---|---|---|
| **Mass transit between cities** | | |
| Intercity railroad | 80 | 469 |
| Intercity bus | 40 | 503 |
| **Mass transit within cities** | | |
| Light rail (within city, including underground) | 55 | 674 |
| City bus | 45 | 728 |
| Rapid rail | 60 | 793 |
| **Automobiles** | | |
| Car pool | 4 | 1,206 |
| Automobile | 1 | 4,823 |

*Worldwatch Institute.

tinations by trucks and vans. Transportation of people is much more varied. Beijing's bicycles, Calcutta's rickshaws, London's buses, Los Angeles's freeways, Moscow's underground, and Venice's canals all make their unmistakable marks on life in these cities. Table 9-2 shows that the more people travel together, the less energy each of them uses.

### Individual Transit

**Individual transit** is people traveling by themselves or in small groups, walking or using automobiles, bicycles, motorcycles, mopeds, rickshaws, or taxis. The larger individual transports, such as cars and taxis, tend to travel slowly in cities, spending much of their time in traffic jams. For this reason, in many areas there is a tendency toward increased use of smaller vehicles such as mopeds, motorcycles, and bicycles. Nearly half of all urban journeys are of less than 8 kilometers and well-suited to bicycles, mopeds, and motorcycles in towns where the land is flat and the weather is not extreme.

Bicycles are the fastest growing means of transportation in the world. Three times as many bicycles as cars are manufactured every year. In many areas, bicycles are encouraged in attempts to reduce air pollution and traffic congestion. The trouble with bicycles and motorcycles that share roads with larger vehicles is that their riders are more likely to be killed or injured than are people in automobiles. The use of bicycles can be increased by providing bicycle paths that segregate cars from bicycles. Davis, California, has constructed many kilometers of bicycle paths, closed some streets to automobiles, and pro-

vided city employees with bicycles. Bicycles now account for about one fourth of all transportation within the city.

### Motor Vehicles

The hallmark of most twentieth-century cities is the curse and convenience of motor vehicles: cars and trucks. There are about 600 million motor vehicles in the world, three quarters of them in North America and Europe. The U.S. economy, more than that of any other country, depends on the automobile. One out of every $6 spent and one out of every six nonfarm jobs are related to automobiles, and automobiles account for 20% of the gross national product.

Cars have the overwhelming advantage of giving people the freedom to go where they want when they want, but they have many disadvantages. Cars are not necessarily fast, they consume enormous amounts of energy and money, and they produce much of the air pollution in developed countries. Each year, the world's trucks and cars kill some 180,000 people and injure 10 million.

Highways occupy millions of hectares of land, much of it prime land for agriculture because highways are easiest to build in flat valley bottoms. Roads and parking spaces occupy more than half the area of many American cities. Because paved areas do not absorb water, this contributes to the flooding of streets and houses that disrupts life in many cities after a heavy rain.

We shall see how motor vehicles pollute the air in Chapter 20. Vehicles use fuel inefficiently, and much

**Figure 9-15**   Light railways are a fast and energy-efficient way of moving people around cities. Here, a train in the Paris Metro system crosses a bridge over the River Seine.
*Thom Smith*

of the fuel is converted into heat and chemicals that pollute the air. In Mexico City, you are not allowed to drive a car on days when pollution is heavy. Denver has experimented with a similar system—permitting people to drive into the city only on certain days of the week. Pollution is one of the main reasons that many towns would like to get rid of most automobiles

from city streets: unless they do so, they cannot meet legislative clean air standards. Mass transit is the main solution to getting around without cars.

> Cities try to encourage forms of transportation other than motor vehicles, which cause pollution and traffic jams.

### Mass Transit Versus Cars

Mass transit comprises all sorts of methods of moving people in masses: regular buses and water buses, trolley cars powered by overhead electric wires, trams running on street railways, trains in subway tunnels, and trains above ground (Figure 9-15). Most of these are **public transport** systems run by states and towns. Many European and Asian countries have made a policy commitment to mass transit. In the United States, public and private efforts have encouraged the use of automobiles at the expense of mass transit (Figure 9-16). Part of the decline of mass transit was the result of a 1930s conspiracy between oil producers and automobile manufacturers who bought up and destroyed the trolley cars and trams that once served many American cities. Federal subsidy of gasoline prices and federal funds to build highways have also supported the use of motor vehicles over mass transit.

In 1993, it cost the average American $1,700 to commute to work by automobile. The average Euro-

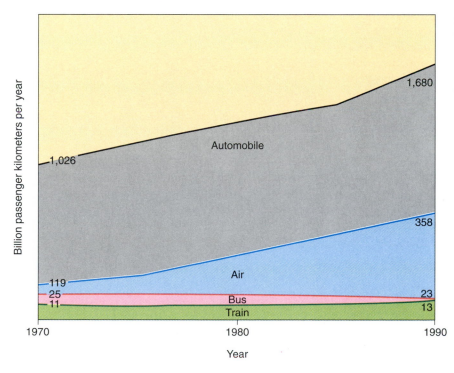

**Figure 9-16**   How Americans traveled between 1970 and 1990. Bus travel declined, while train travel increased slightly. Air travel grew even faster, and travel by automobile grew fastest of all.

**TABLE 9-3**   Urban Densities and Commuting Choice in Selected Cities: More people commute to work by mass transit in cities where jobs and population are concentrated in a small area

| City | Density (population + jobs per hectare) | Private Car (% of commuters using) | Mass Transit (% of commuters using) | Walking and Cycling (% of commuters using) |
|------|------|------|------|------|
| Phoenix, U.S.A. | 13 | 93 | 3 | 3 |
| Perth, Australia | 15 | 84 | 12 | 4 |
| Washington, U.S.A. | 21 | 81 | 14 | 5 |
| Toronto, Canada | 59 | 63 | 31 | 6 |
| Hamburg, Germany | 66 | 44 | 41 | 15 |
| Stockholm, Sweden | 85 | 34 | 46 | 20 |
| Vienna, Austria | 111 | 40 | 45 | 15 |
| Tokyo, Japan | 171 | 16 | 59 | 25 |
| Hong Kong | 403 | 3 | 62 | 35 |

*Newman, P., and J. Kenilworth, *Cities and Automobile Dependence: An International Sourcebook.* Aldershot, U.K.: Gower, 1989.

pean spent less than half of that to get to work for a year by public transport. Because mass transit is a cheap and efficient way to get to work, residential areas cluster near train stations and bus stops. Without mass transit to attract high-density housing, American cities have spread more than cities in any other part of the world. In the 1992 election, for the first time, more Americans lived in suburbs than in cities and the countryside combined. Table 9-3 shows that people are least likely to commute to their jobs by mass transit when jobs and population are spread out over a large area as they are in most parts of the United States.

Those who advocate mass transit argue that this form of transport reduces highway congestion, land destruction, air pollution, and dependence on imported oil. Studies show that mass transit is an effective way to revitalize decaying inner cities, stimulating the development of new businesses in the city. Land values rise and businesses start up wherever mass

transit, such as a city railroad or subway system, is built. The American Public Transit Association estimates that every $100 billion spent on mass transit generates $327 million in urban business revenue and supports 8,000 jobs.

People find cars so convenient that if a city wants to encourage the use of mass transit it has to make cars inconvenient. Many cities are experimenting with ways to stop people using cars by closing some streets to cars, which permits delivery trucks, buses, and taxis to move more rapidly. As the number of cars continues to increase, urban experiments in banning them from the city center can be expected in many places.

Mass transit revitalizes cities and reduces congestion and pollution, but it is expensive to install and commuters have to be encouraged to switch from automobiles to mass transit.

**TABLE 9-4**   LA Grinds to a Halt: Population and transport in Southern California

| In Southern California | 1992 | Forecast for 2010 | Growth (%) |
|------|------|------|------|
| Population (millions) | 13 | 19 | 46 |
| Vehicles in use (millions) | 8 | 10 | 30 |
| Vehicle kilometers driven (millions per day) | 384 | 619 | 61* |
| Average road speed (km/hr) | 56 (35 mph) | 30 (19 mph) | − 46 |

Source: South Coast Air Quality Management Plan.

*Driving distance is increasing faster than population or numbers of vehicles because people are living further and further away from where they work.

### Why Building More Highways Won't Help

Traffic engineers tell us that building more roads or widening existing ones will not reduce traffic congestion in southern California (Table 9-4). They quote the example of the M25, a British freeway built in the 1980s. This road encircles London. It is supposed to carry traffic traveling from one side of London to the other, relieving congestion in the center of the city. Already, the new road is so crowded that it is often faster to drive through the narrow streets and traffic lights of central London than it is to take the new road.

The reason new roads often do not relieve congestion is that new or improved roads may actually generate new traffic. Consider a city where many people work and a small rural town 20 kilometers away. If we widen the country lane between city and town so that rural people can get into the city more easily, people discover that they can live in the town and work in the city without spending much more time traveling. Since life in a small town is often more pleasant than life in the city, people move from the city to the small town and commute to work, creating new traffic on the road.

### Internodal Surface Transportation Efficiency Act of 1991

"Ice tea" is the nickname for the Internodal Surface Transportation Efficiency Act (ISTEA) of 1991, which acknowledges that the United States will have to stop relying so heavily on automobiles for transportation. For the first time, state and local officials can use money from the trust fund created with the federal tax on gasoline for things like bicycle paths and subways as well as for highways. This will permit different areas to come up with suitable solutions to their own traffic problems. One area may decide bike paths would relieve traffic jams, while another may build trains for commuters.

### Trains

Trains are much safer than automobiles. Japanese bullet trains and the French *Trains a Grande Vitesse* (TGV) have carried billions of passengers virtually without accidents. They are also fast. A French TGV did a high-speed run at over 500 kilometers per hour (nearly 320 miles per hour) in 1990, and the trains regularly average almost 200 miles per hour. Europe is planning a network of electric high-speed trains extending from the Atlantic to eastern Europe. The railways already exist, but they are being upgraded to take high-speed trains and to connect new towns and even countries into the system. The "Chunnel"

under the English Channel between France and Britain is a train tunnel and will tie in with this network.

Modern trains are powered by electricity and produce little air pollution. Trains that run on diesel oil, and even on coal, still exist in parts of the world, and even these produce much less air pollution for each person transported than a car does.

It is often faster to travel by train than by airplane from one city to another, especially for tourists and business people who spend much of their time in the center of the city. The train goes from one city center to the other. Although an airplane may be faster in the air, the time taken to reach an airport on the outskirts of a city and delays at the airport often make air travel more time-consuming. Many passenger trains also transport automobiles so that travelers can collect their cars for local use at the end of the trip.

The United States is belatedly grasping the necessity of trains for transport in the twenty-first century. The U.S. passenger network, Amtrak, carries more passengers every year. Since it does not receive the subsidies of a European system, it has not been able to afford to upgrade to a high-speed system. Many states are now interested in trains as an alternative to endless road building. In 1991, Texas granted a franchise to a consortium, which includes the manufacturers of the TGV, to build the first high-speed rail system in the United States, linking five Texas cities by 2000. Even before that, Amtrak is scheduled to start operating a maglev system in Florida in 1996.

### Maglev

A maglev system looks a bit like a train (Figure 9-17a). Although the cars travel along tracks, they do not touch the track when they are in motion. "Maglev" stands for "magnetic levitation" because the cars start their trip by being hoisted a few inches into the air by powerful magnets and are then propelled along the track by electromagnetic motors. Maglevs, in theory, can travel faster than trains. In practice, some people think that developing this new technology is a waste of money when electric trains already travel at such high speeds.

A vehicle can travel faster when suspended in air because contact with the ground (or a rail or water) produces friction that slows the craft down. The principle is rather like that of a Hovercraft (Figure 9-17b), a water transport vehicle that rises from where it is parked by shooting a jet of air downward under its "skirt." Then it moves forward on its bubble of air under the power of aircraft engines. Hovercraft ferries carry cars and people in many parts of Europe. Riding in a Hovercraft is more like a plane

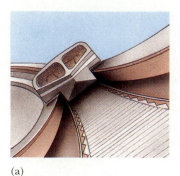

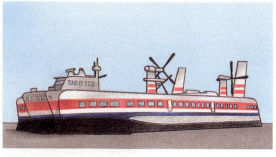

(a)          (b)

**Figure 9-17** Maglev and Hovercraft. (a) A German maglev. Note that the maglev has no wheels and does not touch the tracks as it moves. (b) This big Hovercraft is a ferry between England and France. It carries cars and trucks as well as passengers.

ride than a boat ride. Seat belts are buckled and the craft races across the ocean at up to 100 knots (more than 100 miles per hour). On a rough day, the windows are so covered with spray that you cannot see out and you cannot imagine how the pilot can possibly avoid mowing down motorboats and sailboats as you come whizzing into harbor and up onto dry land—where you drive your car out of the Hovercraft and go on your way.

Maglevs came from American inventors in the 1960s. Even in the 1950s, inventors had imagined putting tracks alongside or above interstate highways to carry railroad cars that would move as many people as ten lanes of a highway. American technology and ingenuity are arguably the best in the world, but we have been cursed with unimaginative governments for much of the twentieth century, at least when it comes to civilian projects. So it should come as no surprise that U.S. government funding for maglev research dried up in 1973 and subsequent development took place in Japan and Germany. The Florida maglev, running between Orlando airport and Walt Disney World is being built by a consortium of U.S., German, and Japanese companies.

## SUMMARY

The industrialized world depends on nonrenewable natural resources—fossil fuels and nuclear power—for the enormous amounts of energy it consumes.

### Renewable and Nonrenewable Energy Sources

Agricultural societies use renewable energy from sources such as wood, sunlight, and muscle power. Some 200 years ago, the industrial revolution was fueled by the much greater amounts of energy available from nonrenewable sources. These are now becoming depleted.

### Electricity

The type of fuel used for energy changes, depending on factors such as cost and availability. Worldwide, less than one quarter of all energy used comes from renewable sources, but that fraction is slowly rising. In developed countries, around one half of all energy reaches the user as electricity, produced in various ways.

### Evaluating Energy Sources

The main types of energy use are to supply buildings, to power transportation, and to provide high temperatures for industry. Different types of energy are

appropriate for these different uses. Enormous sums can be saved by increasing the efficiency with which energy is used.

### Petroleum (Oil)

Oil is the world's largest energy source. Among its advantages, oil is suitable for most types of energy use, is relatively cheap to transport, has a high net useful energy yield, and needs little processing. Its main disadvantages are the pollution caused by oil spills and the emissions from burning it and the fact that it is the most depleted of all major energy sources.

### Natural Gas

Natural gas is replacing oil for many purposes, and its price is rising. It has the advantage that it is relatively easy to process and burns fairly cleanly. Its main disadvantage is that it is expensive to transport except through pipelines.

### Coal

Coal was the first fossil fuel of the industrial revolution and will probably be the last to be used up. Its main advantage is availability, but it has several disadvantages. It must be converted into synfuels or electricity for many uses. Mining and burning coal can cause massive damage to health and the environment unless they are carefully controlled.

### Controlling Damage from Burning Coal

Environmental damage from the use of fossil fuel can be broadly divided into that caused by extracting and processing the fuel and that caused by burning it. Damage such as that from mining and oil spills can be controlled by modern technology and enforcing environmental regulations. Damage from pollutants such as fly ash and sulfur compounds can be reduced by pollution control systems when fossil fuels are burned. The most difficult problem to overcome is the emission of carbon dioxide, which contributes to the greenhouse effect.

### Nuclear Power

Nuclear fission reactors supply about 3% of the world's electricity, and that percentage is increasing.

Nuclear power has the advantage that it produces no carbon dioxide. But no country has yet discovered how to dispose of nuclear waste safely, and nuclear power has proved expensive and occasionally very dangerous. As a result, experts doubt that many more conventional fission reactors will be built. It is possible that breeder reactors or fusion reactors may supply safer nuclear power in the future.

### Transportation

The amount of energy we use has skyrocketed during the last 200 years, but the demand for energy is now slowing down in the industrialized nations. This is partly because population growth is slowing and partly because countries are cutting their energy bills by using energy more efficiently. Up to one half of all energy used is wasted so there is plenty of room for cost-cutting conservation. Conservation also makes depleted energy resources last longer, buying time for us to develop the next generation of energy sources. Using mass transit instead of automobiles is an effective way to save energy.

## DISCUSSION QUESTIONS

1. Consider the future energy needs of the state you live in. List all the factors you can think of that will determine how much energy the state will need in 2020.

2. What evidence of energy conservation can you find in your community?

3. Most people demand to look at the past year's utility bills before they rent or buy a house or apartment. Is there additional information that you will probably want to see before you rent or buy in 2010?

4. Suppose the cost of energy doubles in the next 20 years. What steps might you take to prevent energy gobbling up more and more of your paycheck?

5. Why does the United States use more oil than coal when coal is the country's most abundant fossil fuel?

6. What steps would you advocate if you were on a commission to suggest ways to supply the energy North America will need in 2050?

7. What do you think of the argument that the United States should reduce its dependence on foreign oil (even though this is the cheapest source of energy) by investing in new sources of energy within the country?

8. I am not as strongly opposed to nuclear power as most environmentalists. This is partly because I grew up with it. (My father was a nuclear engineer.) But it is also because the argument against nuclear power relies so heavily on the people a nuclear accident *might* kill, ignoring the hundreds of thousands of people who *are* killed every year in the production and consumption of other forms of energy. What do you think? Is your opinion different if your relatives work in coal mining or health care than if they work in an industry unrelated to our energy supply and the damage it does?

# Chapter 10

# ENERGY CONSERVATION AND RENEWABLE ENERGY

*Renewable energy in sea water, waves, and wood.*

**Key Concepts**

- As fossil fuels are depleted, we shall obtain more and more of our energy from conservation and renewable sources.

- The energy needs of industrialized countries are growing more slowly than before, as population growth slows and conservation is used to reduce energy costs.

- New sources of energy and much more energy are urgently needed by developing countries. Their energy needs are now supplied largely by wood (environmentally destructive) and imported oil (expensive).

- Energy conservation is the most cost-efficient, environmentally safe method of supplying future energy needs.

- We have only just begun to explore the vast amounts of energy available from renewable sources. The technology to exploit these resources will develop rapidly in the future.

Utility companies in the northwestern United States are giving grants to businesses for building alterations that decrease energy use. Homeowners get advice and incentives to induce them to conserve energy. Things have changed since the days when utility companies exhorted us to use more energy. Why are utility companies urging conservation when energy prices at the gasoline pump, adjusted for inflation, are as low as they have been in many years? The answer is that utility companies and energy planners want to avoid the high cost and environmental damage of building large new power stations. They believe that U.S. energy needs will grow only slowly in the next century and can be supplied by a wide mixture of renewable energy sources and by conservation.

At the moment, we are very dependent on fossil fuels. But reserves of fossil fuels are limited, and it is essential to find new sources of energy before they are seriously depleted. The search is made more urgent because people are increasingly aware of the environmental damage caused by burning fossil fuels, which range from local pollution of air and water to global warming. New energy sources should be renewable and less harmful to the environment. As a result, experiments are under way on alternative fuels that reduce air pollution from automobiles and factories and on conservation and solar energy—the two most promising energy sources of the future.

> Conservation and solar energy are the renewable and environmentally friendly energy sources of the future.

## 10-A ENERGY CONSERVATION

In 1992, a consortium of 23 utility companies, headed by Pacific Gas and Electric, announced a $30 million prize to be awarded to the manufacturer that can put a super-efficient refrigerator on the market by 1995. In an attempt to encourage energy conservation, the consortium is hoping for one that uses only half as much electricity as a 1992 model. Refrigerators and freezers use about 20% of the nation's electricity, although modern refrigerators are much more efficient than those made 20 years ago (Figure 10-1). **Conservation** is the reduction of waste and of unnecessary use of a resource. Utility companies sell energy for a living. What are they doing trying to make people use less of it?

Utility companies now look on conservation as a source of energy. California's State Energy Commission estimates that the state will need an additional 11,000 megawatts of electricity a year by 2000. Three quarters of that new demand is supposed to be met by conservation. As demand for energy rises, it is cheaper for utilities to induce their customers to conserve enough energy to cover the new demand than it is to build new power stations to produce the necessary energy. In the Pacific Northwest, most energy is hydroelectric, produced by generators powered by water backed up by dams (Section 10-D). Rivers now have so many dams on them that fish can no longer make their way upriver to spawn, and the important commercial and recreational fishing industries are declining. A proposal to construct a large new hydroelectric dam would cause public outcry, and building a new hydroelectric plant or a conventional

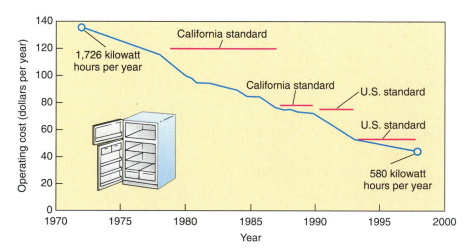

**Figure 10-1**  The increasing energy efficiency of American refrigerators. The cost of running a refrigerator for a year has fallen as refrigerators have become more energy-efficient. In the early 1970s, energy prices rose and consumers began demanding more efficient appliances. In the 1980s, state and federal governments, beginning with California, have encouraged energy efficiency by setting standards for energy use by appliances. *Data from: Association of Home Appliance Manufacturers, Lawrence Berkeley Laboratory*

power station to produce electricity from oil or coal is enormously expensive.

Although we do not usually think of it in this way, conservation is the perfect renewable energy source. Between 1975 and 1985, conservation made more energy available than all energy alternatives combined. A dollar of gross national product uses less energy to produce now than it did in 1977, and American houses cost less to heat than they did then. The potential for saving energy by conserving it is obviously enormous (Lifestyle 10-1).

Western nations still use energy so inefficiently that even tough conservation measures need not interfere with economic growth. Conservation is attractive to industries and to individuals because it not only saves them money but it also gives them more control over where their money goes and when it is spent. I can decide to invest in insulation or a new furnace for my house this year to reduce utility costs in the future or, if I am short of cash today, I can pay higher utility bills at the moment and postpone the investment. There are two methods of conservation: changing energy-wasting habits and investing in energy savings.

## Light Bulbs

Lighting consumes 25% of the electricity used in the United States, and most of that energy is used inefficiently. In an ordinary incandescent bulb, 95% of the electricity is converted into heat, leaving only 5% to produce light. Fluorescent bulbs are more efficient, but they contain filaments that eventually burn out.

In 1992, a company in Sunnyvale, California, announced production of a super-efficient light bulb that consumes less than half the energy of an incandescent bulb to produce the same amount of light. The bulb is called an electronic light or "E-lamp." Figure 10-2 shows how it works. The bulbs costs about the same as compact fluorescent bulbs, but the makers project that the bulbs will last for about 20,000 hours (as opposed to 1,500 hours for the best incandescents). Using an E-lamp would save the consumer up to $100 over the lifetime of the bulb.

## More Efficient Buildings

Residents of Osage, Iowa, adopted an energy conservation plan that saves this town of 3,600 people more than $1 million in heating costs each year. They plugged the leaks around windows and doors, through which much of the heat escapes from a house. In addition, they replaced inefficient furnaces and insulated their hot-water heaters. Adding insulation can greatly reduce the cost of heating and cooling a house. The insulating effect of a substance is described by a number known as its R-value. A substance with a high R-value is a good insulator. Figure 10-3 shows the different heating zones in the United States, with recommended R-values for insulation in each zone and the R-values of different types of insulation.

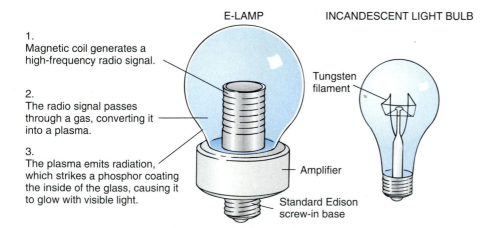

**Figure 10-2**   How an E-lamp and a traditional incandescent light bulb work. In an incandescent bulb, electricity passes through the filament, heating it until it glows with visible light (white hot). The E-lamp has no filament and produces light somewhat like a television screen, by inducing a phosphor coating on the inside of the bulb to glow with visible light. (A problem with this lamp is that since it produces a radio signal, it may have the unfortunate effect of interfering with shortwave radio frequencies such as those used by old cordless telephones.)

HEATING ZONES

### Recommended R-Value

| Heating Zone | Attic Floors* | Exterior Walls | Ceilings Over Unheated Crawl Space or Basement |
|---|---|---|---|
| 1 | R-26 | R-value of full wall insulation, which is 3½″ thick, will depend on material used. Range is R-11 to R-13. | R-11 |
| 2 | R-26 | | R-13 |
| 3 | R-30 | | R-19 |
| 4 | R-33 | | R-22 |
| 5 | R-38 | | R-22 |

*If attic floors already contain R-11 or R-19 insulation, it may not be cost-effective to increase the insulation to the values given here.

### R-Values Chart

| | Batts, Blankets, Boards | | †Loose Fill (Poured In) | | |
|---|---|---|---|---|---|
| | Glass fiber | Rock wool | Glass fiber | Rock wool | Cellulosic fibers |
| R-11 | 3½″– 4″ | 3″ | 5″ | 4″ | 3″ |
| R-13 | 4″ | 4½″ | 6″ | 4½″ | 3½″ |
| R-19 | 6″–6½″ | 5¼″ | 8″–9″ | 6″–7″ | 5″ |
| R-22 | 6½″ | 6″ | 10″ | 7″–8″ | 6″ |
| R-26 | 8″ | 8½″ | 12″ | 9″ | 7″–7½″ |
| R-30 | 9½″–10½″ | 9″ | 13″–14″ | 10″–11″ | 8″ |
| R-33 | 11″ | 10″ | 15″ | 11″–12″ | 9″ |
| R-38 | 12″–13″ | 10½″ | 17″–18″ | 13″–14″ | 10″–11″ |

†R-value of rigid foamed boards is 5.2 per inch when new.

**Figure 10-3**   Heating zones in the United States with the insulation recommended for each zone to reduce the energy used for heating.

Most buildings in the United States use almost twice as much energy as they need. One surprising example—surprising because it was built during the energy shortages of the 1970s—is the twin-towered World Trade Center building in Manhattan. This 110-story monument to our extravagant way of life uses as much energy as a city of 100,000 people. It has walls of glass, which soak up heat in the summer and lose it in the winter. None of the windows opens to permit natural cooling or heating, and the building's air conditioning and heating systems run all the time, even when no one is there. Needless to say, the high utility bills make it hard to find tenants. In contrast, the National Audubon Society's renovated headquarters in New York uses half as much energy as conventional buildings the same size (Figure 10-4). Windows can be opened, and daylight is used for lighting wherever possible. Energy-efficient lights focus on desks instead of lighting entire rooms. As a result, lighting costs are about one third those for a conventional office, where 80% of the money spent on light is wasted. The building is heavily insulated,

# The Energy-Efficient House of the Future

Life would be very different for us all if we lived in houses with utility bills of only a few hundred dollars a year. Such houses can be built.

The house's construction emphasizes insulation because much of the energy we use to heat and cool the air in an ordinary house is lost through the windows and walls. New insulating material is thin so that the roof of our energy-efficient house is insulated to R-100 and the walls to R-40 without being thick. In an ordinary house, large amounts of heat seep through the wood studs between the outer and inner walls. Our energy-efficient house avoids this problem. It's frame is built of metal bars thinner than the usual wooden timbers and studs, and the outer and inner walls are held together by cross beams (Figure 10-A).

Heat loss through new double-glazed windows is half that in a conventional house, partly because the gap between the two panes is filled with a chemically unreactive gas such as argon or xenon instead of with air. In addition, one of the inner surfaces of the window is coated with tin oxide, which reduces the transmission of infrared radiation (heat) through the window while letting visible light pass through (see Figure 10-A). These ''low-emissivity'' windows will be put into all new houses shortly after 2000.

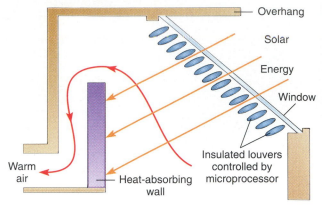

**Figure 10-B** A window that collects solar heat, with insulated blinds to slow heat transfer at night and during the summer.

Compared with today's windows, they will save the equivalent of one sixth of all the energy in the Alaska oil fields.

The way the sun strikes the windows of our new house is carefully controlled. Several windows are angled toward the winter sun and backed by a heat-absorbing wall (Figure 10-B). Fans draw the hot air into the house. A roof overhang lets the winter sun shine in but excludes the higher summer sun. When the temperature in the house rises, insulated louvers (like those of a venetian blind) automatically turn to exclude the sun. Like all the heating and cooling devices in the house, the louvers are controlled by a microprocessor. Other microprocessors are programmed to control the temperature by opening and closing windows and by turning fans on and off.

In the ground underneath the house is a water tank connected to a heat pump. This is the heart of the heat exchange system. It acts as a central heating and air conditioning unit. The water tank acts as a heat (or cold) store. The heat pump passes heat to it in summer and cold to it in winter. The heat pump is powered by electricity from a small cogeneration unit. This unit produces hot water and steam as needed and can also use the steam to generate electricity. It is about the size of a refrigerator and runs on natural gas or synthetic gas (made from coal). The house is not even connected to the utility company's electrical supply, although it contains all the latest energy-efficient appliances.

This house costs about 10% more than a conventional energy-efficient house, but it will save at least ten times that amount over a 40-year period by almost halving lifetime heating and cooling bills.

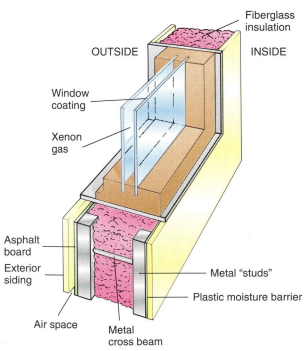

**Figure 10-A** Part of a wall and window of a superinsulated house to show the structures that make the house so efficient at retaining hot and cold air.

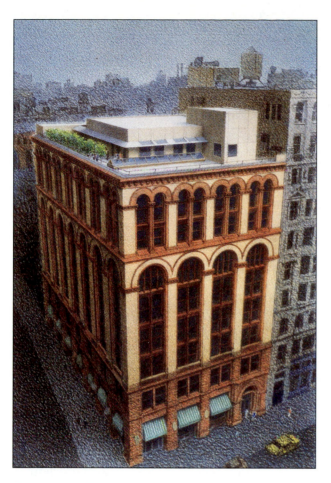

**Figure 10-4**   Drawing of the new Audubon Society headquarters in New York by Paul Stevenson Oles. The Society has taken an old building and rebuilt its interior completely. To increase the amount of natural light (and reduce costs of energy used for lights), it has installed skylights on the roof, where they will not alter the building's outside appearance.

and windows and skylights are fitted with low-emissivity glass. The extra insulation permitted the Society to install a heating/cooling unit only half as large as is usually needed for a building of this size, and they chose a highly efficient natural gas unit instead of the more usual electrical model.

Similar savings are possible in the home. When we are thinking of buying almost anything for the house, from the house itself, to windows, carpets, heating systems, or dishwashers, it is worth researching the newest developments. The kind of energy-efficient house that can now be built is described in Close-Up 10-1.

Sometimes little more than common sense is involved in saving energy. For instance, in the United States, the average Southerner spends more on cooling the house during the relatively short summer than the average Northerner spends on heat during a much longer winter. Air conditioning consumes huge amounts of energy, particularly since it is usually powered inefficiently by electricity. Yet ways to cool houses without air conditioning are old and well known. On a sunny day, the air under a tree may be 30° cooler than the air over nearby pavement. Planting trees to shade buildings and walkways can save a lot of energy. Similarly, black objects absorb heat from the sun while white objects reflect it. It seems obvious that houses in the South should be built with white roofs, although they often are not (Figure 10-5).

When we heat or cool buildings, one of the most important appliances is a heat pump. A **heat pump** is a machine that moves heat (or cold) from one place to another. Figure 10-6 shows how a heat pump works. We shall see that heat pumps may one day permit us to generate electricity from heat stored in the ocean (Section 10-F).

**Figure 10-5**   White roofs in Central America save on cooling costs by reflecting solar radiation away from buildings.

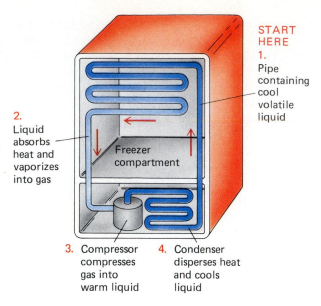

START
HERE
1.
Pipe
containing
cool
volatile
liquid

2.
Liquid
absorbs
heat and
vaporizes
into gas

Freezer
compartment

3. Compressor
compresses
gas into
warm liquid

4. Condenser
disperses heat
and cools
liquid

**Figure 10-6**   How the heat pump in a freezer works. The pump's pipe carries heat away from the freezer compartment. The engine of the heat pump is a compressor (powered by electricity) that compresses the heated gas. It passes the gas to a condenser, where the gas is cooled and its heat dissipated. You can find the warm pipes of the condenser on the outside of many older freezers or refrigerators.

### Energy-Efficient Transportation

Most transportation is powered by oil, which most countries have to import, creating a heavy economic burden. If we could reduce the amount of energy used in transportation, we would save large sums of money and reduce the air pollution caused by vehicle engines. In many cities, such as Los Angeles and Mexico City, trucks and automobiles are by far the main source of air pollution.

Inhabitants of many European countries, with a standard of living at least as high as that in the United States, manage to survive on about half the energy used by each American in a year. Part of the reason is widespread use of mass transit. Western Europeans use automobiles much less than North Americans. They own almost as many cars per person, but the cars are more fuel-efficient and are used less often. The fuel efficiency of vehicles in the United States is determined by government regulation, which, as we shall see (Section 22-C), is an expensive and inefficient method of control. Economists point out that it would be more cost-effective to raise the tax on gasoline. The tax would help pay for the externalities of using gasoline (Section 9-A). With higher gasoline

prices, consumers demand more fuel-efficient vehicles and manufacturers supply them.

There is much we can do to improve the energy efficiency of American transport. More than half of the money Americans saved by energy conservation between 1975 and 1985 resulted from improved fuel economy in trucks and cars. Still, Japan has the highest average fuel economy of any country for new and existing cars, whereas the United States and Canada have the lowest. Experimental vehicles have shown that fuel economy for American light cars and trucks could reach between 30 and 45 kilometers per liter (70 to 100 miles per gallon). This is made possible by lighter vehicles, which are more aerodynamic, with less wind resistance, combined with more efficient engines and drive trains.

Fuel makes up a high proportion of the cost of running an airplane, and airlines have been investing in new fuel-efficient airplanes since the 1970s. However, shifting some of our freight from airplanes and trucks to trains and ships would save a lot of fuel. This is happening in some places. For instance, ships coming into the port of Savannah, on the southeastern coast of the United States, sometimes bring imported cargo in lighters, containers that float. The lighters are unloaded by hoisting them straight off the ship into the river. Then a tugboat tows or pushes a string of lighters through the Intracoastal Waterway or the South's network of rivers, to their destination—an energy-efficient means of moving cargo.

> The area of energy conservation in which the United States lags farthest behind the rest of the world is encouraging the production and use of fuel-efficient transportation.

## 10-B  SOLAR ENERGY

Nearly all renewable energy comes, directly or indirectly, from the sun, which powers plant growth and the water cycle (see Figure 10-14). When we think of using solar energy, we usually think of ways of capturing it directly.

There are three main types of energy systems that use the inexhaustible fuel of the sun's energy directly: passive, active, and photovoltaic.

### Passive Heating and Cooling

Passive solar heating is the most cost-effective way of warming buildings. Figure 10-7 shows the main features of passive solar heating and cooling systems. Putting all or part of a building underground so that

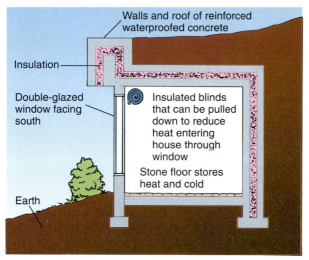

(a) Underground house

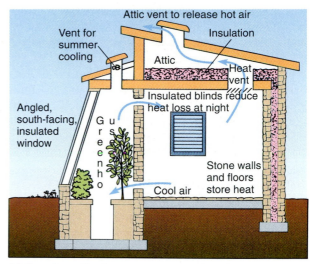

(b) Greenhouse heat

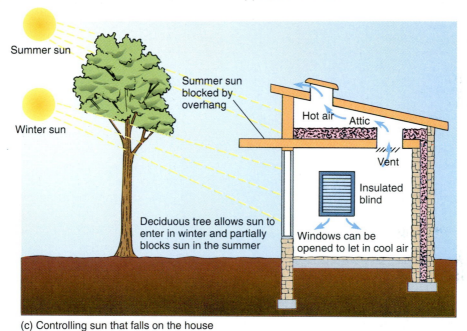

(c) Controlling sun that falls on the house

**Figure 10-7** Passive heating and cooling using the sun's heat and the cool, stable temperature of the soil. The features shown here in three different houses can all be incorporated into one house. All three have thick insulation, energy-efficient windows, and make use of the ability of stone and earth to store heat and cold. (a) The cool, stable temperature of the earth keeps an underground house at roughly the same temperature year round. Sunlight striking the window heats the house, but heating can be reduced by lowering insulated blinds across the window. (b) Greenhouse windows and stone walls and floors capture and retain as much heat as possible. The windows are angled for maximum exposure to the sun. Stone absorbs heat and releases it much more slowly than wood. Vents and windows can be opened to reduce the temperature. (c) The main feature of this house is the use of overhang and tree to control the amount of sunlight that falls on the window and heats the house. The sun is higher in the sky in summer than in winter so a large overhang sticking out above a window permits sun to fall on the window in winter and blocks it in summer. A deciduous tree has a similar effect. The tree loses its leaves in winter, permitting sunlight to reach the window. In summer, the tree's leaves keep most of the sunlight from striking the window.

its walls are surrounded by earth is one way of insulating a building, and this can also help cool the air. The soil a few feet below the surface remains at a constant temperature of 5 to 10°C all year in most parts of the United States. Taking advantage of this can reduce the energy spent on cooling. The house can be built either partly or completely underground, or it can be cooled with cold air from a water tank, tunnel, or cellar in the ground beneath the house, using a heat pump to transfer heat from house to ground.

Houses heated by passive solar energy have energy-efficient windows that face south so they absorb as much heat as possible from the sun. They also have large areas of stone or tile in walls and floors because stone, like earth, stores heat and cold better than the wood or plaster walls of a conventional house. The

body heat of people in the house and waste heat from appliances supply part of the heat. The only energy input involved is the human muscle power used to pull blinds over the windows to prevent heat loss or gain or to open windows and vents when necessary to release hot air or recycle cold air. Such a system is 90% efficient (90% of the energy that reaches the house heats the living space) and can heat a properly built house even in winters where temperatures drop to −30°C—as long as it is in an area that gets plenty of winter sunshine.

### Active Solar Heating

Active solar heating systems have heat-collecting boxes (Figure 10-8). Most are insulated boxes with a layer of glass or plastic polymer on one side. The

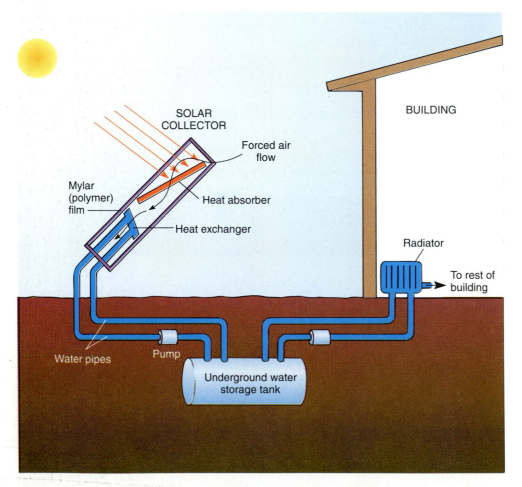

**Figure 10-8**   How a solar collector warms water used to heat a building. In this example, the water is heated by air that has passed over a heat absorber exposed to solar radiation.

inside of the box is dark to absorb as much solar radiation as possible and convert it into heat. The heat is carried away from the box by water flowing through pipes or by air circulated by a fan. The heat is used to warm an insulated storage tank of water, and the air or water is then returned to the collector. The storage tank provides hot water and heating for the building, and it can also be used to run a cooling system (by way of a heat pump).

Solar heat can also be used to generate electricity. One solar plant in the Mojave Desert produces electricity that is used in California. Curved reflectors focus the sun's rays on pipes containing oil. The hot oil heats water to produce steam that drives a steam turbine to generate electricity.

### Solar Cells (Photovoltaics)

**Solar cells**, or **photovoltaics**, are devices that convert solar energy directly into electricity. Photovoltaics are most familiar in small solar cell calculators, but the electricity can be used for any purpose and can be stored in a battery until it is needed. A solar cell consists of thin wafers of semiconductors, substances that give out electrons when struck by light energy. The electrons flow out of the wafer as an electric current (Figure 10-9). Solar cells were expensive and inefficient in the 1970s because photovoltaics contain small amounts of rare or expensive elements such as gallium and cadmium. However, new semiconductors, used in exceedingly thin films, are rapidly reducing the price and increasing the efficiency, making solar power more attractive (Figure 10-10).

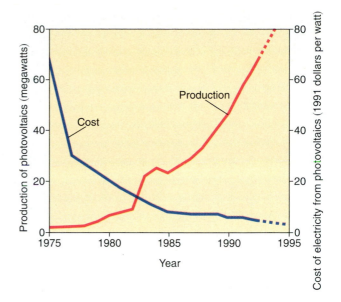

**Figure 10-10**   The falling cost and rising production of photovoltaics worldwide since 1975. *Data from: Paul Mycock, PV Energy Systems*

Solar cells have been used to generate electricity in space, for systems such as satellites, since 1958 (Figure 10-11). Now they are used in remote areas such as in lighthouses, fire lookouts, and roadside shelters, a variety of places where it is expensive or impossible to plug into the main power supply, which still provides cheaper electricity for most purposes.

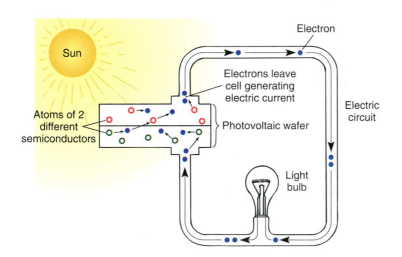

**Figure 10-9**   How solar cells work. Sunlight falls on a semiconductor in the photovoltaic wafer, causing it to release electrons that generate an electric current. The circuit is completed when another semiconductor in the wafer absorbs electrons, passing them on to the first semiconductor.

**Figure 10-11** An artist's impression of a solar power station in space in the future. The solar cells in this example are recharging portable batteries (the barrel-like objects). The spacecraft is swapping an exhausted battery for a new one. *NASA*

Solar cells, in panels on the roof, also supply electricity to some 12,000 homes worldwide, particularly in isolated areas where it would be very expensive to run electric lines to each house. Solar cells are also useful in many situations where conventional batteries are often used at the present time. A solar cell can be used as a battery charger that works merely by exposing it to light. Thus, solar cells are common and convenient as battery chargers on sailboats, in recreational vehicles, and in outdoor lighting.

In developing countries, more than 1 billion people do not have main electric power. Many rely on wood for cooking, kerosene lamps for light, and car batteries for electricity. The batteries have to be hauled long distances for recharging. As an alternative, people can spend about $500 on a photovoltaic system that provides enough power (about 40 watts) to run a television and five light bulbs for 3 hours each day. The Dominican Republic, India, Zimbabwe, and Sri Lanka all have programs that bring photovoltaic electricity to isolated villages.

> Solar energy can be used for passive and active heating and cooling, or it can power solar cells that generate electricity.

## Converting Solar Energy into Chemical Energy

There are problems with converting solar energy directly into electricity by solar cells. First, solar cells are a reliable source of energy only where a lot of sunshine falls on Earth (Figure 10-12). Most of these areas are deserts, where few people live, so the energy must be stored and transported. The only way to store electricity is in batteries, which are heavy and inefficient. Transporting electricity also has disadvantages, involving ugly, potentially dangerous power lines and pylons stretching across the landscape. Scientists are trying to get around these problems by capturing solar energy in the form of chemical instead of electrical energy. Chemical energy sources can be piped and stored as gases, liquids, or solids, which are then used for fuel.

Many of the experiments in producing chemical energy from solar energy produce hydrogen fuel. Hydrogen can be made from water by **electrolysis**, which means passing an electric current through a substance to separate it into positively and negatively charged components:

$$2H_2O \rightarrow \underset{\text{hydrogen}}{4H^+} + \underset{\text{oxygen}}{2O^-} \longrightarrow \underset{\text{hydrogen}}{2H_2} + \underset{\text{oxygen}}{O_2}$$
$$\underset{\text{water}}{}$$

In the Hysolar project in Saudi Arabia, Saudi and German engineers are converting solar energy to electrical energy by photovoltaics. The plant then uses the electricity to produce hydrogen fuel from water by electrolysis.

Other experiments focus on producing gas called **syngas**, a mixture of carbon monoxide and hydrogen. Syngas can be burned as a fuel, or it can be used to manufacture other fuels, such as hydrogen, methanol, and gasoline. Syngas is usually made from biomass, any kind of plant material. When biomass is heated to about 800°C, in the absence of air and presence of steam, it is converted into syngas. Israeli and German scientists are experimenting with solar energy, collected by a series of mirrors, to heat the biomass to a high enough temperature to produce syngas.

All of these processes for turning solar energy into chemical energy are still fairly inefficient, but as techniques improve, we are sure to see more and more chemical fuel produced using solar energy.

## Hydrogen Fuel

In 1985, a small car fueled by aluminum wire and water left Paris on a 50,000-kilometer demonstration trip around Africa. The car, with an ordinary gaso-

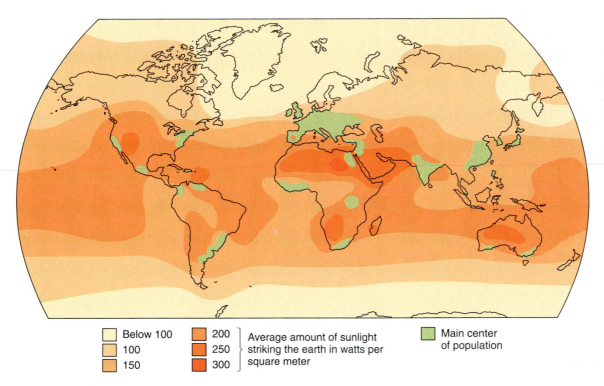

**Figure 10-12**   Where most sunlight falls on the earth and where people live. The darkest orange areas are most suitable for solar power plants, but they lie in desert areas some distance from populated centers where most energy is needed.

Legend:

- Below 100
- 100
- 150
- 200
- 250
- 300

} Average amount of sunlight striking the earth in watts per square meter

- Main center of population

line engine, burned hydrogen gas produced from the water by electrolysis. Russia has built airplanes that run on hydrogen, and several types of vehicles that run on hydrogen have been built in the United States.

Hydrogen is easy to transport, and it burns to produce water so it is a nonpolluting fuel, as long as it is made by processes that do not cause pollution. Gaseous fuels to power motor vehicles may be the wave of the future. A conventional car converts its gasoline into a gas before burning it, so gasoline engines do not need major modification to run on many other gases.

> Hydrogen gas is likely to become more widely used in vehicles.

## 10-C  ENERGY FROM WIND

Some of the sun's energy that strikes Earth is converted into wind energy. There is enough energy in the windiest spots on Earth to generate more than ten times the electricity now used worldwide. Today, wind-generated energy fills only a tiny portion of the world's energy needs. However, the use of wind energy is increasing as the technology improves and costs fall.

Wind can be used to do work directly—for instance, by powering a water pump—or it can be used to generate electricity. Small windmills and generators are relatively cheaper and less subject to breakdown than large ones, and they can be powered by lighter winds. They have been used as a source of power for generations in areas without electricity. In the 1930s, for instance, Zenith Radio sold thousands of "Windchargers" to people living in remote parts of the United States. These were windmills whose spinning powered automobile generators to recharge batteries. The batteries were used to run radios, the only contact with the outside world for isolated farms without electricity or telephones.

The technology of windmill generators is well developed, and the cost of the electricity they produce is lower than that from many energy sources. The most important restriction on wind energy is that windmills must be sited where there is enough wind to make them economical. Wind generators have disadvantages. They are large, often unattractive additions to the landscape, and the blades of large gen-

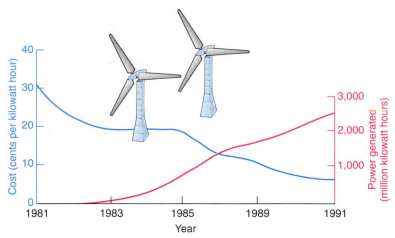

**Figure 10-13**    The output and cost of electricity from California wind farms. During the 1980s, wind technology improved markedly, and the cost has fallen as output has increased.

erators may interfere with microwave communications. Despite their disadvantages, wind generators are popular in windy areas, such as some of the Caribbean islands and parts of the United States, where they provide electricity at a reasonable cost (Figure 10-13).

> Wind generators are a cost-efficient method of producing electricity in windy parts of the world.

## 10-D  POWER FROM WATER

Hydropower, power from water, is driven by solar energy, which causes evaporation and rainfall (Figure 10-14). The energy in waterfalls has provided people with energy to power things like water pumps and mills to grind grain for thousands of years. Nowadays its most important use is driving turbines that generate electricity, called **hydroelectric power (hydropower)**. The technology needed to produce hydroelectric power is well developed, and hydropower produces no direct air or water pollution (Figure 10-15).

The main problem with hydropower is the dams that are usually built to produce the waterfalls that power turbines. Although hydropower is renewable, the dams and reservoirs needed to capture this energy have limited life spans. The reservoirs behind dams invariably fill with sediment, giving the typical dam a life span of 20 to 100 years. Once a hydropower reservoir is filled with sediment, it is gone forever. Dams also create a number of other environmental problems (Section 11-G).

Dams have adverse effects on the ecosystems of rivers that are dammed. The salmon fisheries of the northwestern United States have been destroyed because dams prevented the salmon from swimming upriver to the streams where they spawned (laid their eggs). The ecosystems of large parts of the Northwest depended on the salmon, and many habitats have been degraded now that they are gone. For instance, the loss of the salmon means that animals such as bald eagles and bears that used to feed on the salmon starve and their populations fall. Then organisms that depend on bears and eagles also die, and the forest ecosystem has been damaged beyond repair.

The cost of building dams has skyrocketed in recent years, and any proposal for a large new dam today meets with great resistance from environmental groups. However, even if no new large projects are undertaken, the amount of energy generated by hydropower in the United States will continue to increase for some years by the expansion of existing plants and the construction of small new plants. Corporations, municipalities, and utility companies continue to install turbines on some of the 50,000 or more dams built to store drinking water or to provide flood control. In addition, small new dams can be built in some locations to provide local power without major damage to the environment.

Hydroelectric power is much less well developed in many other countries, and here it will be tapped to provide energy in the future. Many South American and African countries and other countries with mountainous areas, such as China and Nepal, have vast untapped sources of hydroelectric power. Because of the escalating cost and environmental dam-

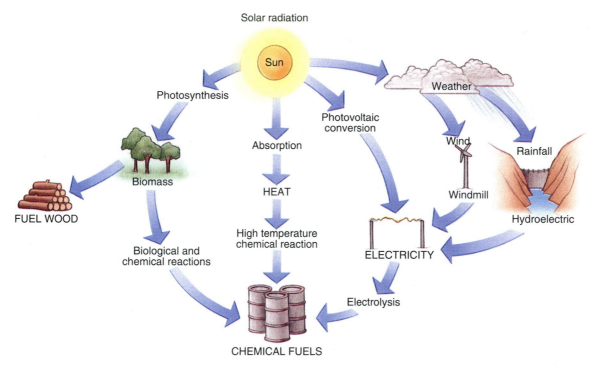

**Figure 10-14**   How solar energy powers renewable energy sources. Fuelwood, heat, liquid and gaseous fuels, and electricity can all be produced indirectly from the sun's energy.

age from building large dams, many of these will probably be developed as small plants to supply local needs. If this is the case, hydropower will not solve the problem of how to supply energy to the huge and rapidly growing cities of less developed countries.

For instance, the energy needs of Mexico City, one of the largest cities in the world, can never be supplied by small hydropower projects. At best, hydropower can provide only part of the energy for such large concentrations of people.

**Figure 10-15**   Hoover Dam. The dam rises 221 meters (726 feet) above the bed of the Colorado River between Nevada and Arizona and was the world's tallest dam at the time of its completion in 1936. Hoover Dam was built to provide flood control, hydroelectric power, and drinking and irrigation water. The 185-kilometer-long reservoir behind the dam, Lake Mead, is a recreation area and supplies water to Las Vegas and parts of California. *Richard Feeny*

Many areas with large amounts of running water generate much of their electricity cheaply from hydroelectric plants. However, the many problems caused by dams mean that hydroelectric power is unlikely to expand much further in developed countries.

## 10-E  BIOMASS

Biomass is the organic matter contained in plants and produced by photosynthesis. For most of the world's population, biomass, in the form of wood used for fuel, is the main energy supply. In industrialized countries, biomass supplies only a small proportion of energy needs, but developed countries have shown considerable interest in biomass energy in recent years for several reasons:

1. Biomass may provide cheap replacements for vanishing oil supplies.
2. Many countries have agricultural surpluses; perhaps it would be economical to convert these into fuel.
3. Disposing of waste is a growing problem in all developed countries, and biomass makes up a large part of that waste.

### Burning Biomass

Most biomass is burned for heat. Wood is the primary source of energy for cooking and heating for half the world's population and for 80% of people in developing countries. The wood is either burned directly or is made into charcoal, which is sold in the cities.

Fuelwood is being harvested faster than it can grow in many countries and is causing the destruction of many tropical forests. More than 1 billion people do not have access to enough wood to supply their energy needs or to alternative sources of energy (Figure 10-16). Possibly the most pressing item on the world's energy agenda is finding a substitute for biomass energy in developing countries (Section 17-B).

In countries where wood reserves are adequate, the use of wood to heat homes and to produce steam and electricity in industrial processes has increased in the last 30 years. For instance, the use of wood-burning stoves to heat homes has increased enormously in the United States since 1970 (Figure 10-17). Wood now heats as many single-family homes as oil does. Wood is bulky and heavy for its energy content, so it is inefficient to transport it over large distances. Thus, most of the fuelwood used in developed countries is cut by people in rural areas who pay little or nothing for it and do not carry it far.

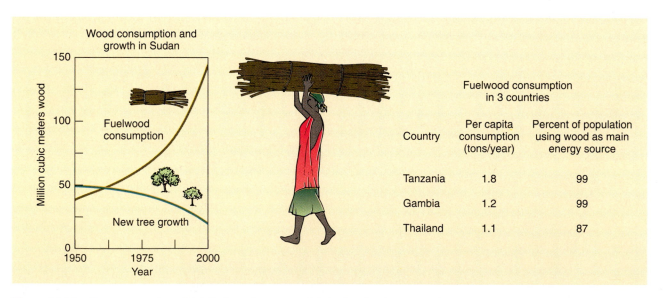

**Figure 10-16**  Some examples of the fuelwood energy crisis in developing countries. Most developing countries are almost completely dependent on fuelwood supplies that are highly depleted. The graph on the left, with projections to 2000, shows that fuelwood use has increased much faster than wood has grown in Sudan (Africa). The table on the right shows that each person in Tanzania and Gambia (Africa) and in Thailand (southeastern Asia) uses more than a ton of wood each year and that nearly everyone in these countries is completely dependent on wood for energy.

**Figure 10-17** The increase in use of wood for energy in the United States during the 1980s. These figures include domestic heating, the use of wood by industry, and electricity generation powered by wood.

Wood's main disadvantage as a fuel is the particles and carbon dioxide added to the air when it burns. Several states with pollution problems from burning wood passed laws requiring that wood burners be fitted with catalytic combusters, which curb air pollution and also make wood-burning stoves more efficient. In 1992, the U.S. Environmental Protection Agency established national emissions standards for stoves. Manufacturers have responded by producing several varieties of clean-burning stoves:

1. **Pellet stoves.** New, popular pellet stoves burn waste biomass compressed into slow-burning pellets. You just load the pellets into a hopper behind the stove to feed a fire that burns for about two days. The pellets are made from things such as sawdust, sunflower seed hulls, and recycled cardboard.

2. **Catalytic stoves.** Catalytic combusters, rather like those that reduce automobile emissions, lower the temperature of the smoke so that more of it is burned and less goes up the chimney.

3. **Noncatalytic stoves.** These stoves route the smoke through a series of chambers around the

fire, heating it so that most of the particles in the smoke are incinerated.

> Finding replacements for biomass energy in developing countries would slow the destruction of tropical forests. In developed countries, the use of biomass energy is increasing.

## Biomass in the Waste We Produce

When we produce energy from biomass, it is more economical to use biomass that is formed as a waste product of other processes than it is to grow plants especially for energy production. One study showed that Virginia produces enough sawdust, logging residues, and unsalable trees each year to replace nearly half of the gas and oil consumed by its industries. It would make sense for the state to generate energy from this waste.

Biomass is used in many American factories. For instance, the U.S. pulp and paper industry meets more than half of its energy needs by burning its own waste products. This also solves part of the industry's waste disposal problem. The Hawaiian sugar-growing industry started selling electricity in the late 1970s. It uses bagasse, the plant residue left after juice is extracted from sugarcane, to power its generating plants. Some examples of companies generating electricity from waste biomass are shown in Table 10-1.

The most efficient way for an industry to make use of energy from waste biomass is to install cogeneration units. **Cogeneration** means electricity generation together with some other use of the energy. The industry burns the biomass for purposes such as heating buildings or boilers. In addition, it uses any excess heat to generate electricity. It may use the electricity itself or sell it to the local utility company.

## Gas Turbines

The simplest way to get energy from biomass is to burn it. To generate electricity, the heat has traditionally been used to produce steam that powers a steam turbine. However, it is more efficient to use the hot gases from burning biomass directly to power a gas turbine. Improved gas turbines are now being produced, and some experts believe that they will revolutionize the production of energy from biomass. One study showed that developing countries that grow sugarcane could use this biomass (with gas turbines) to produce as much electricity, at a lower cost, as they now produce from oil.

| TABLE 10-1   Some Plants Generating Electricity from Waste Biomass in the United States | | |
|---|---|---|
| **Plant Location** | **Fuel** | **Start-Up Date** |
| Union Camp (Virginia) | Pulp waste, peanut shells | 1937 |
| Champion Intl. (Florida) | Pulp waste, bark | 1961 |
| Lihue Plantation (Hawaii) | Bagasse | 1980 |
| Louisiana Pacific (California) | Wood waste | 1983 |
| Farmers Rice Milling (Louisiana) | Rice husks | 1984 |
| Wheelabrator (California) | Orchard prunings | 1989 |

U.S. government agencies estimate that biomass could supply up to 18% of this country's energy with adequate research and development, tax, and other incentives. The Department of Energy has awarded a number of grants for research in this area in recent years.

## Biomass into Biofuels

Sometimes it makes sense to convert biomass into liquid or gaseous fuel instead of electricity. Organic matter can be converted into fuels by decomposer organisms with the assistance of various chemical processes. The main liquid fuels produced from biomass are methanol (methyl, or wood, alcohol) and ethanol (ethyl, or grain, alcohol). The main gaseous fuel is **biogas**, a mixture of methane and carbon dioxide. Biogas digesters are fermenting vats where bacteria convert plant and animal waste into biogas used for heating and cooking. They have been widely used in China to recycle village waste. When the bacteria have released all the methane from the organic matter, the waste is removed and used to fertilize crops. Biogas digesters are vulnerable to the same problems that may beset a garden compost heap or anything else powered by living organisms: they work best at particular pH values and temperatures and can be poisoned by detergents, heavy metals, pesticides, and industrial wastes. These factors mean that gas production is unpredictable, and the process will probably continue to be used only in small-scale local plants.

Methane is produced naturally by anaerobic decomposers in any pile of organic matter not exposed to air. As a result, garbage dumps invariably contain methane, which can be collected merely by inserting pipes. Hundreds of plants for recovering gas from landfills are now in operation in the United States, many of them in California. Methane could be recovered from thousands of other landfills.

Methane is also being produced by the anaerobic decomposition of sludge produced by sewage treatment plants and manure from feedlots. Part of Chicago's gas supply comes from this source. Again, this tends to be a local solution, since it takes a lot of energy to collect and transport sludge and manure any distance.

Some analysts see ethanol and methanol as the liquid fuels that will replace gasoline in transportation. Ethanol is produced from crops that contain sugar, such as sugarcane, sugar beets, sorghum, and maize, in the same way that these are fermented and distilled to produce alcoholic beverages. One advantage of ethanol and methanol is that they have high octane ratings without the need for additives and can be used in existing engines. Both fuels cause a lot of corrosion if they are burned in today's cars, and they are best used in engines specially made to resist the corrosion. However, mixtures of gasoline and alcohol do little damage, and much of the gasoline purchased in the United States is actually **gasohol**, gasoline containing up to 23% ethanol.

One major advantage of alcohol fuels in vehicles is that they produce less air pollution. Colorado has tackled its air pollution problems partly by requiring motorists in major cities to use gasohol during winter months when air pollution is worst. As a result, carbon monoxide emissions have been reduced. An even cleaner solution is vehicles powered by electricity from fuel cells that run on methanol. The U.S. Department of Energy has awarded contracts to two teams that will build buses powered by methanol fuel cells.

Brazil has experimental airplanes flying on a fuel called prosene, a vegetable oil derivative. Biologists have thought for years that fuel for transportation might be made from various plants that produce high yields of oil. Vegetable oils similar to prosene have already been developed for tractors, where they substitute for diesel fuel made from petroleum. Prosene technology may prove a valuable substitute for expensive imported petroleum, particularly in tropical countries that do not have their own oil reserves.

A major problem with biofuel production in developed countries is that so much energy goes into the methods we use to grow maize, sugarcane, and the other crops that are made into biogas. As a result, the energy efficiency of production is close to zero, or even negative, and has to be supported by subsidies from taxpayers. This does not apply in developing countries, where agriculture uses more human power and less tractor fuel. A further problem is that distillation produces large volumes of toxic waste known as swill, which kills algae, fish, and plants if it leaks into waterways.

> Biofuels of many kinds are becoming widespread. Their use will undoubtedly continue to increase.

## 10-F  ALTERNATIVE ENERGY SOURCES

Aside from the major sources of renewable energy discussed in this chapter, there are a number of sources that are not of great importance at the moment. Some of them, such as geothermal energy, are becoming too expensive to use. Others, like wave power, still face major technical problems, although they may become more important in the future.

### Geothermal Energy

Earth contains large amounts of **geothermal energy**, which is heat from molten rock or the decay of radioactive elements, some of which can be tapped for human use (Figure 10-18). Some forms of geothermal energy are renewed by geological processes so slowly that they are essentially nonrenewable. **Dry steam deposits** are the most valuable sources of geothermal energy but also the rarest and most depleted. They are exploited merely by drilling a well. This releases superheated steam, which is used to drive a turbine. A dry steam well in Italy powers that country's electric railroads. The Geyser Field north of San Francisco will supply a portion of California's electricity by 2000.

**Wet steam deposits** contain drops of salty water and are more expensive to convert into electricity because salt water corrodes metal fittings. New Zealand, which gets much of its energy from geothermal sources, has a large wet steam plant. New Zealand's

(a)

(b)

**Figure 10-18**   Geothermal power. (a) A geothermal power station in New Zealand, a country that once generated most of its electricity in this way. Geothermal sources in New Zealand are depleted, and the country is beginning to turn to other sources of energy. (b) The red patches show those parts of the continental United States where geothermal energy lies close enough beneath the surface to be usable.

geothermal sources are already depleted. New Zealand used to have 150 geothermal power plants; now it has 20.

**Hot water deposits** occur in several parts of the world. The most economical way to use their energy is directly: the water is merely pumped to the surface and piped to where heat is needed. Nearly all the buildings in Reykjavík, capital of Iceland, are heated by this means, and nearly all that country's vegetables are grown in geothermally heated greenhouses.

**Hot-rock zones** are the most widespread of all geothermal energy sources. These are areas where magma penetrates Earth's crust and heats up rocks relatively near the surface. Research into the best way to use this energy is under way. One system, used at a small plant in New Mexico, pumps water down to the hot rock zone, where the heat converts it into steam. The steam is then brought to the surface and used directly for heat or used to generate electricity by way of a steam turbine. The U.S. Geological Survey estimates that molten rock lying no more than 6 miles below Earth's surface could easily supply all the energy used by the United States. Extracting the energy at an affordable price is the problem.

The use of geothermal energy can cause environmental problems. For instance, in many parts of Idaho, Montana, and Wyoming, hot water rises to the surface where it is used for purposes such as swimming pools and to heat greenhouses. The geysers in Yellowstone National Park come from the same source. As geothermal energy in the area is used up, the level of underground hot water sinks and the geysers and swimming pools will eventually dry up. For this reason, environmentalists are fighting the use of geothermal energy to produce electricity in areas around Yellowstone.

> A large amount of geothermal energy exists as hot rocks and water beneath the surface of Earth.

## Tidal and Wave Power

Ocean waves and tides contain enormous amounts of energy, but we still have few efficient ways of tapping these energy sources. The tides are daily movements of water driven by gravitational attractions between the sun, Earth, and moon. Twice a day, water flows in and out of bays and river openings to produce high and low tides. However, there are not more than a few dozen sites in the world that are suitable for tide-driven electric generators. What is needed is a bay where the tides are large and with an opening to the sea narrow enough to be closed by a dam, with gates that can be opened and closed. The bay fills as the tide comes in through the open gates. Then the gates are closed, and the water flows out through a water turbine that generates electricity.

Since 1968, a small commercial tide power plant has operated near Saint Malo in the north of France, where the tide rises and falls some 12 meters. The plant generates electricity cheaply, but it cost more than twice as much to build as a conventional hydroelectric plant farther up the same river. A small tide power plant operates in Russia, and an experimental plant has been operating in Nova Scotia since 1984.

This plant is on the Bay of Fundy, which has the biggest tides in the world—16 meters. Canada plans to expand this plant in the future if it is successful.

## Wave Power Plants

Tidal power plants are threatened by corrosion from sea water and also by damage from storms—a problem that is even worse in the case of energy from waves. In 1986, the United Kingdom abandoned its experiments on energy from waves for this reason. Also in 1986, Norway brought the world's first wave power plant on line. The plant consists of a reservoir, connected to the sea by a channel that narrows as it approaches the reservoir. As waves approach the top of the channel, they spill over the sides, keeping the reservoir 3 meters above sea level. As it returns to the ocean, the water passes through a turbine, generating electricity. The success of this plant has produced orders for wave plants from Indonesia, Portugal, and Puerto Rico.

## Ocean Thermal Energy Conversion

A vast amount of solar energy is stored in the oceans as heat. Experiments focus on generating electricity from the large temperature difference that occurs in tropical seas between surface waters and the depths of the ocean. **Ocean thermal energy conversion (OTEC)** works on the same principle as a heat pump (see Figure 10-6). Warm water from the ocean surface is used to evaporate a fluid that boils at a low temperature, such as ammonia. The ammonia gas is compressed and drives a turbine that generates electricity. Then cold water from the ocean floor is pumped to the surface and used to cool the ammonia, converting it back into a liquid that can be used to start the cycle again.

Ocean thermal energy conversion plants would be located on platforms anchored near shore. The electricity generated could be taken ashore by way of a cable. More economically, the electricity could be used on the spot to desalinate sea water into fresh water, to extract minerals from the sea, or to break down water into hydrogen gas that could be pumped ashore for use as a fuel. The main problem with such a plant is that the power required to pump cold water up from the sea floor uses about one third of the electricity that the plant produces, reducing its energy efficiency. Japan and the United States are experimenting with this source of power, although some experts believe that high costs and low efficiency mean that these plants may never compete with land-based sources except when the electricity they generate is going to be used at sea.

### Inland Solar Ponds

Anyone who has ever felt the temperature of water in a garden hose left lying in the sun can imagine how a solar pond works. A pond is enclosed in plastic bags, which act like greenhouses, increasing the sun's power to heat the water. Freshwater solar ponds are used as a source of hot water. When the water reaches a preset temperature, sometime in the afternoon, it is pumped into insulated storage tanks to be distributed as hot water or to heat buildings.

Solar-heated ponds containing salt water are used to generate electricity in a way similar to that used in ocean thermal energy conversion. The bottom layer remains cool, while the sun heats the surface layer. The temperature difference between top and bottom is used to generate electricity. An experimental pond of this type has been operating on the Dead Sea in Israel for several years, and more are planned. More than a dozen experimental ponds of this kind have also been built in the United States, most in desert areas near naturally salty bodies of water, such as the Salton Sea in California and the Great Salt Lake in Utah.

> Several countries are experimenting with energy from tides, waves, and heat in the ocean and in inland solar ponds. It is not yet clear whether these will ever become major sources of energy.

## 10-G  CONCLUSIONS

This chapter contains a long list of alternative energy sources that are in use somewhere in the world or are in the experimental stage. Some scientists and governments are planning now for the time when fossil fuels will be depleted and we shall have to turn to other energy that will cause less environmental damage and be as inexpensive as possible. It appears that many sources of renewable energy will become more important in the next 50 years, especially conservation and solar energy.

Table 10-2 shows that different countries are taking different approaches to the future. Brazil plans to supply two thirds of its energy from renewable sources by 2000. In contrast, the United States and Germany will still rely on oil and coal by that date, with less than one tenth of their energy from renewable sources. These differences are partly a matter of geography: a country can develop geothermal energy only if it is sitting on the right kind of geological formations. But the differences also result from different priorities and planning. Brazil and Norway

**TABLE 10-2   Share of Total Energy Supplied from Renewable Sources in Selected Countries in 1985, with Projections for 2000.**

| Country | % in 1985 | % in 2000 |
|---|---|---|
| Brazil | 59 | 65 |
| Norway | 61 | 63 |
| Japan | 5 | 14 |
| Australia | 10 | 13 |
| Denmark | 2 | 10 |
| United States | 8 | 9 |
| Germany | 3 | 6 |

give high priority to energy independence, generating their own energy so they are not dependent on foreign producers. The United States gives higher priority to low cost—as long as oil can be imported cheaply, it will obtain energy in that way and worry about the future later (Equal Time 10-1).

Despite arguments over priorities, we can make a list of goals for energy planning. Conservation and environmental protection must head the list. Conservation is the most effective way to stretch existing energy resources and to save money. It also reduces the pollution caused by energy use. Within the next 50 years, the reasonable lifetime of cheap petroleum, we must also find a replacement for oil. Transportation fuel may well come from synthetic fuels made from biomass and coal instead of from oil. Our dwindling reserves of petroleum should probably be saved to make plastics. Sometime in the twenty-first century we shall also need to start replacing natural gas, which is used mainly for heat—in buildings and industrial processes. Synthetic gas from coal will probably produce most of the high-temperature industrial heat, and passive and active solar energy will heat most buildings (Figure 10-19). Eventually we shall also need a replacement for coal, which is used mainly to generate electricity. Nuclear power plants will fill part of the gap as will plants using biomass and photovoltaics.

It seems likely that we shall never need as much energy from central sources, such as power plants, as was once believed. Since about 1970, industry, communities, and individuals have experimented with supplying their own energy needs on a small scale. They gain a measure of control over the cost of the energy they use and may solve other local problems such as garbage disposal and pollution. They will probably never go back to relying completely on energy from a utility or an oil company.

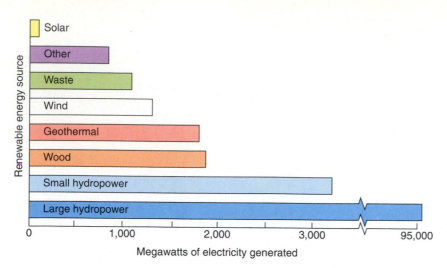

Renewable energy source

Solar
Other
Waste
Wind
Geothermal
Wood
Small hydropower
Large hydropower

0 · 1,000 · 2,000 · 3,000 · 95,000

Megawatts of electricity generated

**Figure 10-19** The amount of electricity generated from different renewable sources in the United States—which has changed very little since the mid-1980s. Electricity from large hydroelectric projects still contributes the vast majority of energy from renewable sources. "Other" includes sources such as biogas and wave power. "Waste" includes incinerated agricultural and municipal waste. All these sources together supply about 14% of the United States's electricity. The rest comes mainly from coal, oil, natural gas, and nuclear power plants.

# EQUAL TIME 10-1

## Energy Policies

The United States developed a coherent national energy policy for the first time in 1992, whereas other developed countries have had such policies since at least 1970. The U.S. policy encourages conservation and supplies tax credits for research on renewable energy. The long delay in producing the policy resulted from conflicts between two main strategies for supplying energy in the future. Despite the passage of an energy policy, the disagreements remain:

**On one side** are supporters of the policies of the 1980s. These people propose to increase energy production largely by using coal and oil. They want to continue to subsidize the price of imported oil (Section 9-A) and to increase U.S. oil production. This would involve drilling for oil in protected areas, such as more of the coastline and the Alaska Arctic Wildlife Refuge.

### Who is on this side?

1. Politicians who believe that cheap energy is essential to a strong U.S. economy.
2. Businesses that depend on subsidized petroleum prices to keep their products cheap enough to compete in world markets, including farmers and plastics manufacturers.

### On the other side

Critics argue that environmentally sensitive areas must be preserved from oil drilling and oil spills. They say that:

1. Drilling in new areas in the United States will not produce much new oil as long as the price

of oil is low because the remaining North American oil is relatively expensive to extract.

2. Taxes on oil, and particularly on gasoline, should be raised until the price of oil reflects its hidden costs. Then alternative energy sources, such as synfuels from coal and biogas, and various forms of solar energy, could be produced at prices that would compete with oil.

3. Conservation can supply much of our future energy needs and should be encouraged by tax deductions for passive solar buildings and research and development on efficient appliances (and light bulbs).

### Who is in favor of higher gas taxes?

1. American oil producers. They point out that subsidies, such as fighting the Gulf War and low gas taxes, keep oil too cheap for them to produce domestic oil profitably.
2. Young people who understand that the cost of importing oil adds to the federal deficit, which they will have to pay off—by higher taxes and lower standards of living in the twenty-first century.
3. Environmentalists concerned about the damage done by fossil fuels, who would like to see us move rapidly to greater use of renewable energy before much more damage is done.
4. The international community, represented at the Rio conference, which is trying to reduce carbon dioxide emissions to slow global warming.

# LIFESTYLE 10-1

## Saving Energy

**On Campus** Students at Connecticut College found that reducing room temperatures by 1° in every campus building would save 20,000 gallons of fuel and $8,000 annually. Temperatures above 68°F use 3% more fuel oil for each degree increase.

Students at Brown University found that the University could save $65,000 if it replaced 2,200 "EXIT" signs on campus with fluorescent fixtures.

**Dressing Warmly** Long-sleeved sweaters add 2 to 4° to the body's warmth. Two lightweight sweaters add even more because the air gap between the sweaters insulates the body, preventing heat loss.

**Energy Audits** Most utility companies will send an auditor to examine your home and recommend inexpensive steps you can take to save energy immediately as well as the potential for using more renewable energy for heating and cooling. They can also advise on insulation.

Simple steps include installing storm doors and windows, changing filters on heaters and air conditioners regularly, and blocking leaks around doors and windows (Figure 10-C). You can often find leaks by using a candle flame. If the flame flickers when you hold the candle still, it is in a draft from a leaky door or window. Windows lose heat through the glass, and you can reduce this by hanging heavy or insulated curtains and closing them promptly at sunset.

**Figure 10-C** The main places where heat leaks out of an American wood-frame house. Loss through the ceiling and roof would be a lower percentage in a two-story house. Leaks through windows and cracks are the easiest to reduce.

**Timers and Clock Thermostats** It is a waste of energy to continue heating or cooling the house after you go to bed or to work. Inexpensive timers for heating and air conditioning systems can solve this problem. *Consumer Reports* recommends the Hunter Energy Monitor, which plugs into a wall socket and accepts the plug from an air conditioner. You can set it for up to four temperature changes a day.

**Lighting** It costs about $80 to burn a 100-watt incandescent bulb day and night for a year. Turn off lights when you leave a room. Replace incandescent bulbs by compact fluorescents as the bulbs burn out. A single 100-watt bulb gives off 20% more light than two 60-watt bulbs and uses less energy. For greater efficiency, use one large bulb in place of two small ones (but don't exceed the acceptable wattage for a light fixture or you may set the house on fire). Dust light bulbs; dust reduces the light from a bulb.

Dark colors absorb light and heat; light colors reflect them. If you have ever been in a room with black or purple walls, you may have noticed how many lights are needed before you can see anything. Save energy by painting walls white and using light-colored carpets and furniture—unless you have pets or small children, in which case your white sofa may soon be mud-colored.

**Hot Water** Heating water can use up to 70% of the energy in a house. You can reduce the energy used to heat water in the following ways:

1. Insulate the hot water tank.
2. Repair leaky hot water faucets.
3. Install water-saver shower heads so that a shower uses less hot water.
4. Lower the thermostat on the hot water heater.

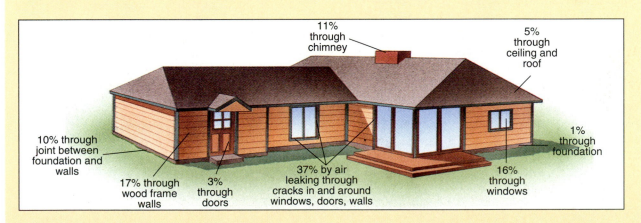

## SUMMARY

As fossil fuels are depleted, we shall switch to energy conservation and renewable sources of energy.

### Energy Conservation

Many utility companies are preparing for increased demand for energy by encouraging conservation among existing customers instead of by building new power stations. The 1970s showed that the developed nations can reduce their energy use by more than one third by conservation, which has the economic advantages of lowering cost and providing more control over how money is spent on energy. Conservation involves changing wasteful habits and investing in energy-efficient buildings, vehicles, and appliances.

### Solar Energy

Nearly all renewable energy comes indirectly from the sun, but there are also several ways of capturing solar energy directly. Passive use of the sun's energy is the most cost-effective way of heating a building. Active solar heating is achieved by using the sun to heat water, which is then used to heat a building. Photovoltaics convert solar energy directly into electricity. The price of photovoltaics is falling rapidly and their use is expanding rapidly—as the electricity supply for small or portable devices and to supply electricity in isolated areas without main electricity. The electricity produced by photovoltaics (or windmills) is also used to produce chemical fuels such as hydrogen gas from water. We can expect more vehicles to be powered by hydrogen in the future.

### Energy from Wind

Wind generators have become cheaper and more efficient over the last 20 years. Considerable electricity is now produced by wind generators in windy parts of the world, and their use is spreading rapidly. One difficulty with wind generators, as with all small-scale methods of generating electricity, is that the batteries for storing electricity are inefficient.

### Power from Water

Hydroelectric power is well developed in industrialized countries where dams cause many environmental problems. More hydroelectric power will be generated in some developing nations in the future. Hydrogen gas, produced by the electrochemical splitting of water, is a potential source of fuel for vehicles.

### Biomass

Most people in developing countries get most of their energy from burning wood and face a worsening shortage of fuelwood. Burning wood causes air pollution, which may be controlled by catalytic combusters fitted to wood-burning stoves. Waste biomass can be burned as a fuel, and organic matter can also be converted into liquid and gaseous biofuels, used to power electricity generation or vehicles. Crops may be grown specifically for conversion into fuel as crops in the United States and Brazil are grown to produce alcohol for vehicle fuel.

### Alternative Energy Sources

There are many sources of renewable energy that are not widely used because they are not as convenient or cheap as other renewable sources. For instance, the enormous store of heat beneath Earth's surface is being used to heat buildings and to generate electricity, but many sources of cheap geothermal energy are depleted. Experimental plants have been built to harness the energy from waves, the tides, and the heat stored in the ocean.

### Conclusions

The use of renewable energy is increasing. The experiments now in progress show that many people believe renewable energy sources are or will become cost-effective as fossil fuel supplies are depleted. Some renewable energy sources have the added advantage of reducing air pollution and disposing of waste. It is still not clear which energy sources will prove most cost-effective for various purposes in the twenty-first century, but it appears that we shall increasingly use a wide variety of sources.

## DISCUSSION QUESTIONS

1. We must not forget that petroleum is used for things other than energy. Its most widespread use other than energy is as the raw material for artificial polymers. More than one quarter of a million barrels of oil every year were once used to make the polystyrene in which McDonald's sold its hamburgers. They have now switched back to paper for environmental reasons. When the price of oil goes up, so does the price of making plastics. The price of fiberglass boats, for instance, rose much more rapidly than inflation during the 1970s. What is going to happen to plastics manufacture and use as our oil supply starts to run out? What possible solutions can you think of?

2. List ways in which energy consumption could be reduced in the buildings where you spend the most time.

3. Do you think that federal subsidies for all energy sources should be abolished so that all alternatives can compete in a free market? Why or why not?

4. Some people believe that the United States should fill its future energy needs by exploiting its coal, oil, and natural gas supplies to the last, building nuclear power plants, and importing oil, in various combinations depending on their price at the time. List all the disadvantages of this strategy that you can think of.

5. Many countries provide grants and tax incentives to stimulate research and development of renewable energy sources. Why do you think they do this while oil is still a cheap and convenient source of energy? Why not wait until the oil runs out?

6. Explain how the U.S. policy of subsidizing fossil fuels and nuclear power can discourage exploration for domestic coal and oil, increase unemployment, increase dependency on foreign oil, discourage energy conservation, and discourage research into alternative energy sources.

# FRESH WATER

**Key Concepts**

- Clean, fresh water for drinking, cooking, washing, sewage disposal, and agriculture is vital to healthy human life.
- Although ample fresh water is available on Earth, billions of people do not have the water they need because their local water supply is polluted or depleted.
- Our water supply is purified by the water cycle: solar heat evaporates water, which then falls as rain or snow.
- When water from a river or an aquifer is removed faster than the water cycle replaces it, local shortages develop.
- Developed nations waste much water by inefficient irrigation, rapid runoff from urban and deforested areas, and waste.
- Conservation and reducing runoff can make more water available, reduce water pollution, and recharge aquifers.
- Many experts believe that in the future wars will be fought over access to water.
- Human interference with waterways, by building dams or channelization, destroys the waterways' ecosystems and often merely moves flooding and other problems elsewhere.

*Irrigating cattle pasture.*

*The number of water taps per thousand persons is a better indication of health than the number of hospital beds.*

World Health Organization

In many parts of the world, a growing shortage of clean, fresh water is the most pressing environmental problem. Lack of water hampers economic development in Los Angeles, Salt Lake City, and much of Florida. It prevents many African nations from producing enough food to feed their populations, and it causes conflicts that threaten to lead to war in the Middle East. When I was a teenager, camping or backpacking in England, France, Idaho, and Montana, we took our drinking water from rivers and streams. It never occurred to us that the water might need to be boiled or purified before we drank it. Today streams and rivers everywhere are so polluted that it would be foolish to drink from such a source.

Clean, fresh water is the substance most essential to life. People can survive for more than a month without food but for only a few days without water. The main reason people live longer today than they did 200 years ago is clean water: water to drink; water to wash dishes, clothes, hands, and babies; water to flush away sewage; and water to irrigate agriculture.

Fresh water is a renewable resource. The water available to us is constantly replenished and cannot be depleted unless we use and contaminate it faster than it is cleaned up by human or natural processes. Water pollution and the way in which it alters ecosystems are covered in Chapter 19. In this chapter, we consider the uses and distribution of water, some freshwater habitats, and why many communities are running short of water.

## 11-A  PROPERTIES OF WATER

Living organisms contain more water than any other substance. Every day, we drink water, wash in it, dispose of wastes in it, and never give it a second thought. When we take a closer look, however, we realize that this common substance has several uncommon properties. These properties are so important to organisms that we cannot imagine life existing on any planet that does not have an abundant supply of water.

The unique properties of water result from the structure of its molecules. A water molecule contains an atom of oxygen bonded to two atoms of hydrogen (Figure 11-1). The molecule is electrically charged, and water molecules bond to each other. This structure endows water with a number of properties vital to life.

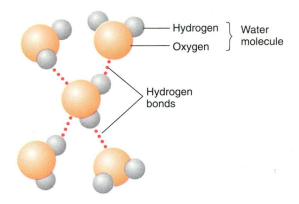

**Figure 11-1**  Five water molecules held together by bonds (red dots). These bonds are easily broken and new ones formed to other, dissolved, molecules as the molecules tumble about in liquid water.

1. **Water sticks to itself and to other substances.** You can fill a glass of water slightly above its brim, and a mosquito larva can hang from the surface of a pond. Both are possible because of the **surface tension** of water, which makes the surface of water act as if it were covered by a skin. The molecules of water also stick to any electrically charged surface. This accounts for the **capillarity** of water—its ability to move upward, against the force of gravity, in small spaces such as the pores in paper or soil (Section 14-A).

2. **Water is a solvent.** More things dissolve in water than in any other liquid. When a substance dissolves, its molecules separate from one another and mingle with molecules of the solvent. Because water is electrically charged, it dissolves other charged molecules such as salts, but it does not dissolve uncharged molecules such as oil. In practice, this means that water dissolves sulfur compounds in the air (to form acid rain) and salts in the soil, but an oil spill floats on water and does not mix with it.

3. **Water is a good evaporative coolant.** It takes a lot of energy to make water molecules move fast enough to break the bonds holding them together. When this does occur, liquid water becomes a gas, called **water vapor**, in which each molecule is separate. Water vapor molecules carry a lot of heat energy away with them. So when water evaporates from a body, it cools the body. This is why we sweat when we are hot and why we use large quantities of water as coolants in power stations and industrial processes.

**4. Water is the only common substance that expands rather than contracts when it freezes.** The low density of ice is the reason icebergs float. In temperate climates, this means that ice floats on top of water in winter, forming a blanket of insulation between the water and the cold air above. This slows the formation of more ice and protects organisms living below the ice from freezing. Water's peculiar thermal properties also account for the fact that lakes in temperate areas are mixed by natural processes twice a year, an important part of their self-cleaning systems (Section 19-A). Because water expands when it freezes, it also breaks pipes, cracks streets and rocks, and requires us to put antifreeze in our cars.

## 11-B  AVAILABILITY AND USES OF FRESH WATER

There is water, water, everywhere, but most of it is of no use to human populations. Most of the water on Earth is salt water in the ocean. Of the 3% of water on Earth that is fresh, most is locked up in polar ice caps and glaciers and most of the rest occurs underground. Less than one hundredth of 1% of the water on Earth exists in the atmosphere, rivers, or lakes where it is easy to get at (Figure 11-2). Nevertheless, this is about 5 million liters of water per person, which is more than we need. The problem, as with many natural resources, is distribution. Some people

suffer from water shortages and some from floods. Other unfortunates suffer both in swift succession.

> Only a tiny fraction of the water on Earth is fresh water in rivers, lakes, accessible aquifers, and the atmosphere, where we can reach it.

How much fresh water do people need? We must drink about 2 liters of water every day, and we need another 3 liters for cooking. In addition, unless there is enough water for proper sanitation, human waste washes into the water supply and makes the drinking water unsafe. In developed countries, most people have ample, clean piped water and safe sewage systems. However, even in the United States as much as 20% of the "drinking water" is considered unsafe for human consumption according to federal guidelines. The increase in purchases of bottled water in recent years reflects the number of people who are not convinced that their tap water is safe to drink (Equal Time 11-1). In developing countries, fewer than half the people have easy access to safe water and fewer still to adequate sanitation. Even in the United States, thousands of people in southern Texas and New Mexico live without running water and suffer from high rates of hepatitis as a result.

> About half the world's people do not have easy access to all the clean water they need.

**Figure 11-2**  Water in lakes and rivers is the fresh water that is most useful to humans because it is easy to get at. This is the Salmon River in Idaho.

# EQUAL TIME 11-1

## Bottled Water

Sales of bottled water soared by 400% in the 1980s as more and more people decided their tap water was not fit to drink. True? Or merely good advertising by the people who sell bottled water?

Most people assume that bottled water comes straight from some pure mountain stream. Much more often, bottled water is simply tap water, filtered and treated with various chemicals. After all, you can honestly say the water is "mountain spring water" if it comes from a river that starts as a mountain spring, which most rivers do. If you call it "natural spring water," however, you are guilty of misleading advertising if the water doesn't come straight from a spring.

The table compares tap water with various brands of bottled water. Trihalomethanes are compounds that form when organic molecules combine with chlorine and fluorine that are added to water to kill bacteria and prevent tooth decay. Calcium carbonate dissolves in water that runs over limestone rock. Water containing a lot of carbonates is said to be "hard." The best-tasting water is hard water containing considerable dissolved oxygen.

Consumers have reported finding pine needles, cigarette butts, and even a dead frog in sealed jugs of bottled water. Bottled water plants are regulated by government, but bottled water is not tested for pollutants as often as the public water supply is.

| Tap Water and Bottled Water Compared* | | |
|---|---|---|
| **Brand of Water** | **Trihalo-methanes (parts per billion)** | **Calcium carbonate** (parts per million) |
| South Florida tap water | 18 | 76 |
| Deer Park Spring Water | 0.2 | 138 |
| Evian Natural Spring Water | 0 | 306 |
| Glacier Spring 100% Purified Water | 5 | 2 |
| Perrier Mineral Sparkling Water | 0 | 371 |
| Publix Drinking Water | 27 | 96 |
| Ritz (Seltzer) Sparkling Water | 11 | 7 |
| Triton Water (natural mountain spring water) | 0 | 6 |
| Zephyrhills Natural Spring Water | 0 | 164 |

*Source: Florida International University Drinking Water Research Center.

**Calcium carbonate makes water hard, which makes it taste better but tends to block pipes and is not as good for washing. It is not surprising, therefore, that many bottled waters contain more carbonate than does tap water.

Most of the water used by households and industry is eventually returned to the water supply. It is said to be **withdrawn** but not **consumed**. In all countries, however, much more water is used to irrigate farmland than goes to households, industry, and all other uses combined. And much less of the water withdrawn for irrigation returns to the water supply. Less than one sixth of water withdrawn for households and industry is consumed, whereas more than half of water withdrawn for irrigation never returns to the water supply because it evaporates into the air (Figure 11-3).

> Most of the water used in the world goes to irrigate crops.

## 11-C  THE WATER CYCLE

We continually pour pollutants into water, so how is it that all the fresh water on Earth is not already too polluted to drink? The answer is that water vapor, the gas that evaporates from a damp surface into the air, is pure water. The pollutants are left behind as the water evaporates.

Water is continuously purified by the **water cycle**, in which water moves between the atmosphere, the land, and the oceans. As water vapor in the air rises, it eventually reaches altitudes so cold that it condenses on dust particles in the atmosphere into liquid water or ice to form clouds. When the water is liquid, it can dissolve gases and other substances as it travels through the atmosphere, so rain is a dilute solution of various chemicals. We call this "fresh wa-

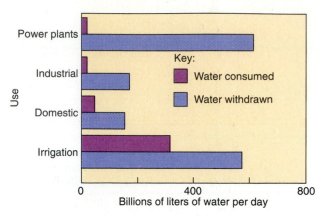

**Figure 11-3** Water withdrawn from the water system and water consumed for different uses in the United States. Power plants withdraw most water (for cooling), but they consume very little of it. The vast majority of water consumed every day evaporates from irrigated fields.

ter'' to distinguish it from ''salt water,'' the fairly concentrated salt solution that makes up the oceans.

The water cycle is driven by the sun's heat energy, which causes water to evaporate (Figure 11-4). Some of the water evaporates from soil, roads, and lakes. A lot comes from plants, during **transpiration**, evaporation of water from a plant, most of it during photosynthesis. Seventy percent of Earth's surface is ocean, and this is where most water evaporates. Much of the water that evaporates eventually descends on the ocean as rain or snow. Only about 30% falls on land, where it drains slowly back toward the ocean and is known as **runoff**. This runoff makes up our ''water income,'' the recycling supply of purified water on which life depends. All of the runoff eventually runs from the land back into the oceans.

Water is continuously purified in the water cycle, powered by the sun's energy.

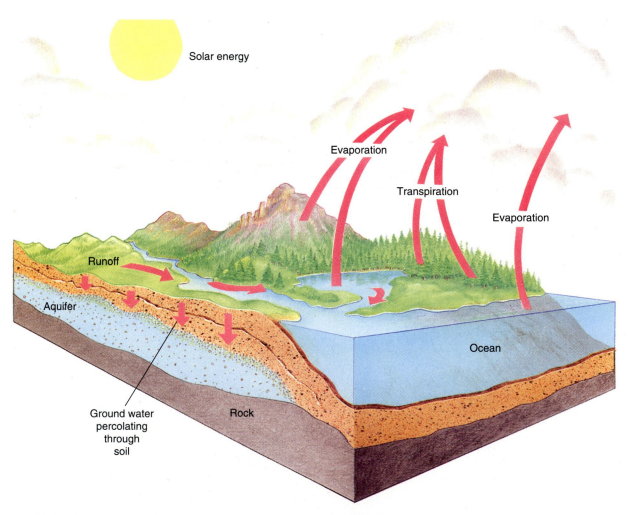

**Figure 11-4** The water cycle. Water that reaches the atmosphere by evaporation or transpiration (from plants) is pure fresh water. (Ground water and aquifers are discussed in Section 11-D.)

## 11-D GROUND WATER

Of all the water that falls on land as rain or snow, about one third runs off the surface into streams and rivers, and another one third evaporates or is absorbed by plants. The rest drips or **percolates** down through soil and rocks and becomes part of the **ground water**, water below Earth's surface. Ground water may flow vast distances underground and rise to the surface again or be pumped up, far from where it entered the earth. In very dry regions, such as the Middle East, ground water flowing from other parts of the world is the only source of fresh water.

### Aquifers

Much of the world's unfrozen fresh water exists as ground water, but there are two main problems with using this water supply. First, the water must usually be pumped to the surface, which is expensive. Second, ground water is replaced very slowly. This means that ground water pollution is often irreversible and that the supply can be depleted.

An **aquifer** is any natural material that contains water that can be brought to the surface in useful amounts by means of a well (Figure 11-5). Aquifers are usually rocks with a lot of air spaces in which water can accumulate, such as sand and gravel. Oc-casionally, aquifers contain large areas of water without any rocks in them. Limestone ($CaCO_3$), for instance, may dissolve to leave large caves full of water. Similar underground lakes may form in basalt. (Basalt is one kind of rock that may form when the lava that flows from an erupting volcano hardens. There is a lot of it around. It underlies most of the oceans and is the rock from which the Hawaiian Islands and Iceland are formed.)

An aquifer constantly gains water that percolates down from the surface. Its **recharge** area is the area of land from which the aquifer is filled. This often lies hundreds of miles from the main body of the aquifer. For instance, the Floridan Aquifer under south Georgia and north Florida is recharged from rain that falls in the hills of western Georgia and eastern Alabama. Some of this rain percolates down through the soil into layers of rock and gravel beneath. In the rock layers, some of the spaces contain water and some air. This area is called the **zone of aeration**. Further down, the rock spaces are all filled with water in the **zone of saturation**. The upper surface of the zone of saturation is called the **water table**. The zone of saturation loses water slowly to the surface through springs, or to rivers, lakes, and the ocean.

When water is withdrawn from an aquifer faster than the aquifer is recharged, the aquifer is said to

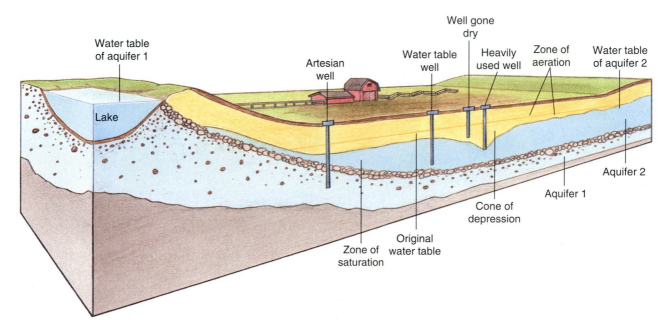

**Figure 11-5** Aquifers with an artesian well and two wells drilled into the aquifer to bring water to the surface.

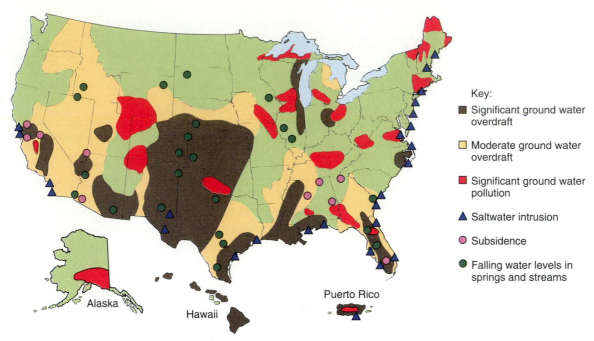

**Figure 11-6** Areas in the United States with ground water problems. Much of the Midwest, Gulf Coast, and Florida suffer from ground water overdrafts that threaten the water supply. Nearly all western and southeastern states and the Mississippi Valley have moderate ground water overdrafts. Ground water pollution, areas where the ground has subsided, salt water intrusion, and falling water levels in waterways are also found scattered in most parts of the country.

Key:
- Significant ground water overdraft
- Moderate ground water overdraft
- Significant ground water pollution
- ▲ Saltwater intrusion
- Subsidence
- Falling water levels in springs and streams

have an **overdraft**, meaning it loses water faster than it gains it. Ground water overdrafts exist in many parts of the United States (Figure 11-6). They cause numerous problems. Wells to reach the falling ground water have to be dug deeper, making them more expensive. The amount of water that naturally flows from the aquifer into streams and rivers is reduced so that the water level in the waterway falls, and the ground above the aquifer may subside as the ground, no longer full of water, collapses. Dramatic examples of subsidence are the sinkholes that have developed in Florida and other southern states in recent years, consisting of holes as much as 100 meters across and 50 meters deep. Venice, on the Adriatic coast of Italy, has suffered more gradual subsidence. Venice has sunk more than 3 meters since it was founded 1,500 years ago. The city has recently attempted to recharge the aquifer beneath the city, partly by rationing the withdrawal of water, and the city is now subsiding more slowly. Similarly, parts of Mexico City have subsided up to 8 meters, so that houses have to be entered by second-floor windows. San Jose, California, has sunk more than 3 meters in the last 50 years.

> About one third of the water that falls on land percolates down through the soil and rocks to become part of the ground water.

## Wells and Springs

Water at the bottom of an aquifer is under pressure and will move toward an area of lower pressure, such as an outlet to the surface of the land, even if it must flow upward to do so. (The water cannot rise to a level higher than the highest point of the water table unless it is pumped up.) A **spring** is a natural flow of ground water onto the surface of the earth. The differences in water table height in different parts of an aquifer generate hydraulic pressure at the lower points. This may produce an **artesian well**, a spring that spouts out of the ground.

A deep enough well will strike an aquifer in most parts of the world. But deep wells are expensive to dig and then require powerful pumps to bring the water to the surface. In most aquifers, ground water moves at only about 1 or 2 meters per day. A well provides most water if it is drilled into material with plenty of holes in it so that water removed from the well is replaced as fast as possible by percolating in from the sides. Even in fairly permeable material, a conical depression in the water table forms in the area surrounding the well. This **cone of depression** may eventually reach the level of the well intake, causing the well to run dry. To the distress of neighboring users, it may also cause nearby wells to dry up (see Figure 11-5).

### Contamination of Ground Water

We deal with water pollution in general in Chapter 19. Until the 1970s, few people realized that ground water could be polluted at all. Ground water was viewed as protected by the soil and rocks above it from the fertilizer runoff, industrial chemicals, pesticides, sewage, acid rain, and other pollutants that threaten surface water. We now know that ground water can be contaminated by many human practices.

**Sea Water Intrusion**  Fresh water is less dense than salt water, so in an aquifer on the coast, fresh water from rainfall lies on top of salt water that percolates in from the sea. The density difference between fresh and salt water is such that a balanced state develops in which fresh water extends about 40 meters below sea level for every meter that the water table on land extends above sea level (Figure 11-7). This means that if the freshwater table is lowered by only 2 meters, the salt water underneath will rise as much as 80 meters and may easily contaminate wells and irrigation systems. Long Island, New York, has suffered extensively from this problem. In the industrialized western part of the island, so much ground water was pumped out that the salt water level rose and contaminated the wells. Today water for this area has to be supplied by pipes from the mainland.

**Pollutants**  Ground water from fine, sandy rock is often very clean. This is because the sand has an enormous surface area on which impurities are absorbed. This kind of filtration works particularly well for biological impurities such as bacteria, which stick to the rock grains and then act as decomposers to break down any organic material in the water. However, many chemical pollutants, such as industrial wastes, are not removed from ground water by filtration of this kind. Also, not all aquifers are made of fine-grained material. Large channels, such as those often found in limestone or basalt, provide little filtration.

In Florida, where aquifers are of sand and limestone, more than 1,000 wells have been shut down as drinking water sources because they are contaminated with a pesticide used to kill nematode worms. The state and Dow Chemical, which made the pesticide, have spent more than $3 million supplying drinking water to communities that depended on these wells. The cleanup after Hurricane Andrew in 1992 made the situation worse. The Biscayne Aquifer, which supplies most of south Florida's water, is separated from the surface by only a foot or two of loose rock so it is easily contaminated. After the hurricane, military and local officials bulldozed waste into heaps and set fire to them. Toxic waste and pollutants seeped down into the aquifer (Close-Up 11-1).

When sewage or chemical wastes are dumped untreated on the ground, particularly if the underlying rock is permeable, there is always danger that ground water will be contaminated. The problem is widespread. Some of the United States's ground water is contaminated from individual sources such as leaking landfills and hazardous waste dumps. California estimates that pollutants in one fifth of the state's large drinking water wells exceed state pollution limits. Industrial solvents and gasoline from leaks at gas stations are the most common contaminants, especially around Los Angeles and San Francisco. Pesticides and other synthetic chemicals have been detected in more than half of Iowa's city wells. Treating or cleaning up contaminated ground water is costly,

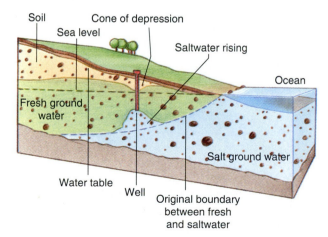

**Figure 11-7**  Sea water penetration into an aquifer. On the coast, the line between fresh ground water and salt ground water is determined by the relative densities of salt and fresh water and by the slow flow of fresh water toward the ocean. Pumping from the well lowers the water table. Salt ground water from the ocean rises in response, and the well pumps salty water.

# CLOSE-UP 11-1

## How Water Reaches Your Tap

What happens to water between the time it leaves the aquifer, river, lake, reservoir, or wherever it comes from and when you turn on the tap? The steps vary in different places, but here is what happens at the Pembroke Pines water plant in south Florida:

The Pines draws water from 35-meter wells into the Biscayne Aquifer, which supplies most of south Florida's water, and pumps it into treatment tanks. The raw water is tea-colored and exceedingly hard (full of minerals). So workers add lime and coagulants that cause most of the minerals to precipitate out of the water and fall to the bottom of the tank.

The softened water flows twice through eight filters, which remove any remaining lime and also add a phosphate compound. The phosphate helps to prevent remaining minerals in the water from precipitating on water pipes and clogging them.

The demineralized water is pumped into storage tanks where fluoride and chlorine are added. The fluoride combats tooth decay very effectively. The chlorine kills any bacteria and makes the water less yellow. The main disadvantage of chlorine is that it breaks down organic matter to form trihalomethanes (THMs), which are carcinogenic (cancer-causing). By federal standards, drinking water must contain no more than 100 parts per billion of THMs, so the water's content of THMs and fluoride is tested frequently. Experiments have shown that adding ammonia with the chlorine produces a slower-acting disinfectant that reduces formation of THMs, so now ammonia is also added.

The treated water is pumped into storage tanks where it awaits supply to households through 175 miles of steel pipes. There are only two major threats to the water supply: a broken water main or a lack of electricity, which is needed to power the wells and pumps.

**Figure 11-A**  Surface storage of drinking water: a reservoir.

and numerous experts have urged that laws to protect our underground water supplies be better enforced.

> Because ground water may take hundreds of years to replace, poor management of ground water leads to many problems, such as ground water pollution, aquifer depletion, sinkhole formation, and saltwater intrusion.

## 11-E THE WATER SHORTAGE

Although the average world water supply appears sufficient, local shortages exist in many places.

### Water Use in the United States

For every person living in the United States, about 7,500 liters of water per day are withdrawn from the freshwater supply. Most of this is used as cooling water in electric power plants. The water is heated and some of it evaporates, but most of it is returned to a lake or river unaltered. The other uses of water, for industry, irrigation, or the home, degrade the water to a greater or lesser extent. Most water taken in by industrial plants is used several times before it is discharged, but much of it is returned to lakes and rivers somewhat polluted. In 1984, the U.S. Water Resources Council listed only eight states as having no

serious problem with pollution of their surface or ground water. The worst water shortages in the United States exist in dry western states, where large volumes of water are consumed by agriculture (Close-Up 11-2; Figure 11-8).

> Most water used in the United States goes to cooling systems in power plants, but most water disappears when it is consumed in inefficient agricultural irrigation.

Between 1940 and 1980, the U.S. population almost doubled, but water use more than tripled. In many places, water is too scarce to permit the community to welcome water-using industries, however much it needs the jobs.

Most people have experienced water shortages caused by droughts (Figure 11-9). Meteorologists define a drought as existing when rainfall is 70% below average for 21 days or longer. Water levels in rivers and reservoirs fall, and water may be rationed to conserve it. More important than droughts in the long run, however, are shortages caused by the depletion of aquifers or overuse of a particular source of water. Potato production in Idaho has declined by 10% since 1977 because less water is available for irrigation. During the same period, 80,000 hectares of corn-producing land in the Midwest was taken out of production because of water shortages (Figure 11-10).

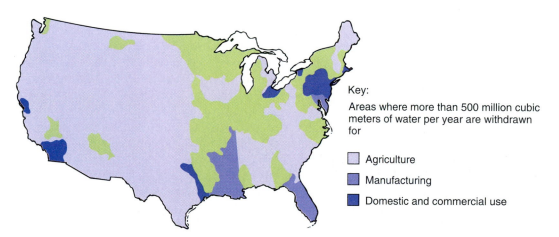

Key:
Areas where more than 500 million cubic meters of water per year are withdrawn for

☐ Agriculture

☐ Manufacturing

☐ Domestic and commercial use

**Figure 11-8**   Where most of the water withdrawn from the U.S. water supply is used. Most agricultural use is in the West. Centers of municipal use, for households and businesses, are clustered in the Northeast, coastal California, and the Gulf Coast in large centers of population. Heavy water use for manufacturing is largely restricted to the eastern half of the country and the Northwest.

# Drought in California

In February 1991, flooding caused some residents of Ohio, Indiana, Georgia, and Pennsylvania to take to their boats and head for the high ground. At the same time, California's governor Pete Wilson announced a $100 million "battle plan" to deal with that state's drought. A drought action team recommended spending $100 million to improve firefighting ability in what was expected to be the worst year for fires in California's history. And it was. The chaparral biome is particularly susceptible to fires (Section 4-K). California had received less than average rainfall for six years in a row and was suffering from an acute water shortage. Especially hard hit were cities on the state's central coast that depend largely on depleted aquifers and nearly empty reservoirs (Figure 11-B).

Even small decreases in rainfall may produce large reductions in the amount of available water. For instance, one calculation on the effects of global warming showed that in the Colorado River Basin, a 1.5 °C temperature rise coupled with a 10% decrease in precipitation would reduce water runoff into streams and lakes by a massive 40 to 70%. This is partly because water resources are the difference

**Figure 11-C** A southern California avocado orchard, the trees dead from lack of water. Water was so scarce and expensive in parts of California in the 1980s that even some valuable crops such as fruit and vegetables did not receive enough water to survive. *(Richard Feeny)*

between the water that falls and the water that evaporates (or is transpired by plants). Lower rainfall means fewer clouds and higher ground temperatures so evaporation increases.

Because of its massive areas of irrigated agriculture, California both withdraws and consumes more water per capita than any other state. In 1985, the U.S. Department of Commerce reported that 43% of the water withdrawn from California's water supplies was consumed, compared with 23% for the nation as a whole. Of the water withdrawn, 4% went to industry, 11% into the public water supply, and most—62%—to agricultural irrigation. Some of the water is used to grow California's famous fruit and vegetables (Figure 11-C), but much of it is used to grow fodder for cattle. Cheap irrigation water throughout the West is heavily subsidized by the American taxpayer. It seems foolish for taxpayer money to subsidize cheap beef while citrus orchards and home vegetable gardens die from lack of water. However, changing the distribution of water takes time and legal changes.

**Figure 11-B** A California reservoir in 1992, nearly empty after eight years of drought. *(Richard Feeny)*

**Figure 11-9** When water shortages occur, local authorities often prohibit needless use of water for purposes such as washing cars and watering lawns. This lawn in California has just about dried up after water restrictions were imposed in 1991. *(Richard Feeny)*

## Ogallala Aquifer

The Ogallala Aquifer is a huge body of ground water formed from melting glaciers about 2 million years ago. It lies under seven states, which have used its water for agriculture since about 1950 (Figure 11-11). Today it supplies the water for one fifth of the United States' irrigated farmland, supporting more than $30 billion worth of agriculture every year, including the production of maize, cotton, sorghum, and beef.

**Figure 11-10** Farm soil for which irrigation water is no longer available. Irrigated farmland often has soil of poor quality, like the clay shown here, which is easily eroded if it is not covered with vegetation. *(Richard Feeny)*

**Figure 11-11** The huge Ogallala aquifer (blue) lies under eight states in the Midwest.

The aquifer has a low recharge rate because it lies under an area that receives little rain. As a result, the aquifer is being emptied eight times as fast as natural processes can refill it and up to 100 times as fast in parts of Texas, New Mexico, Oklahoma, and Colorado. The agriculture of the Texas High Plains depends on withdrawing from the Ogallala aquifer each year a volume of water almost equal to the entire annual flow of the Colorado River. Obviously this rate of depletion cannot continue forever. Water experts predict that the aquifer will be dry in most areas by 2020. Before the aquifer dries up, however, the cost of its water is rising rapidly as wells have to pump water up from further and further below the surface. Farmers are giving up farming, or switching to crops that can be grown without irrigation. The amount of irrigated land is already declining because of the rising cost of water in five of the states that use water from the aquifer.

## Who Gets the Water?

The Salmon River in Idaho provides an example of what happens when demand for water is so high that people compete for it. This lovely river is used for white-water rafting and canoeing as well as for industry, fishing, fish farming, hydroelectric plants, domestic purposes, and irrigation. If everyone took all they wanted from the Salmon River, the river would be dry (Figure 11-12). So the river water is rationed to each user according to antique riparian rights laws

(a)

**Figure 11-12** Water has many functions other than domestic, industrial, and agricultural use. (a) This sailboat race on a New York lake represents the many recreational uses of fresh water, which include fishing, duck-hunting, canoeing, motor-boating, and bird-watching. (b) This trout farm in Idaho produces much of the frozen trout sold in the United States. The use of fresh water for fish-farming is growing rapidly.

(b)

and nobody is satisfied. **Riparian** means relating to the banks of a river. Riparian laws govern the land and water rights of people who live along a river.

Hundreds of lawsuits over water occupy the courts nowadays, and the battle lines are being drawn for an even bigger fight. The arid western states want to pipe water from the Great Lakes, the biggest body of fresh water in the world. Cheap water was a major reason for the early industrial development of the Northeast, and the Canadian provinces and American states that already rely on this water are organizing to defeat any plan to take it elsewhere.

## The Western Water Mess

The history of the West can be written in endless battles over water, which is the limiting natural resource in most of the western half of the United

States. Western waterways have been completely altered in little more than 100 years. In 1900, California's largest body of water was Lake Tulare. The Salton Sea was sand, and most rivers flowed from the Sierra Nevada to the sea. Today Lake Tulare is a cotton field, the Salton Sea (formed by runoff from irrigated fields) is the largest body of water, and all major rivers, except the Eel, flow mainly to the fields, swimming pools, and lawns of southern California.

One of the biggest bodies of water in the West is the Colorado River, which flows through Grand Canyon and Glen Canyon. The river's history is one of endless disputes over who can use its water. In 1985, the Central Arizona Project, funded by $3.9 billion of taxpayer money, began pumping water from the Colorado uphill to Phoenix and Tucson (Figure 11-13). Southern California immediately lost

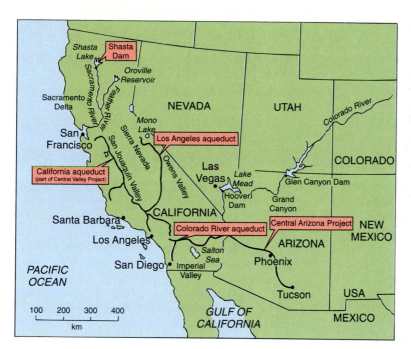

**Figure 11-13** Diverting the Colorado River to Phoenix, Tucson, and southern California. This map shows some of the waterways and huge water projects in the southwestern United States. Most of the water for southern California comes from lakes and reservoirs in northern California and from the Colorado River.

about one fifth of its water supply, which is diverted from the river through the Colorado River Aqueduct (Figure 11-14). As a result of these and many other diversions of water along the length of the Colorado, the river is nothing but a muddy, polluted trickle when it reaches Mexico and empties into the Gulf of California.

In the West, the water in major bodies of water, such as lakes and rivers, belongs to the federal government and is controlled by the Department of the Interior's Bureau of Reclamation. The Bureau was set up in 1902 to build 600 dams and 53,000 miles of canals that supplied water to the farmers and ranchers who occupied most of the land. Because the Bureau was formed primarily to supply water for agriculture, farmers are supplied with water at low, highly subsidized prices, while all other users suffer from water shortages, even if they would be happy to pay much more for their water.

Almost no one is happy with the way in which the Bureau handles water rights or the way water is priced. The problem can be traced back to the Anglo settlers of the West, who came from northern Europe and the East where water is plentiful. The laws these settlers brought with them treat water differently from any other commodity: water in streams and lakes is not privately owned; only the rights to use it are owned. This works fine in areas where there is plenty of water, but it has several fatal flaws when applied to dry areas:

1. **Conservationists may save wetlands but lose all the water in them.** States and other organizations sometimes buy wetlands, for instance to preserve the habitat of endangered birds, but in the West they also need to buy rights to the water in the wetland if the supply is to be ensured and they often cannot do so because the Bureau will not sell it. In Stillwater National Wildlife Refuge in Nevada, for example, thousands of birds have

**Figure 11-14** Part of the Colorado River Aqueduct that brings water from the Colorado River to Los Angeles, seen here in a dry year. *(Richard Feeny)*

died because the water that used to flow into the wetland refuge has been diverted to agriculture. The problem is particularly acute in California's Central Valley, part of the Pacific Flyway. The **Pacific Flyway** is the most westerly of four main "flight lanes," used by millions of waterfowl, shore birds, and other migrants as they travel between their nesting grounds in the far north and regions farther south where they spend the winter, stopping in ponds and wetlands on their way. In 1970, more than 10 million ducks using the Flyway wintered in the Central Valley. By 1985, so much water had been diverted from Valley wetlands to agriculture that the number had fallen to less than 2 million.

2. **The Bureau of Reclamation has a "use it or lose it" rule.** If farmers do not use all the water they are allotted in one year, the Bureau can allow them less water the following year. So farmers have no incentive to conserve water, and more than 50% of irrigation water is wasted by evaporation and leakage. Compare this with Israel, which practices aggressive conservation. In 20 years, Israel has doubled the amount of food it can produce using 1 acre-foot of water (defined in Facts and Figure 11-1).

3. **Owners cannot transfer the right to use Bureau water to anyone else without the Bureau's permission.** This is the most crippling regulation of all. In 1989, a Nevada water company agreed to sell Colorado River water that it did not need to the Las Vegas suburb of Henderson. The contract was already signed when Bureau lawyers stopped the sale. Thus, a farmer who pays the Bureau $5 an acre-foot for water usually cannot get permission to sell that water to a town that would be happy to pay $500 for it.

### Progress on Western Water Rights

Slowly the ridiculous rules that cover western water rights are changing. Cities like Tucson, Phoenix, and Albuquerque now buy water rights from nearby farmers for up to $2,000 an acre-foot. Since about 1988, even the Bureau has permitted water to be transferred from agricultural areas such as Imperial Valley to Californian cities, but these transfers are still controlled by the Bureau and do not reflect the real price of water.

A minor breakthrough occurred in 1992, when the rules for California's Central Valley Project (CVP) were changed. The CVP is the largest federal water system and controls more than one fifth of California's water. For the past 50 years, it has sold 90%

### FACTS AND FIGURES 11-1

#### Liters, Gallons, and Acre-Feet

- An **acre-foot** is the amount of water needed to cover an acre of land 1 foot deep in water.

- An acre-foot is about 326,000 **gallons**, enough to supply two households for about a year.

- The average U.S. household uses about 150,000 gallons of water a year. This is about 550,000 **liters**.

- International agencies put the amount of water a country needs at 600 **cubic meters** for each person every year. This is 600,000 liters, slightly more than the average American household uses in a year. It includes water for agriculture and industry as well as household use and is not a very generous allowance.

- The United States uses more than three times as much as the standard 600 cubic meters. We **withdraw** from the water supply nearly 2,000 cubic meters per person per year, and we **consume** about one fifth of this—meaning that more than 400 cubic meters of water per person per year are not returned to the water supply.

of its water to agriculture at very low prices, while cities, industry, and the environment suffered from water shortages. Finally the business community rebelled, arguing that California had little chance of further economic growth as long as industries could not move into southern California and towns could not expand for lack of water. Over the objections of congressional representatives from the Central Valley, Congress voted to allot more CVP water to towns and to wildlife habitats such as rivers and wetlands. The wetlands of the Pacific Flyway will get some of their water back, and some rivers will lose less water.

The United States could save itself enormous sums by abolishing the Bureau, turning its assets (mainly dams) over to local governments, and permitting people to buy and sell water freely. Zach Willey, of the Environmental Defense Fund, says, "If each Las Vegas homeowner could sell the water used on their lawns back to the Water District for the $1,000 a year it is worth, you would see desert gardens replacing lawns overnight in Las Vegas and a huge saving of a precious resource."

> Water shortages are reducing production and restricting development in many parts of the United States.

### World Picture

Europe is the only continent with all the water it needs for the moment. In many developing countries the most pressing problem is lack of clean drinking water. The World Health Organization estimates that 80% of disease in the Third World is caused by the lack of sufficient clean water for drinking, washing, and sewage disposal. Every hour more than 1,000 children die of water-borne diarrhea (Section 8-F).

Inadequate drinking water means no source within several hundred meters. Inadequate sanitation means no pit privy or bucket latrine, let alone a sewage system. In some areas, women and children (who usually perform this chore) spend as much as 6 hours a day walking to bring home adequate drinking water, depriving the children of time to go to school and the women of time to do more useful work.

Complete lack of water is often not the problem. In many places there is plenty of water in the local irrigation system. But purification plants and hundreds of kilometers of pipes are needed to deliver clean water to the people who need it. Some sort of sanitation is then vital if human waste is not to end up back in the drinking water supply.

The problem is being attacked vigorously in many countries by aid organizations and governments, and some progress has been made. The trouble is that populations, especially of many towns, are growing faster than engineers can add to the clean water supply.

> Supplying people with clean water is one of the highest priorities in many developing countries.

### International Law and the Euphrates River

International agencies put the amount of fresh water that a country needs at 600 cubic meters for each person every year. This is about 150,000 gallons, the amount that the average American household uses in a year, but it includes amounts for irrigation and industry. In 1993, 26 nations had less than this minimum amount. Nine of these countries are in the Middle East (Figure 11-15). Sandra Postel of the Worldwatch Institute warns that "By the end of the 1990s, water problems in the Middle East will lead either to an unprecedented degree of cooperation or a combustible level of conflict." Worldwatch says that no amount of conservation or technology can produce adequate water supplies for these countries unless population growth ceases almost immediately—which will not happen. For an example of the international tensions generated by struggles over water rights, consider the case of the Euphrates

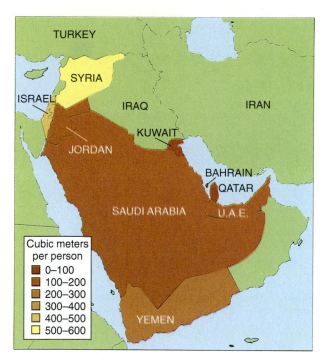

**Figure 11-15** Severe water shortages in the Middle East. All the countries colored yellow, orange, and brown on this map have access to less than the minimum amount of water the United Nations estimates is needed to supply domestic and agricultural needs.

River, which starts in Turkey and flows through Syria and Iraq to the Persian Gulf.

In 1989, Turkey announced that it would hold back the Euphrates's flow for one month to fill the reservoir behind its new Ataturk Dam (Figure 11-16). To compensate, more water than usual would flow during the preceding month. Turkey has plans for 21 dams on the Tigris and Euphrates rivers. If these are all completed, the amount of water in the Euphrates in Syria would be little more than half what it is now. Syria relies on the Euphrates River for almost all its water and already has too little irrigation water to feed a rapidly growing population. It faces complete catastrophe if it loses half its water.

What does international law have to say about control of a river that runs through three countries? The answer is not much. International struggles over water are relatively new, and there is little law on the subject. Turkey, Iraq, and Syria belong to a Trilateral Commission, which is supposed to regulate water use, but the commission is ineffective. In 1974, Syria finished the Al-Thawra dam on the Euphrates. Iraq complained about the reduced water flow, threatened to bomb the dam, and massed troops on the frontier. The World Bank is so worried about the in-

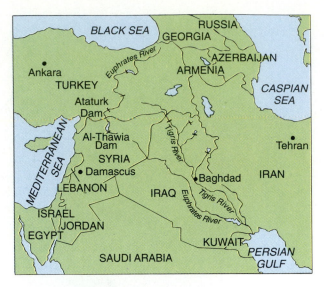

**Figure 11-16**    Position of the Ataturk Dam and the many other dams along the length of the Tigris and Euphrates rivers, which rise in Turkey and flow into the Persian Gulf.

ternational tensions created by water projects in the Middle East that it has withheld money that was to be used for the Al-Thawra project.

> Wars over water seem likely in the Middle East within the next decade or two.

## 11-F    SOLUTIONS TO WATER SHORTAGES

There are numerous potential solutions to local water shortages, but it is probable that none of them alone is sufficient. We need to develop new sources of fresh water, to use less by practicing conservation, and to minimize pollution so that we do not have to spend vast sums cleaning up or replacing the water we already have. Here are some of the relatively small-scale solutions that may, nevertheless, one day add up to enough to prevent us from running out of fresh water.

### Wasting Less Water

An estimated 30 to 50% of water used in the United States is wasted (Figure 11-17). This could be drastically reduced. Between 1950 and 1980, Israel reduced its water wastage from 83% to a remarkable 5% by changing to improved irrigation methods and conservation. The most important of these changes was from spray to drip irrigation. When water is

sprayed up into the air to spread it over a crop, more than half the water evaporates and never reaches the plants. Modern farmers and gardeners reduce evaporation by using drip irrigation, in which water reaches the plants by way of pipes along the ground. The pipes are punctured by tiny holes through which water leaks into the soil. Since the price of water started to rise in many parts of the United States during the 1980s, drip irrigation systems have become available in most American farm and garden shops.

Conservation is much easier than most people imagine. During droughts in California, residents have reduced their water consumption by 65% or more. Tactics include using water for more than one purpose, for instance, by siphoning bathtub water out to water the lawn. **Gray water** is household water not used in toilets, which is reused for watering gardens or flushing toilets. Many health codes, ridiculously, forbid setting up gray water systems, for fear that disease-causing organisms from washing clothes and hands will reach lawns and gardens and infect others.

### Pricing Water

One of the main reasons for water shortages in the United States is that the government has kept the price of water to farmers artificially low. As a result, vast amounts of water are used to produce crops with little economic value instead of for more valuable uses such as industrial development and conservation of wildlife habitat (Section 11-H). For instance, 1 acre-foot of water in California may cost a water board $250 to produce and be sold to farmers for $3.

**Figure 11-17**    Wasting water: this is a Los Angeles street in 1991, a drought year. Although the house is unoccupied, an automated irrigation system supplies this strip of lawn with water at regular intervals, spilling most of the water into the street. *(Richard Feeny)*

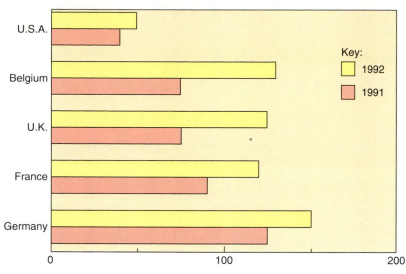

Price of water to consumer (cents per cubic meter)

**Figure 11-18**   Cheap U.S. water. This graph shows the price of water to consumers in various countries. Although the cost of water is rising, it is rising more slowly in the United States than in European countries, although Europe has much more water. *(Data from National Utility Services)*

In 1980, the irrigation of rice, alfalfa, cotton, and pasture consumed more than one third of California's water, although these crops made up less than 1% of its economy.

The average cost of water to the U.S. consumer is rising. In the summer of 1990, some customers in Lafayette, Colorado, had water bills of more than $1,000 a month! Nevertheless, the price of water in the United States in 1992 was still much lower than the price in European countries, although Europe has more water than the United States (Figure 11-18). Low prices give American consumers little incentive to conserve water.

When water prices rise, people rapidly learn to use less (Lifestyle 11-1). For instance, Las Vegas uses

475,000 liters per person each year for which the consumer pays about $300, about one third of what the water costs the city. In Tucson, consumers pay more than twice as much each year for their water and, not surprisingly, use about half as much. At 1992 rates of use, Las Vegas would run out of water in 1997. Half the water Las Vegas consumes irrigates landscaping (Figure 11-19). The inhabitants of Tucson conserve water partly by not using water to maintain lawns around their houses. More and more Americans are learning how to create attractive home landscapes using plants that do not need watering. Dry landscaping does not have to consist of gravel decorated with scattered rocks and cactuses. It can be full of greenery and flowering plants (Figure 11-20).

**Figure 11-19**   Las Vegas. The green area (middle left) is decorative landscaping, a wildly extravagant use of water in a desert. Irrigated landscaping would disappear from Las Vegas if the price of water to the consumer reflected its real cost more accurately. Instead, Las Vegas's water is subsidized by all American taxpayers.

**Figure 11-20**   A dry garden in Boise, Idaho. The garden contains plants that are native to areas with little rainfall. When they are first planted, the plants need to be watered. But after a few months, their root systems grow large enough to collect all the water they need and the garden never again needs watering. Many plants adapted to dry conditions are succulents, which store water in fleshy leaves or stems.

## Legal and Economic Changes Needed

At the moment, the crazy quilt of laws and regulations governing water in the United States provides little incentive for people to change their wasteful ways. For instance, leaky pipes and faucets waste an estimated 30% of piped water. There is little incentive for anyone to fix the leaks, however, where water is very cheap or, as in many cities, not metered at all.

Water resource management needs to be rationalized. Chicago, for instance, is supplied by 349 water supply systems and 135 waste treatment plants! One authority's waste treatment plant may discharge its waste water into a different authority's water supply.

A better way to organize water resources would be to consider each watershed or aquifer a management unit. A **watershed** is the land area drained by a particular lake or river. In such a unit, the people who consume the water will, ideally, pay for the resource

development. They will refrain from polluting their own water supply because they will have to clean it up to make it fit to drink. They may be encouraged to reduce costs by producing a dual water system: highly purified water for drinking and less purified water for other purposes. Lightly contaminated water from waste treatment plants and industry could be used for recharging aquifers and for irrigation.

One environmentalist has suggested that crop yields should be measured not in terms of the land needed to produce them but in terms of the water they require, since water is often a more valuable resource than land.

## Towing Water

The world's icebergs contain about five times as much fresh water as we use for domestic purposes worldwide today. People have toyed for years with the idea of towing flat-topped Antarctic icebergs (less prone to roll over than spiky Arctic ones) from where they are to where they would be more useful. Indeed, the cost of water is rising so fast in places that some experts think the technical problems may soon be worth solving. Saudi Arabia, where water costs more than oil, recently commissioned a series of experiments on towing icebergs. Some of the snags are obvious: icebergs are hard to tow, they melt rapidly when you get them near the tropics, and how do you get the water ashore at the other end?

Alaska is experimenting with towing water in huge plastic bags. Alaska contains more than 40% of the United States's fresh water. Therefore, says the state's water chief, Ric Davidge, "Let's bag it up and sell it south." The Medusa Corporation of Calgary has designed bags of polyvinyl chloride, the smallest of which holds 275 million liters of water. The idea is to float these bags of fresh water down the Pacific coast, probably to California, which is perennially short of water. Some towing company executives have their doubts about the project. They believe that logs and snags in rivers will puncture the bags and that rough weather at sea may tear the bags from the tugs that tow them. Water delivered in this manner will not be cheap, but the price of water in the United States is, in any case, bound to rise rapidly in coming years.

## Desalination

**Saline water** is any water containing more than one part per thousand (ppt) of dissolved solids. It includes brackish water (1 to 4 ppt of dissolved solids) and sea water (18 to 35 ppt of dissolved solids). Saline water is too salty to drink and ruins the soil if used for irrigation. The process of removing dissolved salts from water is called **desalination**. California cities such as Santa Barbara built desalination plants after suffering from drought during the 1980s. Nearly all the water in desert countries such as Saudi Arabia is produced by desalination.

The two main methods of desalination are distillation and osmosis. In distillation, heat is used to evaporate fresh water from salt water, leaving the salts behind. With osmosis, pressure is used to push the water through a semipermeable membrane that will not permit the salts to pass. Small desalination units that use solar heat to distill water are common as emergency and backup systems on boats. Only a tiny fraction of the world's water is produced in this way, usually in coastal cities in arid areas. The reason is cost. Desalination uses a lot of energy, and more may be needed to pump the desalinated water uphill from the coast where it is produced to where it is needed. Water produced in this way will probably never be cheap enough to use for purposes such as irrigation.

### Cloud Seeding

Raindrops and snowflakes do not form spontaneously in air saturated with water vapor. They form when water molecules are absorbed onto **condensation nuclei**, dust particles in the atmosphere. Cloud seeding is an attempt to increase precipitation by adding condensation nuclei (often silver iodide dust) to a cloud. In the United States, ski resorts and western farmers have commissioned cloud seeding. Seeding is not really a solution to the water problem. It does not increase the amount of water that falls on the ground (although it doubtless increases the precipitation of silver iodide). At best, it induces rain or snow to fall on you rather than on me. And no one has ever provided convincing evidence that it even does that. In fact, cloud seeding is probably a nonstarter as a solution to water problems.

> Many approaches are needed to do away with water shortages. The most important are wasting less water, reducing evaporation, and legal and economic reforms.

## 11-G WATERWAYS

Most of our water comes from lakes and streams, waterways that collect surface runoff. We manipulate our waterways in many ways and for many purposes. The two most usual reasons are to increase the water supply in an area and to control floods.

### Rivers

Streams and rivers provide us with water, electrical power, agricultural soil, waste disposal, and valleys that are convenient routes for highways. A stream is fed by ground water and by the water that runs off its watershed. The rate at which water is supplied to a stream determines its discharge, or volume of flow. Discharge, in turn, affects the stream's velocity, width, depth, and amount of sediment or silt (soil particles) it can transport.

### Floods

A flood is any relatively high stream flow that spills out of its channel. The banks of a stream are shaped largely by its usual amount of discharge. When the stream receives an unusual volume of water, the surplus spills over the banks and flows out onto the floodplain. Floodplains are natural safety valves, areas where a river has overflowed its banks during every flood for thousands of years, depositing silt, which often creates fine agricultural land.

Floods can be described by their **recurrence intervals**, how often they occur in an area, and their **peak discharge**, the unusually high water discharge during a flood. Flooding gets worse in an area when the recurrence interval falls or the peak discharge increases.

### Deforestation

Trees hold large volumes of water. When they are cut down, water runs off the land whenever it rains, carrying soil away and causing floods. The observation that deforestation could cause floods was one of the first indications of major environmental damage from human interference.

The first experiments on how changing land use affected stream flow began at Wagon Wheel Gap, Colorado, in 1910. Stream flow from two similar watersheds of about 80 hectares each was compared for eight years. All the trees in one valley were then cut down and the measurements continued. The results from a similar experiment in North Carolina are shown in Figure 11-21. Deforestation always produces an increase in peak discharge. As vegetation regenerates, runoff steadily decreases again. Replacing forest with crops also increases runoff because crops tend to be shorter than trees, with less biomass to hold water. The greatest immediate danger from

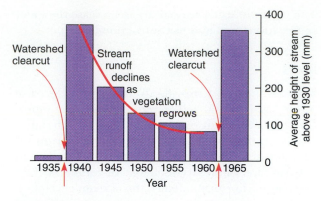

**Figure 11-21** Increase in water runoff to a stream as a result of clear-cutting the forest on a watershed in Coweeta, North Carolina. These results show how much precipitation vegetation absorbs. As the forest regrew (1940 to 1960), the stream's volume of flow declined because large trees and shrubs absorb more water than small ones. Deforestation is a major cause of floods.

| TABLE 11-1    Some Effects of Dams | |
| --- | --- |
| **What Is Affected** | **Effect** |
| Water | Stream flow controlled |
| | Evaporation increases (from surface of reservoir) |
| | Water temperature becomes less variable |
| Organisms | Habitats flooded |
| | Fish prevented from migrating |
| | Creates still-water habitat for flies and snails that transmit disease |
| | Changes the species that can survive in the area |
| Geology | Soil (and nutrients) trapped behind dam; fields and habitats downstream receive fewer nutrients, and coast erodes faster |
| | Water table rises above dam: may waterlog the soil and cause landslides |
| | Changes in water pressure above the dam and downstream may cause earthquakes |
| Local climate | Humidity increases above dam |
| | Precipitation increases above dam and decreases downstream |
| | Temperature falls upstream of dam |

deforestation is flooding and soil erosion (Section 14-G). But when the forest regrows, it may also cause problems. In the eastern United States, many farms have been abandoned and forest permitted to regrow. The increased vegetation cover has reduced runoff and river flow at a time when water supplies for some eastern cities are becoming critically short.

### Dams

Building dams and ditches is a widespread human behavior pattern. From a child constructing waterworks on the beach to a peasant diverting irrigation trickles into a rice paddy, from 2,000-year-old Roman aqueducts carrying water across Europe to the Hoover and Aswân dams holding back great rivers, changing the natural course of a waterway has a long and venerable history. Dams are built for various purposes: to prevent floods below the dam, to generate power, and to store water. More than 15% of the world's surface runoff ends up in the reservoir behind a dam.

Dams have had bad publicity recently. While reviewing the problems, we should, nonetheless, remember that most dams achieve their aim of controlling river flow, and that millions of people depend on them for survival, employment, and welfare. The snag is that dams tend to have a whole series of environmental effects that their designers may or may not have predicted (Table 11-1). Some of these effects, such as sediment retention, soil salinization, and waterlogging, are considered in other chapters. Dams tend to precipitate bitter fights be-

tween conservationists and pro-dam forces. They reduce the river's velocity and are often built to flood wild and lovely valleys, ruining sports such as whitewater canoeing and destroying wildlife habitat.

The reservoir behind a dam also provides a vast surface from which water is lost by evaporation. Lake Nasser, behind the Aswân High Dam in Egypt, loses an incredible 10% of its water by evaporation. Because of this and because it is fast filling with silt, Lake Nasser holds much less water than its builders anticipated. Possibly the worst problem with dams, however, is that they may burst and kill people, usually because silt has built up behind the dam faster than anticipated. Once the dam is built and the river controlled, people tend to settle on the flood plain below. Thousands of people in India and Pakistan and hundreds in North America have been killed by bursting dams. It is alarming to realize that 90% of the dams in the world are less than 40 years old and have not yet had time to silt up and become unsafe.

Because of the environmental problems they cause, the era of big dams in the United States is probably over. All countries, however, will continue to build some dams to provide reservoirs of drinking water and supply small hydroelectric plants with energy.

## Channelization

Channelization is the practice of altering the banks of a stream by constructing levees, dikes, floodwalls, and embankments or of increasing its discharge capacity by straightening, widening, or deepening it. The practice may have various goals, such as increasing the stream's capacity to hold floodwater or guiding the stream in particular directions (Figure 11-22). Some of the largest rivers in the world are now lined by huge embankment systems like those that run 1,000 kilometers beside the Nile, 1,400 kilometers by the Red River in Vietnam, and more than 4,500 kilometers in the Mississippi Valley. Like dams, artificial channels often achieve what they are meant to achieve, but they also have undesirable effects and may create new problems. One universal effect is that channelization reduces the number of habitats in the stream and on its banks, reducing the variety of organisms that live there, often ruining the fishing, and reducing the ability of the water to purify itself.

The water has to go somewhere: when dikes and levees are built to prevent flooding, they may contribute to flooding upstream. In addition, when water spreads out over a floodplain, it slows down. By forcing it back into a channel, an embankment speeds it up again and delivers water faster to areas downstream. These may flood after channelization, although they never did before the embankment was built.

In the United States, the Flood Control Act of 1938 provided federal money for waterway projects. From the 1940s to the 1970s, "channel improvement" projects were constructed on nearly 65,000 kilometers of waterways by the U.S. Army Corps of Engineers and the Soil Conservation Service. Another 300,000 kilometers were modified by local governments.

> Altering natural waterways often has unexpected results and adverse environmental effects.

## Undoing the Damage

People are beginning to rebel at their streams being straightened, encased in concrete, or buried (Figure 11-23). In Snohomish County, Washington, community groups have begun to restore streams under the direction of a group called Adopt-A-Stream.

In California, an estimated 95% of streamside habitat has been destroyed, but the state now has almost 300 private and public projects aimed at preserving or restoring urban streams. Citizens of Richmond, California, were angered by plans to channelize Wildcat Creek, one of the last natural streams in the Bay Area. They blocked a plan to bury the creek and suggested an alternative plan: use nature's flood control methods by leaving a meandering streambed to carry storm runoff, with higher flood terraces to allow for higher flow, and plant vegetation to prevent erosion of the stream banks. This plan was adopted in 1984.

**Figure 11-22**   Rivers are usually channelized where they run through cities so that roads and parks can be built right to the edge of the river without fear of them collapsing into the river. This is the River Seine as it runs through Paris. Its banks were channelized hundreds of years ago to support bridges carrying traffic over the river.

(a)

(b)

**Figure 11-23**   Which would you rather have in your neighborhood: (a) a natural stream like this one near Denver or (b) a channelized stream like this one in Sacramento? There are no prizes for guessing which of these streams contains more plants, fish, and other wildlife. People are beginning to protest the conversion of their streams into ditches lined with concrete. *(b, Richard Feeny)*

In Florida, the most ambitious U.S. project of all is under way. Flood control projects in the 1960s destroyed 80,000 hectares of wetlands at the headwaters of the Everglades. The meandering 98-mile-long Kissimmee River was turned into a 52-mile concrete ditch, and the flood plain was opened to farming and housing. The fish population dropped by 75%, the waterfowl almost disappeared, and pollutants from farms ran into the river. The Corps of Engineers is now embarked on an ambitious plan to restore part of the river to its natural flood plain. The cost of restoring about 4,000 hectares of flood plain will be about $100 million, most of it spent to buy back land from farmers.

## 11-H  FRESHWATER WETLANDS

Freshwater wetlands are among the most threatened of habitats in the United States and indeed all over the world. They range from peat bogs and prairie potholes in the Midwest, where waterfowl nest or feed on migration, to the Pantanal in Brazil, the world's largest freshwater wetland, covering an area greater than Florida and home to endangered jaguars, anteaters, caiman, and waterfowl (Figure 11-24).

Whatever their makeup, wetlands serves functions that cannot be achieved in any other habitat:

1. **Wildlife habitat.**   Wetlands are home to hundreds of species of plants and animals that cannot survive elsewhere (Figure 11-25).
2. **Flood control.**   Wetlands are able to absorb large amounts of water when it rains and rivers rise, so they provide flood control.
3. **Water purification.**   Wetland plants absorb enormous quantities of pollutants, which is why artificial wetlands are being built in many areas to clean up the water from sewage plants and feedlots before it is returned to a river (Section 19-F).

**Figure 11-24** Location of the huge Pantanal freshwater wetland in southern Brazil, home to a myriad of interesting organisms.

**4. Nutrient recycling** When water flows through a marsh, plants remove most of the nutrients in the water, preventing them from washing out at the far side and eventually being washed out to sea.

One tenth of the continental United States was once wetland (Figure 11-26). More than half of that

area has been destroyed. The Swamp Land Act of 1849 essentially gave away thousands of hectares of wetlands to anyone who would drain them and turn them into farmland. California once contained 2 million hectares of wetland; today it contains less than one tenth of that. To land-hungry developers and farmers, wetlands look like just another difficult bit of land to be filled and farmed. Wetlands sometimes don't even look wet. They may consist of pine forests with water an inch or two beneath the soil surface for most of the year. Or, like the Pantanal, they may be flooded during the wet season and consist of grassland dotted with water holes during the dry season. The difficulty of deciding precisely whether an area is wetland or not allowed the Bush administration to declassify about half of all federally protected U.S. wetland so that it could be built on and farmed.

Wetlands vary in the amount of water they contain. **Marshes** contain low-growing plants, such as pickerel weed and cattails, which stick up out of the water, and areas of open water (Figure 11-27). The Everglades is the largest freshwater marsh in the United States. **Swamps** are wetlands dominated by trees and shrubs that can survive floods. In northeastern swamps, red maples are the usual trees. Throughout the Southeast, bald cypress and oaks stand in water year round. Big Cypress National Preserve in western Florida contains immense bald cypresses, festooned with orchids and other epiphytes

**Figure 11-25** Common moorhens in the Savannah National Wildlife Refuge. The refuge is a freshwater swamp that was damaged by an Army Corps of Engineers project, which flooded it with salt water. The Corps built a tide gate that deflected sea water that runs up the Savannah River when the tide comes in. The idea was to increase the river's rate of flow so that the water would wash more mud out to sea and save the cost of dredging the river bottom every few years. Instead, the tide gate diverted salt water into the refuge, killing freshwater plants on which ducks and other wildlife depend. In 1992, the Corps dismantled the tide gate and the refuge began to recover. Since common moorhens can live only in fresh water, this photograph taken about 6 months after the tide gate was abandoned shows that fresh water and freshwater vegetation are returning to the refuge. *(Steve Bisson)*

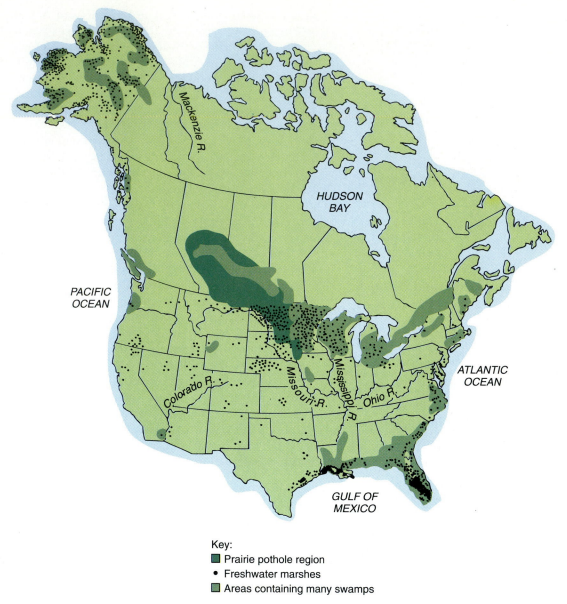

Key:
- ■ Prairie pothole region
- • Freshwater marshes
- ■ Areas containing many swamps

**Figure 11-26**   The main areas of wetland in the United States.

(Figure 11-28). Within the preserve lies the Corkscrew Swamp, once home to a huge rookery of egrets and wood storks, which build their nests in the tops of large trees.

**Bogs** and **fens** are wetlands found in colder parts of the world, where decomposition is so slow that dead plants are not recycled into nutrients that living plants absorb. Instead, the dead plants accumulate as peat. Northern Maine, Minnesota, Canada, and Alaska contain about 35% of the world's peat bogs. These areas are very acidic, low in nutrients, and contain specialized plants, such as sphagnum moss,

heathers, pitcher plants, and black spruce, which are adapted to these unusual conditions.

### The Everglades

Almost the whole of southern Florida was once freshwater wetland: the Florida Everglades. This was essentially a huge, slow-flowing, wide, shallow river that flowed from Lake Okeechobee to Florida Bay (Figure 11-29). In some places, it flowed a foot or two beneath the ground (forming the Biscayne Aquifer); in others, it flowed on the surface, partly filled with

**Figure 11-27** Pickerel weed, *Pontederia cordata*, a common flowering plant of wetlands, with a monarch butterfly collecting nectar. *(Steve Bisson)*

saw grass and other water-loving plants. Near the coast, salt-tolerant mangrove trees grew in a dense tangle. Their above-ground prop roots were the nurseries where more than 40 species of juvenile fish grew to maturity. In the dry season, from December to April, animals collected in huge numbers around the remaining pools of water where herons, egrets, roseate spoonbills, and anhingas fished. The Everglades were home to hundreds of species of plants and animals, hundreds of which are now extinct and 400 of which are endangered.

Agriculture destroyed the Everglades. More than 50 years ago, settlers started to drain the land around Lake Okeechobee to grow crops on soil that was fertile because it contained the partly decomposed remains of saw grass. A dike now rings the lake, preventing it from flowing south on its natural course. In times of drought, Everglades water is contained in the lake and in three "conservation areas," which can release water into the Biscayne Aquifer or for irrigation. The Everglades National Park, which is all that remains of the Everglades, receives water only when it is released from a conservation area. As a result, the park is chronically short of water.

The worst damage to the Everglades water comes from sugarcane farms around Lake Okeechobee,

(a)

**Figure 11-28**  Plants and animals of American wetlands. (a) The bulging "knees" of bald cypress, *Taxodium distichum*, rise from the water of a southern swamp. (b) Egrets, which are herons, nest in colonies, known as rookeries, in the tops of trees. (c) In autumn, the delicate flowers of marsh lily, *Crinum americanum*, brighten the water's edge. (d) Southern swamps host several species of epiphytes, plants with their roots in tree bark instead of the soil. These are grey strings of Spanish moss, *Tillandsia usneoides*, the most common epiphyte in North America. (e) Stalking dinner in the swamp, a turtle and an egret wade and swim through duckweed, *Lemna*, a plant that grows floating on fresh water. (a, *Steve Bisson*; d, *Thom Smith*)

(b)

(c)

(d)

(e)

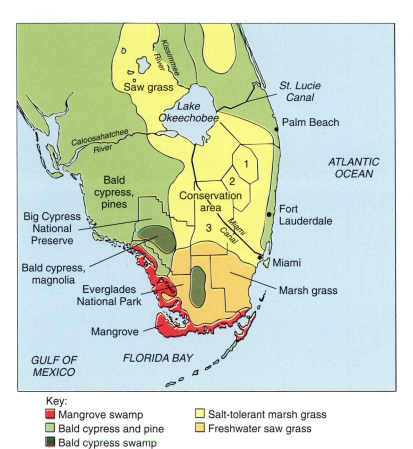

**Figure 11-29** The Everglades. Colors show the natural vegetation before human interference. Pale yellow saw grass (*Cladium*) is a freshwater marsh grass. In the south, cordgrass (*Spartina*), shown in gold, tolerates salt and brackish water. Bald cypress is an adaptable tree that can live in water or on drier land. Loblolly pine trees can also tolerate soil that is flooded occasionally by fresh or salt water.

which use large amounts of water and release fertilizer into the water. The polluted water supports the growth of huge stands of European cattails, which crowd out native water plants (Figure 11-30). The strange thing about this situation is that the taxpayer subsidizes the corporations that own the sugar farms. Although the Reagan and Bush administrations professed to believe in free trade, they maintained quotas on imported sugar. This kept the price of sugar artificially high, which was tough on American consumers and food companies but permitted Florida sugar farmers to make a living. Another peculiar feature of this situation is that many of the sugar corporations are owned by Central Americans, not by U.S. citizens.

Marjory Stoneham Douglas, 100 years old in 1990, started the Friends of the Everglades to fight this situation, with the rallying cry, "Sugar does not belong in the Everglades." Politics makes strange bedfellows, and Douglas's pleas were eventually heard not only by environmentalists but also by the South Florida Water Management District, which woke up to

the fact that farm fertilizer was polluting the Biscayne Aquifer and threatening Miami's water supply. This coalition eventually led to the Kissimmee River restoration project (Section 11-G), which includes plans to turn thousands of hectares of farmland near the lake back into marsh that will absorb much of the fertilizer and pesticide pollution and send clean water down to the conservation areas. This land is not much loss to the sugar corporations, which will be well paid for their land. Anyway, farming has depleted the soil's fertility to the point that agricultural experts estimate that it would be little use for farmland after 2000. Nevertheless, the project is opposed by 500 Ladies of the Lake United to Save Agriculture, most of them employees or wives of employees of the sugar corporations.

## 11-I URBANIZATION

When houses replace natural habitat in many parts of the Southeast, the Midwest, and California, wetlands are lost. Building towns also has many other

**Figure 11-30**   Cattails (*Typha*). *(Steve Bisson)*

**TABLE 11-2**   Stages of Urban Growth and Their Effect on the Water Cycle

| Stage | Effect |
|---|---|
| **Scattered Buildings and Roads Built** | |
| Removal of vegetation | Decreased transpiration, increased runoff in storms |
| Scattered houses with wells and septic tanks | Lower water table, increase in soil moisture, some pollution from septic tanks |
| **Widespread Development** | |
| Bulldozing land | Soil erosion |
| Wells abandoned | Rise in water table |
| Diversion of streams to public water supply | Decrease in stream flow |
| Increase in sewage | Pollution of streams and wells |
| **Urban Modernization** | |
| Sewers and treatment plant installed | Reduced pollution; more water removed from area |
| Improved storm drainage system | Oil, pesticides, and other substances from town into local waterway |
| Deep, high-capacity wells | Lower water pressure, subsidence, salt water intrusion, aquifer depletion |
| Increased water use for air conditioning | Overloading of sewers and other drainage systems |
| Conservation, waste water reclamation | Recharge of aquifer, more efficient use of water |

effects on the water supply (Table 11-2). One is to increase the danger of floods. When vegetation is replaced by streets and houses, the soil surface is replaced by impermeable tarmac, tiles, and concrete. Much of the rain that sinks into the soil in rural areas runs off the streets in towns. Installing sewers and storm drains accelerates this trend. Studies in Jackson, Mississippi, showed that the peak discharge in floods of an urban watershed was three times that in a similar rural area. The water that runs off city streets is full of pollutants: plastic and paper litter, dog and cat feces, oil and gasoline from the streets, pesticides and fertilizers from lawns, and any chem-

icals spilled in the streets. If storm drains run straight into the local waterway, they invariably pollute it.

Urban planners are aware of this problem and encourage practices that reduce the rate at which water leaves an urban area after heavy rain. Vegetation is one of the most important of these because plants and soil absorb water. Trees along the streets, parks, and grassy strips beside parking lots are all helpful. Many artificial structures also hold a little water. Flat roofs with small drainpipes, gravel surfaces and trenches, ponds, and roughened road surfaces can all help to prevent polluted water draining into the waterways (Table 11-3).

**TABLE 11-3** Means for Reducing and Delaying Runoff from Urban Areas after Heavy Precipitation

| Area | Reducing Runoff | Delaying Runoff |
|------|-----------------|-----------------|
| Flat roof | Cisterns to store water<br>Rooftop gardens | Drain roof by small drainpipes<br>Rough surface on roof |
| Parking lot | Porous pavement<br>Cisterns beneath lot<br>Surrounding vegetation | Grass strips and drainage ditches<br>Rough surface with depressions |
| Residential | Gravel (porous) drives<br>Build small hills<br>Aquifer recharge via perforated<br>pipe, sand, trenches, dry wells | Small ponds or lagoons<br>Rough grass that delays runoff<br>Slow runoff by gutters, diversions<br>Vegetation everywhere possible |
| General | Plant trees<br>Gravel alleys<br>Porous pavements | Rough surfaces<br>Parks and ponds<br>Berms (small hills) |

# LIFESTYLE: 11-1

## Indoor Water Use

In most U.S. households, more water is used outdoors to water the lawn than is used indoors. To save water, stop watering the lawn and then consider these facts:

| Typical Daily Water Use Per Person (total: about 300 liters) | | |
|-----|-----|-----|
| Use | Percent of Daily Use | Liters of Water |
| Toilets | 39 | 115 |
| Showers | 33 | 100 |
| Laundry | 10 | 30 |
| Washing dishes | 5 | 15 |
| Cooking and drinking | 5 | 15 |
| Brushing teeth | 4 | 10 |
| Cleaning | 4 | 10 |

### Facts about Toilets

A standard toilet uses 13 to more than 25 liters per flush.

An average adult flushes the toilet six times a day.

An average household flushes the toilet 15 to 20 times a day.

Ultra low-flow toilets use as little as 4 liters per flush and cost $100 to $300, not counting installation.

For about $30, you can buy devices that you can install yourself in existing toilets that reduce water use to about 6 liters per flush but that also provide big flushes when needed.

### Facts about Staying Clean

Standard shower heads use 20 to 30 liters per minute at full pressure.

Low-flow heads, which are cheap and easily installed by the householder, use 6 to 12 liters per minute at full pressure.

Up to 15 liters of cold water may flow out of the hot water tap before the water runs hot. Consider catching this water in a bucket to use in the garden.

Taking a bath uses less water (20 to 25 liters) than a 10-minute shower (180 to 250 liters).

A washing machine load uses 100 to 150 liters; a dishwasher load uses 35 to 45 liters.

## SUMMARY

Clean, fresh water is a vital natural resource. There is plenty of fresh water on Earth, but the supply is not always available to those who need it.

### Properties of Water

Water is made up of polar molecules held together by hydrogen bonds. As a result, it has several properties vital to life: it moves up through soil because water molecules stick together, it dissolves polar substances but not nonpolar substances such as oil, it is an evaporative coolant, and it expands when it freezes.

### Availability and Uses of Fresh Water

Most of the water on Earth is unavailable for human use. All our fresh water comes from the atmosphere and from water in the ground, lakes, and rivers. More than half the people in the world do not have easy access to the 5 liters of clean water that humans need each day for drinking and cooking. Even fewer have access to water for sanitation.

### The Water Cycle

Water travels through a cycle that purifies it and that is driven by the sun's energy. It evaporates into the atmosphere, mainly from the ocean, and condenses, and about one tenth of it falls as precipitation on the land. Here it percolates into the ground or runs off into rivers and thence into the ocean.

### Ground Water

Much of the water we use comes from wells and springs that bring water to the surface from aquifers. If ground water is removed faster than it is replaced, the water table falls, sea water may intrude into the ground water, or sinkholes may form. Various pollutants may enter ground water and contaminate it for long periods since ground water is usually replaced more slowly than water on the surface.

### The Water Shortage

Local shortages of water, especially for industry and irrigation, exist in many parts of the United States. These shortages are getting worse as aquifers are depleted. Water use has grown faster than population, and water shortages restrict economic development in some areas. Worldwide, Europe is the only continent with all the water it needs. Serious shortages of water for drinking, washing, and sewage disposal exist in much of the world.

### Solutions to Water Shortages

Water conservation is the most effective means of increasing the freshwater supply in many areas. More efficient irrigation, reuse of water in industrial processes, and less domestic use can easily be instituted. Evaporation can be reduced and legal and economic changes made that will encourage conservation. Experiments with towing icebergs, desalination, and cloud seeding appear to be expensive and not particularly effective methods of increasing the water supply.

### Waterways

Our surface water supply is contained in waterways, including rivers, lakes, and reservoirs. The volume of water in waterways varies with precipitation and type of vegetation in the watershed and the amount of water withdrawn from the waterway. Many waterways flood periodically and have created floodplains covered with fertile silt. Human interference with natural patterns is caused by deforestation, dams, urban development, and channelization. These may cause unintended disruptions, such as increased flooding.

### Freshwater Wetlands

Wetlands are important wildlife habitats that also reduce flooding and purify water. More than half the wetland area in the United States has been converted to farmland or urban uses.

### Urbanization

Building towns has many effects on the water supply. It increases the danger of floods and adds pollutants to the water supply. The rate at which water runs out of a town after heavy rain can be reduced by minimizing the area of hard surfaces and by planting vegetation.

## DISCUSSION QUESTIONS

1. Suppose that you are appointed to a panel to study floods and flood control in your state. List the reasons floods occur now and ways to prevent them.

2. Suppose water was metered in your community and the consumer charged $2 for each 100 liters used each month. How would you set about reducing your water bill?

3. You are put in charge of supplying clean water and water for agriculture to an African village. At the present time, animals drink at a waterhole fed by a spring that is on a hill above the permanently flowing stream where the villagers get all their drinking water and wash clothes. The only sanitation is a latrine over a dirt pit. What would you do?

4. Why is average annual precipitation in an area not a measure of the water available for human use?

5. In your community, what are the main sources of water, which are the biggest consumers of water, where does used water go, and what has happened to the price of water in the last ten years?

6. The laws of thermodynamics explain why desalination is unlikely ever to be a major source of water. Why is this so?

7. What is the nearest freshwater wetland to your home? What type of wetland is it? What animals and plants live there? Is the wetland in good shape, or has it been degraded by pollution or filling?

# Chapter 12

# THE OCEANS

*A shrimp boat at its dock.*

**Key Concepts**

- The oceans provide us with fresh water, oxygen, minerals, energy, and food.

- The ocean is most productive where there are nutrients in surface waters to support phytoplankton.

- Estuaries are important areas where nutrients are trapped and wetlands develop. They are also the locations where large harbors and cities are often built.

- Coastal wetlands are productive ecosystems where many marine species spend part of their lives. Large areas of wetland have been lost, but efforts to save or restore them are now under way.

- With better research and management, the sustainable yield of food from the ocean could be higher than it is now.

- Pollution of the ocean is a continuing problem.

- Because every ocean is bordered by many nations, international cooperation is necessary to manage ocean resources successfully.

- The developed nations lay claim to the resources in the ocean depths by virtue of their superior technology despite attempts by the United Nations to declare that the ocean floor is communally owned by all people.

S even tenths of Earth is covered by an ocean about which we still know very little. In this chapter, we consider the ocean's resources and our attempts to manage them in a sustainable fashion.

Businesses the world over work together for their mutual profit. But nations, although they are merely the political entities to which the businesspeople belong, are not famous for their ability to collaborate. In the ocean, we encounter a resource that cannot be managed without some degree of international cooperation. We shall see that attempts at such cooperation have produced some successes that might serve as models for other types of environmental management.

## 12-A  OUR OCEAN RESOURCES

The two main differences between fresh water and ocean water are that the ocean contains more salts and is constantly moved by the tides. The gravitational forces of the sun, moon, and Earth interact to move water back and forth in the ocean, so it rises and falls on the coast. If you live in an area where tides move the water up and down by 2 meters or more every 12 hours, it is a shock to visit an area where there is hardly any tide. On my first visit to Puerto Rico, I was amazed to find that the tide is hardly noticeable. Someone from the coast of northern France or Canada's Bay of Fundy, where the tide rises and falls more than 5 meters, would find the tides around most of the United States fairly trivial. Because of its tides and its high salt content, the ocean's organisms are significantly different from those found in fresh water. Few organisms can survive in both fresh and salt water and move from one to the other.

All of the ocean on Earth is connected. It fills four great basins that contain the Arctic, Atlantic, Pacific, and Indian oceans, joined around the Southern Hemisphere by the vast Southern (Antarctic) Ocean. Smaller seas join the oceans, including the Mediterranean, Caribbean, Philippine, and Baltic seas. The ocean is, perhaps, our last frontier. It is so vast, and so inhospitable for land animals like ourselves, that we have tended to see it as an obstruction to be crossed, a bottomless pit for garbage, and the home of a few useful fish. We now know how shortsighted this view is.

The ocean supplies most of the water to the water cycle, which provides us with fresh water. It supplies oil, minerals, energy, much of our oxygen, and nearly one quarter of the protein we eat. It also acts as a great reservoir of dissolved gases that help to regulate the composition of the air by absorbing much of the carbon dioxide we produce by burning fossil fuels. Ocean currents also affect the climate. Without the Gulf Stream, much of temperate Western Europe would have a very cold climate.

> The ocean acts as a great reservoir of dissolved gases, which helps to regulate the composition of the atmosphere.

The twentieth century has witnessed a surge of interest in the ocean and its resources and in ways to exploit these resources without depleting them. Fisheries management is becoming more precise, and fresh moves are under way to prevent pollution and protect marine environments. The United States has set up a series of national marine sanctuaries, areas preserved as ecologically interesting habitats for marine organisms, to be used only for research and recreation (Figure 12-1). A series of conventions and treaties on the sea is coming into force, and the Law of the Sea acknowledges the philosophy that the ocean is the common heritage of all humanity.

Moves to protect the ocean come none too soon. Throughout human history, coastal communities have relied on food from the sea. A breakfast of seaweed or a stew of rock pool inhabitants are still treats for visitors to the coasts of Wales and Ireland or Japan. The skeletons of many prehistoric Guale Indians on the coast of Georgia have teeth that are worn down almost to the gum by the sand grains in the oysters that provided most of their protein. The Chickasaw of Mississippi drugged fish to make them easier to catch. They made a mild poison out of green walnut husks and threw it into the water, then speared the drugged fish as they floated to the surface (Figure 12-2). Until the nineteenth century, nearly half of the protein consumed in the United States came from the sea. But yields of many species are declining. For instance, the worldwide catch of herring and mackerel averaged 6 million tons per year in the 1960s but was down to 2 million tons in the 1980s. The salmon fisheries of the U.S. Pacific Northwest have been completely destroyed by dams that prevent the salmon from swimming upriver to reproduce.

> The stocks of many fish species are seriously depleted.

Part of the reason for declines in catches is the desire for profits today even if it results in a smaller catch next year. Some of it is due to inadequate re-

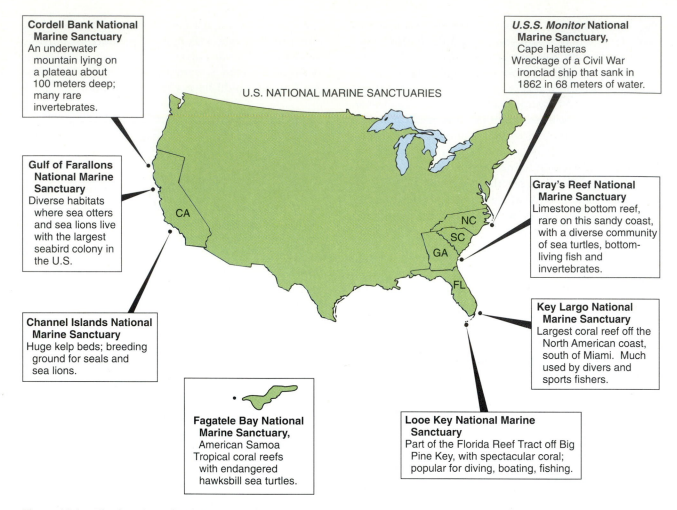

**Cordell Bank National Marine Sanctuary**
An underwater mountain lying on a plateau about 100 meters deep; many rare invertebrates.

**U.S.S. Monitor National Marine Sanctuary,**
Cape Hatteras
Wreckage of a Civil War ironclad ship that sank in 1862 in 68 meters of water.

U.S. NATIONAL MARINE SANCTUARIES

**Gulf of Farallons National Marine Sanctuary**
Diverse habitats where sea otters and sea lions live with the largest seabird colony in the U.S.

**Gray's Reef National Marine Sanctuary**
Limestone bottom reef, rare on this sandy coast, with a diverse community of sea turtles, bottom-living fish and invertebrates.

**Channel Islands National Marine Sanctuary**
Huge kelp beds; breeding ground for seals and sea lions.

**Key Largo National Marine Sanctuary**
Largest coral reef off the North American coast, south of Miami. Much used by divers and sports fishers.

**Fagatele Bay National Marine Sanctuary,**
American Samoa
Tropical coral reefs with endangered hawksbill sea turtles.

**Looe Key National Marine Sanctuary**
Part of the Florida Reef Tract off Big Pine Key, with spectacular coral; popular for diving, boating, fishing.

**Figure 12-1**    The location of U.S. National Marine Sanctuaries.

search. The goal of the world's many fisheries boards is sustainable yield: to control the size and number of animals caught each year so as to maintain or increase the catch in future years. To judge how much seafood should be harvested each year, managers need to know how the animals feed, whether they migrate, where they breed, and how old they must be before they breed. These facts are known for some commercial species, but our ignorance about many of the others demands more research.

> We need to know more about many ocean animals before we can manage them for sustainable yield.

### Recreational Fishing

One of the most significant changes in fisheries management in recent years has been in recreational fishing. Most important economically is the charter boat industry, where guides take vacationers out for a day's fishing. More farsighted than many commercial fisheries, the people who run these enterprises realized in the 1970s that they could not keep expanding their business if the fisheries were depleted. The solution: catch the fish, but throw them back alive. Costa Rica's west coast is famous for its magnificent sport fishing. Sailfish are particularly prized because of their beauty as they leap and run after they are hooked. Stay at a Costa Rican fishing hotel and you can have day after day of wonderful fishing, with photographs to take home and even models of the fish you have caught to decorate your walls. The industry is booming, and the fish stocks are stable. The U.S. industry is beginning to follow suit. Shark fishing is popular in the southeastern states. But shark do not reproduce until they are seven or more years old, and most bear few offspring each year. In 1992, quotas were introduced for some species: contestants in shark-fishing competitions weighed and photo-

**Figure 12-2** An artist's impression of Chicasaws spearing fish after poisoning them on the Gulf Coast. As in most hunter-gatherer societies, men were most responsible for collecting meat, by hunting, fishing, and collecting, while women were responsible for gathering the plant food that made up most of the diet.

graphed their catch and then returned them to the sea instead of killing them as in previous years.

An advantage of increasing recreational fishing is the pressure it puts on commercial fishing to preserve fish. Recreational fishing in Florida now brings more money into Florida than commercial fishing. So when the recreational industry asks government officials to stop commercial fishing from destroying the state's endangered coral reefs, the fishing authorities are beginning to listen for the first time.

## 12-B THE OPEN OCEAN

Along the edges of land masses are shallow, gently sloping **continental shelves**. These shelves are fertile areas for plant growth because they are constantly washed by nutrient-rich sediments from the land. They support dense populations of organisms in and above the water. Beyond the continental shelves, the floor of the open ocean drops away to a vast plain 3,000 meters or more below the surface. The center of this plain is marked by huge ridges. In places, deep canyons and trenches plunge as much as 11,000 meters down. As on land, geothermal activity heats some parts of the sea floor, but most of the deep ocean is very cold and permanently dark because the sun reaches only the top few hundred meters of water. The ocean's depths were once thought to be lifeless deserts, but we now know that they support scattered populations of fish and other organisms, most of which feed on dead organic matter that falls from the surface.

**Figure 12-3**   Black skimmers. The lower bill is shorter than the upper so these birds can fly along just above the surface of the ocean with the lower bill in the water, skimming up plankton and small fish for food. How they manage to fly with part of their bill in the water is a mystery to me. *(Steve Bisson)*

> Most fisheries are on the continental shelves, where nutrients washed from the land keep the waters productive.

## The Ocean Food Chain

Tiny algae and one-celled protist producers are the foundation of the ocean's food chain. These live as **phytoplankton**, producers that float in the water but cannot swim against the ocean currents (Figure 12-3). Because they produce their own food by photosynthesis, phytoplankton need light and mineral nutrients. Enough light for photosynthesis is found only in the top 100 meters or so of the ocean. Minerals are scarce in much of the open ocean, so most of the phytoplankton, and the world's major fisheries, lie on continental shelves that receive minerals washed down rivers. Many fisheries lie where an **upwelling**, a current rising to the surface, carries mineral nutrients up from the bottom (Figure 12-4). Some upwellings occur near land, others in the open ocean.

El Niño, a periodic shift in the currents of the Pacific Ocean, suppresses an upwelling that is normally found along the California coast. This occurred in January 1992. Without the upwelling, populations of plankton and fish were reduced, and those higher up the food chain went short of food. Many seabirds did not nest at all or abandoned their eggs or nestlings. Young sea lions, about half their usual weight, crowded Monterey harbor in search of food, even venturing into parking lots and restrooms near a boat launching area. Some tried to snatch fish and caught themselves on hooks and nets. The Marine Mammal Center bulged with malnourished animals and even treated 50 sea lions with gunshot wounds, presumably inflicted by irate fishermen.

Wherever phytoplankton occur, we also find **zooplankton**, tiny animals and protists that feed on phytoplankton and bacteria. Although plankton are primarily drifters, many can control their vertical position in the water, thereby moving from one current to another, for currents flow in different directions at different depths in many parts of the ocean. Fish and similar large animals in the ocean make up the **nekton**, creatures that can swim in any direction, feeding on plankton and on other fish. The nekton can be divided into bottom-living and pelagic species. **Pelagic species** are independent of the bottom and swim, often in shoals, throughout the ocean. Most commercial fish, such as herring, anchovies, tuna, cod, and menhaden, are pelagic members of the nekton. Fish, in turn, provide the food for birds and for mammals such as seals and some whales.

> Most of the ocean's productivity is in the top 100 meters of water, where there is sufficient light for photosynthesis.

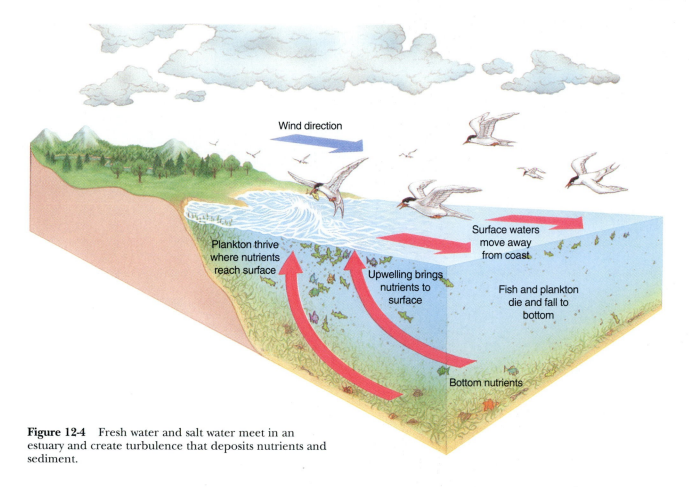

Wind direction

Plankton thrive where nutrients reach surface

Upwelling brings nutrients to surface

Surface waters move away from coast

Fish and plankton die and fall to bottom

Bottom nutrients

**Figure 12-4** Fresh water and salt water meet in an estuary and create turbulence that deposits nutrients and sediment.

## 12-C REEFS

Many prized commercial fish, such as snapper and grouper, are inhabitants not of the open ocean but of reefs. A reef is an area where a rocky outcrop rises from the sea floor (Figure 12-5). The best known reefs are coral reefs, restricted to warm oceans where the water temperature seldom falls below 21 °C. Corals are cnidarians, small animals that live in association with photosynthetic protists. The reef itself is made up of material containing calcium that is secreted by the coral animals and by red and green algae. Because photosynthetic algae are important to their formation, coral reefs are found only in clear, shallow water, where there is enough light for photosynthesis.

A reef contains many different habitats and a much greater variety of organisms than nearby open ocean or sea floor. The reef provides anchorage for algae and animals. Many fish and swimming invertebrates shelter in crevices of the reef. Today many coral reefs are being damaged by waste from nearby tourist areas and by drilling for underwater oil deposits. Most countries have laws protecting reefs be-

cause a coral reef, once destroyed, takes many years to regrow. The most widespread destruction of coral has been by the United States, which has converted many coral islands, with their surrounding reefs, into military bases or nuclear test sites. Diego Garcia, an island in the Indian Ocean, was once the fertile home of 2,000 people. Today, its land and lagoons are paved with concrete, their productivity destroyed.

In recent years, reefs have also suffered more and more from **coral bleaching**, turning white because their algae have died. A bleached reef does not grow. Sometimes the algae return, and a bleached reef recovers. Despite much research, nobody is sure what causes coral bleaching. It appears to occur when the water temperature increases, but there may also be other causes.

## 12-D THE EDGES OF THE OCEAN

Along the sea coasts, many kinds of plants and animals thrive in the **littoral zone**, the area between high and low tide marks, where they are submerged for

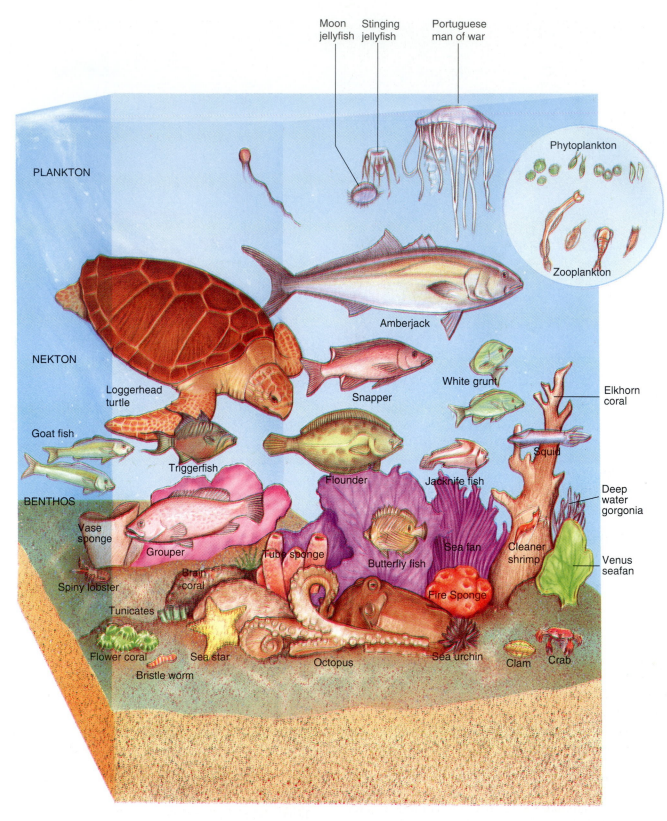

PLANKTON

Moon
jellyfish

Stinging
jellyfish

Portuguese
man of war

Phytoplankton

Zooplankton

Amberjack

NEKTON

Loggerhead
turtle

Snapper

White grunt

Elkhorn
coral

Goat fish

Squid

Triggerfish

Flounder

Jacknife fish

Deep
water
gorgonia

BENTHOS

Vase
sponge

Grouper

Brain
coral

Tube sponge

Butterfly fish

Sea fan

Cleaner
shrimp

Venus
seafan

Spiny lobster

Fire Sponge

Tunicates

Flower coral

Sea star

Octopus

Sea urchin

Clam

Crab

Bristle worm

**Figure 12-5**   A coral reef such as might be found in the Caribbean or off the coast of
Florida.

(a)

(b)

**Figure 12-6** Creatures of mud flats, part of the littoral zone, which is under water at high tide. (a) A juvenile egret hunts for fiddler crabs among the oysters of a salt marsh. (b) A blue crab, the species used to make delicacies such as deviled crab in South Carolina and crab cakes in Maryland. The narrow notch on its abdomen shows this is a male. The female, used to make she-crab soup, has red tips to her claws. Harvests of both sexes are regulated to ensure sustainable yields. The main threat to the crab supply is water pollution in areas such as Chesapeake Bay.

part of the day. There are three main types of littoral zone: rocky, sandy, and muddy shores, which support different communities.

**Mud flats** occur where the water moves slowly enough to deposit sediment. Algae cover the mud and make up the food of numerous burrowing worms, **molluscs**, such as oysters and clams, and **crustaceans**, which are the arthropods such as shrimp and lobsters. Clams and oysters are valuable species from mud flats (Figure 12-6).

**Sandy beaches** are less stable than mud flats, for sand shifts constantly and dries out more rapidly than mud when the tide is out. Most of the tiny protists and crustaceans that live between the sand grains eat plankton stranded when the tide goes out.

Neither muddy nor sandy shores provide much foothold for anchored seaweed or animals that live attached to the ground. These forms are more common on rocky shores, which support a large variety of plant and animal life. Since the water crashes onto the rocks, everything must be firmly anchored. Animals that can move about anchor themselves firmly to rocks or seaweed by their legs or hide in crevices.

The **sublittoral zone** occupies the continental shelves extending from low tide mark to a depth of about 200 meters. In the sublittoral zone, temperature fluctuates less and wave movement is less violent than in the littoral zone. Mineral nutrients are readily available, washed from the land by rivers. Sublittoral areas are among the most densely populated on Earth, crowded with shrimp, lobsters, octopuses, scallops, and hundreds of species of fish.

## Coastal Wetlands

In some parts of the world, the land rises steeply out of the sea. In others, the coast is flat, and the ocean influences the land for some distance inland. The ocean's tide may rise and fall far up a river or in a coastal marsh. These areas where the ocean penetrates the land are the site of **coastal wetlands**. Coastal wetlands include mangrove swamp, found in tropical and subtropical regions, and salt marsh, its temperate equivalent. These wetlands are the nurseries of many marine species. In the United States, it is estimated that half of the commercial harvest of the Pacific Ocean and two thirds of the harvest of the Atlantic Ocean and Gulf of Mexico depend on the coastal wetlands for their existence.

Most coastal wetlands form in estuaries. An **estuary** is a body of water where a river runs into the sea, where salt water mixes with fresh water from the river. Fresh water is less dense than salt water, so river water flows out through an estuary over a wedge of salt water. Currents are set up where the two bodies of water meet and cause sediment to fall to the bottom and slowly fill the estuary. East Coast estuaries average about 4 meters in depth with a deeper channel in the middle. Where the water is shallow

**Figure 12-7**  The coastal wetlands of North America. On the East Coast, most of the area is salt marsh, dominated by smooth cordgrass, *Spartina alterniflora.* On the West Coast, salt marsh is made up mainly of a different cordgrass, *Spartina foliosa.*

enough, rooted plants start to grow in the bottom sediment, and a marsh forms. Estuaries are very productive ecosystems because they constantly receive fresh nutrients from the river and from the ocean, and they are protected from the worst force of ocean waves. Figure 12-7 shows the location of coastal wetlands in North America. Figure 12-8 shows how the collision of fresh and salt water at the mouth of a river produces a sediment trap of fertile mud.

Most estuaries, like Chesapeake Bay, are **drowned valleys**, river valleys that have filled with water because the sea level has risen or the land has subsided (Close-Up 12-1). Figure 12-9 shows how a drowned valley estuary forms on a coast with sand dunes, formed from sand washed up by the ocean. After the estuary has formed, the sand dunes become **barrier islands**, islands off the coast, like those found along most of the East and Gulf coasts of the United States. Barrier islands are battered on the ocean side by wind, waves, and sand, but on the estuary side water moves more slowly and is quite shallow. As a result, barrier islands develop distinctive vegetation, with tough plants that can tolerate salty wind on the seaward side and marsh on the other side (Figure 12-10).

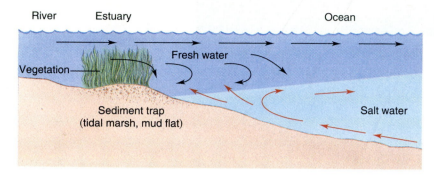

**Figure 12-8**  Currents are set up in an estuary between fresh water from the river and salt water from the ocean. The currents create a sediment trap, where both bodies of water deposit mud, sand, and nutrients. The sediment trap becomes a fertile mud bank and, where the water is shallow enough, permits marsh grasses to grow.

Thanks to federally subsidized flood insurance, people have been able to build houses on the seaward sides of barrier islands and on sandy shores in states such as Florida and New Jersey. These sandy shores are vulnerable to storms, and building on them destroys sand dunes, which normally protect sandy shores from erosion. In recent years, some communities have moved to protect their beaches from erosion by forbidding construction near the ocean, restoring damaged dunes, and building walkways over dunes so the vegetation that holds the dune in place is not destroyed by people walking on it (Figure 12-11). Many of these measures have precipitated lawsuits from land owners who argue that

the authorities have no right to prevent them doing what they like with their own property.

**Salt Marsh**  Salt marsh dominates much of the flat shoreline of the Gulf of Mexico and Atlantic Coast of the United States. Here rivers such as the Mississippi deposit their load of mud and nutrients in estuaries. Smooth cordgrass (*Spartina alterniflora*) covers hundreds of thousands of acres of the marsh, with scattered stands of other plants on higher, drier ground.

Cordgrass is the backbone of an ecosystem that is unusual in two ways. First, it contains few species of plants. Few flowering plants can tolerate wet, salty

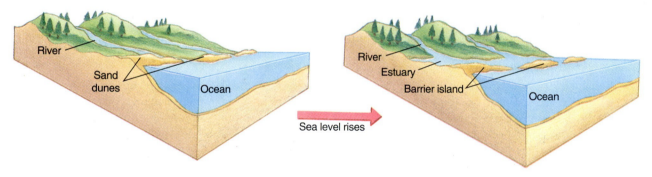

**Figure 12-9**  Formation of a drowned valley estuary on a sandy coast. As the sea level rises, it floods the lower part of the river valley to form an estuary. If several rivers lie close together, their estuaries may run together to form one large estuary.

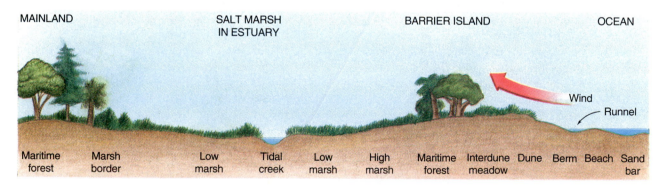

**Figure 12-10**  Profile of the land and vegetation from the mainland, through an estuary and a barrier island to the ocean in the southeastern United States.

# The Rising Sea

The sea level is rising at a worldwide rate of about 2 millimeters a year. This is partly because the ocean is warming up in many places. Studies off the coast of California show that temperatures near the ocean surface have increased by 0.8 °C in the last 50 years. Water expands as it warms up, raising the water level (Figure 12-A). However, the volume of the ocean is not the only thing that determines sea level. We measure sea level by the level of the land, and land rises and falls for several reasons.

The outer shell of Earth is made up of huge **tectonic plates**, that float on the fluid core of Earth (Figure 12-B). These plates move slowly in various directions, sometimes causing earthquakes. On the West Coast of the United States, the Pacific Plate, under the ocean, is sliding under the American Plate on which the land lies, pushing it up. So the West Coast is rising. (The San Andreas Fault marks this collision of plates.) The East Coast, meanwhile, is sinking on the American Plate. So any increase in the volume of the ocean raises the sea level on the east coast of the United States more than on the west. Alaska is rising, possibly because some of the ice that weighs it down has melted and let the land rise.

Many human activities cause land to sink. For instance, as we pump or dig oil, water, or coal out of the ground, the land above may subside. Deltas are peculiarly vulnerable. A **delta** is the land that forms at the mouth of a river as the river deposits sediment that it has picked up farther inland. This sediment is often fertile soil, so many deltas are important agricultural regions. Bangladesh lies almost entirely on the delta of the Ganges River. The dreadful floods in that country in recent years result from deforestation in other countries higher up the river. Because the vegetation that used to absorb some of the heavy monsoon rains is gone, the water now runs down unchecked, flooding Bangladesh and washing much of its soil out to sea.

Another example is the Nile delta, which has produced food and prosperity for Egypt for thousands

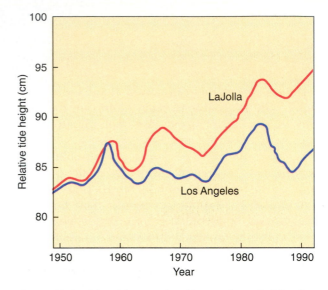

**Figure 12-A** The rising sea level in southern California, measured at La Jolla and Los Angeles since 1948 (*Source: Roemimich, 1992.*)

of years. The delta used to be fertilized by annual floods that overflowed the banks of the river and dropped sediment on the surrounding fields. But now, in the name of flood control, the upper Nile has been dammed. The sediment that once fertilized the delta accumulates behind the dam. Erosion removes soil from the delta, and it is not replaced from upstream. To make matters worse, irrigation projects pump water to the surface, and the delta subsides into the space left beneath it. A slight rise in sea level threatens to flood Egypt's main agricultural area forever, this time with sea water.

Louisiana, which contains most of the delta of the Mississippi River, is losing land faster than either Bangladesh or Egypt. It loses about 70 square kilometers of land every year to the rising sea and sinking land, more than any other state or country in the world.

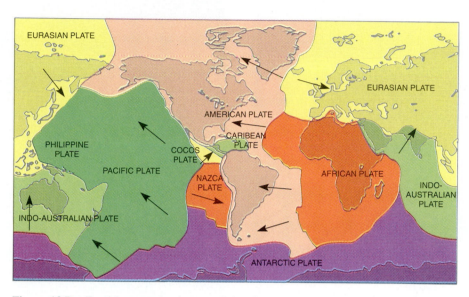

**Figure 12-B**    Earth's tectonic plates and the direction in which they are moving.

(a)

(b)

**Figure 12-11**   Attempts to protect sandy beaches from erosion. (a) The dunes on this barrier island were destroyed by development, and the beach erodes away completely about every 4 years. Then it is replaced, at a cost of millions of dollars, by dredging sand from the bottom of a nearby river. In an attempt to slow this costly cycle, the town is trying to create artificial dunes. Snow fences trap sand that blows inland from the beach and sea oats are planted on the sand to hold it in place. There is so little space between beach and road, however, that the newly formed dunes are very narrow and were easily breached by the waves of Hurricane Hugo in 1989, just before this photograph was taken. (b) This wall of concrete blocks is supposed to act as an artificial dune and prevent erosion of a barrier island beach. It is not very effective. In the background is a walkway to carry pedestrians over a dune so the dune is not destroyed by foot traffic. The dune it was supposed to protect has eroded away.

mud and sand, but cordgrass thrives. Second, few animals eat cordgrass because its leaves contain quantities of glass-like silica.

Although cordgrass is the ecosystem's main producer, its energy and nutrients pass to consumers indirectly. Every year, the dead stalks of last year's cordgrass break off and wash up high into the marsh, where they are broken down by decomposers. The tide then washes the decomposing detritus toward the many creeks that meander through the marsh. The detritus provides nutrients for algae and food for crabs, mussels, worms, snails, and clams that live in the mud and among the cordgrass stems. At the edges of creeks, oysters feed on the detritus and algae they filter out of the water. All these invertebrates, in turn, feed human collectors, raccoons, marsh rats, and a host of birds.

The marsh is the nursery where many animals find food and protection while they are small. For instance, southern commercial shrimp live in the ocean as adults. They lay their eggs on the bottom, and the young move in with the tidal currents until they reach the marsh, where they grow to adulthood. Other commercially important species that spend

(a)

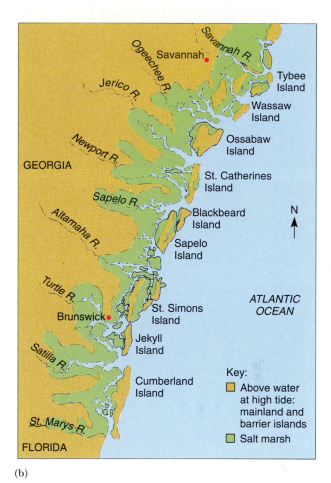

(b)

**Figure 12-12** Salt marsh. (a) Springtime at low tide in a salt marsh. This year's young cordgrass is especially tall and green where nutrients are most plentiful at the edges of creeks in the foreground. The sandy patch in the middle is high marsh, which is seldom flushed by the tide and, therefore, so salty that even cordgrass cannot grow on it. The decaying brown stalks of last year's cordgrass are visible, especially in the background. (b) Salt marsh (green) on the Georgia coast. The marsh forms in the estuaries behind the barrier islands. *(a, Thom Smith)*

part of their lives in the marsh include the young stages of several species of crabs as well as many species of fish. As they grow to maturity, these animals migrate down the creeks and out to sea, where they may end up as seafood on the table or as food for ocean fish such as swordfish, snapper, grouper, and tuna (Figure 12-12).

Salt marsh has a considerable ability to absorb and detoxify pollutants. For instance, in 1987 an oil spill in the Savannah River caused enormous damage in the river itself but closed the oyster beds for only a few days when oil leaked into the nearby salt marsh. Large quantities of inadequately treated sewage, farm fertilizer runoff, and even heavy metal industrial waste, which kill fish and pollute lakes in other places, are rendered harmless by the marsh.

**Destruction of Coastal Wetlands** Estuaries provide protected harbors, access channels to the ocean, and connection with a river system. Not surprisingly, they are the sites of many of the world's major ports. Of the ten largest urban areas in the world, seven, containing more than 50 million people, border estuaries (Tokyo, New York, Shanghai, Calcutta, Buenos Aires, Rio de Janeiro, and Bombay). It is not surprising that nearby wetlands have often been viewed as convenient places to dump waste and then, as the wetland fills in, as potential building sites. About half of all the wetlands in the United States have been destroyed, and the situation is much worse in many other countries. For instance, in the Philippines, coral reefs and mangrove swamps support a huge harvest of fish and shellfish, much of which is ex-

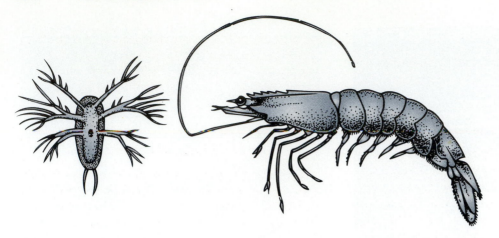

**Figure 12-13**   The southern brown shrimp. *Left*, the nauplius larva; *right*, the adult. The nauplius hatches from an egg on the sea floor and makes the long journey up a river into the marsh. Here it develops into a small shrimp, which returns to the sea. These shrimp grow to about 20 centimeters long. Every year, millions of dollars' worth of them are caught for the dinner table by trawlers. *(Carol Johnson)*

ported to earn foreign currency. Harvests from both ecosystems are decreasing as the habitats are destroyed. Swamps are often converted into agricultural land or cleared for fish farming. Sometimes they are converted into ponds from which sea water is evaporated to produce salt. In Puerto Rico, mining for sand and construction of an airport have destroyed large areas of mangrove. In Hawaii, soil eroded from the land has washed down rivers into the ocean and choked coral reefs.

Along the coast of the United States, landfill operations, together with dredging for sand and gravel, have destroyed 20,000 square kilometers of wetlands. Much of this filling occurred in California, which now has plans to restore some of its wetlands (Close-Up 12-2). And one third of the United States's estuarine and wetland area is closed to shellfish collecting because it is polluted. Figure 12-13 shows one species of shrimp that is becoming less common as U.S. wetlands are destroyed.

The pollutants that damage wetlands are the same ones that pollute lakes, rivers, and the oceans: sewage; industrial waste containing toxic chemicals; and agricultural runoff of soil, pesticides, and fertilizer. Aquatic systems can degrade all of these pollutants except some toxic wastes, but they cannot cope with the quantities produced by dense human populations. Off the coasts of New York and New Jersey, for instance, 9 million tons of sewage are dumped each year, destroying an ecosystem that would normally occupy 12,000 square kilometers of continental shelf: the normal inhabitants of the area are replaced by

organisms that can live in high concentrations of sewage. These states have also dumped much of their garbage at sea for many years, a practice that drew protests in 1991 when medical waste, including syringes and bags that once contained blood, started washing up on New Jersey beaches.

> Coastal wetlands are vital to the survival of many valuable seafood species and absorb large amounts of pollution. Wetlands are disappearing rapidly all over the world.

## 12-E   DEPLETED FISHERIES

In 1966, many of the fish for sale in Spanish markets were too young to have reproduced. The size and quantity of seafood caught in the Mediterranean declined rapidly during the 1960s, and many fishermen went out of business. Today strict limits on catches have restored some of the Mediterranean's productivity.

In 1950, 21 million tons of fish, shellfish, crustaceans, and mammals were harvested from the sea. Thereafter, the harvest increased by about 7% each year to 70 million tons in 1970, a rate of growth faster than that produced by increases in farm production on land. Figure 12-14 shows the composition of the catch. Most experts believe that the oceans could sustain an annual harvest of 100 million tons, but this would require better management than we have so far achieved (Equal Time 12-1). Since 1970, the annual increase in the harvest has been small, and the food value per ton has declined. This is because the fish caught are smaller, and fewer are used directly for human food. Most of the rest are converted into fish meal and oil that is used to feed cattle and chickens or is used for fertilizer. Many large fisheries have declined to a mere fraction of their potential yield. The most notable decline is of the whale fisheries, which were not effectively regulated until the 1980s.

Part of the problem is technological: fishing boats are much more efficient at catching fish than they used to be. Part of the problem lies in our ignorance of ecosystems, particularly in the open sea. The main cause, however, has been an increased demand for fish as food. Island nations like Japan and the United Kingdom have always consumed a lot of fish. The Japanese get 60% of their protein from fish, compared with a global average of 24%, and Americans and Europeans now eat more fish than they used to.

> Although we are not yet harvesting the full sustainable yield of the world's fisheries, catches have hardly increased since 1970, and the yield of many species has declined.

## Large-Scale versus Local Fisheries

Technological advances in fishing have created problems for local labor-intensive fisheries, particularly in developing nations. Fishing in foreign waters by long-distance ships is now declining thanks to international agreements. However, misused foreign aid has caused similar problems. The aid has often been applied to create efficient factory fishing fleets whose catch is used to feed urban populations, resulting in increased poverty for small, traditional fishing communities.

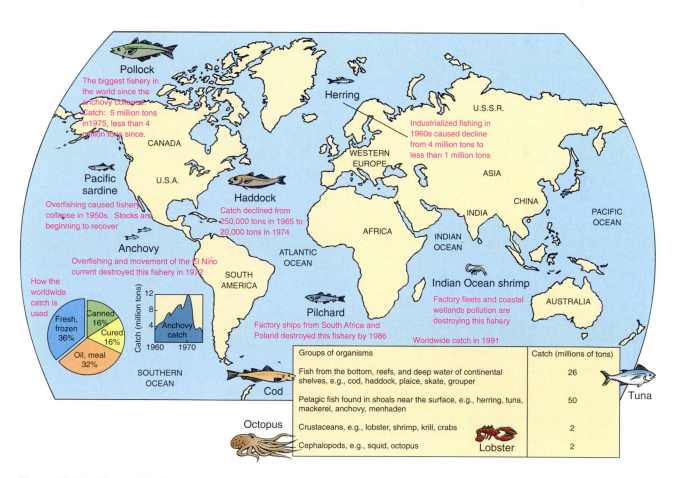

**Figure 12-14** The worldwide seafood catch and the decline of various fisheries.

# California and Georgia: Destroying and Saving Salt Marsh[1]

Until the 1960s, most people considered wetlands a nuisance. They were dredged for harbors, paved over for construction, and filled for farmland. Now the tide has turned, and legislators and the public are pressing for wetlands to be preserved. States such as California and Florida, which started protecting their wetlands late, have a much more difficult and expensive task ahead of them than states that started early.

Georgia, led by ecologist Eugene Odum, was among the first states to try to save its coastal wetlands. A 1970 act protects salt marsh against all destruction except by road building and the federal government. Legislators were particularly concerned to protect the marsh's roles in seafood production and pollution control. When salt marsh is lost, usually to road building, the state offsets the loss with a **marsh loss mitigation** program, by which new marsh is created or degraded marsh restored (Figure 12-C). For instance, one corporation has purchased several hundred hectares of marsh that were converted into rice paddies in the eighteenth century. The paddies can be turned back into marsh by removing dikes around the paddies, dredging creeks, and planting appropriate vegetation. Now when a developer fills an area of marsh to build a shopping center, it pays the corporation to convert a slightly larger area of rice paddy back into functioning marsh.

California, in contrast, was slow to realize the value of its wetlands, with the result that 90% of the coastal marsh of southern California has been destroyed and most of the rest is privately owned, so conservation groups have to purchase the land if they want to preserve it. Court battles between preservationists and developers add to the cost. Figure 12-D shows the work involved in restoring a marsh. Table 1 shows the estimated cost of restoring the 340-hectare Bolsa Chica Wetlands in Orange County—and this does not even include the cost of buying the land, which is incredibly expensive, as Carpinteria has discovered.

Carpinteria is a tiny community near Santa Barbara, so strapped financially that it has dismantled its police force. Nevertheless, Carpinteria has rejected developers' offers to build income-producing marina and condominiums on its beachfront property and wetlands. Instead, city leaders have won a grant to purchase two hectares of the 90-hectare

[1]Source: *Los Angeles Times*.

**Figure 12-C** Marsh loss mitigation. Three hectares of salt marsh were destroyed during the building of this bridge. To compensate, the Georgia Department of Transportation bulldozed an area of young forest near the bridge and removed soil until the tide in the creek under the bridge would flood the area. The marsh is slowly establishing itself without further help. Between the trees and the road, you can see a new stand of cordgrass, the most important part of salt marsh. It could be argued that you haven't achieved much if you cut down forest to create marsh. But ecological succession eventually converts marsh into forest anyway, so there is no particular harm in moving this patch of ground back several steps in succession.

Carpinteria Salt Marsh to preserve it. For its money, the city will get only degraded marsh. Suburban lawn grasses have invaded the marsh and drainage channels from farms, and suburbs carry pesticides into the marsh.

Despite the difficulties and expense, wetlands are being fought for, preserved, and restored along the coast of southern California (Figure 12-E). For instance, nearly 300 hectares of Upper Newport Bay, once a degraded salt marsh, is now a state preserve. It was restored with money that Union Oil Co. paid to the state in compensation for a 1969 oil spill in the Santa Barbara Channel. Money for the restoration of wetlands along the San Dieguito River will come from Southern California Edison Co. in compensation for fish killed at its San Onofre nuclear power plant.

Ecologists tend to be doubtful about the value of wetland loss mitigation. Since this is a new area of research, we have only limited understanding of how to create wetlands. Some fear artificial marshes will not perform all the functions of natural marshes. It is obviously best to preserve natural wetland whenever possible. Second best is to take an

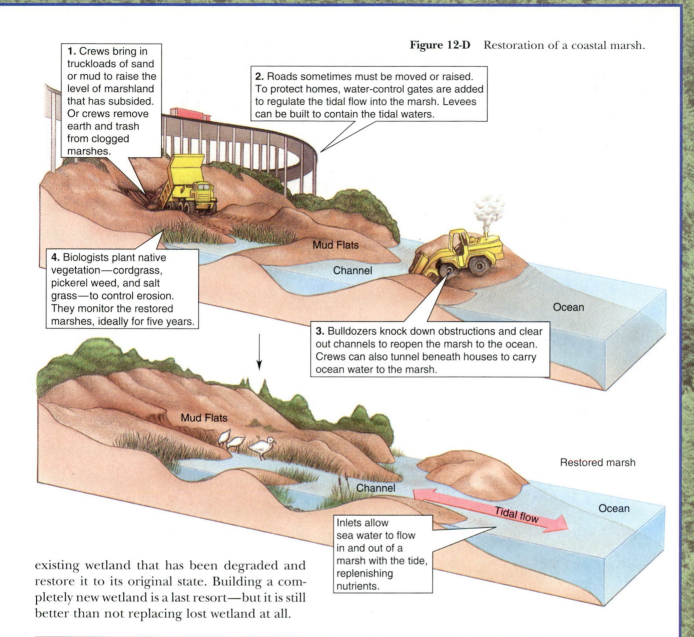

**Figure 12-D** Restoration of a coastal marsh.

1. Crews bring in truckloads of sand or mud to raise the level of marshland that has subsided. Or crews remove earth and trash from clogged marshes.

2. Roads sometimes must be moved or raised. To protect homes, water-control gates are added to regulate the tidal flow into the marsh. Levees can be built to contain the tidal waters.

4. Biologists plant native vegetation—cordgrass, pickerel weed, and salt grass—to control erosion. They monitor the restored marshes, ideally for five years.

3. Bulldozers knock down obstructions and clear out channels to reopen the marsh to the ocean. Crews can also tunnel beneath houses to carry ocean water to the marsh.

Mud Flats

Channel

Ocean

Mud Flats

Channel

Restored marsh

Tidal flow

Ocean

Inlets allow sea water to flow in and out of a marsh with the tide, replenishing nutrients.

existing wetland that has been degraded and restore it to its original state. Building a completely new wetland is a last resort—but it is still better than not replacing lost wetland at all.

| **TABLE 1** Estimate of Cost of the 20-Year Plan to Restore Bolsa Chica Wetlands* | |
|---|---|
| **How the Money Will Be Used** | **Price Tag** |
| Buying oil rights to the land and removing wells, tanks, pipelines | $29.7 million |
| Moving earth, constructing berms, cutting channels to bring in fresh water | $43.3 million |
| Planting native vegetation | $9.4 million |
| Cutting inlet to sea, raising part of a road, building pedestrian bridge | $15.7 million |
| Monitoring and maintenance for about 5 years | $4 million |
| Total | $102.1 million |

*The Koll Co.

Santa Barbara
Lancaster
Ventura
Carpinteria Marsh
Los Angeles
Ballona Wetlands and Lagoon
Bolsa Chica Wetlands
Upper Newport Bay Ecological Reserve
Batiquitos Lagoon
Tijuana River National Estuarine Sanctuary
● Marsh
Palm Springs
San Diego
MEXICO

**Figure 12-E** Some of southern California's most important marsh sites.

# EQUAL TIME 12-1

## U.S. and Canadian Management of Alaskan Fisheries

The story of the Alaskan halibut fishery illustrates the difficulties of regulating fishing intelligently.

On June 8, 1992, nearly 10,000 boats set out from Alaska to catch as much halibut as they could before the "season" closed 24 hours later. When they turned for home, the boats were dangerously overloaded with 13 million kilograms of halibut. "We were lucky with the weather this year," said a Coast Guard spokesman. "For the first time in 20 years, nobody drowned." Fishing frantically to beat the time limit, the fishermen did not gut the halibut or store them properly. The *Los Angeles Times* reported on one returning boat: "They brought in 70,000 pounds of fish. 15,000 pounds had gone bad and had to be thrown away. The rest was of poor quality. But it was cleaned and sold to American consumers as frozen fish." None of the catch was fit to be sold as fresh fish. The governor of Alaska was infuriated by the poor quality. He said, "I can't get fresh Alaskan halibut for my restaurant. I have to buy it from Canada."

What does Canada do differently? After determining the sustainable harvest of halibut (and other fish) in West Coast waters, Canada sells fishing licenses to about 100 boats. Each boat can fish all year long, permitting it to avoid stormy weather and the high death rate among U.S. fishermen. With plenty of time, each fish is cleaned and iced as it is caught so the fish reach market fresh and command high prices.

In contrast, the United States lets anyone who wants go fishing, and more and more boats turn out each year. Authorities react by making the fishing season shorter and shorter until it now consists of only two 24-hour periods. Steve Hoag, of the U.S.–Canada Halibut Commission says, "Imagine the waste and chaos if the nation frantically baked all its bread on just two days of the year. It's crazy, it's blood and guts. And the real loser is the consuming public."

Why doesn't the United States solve the problem by licensing fewer boats for the whole year? Asked to change the system in 1982, aides to President Reagan replied, "We're opposed to limiting fishing by licenses. It would interfere with basic economic liberties." The fisheries are the last frontier, where anyone with a boat can go out twice a year and catch a few fish to supplement the family income. The local schoolteacher objects to licensing, as do fish processors, who would have to hire workers year round if the catch to be sold and frozen arrived throughout the year.

Sixty percent of America's domestic fish comes from Alaskan waters, although the quality is so poor that this accounts for only 50% of the value of fish caught in the country. The fisheries are managed under the Magnuson Act of 1976 by the North Pacific Fishery Management Council. The Council is made up of people from the fishing industry and government. Consumers and environmentalists are not represented. In 1992, the Council finally voted to recommend restricting those who could fish for halibut to those who have been in the business for at least three years, but even this modest suggestion for improvement had not been approved by the Secretary of Commerce by 1993.

---

Consider one ton of fish caught in the rich fishery off the west coast of Africa by a European factory ship. Converted to fish meal, this ton of fish produces less than a quarter of a ton of meat after it has been fed to livestock in Europe. It produces still less if it is used as fertilizer to grow grain for animal feed. Were the same ton of fish fed directly to people on the coast where it was caught, it would increase the protein content of their diet by 50%—which would make the difference between health and malnutrition for most of them. It is no wonder that the developing nations have been stubborn in their demand that any future international treaties on the use of ocean resources must guarantee their fair share (Section 12-H).

### Managing Our Fisheries

Thirteen major international commissions monitor the world's fisheries. Some merely advise national control agencies, others have power (through treaties) to enforce their recommendations. Local entities, such as states, may then add their requirements to the national regulations.

Fishery commissions aim to manage the world's fisheries for sustained harvests. They control such things as the size of mesh in fish nets or the size of caught fish. Most of these regulations are designed to prevent fish from being caught before they reach reproductive size. Regulations also control what fish may be caught where and when, and these rules change from year to year as managers monitor the fish stocks.

Despite some successes, we still have a long way to go. For instance, the weight of fish caught each year in the North Atlantic stopped declining in the mid-1960s. But the species caught have changed since then. Mackerel and herring were once the main catch. Today smaller species, such as hake, sand eels, and mullet, with occasional larger cod and haddock, make up the bulk of the harvest. Apparently heavy fishing of mackerel and herring has altered the ecosystem. Now that fewer of these fish are present, other fish, whose larvae were once eaten by the mackerel and herring, are surviving in larger numbers.

This story highlights a major problem in fishery management: only a few species are commercially valuable. For instance, when anchovy populations in the eastern Pacific declined, the anchovy fishery collapsed. Hundreds of fishing boats from California and western South America went out of business, although there were still plenty of fish—of other species—to be caught. The high price of a few species also leads to appalling waste. A shrimp boat catches dozens of species other than shrimp. All these animals, many of them dead, are shoveled back into the sea from the shrimper's deck when the net is pulled up and the catch sorted. One of the problems here is that we are so conservative in what we eat, largely as a result of ignorance. Grouper may cost $15 a kilogram in the American market, whereas equally tasty squid, tilefish, and triggerfish, caught in the same net, sell for pennies as fish bait or fertilizer. (The species mentioned in this chapter are illustrated in Figures 12-5 and 12-15.)

## Reducing Waste

There are several ways to prevent so much waste of ocean resources. Educating public taste is one of them. If the public would eat more different species, a fishery could be managed by taking into account the fact that as one species declines, the species it feeds on will become more numerous. Fish farming is also beginning to reduce pressure on the ocean's fisheries (Section 15-E). Many fish, shellfish, and crustaceans are now being produced in fish farms more cheaply than they can be harvested from the ocean.

Another part of the solution is to continue to design nets that do not kill species caught accidentally that are not destined for market. Shrimp nets that cannot trap porpoises and **turtle excluder devices** that push sea turtles away from the mouth of a net are recent steps in this direction in the U.S. shrimp fisheries (Figure 12-16).

In the Pacific, fishing boats from Japan, South Korea, and Taiwan have long used huge nylon-mesh **drift nets**, nets up to 50 kilometers long that hang like a curtain to a depth of 15 meters from floats on the surface. Drift nets have been dubbed "walls of death" because thousands of dolphins, whales, turtles, and birds drown every year after being trapped in them. Drift nets that are lost or abandoned become "ghost nets," which float in the ocean for years, trapping and killing fish and other animals. Drift nets are being phased out slowly under pressure from the small island nations of the Pacific, whose fisheries have been decimated by these nets.

## Monitoring Ocean Productivity

Phytoplankton, the ocean's producers, are not evenly distributed. Population explosions occur at different times in different places and are not even constant from year to year. This has been known for years and has always made it difficult to predict where the fish that feed on the phytoplankton will be found. Now photographs taken from satellites are improving our ability to track plankton populations. An ocean color sensor can detect the concentration of the pigment chlorophyll $a$, found in phytoplankton, by the wavelength of light that the chlorophyll absorbs.

Although oceanographers have known for years that plankton is patchily distributed, the satellite pictures reveal that the ocean's productivity varies enormously within short distances, even on the generally productive continental shelves. In the early 1980s, NASA distributed maps showing the boundaries between plankton-rich and plankton-poor areas to fishermen on the U.S. West Coast. They found that albacore catches were highest along these boundaries, suggesting that albacore are attracted to the leading edges of currents dense with phytoplankton, where

**Figure 12-15** Some of the commercially important species mentioned in this chapter.

they feed on the zooplankton and small fish that eat the phytoplankton. Techniques such as this are enlarging our understanding of ocean productivity and increasing our ability to manage fisheries wisely.

## 12-F MINERAL AND ENERGY RESOURCES

Until the mid-twentieth century, international law defined the jurisdiction of each country as extending 3 miles from its shore. (Distances at sea are usually expressed in **nautical miles**; see Appendix A.) Anything beyond this 3-mile limit was international water and governed by a different body of law. Until 1946, most of the continental shelf was of little interest to any-

one except fishermen and oceanographers. In that year, the United States claimed the oil and other mineral resources in the shelf around its coast. For legal purposes, it defined the continental shelf as extending to the point where the sea was 600 feet deep.

Since that time, exploration of the sea floor has proceeded apace. Since the oceans cover most of the globe, it is not surprising that reserves of fossil fuel and minerals as large as any found under the land have been found there. The difficulty, of course, is exploiting these reserves. Established technology permits oil rigs to extract oil from beneath the continental shelf at depths of up to about 150 meters. Dredging is also an important technique, used to mine many minerals, from sand and gravel to dia-

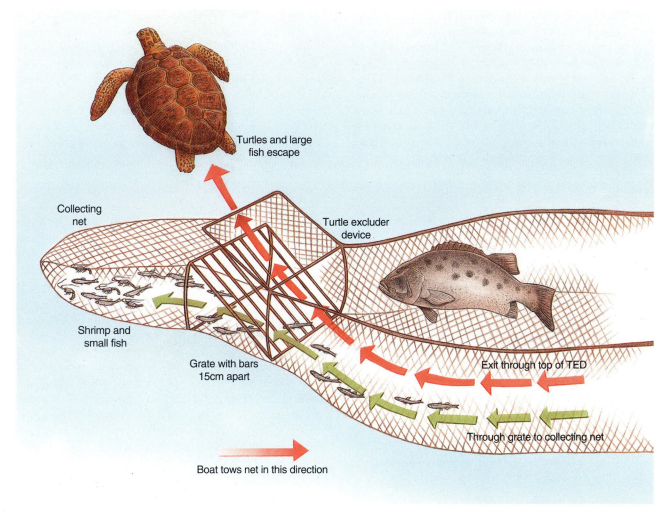

Turtles and large
fish escape

Collecting
net

Turtle excluder
device

Shrimp and
small fish

Grate with bars
15cm apart

Exit through top of TED

Through grate to collecting net

Boat tows net in this direction

**Figure 12-16**   One type of turtle excluder device (TED) used in shrimp nets to prevent them from capturing and drowning endangered sea turtles. The TED places a grate over the collecting net through which shrimp can pass. Above the grating is a hatch through which large objects such as turtles, large jellyfish, and fish can escape. TEDs work. The number of turtles captured in nets has fallen dramatically since their use was mandated in 1992.

monds, from depths averaging only about 30 meters. Exploration and extraction in deeper waters are still in their infancy.

> The ocean is a potential source of minerals and energy. The technology to exploit these resources is only now being developed.

## 12-G POLLUTION

"A river of polystyrene cups and bits of plastic stretch across the ocean. There isn't a clean spot in the At-

lantic from Bermuda to the African coast," reported a small-boat sailor who made the journey.

The ocean has a vast capacity to receive waste, but waste disposal has to be carefully controlled if it is not to damage ocean ecosystems. The questions are: how much waste can the oceans absorb, where is it best deposited, and how long will the waste take to decompose? Little research is being done on these vital questions. We find out the answers only piecemeal, often after it is too late. The most important facts we have discovered are that at least 85% of ocean pollution arises from activities on land and that 90% of these pollutants do not wash into the open sea but remain near land. The pollutants are the same ones that cause problems on land. For instance, excessive plant nutrients washing off the land as runoff from fertilized fields may cause blooms of algae, some of which are poisonous (Close-Up 12-3).

A modern addition to the list of substances that pollute the oceans is radioactive waste, particularly from nuclear power stations. Until 1970, much of this waste was dumped in deep water in drums and glass containers that have subsequently leaked.

Each country is responsible for controlling the pollution that reaches the seas around its coast. Even in developed countries, this control is often inadequate. However, when tourists and surfers find they are swimming in sewage and abandon the area, coastal towns usually see the light and start to clean up the water they discharge. Industrial waste and sewage discharged into rivers is the biggest source of coastal pollution in most areas of the United States. Shortsighted cities, enticing industry by lax pollution standards, then find the local shrimp or oyster fisheries collapsing and eventually change their ways. Fortunately, rivers and coastal ecosystems have remarkable powers of regeneration, even though they sometimes take a long time to recover.

> Most pollution of the ocean comes from the land and remains in coastal waters. It can be controlled only by preventing pollutants from reaching the ocean.

My cousins and I learned to swim in the Atlantic Ocean off the coast of Wales in the United Kingdom. We always returned home with tar on our towels and swimsuits from the tar-spattered rocks along the coast. People who grew up on the East Coast of the United States remember the same sticky feeling. The tar came from oil tankers, which cleaned out their tanks by washing them with sea water. When the volatile part of oil evaporates, globs of tar are left to wash up on beaches. We hear about birds and other marine creatures killed by the oil, but populations of these creatures usually recover. More insidious is the damage done by toxic and carcinogenic components of oil that persist for long periods of time. International rules have now been written to reduce this pollution, and tar on coastal rocks and beaches is much less common than it once was.

## 12-H  INTERNATIONAL COOPERATION

The International Convention for the Prevention of Pollution from Ships (MARPOL) limits the amount of oil a ship can discharge. In enclosed areas such as the Baltic and Mediterranean seas and the Persian Gulf, discharge is prohibited. To trace illegal dumping, oil cargoes are sometimes "fingerprinted" with additives so authorities know the source of a particular spill. As a result of these regulations, beaches are now largely free from tar, despite the ever-increasing volume of oil carried in tankers. MARPOL's controls over the dumping of wastes at sea are shown in Figure 12-17.

MARPOL has also tackled the problem of the huge amounts of garbage dumped at sea. Better ways to disperse sewage sludge and to keep toxic chemicals out of industrial discharges have all been developed. Incineration tankers offer an alternative to ocean dumping of waste. Polychlorinated biphenyls (PCBs) are toxic compounds once used in the plastics industry that can be disposed of in this way. If heated to high temperatures, PCBs break down into water and hydrochloric acid, which can be disposed of safely. Polystyrene and plastic garbage was once largely a product of commercial shipping. Now, it comes increasingly from recreational boaters. In the sea, polystyrene breaks up into tiny pellets, which are now to be found throughout the world's oceans, from Antarctic icebergs to Pacific deep-sea trenches. Only better public education and the use of degradable plastics can solve this problem.

> International regulations have greatly reduced the amount of oil and other pollutants dumped at sea by commercial ships.

The Helsinki Convention in 1974 was the first to address pollution of the oceans from land-based sources as well as from ships at sea. After this convention, the United Nations Environmental Program set up a Regional Seas Program, supported by 120 nations. This has produced remarkable examples of cooperation between nations, where countries as unfriendly to each other as Israel and Libya, Iran and Iraq, the United States and Cuba have been able to sit down together to resolve common problems.

### Cleaning up the Mediterranean

The first cleanup effort under the Helsinki program was the Mediterranean Action Plan, involving 17 nations and the monitoring efforts of 80 marine research stations. The Mediterranean is particularly fragile because its coasts are home to some 50 million people, and it is almost landlocked so that it is flushed clean very slowly. Its only access to the open

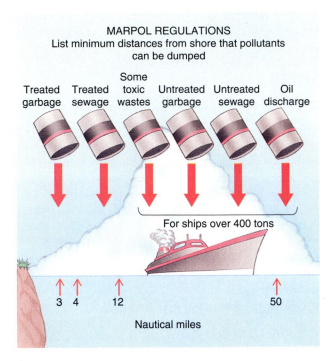

**MARPOL REGULATIONS**
List minimum distances from shore that pollutants can be dumped

Treated garbage | Treated sewage | Some toxic wastes | Untreated garbage | Untreated sewage | Oil discharge

For ships over 400 tons

3 4 12 50
Nautical miles

**Figure 12-17** The International Convention for the Prevention of Pollution from Ships (MARPOL) prescribes the distances from shore at which particular pollutants can be dumped. The distances are shown here. The regulations for ships cover only ships larger than 400 tons. Only 25 nations have signed the MARPOL agreement.

ocean is through the Strait of Gibraltar, which is shallow and only about 20 kilometers wide. For years, the Mediterranean had been becoming filthier and less productive. Now two lists of pollutants have been drawn up. Countries are completely forbidden to dump highly toxic substances, such as DDT, cadmium, and mercury, anywhere they might eventually reach the Mediterranean. Less toxic substances require special licenses before they can be dumped, and their discharge must be limited. A 1982 agreement set up 15 marine refuges to protect endangered species, such as turtles, monk seals, and Aduoin's gull.

> The United Nations has sponsored collaboration between countries that must work together to reduce pollution of the ocean they share.

The problem of ocean pollution once seemed impossible to solve because it was an international problem. The Regional Seas Program has shown that people can cooperate across international lines on difficult problems, particularly if agreement can be

reached before nations with powerful economic interests realize they might be threatened by it. The Law of the Sea is the story of what happens when greed intervenes.

## The Law of the Sea

Since time immemorial, international law has permitted nations to exercise complete control over their **territorial waters**, extending 3 miles from the coast. The rest of the world's oceans were **high seas**, open to everyone. In the twentieth century, some countries claimed to extend their territorial waters to 12, and occasionally to 200, miles from their coasts, usually to protect their fisheries from competition. But the freedom of the high seas remained unchallenged until the 1960s, when it became obvious that technology was permitting some nations to take over more than their fair share of resources in the deeper ocean far from the coast.

The idea evolved that the depths of the oceans belonged to nobody but should be treated as the "common heritage of mankind," administered by an international body such as the United Nations. This philosophy prompted the Third United Nations Conference on the Law of the Sea (UNCLOS), which met from 1973 to 1982 and was attended by more than 150 nations. The hope was to produce a "constitution" covering every conceivable issue involving the oceans, including pollution, fisheries, navigation, the deep-sea bed, and research. Unhappily, this high hope failed. The conference split over the very notion that inspired it, that the deep-sea bed might become the common heritage of humanity. The known resources of the deep-sea bed include such things as nodules rich in manganese, an expensive, nonrenewable resource. We still do not have the techniques to exploit such resources economically. The developed nations are most likely to produce and use such technology and they refused to share the potential wealth, while the developing nations refused to give up their claim to what may one day be an important resource.

UNCLOS was not a complete loss, however. The final agreement was signed by 134 countries. Although the idea was a heroic one, it may be that making UNCLOS an all-or-none affair was a mistake. A nation had to reject or to accept the whole package, although there were many parts of the law to which all parties could have agreed. Although UNCLOS may never become international law, it has clarified a confused situation, and much of the conference's

# Algal Blooms

The Japanese and Irish eat algae; luxury spas wrap it around their customers. But algae also clog water intakes and choke fishing nets. A beach covered with algae is slimy and smelly and sends tourists running.

Some algae are microscopic, but others are quite large, and these are usually known as seaweed. Seaweeds are usually attached to the bottom of the ocean by root-like structures known as holdfasts (Figure 12-F).

Population explosions of algae are known as "blooms," and algal blooms are sometimes signs that the ocean is in trouble. Its temperature or chemical balance has altered. The algal blooms that emptied Italian beaches in 1991 were caused by nutrient pollution: sewage pouring into the sea from rivers provided mineral nutrients that algae need to grow and reproduce. Huge blooms formed and washed up on the beaches.

Every year, more and more algal blooms occur around the coast of Florida (Figure 12-G). Here the probable culprits are nutrient pollution from fertilizer and sewage and the fact that most of Florida's fresh water no longer reaches the sea. South of the Everglades, the ocean is becoming increasingly salty as the fresh water that once flowed through the Everglades to the ocean is diverted to irrigate farmland and supply cities with water. As a result, the coral reefs in Florida Bay are dying, choked by algae, silt, and excess salt.

On Boynton Beach, crews attempted to clean up the algae using tractors with rakes. It did not work. The algae kept washing ashore. The crews described the algae as "looking like globs of slimy worms and smelling like old socks."

Not what the tourists were looking for!

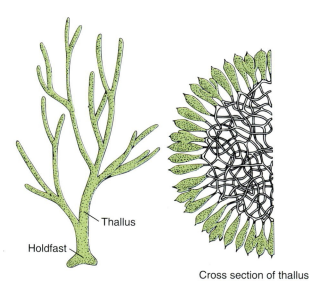

Thallus

Holdfast

Cross section of thallus

**Figure 12-F**  *Codium,* the green alga that is clogging Florida beaches.

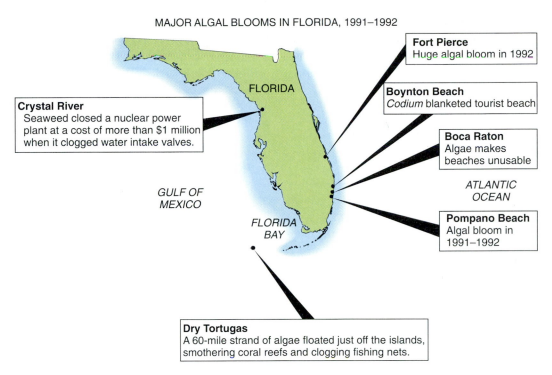

MAJOR ALGAL BLOOMS IN FLORIDA, 1991–1992

**Fort Pierce**
Huge algal bloom in 1992

**Boynton Beach**
*Codium* blanketed tourist beach

**Boca Raton**
Algae makes beaches unusable

**Pompano Beach**
Algal bloom in 1991–1992

**Crystal River**
Seaweed closed a nuclear power plant at a cost of more than $1 million when it clogged water intake valves.

FLORIDA

GULF OF MEXICO

FLORIDA BAY

ATLANTIC OCEAN

**Dry Tortugas**
A 60-mile strand of algae floated just off the islands, smothering coral reefs and clogging fishing nets.

**Figure 12-G**    Sites of major algal blooms in Florida in 1991 and 1992.

work is finding its way into treaties covering various aspects of ocean law.

The UNCLOS agreement placed more than 40% of the oceans under the jurisdiction of the coastal nations, defining four zones of influence. The **territorial sea** extends 12 miles from land. The **contiguous zone** runs to 24 miles out, and the **extended economic zone** to 200 miles or more. The fourth zone is the continental shelf. The territorial sea is completely controlled by the coastal nation, which has limited rights over the contiguous zone. The nation also has certain rights over economic activity, research, and environmental preservation in its extended economic zone. The sea bed of the continental shelf may be exploited as long as this does not infringe on the legal status of the sea and air above it. This means, for instance, that Brazil can prevent other nations from drilling for oil on its continental shelf as long as it does not remove navigation markers at the surface of the sea or try to prevent foreign airplanes from flying in the international airspace above.

UNCLOS retains the traditional freedom of the seas for 40% of the ocean surface waters. But the sea bed under most of this area is designated communal property to be controlled by the International Seabed Authority. The United States, Belgium, Germany, Italy, France, the United Kingdom, and Japan all enacted legislation that conflicts with the authority of the International Seabed Authority, permitting each of them to issue its own licenses to people who want to mine the sea bed.

> UNCLOS modernized many aspects of maritime law but failed to produce agreement on rights to resources on the deep-sea floor.

## Whaling and the Southern Ocean

Most fishing takes place in the Northern Hemisphere, but huge, underexploited resources of fish exist in the stormy and dangerous Southern Ocean that surrounds Antarctica. Although the area contains enormous plankton populations, there is almost no continental shelf, and many species of fish swim too deep for fishing nets to reach them.

The Southern Ocean was the main site of whaling, an industry of romance and infamy. Originally the source of oil for lamps and whalebone for corsets, twentieth-century whaling yielded mainly meat and dog food, from efficient factory ships that brought their beautiful and intelligent prey to the verge of extinction. All species of cetaceans (whales and dol-

phins) are at least somewhat protected from hunting by international agreements. Several species are still declining, however, probably because so few of them are left that some of them cannot find mates and reproduce.

The largest whales feed on **krill**, shrimp-like planktonic crustaceans. Since the whales have declined, krill populations have increased until there may be 600 million tons of these creatures in Antarctic waters at certain times of the year. Several countries now harvest krill for human food, and more plan to do so. However, we have seen in the case of herring and mackerel that it is not enough merely to manage one species. Each species is part of a food chain, so its fate affects many other organisms. Krill are the food for many other species, and their harvesting must be managed with this in mind.

## The Arctic and Antarctic

The ice-covered ends of Earth are very different from each other. The Arctic is an ocean, largely frozen for most of the year. The Antarctic comprises a vast unfrozen ocean surrounding an ice-covered continent twice the size of Western Europe.

Much of the Arctic lies over the continental shelves of the surrounding land masses. It supports some of the richest fishing grounds, accounting for about one tenth of the global catch. Because the Arctic Circle encloses only an ocean and the fringes of populated countries, its ocean resources are controlled by laws that govern oceans and coastal areas elsewhere in the world. The Antarctic is different, largely because it has never been colonized by human populations.

By 1943, seven nations (Argentina, Chile, Australia, Norway, France, New Zealand, and the United Kingdom) claimed parts of Antarctica on the grounds of proximity or because their citizens had explored it. Thirty years later, the Antarctic Treaty was signed by these claimants as well as by Japan, the United States, the Soviet Union, South Africa, and Belgium. Since 1959, the treaty powers have managed Antarctica as an area devoted to peaceful research in an interesting example of international cooperation. Another 13 countries have agreed to abide by the treaty and can observe meetings. While permitting many kinds of research, the treaty powers also administer agreements designed to conserve Antarctica's natural resources. One agreement, for instance, protects Antarctic seals.

The treaty group is a small one and meets in secret. Developing nations charge that the group is a last stand of capitalist colonialism, designed to keep an international resource for the benefit of developed nations. Conservationists promote the idea that Antarctica should be a world wildlife preserve. But nations that covet Antarctica's mineral reserves are unlikely to agree to this. Pending a long-term solution, Antarctica needs some short-term measures to prevent further depletion of its resources: a freeze on fishing quotas and sanctuaries for whales and seals. Above all, many more nations need to participate in the management decisions that affect this great asset.

Many people believe that the treaty powers are not protecting the Antarctic with sufficient care. Scientists complain that tourists disturb the Antarctic wildlife. Tourists retort that research stations in Antarctica have polluted many areas by improper waste disposal. In a 1993 lawsuit brought by the Environmental Defense Fund, a federal court ruled that standards that apply to waste disposal in the United States also apply to U.S. operations in Antarctica and ordered U.S. research stations in Antarctica to clean up their act. Permits to explore for minerals in Antarctica have also recently been given out, and many people believe these do not provide sufficient protection for the environment.

## SUMMARY

The ocean contains resources, many still to be exploited, that can be managed only by international cooperation.

### Our Ocean Resources

The ocean fuels the water cycle and supplies us with minerals, oxygen, recreation, and a garbage dump. But it is so hard to study that less is known about the ocean than about any of our other resources. Marine animals provide nearly one quarter of the protein in the human diet, but most species are declining as a result of overfishing. Major efforts to protect the oceans from pollution, habitat destruction, and overfishing have emerged in recent years, some of the most successful being those of the recreational fishing industry.

### The Open Ocean

Continental shelves are the most fertile part of the ocean. The ocean depths contain fewer organisms. Phytoplankton live on continental shelves and in areas of upwelling, where there are light and nutrients. Phytoplankton provide food for zooplankton and nektonic animals, which, in turn, feed larger fish, marine mammals, and birds.

### Reefs

Reefs are among the most productive ecosystems in the world because of the variety of habitats they provide. Because they lie in relatively shallow water, which is usually near land, reefs have been exposed to massive destruction from nearby human communities.

### The Edges of the Ocean

The edges of the ocean include rocky, sandy, and muddy shores. Coastal wetlands usually form in estuaries. Here fresh water from a river meets salt water, and currents cause sediment to fall to the bottom. On sandy coasts, barrier islands may form when the sea level rises to form estuaries. Salt marsh and mangrove swamps are highly productive ecosystems that develop in estuaries. These wetlands are important as the breeding ground for many marine species. Coastal wetlands are destroyed by dredging, building, dumping of waste, pollution, and conversion into agricultural land. In some areas, they are being preserved, restored, and even created from scratch.

### Depleted Fisheries

Catches of many species have declined because of overfishing, although we are probably still not harvesting the sustainable yield of the sea. One reason is the enormous wastage of marine species for reasons including (1) lack of consumer demand for many species, (2) the use of wasteful, destructive drift nets, and (3) management of many fisheries for the convenience of powerful interests instead of for sustainable yield. Modern long-distance fishing boats have depleted local fisheries that once supported local populations, particularly on the coasts of developing nations. International, national, and local commissions attempt to manage fisheries for sustainable yield in all parts of the world, with varying success.

**Mineral and Energy Resources**

We are only beginning to develop the technology to exploit the minerals and energy that the oceans can supply. Control over these potential resources causes conflict between developed and developing nations and was responsible for their failure to reach agreement at the United Nations Law of the Sea conference.

**Pollution**

People have assumed that the ocean could absorb infinite amounts of waste. The destruction of ocean ecosystems, fisheries, and beaches has shown that this is not true. Efforts are now being made to control the dumping of oil and plastic waste by ships, to remove sewage and industrial chemicals from waste water that reaches the ocean, and to make the disposal of radioactive waste less hazardous.

**International Cooperation**

Under the auspices of the United Nations, great strides have been made toward cooperation between countries that need to work together if they are to clean up and manage the ocean that they share. A program to clean up the Mediterranean has had considerable success. The Law of the Sea conference declared the ocean depths the common heritage of mankind and clarified international maritime law, although the most powerful developed nations did not sign the conference agreement. There is hope that international cooperation will permit multispecies management for sustainable yield of the underexploited fisheries of the Southern Ocean. The Arctic is an ocean. Antarctica is a continent managed by an international agency and used for research. Many think it is not being managed well.

## DISCUSSION QUESTIONS

1. Why can we not use the distant depths of the ocean as a dump for toxic wastes that are difficult to dispose of?

2. Suppose you sit on the commission of a coastal community. Draw up a list of regulations for use of the local beach and sand dunes that will ensure these resources are still usable in the future.

3. Should all ocean dumping of waste be banned? Why or why not? If so, what would you do with the waste that is now dumped at sea (for instance, that from New York City)?

4. Banning ocean dumping will not eliminate pollution of the oceans. What else needs to be done?

5. What would you include in an advertising campaign that would convince people to eat species of seafood to which they are not accustomed?

6. If you were in charge, what steps would you take to reduce the damage done by a particular fishery when the damage includes killing endangered species and dumping garbage and nylon lines and nets in the sea?

# MINERAL RESOURCES

# Chapter 13

**Key Concepts**

- Minerals are nonrenewable resources that are extracted from the ground and processed for their final use.

- Obtaining minerals causes damage from mining and pollution from waste and processing.

- Depletion of minerals leads to the use of more and more energy to process low-grade ores. This raises the cost of minerals, promoting recycling and substitution.

- Plastics, which are petroleum products, have replaced minerals for many purposes, and many functions can be fulfilled by more than one mineral.

*Limestone pinnacles, Badlands National Monument, South Dakota.* (Richard Feeny)

Since the first prehistoric potter collected clay from a riverbank to make a bowl, people have depended on minerals. Earth is pockmarked with ancient mines and heaps of the debris dug out of them. Since the industrial revolution, our use of minerals has increased rapidly. From 1750 to 1900, the world population doubled, while use of minerals increased tenfold. For instance, we now produce 22,000 times as much pig iron, from which steel is made, as in 1700. Instead of scratching shallow holes with stone tools, today we use explosives and massive machines to mine minerals used in almost every aspect of our lives.

A **mineral** is a naturally occurring inorganic solid with a specific chemical composition and crystalline structure, extracted from Earth's crust. The term is also often used to include anything that is extracted from the ground, including organic substances such as fossil fuels. Some minerals, such as sand and gravel, are used as they are found. Others, such as metal ores and raw diamonds, are processed before they are used. An **ore** is the mineral in which a substance occurs naturally.

Mineral resources are generally divided into four groups:

1. Metals, such as iron, aluminum, and copper.

2. Industrial minerals, such as lime and potash, that are valued for their particular chemical properties.

3. Construction materials, such as sand, stone, and gravel, that are used to build such structures as roads and buildings.

4. Energy minerals, such as uranium.

In this chapter, we consider only the first three groups of minerals.

Extracting minerals from the earth and processing them does enormous environmental damage. When we worry about how to dispose of waste, we think of garbage from households and waste from factories. Yet mining produces three quarters of all waste (Figure 13-1). Mining produces as much soil erosion as all Earth's rivers combined. Air pollution from the mineral industry has made tens of thousands of hectares of land uninhabitable and causes thousands of birth deformities and illnesses every year.

When people talk of reusing and recycling minerals such as aluminum, they usually assume that this will prevent us from running short of minerals. In fact, a much more important reason for conserving minerals is that it permits us to reduce environmental damage from extracting and processing minerals.

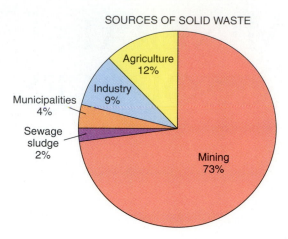

**Figure 13-1**    The sources of all the solid waste produced in the United States. Municipal garbage is mainly waste collected by garbage trucks from homes and businesses in a community.

The environmental effects of mining are poorly reported and regulated. For instance, manufacturers in the United States are required to report how much toxic waste they emit each year (Section 20-C). The mining industry, although it is a huge producer of toxic emissions, is not required to report them.

> Using minerals does less environmental damage than extracting and processing them.

## 13-A  THE MINERALS WE USE

During the course of history, we have discovered and learned to use an increasing number of substances. Today industry depends heavily on approximately 80 minerals. Some of these are listed in Table 13-1.

### Prehistoric Mineral Use

When we think of mineral processing, we usually think of steel furnaces, and indeed many useful substances, including nonmetals, are extracted from their ores by heat. These include silicon, copper, tin, and lead. Many minerals occur near the surface of Earth in many places and are not particularly difficult to extract (Figure 13-2). Even primitive technology can produce furnaces with quite high temperatures, using charcoal for fuel and a bellows to produce a current of air. As a result, many of these substances have been used from early times.

The first metal used was probably "native copper," a metal that occurs naturally. The ancient Su-

**Figure 13-2**  A small gold mine on the Salmon River in Idaho. A pipe draws water out of the river for use in the mine. The wooden chute dumps waste rock into the river. *(Thom Smith)*

**TABLE 13-1**  **Estimated Worldwide Production of Some Minerals in 1991\***

| Mineral | Production (in thousands of tons) | Main Producing Countries |
|---|---|---|
| **Metals** | | |
| Pig iron (for steel) | 531,000 | Soviet Union/CIS†, Brazil |
| Aluminum | 18,500 | Australia, Guinea |
| Copper | 9,100 | Chile, U.S. |
| Zinc | 7,400 | Canada, Australia |
| Manganese | 6,700 | Soviet Union/CIS, China |
| Chromium | 3,800 | South Africa, Zimbabwe, Soviet Union/CIS |
| Lead | 3,370 | Australia, U.S. |
| Nickel | 953 | Soviet Union/CIS, Canada |
| Tin | 210 | China, Brazil |
| Molybdenum | 82 | U.S.A, China |
| Tungsten | 39 | China, Soviet Union/CIS |
| Silver | 14 | Mexico, Peru |
| Mercury | 6 | Algeria, Spain, Mexico |
| Gold | 2 | South Africa, U.S. |
| Platinum-group metals | 0.3 | South Africa, Soviet Union/CIS |
| **Nonmetals** | | |
| Stone | 11,000,000 ⎫ | *Usually* |
| Sand and gravel | 9,000,000 ⎬ | *produced* |
| Clays | 500,000 ⎭ | *locally* |
| Salt (sodium chloride) | 186,000 | Canada, Bahamas, Mexico |
| Phosphate rock | 160,000 | Peru, U.S. |
| Potash (potassium compounds used mainly as fertilizer) | 160,000 | Canada, Israel, Germany |
| Lime | 135,000 | *Most countries* |
| Gypsum | 98,000 | Canada, Mexico, Jamaica |

†CIS = Commonwealth of Independent States, successor to most of the land area of what was the Soviet Union.

\*Some strategic minerals, those that are in relatively short supply but of vital importance for industrial and military uses, are highlighted.

merians, in what is today southern Iran, heated copper ores with wood fires. The ores were not purified, and sometimes a mixture containing both tin and copper was heated. Heating such a mixture produces **bronze**, an alloy of tin and copper that is hard and resists corrosion. Bronze objects from nearly 5,000 years ago have been found in Egypt.

Lead, which can be obtained from its ores by heating them with charcoal, was also known to the ancients. The water systems of the Hanging Gardens of Babylon are said to have been lined with lead, and the Romans used lead for water pipes because it is soft and does not crack easily. We now know that lead is poisonous. Lead poisoning causes mental deficiency, blindness, and many deaths, especially in children. Historians have even suggested that the Roman Empire collapsed because her citizens were poisoned by the lead in their drinking water. Lead is still found in many water pipes and faucets worldwide because we have only recently discovered how dangerous it is, and substances that can be used in place of lead in water systems are still not satisfactory for all purposes.

### Modern Mineral Use

Modern civilization is completely dependent on minerals. We use them to make irrigation pipes and fertilizer, to make machinery, to build transport systems, and to produce household goods (Table 13-2). The importance of minerals in modern life can be

gauged from the amount of energy used to produce them. Some 20% of all energy used in the United States goes to extract and process minerals.

Industry depends heavily on a number of minerals that are relatively plentiful, such as aluminum and iron. However, there are a few minerals that are critically important to industry, and particularly for military uses, but that are in relatively short supply. These are known as **strategic minerals**. Chromium, for instance, is a strategic mineral that not only makes shiny automobile bumpers but also is part of alloys vital to the production of tools and jet engines. Other strategic minerals are manganese, an essential component of high-grade steel; cobalt, used in alloys for the aerospace industry; and platinum, used in communications equipment and catalytic converters for pollution control.

> Strategic minerals are those vital to industry but in relatively short supply.

### 13-B  MINING

The largest excavation in the world is Bingham Canyon copper mine in Utah, which is 775 meters deep. About 3.3 billion tons of material have been removed from this hole in the ground, about seven times as much as was moved to build the Panama Canal.

In surface **strip mining**, the topsoil is first scraped off, then the **overburden**, the rock lying above the mineral, is removed by huge shovels called draglines to expose the mineral (Figure 13-3). This results in huge piles of waste. This waste consists of overburden and **tailings**, the material that is left after the mineral has been extracted. In the United States, surface mining produces about eight times as much waste as underground mining. But deep mining can produce even worse problems. It can even cause earthquakes. When underground mines cave in, not only do they kill miners but they also cause subsidence of the surface, forming holes into which roads and houses may collapse.

### Waste

In 1991, about 990 million tons of ore were mined to produce only 9 million tons of copper—and this does not include the overburden removed from above surface mines. As minerals near the surface are depleted, the amount of waste produced increases as miners have to dig deeper to find the mineral. A study by the National Academy of Sciences predicted

| **TABLE 13-2**    **Some of the Minerals Most Important to Industry** | |
|---|---|
| **Mineral** | **Some Uses** |
| Diamonds | Industrial abrasives, jewelry |
| Chromium | Tool making, jet engines |
| Copper | Electrical wiring, alloys |
| Iron | Steel making |
| Aluminum | Structural material, packaging, fireworks |
| Gold | Jewelry, money, electric circuits |
| Platinum | Electrodes, jewelry, cancer chemotherapy, catalysts in pollution control and fertilizer synthesis |
| Silver | Jewelry, photography |
| Zinc | Metal alloys, electrodes |
| Sulfur | Paper making, photography, food additive, paint, explosives, pesticides, pharmaceuticals, rayon |
| Sand | Concrete, road making, silicon, glass |

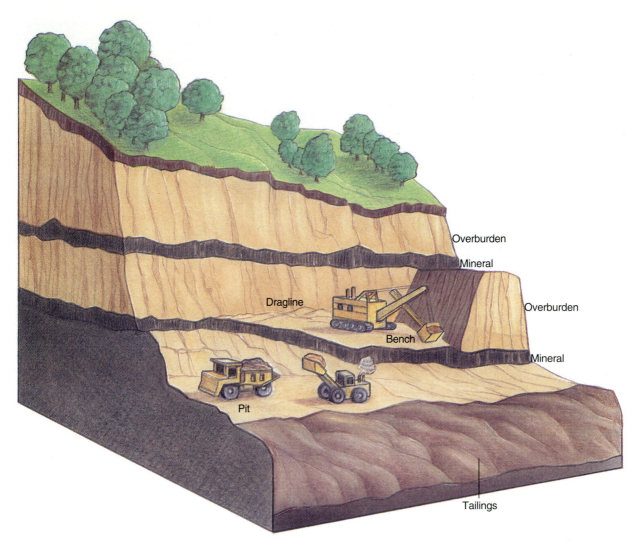

**Figure 13-3**  The structure of a surface mine.

that copper mining in the year 2000 would produce three times as much waste for each ton of copper produced as did copper mining in 1978.

Piles of debris from mines often appear stable for many years, while they are in fact being undermined by streams or by the earth subsiding. Terrible landslides of waterlogged mine waste have wiped out whole communities.

## Pollution

Tailings, in particular, often contain toxic materials, which leach out to pollute nearby waterways. Tailings and mines in Eastern coal mines leak sulfuric acid ($H_2SO_4$). The acid is formed from water, air, and iron sulfides in the mines. The acid runs off into streams, where it destroys wildlife just as acid rain does and corrodes bridges, ships, and locks. It also

dissolves toxic elements such as copper, zinc, arsenic, aluminum, and magnesium from tailings and the soil and carries them into waterways and even ground water, often making the water unfit to drink (Figure 13-4).

## Soil Erosion

In the United States, environmental laws now require that companies mining on public land must restore the land to its previous state when they have finished (Equal Time 13-1). Miners have been heard to complain that this means they have to plant sagebrush instead of more useful vegetation when they fill in a surface mine. Where it applies, the law prevents the devastation that mining has created in, for instance, parts of Pennsylvania and Wales that were mined for coal in the nineteenth century. The trouble is that

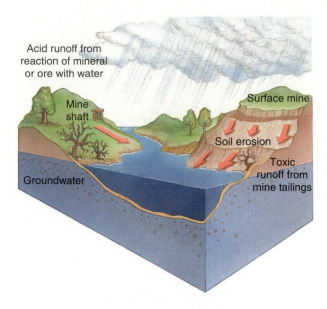

**Figure 13-4** Pollution of water sources by runoff from surface and underground mines.

regulations requiring land reclamation do not apply to all mines, and they are not universally enforced. As a result, mining in many places still does enormous damage to the countryside and destroys the habitat of many animals and plants.

Until recently, the overburden from surface mines was merely dumped downhill, causing landslides and erosion. As with the lumber industry, mines also

| TABLE 13-3 Effect of Building Dirt Roads on Soil Erosion* | |
|---|---|
| **Site** | **Volume of Soil Lost (cu m/sq km/year)** |
| **Alder Creek, Oregon** | |
| Uncut forest | 45 |
| Clear cut area† | 117 |
| Dirt road | 15,565 |
| **Coast Mountains, Southwest British Columbia** | |
| Uncut forest | 11 |
| Clear cut area | 25 |
| Dirt road | 283 |

*Data from Swanston, D.N., and Swanson F. J. "Timber harvesting, mass erosion and steepland forest geomorphology in the Pacific northwest." In Coates, D. R., ed. *Geomorphology and Engineering.* Stroudsburg, PA: Dowden, Hutchinson & Ross, 1976.

†A forest is clear cut when all the trees in an area are felled. The ground still has some vegetation cover, however, because herbs and shrubs in the undergrowth survive.

**Figure 13-5** Erosion on a dirt road: the road has deep gullies caused by water running down the road and carrying the soil into a nearby river. *(Thom Smith)*

cause erosion because dirt roads are often built to reach them. Dirt roads are a major source of soil erosion and landslides (Table 13-3; Figure 13-5).

Many landscapes are dotted with "borrow pits" left after a small deposit of clay, sand, or gravel has been dug out of the surface. These pits often fill with water and are dangerous and unsightly. Some municipalities now make a virtue of necessity by converting these pits into small recreational lakes for canoeing, swimming, sailing, and windsurfing.

Enormous environmental damage has resulted from extracting and processing minerals.

## 13-C PROCESSING MINERALS

Few minerals are found in a form in which we can use them. Most have to be processed in various ways. Although methods vary with the mineral involved,

# EQUAL TIME 13-1

## Oh, Give Me A Home Where the Subsidies Roam?

In 1991, *The Washington Post* published an article by Jessica Mathews that criticized the federal government for subsidizing mining on publicly owned lands, including national parks and national forests. Mines have done great damage to public lands, polluting nearby rivers and destroying ecosystems (Figure 13-A). Mathews blamed the "antiquated" Mining Law of 1872, which allows anyone to claim mining rights on public land for $5 an acre, which is much less than the price for which mineral rights on private land would be sold.

When minerals are found, the mining company pays a royalty to the federal government for minerals produced and receives a tax credit for the royalty payment. The tax credit reduces the amount of tax collected from the company, which then has to be made up by other taxpayers. This is another subsidy of the mining company by the general public. Why, asked Mathews, is the general public paying for its own lands to be mined for the profit of private companies?

- Barbara Vucanovich, Republican representative from Nevada, replied:

- The Mining Act of 1872 sounds out of date, but in fact it has been amended many times. It is designed to encourage private enterprise to search for minerals. For instance, when the federal government tried to find uranium deposits to make bombs after World War II, it failed. When uranium was put under the control of the Mining Law, an army of prospectors

**Figure 13-A**   A gravel pit on National Forest land in Montana. Is this an appropriate use of public land?

combed the West and found hundreds of deposits.

- The law does not guarantee the right to mine public lands. Control of land destruction and pollution control are still in the hands of federal agencies.

- The $5-per-acre fee applies only when a valuable discovery of minerals has been shown. It generally takes large sums of money to prove that you have found a mine that is worth developing.

---

the extraction of two metals illustrates the types of processes involved.

## Aluminum and Gallium

Aluminum and gallium are chemically related metals. Almost 20 million tons of aluminum were used worldwide in 1991, in thousands of products from saucepans and automobiles to deodorants and furniture. Aluminum is used as a structural material because it is light and resists corrosion. The main use

for gallium is to make semiconductors, used in computers and photovoltaics. Computer circuits based on gallium arsenide operate at speeds up to five times as fast as those of the silicon chips generally used today, and they operate at a wider range of temperatures.

Both aluminum and gallium have to be extracted from ores. Aluminum is found combined with silicon and oxygen, often in clay deposits. As these ores weather, they break down to various forms of aluminum oxide ($Al_2O_3H_2O$). A mixture of these oxides forms the ore of aluminum, **bauxite**.

**Figure 13-6**   Energy and the supply of minerals: many small gravel pits dot the countryside because gravel takes a lot of energy to transport and so it is usually dug near where it is used. This machine crushes gravel from a nearby pit for use on a road that is being rebuilt.

Aluminum is extracted from the earth in three stages:

1. Raw bauxite is dug out of the earth, either from a surface mine or an underground mine.

2. The ore is processed to concentrate the aluminum oxide it contains. First the bauxite is treated with hot concentrated caustic soda (NaOH), which dissolves any silicon present as well as aluminum oxide. Any of the bauxite that does not dissolve is filtered out. Then the solution that remains is treated with carbon dioxide, which causes aluminum oxide to precipitate.

3. The aluminum is purified by extracting it from the solid aluminum oxide by **electrolysis**, separation by an electric current. Aluminum oxide dissociates in water to form aluminum ions ($Al^{3+}$) and hydroxide ions ($OH^-$). When these ions are exposed to an electric current, the positively charged aluminum ions are attracted to the negative electrical terminal, where they precipitate as pure aluminum.

Pure aluminum is seldom used to make things because it is soft and weak. To improve its properties, it is mixed with other metals, such as copper and manganese, to form alloys (mixtures of metals). Quite a lot of aluminum is never purified but is used in the form of one of its ores. For instance, crystals of aluminum oxide, known as **corundum**, are very hard. Corundum is used as an abrasive in grinding wheels, sandpaper, and toothpaste. Some precious stones are impure aluminum oxides. Sapphires, valuable blue gems, are aluminum oxide containing iron and titanium as impurities. Rubies are aluminum oxide contaminated with small amounts of the metal chromium. They are prized as the most valuable red gem, are used in some lasers, and form the jewel bearings in watches and instruments. Rubies occur naturally, and they are also made artificially. Some 200,000 kilograms of artificial rubies are produced each year.

Table 13-4 shows the potential environmental damage from extracting and processing a mineral such as aluminum from a large volume of ore.

Production of gallium is entirely different from that of aluminum and has much less effect on the environment. Gallium does not occur as a mineral in large concentrations. It is found in tiny amounts associated with other minerals. It is extracted in small amounts as a by-product during the recovery of metals such as zinc and aluminum. As a result, gallium is expensive. One gram of pure aluminum cost about 20 cents in 1993, whereas 1 gram of gallium cost about $4.

## 13-D  ENERGY AND MINERAL SHORTAGES

Minerals are generally regarded as nonrenewable resources. Mineral deposits that can be extracted economically are formed so slowly that their formation is of no practical use to us.

At the rate we are using minerals, Earth's reserves of aluminum will last for about 200 years, whereas reserves of tin, zinc, and lead will last for only 20 to 30 years. Although 20 to 30 years does not sound long, there is little danger of the world running out of tin, zinc, or lead. All are plentiful in Earth's surface, and new reserves are constantly being discovered whenever the price of a mineral makes exploration economical—which usually means that the mineral is in relatively short supply.

In one sense, the factor limiting our supply of minerals is energy (Figure 13-6). For instance, it has been estimated that any 1 cubic kilometer of rock contains tons of aluminum, zinc, iron, and copper. If we had unlimited energy, we would never run short of these metals because we could extract them from ordinary rocks, from air, soil, water, and even from our own garbage landfills. But we do not have unlimited energy.

It takes more energy to extract a mineral from a low-grade (less concentrated) deposit than from a high-grade (concentrated) deposit of ore. The

**TABLE 13-4** **Environmental Effects of Extracting and Processing Minerals***

| Activity | Potential Effects |
|---|---|
| Digging a mine and removing ore from it | Surface mining: destroying habitat, farmland, and residences<br>Underground mining: sinking of land above the mine (subsidence)<br>Soil erosion caused by disturbing soil; silting of lakes and rivers<br>Production of waste (material removed from mine and not used)<br>Toxic drainage into waterways when water runs through mine waste (usually acid sulfur compounds and toxic metals) |
| Concentrating the ore | Production of piles of tailings, which are the waste materials usually left near a mine<br>Organic waste added to tailings when organic chemicals are used to concentrate ore<br>Toxic drainage from waste piles into waterways |
| Purification (by chemical methods or smelting and other high-temperature processes) | Air pollution by emissions from smelter, including arsenic, lead, sulfur dioxide, cadmium<br>Production of more waste, such as the slag that is left after smelting<br>Energy use (most of the energy used in extracting minerals is used in smelting and refining) |

*Source: Worldwatch Institute.

amount of energy used to extract minerals can only increase as we use up easily mined high-grade ores.

At the moment, no mineral that is industrially important is in critically short supply. Experts foresee future difficulties with the supply of some 18 minerals, even when more efficient recovery and recycling are taken into account. These are the minerals whose use is increasing most rapidly. They include lead, sulfur, tin, tungsten, and zinc. If the amount of a mineral used each year increases, no amount of recycling will provide enough for future use. Recycling can merely slow down the rate at which reserves of the mineral are depleted.

## 13-E WORLDWIDE USE AND DISTRIBUTION OF MINERALS

Reserves of important minerals are not equally distributed around the world. The minerals in an area depend on the rock formations, and particular types of rock are simply absent from Earth's surface in some places. As we might expect, the industrialized countries use more than their fair share of minerals. The 25% of people in developed countries use 75% of the world's mineral resources.

Many developing countries contain reserves of minerals that they export. The United States, Europe, and Japan import most of their strategic min-

erals (such as chromium, manganese, platinum, and cobalt) from developing nations (Figure 13-7). Developing nations need to export more than just minerals, however, to have stable earnings from exports. The price of a mineral can change rapidly as new reserves are discovered, so the income from exporting minerals tends to be erratic. The foreign policies of importing and exporting nations take these facts into account. For instance, in 1966 the United States joined in international economic sanctions against Zimbabwe because of its racist internal policies. Imports from that country were banned. In 1971, when supplies of chromium imported from Zimbabwe ran low, that policy was abolished.

### Stockpiles

Many nations stockpile minerals, particularly those that they must import. The United States imports all the titanium it uses, 97% of its manganese, and more than 90% of its chromium, largely from developing countries. It has accumulated stockpiles of these minerals, as well as of bauxite, tin, palladium, and platinum. Stockpiles are designed to ensure that a country will not run short of minerals, particularly those needed for military purposes. Stockpiles prevent political upheavals in the exporting country and disruptions in the mineral supply by international cartels; they also reduce fluctuations in a mineral's price.

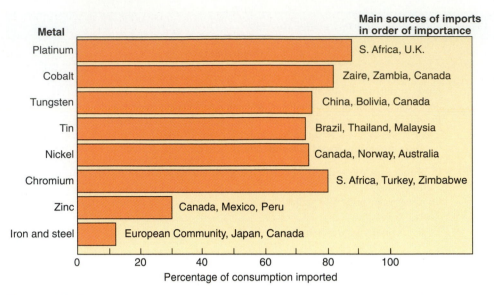

**Metal**

**Main sources of imports in order of importance**

- Platinum — S. Africa, U.K.
- Cobalt — Zaire, Zambia, Canada
- Tungsten — China, Bolivia, Canada
- Tin — Brazil, Thailand, Malaysia
- Nickel — Canada, Norway, Australia
- Chromium — S. Africa, Turkey, Zimbabwe
- Zinc — Canada, Mexico, Peru
- Iron and steel — European Community, Japan, Canada

Percentage of consumption imported

**Figure 13-7** U.S. imports of some minerals. The country depends on imports for many of the strategic minerals it uses. *(U.S. Department of Commerce)*

## 13-F RECYCLING

Every year a city the size of San Francisco throws away more aluminum than is produced by a small bauxite mine and more copper than a medium copper mine. Disposing of solid waste is a worldwide problem (Chapter 18). As it increases, and as the prices of most minerals rise, recycling becomes increasingly attractive. Experts believe that almost half of the world's metal will be recycled by 2010.

**Recycling** is the use of previously used materials for new purposes. Recycling is suitable mainly for "pure" products, those that are not mixed with many other substances in the final product and those that are not degraded when they are used. Thus, aluminum and glass, which are not mixed with other products to make cans and bottles, are easily recycled, whereas paper, which is often printed with ink and coated with plastic, is often more difficult to recycle. In the case of mixed or degraded products, such as things made of several types of plastic, the cost of separation and recovery exceeds the value of the recycled product. Recycling is built into many aspects of the economies of most developed nations.

Recycling is often very cost-efficient. For instance, making new aluminum from scrap aluminum uses 95% less energy than extracting it from the earth and processing it. It also reduces the air and water pollution associated with aluminum production by 95%. Recycling steel saves energy, air pollution, water pollution, and the production of mining wastes that

must be disposed of. Roughly one fourth of the steel used in the world comes from recycled scrap steel. The United States leads the world, with nearly half of the steel it uses made from recycled material (Figure 13-8).

Compared with virgin steel produced by mining and processing, steel produced by recycling uses 40% less water and creates more than 75% less air pollution, water pollution, and mining waste. Recycling steel is big business in the United States. The number of people employed in the industry rose from 87,000

**Figure 13-8** The United States leads the world in steel recycling. This graph shows how the share of steel produced in the United States from scrap metal has risen since 1960. *(U.S. Bureau of Mines, American Iron and Steel Institute)*

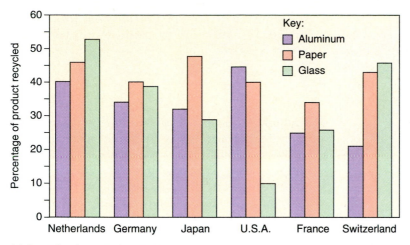

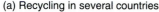

(a) Recycling in several countries

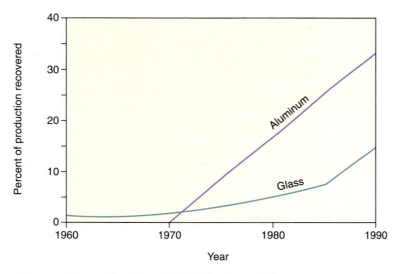

(b) Increase in recycling in the United States since 1960

**Figure 13-9** Recycling from municipal garbage. (a) The percentage of various substances that are recycled in different countries. Paper is not a mineral but a biological resource. It is included here because paper, glass, and aluminum are the substances most commonly recycled by households and municipalities. (b) The enormous increase in recycling in the United States since 1960. These graphs do not include recycling by manufacturers, which adds even more to the total.

in 1986 to nearly 120,000 in 1990. The volume of steel that can be recycled is limited by the fact that the large amounts of steel used in automobile parts cannot be recycled because copper has been added to the steel. If automobiles and other products were built without copper and other elements that limit recycling, even more steel would be recycled. An industrial society of the future, with a stable population, could have an almost sustainable steel industry, with nearly all its steel recycled and enormous reductions in energy use and pollution.

In the United States, many states have **container laws** that encourage the recycling of aluminum and glass beverage containers. Nevertheless, less than half of the aluminum we use is recycled (Figure 13-9). We could save money and resources by increasing this percentage.

Recycling becomes more widespread and cost-effective as reserves of minerals are depleted.

## 13-G CONSERVATION AND SUBSTITUTION

Two ways to make mineral reserves last longer than they otherwise would are conservation and substitution—using less of a mineral and substituting something else for it. Sometimes this happens anyway as technology develops. Small computers with plastic cases have replaced rooms full of metal computers. Plastics in general have replaced metals for many uses in modern society. This increases the reserves of metals (although it depletes the reserves of petroleum used to make plastics).

## Conservation

We could encourage conservation of minerals by producing and using products that last for a long time so they do not need to be replaced often. This is not likely to happen in the case of products made from minerals. Shoddily made products, which will not last for long, are cheaper to produce than sturdily made items with longer lives. But a low price is only one reason that we buy cheap stereo systems, toasters, computers, and other products. A second reason is that technology changes so fast today that we imagine we shall want to replace our present appliances with more advanced models in a few years' time. It seems unlikely, then, that conservation will ever form a major part of our efforts not to run out of mineral resources.

> We are unlikely to practice widespread conservation of minerals.

## Substitution

Manufacturers often stop using a mineral that is depleted and becoming expensive by substituting another mineral in the production of a product. In the nineteenth century, the most important structural materials were steel, brick, and wood. Now these have been replaced by substances as diverse as aluminum, fiberglass, concrete, titanium alloys, ceramics, and plastics. The compounds that most often substitute for inorganic minerals today are plastics and fibers, artificial organic molecules made from the chains of carbon and hydrogen atoms in petroleum.

> Plastics, made from petroleum, have taken the place of other mineral products in many aspects of modern life.

**Synthetic Polymers**   Plastics are manufactured polymers. A **polymer** is any molecule made from repeated subunits called **monomers**. Figure 13-10 shows the repeating structure of cellulose, a natural polymer found in all plant cells, and of polystyrene, a synthetic polymer. Polymers are the molecules of life. The proteins, carbohydrates, and nucleic acids that make up our own bodies are polymers. When chemists set out to make polymers, they start with the natural polymers and monomers in petroleum, the remains of ancient living things. You can see from Figure 13-10 that the backbone of a natural or a synthetic polymer is a string of carbon compounds joined together.

Synthetic polymers include polyvinyl chloride (PVC) used to make records and plumbing pipes; polyethylene for garbage bags and bottles; fiberglass to make cars and boats; and Dacron, nylon, Orlon, and polyester fibers used to make fabrics. Polymers have taken the place of natural substances such as wood, cotton, leather, and bone in thousands of products. The United States now uses more plastic than steel, aluminum, and copper combined. About 80 kilograms of synthetic polymers are made every year for every person in the United States.

Our increasing dependence on plastics makes the fossil fuel shortage look even more critical, for our

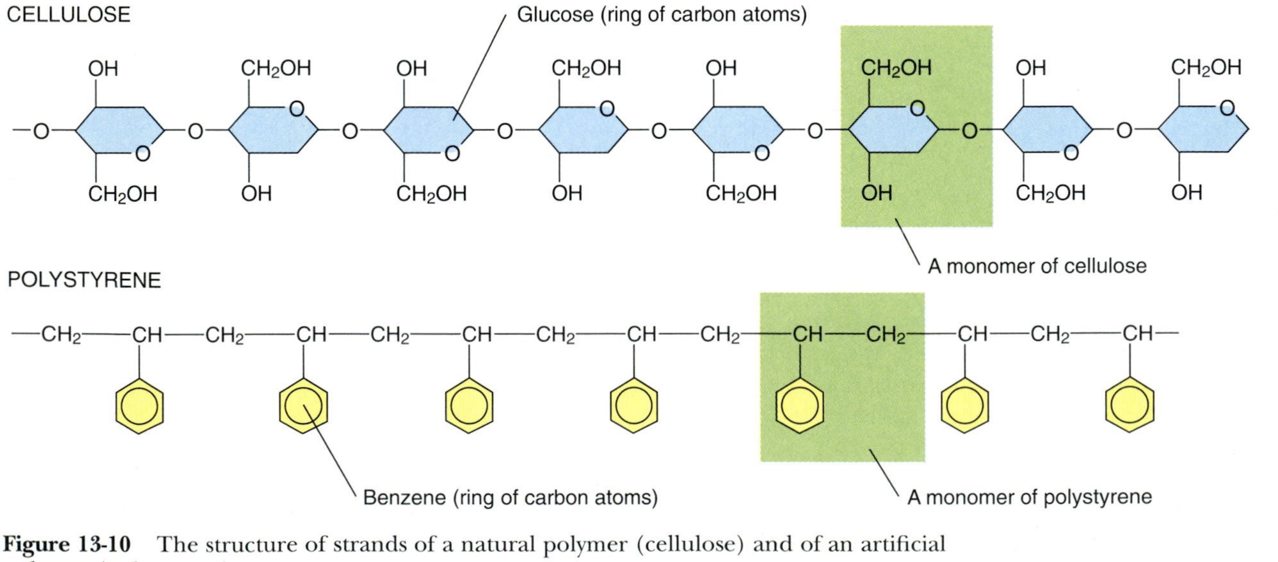

**Figure 13-10**   The structure of strands of a natural polymer (cellulose) and of an artificial polymer (polystyrene).

supplies of petroleum are limited. We should remember, however, that petroleum is a mixture of organic compounds and that similar molecular structures can be obtained from today's plants. We can produce fuel and polymers from various crops.

Although plastics cause environmental problems, the replacement of minerals and natural products by plastics has many benefits. These include protecting endangered species and producing consumer goods that are inexpensive enough for most people to afford. Piano keys were once made of ivory from elephant tusks. African elephants are threatened today by the demand for ivory. They might be extinct by now if plastics had not replaced ivory for piano keys. Similarly, plastics have partly replaced metals in many consumer products. These products are inexpensive enough for most people to buy partly because they contain plastic instead of metal.

Polymer science is beginning to imitate biology in making new materials. Consider tendons, the structures that hold muscles to bones in our bodies. Tendons are made up almost entirely of collagen, a protein. Collagen by itself is not a particularly strong molecule. A tendon gets its strength from the way the collagen is organized into fibers. When a tendon is subject to great stress, individual fibers break into fibrils. This breakage absorbs energy and prevents the whole tendon from failing until it has sustained enormous damage. Polymer scientists are now learning to organize synthetic polymers into materials of this kind, which are stronger than the individual polymer sheets or fibers.

**Ceramics** For thousands of years since the first pot was fired, people have taken natural substances and converted them into other substances with particular uses. Only recently has material science begun to discover why treating materials in certain ways gives them particular properties. Today for the first time, we are learning to work the other way round: to identify a need and then to develop a material that can fill that need. This ability will help free us from dependence on particular minerals as they become depleted. One example is the widespread use of ceramics.

Ceramics are one group of materials that are being designed for specific purposes. **Ceramics** include glass and glass-like materials made from silicon nitride powder (whose raw ingredient is sand). Ceramics are now used for things as diverse as telescopes for astronomy, turbochargers, nose cones on space rockets, mechanical seals, automotive valve guides, and ovenproof dishes.

## SUMMARY

Industry depends on supplies of nonrenewable minerals, which are extracted from Earth's surface.

### The Minerals We Use

The use of metals for tools and clay for pots is thousands of years old. The United States uses so many minerals that about one fifth of all energy used in the country goes to mineral production. Among the most important minerals are strategic minerals, those that are vital for industrial and military purposes.

### Mining

Mining is destructive to the environment. It generates most of the waste produced in the United States, some of it toxic, so runoff from mines often pollutes water supplies. Piles of mine waste, digging underground mines, and building dirt roads to mines cause landslides, subsidence, and erosion. Mining uses large amounts of energy and destroys large areas of land.

### Processing Minerals

After they are extracted, minerals usually need to be processed before they can be used. For instance, metals are usually extracted from ores by heating and treatment with various chemicals or by electrolysis.

### Energy and Mineral Shortages

It takes more energy to extract minerals from low-grade than from high-grade deposits. Therefore, as minerals are depleted, the amount of energy needed for extraction increases. Energy prices determine whether or not it is economical to extract a mineral from a particular deposit.

### Worldwide Use and Distribution of Minerals

Because minerals are not evenly distributed in Earth's crust, countries have to obtain particular minerals from other countries. The international trade in minerals is an important factor in the foreign policies of nations. Many nations stockpile minerals that they must import to cushion their industries against fluctuations in price and supply.

### Recycling

Recycled minerals become increasingly important as resources are depleted. Recycling is cost effective, particularly for many metals. It saves energy, water, and money and reduces pollution caused by mining and processing. Almost half of the world's metal will be recycled by 2010.

### Conservation and Substitution

Conservation of products made from minerals is not a practical method of increasing reserves appreciably, partly because advances in technology make it desirable to replace many products frequently. Industries react to price changes of minerals by substituting a cheaper mineral for a more expensive one whenever possible. Plastics have been widely substituted for several metals in the last 40 years. Ceramics are also often used in place of minerals and are now being designed for specific purposes.

## DISCUSSION QUESTIONS

1. Why is it difficult to estimate the supply of minerals available to us?

2. One way that governments encourage the extraction of minerals by mining is by "depletion allowances" that give the producer a tax reduction as the mineral is depleted. It seems contrary to environmentalist principles to encourage depletion by tax breaks. What do you think?

3. How can the second law of thermodynamics be used to predict (correctly) that it is seldom economical to extract minerals from sea water or to mine increasingly low-grade deposits of minerals?

4. What are the arguments for and against insisting that all strip-mined areas should be restored to their former state when the mine is exhausted? Can you think of any other ways of making a strip-mined area useful and attractive again?

# AGRICULTURAL LAND

# Chapter 14

## Key Concepts

- A shortage of fertile agricultural land threatens our ability to feed the human population.

- Sustainable farming requires that the soil's nutrient level and physical structure be continually restored. The only way we know to achieve this is to add organic matter to the soil.

- Vast areas of land are lost from agricultural production every year because the soil has eroded or become so saline that crops will not grow in it.

- Irrigation is the main cause of soil salinization.

- Desertification of millions of hectares of arid land every year results from deforestation, overgrazing, and overcultivation.

- Degraded land can be restored to productivity, but this takes many years.

- Measures that reduce the rate of topsoil erosion include reforestation, planting windbreaks, efficient irrigation, and no-tillage farming.

*Plowing produces a dust storm over a field and road in California.* (Richard Feeny)

*"When our fore-Fathers settled here...the Land being new, they depended upon the Natural Fertility of the Ground...and when they had worn out one piece they cleared another."*

—Jared Eliot, Connecticut minister, doctor, and farmer, 1748

As this quotation suggests, farming has steadily reduced the fertility of the soil in most parts of the world. Fertile soil is one of our most important natural resources because it grows the food we need.

Only about 15% of Earth's surface can be farmed because the rest is made up of ocean, permanent ice, steep rocky slopes, and deserts. About half of this 15% is already farmed. The rest consists mainly of tropical forest and of land without enough rainfall to grow crops. Fresh land is brought into production every year, but every year an almost equal area is lost to production when it is built on or when its soil becomes so infertile that plants can no longer grow in it. In this chapter, we consider what makes farmland productive and what we must do to ensure that we have enough land to grow the crops we need in the years to come.

## 14-A  FARMLAND AND SOIL

Agricultural land can be divided into **arable land**, land on which crops can be grown, and **rangeland**, land better suited to animal grazing (Figure 14-1). Rangeland is usually land where grass and shrubs grow but where the soil is thin so that it tends to erode, wash or blow away. When we talk of arable land, we generally mean land that can be used to grow our most important food plants, **grains** (also known as **cereals**), which are members of the grass family, such as wheat, rice, and barley.

**Fertile soils** are those in which most plants are healthy and grow rapidly. Fertile soil is composed of rock particles, decaying organic matter, living organisms, water, and air. A plant's roots, buried in the soil, support the plant and draw in water, in which mineral nutrients are dissolved. Most plant roots grow in **topsoil**, the loose surface layer of soil. Topsoil usually lies on top of **subsoil**, which consists largely of tightly packed rock particles that only the toughest roots can penetrate. Subsoil contains much less organic matter and air than topsoil. Under the subsoil lies **bedrock**, solid rock that may break down to form soil over long periods of time (Figure 14-2).

### Soil Structure

Rocks contain most of the minerals that plants need, trapped within the rock crystals. Soil is produced from rock by weathering. **Physical weathering** means the breakdown of rock by erosion and weather, which break the rock into smaller and smaller particles. The largest particles in soil are grains of **sand**, smaller particles are called **silt**, and the smallest of all are **clay** particles. Sandy soil contains large pores so water drains through it rapidly. It tends to contain plenty of oxygen, which plant roots need, but it dries out rapidly. Clay particles are so small that the spaces between them hold water by capillarity (Section

**Figure 14-1**  Arable land and rangeland in the western United States. The tractor is plowing flat arable land for a crop. The hillside behind is rangeland used for grazing cattle. This land has low rainfall, so its soil is thin and easily washes or blows down the hills if the vegetation cover is removed. The soil on the hills would rapidly erode away if it were plowed. *(Thom Smith)*

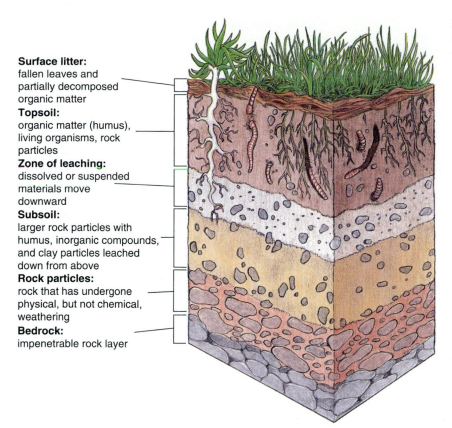

**Surface litter:**
fallen leaves and partially decomposed organic matter

**Topsoil:**
organic matter (humus), living organisms, rock particles

**Zone of leaching:**
dissolved or suspended materials move downward

**Subsoil:**
larger rock particles with humus, inorganic compounds, and clay particles leached down from above

**Rock particles:**
rock that has undergone physical, but not chemical, weathering

**Bedrock:**
impenetrable rock layer

**Figure 14-2** The structure of fertile soil. The number and details of the soil layers may be different in different types of soil.

11-A). As a result, clay soil usually contains plenty of water but little air. The best soil for plant growth contains a mixture of particle sizes. Its clay particles hold water; its larger particles assist drainage and allow air to penetrate the soil.

Fertile soils contain living organisms and a great deal of organic matter: dead plants and animals, animal excrement, and microorganisms (Table 14-1). Many of the soil organisms are decomposers that release nutrients from organic matter as they convert it into **humus**, which is partly decayed organic matter. In addition, the organisms in soil cause **chemical weathering**, reduction in the size of soil particles by chemical reactions. Many organisms excrete acids, and their respiration produces carbon dioxide, which dissolves in water to produce carbonic acid (Section 3-B). Acids attack soil particles, dissolving minerals from their edges. The soil particles get smaller, and the dissolved minerals can be taken up by plant roots.

As well as releasing mineral elements from rock, chemical weathering alters the chemistry of soil particles. As rock dissolves, it releases **cations**, positively charged ions (Section 3-B). This leaves the remaining rock particle with a negative charge. In sandy soils, most of the cations wash away, but in clay, the particles are closer together, and the cations remain in the soil spaces, attracted to the negatively charged particles (Figure 14-3).

Living organisms are so important to the formation of soil that it may take 10 million years for 1 centimeter of soil to form in dry, cold areas where soil organisms grow and reproduce slowly. These include the high slopes of mountains in temperate zones and the prairies of the northern plains states and Canada. In contrast, 1 centimeter of soil may form in a year or less in warm, wet tropical regions where organisms grow rapidly.

| TABLE 14-1 Number of Organisms in Average Farm Soil | |
| --- | --- |
| **Organisms** | **Quantity** |
| Insects | 670 million/hectare |
| All anthropods (including insects) | 1.8 billion/hectare |
| Bacteria | 1 billion/gram |
| Algae | 100,000–800,000/gram |
| Earthworms | 1.8 million/hectare |

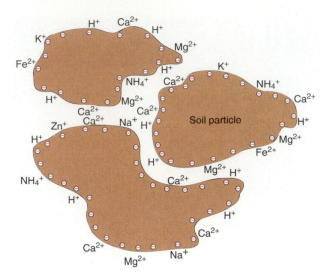

**Figure 14-3** Clay particles, bearing negative charges, with some of the cations they attract.

Organic matter has many functions in soil. When it falls on the soil surface as **litter**, it insulates the soil from the heat of the sun and reduces water loss by evaporation (Figure 14-2). Organisms in soil act as a store of mineral nutrients, and some of them produce specific nutrients that plants need. For instance, some bacteria can fix nitrogen from the air into nitrates that plants can use (Section 5-D). These nitrogen-fixing bacteria grow in nodules on the roots of legumes—the family of plants that includes peas, beans, and acacias. Planting legumes adds nitrogen nutrients to the soil.

Fertile soil is actually a series of complex ecosystems, containing lots of organisms. Although we hardly ever notice them, these soil organisms are essential to agriculture and indeed to the life of most other living things.

> Fertile soil contains a mixture of particles, air, water, and large amounts of organic matter, including live organisms.

### Soil pH

Plants can absorb water and salts through their roots only if the pH of the soil is within a certain range. Recall that an acid releases hydrogen ions ($H^+$) when it dissolves (Section 3-B). Roots add acid to the soil when carbon dioxide from respiration dissolves in the soil water. A hydrogen ion in this acid may take the place of a cation attracted to a soil particle, releasing a cation, such as calcium, magnesium, or

potassium, that the root may absorb. This process is called **cation exchange**. Very acidic soils, however, are not very fertile. Here hydrogen ions displace most of the other cations from soil particles so rapidly that the displaced cations are washed, or **leached**, out of the soil. In contrast, the soil in dry areas tends to be alkaline and contain few hydrogen ions. There are plenty of salts in the soil, but they are not dissolved in water so plants cannot use them. Cation exchange is slow, and the soil is not fertile.

Some mineral nutrients are more soluble in acid and some in alkaline soil water. In acid soil, aluminum and manganese can become so soluble that they reach toxic levels. Most plants do best at a slightly acid soil pH of 6.5 to 7, at which most minerals are fairly soluble. However, because some soils are more acid than others, plants that can grow in both acid and alkaline soil have evolved. An alkaline soil of pH 7 to 7.5 suits apples, asparagus, cabbage, and soybeans. Azaleas, blueberries, cranberries, and potatoes prefer an acid pH of 4.5 to 5.5. In general, acid soils are found in areas with high rainfall, such as the East Coast of the United States, where plants grow rapidly and produce acids by respiration and decomposition. Alkaline soils are common in western areas where there is little moisture.

## 14-B HOW CULTIVATION AFFECTS SOIL

One of the oldest agricultural practices is **tilling** the soil, that is, turning it over before planting seeds and turning over the soil around plants as they grow. Tilling has two main purposes. First, it mixes up nutri-

**TABLE 14-2** Soil Compaction by Animals: Rates at Which Water Filters Through Grazed and Ungrazed Lands*

| Site Location | Median Rate of Water Filtration (ml/hour) | |
| --- | --- | --- |
| | **Ungrazed** | **Heavily grazed** |
| Montana | 31 | 5 |
| Oklahoma | 88 | 62 |
| Colorado | 62 | 22 |
| Wyoming | 34 | 23 |
| Louisiana | 46 | 18 |
| Kansas | 33 | 20 |
| Arizona | 41 | 31 |

*Data from Gifford, G.F., and Hawkins, R.H. "Hydrologic impact of grazing on infiltration: A critical review." *Water Resources Research*, 14:305, 1978.

**TABLE 14-3  Soil Compaction by Recreational Vehicles: Changes in Soil Properties after Passage of 100 Motorcycles***

| Soil Property | Direction of Change | Change (%) |
|---|---|---|
| Penetration by water | Decrease | 78 |
| Weight bearing | Decrease | 19 |
| Moisture in soil | Decrease | 17 |

*From a study in New Zealand by Crozier, M.J., Marx, S.L., and Grant, I.J. "Impact of off-road recreational vehicles on soil and vegetation." *Proceedings of the 9th New Zealand Geography Conference*, Dunedin, 76:9, 1978.

**Figure 14-4**  A drainage ditch carries not only water, but also soil from this irrigated field into a nearby river.

ents and loosens soil particles. Second, it gives crop plants a competitive advantage over weeds, which are deliberately disturbed to damage their roots.

Many human practices have the opposite effect to tilling—they compact the soil by exerting pressure on it. For example, the weight of machinery and animals compresses the soil (Table 14-2). This closes the spaces between soil particles, making it harder for roots, air, and water to penetrate (Table 14-3). One common result is that more water than usual runs off the surface, washing soil particles with it and contributing to soil erosion. Soil flowing into waterways is one of the most widespread forms of water pollution (Figure 14-4). Plowing itself can compact the soil, producing a loose surface layer over dense subsoil where the soil particles have been pressed together by the weight of the plow. Recent methods to reduce and reverse soil compaction include adding municipal wastes and other organic matter to the soil and adopting no-tillage cultivation in which machinery moves over the soil less often (Section 14-H).

When a crop is harvested, soil is lost because the crop contains minerals that were once part of the soil. In addition, some soil is usually lost to erosion when the crop is planted and harvested. The U.S. Comptroller General calculates that in the Great Plains of North America, producing one bushel of maize (corn) consumes two bushels of topsoil. If farmers do not constantly renew the soil, the layer of topsoil gets steadily thinner and less fertile.

Farmers improve the soil in many ways. Since prehistoric times, lime (calcium carbonate [$CaCO_3$]) has been added to acid soils to raise the pH. Fertilizers are added to increase the nutrients available to plants. Organic fertilizers, such as manure (a mixture of excrement and straw), have the advantage of improving the structure of soils as well as providing nutrients. Their disadvantages are that they are heavy and expensive to spread, and their nutrient content varies. For these reasons, inorganic fertilizers are more often used.

The mineral elements plants need in relatively large quantities are nitrogen, phosphorus, and potassium. Commercial fertilizers are labeled with the percentage of each of these elements they contain. A "5-10-5" fertilizer contains 5% nitrogen, 10% phosphorus, and 5% potassium, by weight. Despite their convenience, inorganic fertilizers damage the environment in many ways, and their use is decreasing in the United States and many other countries. Fertilizers are expensive, and when too much fertilizer is used it can "burn" plants and kill them. The concentrated chemicals in a fertilizer kill many soil organisms, slowing down the decomposition of organic matter and the formation of new soil. Also, fertilizers running off into waterways are a significant source of water pollution.

## 14-C  IRRIGATION

Since agriculture began, farmers have helped their plants to grow by **irrigation**, the addition of water. Many ancient civilizations were crisscrossed with irrigation ditches. Ancient Egyptian art shows ox-

**Figure 14-5**   An irrigation system. This strip of pipe is mounted on wheels so that it can be moved across the field. Notice that the water blows up into the air, and a lot of it will be lost by evaporation before it even reaches the ground.

powered pumps raising water from the River Nile into fields. In many places where agriculture was traditionally rain-fed, experiments showed that crop yields could be increased as much as sixfold by irrigation. In addition, the world is full of land that can be farmed only if it is irrigated. Some 13% of Earth's arable land is now irrigated, and irrigation consumes more of our freshwater supply than any other use (Figure 14-5).

Without irrigation we could not feed the human population, but irrigation can also destroy the soil. Worldwide, for every hectare of new land brought into cultivation by irrigation, another hectare of already cultivated land goes out of cultivation, its fertility destroyed. The culprit is usually salinization, the accumulation of salts in the soil (Facts and Figures 14-1).

## Salinization

In California's Central Valley, reports dating back more than a century tell of farmland that had to be abandoned because it became too salty for crops to grow. Salts enter the soil naturally as rock particles dissolve, and salts are also present in all the water that falls on the soil, either from rainfall or from irrigation. Salts leave the soil when they are absorbed by plants and when they are washed out into rivers or ground water. Water flowing off irrigated land is often salty and pollutes nearby waterways.

## FACTS AND FIGURES 14-1

### Salinization Around the World

Around the world, fertile soil is rapidly washing away, turning to dust in the wind, or becoming so saline that it is useless, says a three-year study by the United Nations Environmental Program. The report concludes that more than 1.2 billion hectares of fertile land—an area the size of China and India combined—have been seriously degraded. The main culprits are overgrazing, poor agricultural practices, and deforestation. Severe salinization affects 24% of the world's irrigated soil (Table 1). The study concludes that national programs of soil conservation and water management are needed, in dozens of countries, if soil loss is to be slowed.

**TABLE 1   The Extent of Salinization in Selected Countries in 1988**

| Country | Saline Land in millions of hectares | Saline Soil as a % of all Irrigated Land |
|---|---|---|
| India | 20 | 36 |
| China | 7 | 15 |
| United States | 6 | 20 |
| Pakistan | 3 | 27 |
| Soviet Union | 3 | 12 |
| World | 60 | 24 |

**Figure 14-6** Irrigated areas from the air. In the arid western United States, irrigated fields often show up as green areas. This photograph was taken over New Mexico. The circular green areas are irrigated by a device that moves around a central pivot like the hands on a clock. The salty soil appears white when nothing is growing on it.

Arid lands have naturally salty soil. Land is **arid** if more water leaves the ecosystem (from plants and by evaporation from the soil) than enters it. In practice, land is arid if it receives 20 to 25 centimeters of rainfall a year. The soil water becomes steadily more salty until only salt-tolerant plants can live there. The soil is usually alkaline from a buildup of calcium and magnesium carbonate (Figure 14-6).

Irrigation is the most important cause of rapid salinization. When land is irrigated, much of the water evaporates, leaving the salts it contained behind. Calcium and magnesium tend to precipitate, leaving sodium as the main ion in the soil solution. Sodium is absorbed onto clay particles, turning the clay into a cement-like solid that neither water nor roots can penetrate. Worldwide, about 5,000 square kilometers of irrigated land each year go out of production, their fertility destroyed by salinization.

Human practices that alter the position of the water table can also contribute to salinization. Ground water is often more saline than rainfall because it lies on the underlying bedrock, where it dissolves salts. When ground water comes within 3 meters of the surface in clay soils (less for sandy soils), it is drawn up through the soil by capillary action. When it reaches the surface, it evaporates, adding its salts to the soil.

> Irrigation can increase crop yields enormously, but it can also cause salinization that destroys soil fertility.

Irrigation, irrigation canals, reservoirs, and wells can all raise the water table. In areas where there is a layer of clay not far beneath the surface, replacing native vegetation with crops can also cause this problem. Soil water collects on top of clay. Trees and deep-rooted grasses absorb much more of this water than crop plants, and they absorb it from deep in the soil (Figure 14-7). When the native vegetation is replaced by shallow-rooted crop plants, the water table rises, and the soil becomes waterlogged. Few plants can grow in waterlogged soil because it contains little oxygen, which plant roots need for respiration.

The opposite problem may occur when the water table falls, permitting salt water from the sea to penetrate what was previously a reservoir of fresh ground water (Section 11-D). This has occurred in areas such as the coasts of California, Israel, and Japan. One of

**Figure 14-7** Native vegetation of arid land in the western United States. The yellow-flowered shrub is rabbit brush. The plants grow some distance from each other, allowing each plant a larger area from which to collect water. The roots go deep into the soil to reach any water that may be available. If native vegetation is replaced by irrigated crops with shallow roots, the water table may rise.

**TABLE 14-4** Some Causes of Soil Salinization and Waterlogging

| Cause | Effect |
|---|---|
| Irrigation in dry areas | Water evaporates, leaving salts in soil |
| Rising water table as water is added to soil by irrigation or by leaking from reservoirs or irrigation ditches | Water comes close enough to the soil surface to evaporate, leaving salts behind |
| Depletion of ground water near the ocean | Falling water table permits salt water from sea to flow into ground water |
| Replacing native trees and shrubs with crops in areas where water table lies not far below the surface | Crop plants cannot use soil water because their short roots do not reach it, so the water table rises |

the best-known irrigation projects in the world is the Aswân Dam across the River Nile in Egypt. This dam controls the river's annual floods, preventing fresh water from being lost to the sea. The dam, however, has lowered the water table downstream, permitting salt water from the sea to seep into arable land.

Soil salinization has destroyed civilizations. Grain records in southern Iraq show that equal amounts of wheat and barley were grown there in 3500 B.C. Fifteen hundred years later, barley, which can grow in soil containing a fairly high salt concentration, accounted for almost all the crop. Temple records show that the yield per hectare had been more than halved. Cities such as Ur and Uruk, where writing and mathematics were invented, dwindled to mere villages or were abandoned as the soil deteriorated. The main causes of salinization and waterlogging are summarized in Table 14-4.

## Solutions to Salinization

Australian Girl Guides sing as they scramble over the hills of southeastern Australia planting tree seedlings: "We can halt the salt, help heal the land." And so they can. Land that has become so saline that crops will not grow on it can be reclaimed, and planting trees is an important part of the solution, especially if large amounts of water cannot be spared for the project.

**More Water** Salinization can be slowed, if not reversed, by measures such as lining irrigation canals to reduce water seepage into the soil and using salt-tolerant crops. Some of these may even accumulate salts from the soil so it is removed when they are harvested.

Young seedlings are much more susceptible to salt damage than are older plants, and irrigation methods can be improved to take account of this. If the ground is watered heavily before seeds are planted, most of the salts that might damage the seedlings are removed, and then the crop can be lightly watered for the rest of the season so as little salt as possible is added back to the soil. In this context, sprinkler or drip systems, applying only as much water as the plants need, are better than irrigation methods in which water is led to the plants through open channels from which much of the water evaporates.

Another way to reverse salinization is to add more water. When water is added to soil faster than it can evaporate or be absorbed by plants, some of it runs through the soil into the nearest river, leaching salts out as it goes. This solution is expensive: it requires more irrigation water and, usually, drainage ditches so the water will run off the land. It may also pollute the local water supply. Water running into rivers from irrigated land may become so salty that it is unfit for human consumption, for instance, when it contains large quantities of nitrates (Table 14-5).

**TABLE 14-5** Increase in Salinity of River Water as a Result of Irrigation*†

| Location | Salinity Upstream of Irrigation | Salinity Downstream of Irrigation | Increase in Salinity Caused by Irrigation (%) |
|---|---|---|---|
| Rio Grande, Texas | 111 | 631 | 570 |
| Sunnyside, Washington | 40 | 299 | 750 |
| Arkansas River | 212 | 890 | 420 |
| Sutter Basin, California | 72 | 480 | 670 |

*The salinity of the river water is reported as milligrams per liter of $CaCO_3$ upstream and downstream of an area where irrigation water is applied to cultivated fields.

†Data from Hotes, F.L., and Pearson, E.A. "Effects of irrigation on water quality." In Worthington, E.B., ed. *Arid Land Irrigation in Developing Countries: Environmental Problems and Effects.* Oxford: Pergamon Press, 1977.

**Drainage** On the west side of California's San Joaquin Valley, irrigation has raised the water table to within 2 meters of the surface beneath more than 300,000 hectares of fields. Farmers cope with the saline soil and high water table by digging ditches that drain the salty water away into ponds that cover nearly 1,500 hectares of land. As the water evaporates from the ponds, substances drained from the soil become concentrated, and the water becomes toxic, high in poisonous arsenic and selenium. Both are naturally present in the soil at low, nontoxic levels. Before the San Joaquin Valley was turned into farmland, it contained hundreds of thousands of hectares of freshwater lakes and swamps. These lakes were home to millions of birds. Now that the lakes are gone, the birds visit the artificial ponds instead—where many of them die from selenium and arsenic poisoning. Officials say that some of the ponds will have to be closed and filled in eventually, causing the toxic water to back up in the fields and ruin them as farmland. Experts warn that more than 30,000 hectares of land is so full of salts and selenium that it will probably have to be removed from production in the next 50 years.

The solution that farmers favor is a massive drainage scheme that would drain the polluted water into "the ultimate evaporation pond," the ocean. This would be enormously costly and would pollute the ocean, at least locally.

> Soil salinization can be prevented and reversed by careful irrigation, drainage, and reforestation.

### Kesterton National Wildlife Refuge

The most dramatic disaster resulting from saline soil drainage systems occurred at Kesterton National Wildlife Refuge in northern California. Kesterton was polluted by a drain from the San Joaquin Valley that ended in a pond at Kesterton. In 1981, Gary Zahm, Fish and Wildlife manager at Kesterton, noticed that there were no crayfish or frogs in the Kesterton marshes. He called in experts who discovered that the wildlife was being poisoned by selenium and arsenic that had entered many of the plants in the refuge. Birds that ate the seeds of the bulrushes and cattails produced deformed offspring, with missing wings, swollen heads, and misshapen beaks. At first, Californian authorities fired off explosions to keep wildlife away from the toxic ponds and marshes. The final solution? At a cost of more than $25 million, the state blocked the drain and filled the pond with soil. Today, cottontail rabbits and coyotes have moved back to Kesterton and hawks and blackbirds wheel above the marsh.

## 14-D SOIL DESTRUCTION IN AFRICA

There is not enough naturally good arable land in the world to support the human population. So crops are grown and livestock raised on poor farmland, where the soil or rainfall is less than ideal, as in the arid lands and tropical forest that cover a large part of the world. The topsoil of arid scrubland is held in place mainly by the roots of small trees and shrubs, such as the brush of the western United States. These plants grow slowly because the rainfall is low. Nevertheless, arid lands have supported human populations for thousands of years. The practices in the Sahel region of northern Africa are typical of aridland agriculture.

> Soil in arid regions is easily destroyed because it is thin and naturally saline.

In the northern Sahel, less than 35 centimeters of rain fall every year, and the rainfall is erratic in place and time (Figure 14-8). This climate supports shrubs that animals can eat, but there is too little rain for crops. The area has long been the home of nomadic people who travel with their flocks of cattle, camels, goats, and sheep to wherever rain causes young leaves to grow on the native shrubs. This was how the population of northern Somalia and much of Ethiopia traditionally made a living.

In the less arid region farther south, farmers live in permanent villages and plant crops. A number of crops are grown to reduce the risk of one crop failing completely in a dry year. The main crops are drought-tolerant sorghum and millet. More demanding, but more profitable crops such as peanuts (groundnuts) and cotton may be grown as cash crops where rainfall permits. A **cash crop** is a crop that is sold rather than consumed by the grower.

**Fallow periods**, periods when crops are not grown, are a vital part of this type of agriculture. After four or five years of continuous cropping, the land is left idle, or lightly grazed as pasture, for several years to allow the soil to regenerate. During the fallow period, the soil is protected by a covering of vegetation so it will not erode and so nutrients and humus can accumulate. In some areas, *Acacia senegal* shrubs were allowed to invade the land (Figure 14-9). After five years, these shrubs could be tapped for a cash crop of gum made from their sap. Nomads and village

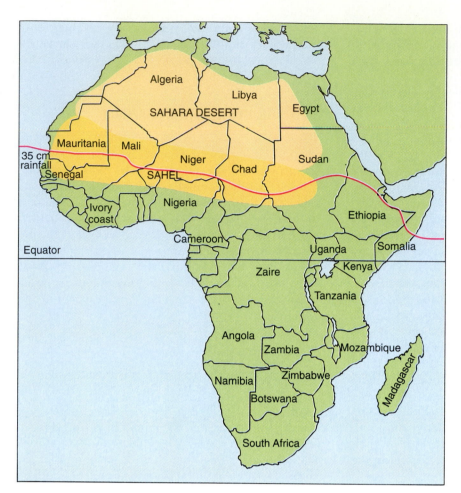

**Figure 14-8**   A map of Africa to show the Sahel region south of the Sahara Desert. South of the red line, annual rainfall averages 35 centimeters, and crops can be grown as long as precautions are taken to prevent soil erosion. In the arid land north of this line, livestock grazing has been the traditional type of agriculture.

farmers cooperated with each other. The nomads brought meat, milk, and camel-loads of salt and other products to the village. Here they traded them for cereals, cash crops, and the right to graze their animals on village fallow land when their usual pastures were exhausted.

## Desertification

The topsoil of arid scrubland is shallow, and the land can support only a low-density human population. As the population density increased, the land was cultivated, grazed, and deforested faster than it could recover. Several crops a year may now be planted until the soil's fertility is exhausted. Trees and shrubs are cut and used as fuelwood or animal fodder faster than they can regrow. More and more animals are put out to graze. The result is that the land becomes denuded of vegetation, leaving nothing to hold the soil in place or to retain any rain that falls. The topsoil blows away or is washed away by the occasional heavy rain and the land turns into desert. The surviving inhabitants become environmental refugees. They move on to add to the population of nearby areas, increasing the chances that the whole cycle will be repeated there.

Although grazing and farming on arid land have supported a modest number of people since prehistoric times, this use has almost always been unsustainable. In northern Africa, even traditional land use with small populations added many hectares each year to the Sahara Desert. Twentieth-century conditions, however, have increased the rate at which land is converted into desert, a process known as **desertification**. There are two main reasons. First, the populations of both humans and livestock have grown rapidly. Second, political and social conditions have led to a call for more cash crops, to be sold abroad for foreign exchange and to feed rapidly growing urban populations. These changes have had numerous results, including the following:

1. *Overcultivation.* Land is farmed so intensively that its productivity declines. Lands traditionally used

**Figure 14-9**  *Acacia senegal.* This is the small spiny tree that produces gum arabic, used to make glues and other products. Its nutritious seed pods are used for animal feed.

for grazing have been brought under cultivation, and fallow periods have been abandoned or shortened. More intensive cultivation may even mean that crop yields fall. In central Sudan, 5 hectares were needed in 1973 to produce the same peanut crop produced by 1 hectare in 1961. In the same time, the number of animals in the region increased almost sixfold, decreasing the area available as arable land.

Overcultivation may exhaust soil nutrients, or soil exposed by plowing may become so crusted, under the influence of rain and sun, that plant roots cannot penetrate it. Exposed topsoil may also wash or blow away. Where native shrubs have been chopped down to expose more land to the plow, there may be nothing to stop sand dunes blowing onto the land and destroying crops. Although it is easy to make laws against cultivating grazing land, it is difficult to enforce them.

**2.** *Cash crops.* Much of the new arable land in the Sahel has not been used to grow food for the local population but to produce cash crops, such as cot-

ton. The infamous peanut scheme is an example. After World War II, France set a high guaranteed price for peanuts, encouraging its former colonies in West Africa to grow this plant. The aim was to produce peanut oil and combat an American campaign to dominate the European market for vegetable oils with soybean oil. Because farmers were reluctant to replace all their food crops with peanuts, peanuts were grown largely on previously fallow land. This forced grazing animals from this land and led to overgrazing in other areas.

There are signs that cash cropping in the Sahel is declining, probably as a result of the recent famines. Some of the land taken out of peanut cultivation is now used to grow millet and sorghum, and some is used for niebe, a valuable bean that enriches the soil with nitrogen.

**3.** *Overgrazing.* In parts of the Sudan, the number of cattle more than quadrupled between 1957 and 1977, while the pasture to feed them decreased. Well-intentioned governments seem to have compounded the problem by digging thousands of wells and water holes since World War II. In places, animals have trampled the ground and denuded it of vegetation for up to 10 kilometers around a water hole. Overgrazing frequently compacts the soil and permits plants that animals cannot eat to replace more palatable species. This makes the land essentially useless for farming.

Overgrazing does not always produce permanent desertification. The soil is usually still there and may even be enriched by animal manure. A sufficient period without grazing restores the productivity of much overgrazed land. This simple method has been used by the U.S. government to restore some overgrazed federal rangeland. The trouble is that, in most parts of the world, this land is never permitted to lie fallow.

**4.** *Deforestation.* Woodlands and forest in the tropics are being cut down rapidly. One consequence may be that people burn animal dung because they have no wood to use for fuel. This dung would otherwise have fertilized the land. One expert estimates that, worldwide, burning dung reduces grain production by about 20 million tons a year, enough to feed 100 million people.

Deforestation also causes floods, which erode soil. The flood waters carry the soil they have washed off the land and often deposit it as the water slows down behind irrigation and flood-control dams. In 1979, a silted-up dam in India burst, unleashing a 5-meter wall of water and mud on the town of Morvi and killing 15,000 people.

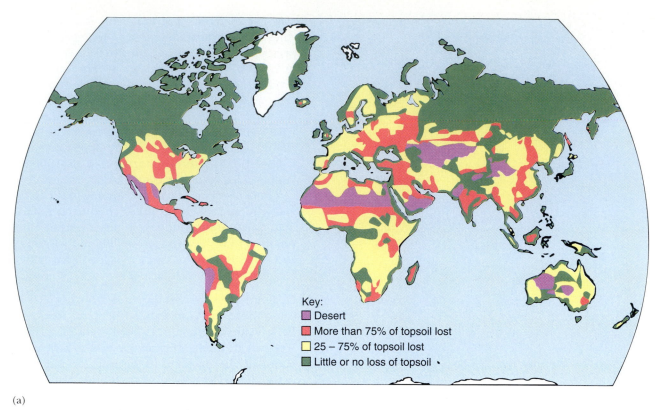

(a)

**Figure 14-10**    Soil destruction around the world. (a) The worst soil loss occurs in arid areas, which lie on either side of the world's large deserts (shown in purple). The yellow area in the Northern Hemisphere shows that much topsoil has also been lost from the world's most important food-producing regions in temperate parts of North America and Europe.

### Drought and Famine

People often talk as though droughts cause the periodic famines in northern Africa, but they do not. Famines are caused by wars and inadequate food reserves and food distribution. They are made worse by overpopulation and poor agricultural practices. **Droughts** are periods when less rain falls than in the average year, and they are part of the natural weather cycle in many parts of the world. If agriculture is to be sustained in a region, it must adapt to the erratic rainfall that is common over much of the globe. In 1985, people collected millions of dollars to feed the thousands of people starving in Ethiopia and neighboring countries. A drought, combined with soil degradation and war, had caused a major crop failure for the second time in 15 years. By the end of the year, the rains finally came, and thousands of people watched, with tears in their eyes, as millions of tons of topsoil were washed by the rains off the land and into the rivers. That year, another 15,000 square kil-

ometers had been stripped clear of soil and added to the north African desert, and, barring miracles, nothing will ever grow there again.

### 14-E    PREVENTING AND REVERSING SOIL DESTRUCTION

Fertile soil is disappearing all over the world to salinization and erosion (Figure 14-10). The loss of soil is particularly high in arid lands, which lie on either side of the world's great deserts. The scale of this problem is hard to imagine. Two thirds of the world's 150 nations are affected. More than 20% of Earth's land surface, housing some 80 million people, is directly threatened with becoming desert in our lifetimes. Every year an area larger than Nebraska deteriorates to the point where the land no longer produces an economic yield—not food crops, nor wood, nor grazing for cattle. The agricultural production lost each year is worth at least $80 billion.

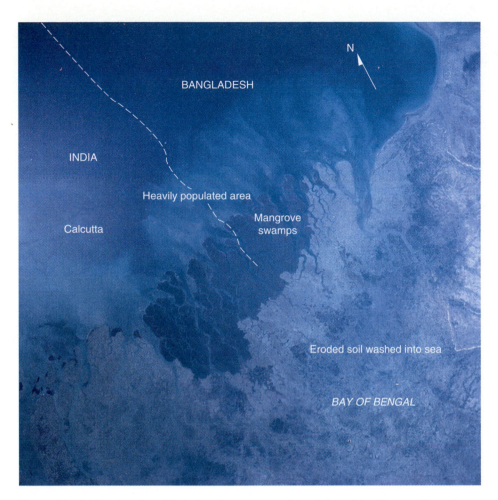

**Figure 14-10** (*Continued*)    (b) A view from space of the delta of the Ganges River shows topsoil washing through mangrove swamps into the Bay of Bengal. Some of the topsoil comes from hillsides in the Himalayas, where the forest is being cut down to produce farmland. The rest comes from farmland in Bangladesh, one of the most overpopulated countries in the world. (b, *NASA*)

> Soil destruction adds an area the size of Nebraska to the world's deserts every year.

The loss of topsoil is an ancient problem. Early civilizations developed where fertile soils permitted the agriculture that would support large populations. As soil was allowed to deteriorate, the civilization declined or moved to other areas. Ruins of great civilizations are found in many sandy wastelands. The fertile crescent of Mesopotamia, sometimes known as the cradle of civilization, is the floodplain of the Tigris and Euphrates rivers, now part of Turkey, Syria, and Iraq. Salinization came first; the desert followed. Crop yields are now among the lowest in the world. The Mali and Ghana empires flourished some

800 years ago in the northern Sahel, now an area of desperate poverty containing hardly any arable land. Soil erosion probably contributed to the decline of the Incan and Mayan empires in Central and South America.

We know enough about its causes to be able to prevent soil loss. Obviously we could stop the practices that cause deserts to spread, but there are also positive ways of tackling the problem.

## Tree Planting

Aside from reducing overgrazing and overcultivation, planting trees and shrubs is the single most useful way of improving degraded arid land and reclaiming land that has become salinized by irrigation.

Trees hold the soil in place and act as windbreaks that reduce the amount of soil blown away by the wind. They also provide fuelwood for people and food for animals. The International Livestock Center for Africa encourages the planting of *Prosopis*, a legume tree related to American mesquite, that tolerates drought and saline soil and produces pods that can be stored for animal food in winter. *Acacia albida* has similar virtues and also has a relatively open canopy so millet can be grown underneath it. Where this tree is planted, 20 cattle per square kilometer can be grazed on land where only 10 cattle could be supported without the trees.

In Niger, West Africa, a windbreak project by the U.S. voluntary agency CARE has planted 250 kilometers of trees since 1975, protecting about 3,000 hectares of land from soil erosion. Cereal production on this land has increased by more than 15%. Wood from the trees is also harvested for fuelwood without destroying the windbreak.

**Trees for Saline Soil** In the 1980s, the government of Victoria, Australia, decided on a grassroots campaign to slow the salinization that has reduced or destroyed the productivity of 400,000 hectares of farmland in the state. The salinized land looks like desert, with patches of white salt crust, but underneath the soil is waterlogged. This is where the tree seedlings come in. First the soil is treated to make it less alkaline, and then salt-tolerant trees are planted. Time does the rest. Water leaves the land more rapidly now that trees are absorbing it. As a result, the water table will fall if it is high. Shade from the trees reduces evaporation from the soil surface, which slows the upward movement of ground water. Tree roots penetrate the soil, making it more permeable so salts wash out more rapidly, and fallen leaves add humus, which improves the soil. The water table begins to fall within a year or two, although the complete restoration of salinized land takes decades.

Planting trees is the most important step in slowing and reversing desertification.

**Reclaiming Desert**

Some desert can be made to flower again, but the process is not nearly as easy or inexpensive as Israel hoped when it proclaimed this goal for the Negev Desert in 1968. China has probably the most impressive record of reclaiming desert, using its vast resource of cheap labor. In the 1950s, a basket brigade bodily removed 2 million cubic meters of sand that had blown over farmland in Chifeng province. The land was then flooded with muddy water to replace some of the soil, and rice paddy fields were built, protected by windbreaks to prevent the sand blowing back over the fields.

Sand dunes themselves can be reclaimed. The area is first fenced to keep livestock out, and then strips of brushwood are stuck into the dunes as windbreaks. If the soil is acid, lime may be added at this point. The next step is to plant hardy trees and grasses or merely to wait long enough for these plants to seed themselves on the dunes. Some of the plants can be cut for animal fodder after about five years, and the land must always be protected from overgrazing if it is not to revert to desert.

## 14-F  TROPICAL FOREST AND LATERIZATION

A unique soil problem arises when agriculture is practiced in parts of the tropical forest biome where the soil contains **laterite**, particles containing iron or aluminum oxides. When lateritic soil dries out or is exposed to air, it **laterizes**, hardening into a concrete-like substance that plant roots have difficulty penetrating. Deforestation promotes laterization. In parts of Cameroon, West Africa, laterized soil has formed to a depth of 2 meters in less than 100 years. Crops cannot be grown on highly laterized soil, and even the native forest is slow to reestablish itself. Indian foresters have long known that teak plantations may be affected by laterization. Teak trees have to be planted a long way apart, and so the soil in a teak plantation is more exposed than that in the native evergreen forest, and laterization proceeds more rapidly.

Natives in many lateritic areas practice **swidden**, or **slash-and-burn agriculture**. An area of forest is felled or burned and the land cultivated for a few years. The shallow tropical soil is then exhausted, and the farmers move on to another area, leaving the cultivated patch to be overrun by forest plants again, allowing its soil to regenerate (Figure 14-11). The soil is not exposed long enough for severe laterization to occur.

Swidden agriculture does not suit modern farmers, who want to cultivate large areas that make heavy machinery worthwhile. The failure to understand tropical soils and laterization has caused some spectacular agricultural disasters, including the loss of millions of American dollars invested in failed banana farms and similar projects in what was once Brazilian rain forest.

**Figure 14-11**   Tropical topsoil is often very thin. This road cut in Central America shows the deep layer of subsoil with plants growing in the thin layer of topsoil above it.

> Up to one third of all land in the tropical forest biome is threatened by laterization when it is deforested for agriculture.

## 14-G  SOIL EROSION IN NORTH AMERICA

[*T*] *here was something fantastic about a dust cloud that covered 1.35 million square miles, stood 3 miles high, and stretched from Canada to Texas, from Montana to Ohio—a cloud so colossal it obliterated the sun . . . a four-day storm in May 1934 . . . transported some 300 million tons of dirt 1,500 miles, darkened New York, Baltimore, and Washington for 5 hours, and dropped dust not only on the President's desk in the White House, but also on the decks of ships some 300 miles out in the Atlantic . . . ducks and geese suffocated . . . the sky so black that chickens, thinking it night, would roost. Oklahoma counted 102 storms in the space of one year; North Dakota reported 300 in eight months.*

(a)

(b)

**Figure 14-12**   Wind erosion. (a) A car drives into a cloud of soil thrown up by a plow and blown by the wind. (b) This sandy soil was left without vegetation and has blown into the road from the neighboring field. (a, *Richard Feeny*)

Thus one writer reported the American Dust Bowl of the 1930s. The Dust Bowl was caused by poor farming, followed by several hot, dry years that killed vegetation, leaving loose, dry soil waiting for a wind to blow it away. After World War I, tractors were introduced to the Great Plains and used to plow up the native grassland. Combine harvesters began to speed the harvest. Farmers planted yet more wheat in attempts to keep up the payments on all this machinery. They more than doubled the area planted to wheat in less than ten years. Early settlers struggled with the tough prairie sod, but this was now replaced by light, dry soils. Dry weather and high winds did the rest (Figure 14-12).

**TABLE 14-6   Sources and Rates of Soil Erosion in the United States**

| Source | Percent of Total Erosion |
|---|---|
| **Erosion by water** | |
| Roads and construction | 5 |
| Gullies | 6 |
| Stream banks | 11 |
| Cropland | 38 |
| Rangeland | 8 |
| Pastureland | 3 |
| Forests | 29 |
| **Erosion by wind** | |
| Rangeland | 45 |
| Cropland | 55 |

Dust storms did not disappear with the 1930s. In 1977, a storm stripped more than 25 million tons of soil from California's San Joaquin Valley in one day. Erosion by wind and erosion by water are intimately linked. Water runs rapidly off land that has been stripped of its loose topsoil by wind erosion. In the months after a dust storm, water runoff causes floods and carries away yet more soil. Topsoil is lost from American agricultural land at the rate of about 30 tons per hectare per year, approximately eight times the rate at which topsoil is formed (Table 14-6). About half of the country's topsoil has now been lost. Every year, about 1 billion tons of soil are eroded by wind, and water deposits another 4 billion tons of soil in our rivers (Figure 14-13).

The main reason for soil loss is that most farmers give short-term crop yields and profits precedence over the longer term advantages of conserving soil. Even when soil begins to lose its fertility, applying yet more fertilizer and pesticides may keep crop yields high for a while, hiding the real problem.

Deforestation is another important cause of erosion. Sometimes deforestation permits the topsoil to wash away slowly. Often it creates more spectacular erosion in the form of landslides and avalanches. Trees and shrubs absorb large quantities of water so deforestation increases the water runoff from an area. Experiments in Arizona showed that when the sparse chaparral scrub was burned, water runoff increased tenfold. A lot of the erosion caused by logging results from the construction of dirt roads and could easily be prevented with greater care.

> The United States has lost about half its topsoil to erosion caused by farming and deforestation.

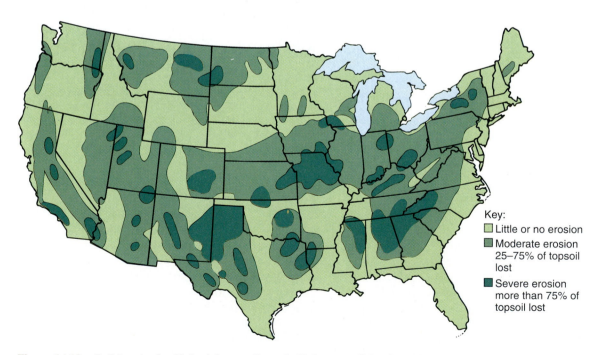

**Figure 14-13**   Soil loss in the United States. About half the topsoil has been lost in just 200 years.

Key:
- Little or no erosion
- Moderate erosion 25–75% of topsoil lost
- Severe erosion more than 75% of topsoil lost

## 14-H SOIL CONSERVATION

Every year, farming causes large areas of land to lose their fertility to the point where the soil is so poor or so completely eroded that it can no longer be farmed. Yet the world contains land, notably in Europe and Asia, that has produced crops every year for thousands of years. This land has obviously been farmed in a sustainable manner so that it has remained fertile. How is this achieved? First, most of this long-term agriculture is rain-fed, watered by rain and not by irrigation, so salinization is not a problem. Second, farmers replace the nutrients removed when they harvest crops. They do this by what we should now call **organic farming**, farming without the use of inorganic fertilizers and pesticides. Instead, organic matter is added to the soil each year to replace nutrients and improve soil structure.

### Managing Hillsides

There are many ways of conserving soil by reducing erosion. Adding organic matter to the soil, crop rotation, and methods of tilling the soil can help. So can planting windbreaks or terracing the land.

In hilly country, building terraces is an age-old practice to prevent soil washing and blowing down the hill. Terracing takes a lot of labor and tends to be most suitable in places such as China, where labor is cheap and plentiful, or in areas where the crop is sufficiently valuable to justify the expense, as it is in wine-growing areas. Terracing is not without its own risks. Terraces have been known to hold back so much water that the soil becomes saturated and mud slides wash the hillside away.

On gentle slopes, contour plowing is the traditional method of reducing soil loss. **Contour plowing** means plowing parallel to the slope of a hill instead of up and down it (Figure 14-14). The plow furrows form tiny terraces that help prevent the soil from washing or blowing down the hill. A variation is **contour stripcropping**, more effective than pure contour plowing in reducing erosion. In this method, the whole of a hillside is never plowed at once. Instead, strips of vegetation are left to catch soil and water that run down the hill.

In many areas, steep hills are best used as forest. The economic yield is not as rapid as with farmland, but it can be sustained indefinitely if managed properly, whereas eroding agricultural land has only a limited productive life. I used to live on a hill in upstate New York, on soil described by the county soil map as "Volusia Channery Silt Loam, 90% eroded." The early settlers cleared the forest and plowed up

**Figure 14-14** Contour farming. This field in Washington is not on a steep hillside, nevertheless; erosion has been minimized by plowing and planting along the slope of the hill rather than up and down it. Here a crop of hay has been cut and lies drying before it is collected. *(Richard Feeny)*

and down hills. In less than 80 years, nearly all the topsoil eroded from the hills into the valleys. Most of the hill farmers were bankrupt by the 1930s. Then during the Great Depression, a farsighted state government bought up the mortgages on millions of hectares of hillside land, and now almost every hill is crowned with state forest. Farms survive only in the valley bottoms where the eroded soil collected (Figure 14-15).

In the 1980s, the United States finally instituted a reasonably effective program for reducing soil erosion. In 1987, its second year, the Conservation Reserve Program helped to reduce soil losses by 400 million tons, the greatest year-to-year reduction in American history. The program involves converting easily eroded land to grassland or woodland. It penalizes farmers who do not practice soil conservation by denying them benefits from farm programs, including price supports, crop insurance, and low-interest loans. Note, however, that this program merely reduces the rate of soil erosion. It does not prevent it or rebuild the soil.

### Organic Farming

**Organic farming or gardening** is growing plants without the use of pesticides and inorganic fertilizers. Until the last hundred years, most farms were organic because chemicals were expensive or unavailable. Then artificial fertilizers and pesticides came

**Figure 14-15**   The result of soil erosion in New York. State forest covers the tops of hills from which nearly all topsoil has eroded. Farming is confined to the valley bottoms.

into wide use, increasing agricultural productivity enormously. From about 1920 to 1980, anyone who advocated a return to organic farming was ridiculed as an "ecofreak."

Plenty of people did advocate organic farming. Agronomists F. H. King (1927) and Albert Howard (1940) studied humus and compost and concluded that farm soil was degraded if organic matter was not added back into the soil. **Compost** is waste plant material, such as grass clippings, weeds, and other yard waste, that gardeners place in piles to decompose before they add it to the soil. Other sources of organic matter that are often added to soil are animal manure and sewage sludge, the organic material from a sewage works (Section 19-E).

Philosopher Rudolf Steiner (1970s) argued from studies of plant interactions that planting just one species of plant was bad for the soil. Farmers and gardeners have known for centuries that one patch of soil should be planted with different plants in different years to increase the yield and reduce pest damage. This practice is known as **crop rotation**. It works because different plants require different nutrients and have different pests. After a crop of tomatoes has been grown, the soil contains little nitrogen and is full of pests that feed on tomatoes. If tomatoes are planted next, they will not do well. But if beans are planted, the tomato pests will not attack them, and beans are legumes, which add more nitrogen to the soil than they remove.

Throughout this period, Robert Rodale, founder of *Organic Gardening* magazine, advocated organic

practices and was widely read by organic gardeners and farmers. In the 1970s and 1980s, farmers and the U.S. Department of Agriculture, faced with degraded farmland and regional drought, at last began to listen to what ecologists and organic farmers had been saying for a long time. New approaches to farming influenced by organic methods are usually described as **low-input sustainable agriculture**, which emphasizes reducing farmers' costs and reducing environmental damage.

Studies show that organic farming works just as well now as it has for thousands of years. Researchers from Washington State University studied two farms next door to each other near Spokane, Washington. One of the farms has been managed with applications of inorganic fertilizer and pesticides, as recommended by the state, since 1948. Crops of winter wheat and spring peas are grown in succession in each field. The other, organic, farm has never been treated with inorganic fertilizer. It too grows wheat and peas, but it also grows a crop of green manure on each field every third year.

**Green manure** is a crop that is not harvested but is plowed back into the soil to add organic matter. Topsoil on the organic farm is now 15 centimeters deeper than on the conventional farm, and both farms have provided similar profits for their owners over the years. How can this be? The organic farm sells only about two thirds the harvest of the other farm because its fields are devoted to green manure for one third of the time (and it also probably loses more crops to pests). The secret lies in the higher operating cost of the conventional farm. Inorganic fertilizers and pesticides are expensive. In this case, it takes about one third of the conventional farm's total production to cover their cost.

> The most important practices that conserve soil are preventing erosion and adding organic matter to the soil.

### No-Tillage Farming

The fastest growing method of soil conservation in the United States is simply to stop plowing the soil. Consider a wheat field. Traditionally the wheat has been cut down and collected by a combine harvester. Then the soil is plowed to turn it over and dig in the remains of the wheat. The soil is raked and the seeds of the next crop planted. With no-till farming, the plowing and raking are omitted. The seeds of the next crop are planted in holes drilled in the soil straight through the remains of the previous crop. The roots of the first crop hold the soil in place while the new crop develops. The remains of the first crop decay on the surface and in the soil, adding organic matter to the soil. No-tillage farming uses less water and also uses less energy—by farmer and tractor—than does traditional cultivation. It reduces soil erosion by up to 90%.

No-till methods also save time. They may permit three crops a year to be grown in areas with long growing seasons. For instance, in the southeastern United States, barley, sweet corn, and soybeans may be grown one after another in one field in the same year. Barley, maize, and snapbeans is another common combination. Note that both rotations include legumes (snapbeans and soybeans), which add nitrogen to the soil. In colder areas, where crops are not grown in the winter, soil is conserved by planting a green manure crop, preferably a legume, in the fall. This produces a layer of organic matter on the soil the following year and often prevents soil erosion completely.

The same effect can be duplicated in the home garden. Experiments at the University of Florida showed that the highest yields from tomato plants came when the baby plants were planted in a plot of green manure. In the fall, you plant green manure, such as mustard, annual rye, or beans. In the spring, you mow the green manure so as not to disturb the roots and then you dig holes and plant the tomatoes. Presto! A bumper crop. Methods that reduce soil erosion and restore fertility to the soil are summarized in Table 14-7.

## 14-I ROADS AND URBANIZATION

U.S. Assistant Secretary of Agriculture Robert Cutler once observed that "asphalt is the land's last crop." When productive cropland is lost to nonfarm use, it is hardly ever used again to produce food. Worldwide, about 6 million hectares of farmland disappear under houses, mines, roads, reservoirs, factories, and power plants every year (Figure 14-16; Facts and Figures 14-2). The farmland lost in this way is often the most valuable because we usually build in valleys, which contain the most productive farmland. Can-

| **TABLE 14-7**  Farming Methods that Reduce Erosion and Restore Soil Fertility | |
|---|---|
| **Practice** | **Effect** |
| **Methods of Slowing Erosion** | |
| Terracing | Holds back soil and water that would otherwise flow downhill |
| Contour farming | Prevents soil washing and blowing down hill |
| Contour strip cropping | At all times keeps strips of vegetation across the hill to catch any soil that washes downhill |
| Reforestation (in areas with enough rainfall to support trees) | If even a little soil is left, trees will grow, holding the remaining soil in place and contributing organic matter to rebuild the soil |
| **Methods of Increasing Soil Fertility** | |
| Spreading organic matter such as compost, manure, and sewage sludge on the soil | Adds nutrients to soil, improving conditions for crop plants and for soil-forming organisms |
| Planting and plowing in green manure | Green manure plants hold soil in place while growing and add organic matter when plowed in |
| No-tillage farming | Keeps soil covered with vegetation to prevent erosion; remains of previous crop add organic matter to soil as they decompose |
| Crop rotation including legumes | Adds nitrogen to soil |

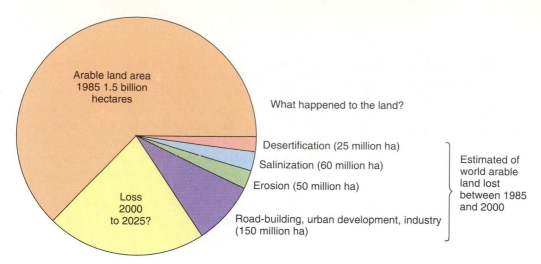

Figure 14-16    Our disappearing cropland: a summary of the problems discussed in this chapter. This graph shows the fraction of the land farmed in 1985 that experts estimate will have been lost from production by 2000. Eleven million hectares (ha) of land become unfarmable each year as a result of erosion, desertification, salinization (and other forms of poisoning), and conversion to nonagricultural uses. (This does not count the 7 million hectares of rangeland lost each year to desertification.) At the present rate, we shall have lost more than one third of the world's cropland from production by 2025.

ada reports that half of the farmland lost to urban expansion every year comes from the most productive one twentieth of cropland.

Few things will affect our future well-being more directly than the balance between human numbers and area of farmland. If the price of food is not to rise rapidly in the near future, we have to stop the further loss of fertile land. This requires long-term land-use planning. A few countries, such as Japan, have long-term plans designed to protect their cropland. In 1968, Japan produced a simple nationwide zoning plan that puts all land into one of three categories: agricultural, industrial, or other. The plan causes various problems as towns grow outward around agricultural land, but it ensures that cropland is preserved. Another part of the Japanese scheme is the use of mass transportation systems. Roads eat up much more land than any other method of transportation. Several European countries have established less effective national land-use guidelines. National land-use planning is still at a rudimentary level in the United States, with most planning on a local level.

## 14-J  CONCLUSION

Although most of this book concentrates on the environmental changes wrought by industrial societies, much of the destruction of agricultural land described in this chapter was and is brought about by nonindustrial ways of life. It can be argued that de-

### FACTS AND FIGURES 14-2

#### Where Does the Agricultural Land Go?

Grains are our most important crops, providing most of the world's food. In 1991, the area of land from which grains were harvested shrank from 695 million hectares to 693 million hectares. Where does the land go? Here are a few of the answers:

- Soil erosion. In the republics of the former Soviet Union, 20% less land is now used to grow grain than in 1970. Essentially all the soil has eroded from the abandoned land, and plants will no longer grow on it.

- Water shortages. In the American West, about 10,000 hectares of irrigated farmland were permitted to return to desert in 1992 because the water was needed to supply the growing needs of the Los Angeles area.

- Fertilizer poisoning. In Ukraine, thousands of hectares of farmland have been abandoned. So much fertilizer had been applied that the soil was salty and nothing would grow in it.

- Urban sprawl. In Thailand, the capital city, Bangkok, is expanding outward so rapidly that each year it covers more than 3,000 hectares of what was once farmland.

- Industrial growth. In China, rapid economic growth means that factories are expanding over farmland. The area planted to grain has fallen by 10% since 1976.

- Pollution from radioactivity. The 1986 nuclear explosion at Chernobyl caused half a million hectares of land to be evacuated because they were no longer safe to live on or farm.

spite the rapid industrialization and urbanization of the last century, many of our most serious problems are caused by advanced and primitive agriculture. Thus, soil salinization and desertification are two of the most serious problems we face, and the erosion of soil, fertilizer, and pesticides into rivers probably causes more pollution of the world's waters than does industrial waste. The habitat destruction that causes most species extinction results from the expansion of farmland, not of industry. When we think of human activities that alter Earth's atmosphere and climate, burning fossil fuels springs to mind. But the release of carbon dioxide and methane from agri-

cultural machinery and livestock and dust from soil erosion are at least as important.

The other point that should be made is that although many adverse human effects on the environment can easily be stopped or reversed, many of the changes caused by agriculture cannot. Once the soil has eroded from an area, it may take centuries to replace; a layer of laterite is almost impossible to break up; and an extinct plant or animal can never be brought back. It is ironic to realize that if bigger and better bombs do not destroy our civilization, our attempts to feed ourselves well may.

## SUMMARY

A shortage of fertile land on which to grow crops threatens our ability to provide food for the world's growing population.

### Farmland and Soil

Farmland is composed mainly of arable land and rangeland. Good arable land has fertile soil that supplies plants with mineral nutrients, water, oxygen, and support. Soil forms slowly as rock particles are weathered by the elements and by soil organisms. Fertile soil contains much organic matter and has a slightly acid pH.

### How Cultivation Affects Soil

Farming affects the physical and chemical structure of soil. Heavy machinery compacts soil, making it more difficult for water to penetrate. Fertilizers, lime, and other additives affect the physical structure of the soil, alter its pH, and supply plant nutrients. Inorganic fertilizers kill soil-forming organisms and reduce fertility in the long term.

### Irrigation

Agricultural yields in many parts of the world can be increased by irrigation. Irrigation can cause many problems, such as making the soil too salty for crops and raising the water table. The problem is particularly acute in the naturally saline soils of arid regions. Salinization can be prevented and reversed by better irrigation techniques and drainage and by planting trees.

### Soil Destruction in Africa

Arid regions make up much of the world's farmland. They have thin soil that is easily destroyed. Arid land is best used as range for livestock because its soil erodes rapidly if plowed. Somewhat less arid land can be used for crops as long as its content of organic matter is maintained by practices such as fallow periods and preservation of the vegetation cover. Overgrazing, overcultivation, the growing of cash crops, and deforestation have completely destroyed the soil on millions of hectares of land in Africa that have become desert.

### Preventing and Reversing Soil Destruction

Every year, poor farming practices convert an area the size of Nebraska into land so poor that it can no longer be farmed. Soil loss destroyed ancient civilizations and is destroying countries today. It is most effectively combated by planting trees and leaving the land fallow. Modern research has identified trees that are highly effective in preserving the soil and reversing salinization in arid regions. The trees may also provide shade, windbreaks, animal food, and fuelwood. However, in overpopulated countries where famine is a constant threat, land is seldom left unfarmed long enough to recover its fertility.

### Tropical Forest and Laterization

Aside from arid lands, the world's largest area of poor farmland lies in the tropical forest biome. Soil in this biome is subject to erosion and degradation just like any other farmland, but much of it is also lateritic, turning into a rock-like substance when exposed to air and light. Laterized soil may take centuries to recover.

### Soil Erosion in North America

Soil erosion by wind and water is a massive problem in North America. Although most of this continent has been cultivated for less than 100 years, very little land in the United States now has more than half the topsoil it started with. The main causes of erosion are deforestation and intensive cultivation without adding organic matter to the soil.

### Soil Conservation

The most important practices that conserve soil are preventing erosion and adding organic matter to the soil. Erosion is prevented by practices such as building terraces on hillsides, maintaining windbreaks, and plowing across slopes rather than up and down them. No-tillage farming prevents erosion by never plowing the soil so that it is covered with vegetation at all times. Organic matter can be added to soil as manure, sewage sludge, green manure, or compost or by letting part of the previous crop decay into the soil.

### Roads and Urbanization

A large area of the world's best farmland is lost every year to transportation, urbanization, and industry. Land-use planning to slow this loss is important if many countries are not to face food shortages in the near future.

### Conclusion

Agriculture is a major cause of the world's worst environmental problems, such as species extinction, deforestation, and water pollution.

## DISCUSSION QUESTIONS

1. Most suburban vegetable gardens produce higher crop yields per hectare than do farms in the same area, with the use of much less inorganic fertilizer. Why do you suppose this is?

2. During the drought that affected much of the United States in 1988, soil erosion was worse than usual in the southeastern states. Why do you suppose drought has this effect? If you were an agricultural extension agent in the area, what would you recommend to prevent a similar loss of soil in the next drought?

3. The populations of developed countries are stabilizing or declining. Why, then, has the damage done by intensive agriculture not declined?

4. Should tax breaks and subsidies be used to encourage more farmers to switch to low-input or organic farming? Why or why not?

5. Why do agriculturalists believe that the price of food will increase considerably within the next ten years?

# FOOD AND HUNGER

# Chapter 15

## Key Concepts

- World food production increased rapidly between 1950 and 1980.

- Food production is no longer increasing rapidly, partly because the green revolution is essentially complete and partly because the fertility of much farm soil is falling.

- The world food supply is dangerously dependent on only a few species of plants.

- If we are to feed a much larger population in the twenty-first century, the next step should be to improve the productivity of low-input and subsistence farming.

- Most hunger in the world today is a result of population increases, unemployment, and landlessness in rural areas, coupled with the greater political power of people who live in cities.

*A fruit stall in a Nicaraguan market.*

In 1988, the United Nations warned that for the first time in history, world food production was lower than in the previous year. Also in 1988, as in every other year at the end of the twentieth century, millions of people starved to death, and millions more suffered from diseases caused by malnutrition (poor nutrition). Even while people are starving, our attempts to produce enough food have caused many of our most pressing environmental problems, as forests and wildlife habitats have been converted into agricultural land, and agriculture has caused soil erosion and water pollution.

The rapid growth of the human population has strained our ability to supply people with food. In one way, this is a success story: today's farmers are producing food faster than ever before, and this represents agricultural productivity that few would have believed possible 30 years ago. In another way, it is a failure: we are not producing enough to feed everyone on Earth.

You will sometimes read that starvation is caused by poverty and distribution problems, not by a shortage of food. This is both true and false. Civil wars, government corruption, and poor transport systems do prevent food from reaching many people, but even if these problems were solved, there is still not enough food in the world for everyone. A nutrition expert from Brown University expressed it this way: "If the world's food were equally distributed, we are producing enough food to keep 120% of the world's present population alive but too malnourished to work; enough to keep 100% of the population alive and fit enough to do a little work; but only enough to keep 80% of the population on the diet recommended by the U.S. Department of Agriculture as an ideal human diet." If everyone in the world today received an equal share of all the food produced, few people would be healthy enough to do a full day's work. The situation can only get worse as population continues to outstrip increases in food production (Figure 15-1).

Food and farming vary enormously in different parts of the world. A trip to the Caribbean or to Central America introduces a North American to dozens of new foods. Breadfruit and soursop, shaddock, and goat on the menu make us realize that the foods we usually eat are only a tiny proportion of those available. People in various parts of the world eat 3,000 or more plant species and 60 or so different animals. But nearly all the increases in food production during the last century have come from half a dozen plant species.

## 15-A NUTRITION

Animals are heterotrophs; we cannot make our own food from inorganic substances but must take in organic molecules from the environment. The nutrients that any animal must ingest as food may be divided, for convenience, into **macronutrients**— nutrients that are needed in large quantities—and **micronutrients**—nutrients that are required in lesser amounts.

### Macronutrients

The macronutrients are the lipids, carbohydrates, and proteins. All are polymers (Section 13-G), and all can be used to build and repair the body as well as serving as sources of energy. The amount of energy available from a given amount of a macronutrient is commonly measured as the number of kilocalories of heat it yields when fully used (Table 15-1).

Nutritionally the most important macronutrient is **starch**, the carbohydrate that plants make and store. All the most important food plants, such as grains, potatoes, and beans, contain high concentrations of

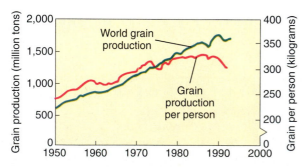

**Figure 15-1**  World food production falls behind population growth. Grains make up most of the world's food. From 1950 to about 1975, grain production grew faster than population. Since 1975, the amount of grain produced for each person on Earth has fallen as the population has grown faster than grain production.
*(Data: U.S. Department of Agriculture)*

| TABLE 15-1 | Energy Yields of Macronutrients | |
|---|---|---|
| **Class of Macronutrient** | **Composition** | **Energy Yield** |
| Carbohydrates | Polymers made of sugar monomers | 4 kilocalories/gram |
| Lipids | Fatty acids plus alcohols | 9 kilocalories/gram |
| Proteins | Polymers made of amino acid monomers | Varies: about 4 kilocalories/gram |

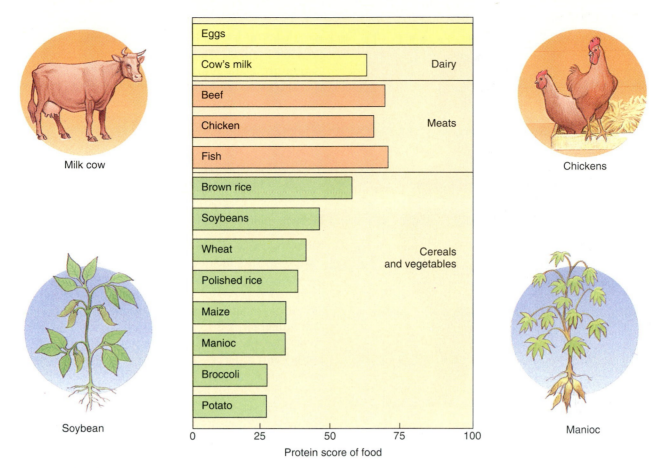

Milk cow

Chickens

Soybean

Manioc

**Figure 15-2** Protein quality of various foods: their content of essential amino acids. Egg is considered to have an almost ideal amino acid content for human nutrition, and the other foods are rated in comparison with egg to give a protein score (which is not the same as the amount of protein).

starch. In the body, starch and other carbohydrates are usually broken down to supply energy. The body breaks proteins down into their monomers, called **amino acids**, most of which are used to build the body's own proteins. Lipids are fats and oils. They are important as the molecules that make up some hormones and all membranes. Every cell and every organ in the body contains large areas of membranes, which are built and repaired with lipid molecules.

Carbohydrates and lipids can be stored in various ways until the body needs them. Proteins, however, cannot be stored. Protein that the body does not use immediately is broken down into fats or carbohydrates.

**Problems with Macronutrients** The most common dietary problem for people in industrialized countries is obesity. This is especially true in the United States, where food is very cheap by world standards. If more calories are ingested in the food than are used up, the excess is stored as fat. Women have a greater tendency than men to store fat. This is bio-

logically valuable, if currently unfashionable, because it provides food reserves that can carry not just the woman but also her fetus or nursing infant through times of shortage. A woman without sufficient fat in her body ceases to ovulate so that she cannot conceive a fetus that her body could not support. This is the reason that women athletes with very low body fat may have difficulty becoming pregnant.

Much more common in the world than obesity is lack of food. Thousands of people die of starvation every day, and at least 10 million children in the world are so malnourished that their lives are in danger. **Starvation** means death from lack of food. However, most people who are inadequately fed do not actually die because they take in too few calories to survive but because their malnourished bodies have little resistance to diseases that would not be fatal to the properly fed.

Probably the most common dietary problem in the world is protein deficiency. This occurs when the diet contains too little total protein or when the protein is not of high quality because it does not contain enough of some essential amino acid (Figure 15-2).

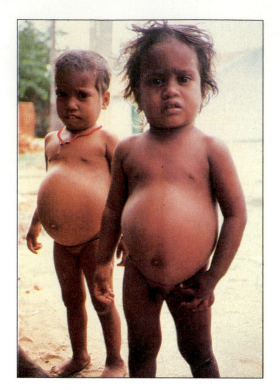

**Figure 15-3** Children suffering from kwashiorkor, a condition caused by severe protein deficiency. The swollen belly results from fluid retention. *(Food and Agriculture Organization of the United Nations)*

Essential amino acids must be supplied in the diet because the body needs them but cannot make them from other amino acids. One of the best-known protein deficiency diseases is **kwashiorkor**, often found in African populations where the diet consists primarily of maize ground into cornmeal. Cornmeal contains very little of the essential amino acids tryptophan and lysine. Most victims of kwashiorkor are growing children, who need more protein than adults (Figure 15-3). They show symptoms such as skin problems, failure to grow normally, lethargy, and edema (swelling because fluid is retained in parts of the body).

**Marasmus** is the other common disease of malnutrition. It occurs when the diet is deficient in both total calories and essential amino acids. Most victims are infants that are not breast-fed or young children in poor families where there is not enough food. A child suffering from marasmus has a bloated belly, shriveled skin, wide eyes, and an aged-looking face. This is the face of poverty with which we are so familiar from news reports on famine. Both kwashiorkor and marasmus cause brain damage unless they

are cured rapidly. This is why commentators say that "an entire generation of Somalis" has been lost to the widespread starvation of 1991 and 1992, not because an entire generation has died, but because many of those who survive will grow up to be mentally handicapped adults.

> Starvation is usually due to death from diseases that would not kill the properly nourished.

### Micronutrients

Micronutrients are the substances an organism must have in its diet in small quantities because it cannot make them for itself or because it cannot make them as fast as it needs them. Micronutrients can be divided into **vitamins**, which are organic compounds, and **minerals**, which are inorganic. Various deficiency diseases result from shortages of particular vitamins and minerals in the diet. Vitamin-rich foods include fresh fruits and vegetables and whole-grain or enriched breads as well as milk, liver, and fish oil.

We need some minerals in relatively large amounts. Sodium and potassium, for instance, are vital to the working of every nerve and muscle in the body. Large quantities of these minerals (particularly sodium) are excreted in the urine every day. Sodium excretion is also a vital part of sweating, which is necessary to the regulation of human body temperature. Calcium is required for muscular activity and, with phosphorus, is needed in large amounts for bone formation. Foods rich in minerals include meats, milk and cheese, nuts, legumes, and spinach.

> A healthy human diet contains sufficient carbohydrates, proteins, and lipids to provide energy and to build body molecules, plus smaller amounts of minerals and vitamins.

### 15-B THE FOODS WE EAT

Agricultural societies invariably have much more limited diets than do hunter-gatherers. Native Americans in sixteenth-century North America ate more than 400 species of plants and about 35 species of animals. A modern American is unlikely ever to eat more than about 40 species of plants and 15 species of animals. Without seafood, most of us would eat no more than about four species of animals. This dependence on a few species poses many dangers. A pest outbreak that severely reduces the yield of any one crop lays societies open to famine on a scale unimaginable in earlier times. For instance, in some years during the 1980s, population explosions of the corn borer, an insect pest of maize, reduced the amount of grain produced in the United States by 25%.

## Diets

If you think about the diet in most parts of the world, you will realize that it consists of large amounts of carbohydrate, such as bread, rice, and potatoes, with smaller amounts of protein and fat, plus helpings of fruit and vegetables. Fruit and vegetables contain minerals and vitamins but less starch than staple food plants. For instance, Italian cuisine is typically pasta with a sauce of vegetables, cheese, or meat. The Asian diet consists mainly of rice, with a little meat and vegetables. American pizzas and hamburgers are basically bread with protein and vegetables.

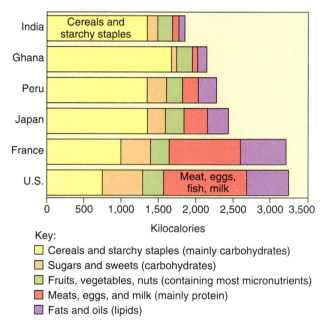

Key:
- ☐ Cereals and starchy staples (mainly carbohydrates)
- ☐ Sugars and sweets (carbohydrates)
- ☐ Fruits, vegetables, nuts (containing most micronutrients)
- ☐ Meats, eggs, and milk (mainly protein)
- ☐ Fats and oils (lipids)

**Figure 15-4** People in wealthy countries eat more food and more protein and lipids than people in poor countries. This graph shows the number of kilocalories of energy from different types of food in the average diet in various countries. Cereals and starchy staples include bread, rice, and potatoes. The U.S. National Academy of Sciences recommends a daily intake of 2,700 calories for an active man weighing 70 kg. (The figure is less for women, unless they are pregnant.) We can conclude that most Americans and French people eat more than they need and that in India, Ghana, and Peru the average person does not consume enough calories to be healthy. What conclusions can you draw from the high consumption of lipids and proteins in France and the United States? Why is the Japanese diet closer to that of a developing nation?

Wealthy people tend to eat more total food and more meat and fats than poor people, and this is reflected in the food consumed in a country (Figure 15-4). Even the Japanese have started to consume large quantities of beef in recent decades, although the traditional Japanese diet contains a healthy mix of carbohydrates, proteins, and fats. This trend is most extreme in the United States, where up to half of all calories come from fats and proteins. We are now told that this is unhealthy, and the consumption of high-fat, high-protein foods such as beef has declined in recent years. Beef, however, remains the typically American food, which is one of the reasons that our agriculture consumes such staggering quantities of natural resources. (See Close-Up 15-2.)

Meat provides more nutrients per gram than does plant food. But agriculture can feed many more people from a given area of land when people eat plants instead of meat. If we eat animals, the animals must eat plants first, and animals convert only about 10% of the plant calories they eat into calories in their own bodies, much of it in the form of bone and other inedible body parts. In contrast, about one third of a plant's productivity can usually be harvested for use as human food. This is why meat is more expensive to buy than the same weight of plant foods: it takes more energy, water, and land to produce a kilogram of meat than a kilogram of bread or vegetables. The highest efficiencies with which animals convert plant calories into calories in animal products is about 25%, achieved in milk and egg production. Since milk and eggs are excellent sources of fat and protein, chickens to lay eggs and goats and cattle to give milk are important for food, even in poor societies.

## Importance of Legumes

Some people eat only plants, sometimes because of preference or tradition, more often because they are too poor to buy meat. It is more difficult to obtain adequate nutrition with a vegetarian than with a mixed diet because plants tend to provide plenty of carbohydrates but few vitamins and minerals and, even more important, little protein. Plants such as wheat and rice are about 10% protein. This is enough for an adult human, although growing children need a higher percentage of protein. The protein content of plants varies from about 3% in leafy vegetables to 34% in soybeans.

Soybeans are legumes, with nitrogen-fixing bacteria on their roots. The main use of nitrogen in the body is to build proteins, and legumes have plenty of nitrogen, making more of their body parts from proteins than other plants do. As a result, legumes

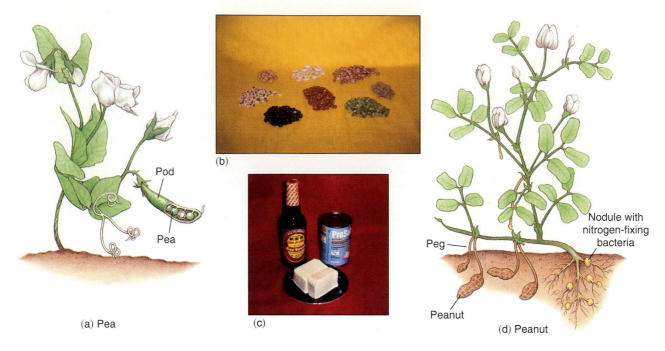

(b)

(c)

(a) Pea

Pod

Pea

Peg

Peanut

Nodule with
nitrogen-fixing
bacteria

(d) Peanut

**Figure 15-5**   Legumes, important sources of protein. (a) A pea plant, showing the type of
flower common to most legumes and the pod, also characteristic of legumes, in which the
seeds that we eat develop. (b) A variety of dried legumes from around the world. The part
of a legume that we eat is usually the seed. Like grains, legume seeds have the great
advantage that they can be stored for long periods without deteriorating, even without
refrigeration. (c) Some of the food products that are made from soybeans. The white
substance is tofu. (d) A peanut plant. You can see the nodules found on the roots of all
legumes that contain nitrogen-fixing bacteria. Peanut ''nuts'' develop underground. After
the flower is fertilized, the flower stalks grow down into the ground, where the peanuts
develop.

are high in protein and form a vital part of the human diet in many parts of the world. Soybeans are
known in China as ''poor man's meat,'' and they are
used to make *tofu*, common in Japanese recipes. *Tofu*
consists of protein concentrated from soybeans by
boiling them with magnesium compounds. The tasty
Arab dish called *hummus* is made from chickpeas,
another legume, and peanuts are widely used in
Indian dishes. Legumes are often combined with
starchy foods, which provide extra calories and
amino acids. Costa Rica's national dish is *gallo pinto*,
black beans and rice. Tortillas and beans are characteristic of Mexican cooking. Soybeans, peanuts,
beans, and peas are the most important legumes
worldwide (Figure 15-5).

> The most cost-effective human diet is plants
> that are high in starch supplemented by small
> amounts of legumes and other species that pro-
> vide amino acids and micronutrients that the
> main food plants lack.

## Energy and Food Production

Food takes a lot of energy to produce but, perhaps
surprisingly, in both traditional and industrial societies, the energy used to process, distribute, and cook
food is greater than the energy used to produce the
food in the first place. In the United States, it has
been estimated that each calorie on our dinner tables has cost 9 calories to put there. Half a calorie
represents investment on the farm; the rest represents the cost of processing, packaging, distribution,
and cooking. Packaging was the single largest contributor to the increase in retail food prices in the
United States between 1967 and 1987. It is obvious
that food production and distribution are vast consumers of energy, with animal products consuming
much more of this energy than plant foods.

In rural India, about twice as much energy goes
into cooking a kilogram of rice as was invested in
producing it. Energy shortages, especially for cooking, have caused environmental problems, such as
deforestation, in many parts of the world. In many

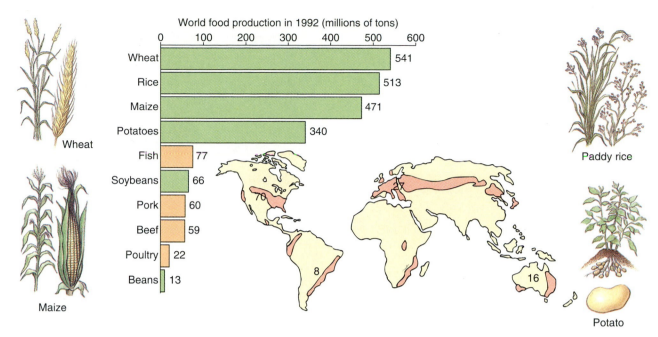

**Figure 15-6** World food production in 1992 (plants in green, animals in red) and the world's four most important food plants. Red areas on the map show the main grain-growing regions of the world. The numbers on the map show 1992 grain exports in millions of tons. Several grain-growing areas have no numbers, meaning either that they produce just enough grain for their own use or that they import grain. Note that North America produces most of the grain that goes to food-importing nations.

poor countries, cooking consumes more energy than transportation, heating, industry, and all other uses combined.

> Cooking and distributing food use more energy than producing it.

## 15-C GRAINS AND POTATOES

Nowadays the vast majority of human food comes from only four plant species: wheat, rice, maize (corn), and potatoes. But nobody can live on one of these plants alone because none of them contains all the essential amino acids. Here we take a closer look at the plant and animal species that supply most of our food.

Grasses, flowering plants of the family Gramineae, have been cultivated since the beginning of agriculture. Today they occupy more than 70% of the world's farmland (Figure 15-6). The grasses include the grains or cereals: among them, wheat, rice, maize, barley, rye, and oats. The part of a grass that we eat is its fruit (grain), which develops from an ovary of one of the many flowers in the compound flower head. The structure of a grain is shown in Figure 15-7. Each fruit develops a seed, containing a layer of stored food (aleurone) inside the seed coat. The embryo (germ) of the seed also contains fats, proteins, minerals, and vitamins. The endosperm, which makes up most of the grain, is mainly starch. When grains are polished to make white flour or white rice, the bran, aleurone layer, and embryo are removed and with them most of the grain's nutritive value except its starch.

Grains produce high yields, are easy to collect, and may be stored for long periods without spoiling, which is not true of plants such as vegetables. It is no wonder that civilizations have become so dependent on grains. Wheat and barley were the staple foods of early empires in the Middle East, rice was the main food of the ancient Asian civilizations, and maize fed the Incan and Mayan empires of Central and South America. In Africa, the most important grains are maize and sorghum. Of these grains, only maize is native to America.

> Since the start of agriculture, grains have provided most of the world's food.

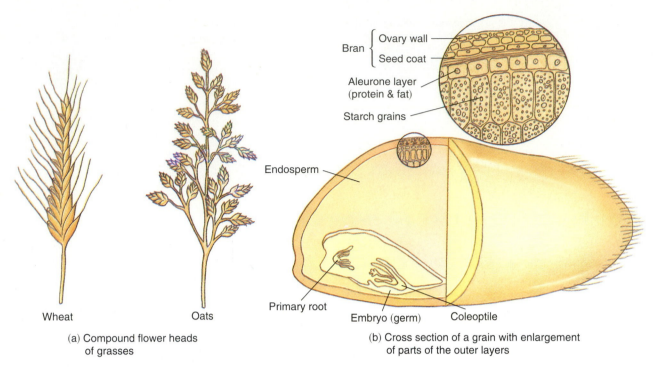

(a) Compound flower heads
of grasses

(b) Cross section of a grain with enlargement
of parts of the outer layers

**Figure 15-7**    The structure of a grain. (a) The fruits (grains) of all grasses develop from
compound flowers. (b) The endosperm (food for the growing embryo) contains most of
the carbohydrate in the grain in the form of starch. When grains are processed, the
nutritive bran, aleurone layer, and germ (embryo) are usually removed. The bran contains
fiber, and the other two parts contain most of the protein and fat in a grain.

## Wheat

Today's most widely grown plant, wheat, was also one
of the first two grains to be cultivated—in the Middle
East. The other ancient grain, barley, is today used
mainly as animal food and as the source of malt for
making beer. A number of other important food
plants are thought to have been domesticated in the
same area about 10,000 years ago at the same time
as wheat and barley.

Wheat grows best in cool climates with only moderate amounts of rain, and many parts of the world
are suited to its culture (Figure 15-8). Two species of
wheat are particularly important today: *Triticum aestivum*, used primarily for flour for bread and pastries,
and *Triticum durum*, used mainly for pasta. The wheat
that is grown today bears little resemblance to its wild
ancestors. Over the years, farmers and plant breeders

**Figure 15-8**    A wheat field in England. Because wheat is
adapted to cool weather and does not require large
quantities of water, it is grown without irrigation in much
of Northern Europe.

**Figure 15-9**  Some of the many products made from wheat. From left: whole wheat flour, bleached white flour, crackers, pasta, bread, cookies, and breakfast cereal.

the seeds are removed by milling. The seeds can then be ground into whole-wheat flour. Producers satisfy a demand for white flour by removing the bran and germ from the wheat seed and bleaching it with various chemicals (Figure 15-9). Although white flours have superior baking and keeping qualities, they are less nutritious than whole wheat. Therefore, laws in many countries require that producers enrich white flour with vitamins and minerals, expensively replacing some, although not all, of the nutrients they have removed during processing.

### Rice

In Japan each year, the emperor himself joins in the ritual harvest of rice on a paddy field within the palace grounds. Rice is revered in many parts of the East for its vital role in human life (Figure 15-10). Although it is not planted as widely as wheat, rice feeds more people. It constitutes the basic food of half of humankind, particularly those in heavily populated Asia. Rice is an Asian native, dating back to about 4000 B.C. in Thailand. Several other important food plants were first cultivated in Asia (Figure 15-11). Rice was introduced into the Carolinas in 1647 but was banned from much of the southeastern United States during the campaign to eliminate malaria. Mosquitoes that transmit malaria need standing fresh water, like that in rice paddies, to reproduce. Today California, Arkansas, Louisiana, and Texas are

have selected varieties with particular genetic characteristics: resistance to disease, rapid germination, short stalks so the plant does not fall over, and grains loosely attached to the plant so they fall from the plant when harvested. Hybridization, the crossbreeding of one variety with another, has produced more than a thousand varieties of bread wheat alone.

When a field of wheat is ripe, it is harvested, and the seed heads are separated from the stalks and leaves (wheat straw). Then the bracts surrounding

**Figure 15-10**  A rice field near Matsumoto, Japan. Little land is wasted in this small country, and rice is planted on nearly every available patch of ground. This temple has rice growing practically to its doorstep. *(Lee Marcott)*

**Figure 15-11**   Food plants thought to have been cultivated first in Asia. Left: onions, banana, garlic; bottom from left: grapes, bitter melon, bowl of chick peas, star fruit, carrot; middle from left: pea plant, bowl of rice, bowl of black peppercorns, ginger; top from left: branch of tea, lemon, mango, pear, orange, apple.

the rice-growing areas, and more than half of the rice crop is exported.

Much rice nowadays is grown on well-drained soil in areas with sufficient rainfall, but the yield is greater from traditional flooded paddies. This is because paddies contain nitrogen-fixing bacteria, which grow symbiotically with a water fern that lives in the paddies. As a result, heavy yields of rice are possible even with no added fertilizer. Typically on subsistence farms in Asia, a paddy is plowed using a water buffalo or ox. The dikes built around the paddy to hold water are built so that the paddy can be flooded with water, usually diverted from an irrigation ditch. The paddy is then drained and the rice planted, either by sowing seed or by transplanting seedlings, backbreaking work that produces a higher yield and is usually performed by women and children. The paddy is flooded several times to supply the plants with water, and the rice is weeded by hand, another exhausting chore.

Rice is harvested with sickles or knives. Then the rice is threshed to remove the seed heads from the stalk by beating it against a log or by having animals or humans trample it. Winnowing, often by tossing the rice in a flat basket, permits the lighter chaff to blow away from the seed. The hull is then removed using a mortar and pestle. This process leaves brown rice that is usually consumed locally. On modern farms, machinery is used to perform all these steps. The thousands of varieties of rice are generally divided into two major groups. Japonica types have

short grains and are sticky when cooked, indica types have long grains and are drier.

When it is prepared industrially, brown rice is whitened by removing its outer layers. As with wheat, this removes much of the nutritional value. The loss of nutrients, particularly vitamin B, during milling and cooking has been responsible for much malnutrition among people whose diet consists almost entirely of white rice. The most widespread deficiency disease is **beriberi**, caused by a lack of B vitamins and leading to blindness and death. In traditional societies where people live mainly on rice, supplements that increase the amino acid and vitamin content of the diet are common. Examples are the fermented fish sauce of Vietnamese cooking, tofu, and soy sauce. Most modern soy sauce is produced by the action of hydrochloric acid on soybeans and has less nutritional value, but traditional shoyu, produced by fermenting soy with bacteria and fungi, is very nutritious. Both of these sauces add vitamins and amino acids, produced by bacteria and fungi, to the diet. Fish sauce also provides B vitamins, which are present in all meat.

## Maize (Indian Corn)

Maize, native to Central and South America, was the most important cultivated plant of prehistoric America (Figure 15-12). Native North Americans were introduced to maize and beans by traders who brought the seeds north from Mexico. Maize is not as good a

**Figure 15-12**   Food plants first brought into cultivation in the Americas. Back row from right: shell beans, butternut squash, papaya, pineapple; middle from right: avocado, potato, sweet potato; front: green beans, wild rice, tomato, maize, peppers, taro.

human food as wheat. Like rice, it has a lower protein content, and it also contains too little of the amino acids tryptophan and lysine and of the vitamin niacin. Deficiencies of these nutrients cause the disease **pellagra**. Pellagra, characterized by weakness so acute it is often diagnosed as mental deficiency, was common in the United States among the rural poor until the 1950s, and it is still occasionally seen. Plant breeders continue the search for high-vitamin maize, with some recent success. In Central and South America, where maize is a staple food, it is usually eaten with legumes, which make up for its nutritional deficiencies.

In the United States, about 80% of the maize grown is fed directly to animals, but in Africa and Central and South America, nearly all of it is used directly as human food. Maize is also used in some 200 products ranging from ethanol and cough syrups to tires. Since it grows so well in the United States, researchers are constantly looking for new uses for this plant.

### Potatoes

In areas unsuited to growing grains, plants with a high starch content, such as potatoes, yams, sweet potatoes, manioc, taro, and bananas, have long formed the staple foods. Although not closely related, all these plants are of tropical origin and are high in calories, although not in protein. Potato-eating Irish peasants always kept a cow if they could possibly afford one: milk supplied the amino acids that potatoes could not.

Potatoes apparently originated in the Andean mountains of South America. They are a very efficient crop, producing more calories per hectare than wheat, rice, or maize. Today scientists at the Potato Research Institute in Chile are searching for varieties used by the local Indians that can be used to breed improved varieties. Potatoes do not grow well in lowland tropical areas, where their place is taken by manioc (also called cassava and yuca).

> Wheat, rice, maize, and potatoes supply most of the world's food. They are nutritionally adequate when supplemented by small amounts of essential amino acids and vitamins.

## 15-D LIVESTOCK

As well as feeding humans directly, the grain and leaves of grasses are the main food of most domestic animals. Although plants are much more important than animals as human food, there are places in the world where animals can be grown more efficiently. Livestock herds have long been the basis of life for populations in North Africa, Lapland, and parts of South America. People who rely heavily on grazing livestock must have access to large areas of land. Thus, many North American cattle farmers graze their cattle partly on their own fenced fields and partly on federal land where they have rights to graze a certain number of cattle. North African herders are largely nomadic, following the rainfall with their herds in search of fresh grass.

The use of grazing land often comes into conflict with settled agriculture, as it did in the "range wars" of the American West. As more and more land is fenced and plowed for crops, the area of grazing land decreases. We shall probably reach the point at which nearly all animals grown for food are not allowed to graze at all but are raised in feedlots and modern chicken-farming operations, with all their food delivered from elsewhere.

Only about 50 animal species have ever been domesticated, meaning that their reproduction is controlled by humans, and this figure includes honey bees, silkworms, fish, and shellfish. Only ten or so of these species are of any importance. If we omit cats and dogs, we find that chickens are the most common domesticated animals; then come sheep; and cattle are third. Goats, pigs, and water buffalo are also important in many parts of the world.

One group of animals has furnished us with 15 of the 22 most important domesticated animals: the artiodactyl order of mammals, the even-toed hoofed mammals. Of these, the **ruminants**, the cud-chewing animals, include three of the most important, as well as the first to be domesticated: cattle, sheep, and goats. Ruminants contain microorganisms in their guts that permit them to digest the cellulose cell walls of plants, which humans cannot digest. People can therefore use ruminants to convert plant material, such as grass stems and woody shrubs, into food that they can digest: beef, mutton, or goat.

Sheep were the first ruminants to be domesticated, valued for the wool of their fleeces as well as for their meat. Sheep became widespread because they can tolerate a wider range of temperatures than goats. They have been replaced by goats in some poor farming areas today because goats will eat woody shrubs and other vegetation that sheep will not eat. As a result, overgrazing by goats does more damage to pasture than overgrazing by sheep. Cattle are particularly widespread in North America, India, and Africa. They are not always slaughtered for meat. In Africa, for instance, the Masai eat milk and blood from their cattle but seldom kill them for their meat.

India contains almost one quarter of the world's cattle. Cattle are sacred among the Hindu religious majority, who will not kill them. Hindu failure to use cattle for food has been much criticized by outsiders because the cattle eat vegetation that would otherwise feed people or at least prevent soil erosion. The cows provide milk, and the bullocks are used to pull plows and carts; the dung is used for fertilizer and fuel. It nevertheless seems clear that India would be better off with fewer, more productive, animals.

> Chickens are the animals most widely used for food. Ruminants are valuable because they can convert plant material that humans cannot digest into meat and milk.

## 15-E  AQUACULTURE

Catfish, shrimp, oysters, salmon—if you buy them in a store, they may well have been produced in a fish farm, not in the ocean. Fish and shellfish are important sources of protein, and most of them still reach the table from hunter-gatherer activities with fishing rods, nets, or spades to dig up clams. But growing amounts come from **aquaculture**, the cultivation of organisms in water. The main species produced by aquaculture in the United States are catfish, trout, salmon, crawfish, and oysters. In the 1980s, production of these species quadrupled, and imports grew even faster (Figure 15-13). The U. N. Food and Agriculture Organization projects a worldwide increase of 500% in food from aquaculture within the next 30 years. Competition with fish produced cheaply by aquaculture is putting fishing boats out of work in many areas.

Although it is a valuable source of relatively cheap protein, aquaculture can cause environmental problems. The fish produce large quantities of nutrient-rich feces, which must be disposed of properly if they are not to pollute waterways. China has taken the lead in developing solutions to this problem in the form of **integrated freshwater systems**, ponds and wetlands that perform more than one function. For instance, one village raises ducks in its rice paddies. Duck droppings fertilize the rice and the ducklings eat insects, reducing the need for chemical insecticides. The rice paddies are flooded several times a year from nearby ponds. Waste plant material and filtered sewage are dumped into the ponds, where they serve as food for carp and other fish. Mulberry trees are planted around the ponds, and silkworms feed on their leaves. The trees get all the water they need from the ponds, and the silkworm droppings

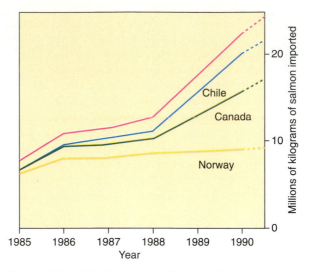

**Figure 15-13**   The increase in food produced by aquaculture imported by the United States between 1985 and 1990. Salmon imported from Norway and Canada and shrimp imported from Chile are considerably cheaper than the same species produced by fishing for them in the United States. As a result, many U.S. fishing boats are put out of business each year.

fall into the pond and fertilize it. The whole system recycles waste; produces silk for sale; produces rice, ducks, and fish for food; and controls insects and other pests. In addition, the system uses little fresh water and recycles that water many times.

> More and more of the world's protein comes from aquaculture. Integrated freshwater systems like those developed in China are cost-effective ways of producing protein.

## 15-F  THE GREEN REVOLUTION

The production of wheat in Mexico increased more than eightfold between 1950 and 1970. The land planted to crops doubled, and the yield per hectare quadrupled. In the same period, India doubled its production of grain and now has food reserves, a future that appeared impossible 20 years ago. These spectacular increases, hailed as the "green revolution," resulted from intensive efforts to breed new strains of grains. Unlike traditional crops, these new varieties produce enormous yields if, and only if, they are supplied with large quantities of water, fertilizer, and pesticides (Figure 15-14). A vital part of the green revolution, therefore, was educating farmers in "high-input" agriculture: the use of chemicals, irrigation, and farm equipment. The economic struc-

**Figure 15-14** A field of poor-quality maize in Central America. Maize requires large quantities of nutrients and quite a lot of water. When it is not adequately fertilized, it produces poor growth like this.

ture of agriculture in developing nations also had to be changed so farmers could borrow money to buy these things. The green revolution has transformed the lives and prospects of hundreds of millions of people and is one of the most successful achievements of international development since World War II.

> The development of new grain varieties suited to high-input farming led to the huge increases in farm productivity known as the green revolution.

However, the green revolution has limitations and drawbacks. By the late 1960s, food production in Mexico began to level off as the green revolution ceased to recruit new farmers. The revolution in agriculture had reached large farms but had no effect on less prosperous farmers, who work 80% of Mexico's agricultural land but produce only 55% of its crops. These poorer farmers occupy the worst land, are often illiterate so they cannot read agricultural literature, and have less money and are less able to borrow it. As a result, the green revolution has had little effect on small farmers in most countries. High-yielding varieties of rice and wheat are planted on less than a third of the grain-growing land in developing countries. More than 1 billion small farmers still raise crops at yields that have hardly increased during the twentieth century.

In addition to its failure to reach many small farmers, the methods of the green revolution caused environmental problems that are now reducing food production in many areas. Grain production in the United States declined during the 1980s, with 1988 becoming the first year in history when the United States produced less grain than it consumed (Table 15-2). All over the world, crop yields per hectare are falling or rising very slowly—too slowly to keep up with population growth. The main reasons for the decline in production are shortages of land, fresh water, and fertile soil. We have seen that irrigation, pesticides, and chemical fertilizers eventually destroy the fertility of soil (Chapter 14). Crop yields are usually higher on irrigated lands, but we are running short of fresh water and the area of land that is irrigated is falling. In 1982, 5% of U.S. farmland was irrigated. By 1992, the area had fallen to 4%. The

**TABLE 15-2** U.S. Grain Production, Consumption, and Surplus Available for Export, 1984–1991*

| Year | Production | Consumption | Exportable surplus |
|---|---|---|---|
| | (In millions of tons) | | |
| 1984 | 313 | 197 | 116 |
| 1985 | 345 | 201 | 144 |
| 1986 | 314 | 216 | 97 |
| 1987 | 277 | 211 | 66 |
| 1988 | 190 | 202 | −12 |
| 1989 | 243 | 216 | 27 |
| 1990 | 267 | 223 | 44 |
| 1991 | 238 | 229 | 9 |

*Data from the U.S. Department of Agriculture, Economic Research Service.

(a)

(b)

(c)

(d)

**Figure 15-15**   Some of the main plants grown by tropical countries for export to temperate countries where the weather is too cold to grow them. (a) A banana tree. Bananas are easy to grow. Each leafy stem grows, flowers, and produces fruit in about 18 months. After the fruit is harvested, the stem is cut to the ground, and new stems grow from the roots. (b) Cacao. Cocoa and chocolate are made from the large colorful fruits, which develop from flowers that grow directly from the stem of a small tree. (c) Sugarcane, with sugar in its stems. Sugarcane is very easy to grow, since it grows again from the roots when the stems are harvested. Most of the world's sugar, however, comes from sugar beets (bred from ordinary beets), which are more productive, although the sugar is not as tasty as that made from sugarcane. (d) Coffee. This coffee bush is in Costa Rica, which grows wonderful coffee, nearly all for export to Finland. Coffee needs shade and is usually grown with bananas or macadamia nuts, which provide that shade. Costa Rican research stations have now developed varieties that will tolerate full sun, which makes it possible to harvest coffee by machinery, which cannot be used when shade trees are interplanted with the coffee. Coffee, like maize, needs many nutrients, so it grows well in fertile volcanic soils in Central and South America.

decline will continue as more and more water is diverted from agriculture to industry and cities (Section 11-E).

> The green revolution produced enormous increases in food production, but new methods will be needed to increase production in the future.

## 15-G  PRODUCING MORE FOOD

North America is the world's largest producer of food, and this will probably always be the case because this continent has more areas with soil and weather suitable for agriculture than any other. Many countries will never be able to produce enough food to feed their people and will have to import some food. In some tropical countries, fertile soil is used to produce high-value crops for export rather than food for local consumption (Figure 15-15). This works well when the exported crops fetch high prices but can lead to famine in years when prices are low and exports produce barely enough to pay for food that must be imported. Most food-importing countries could provide more of their own food than they do now with suitable research and development. Encouraging such research and development is an important goal of the world's agriculture research stations and development banks.

The land that the green revolution reached is now producing close to the maximum possible crop yields. Because of this, agricultural experts predict that further increases in food production will come largely from small farmers who have been little affected by the green revolution. Let us first consider how these small farms compare with modern, mechanized farms.

The world's farms can be loosely divided into **subsistence farms**, which provide the family with food and sometimes a small crop that can be sold for cash, and **modern farms**, where crops are raised to be sold. Modern farms are much more productive in terms of output per person per year, but they often produce less food per hectare, often because the plants must be far enough apart to permit machinery to move between the rows. In addition, mechanical harvesters, which are used to harvest the crop, waste an enormous amount of food that is damaged by the machinery. A subsistence farmer, or a vegetable gardener, usually produces much more per year from the same area of soil than a modern farm, by planting one crop between the rows of another and by planting another crop in places where the first is harvested (Figure 15-16).

Because small farms can produce more food on a given area of land than large farms, placing more land in the hands of small farmers would increase food production. In China, small farms make up much of the world's most efficient agricultural system, which feeds more than 20% of the world population on less than 10% of the arable land. In other parts of the world, land has been subdivided by inheritance to the point where each family has a small-

**Figure 15-16**  A vegetable garden, much more productive than the same area of farmland. Look at the newly planted rows in the background, which are only just far enough apart for someone to get between them to weed and harvest. The sticks are for peas and beans to grow up. Peas and beans are much more productive when grown up sticks or trellises in this way, but they cannot be harvested by machinery, so less productive, bush varieties are grown in commercial vegetable-growing fields. *(Thom Smith)*

**Figure 15-17** The falling number of small farms in the United States. When a small farm goes out of business, the farmhouse is usually abandoned and the land added to a neighboring large farm.

holding of less than half a hectare. Merging these smallholdings into large modern farms reduces food production and throws even more of the rural poor out of work. However, large farms produce crops more cheaply than small farms, partly because they employ relatively fewer people and partly as a result of higher government subsidies for large farms in most countries. As a result, farms in most of the world are getting larger and larger. There were more than 4 million farms in the United States in 1960. Today there are half as many, covering nearly 85% as much land. Half a million small American farmers went out of business between 1975 and 1990, and their land was added to large farms (Figure 15-17).

## Low-Input Farming

The world's remaining small farms must become more profitable if we are to increase the food supply by very much. Agriculturalists believe that any increases in food production will have to come mainly from low-input farming (Section 14-H), which requires less financial investment from the farmer than the methods of the green revolution and causes less environmental damage. Organic and no-till farming save money by avoiding the use of large quantities of energy, pesticides, fertilizer, and water. Consider the use of water for irrigation, which is becoming more and more expensive as a result of shortages all over

the world. Any farmer who can produce the same amount of food, while using less water, can make the farm more profitable. Organic farmers emphasize keeping the soil as moist and fertile as possible by spreading manure, compost (Section 14-H), and other organic matter; keeping the soil planted at all times to avoid erosion; and interspersing different crops to reduce pest infestations. Experiments in Africa showed that maximum yields were obtained when fields were treated with compost with a small amount of inorganic nitrogen fertilizer instead of with large quantities of inorganic fertilizer (Table 15-3).

**TABLE 15-3** Effects of Organic and Inorganic Fertilizers on Sorghum Yields in Burkino Faso (North Africa)

| | Sorghum Yield (Tons/Hectare) | |
| --- | --- | --- |
| Treatment | Without Inorganic Fertilizer | With 60 kg Nitrogen per Hectare |
| No organic treatment | 1.8 | 2.8 |
| 10 tons/hectare manure | 2.4 | 2.4 |
| 10 tons/hectare compost | 2.5 | 3.7 |

Plant breeders are beginning to produce the crop varieties needed by organic farmers—varieties that produce high yields without irrigation, pesticides, or fertilizers. Agriculture specialists have also realized that the most efficient sustainable agriculture is that suited to a particular area. The agriculture of developed nations often cannot be exported efficiently to developing countries where the soil and climate are different. This realization has led to the establishment of agricultural research stations in many developing nations, and particularly in tropical areas, in attempts to find efficient, sustainable agricultural methods for tropical countries with skyrocketing populations.

## New Uses for Old Plants

In a Mexican trial plot, researchers pore over the scarlet heads of amaranth, sacred food of the ancient Aztecs (Figure 15-18). Quinoa, an annual herb with grain-like seeds, grows nearby, as does a new variety of buckwheat. Quinoa thrives with low rainfall and at high altitudes. Buckwheat is adapted to arid land and cool climates. Amaranth is drought tolerant and contains more of the amino acid lysine than do grains. Researchers around the world are taking a new look at some of the food plants eaten by our ancestors, which are seldom planted today as the world has become more and more dependent on the crops developed during the green revolution. They are breeding improved varieties of plants such as jojoba for the oil it produces; this shrub, native to arid lands, is grown in arid areas without irrigation. It would also be very advantageous to develop crops adapted to the growing area of land that is naturally saline or has been salinized by irrigation (Close-Up 15-1).

**Figure 15-18**  Grain amaranth, a productive plant that is the subject of considerable agricultural research. Amaranth is also increasingly popular with home gardeners in the United States as a substitute for spinach in areas with hot summers.

One of the most valuable of the rediscovered plants may prove to be a hybrid between wheat and rye called triticale. Triticale was hybridized a century ago and has now been bred so that it is disease-resistant. More important, it produces much higher yields than wheat on poor soils. These include soils containing levels of salt, acid, and boron that are toxic to most plants, as well as soils that are sandy, acidic, alkaline, and mineral-deficient. We also know of varieties of maize and wheat that will grow with little water or fertilizer. Plants such as these may well be the crops of the future for low-input and subsistence farming.

## The Chinese Model

In 1992, a United Nations study reported that fewer people are malnourished now than 30 years ago. This is almost entirely a result of the spectacular success of the Chinese in eliminating malnutrition among their people, who make up almost one quarter of the whole human population. This is a remarkable feat in a poor country with a population density four times that of the United States. In the early twentieth century, China contained more starving people than all the rest of the world. Today the Chinese are well fed. The Chinese not only produce massive amounts of food but, equally important, ensure that it is shared by everybody.

The Chinese have led the way with organic farming, which uses land as efficiently as possible and recycles wastes. Labor is cheap, and agriculture is labor-intensive, offering jobs to millions of rural people. The country contains the largest irrigation network in the world, enabling it to grow more than one third of the world's rice. Abundant labor permits intensive cropping in one field. Here beans and other legumes, which supply nitrogen to the soil, are grown in alternate rows with crops such as wheat that require a lot of nitrogen. Labor-intensive pest control permits farmers to spray only against outbreaks of specific pests—a method that is much more cost-effective and less damaging to the environment than the broadscale spraying practiced by most farmers in developed countries.

Waste recycling is illustrated by Chinese aquaculture. In a typical pond, carp occupy the top layers of water, feeding on waste grass, sugarcane fiber, and banana leaves that are thrown into the pond. Fish called dace feed on scum near the bottom. The fertile sludge at the bottom of the pond is removed periodically and used to fertilize fields.

China has a massive reforestation program, but it still faces a fuelwood shortage. To ease this, bio-gas

## Water Shortages and New Crops

The world's increasing shortage of fresh water for irrigation has led plant breeders to seek out drought-resistant plants and even plants that can be irrigated with sea water. Now, they are beginning to succeed. In the 1980s, field trials in Mexico confirmed that a useful oil-producing crop can be irrigated with salt water.

*Salicornia bigelovii* is a succulent adapted to saline soils (Figure 15-A). It grows naturally in higher and drier parts of salt marshes and on saline soils inland. Its common name, glasswort, reflects the crisp texture of its leaves. With crisp texture and a salty taste, glasswort leaves are a prized salad ingredient in Europe. They sell for about $25 a kilogram in Paris markets.

Glasswort seeds can be crushed to yield an oil that is high in unsaturated fatty acids and in protein. In the Mexican trials, researchers grew glasswort in coastal desert and irrigated it with sea water. Seed production equaled or exceeded that of sunflower and soybean plants irrigated with fresh water. Humans may not like the taste of glasswort oil, but both the oil and meal that is made from the crushed seeds can be used as chicken feed, saving for human food other oils and wheat now used for chicken feed.

Another use for plants like glasswort is to produce useful yields from saline soils. Most conventional

**Figure 15-A** Glasswort, *Salicornia bigelovii*. This plant is native to salt marshes and is a tasty addition to salads; it also produces seeds from which oil is extracted.

crops can be grown in saline soils only if they are irrigated with large quantities of fresh water. However, most parts of the world with saline soils are in arid areas and also suffer from shortages of fresh water. Crops such as glasswort can be grown on soils that are either naturally saline or have been salinized by years of irrigation. Freshwater shortages and saline soils are now widespread, and we can expect the use of plants adapted to salty conditions to increase rapidly in the future.

**Figure 15-B** The dry, salty boundary between salt marsh and sand dunes: the home of glasswort.

**Figure 15-19** A food market in China. Since World War II, China, with less than 10% of the world's arable land, has done a spectacular job of producing and distributing food to the world's largest population. *(Tricia Smith)*

stoves and digesters, fueled by waste from humans, livestock, and plants, may supply up to half of a village's electricity needs. The digesters also take care of sewage disposal. The Chinese consider human and animal excrement as fuel or fertilizer and view Western sewage disposal systems as very wasteful.

The Chinese system is a model of the kind of sustainable agriculture that could feed and provide employment for most of those who are starving in the world today (Figure 15-19).

## 15-H THE POLITICS AND ECONOMICS OF HUNGER

Unfortunately, for political and economic reasons, few countries are likely to follow the Chinese model and increase their food production much in the future. Chinese communism is a political system that can thrust centrally planned agriculture on people whether they like it or not. Such coercion would not be possible in democratic countries such as India and Bangladesh. Nevertheless, we may note that the greatest gains in Chinese food production have come since 1979, when farmers were first permitted to sell their products in the open market. China's present delicate balance between central planning and individual initiative could, in theory, be imitated by other countries with strong governments. But the governments of most developing countries are not strong. Most of the world's malnutrition occurs in

countries in southern Asia (such as Java, Philippines, Bangladesh, India, Myanmar) and Africa (such as Sudan, Somalia, Nigeria, Mozambique, Kenya) in countries with ineffective governments.

The food situation in these countries will get worse in the future. This is partly because the industrialized nations are putting few resources into helping to prevent starvation. There is little money to be made from providing family planning and new crops to very poor countries, and most research in developed countries is designed to produce a profit. Consider the case of biotechnology.

### The Role of Biotechnology

Most biotechnology companies are involved in **genetic engineering**, the technology used in transplanting genes from one organism to another. Until 1975, the only way of improving crop varieties was by selective breeding. This means collecting seeds from plants with superior genetic characteristics and planting them next year, while preventing inferior plants from breeding. Superior varieties are also crossed with one another to produce hybrids that may have the desirable characteristics of both parents. Plant breeding programs of this type have produced all of our crop plants. For instance, the plants with large tomatoes that we eat today have been developed from a plant species with tiny, almost tasteless fruits that grows in South America. Genetic en-

# Beef: Unsustainable Agriculture

America's love affair with beef has caused massive environmental damage. Cattle are natives of the world's grasslands, but grasslands make the best arable land, so cattle are usually relegated to areas to which they are not well adapted. These include the semiarid scrub region of the western United States, where cattle grazing has degraded thousands of square kilometers of land. Several tropical countries have felled large areas of forest and planted grass instead to produce cheap beef for the American market.

Beef cattle may be raised on pasture and then transported to feedlots where they are fed prepared food to fatten them for market. Alternatively, they may spend their whole lives in feedlots. Either way, it takes more food to produce beef than any other meat (Figure 15-C). On dairy farms, the manure produced by the cows is usually spread on the fields, restoring nutrients and organic matter to the soil. In feedlots, more manure is produced than can be spread on local fields, and it is so heavy that it is uneconomical to transport to distant fields. Some of the spare manure is dried and sold to home gardeners, but much of it is never returned to the soil and causes a huge waste disposal problem, frequently polluting local waterways.

Raising beef also consumes vast quantities of fossil fuel (Figure 15-D). The fuel is used for transportation, to fertilize and irrigate pasture, and to grow crops used to feed the cattle. Finally, enormous amounts of water are used in beef production (Figure 15-E). This is largely because most beef cattle are raised in Texas, California, and other western states where rainfall is low. As a result, crops raised

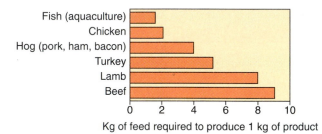

**Figure 15-C**   The amount of animal feed required to produce various forms of meat in the United States *(Data: M. E. Ensinger)*

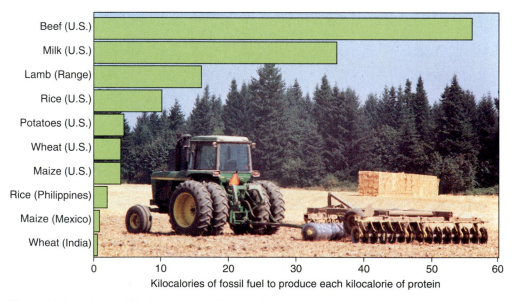

**Figure 15-D**   The fossil fuel energy used to produce food protein in various types of farming. Most of the fossil fuel powers farm machinery and the production, distribution, and spreading of fertilizer, pesticides, and food for livestock.

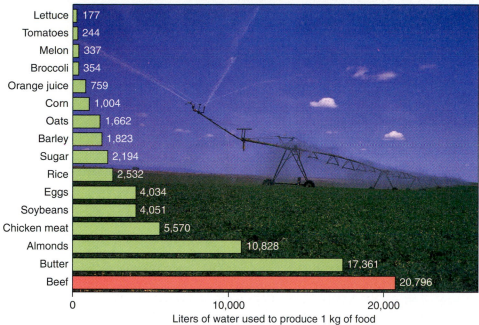

| Food | Liters |
|---|---|
| Lettuce | 177 |
| Tomatoes | 244 |
| Melon | 337 |
| Broccoli | 354 |
| Orange juice | 759 |
| Corn | 1,004 |
| Oats | 1,662 |
| Barley | 1,823 |
| Sugar | 2,194 |
| Rice | 2,532 |
| Eggs | 4,034 |
| Soybeans | 4,051 |
| Chicken meat | 5,570 |
| Almonds | 10,828 |
| Butter | 17,361 |
| Beef | 20,796 |

Liters of water used to produce 1 kg of food

**Figure 15-E** The volume of water used to produce various foods in the United States. *(Water Education Foundation)*

as cattle feed are nearly always irrigated, and even pasture grasses are often irrigated (Figure 15-F). This irrigation is heavily subsidized. In 1991, city dwellers in California were paying the federal government up to $500 an acre-foot for water, while farmers growing alfalfa for cattle feed were paying $5.

As water shortages get worse, these huge subsidies for irrigation water are being reduced. Since crops for cattle feed do not sell for high prices, it will be-

come less worthwhile to spend the money to irrigate them. Water will be saved for high-value crops such as vegetables and fruit, and we can expect beef to become more expensive as production declines. We can also probably expect more of our meat to come from animals such as bison (buffalo), which are well adapted to western rangelands, use much less in the way of natural resources, and do much less damage to the environment.

**Figure 15-F** Cattle on irrigated pasture in Montana. The hillside behind the field contains gray sagebrush. This is arid land where grass is sparse unless the land is irrigated.

gineering is a much faster way of developing new plant varieties than traditional plant breeding.

**Herbicide Resistance** Biotechnology certainly could be used to produce crop varieties that would help to feed the starving, but that is not the goal of most research at the moment. The Monsanto corporation advertised its biotechnology program with a photograph of a single corn plant growing in a desert, above the words "Will it take a miracle to feed the world?" The message was that genetic engineering will produce crops that will grow even in the desert. Is Monsanto really about to release drought-resistant crop varieties? No. Its first genetically engineered products are varieties of soybeans and canola (rape) that can withstand high doses of Roundup, Monsanto's best-selling herbicide (weed-killer). Monsanto's goal: to increase sales of Roundup.

The owners of mechanized farms want herbicide-resistant crops so that they can weed their fields by simply spraying them with herbicides. Plant breeders have been trying to produce herbicide-resistant crops for decades, and genetic engineering has now achieved that goal much more rapidly. However useful they may be in North America and Europe, herbicide-resistant crops pose great dangers to agriculture in developing countries, where they will certainly be sold. Major crops such as wheat, rice, potatoes, and soybeans originated in South America and Asia. Wild relatives of these plants often grow in fields as weeds alongside cultivated crops. Scientists worry that herbicide-resistant crop plants might breed with their wild relatives, producing hybrid weeds that are resistant to herbicides. Then herbicides would be essentially useless to farmers in these countries.

The other major disadvantage of genetically engineered crops is that they make farming even less sustainable than it is now (Close-Up 15-2). Agriculture that depends on large amounts of herbicides destroys the fertility of the soil and requires large amounts of inorganic fertilizers to compensate. Farming becomes ever more concentrated in the hands of large corporations that have the money to buy all these chemicals. This is precisely the wrong direction for agriculture to take if we hope to increase the world food supply and reduce environmental damage.

**Sustainable Agriculture** On a more hopeful note, in universities and government research stations, where making money is not the only goal, scientists are working on biotechnology projects that will contribute to sustainable agriculture and food production.

Recall that legumes are valuable as food plants and soil improvers because *Rhizobium* bacteria in their roots fix nitrogen. Scientists are attempting to produce *Rhizobium* associations with plants other than legumes. Agriculture would benefit if more plants could be grown without nitrogen fertilizer—reducing pollution from fertilizer runoff and the use of fossil fuels. In addition, breeders aim to produce a human food plant that contains "perfect" protein. Legumes contain a lot of protein, but they contain little of several essential amino acids. These amino acids are produced in large amounts by plants such as potatoes, maize, and wheat. If breeders could produce potato, maize, or wheat varieties with *Rhizobium* associations, these plants might be a perfect human food and enormously helpful in feeding the human population.

Despite the great potential value of this research, the major seed or chemical companies that might bring such crops to market show little interest in this work because they do not see any profit in it for many years to come.

## Poverty and Urbanization

Unless caught up in a disaster such as a war, no one with enough money need lack food anywhere in the world, so the problem of hunger is largely the problem of poverty. Malnutrition in the United States, for instance, increased rapidly during the 1980s as federal and state governments reduced payments for programs such as food stamps for needy mothers and infants.

Strange as it may seem, most of the poverty and hunger in the world exists among rural peoples, those accustomed to working the land and growing their own food. Population increases have left more and more of these people without land, and the rural poor have less political power than any other segment of the population. Hard-pressed governments focus their attention on feeding the people of the rapidly growing cities. These are the people who can overthrow a government, who participate in food riots, like those that occurred in the 1980s in Warsaw and Cairo, and who can command the attention of the press. An extraordinary example of the political power of the cities comes from the Soviet Union under Stalin in the 1930s. The government shipped so much food from rural areas to the cities that hundreds of thousands of people starved to death in the villages of one of the world's most productive agricultural areas.

To make matters worse for the rural poor, in times of food shortage urban populations are easier to feed. Food aid from abroad can be distributed to an urban population by a few helpers. When U.S. troops arrived in Somalia in 1993 to help relief workers feed the starving, they were amazed to find most of the urban population well fed and healthy. The starving were in the country. Feeding the rural poor may involve, as it does in Ethiopia, Somalia, and Sudan, sending trucks through countryside torn by civil war to widely scattered groups of people. The only way to feed these people on a permanent basis is to improve rural agriculture so they can feed themselves. Agricultural improvement in the countryside has other benefits: a prosperous rural population may produce surplus food that can be used to feed those in the towns. But only a few developing countries have given much priority to agricultural development. Even where agriculture has been emphasized

and improved, if the rural population is still growing rapidly, as it is in Kenya, improvements in agriculture usually cannot keep up with population growth.

One unknown factor in the world's food equation is the former Soviet Union. Countries such as Ukraine and Russia have large temperate land areas and fairly small populations. They might become major food producers if and when they sort out their economic and political problems. Agriculture has long been unproductive in the Soviet Union, where land was owned by the state and workers were not rewarded for productivity.

The solution to the world's hunger must involve political action on many fronts, including population control, land reform, agricultural improvements, and the development of nonagricultural jobs for the rural unemployed.

## SUMMARY

Food production faces many problems that threaten our ability to feed the growing human population.

### Nutrition

A healthy human diet includes sufficient macronutrients to supply energy, lipids, and essential amino acids and small quantities of micronutrients to supply minerals and vitamins. Lack of essential amino acids is the most widespread dietary deficiency. Most of the millions of people who starve every year die of diseases that would not prove fatal to the well nourished.

### The Foods We Eat

We eat very few of the available species of plants and animals. Our food supply is dangerously dependent on a few species of plants. In most of the world, it is much more cost efficient for people to get most of their food from plants with an adequate protein content and from eggs and milk than to eat meat.

### Grains and Potatoes

The bulk of the world's food consists of wheat, rice, maize, and potatoes. These are adapted to different soil and climates. Grains are particularly important because they are nutritious, productive, and cheap to store. Processing grains by removing the seed's outer layer reduces their nutritional value. Diets high in grain are traditionally supplemented by legumes and by sauces high in amino acids and vitamins to

provide enough essential amino acids and micronutrients.

### Livestock

Chickens and ruminants are the most widely grown livestock. Animals form a large part of the human diet mainly in affluent societies. They also serve to convert indigestible plant parts into milk and meat, particularly on land unsuitable for crops.

### Aquaculture

Aquaculture provides an increasing share of the world's protein. Particularly cost-effective are Chinese integrated freshwater systems, which produce fish and other protein, grow food plants, dispose of waste, and reduce pests all within one recycling system.

### The Green Revolution

We have managed to grow enough food for the exploding human population since World War II only because seed companies have spread high-input agriculture and new varieties of crops from the developed nations to the rest of the world. New varieties of grains produce very high yields when grown by high-input methods with large amounts of fertilizer, pesticides, and water. They account for most of the increase in food production. The green revolution has reached most of the farmland where it can be used. Food production is slowing down as productivity is reduced by water shortages and by adverse effects of high-input farming, such as soil destruction.

## Producing More Food

Most agriculturists think that future increases in food production will come mainly with greater productivity from low-input organic and small farms, which may prove more sustainable than high-input farms. Breeders are searching for plant varieties that produce high yields with low input. The Chinese model of labor-intensive, highly productive agriculture is one that might be imitated in many developing countries.

## The Politics and Economics of Hunger

Politics and economics dictate that starvation will increase. Few developing countries have the kind of strong government that was needed to permit the Chinese to increase that country's food production enormously in the last 50 years and to ensure that enough food reaches all members of the population. Expensive research in biotechnology is not being used to produce the plants needed to feed the starving. It is being used mainly to produce products such as herbicide-resistant crops, which increase the profits of corporations in industrialized nations. Hunger is most widespread, in every country, among the rural poor, who have little political power.

## DISCUSSION QUESTIONS

1. Suppose you were charged with designing an integrated freshwater system for your local area. What would you need to know to set up a cost-effective system?

2. Most of the food production gain of the last 40 years has come from increased yield per hectare. (Although much previously unfarmed land has been brought into production, an approximately equal area has been lost from production.) Do you think that productivity per hectare is likely to increase in the next 20 years in intensively farmed areas such as the American grain-growing area? Why or why not? Under what conditions might your answer be wrong?

3. List as many areas as possible in which food production will increase in the future (for example, more fishing, converting more forest to farmland).

4. The green revolution increased deaths from starvation in several developing countries. How?

5. In attempts to feed the world's population in the twenty-first century, will it be more cost-effective to invest in food production or in family planning?

6. One of the problems with introducing new food crops is that people are very conservative in what they will eat. For instance, in Europe, people consider the brains, pancreas, and other organs of livestock as delicacies, whereas in the United States such meats are considered fit only to be turned into pet food. How conservative are you and your friends? When you see an unknown vegetable in a supermarket, are you likely to buy it? Can you think of ways of encouraging people to try new and unusual food plants?

7. Look at Figure 15-C. Why do you suppose it takes so much more fossil fuel to produce rice in the United States than in the Philippines?

8. Look at Figure 15-D. If you reduced your consumption of beef, would you feel that you were contributing to overcoming our shortage of fresh water? Why or why not?

9. Figure 15-4 suggests that people in North America and Europe are eating considerably more food and more protein and lipids than they need. How might we encourage people to reduce their food consumption?

10. American troops were used to help combat famine in Somalia. Most world leaders viewed this an merely a temporary situation. Looking at Figure 15-1, do you think that they were right? If they understood the true state of the world's food supply, what effective actions might leaders take to reduce the likelihood that civil wars will be fought over food in the twenty-first century?

# SPECIES AND EXTINCTION

# Chapter 16

## Key Concepts

- Human activities during the twentieth century have caused the extinction of thousands of species.

- The main cause of extinction is the destruction of the organism's habitat—usually when it is converted into agricultural land.

- Tropical forest probably contains more species than all other parts of the world put together. Its destruction is causing the extinction of species we know nothing about.

- When a species becomes extinct, we may lose valuable genetic resources or an important link in an ecosystem.

- Some species contain few individuals or have evolved in the absence of competition and are more likely to become extinct than other species.

- People are trying to preserve some endangered species by habitat preservation, laws to prevent the killing of endangered species, and captive breeding programs.

*A shop in Costa Rica that sells herbal medicines. Don't knock it till you've tried it: Costa Ricans live longer, on average, than Americans.*

**Figure 16-1** A leatherback, *Dermochelys coriacea*, largest of the sea turtles, lumbers ashore to lay her eggs. All sea turtles are endangered because they and their eggs are good to eat, and the shells of many species are used to make jewelry. Turtles lay their eggs on beaches, which are now often crowded with people. In addition, lights on shore disorient the hatchling turtles so that they crawl inland instead of toward the sea. *(Fundación Neotrópica)*

*In the end*
*We will conserve only what we love*
*We will love only what we understand*
*We will understand only what we are taught.*

Baba Dioum, Senegalese conservationist

The human population explosion is wiping out species at a record rate, mainly because people are destroying habitat as they convert natural ecosystems into farmland and urban areas. In this chapter, we consider how species form, why they become extinct, and how we can save endangered species.

The battle to save species is often portrayed as a fight between conservation and jobs: either we cut down old growth forest in the Pacific Northwest and wipe out the spotted owl, or thousands of loggers lose their jobs. In this chapter, I argue that this dilemma does not really exist. Conservationists need to do a better job of pointing out the advantages of saving species. And anticonservation forces are often fighting battles that do not need to be fought. Consider the case of turtle excluder devices.

In 1988, the U. S. Endangered Species Act was up for reauthorization. The Senate held up the Act for years while members introduced amendments that their colleagues could not accept. The biggest impediment was resistance to regulations requiring shrimpers to use turtle excluder devices (TEDs), which keep endangered sea turtles out of shrimp nets (Section 12-E). Every year, shrimp nets capture thousands of turtles, most of which used to drown

before the shrimpers could free them or could hang them upside down to drain water out of their lungs, which often permitted them to recover. Vocal shrimpers from Alabama convinced Senator Howell Heflin that TEDs are expensive and decrease the shrimp catch. Senator Heflin removed his objection to the Endangered Species Act only when its sponsors agreed not to enforce TED use until 1991. So what actually happened when shrimpers started to use TEDs?

In April 1991, before TEDs were required, 91 turtles that had drowned in shrimp nets washed up on Georgia beaches. They included 30 leatherbacks and 12 Kemp's Ridleys, the largest and the most endangered of sea turtles (Figure 16-1). The following month, with TEDs in use, only half as many dead turtles washed up and the shrimp catch was more than double that of April. By 1992, the Center for Marine Conservation reported that TEDs were saving 97% of turtles. Are the shrimpers grumbling about having to use TEDs? Not in Georgia. In addition to keeping turtles out of nets, TEDs deflect large stinging jellyfish, which are the curse of the striker who beheads the shrimp on deck, as well as stingrays and horseshoe crabs, which often weigh more than 20 kilograms. Dragging all these heavy and unwanted animals in a net decreases the efficiency of a shrimp boat's diesel engines. Diesel fuel is the costliest item in a shrimper's budget, and the increased fuel efficiency provided by TEDs may make the difference between a profit and a loss. "Most people I know tow

TEDs even if they don't have to," said one shrimper. "We'd catch a lot of horseshoe crabs if we didn't have them."

People always resist having to change their habits, but the TED story shows that this is sometimes mere stubbornness, and the effort to save an endangered species may have unexpected benefits.

## 16-A WHAT ARE SPECIES, AND HOW DO THEY FORM?

We divide the different kinds of organisms in the world into species. The members of a species tend to look alike, to live in similar places, to have the same way of life, and to breed with each other. In biological terms, a **species** is a group of organisms that share many of the same genes. Each individual contains part of the species' **gene pool**, all the genes carried by the species. Members of a species may share a common gene pool because they breed with one another, passing part of the gene pool on to members of the next generation. There are other species whose members never reproduce sexually. These share a common gene pool because they have inherited copies of the same genes from common ancestors.

Probably 99% of all the species that have ever lived are now extinct. Species originate and then they become extinct, a few days or millions of years later. So why do we worry about extinction, which is, after all, a natural process? Ecologists point out that any species may be a vital link in a food web. Its extinction may disrupt an ecosystem or cause the extinction of other species. Agricultural and medical researchers tell us that extinctions deprive us of irreplaceable drugs and genetic resources. They think we should stop our heedless extermination of other species, at least until we have assessed their possible usefulness. Others think that there are more important considerations than the economics of extinction: the aesthetic loss we suffer when a species disappears.

> Almost all the species that have ever lived are now extinct, but the high rate of extinction nowadays is depriving us of valuable organisms and disrupting ecosystems.

### Speciation After Isolation

New species form from existing ones, usually after a species has split into two or more populations that then become separated (Figure 16-2). For example, the original population may be divided into two by

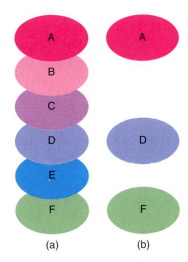

**Figure 16-2**  How species may form from a series of populations of a species. (a) All populations belong to one species as long as continuous, interbreeding populations link them. (b) If intermediate populations B, C, and E are eliminated, then populations A, D, and F are isolated and may evolve into distinct species.

the movement of a glacier, a new road, or a landslide, or a few individuals may colonize a new area and become the founders of a new population. The new population will tend to become genetically distinct from the rest of the species for two main reasons. First, the new population usually contains only some of the genes found in the parent population. Second, the new population will be subject to a new set of selection pressures as it evolves adaptations to its new home.

Most new species probably form like this, when a population becomes isolated from the rest of the species. Suppose a few birds are blown onto an island and breed there. This new population will usually become extinct because its members are poorly adapted to their new home. But sometimes a few individuals will survive and found a new, isolated population. For instance, in Hawaii, there is a species of bird found nowhere else: the Hawaiian goose. This goose closely resembles the Canada goose found on the mainland of North America and almost certainly evolved from a population of Canada geese that migrated to Hawaii. Because islands are isolated, new species tend to form on them more rapidly than on continents.

> Species are groups of organisms that share a common gene pool. They usually originate when a population evolves in isolation, cut off from other members of the species.

## Speciation without Isolation

Although it is less common, there is another way in which plant species sometimes form, by polyploidy. **Polyploidy** is multiplication of the normal number of chromosomes, the structures that carry an organism's genes. The evolution of a polyploid species begins when two different species interbreed to produce hybrid offspring. Hybrids are usually sterile because the chromosomes of the two parent species do not match up so the plant cannot produce eggs and sperm. But most plants can reproduce asexually, so the hybrid may give rise to a group of plants. Then one of these plants undergoes a mutation that doubles its chromosome number. Now that it has matching chromosomes, this individual can produce eggs and sperm. Because many plants can fertilize themselves, the plant can reproduce, giving rise to a new species that cannot breed with its parents.

The origin of the coffee grown today in Central and South America is an interesting example. This species was given the name *Coffea arabica* because coffee was first cultivated in Arabia. Coffee is the ground-up seeds of the plant. It became a popular drink throughout Europe, and by 1675 there were 3,000 coffeehouses in England alone. The Arabs wanted to keep control of coffee production so they dipped the seeds in boiling water before exporting them so that the seeds would not germinate and grow into coffee trees. Eventually the Dutch managed to smuggle live seeds to Sri Lanka, where they started coffee plantations and broke the Arabian monopoly. In 1723, a Frenchman arrived in Martinique in the Caribbean with a single plant of Arabian coffee. Most species of coffee produce seeds only if pollinated by another plant, but *Coffea arabica* is polyploid and can fertilize itself. Even with only one plant, growers in Martinique were able to produce seeds, and these seeds were distributed to Central and South America, where most of the world's coffee is now produced.

Hybrid plants are often larger than their parents and better able to tolerate adverse environmental conditions, such as cold or drought. As a result, farmers and plant breeders often select polyploids as crop plants. Modern varieties of cotton, sugarcane, wheat, potatoes, maize, and barley are all natural polyploids. Researchers have also discovered how to produce polyploids artificially, by treating plants with chemicals such as vincristine, extracted from the Madagascar periwinkle (Close-Up 4-1).

Polyploidy occurs occasionally in animals. Many animal species that are parthenogenetic (reproducing by means of unfertilized eggs) are polyploid: various beetles, moths, shrimp, goldfish, and lizards fall into this category.

## How Quickly Do New Species Form?

The only way to answer this question is to measure the rate at which new species appear in the fossil record. The fossil evidence indicates that species may form rapidly, slowly, or anywhere in between.

Many fossil species existed for millions of years with very little change and without giving rise to new species. Then, in a "short" period of time (which may mean thousands of years), related, but different, species appeared. Sometimes these new species replaced the older one, and sometimes the old and new species coexisted. In many cases, fossils millions of years old are strikingly similar to species alive today. For example, half of the fossil seashells from 7 million years ago apparently belonged to species still alive today. At other times, many new species have evolved in a relatively short time.

Today species continue to form as they always have, usually when populations are isolated from other members of their species.

## 16-B  EXTINCTION

A species becomes extinct when its last member dies. Populations often become extinct, which is a different thing. For instance, in 1953 California adopted the grizzly bear as its state animal—40 years after the last grizzly bear had vanished from California! But the species was not extinct. Populations of grizzly bears survive in other areas.

The biosphere sometimes goes through periods when large numbers of species become extinct. Some 65 million years ago, about half the species on Earth, including the dinosaurs, became extinct in a relatively short period. Biologists are still arguing about the reasons. One theory popular at the moment is that a large meteor hit the Caribbean area, throwing into the atmosphere a cloud of debris that blocked out the sun and caused a "nuclear winter" effect (Section 24-A). This caused global temperatures to fall rapidly and favored the survival of warm-blooded birds and mammals in place of reptiles that had little control over their body temperature.

Some 50,000 years ago, the rate of extinctions again started to increase. A large number of mammals, including many African game species, the Irish elk, the mammoth, steppe bison, wooly rhinoceros, and mastodon, became extinct. The dates of the extinctions coincide with the spread of human populations into new areas. It is likely that early humans were responsible for these extinctions: they hunted these animals and destroyed habitats on a large scale by starting fires. Some ancient human sites contain the remains of extraordinary numbers of animals slaughtered at one time. A site at Solûtre in France contains the remains of more than 100,000 horses. Native Americans slaughtered bison by the thousands by driving them into pits. However, there were still about 60 million bison in North America when white settlers arrived. It took less than 200 years of hunting with firearms to reduce the bison almost to extinction in 1880.

The extinctions since 1900, however, dwarf anything in history. We do not know how many species exist, but biologists think that about 1 million species will be exterminated during this century (Equal Time 16-1). This is mainly because of the rate at which tropical forest is being destroyed. We now realize that tropical forest contains millions of species that have never been described, most of which will be extinct before they are even named (Figure 16-3). The number of species in a tropical forest is hard for people who live in less diverse areas to imagine. In some places, 1 hectare of forest contains more than 100 species of trees, and 10 square kilometers of forest may contain as many species of birds as the whole of Western Europe.

> Biologists believe that the twentieth century will witness the extinction by humans of about 1 million species.

## 16-C CLASSIFYING ENDANGERED SPECIES

The International Union for the Protection of Nature and Natural Resources (IUCN) lists endangered animals and works to save them. In 1992, the IUCN listed more than 5,000 endangered animals, including 698 species of mammals, 1,047 birds, 191 reptiles, 63 amphibians, and 762 fishes. Each species that

# EQUAL TIME 16-1

### Are Ecologists Crying Wolf?

In 1979, Normal Myers published *The Sinking Ark*, a book warning that the world stood to lose "one quarter of all species by 2000," mainly because of tropical deforestation. Within a decade, the world was listening. Madonna held a rock benefit entitled "Don't Bungle the Jungle," and Rainforest Ice Cream showed up on many menus. But several critics say that the suggestion that millions of species are endangered is overstated. They say that:

- Predictions of how many species are lost when habitat is destroyed are based on studies of islands. Mainland areas are almost certainly different because organisms can escape to nearby areas when their own habitat is destroyed—which they cannot do on islands.

- Estimates of the number of species on Earth are exaggerated. In the 1960s, researchers began to understand the amazing species diversity of tropical forests, and estimates of the number of species shot up. This led ecologists such as Paul Ehrlich and Edward O. Wilson to suggest there might be as many as 100 million species on Earth. But fewer than 2 million species have actually been described. Estimates of catastrophic extinction are based in large part on species no scientist has ever seen.

Normal Myers replied to these criticisms:

- There is plenty of evidence of massive extinctions impending in many areas where all the plant species are known and described. These are places where there are unusual concentrations of plant species found nowhere else on Earth and where most of the habitat has already been destroyed. In 14 areas, including parts of Madagascar, Ecuador, Brazil, Borneo, and the Philippines, there are nearly 50,000 endemic plant species (20% of the known species of plants). These areas have all lost 90 to 97% of their natural vegetation, so these 50,000 plant species are all endangered.

- Plant species are better described than animals, many of which are species of ants and beetles. Where animals and plants have been studied in detail, we find that each tropical plant species supports between 20 and 50 species of animals. Taking the lower estimate, it is safe to say that the extinction of 50,000 species of plants will cause the extinction of 1 million species of animals.

(a)

(b)

**Figure 16-3**   Endangered species from tropical forest. (a) A golden beetle, one of a large number of similar-looking species of the genus *Plusiotus*, found in Central America. This beetle lays its eggs in rotting oak branches and is another example of species such as the spotted owl, pileated woodpecker, and many bats that will not survive the loss of dead wood that results from logging old growth forest. Many of the endangered species of beetles in tropical forests are essential to the pollination and seed-distribution of plants. So when the beetles go, the plants go too. (b) An endangered silky anteater from Central America. Anteaters eat ants, which are believed to be the most common group of animals in tropical forests and possibly on Earth. This is the smallest American anteater. As tropical forest is destroyed, ants that live in the forest disappear and so do the anteaters that feed on particular species of ants. *(Fundación Neotrópica)*

Union scientists study is assigned to one of the following categories:

1. **Endangered.**   Species likely to become extinct if nothing changes.

2. **Vulnerable.**   Species that will become endangered if nothing changes. These are mainly species whose population sizes are decreasing or whose habitat is being steadily lost in all the places where the species occurs.

3. **Rare.**   These are species in no particular danger but containing few individuals. They are usually species found only in local populations in particular habitats or thinly scattered over larger areas. Nearly all species found only on small islands fall into this category.

4. **Out of danger.**   This is the heartening place for species that once belonged in the first two categories but have been rescued from risk. The American alligator is an outstanding example. Another is the tuatara, a scientifically important primitive reptile found only on islands in New

Zealand. To preserve the tuatara and other species, the New Zealand government has made many of its islands into wildlife preserves and killed the introduced rats and cats that once competed with the native species.

5. **Indeterminate.**   These are species about which we do not have enough information but that are suspected to be in danger.

## 16-D  WHY ARE SOME SPECIES ENDANGERED?

We live in a world where hundreds of species are disappearing and others have growing populations. As ivory-billed woodpeckers and various whales faced extinction, European sparrows and starlings, African "killer" bees, fire ants, gypsy moths, and water hyacinths were spreading uncontrollably. These are all species introduced into the United States by humans, but many native species are flourishing as well. Coyotes and white-tailed deer are now common in suburbia, and opossums spread farther every year. Why do some species become more common and others become rarer? The fate of a species in the twentieth century usually depends on how compatible it is with growing human populations.

Europe has been densely settled and farmed for thousands of years, and one might imagine that organisms that cannot live with humans would have disappeared long ago. Many European species have indeed been extinct for many years, but it turns out that modern agriculture threatens hundreds more. In the 1950s, for instance, much of Britain and France was a maze of hedgerows, hedges of trees and shrubs that separated fields (Figure 16-4). The hedgerows were home to dozens of species of wildflowers, birds, and mammals that had adapted to these artificial habitats, although their natural habitats had disappeared. Then in the 1960s, agricultural land became very valuable, and farmers started to destroy the hedgerows so they could farm every last inch of land. Primroses, orchids, sparrows, and polecats that survived the destruction of their native habitat are endangered once again by destruction of the hedgerows.

If we look at the various causes of extinction, we can see that some types of species are more prone to extinction than others. Some species have characteristics that permit them to coexist with human expansion. Some do not. The types of species that are particularly vulnerable to extinction are these:

**1. Species with Few Individuals** The fewer individuals a species contains, the more likely it is that an increase in the death rate will wipe out all of them. In addition, a certain population density is often necessary for organisms to find mates or to breed if they do find mates. This is especially true of social animals, which can breed only in a group. We know that even fertile adults often will not breed in zoos, sometimes because they will not mate with the available animals. This is why conservationists are so pleased to find calves being born among the 200 or so surviving right whales. It is still not certain that 200 individuals is a large enough number to permit the species to survive (Section 16-H). The 1,000 individuals that remain may not be enough to save giant pandas.

If the number of individuals falls to less than several hundred, animal species tend to suffer from genetic defects caused by inbreeding. Cheetahs and Florida panthers suffer from this problem. Genetic analysis shows that all the cheetahs that remain are very similar genetically, and their offspring show many genetic defects. Male Florida panthers produce hardly any normal sperm. These defects alone may be enough to doom these species to extinction.

Small populations are especially prone to extinction if they reproduce slowly or if they exist in restricted patches of habitat. Animals such as whales, cheetahs, and albatrosses produce one or fewer offspring each year, and the young take many years to reach sexual maturity. In small populations of such animals, even a few premature deaths may mean the population does not survive.

**Figure 16-4** A country lane in southern England, bordered by hedgerows of trees, shrubs, and wild flowers, which are home to many species of plants and animals.

When an entire species consists only of one small, restricted population, it is especially vulnerable. Such species are common on islands. Here a natural disaster or destruction of the habitat may wipe out the species in one blow. On the Caribbean island of Martinique, the Martinique rice rat was exterminated by a single volcanic explosion about 100 years ago.

**2. Species Unused to Competition** Competition sometimes causes extinction, especially when new species are introduced into an area. At least one species of giant tortoise in the Galápagos Islands has become extinct because of competition from goats introduced to the islands. The goats exterminated the tortoises' food plant (an example of one extinction causing another).

Many of the world's endangered species are **endemic** to islands, meaning they originated on the islands. The opposite of endemic is **exotic**, used to describe a species that has invaded or been introduced to an area. Sometimes endemic island species consist of few individuals, but even where populations are large, island species often become extinct when competitors from other areas are introduced. Species endemic to islands have often faced little competition for food or space and have evolved unusual adaptations. For instance, birds that evolve on islands may lose the ability to fly. This adaptation prevents them from being blown out to sea, but it also prevents them from escaping from some types of predator.

The organisms that usually cause the extinction of island species are those that accompany human settlement: goats, pigs, dogs, cats, rats, mice, and various weeds, such as dandelions and goldenrod. These are all organisms that have evolved on continents in competition with many other species and that adapt easily to life in new areas and among competing species.

An extreme example of extinctions caused largely by competition comes from the Hawaiian islands, where about half of the endemic species are now extinct. The Polynesian discoverers of Hawaii sailed their slender-hulled outrigger canoes to Hawaii from the South Pacific 1,600 years ago. They brought with them dogs, pigs, and chickens and stowaways such as rats, geckos, and skinks. They also imported some 30 species of plants, including yams, taro, bananas, and breadfruit. Centuries later, European settlers brought more new species to the islands. Today 4,000 exotic plants occur in Hawaii. Competition with these imported species caused the extinction of dozens of Hawaiian species. Since 1780, 27 of the 70 endemic species of Hawaiian birds have become extinct, and another 30 are classified as threatened or endangered.

**3. Species That Need Large Areas or Limited Habitats** Most species that become extinct today disappear because humans destroy their habitats. Either they cannot live anywhere else, or they cannot reach alternative habitats when their own is destroyed.

As humans occupy more and more of Earth, animals that need a lot of space are particularly vulnerable. Many of these are large carnivores occupying the top trophic level of a food chain. Bald eagles and California condors are examples of animals that can find enough food only if they can hunt in large areas. Their habitat has steadily disappeared as the human population has increased.

Species with narrow niches that need specific habitats or types of food are also vulnerable to extinction. One example is the Chinese giant panda, which gets 99% of its food from a few species of bamboo. Anything that threatens the bamboo threatens the pandas. In the 1980s, the pandas faced a crisis when their bamboo flowered. Bamboo species have the odd property of flowering all at once, as infrequently as once every 100 years. When they have set seed, the bamboos die. While the new population of bamboo was growing from seed, there was very little bamboo for the pandas to eat and many of them died of starvation, reducing the numbers of an endangered species to a level from which they may not recover.

The wood stork, native mainly to Florida, is endangered in the United States largely because of its peculiar feeding habits (Figure 16-5). The birds feed on fish but, unlike herons, they feel for fish instead of looking for them. This is an adaptation to nesting in trees in swamps. In spring, when the young birds need food, parts of the swamp dry up, trapping large numbers of fish in muddy pools where the fish are invisible. Wood storks feel for these fish and snap them up to feed their young. Wood storks have been driven out of much of Florida as that state's wetlands have been dried up. Georgia conservationists are building shallow, fish-trapping pools near wood stork rookeries, hoping to stabilize the wood stork population in that state.

**4. Species Hunted by Humans** Hunting has exterminated many species of large mammals and left nearly all large carnivores endangered. One odd example is the passenger pigeon, a species endemic to the United States. In 1871, an estimated 140 million of these birds nested in one of their breeding grounds in Wisconsin. Because they lived and bred

**Figure 16-5** Wood storks on their untidy twig nests in a treetop rookery. Rookery is the name for a group of nests of birds that nest in colonies. Wood storks prefer to nest in the tops of bald cypress trees in swamps. *(Steve Bisson)*

in huge flocks, the birds were easy to shoot for food and sport. A mere seven years later, in 1878, conservationists realized that hunting had endangered the species even though thousands of the birds remained. But it was too late. The pigeons were social breeders. Reduced populations, coupled with parasitic infections, had brought their reproduction almost to a standstill. The species was extinct by 1915.

Nowadays in developed countries, groups representing hunting and fishing interests are often among the most active conservationists. They understand that their sport and, sometimes, their livelihood depend on sustained or increasing populations of the organisms they hunt or fish. But in developing countries, where rhino horns for cash or an antelope for meat may mean the difference between starvation and survival, hunting still threatens many species.

> We can identify many of the species most likely to be endangered because they have particular characteristics that make them vulnerable to extinction.

## 16-E GENETIC RESOURCES

Much of the outcry over endangered species today is over the loss of genetic resources. As species become extinct, they often carry with them genes that we need. There are two main ways in which the genetic resources of wild species are used. First, new uses may be discovered for a wild species that will then be farmed for the first time. Second, genes from wild populations may be used to upgrade the gene pools of species that are already domesticated.

### New Uses for Old Species

As a result of recent research, many species are finding new uses in modern economies. Jojoba, long considered a useless desert weed, has been found to produce a wax that retails in Japan for about $3,000 a barrel as a substitute for sperm whale oil. Jojoba plantations are beginning to produce the wax in bulk. Investigating the substance from a tropical tree frog used to tip poison arrows and darts, researchers discovered several valuable substances, including batrachotoxin, now a vital tool of biologists studying how the nervous system works (Figure 16-6).

In attempts to rescue some of the secrets of the wild before they disappear forever, many countries are sending botanists into the forests and savannas to tap the knowledge of tribes whose ways of life are threatened by destruction of their habitats. One team in the Amazon rain forest cataloged more than 1,000 plants used by rainforest Indians that have potential as food, medicines, or industrial raw materials. A few examples are tubocurarine, a muscle relaxant from the pareira plant, a yet-unnamed cure for fungal skin infections, and a high-protein coconut.

**Figure 16-6** A poison dart frog, *Dendrobates granuliferus.* *(Fundación Neotrópica)*

| Compound | Plant Source | Use in medicine | |
| --- | --- | --- | --- |
| | | Modern | Herbal |
| Agrimophol | Common agrimony (*Agrimonia eupatoria*) | Parasitic worm infections | Same |
| Anabasine | Tumbleweed (*Anabasis aphylla*) | Skeletal muscle relaxant | Not used |
| Codeine | Opium poppy (*Papaver somniferum*) | Pain-killer; prevents coughing | Pain-killer. sedative |
| Danthron | Senna (*Cassia*) | Laxative | Same |
| Digitalin | Foxglove (*Digitalis purpurea*) | Heart stimulant | Same |
| Etoposide | May apple (*Podophyllum peltatum*) | Antitumor agent | Cancer |
| Gossypol | Cotton (*Gossypium*) | Male contraceptive | Decreased fertility noted |
| Lobeline | Indian tobacco (*Lobelia inflata*) | Respiratory stimulant | Expectorant |
| Papain | Papaya (*Carica papaya*) | Digesting protein, mucus | Digestive |
| Pseudephoedrine | Ma-Huang (*Ephedra sinica*) | Bronchodilator | Bronchitis |
| Quinidine | Cinchona (*Cinchona*) | Controls heart arrhythmia | Malaria |
| Quinine | Cinchona (*Cinchona*) | Antimalaria, reduce fever | Malaria |
| Salicin | White willow (*Salix alba*) | Pain-killer (in aspirin) | Pain-killer |
| Sanguinarine | Bloodroot (*Sanguinaria canadensis*) | Plaque-inhibitor (red) | No medical use; red dye |
| Scopolamine | Thornapple (*Datura metel*) | Sedative | Same |
| Theobromine | Cocoa, cacao (*Theobroma cacao*) | Diuretic, dilate blood vessels | Diuretic (increase urine production) |
| Trichosanthin | Chinese snake gourd (*Tricosanthes*) | Abortifacient (induces abortion) | Induce abortion |
| Vincristine | Madagascar periwinkle (*Catharanthus roseus*) | Antitumor | Not used |

*Data from Farnsworth, N.R., "Screening Plants for New Medicines." *Biodiversity*, E.O. Wilson, ed., Washington DC: National Academy Press, 1988.

Senna                Opium poppy                Thornapple

**Figure 16-7**   Some compounds found in plants and their uses in modern medicine and traditional herbal medicine.

The World Health Organization estimates that 80% of the world's people depend on compounds from plants for medicine. Norman Farnsworth of the University of Illinois collected a list of 120 chemicals that are extracted directly from plants and used in medicines in various parts of the world. At least 46 of these drugs have never been used in the United States. Until recently, the drug industry showed little interest in traditional plant medicines. Yet Figure 16-7 shows that the uses of substances in traditional medicine and in modern medicine are often very similar. This suggests that many other plants used in traditional medicine contain active ingredients that would be useful in a modern pharmacy.

Native North Americans were seldom consulted by early settlers and much of their knowledge has been lost, but they provided the keys to some important species. The mayapple was used by the Cherokee to kill parasitic worms, and they passed this knowledge on to the early colonists. Research on the mayapple has led to antiviral agents, a drug used to treat testicular cancer, and a means of protecting crops from potato beetles.

Wild rice is an economically important species that grows in shallow fresh water and that was only just saved from extinction. This native of eastern North America is an important food and cash crop for Indians, particularly in the Great Lakes region. Traditional methods of collecting the grain left plenty of seeds to provide next year's crop. Europeans introduced efficient mechanical harvesters and then wondered why the rice was disappearing. Conservation laws were passed in time to save the rice from extinction. Wild rice is now threatened by habitat destruction and by water pollution, which favors the growth of other aquatic plants that compete with the rice.

> The extinction of species deprives us of new additions to the thousands of plant products we use.

**Taxol and the Pacific Yew** In the 1960s, scientists screening plants for the National Cancer Institute discovered a drug they named taxol in the bark of the Pacific yew tree, a plant native to old-growth forests in the Northwest. The drug proved remarkably effective at suppressing tumors, particularly in cases of cancer of the ovaries. To get a drug approved for general use by the Food and Drug Administration, a drug company has to perform a large number of tests to demonstrate that the drug is safe and effective. This required the harvest of enormous numbers of yew trees. Pacific yews are one of the nesting sites of the spotted owl. Both yew tree and owl populations are being dangerously reduced by logging in the national forests of the Pacific Northwest, to produce logs that are nearly all shipped abroad to Japan. So now we have an endangered owl, loggers, an export industry, cancer patients, and drug firms all competing for old-growth forest.

Conservationists insist that we cannot simply harvest every last Pacific yew for taxol. Cancer patients ask how society can value the life of a tree above a human life. Conservationists retort that if we use up all the yew trees, no one will benefit. As our natural resources dwindle and demand for them increases, struggles like this one are likely to be played out more often. "We are going to see a lot more of these debates over how to share the pain," says Henry Shue, director of Cornell University's Program on Ethics and Public Life.

The spotted owl received some help when areas of national old-growth forest were protected from logging in 1993. The ultimate solution as far as cancer patients are concerned will have many facets: timber companies are planting yews in plantations, drug companies are testing taxol-like extracts from other species of yew and from other parts of the yew tree, and chemists are attempting to synthesize artificial taxol. But it will be many many years before there is enough taxol for every cancer patient who might benefit from it.

## Preserving Germ Plasm

As new varieties of wheat, cattle, or any other agricultural species become popular, older ones tend to disappear as farmers cease to use them. In the 1960s, people started to realize that this might be a great loss to agriculture. The dream of turning the drier parts of Asia, for instance, over to modern rice was unrealistic because these rice varieties need lots of water. It would be much better in these areas to plant cereal varieties that produce as high a yield as possible using little water. The genes for drought resistance were more likely to be found in grasses native to dry areas than in imported grains. Plant breeders wanted to search for these genes in the plants that local farmers had been growing before the green revolution. To their horror, they found that many of the old varieties had completely disappeared. Experiences such as this led to the founding of international germ plasm banks. **Germ plasm** is the genetic material contained within reproductive (germ) cells. Germ plasm banks store samples of plants so that if their genes are ever needed in the future, they will be there.

In addition, agricultural expeditions are sent to areas where populations of interest to agriculture might be collected. In the 1970s, a strain of disease-resistant maize was discovered in a Mexican forest. The hardy hybrids produced from breeding this plant are already worth billions of dollars to farmers.

**An Orchid Seed Bank** Orchids are the world's largest family of flowering plants. While most species are tropical, hundreds of lesser-known species are native to the United States, where they are endangered in all but seven states. Orchids reproduce very slowly from seed, and it has only recently become illegal to collect them from the wild. Many orchid species also hybridize with each other readily, and commercial breeders cross one species with another to produce varieties with bigger flowers. As a result, our native orchid species are disappearing rapidly, both in the

**Figure 16-8** An orchid nursery in Homestead, Florida. (*Blue Tahourdin*)

wild and in the hands of growers. To remedy this situation, Derrill Fussell incorporated the International Orchid Seed Bank in Florida in 1992 because Florida and Hawaii are the main orchid-growing centers in the United States (Figure 16-8). With seed storage facilities in Florida, Texas, California, and Arizona, the seed bank collects and stores the seeds of as many orchid species as possible, concentrating on those that are rare and endangered. "As many as 6,000 species of orchids may become extinct within our lifetime," says Fussell. "Anything we can do will be too little, but we can't afford not to try."

> Germ plasm banks and seed banks around the world are hurrying to preserve samples of potentially useful plants before they disappear.

## 16-F LEGISLATION

Most countries have enacted laws and regulations designed to prevent more species from becoming extinct. The U.S. Endangered Species Act is considered one of the strongest of these laws. The Act obligates all agencies of the government to take steps to protect the remaining populations of endangered species. Only a few of the decisions made under the Act have led to major fights of the jobs-versus-wildlife variety (Close-Up 16-1). The snail darter, the Mount

Graham red squirrel, the dune mouse, and the spotted owl made headlines, but thousands of other decisions protecting endangered species have been implemented without controversy. But this does not mean that the Act does the job it was designed to do. People hear well-publicized stories such as the recovery of American alligators and do not hear of the Act's much more numerous failures.

Since the Pilgrims landed at Plymouth Rock in 1620, more than 500 species of plants and animals have disappeared from North America. In an attempt to stem the tide, Congress passed the Endangered Species Act in 1972. The Act covers native American species and foreign species that have been imported at various times. The Act requires that once a species has been officially listed as endangered, a plan to save it must be developed and implemented. The plan usually includes fines for killing a member of the species and protection for its habitat. Critics maintain that although the law is a good one, neither Congress nor successive administrations have ever really wanted it implemented, preferring development of natural habitats to their conservation. As a result, the U.S. Fish and Wildlife Service, which is charged with the Act's enforcement, has never been provided with enough money or power to do the job. Table 16-1 shows the money spent by federal and state agencies in 1991 on some of the most endangered species.

# Florida Keys and Flood Insurance

In 1992, the National Wildlife Federation (NWF) and Florida wildlife groups filed suit against the Federal Emergency Management Administration (FEMA), in an attempt to protect the endangered Key deer, a small deer found on the Florida Keys. FEMA administers the federal flood insurance program, which provides insurance for houses built in areas prone to flooding. This subsidized flood insurance encourages development on land that would otherwise seldom be developed because without it homeowners could not get insurance or home mortgage loans.

The NWF claimed that by providing insurance that permits development on Big Pine Key, FEMA violates the Endangered Species Act by destroying the Key deer's remaining habitat. FEMA admits that development hurts the deer but denies that flood insurance has any direct impact on the deer. The NWF's main witnesses will be from the U.S. Fish and Wildlife Service, which administers the Endangered Species Act, thus pitting one federal agency against another. The result of the case will not be known for some years, but the case will be watched with interest by conservationists. If the court finds that flood insurance harms the deer, this case might be the end of federal flood insurance, which would save taxpayers a lot of money. It would also slow development in flood-prone areas, which include millions of hectares of threatened wetland habitat. For instance, one third of the 500 endangered and threatened species in Florida's Monroe County live at least part of their lives in wetlands and coastal areas where building is usually possible only with FEMA insurance.

Elsewhere in the Keys, biologists report that populations of fish-eating birds are beginning to rebound in the 1990s, thanks to conservation efforts that have increased the flow of water through the Everglades. Although bald eagle nests are still few,

**Figure 16-A** A pair of ospreys with their nest on a navigation mark. Ospreys are large, fish-eating birds that often nest on artificial structures such as bridges and utility poles. The Army Corps of Engineers now puts platforms on top of as many navigation marks as possible to provide nest space for the birds far away from the many people who delight in killing large animals.

the number of Roseate spoonbills in Upper Florida Bay has doubled since 1985, and great white herons and ospreys are also recovering (Figure 16-A). In 1992, one osprey nest on a utility pole right beside the highway on Little Torch Key contained three fledglings. It was attacked in March by two men who climbed the pole and threw the fledglings to the ground. Luckily they were seen. The fledglings were returned to the nest and the men went to jail.

**TABLE 16-1**   1991 Expenditures by U.S. State and Federal Agencies to Preserve Some of the World's Most Endangered Animals*

| Animal | Expenditure |
|---|---|
| Alabama cavefish | $200 |
| Florida panther | $4,113,900 |
| Mexican wolf | $0 |
| California condor | $892,700 |
| Kemp's Ridley sea turtle | $866,200 |
| Black rhinoceros | $0 |
| Gulf of California harbor porpoise | $0 |
| Black-footed ferret | $1,126,700 |
| Green pitcher plants | $42,300 |
| Philippines eagle | $0 |

*Source: Natural Resource Defense Council.

Some 3,900 species of American plants and animals have been declared endangered by biologists but have not yet been added to the Act's endangered list. Only 1,167 species are listed as protected by the Endangered Species Act, 639 from the United States and 528 foreign species. Only 16 species have ever been removed from the list, four because additional populations of the species were discovered so that they are no longer considered endangered, seven because they are now extinct, and five because their populations have grown to a safe level.

Recovery plans have been developed for only half the species on the list, and only a tiny proportion of the existing recovery plans has ever been put into effect. At least 80 species, including the Texas Henslow's sparrow, the dusky seaside sparrow, and the eastern cougar, have become extinct while awaiting protection from the Act. Every time the Act comes up for reauthorization, dozens of legislators introduce amendments that would remove particular species from the list or blunt the protections afforded them. These legislators represent economic interests such as developers who want to build over vanishing habitats or ranchers who want to be able to kill wolves that might kill their livestock.

An example of the Act's failures comes from the Southwest's Mojave Desert, a desolate stony area of searing heat and bitter cold studded with Joshua trees and creosote bushes where the desert tortoise has lived for 150 million years. The tortoises survived the Ice Ages and the heat but will not survive ranching, housing developments, and off-road recreational vehicles. The tortoise has disappeared from the California desert and has declined by 60% in the last five years in Utah. In 1989, scientific documen-

tation of the tortoises' decline was complete, but the Fish and Wildlife Service said that even if the tortoise was listed as endangered, the money to develop a plan to save it was not available.

## The 1988 Endangered Species Act

The 1988 revision of the Endangered Species Act had several useful new provisions. For instance, states were encouraged to monitor endangered species and thus become eligible for a share of an $18 million federal fund. The Act finally made it illegal to dig up or damage protected plants. The fine for violating any part of the Act was increased to a substantial $50,000.

Heavy fines are needed because trade in rare and endangered species generates large sums of money. The collection of large cacti by landscaping services has decimated the vegetation of America's deserts. The cacti are sold to homeowners for thousands of dollars each. Smugglers import parts of dead animals (furs, ivory, rhinoceros horn "aphrodisiac") and an-

**Figure 16-9**   Costa Rican park ranger, Carlos Acevedo, with an injured macaw. The demand for these attractive birds as pets in North America has brought them to the verge of extinction in their native Central and South American forest.

imals such as lemurs, macaws, cheetahs, and tropical fish that Americans want as pets (Figure 16-9). They may often import endangered species legally if they claim they are to be used for research, education, or conservation.

## State Laws

Since 1980, several states have passed protective legislation that works better than the federal law. Consider the problem of protecting endangered plants. Species are threatened because their habitats are destroyed and because people collect them to plant around their houses. The Ohio Endangered Plant Law is relatively weak, but the state's Department of Natural Resources has used it to acquire plant habitats and to educate the public about the 350 species of plants endangered within the state. Increased public understanding of the problem has led to more funding for the state's conservation efforts through a checkoff system whereby the public can contribute when filing state income tax returns.

North Carolina regulates trade in endangered plants. The state's Department of Agriculture has built a network of nursery owners who sell only plants grown in nurseries and not those collected from the wild. A state botanist keeps in touch with regional networks of amateur and professional botanists who monitor the progress of rare plants.

> In the United States, federal legislation has not been very effective at preserving endangered species, but many states are doing an effective job.

## International Trade in Endangered Species

In 1990, the 103 nations belonging to the Convention on International Trade in Endangered Species (CITES) added many species to the list of those whose international trade is prohibited. All spotted cats are now on the list, including ocelots, margays, and the Iberian lynx, once popular as pets of wealthy Americans and all threatened in their native habitats. Macaws and parrots from the Amazon (traded as pets) also joined the list. All hard corals are also now listed because coral reefs were being destroyed to make jewelry. Indonesia wanted to move the green and hawksbill sea turtles off the list, but the delegates prevented this since there was no evidence that Indonesia's turtle populations were large enough to withstand commercial exploitation.

International cooperation is essential to preserving many species. For instance, if a developed country permits citizens to import ivory from African elephants, it will be exceedingly difficult for African countries to prevent poachers from killing elephants. In recent years, international efforts have begun to bear fruit.

## Trade in Small Bulbs

Small bulbs that flower in spring are popular in gardens (Figure 16-10). For many years, most of these bulbs were dug up in the wild and trucked by the millions to Holland to be marketed under "Product of Holland" labels. Few of the buyers realized that these bulbs were dug from the wild. In Turkey alone, during the 1980s, 71 million anemones, 20 million cyclamen, 62 million leucojums, 111 million winter aconites, and 200 million snowdrops were dug and exported, with no attempt to leave behind populations large enough to sustain future harvests. As a result, many species in the Mediterranean and Middle East are endangered.

Steps are now being taken to ensure that at least consumers know when they are plundering wild populations. In 1991, the Dutch bulb industry, the World Wildlife Fund, and the British Flora and Fauna Preservation Society signed an agreement protecting 31 species of bulbs usually dug from the wild, packaged, and sold as Dutch. These bulbs will now be labeled "Bulbs from Wild Source." By 1995, all bulbs will carry labels describing their origins.

## Ivory Trade

The battle over the trade in ivory has been the most visible of CITES's conservation efforts. Tanzania wanted a ban on trade in ivory. Along with other African nations, it has large revenues from tourists who come to the country's wildlife parks to see the elephants and other large mammals. Meanwhile, the elephants were disappearing, killed by poachers for their tusks, which are carved into ivory figures and jewelry that are popular particularly in Japan, the largest consumer of ivory. Tanzania found it impossible to prevent poaching as long as ivory was selling for $300 per kilogram as it was in 1989. The price had risen steadily as elephants became scarcer during the 1980s. CITES agreed with the African nations, the ivory trade was banned, and Japan agreed to enforce the ban on imports.

At the time of the ban, I doubted whether it would help to save the elephants. I feared that it might act like the ban on cocaine imports, merely to drive the price higher, making smuggling more attractive financially. However, that did not happen. As Figure

(a)

(b)

**Figure 16-10**   Small bulbs popular in the garden. (a) A cyclamen. (b) A snowflake, *Leucojum.*

16-11 shows, the price of ivory fell steeply after the ban.

In 1990, both Burundi (Central Africa) and Hong Kong objected to the ivory ban because they had warehouses containing millions of dollars worth of ivory that would become worthless. Ivory need not come from elephants that have been shot. It may be

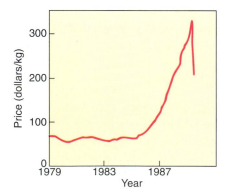

**Figure 16-11**   Changes in the price of ivory on the world market during the 1980s.

removed from an animal that has died of natural causes, but the ban on trading ivory covers ivory from all sources. Burundi pointed out that its elephants were well protected and the populations stable: the ban should apply only to ivory from countries where elephant populations were declining. The trouble was that it was impossible to tell which country any piece of ivory came from. Now scientists seem to have overcome this problem. New tests permit technicians to determine which elephant population a tusk comes from, which may permit Burundi to sell its ivory legally.

## 16-G  ZOOS AND CAPTIVE BREEDING

Zoos are places where the public can see animals that they might never see in the wild. Once viewed as places of entertainment, zoos are now in the thick of the struggle to preserve endangered species.

In zoos, people see animals as attractive individuals instead of merely names in a newspaper, and they may be galvanized into action when they hear that a favorite species is endangered. Aquariums where fish and marine mammals are displayed undoubtedly generate a lot of support for laws protecting such endangered animals as whales and manatees. The role of zoos has expanded as concern over endangered species has mounted.

Conservationists pointed out that one reason some species were endangered in the first place was that animal collectors capturing animals for zoos and for private collectors and research had reduced many natural populations to the verge of extinction. In the 1950s, the more responsible zoos responded to these and other complaints by restricting captures from the wild and by building bigger and more natural enclosures. Some species responded to these better conditions by breeding in captivity for the first time. Zoos realized that they might serve a valuable role in preserving endangered species and even in restoring populations that had become extinct in the wild.

## Captive Breeding Programs

One of the first attempts to restore natural populations by captive breeding was the peregrine falcon program. Peregrines are predatory birds, and many populations became extinct as a result of DDT use (Section 21-B). After many years of study, biologists discovered how to breed these birds in captivity, and in the 1980's they started releasing captive-bred birds into the wild. They hope to reestablish extinct populations. Early results from this program look hopeful. The number of birds raising offspring has been doubling every two years. Some peregrines have even nested on tall buildings in cities, where they feed on pigeons.

The peregrine program nevertheless points up the disadvantages of this method of restoring an endangered species. Inducing wild animals to breed in captivity is a long-term, expensive undertaking even when it is successful. It is then difficult to release animals into the wild in a manner that gives them a reasonable chance of survival. Nevertheless, there are a number of species that are either already extinct in the wild or definitely headed for extinction, and they can only be saved if attempts to breed them in captivity succeed.

Captive breeding was given a vote of approval when the last few wild California condors were caught to be used for breeding, in a last-ditch attempt to save these big scavengers from extinction. One biologist pointed out that the condor breeding program is philosophically futile if there is no habitat left into which these animals can be released safely. Is there any point in preserving a species that must live forever in captivity?

> Zoos tread a fine line between endangering a species further by collecting it and preserving it by captive breeding.

## The Panda Problem

In May 1988, officials from Toledo Zoo flew to Shanghai to collect two giant pandas, found only in China, and bring them back to exhibit. They came home to face a lawsuit by the World Wildlife Fund against the Fish and Wildlife Service for issuing the import permit for the pandas. The zoo claimed that the Convention on International Trade in Endangered Species did not prohibit the pandas' import because the animals would be used for educational purposes and to further conservation. The justification for exhibiting pandas is that it raises public awareness of the panda's plight and provides money for China's conservation efforts. The World Wildlife Fund claimed that the "rent-a-panda" deal was motivated by plain old greed. Not only do zoos make large sums of money from panda loans, but so does the Chinese government—up to half a million dollars in hard currency for a three-month loan.

There have been many short-term loans of pandas to western zoos that are of questionable value to the pandas. The pandas that went to Toledo were two animals of prime breeding age from a captive breeding facility—in violation of China's stated policy of lending only nonbreeding animals to zoos. A judge barred Toledo Zoo from charging a special fee for the panda exhibit, and the Fish and Wildlife Service turned down the next request for a panda import permit—from the Michigan Department of Natural Resources, which wanted two breeding-age pandas for exhibit at the state fair.

Improvements in the panda loan policy may come too late. There are only about 1,000 giant pandas left in the wild. They once roamed much of southeast Asia, but habitat destruction has done them in. Nearly all the bamboo forest that is their home has been destroyed, and the remaining pandas have been pushed into bamboo forests in China, high up against the dry plateau of Tibet, with nowhere else to go. They survive in groups of ten or less (some groups probably too small for breeding) in 12 small reserves, surrounded by roads, rivers, fields, and human settlements. No one knows if the existing populations are large enough to keep the species going. The Chinese government is undoubtedly sincere in its desire to preserve the pandas. It has spent more than $25 million on the animals. Conservation fails at the local level, where penalties for killing the animals are not enforced and a few poachers still roam the preserves. The captive breeding program has not yet succeeded. To date, there have been more deaths than births of animals in the program.

In the wild, breeding can be encouraged only by the difficult task of linking the pandas' scattered patches of habitat by **wildlife corridors** so that different panda groups that are now isolated can interact. Whether Chinese authorities and Western advisers can organize these improvements in time to save the pandas is not clear. Western officials fear that panda loans to zoos may encourage the Chinese to capture even more of the animals from the wild. Some 30 western zoos and parks are now negotiating with the Chinese for loans. It is extremely difficult to decide whether these loans improve the pandas' chances of survival or not.

### Endangered Plants and Gardens

We do not usually think of propagating plants in the garden as captive breeding, but it often is. Homeowners and gardeners are responsible for much of the destruction of wild plants when they dig them from their natural habitats for use in the home landscape. But, as people everywhere become more aware of environmental destruction, gardeners are putting their talents to use by helping to preserve plant species. For instance, the fringed campion is native to woodland in north Florida and western Georgia (Figure 16-12). It is listed as endangered in both states because it was always rare and has become rarer as much of its habitat has been destroyed. Now gardeners in northern Florida have discovered that the campion grows well there in garden soil enriched with compost, and they are growing large numbers of the plants for distribution to nurseries. The species can also be reintroduced to areas where it once grew naturally and that are now being preserved and restored (Close-Up 16-2). Conservation groups in many countries are now organizing gardeners to take responsibility for propagating particular endangered species and varieties so that these will never be lost as their native habitat continues to be destroyed.

## 16-H   HARVESTING WILDLIFE

We think of modern society as fed by agriculture, and so it is. But quite a lot of our food and various other natural resources are acquired as our ancestors acquired them—by hunting and gathering.

Examples of natural resources that reach us by hunting and gathering include most fish, firewood,

**Figure 16-12**   A fringed campion, *Silene polypetala.*

and furs obtained by killing foxes, seals, or beaver. Even though the hunters and gatherers know that overexploitation will deplete the resource, profit or firewood for the moment often takes priority over worrying about the future. The whale fisheries are a good example.

### Saving the Whales

In February 1988, dredgers under contract to the U.S. Army Corps of Engineers started to deepen the entrance to the Kings Bay Submarine Base in Georgia. The dredgers speeding out to sea to dump their waste ran the risk of hitting right whales that had been spotted near the coast (Figure 16-13). Right whales are nearly extinct. Only about 200 remain. The Georgia Conservancy, the New England Aquarium, the University of Georgia, the National Marine Fisheries Service, and local newspapers were up in arms. The New England Aquarium had identified some of the whales spotted in southern waters as those that had earlier moved south past the New England coast. The whales apparently migrate to warmer waters in winter to give birth and nurse their calves. Visions of boatloads of tourists paying for trips to see the whales danced before the eyes of local entrepreneurs.

The filter-feeding whales and their calves bask in shallow water just below the surface. Submarines de-

# Biocultural Restoration at Guanacaste

Ecosystems from the Everglades to California's wetlands have been degraded by introduced species, loss of their water supply, timber-cutting, and dozens of other changes brought about by human interference. Now conservation groups are trying to restore some of these ecosystems to their state before human interference. The most notable example is the biocultural restoration of Guanacaste National Park in Costa Rica (Figure 16-B). Biocultural restoration is the name given to attempts to restore damaged ecosystems with the assistance of their human inhabitants and neighbors. This park, masterminded by ecologist Daniel Janzen, is being pieced together out of bits of existing forest and land that had been deforested for agriculture.

The ultimate aim is to restore Guanacaste's ecosystems to their state before the arrival of Spaniards in the sixteenth century. Guanacaste contains one of America's last remaining tropical dry forests (Figure 16-C). It also includes rain forest, beaches, islands, volcanoes, and rivers. How does one restore a forest? In Guanacaste, trees are being planted and fires controlled. Hunting is forbidden because the animals carry and process tree seeds. Some livestock are allowed to graze on the land to control jaragua,

**Figure 16-C** Looking up into the canopy of part of the remaining tropical dry forest at Guanacaste. If you look carefully in the middle of the picture, you can see the sprawling shape of a howler monkey, asleep at the top of a tree.

**Figure 16-B** Guanacaste: the entrance to the biological research station, which was once a cattle ranch. The scattered trees are the remains of the forest that once occupied this area. The grass in the foreground was introduced to feed cattle.

a grass introduced from Africa that blocks the growth of young trees, shelters mice that eat tree seeds, and fuels forest fires. Within 100 years, the park will be largely free of grass, and within 300 years, it will be climax forest again.

Costa Rica is overpopulated. The first goal at Guanacaste is to teach the park's 40,000 human neighbors to value and sustain it. The educational program, aimed primarily at Costa Rican schoolchildren, is probably the most ambitious in the world. The authorities believe that environmental education is vital to the park's future and that understanding the biological lessons of the park provides an ideal intellectual challenge for a literate population such as Costa Rica's. Teaching environmental biology to schoolchildren may well be the most important step toward a sustainable existence not just for the people of Guanacaste, but for everybody in the world.

**Figure 16-13**   A right whale, *Eubalaena glacialis. (Thom Smith)*

tect and avoid them using sonar equipment. The dredgers had no sonar. Whales are visible to a lookout on a dredger by day, but the dredgers were working around the clock. They were preparing Kings Bay for the arrival of the first new Trident submarine, designed to fire nuclear missiles. At night, the dredgers had little chance of avoiding a whale in their path. The conservancy and the fisheries service called for a halt to nighttime dredging, but the Corps of Engineers said that this would be too expensive. So the contractor and the New England Aquarium used planes to spot the whales. The conservancy pointed out that if a dredger injured or killed a right whale, the fisheries service could sue those responsible for large sums under the Endangered Species Act. A fairly typical battle over an endangered species was being enacted, with conservation, administrative convenience, cost, commercial potential, and even national security lined up in various positions.

Whales are fascinating creatures, and people want to see them. Humpback whales off the coast of California are so well adapted to boats full of whale-watchers that they often come to the surface and make themselves visible when they see tour boats coming. Economists estimate that whale-watching in just the last 20 years has generated more revenue than whale killing has produced in hundreds of years. Why then, is it not better to save the whales than to kill them? The answer is that the people who make money from whale-watching are not the same as those who make money from killing whales. The two groups battle for rights to the whales.

Blue, fin, humpback, sperm, and right whales were hunted to the verge of extinction in the 1960s and 1970s, especially by Japanese and Russian whaling ships. These nations knew that their activities were destroying the fisheries. As one species of whale became too rare to be worth catching, the whalers went on to the next species. In the 1980s, Russia and Japan bowed to public pressure and joined the international ban on whaling. For a while, a complete moratorium on hunting all whale species was in effect, but Japan and Norway are again hunting unprotected sei, minke, and killer whales.

## 16-I   THE NEW WILDLIFE PRESERVES

The usual reason that a species becomes extinct today is that its habitat has been destroyed. Therefore, the best way to save an endangered species is usually to set up a preserve, an area where the species' habitat is saved from destruction and the species can breed and endure. The world contains millions of wildlife preserves of various sorts, including over 1,000 national parks in more than 120 countries. During the 1980s, the area of land covered by these parks almost doubled. The parks attract tourist revenue and are a source of national pride.

One problem with parks is that their boundaries are usually determined by political rather than ecological considerations. As a result, very few of them enclose all the ecosystems many species need to survive, and they cannot hope to preserve all the species in an ecosystem. A worse problem is that growing human populations encroach on the parks in various ways. In the United States, the National Park Service identifies 73 types of threat to our national parks.

They include roads built for access to the parks, garbage left by visitors, industrial and urban development nearby, air and water pollution, invasion by introduced species, illegal planting (usually of marijuana), and damage to soil and vegetation from overuse. The problems are even worse in developing countries, where local people who desperately need wood, land, and other resources within the park cannot be effectively excluded. Poaching, tree cutting, and inadequate staff all threaten endangered species that are, in theory, preserved.

We generally think of national parks as free from human interference, but this is not the case even in developed countries. Lumbering, mining, grazing, and even hunting are all permitted in many parks. Many people now think that the best way to run wildlife preserves, particularly in developing countries, is by a system of **integrated management**, management that integrates conservation with sustainable economic development. Integrated management aims to involve the local population in the fate of large areas of land that stretch beyond the bounds of traditional parks. It also helps to combat another problem with traditional parks—many parks are too small, or are the wrong shape, to ensure the preservation of some large animals.

### Pilanesburg National Park

Pilanesburg is a park in Bophuthatswana in South Africa, created on reclaimed farmland and ringed with volcanoes. It is known for innovative conservation in that it treats the game park as a local industry. Its supporters argue that wildlife parks can never succeed in this overpopulated world unless they are worth something to the people who live near them.

In 1982, Pilanesburg was the first park to sell licenses to hunt the white rhinoceros, once on the verge of extinction but now thriving in the park. Five or six times a year, an American or German hunter pays $25,000 for the privilege of shooting one. When the park must tranquilize a rhino to transport it to a new breeding area, hunters pay up to $10,000 for the right to shoot the tranquilizer dart and pose with the drugged animal for photographs. The park puts 10% of its proceeds into projects for neighboring villages, who now regard the park as an important resource. Poaching in the park has just about disappeared.

Managing endangered species so that they have a commercial value is accepted in Africa as the best method of conservation, but it is not accepted everywhere, particularly among Western conservationists. In 1992, Pilanesburg considered letting hunters kill Van Gogh, an aged black rhino, past reproductive

**Figure 16-14**  A black rhinoceros. *(John Schmidt)*

age and near death (Figure 16-14). The $100,000 they could charge would buy a lot of protection for the world's surviving 2,000 black rhinos. The park decided against the hunt, fearing adverse publicity from animal rights activists in the United States. Instead, Van Gogh will become, as biologist Keryn Adcock lamented, "expensive hyena food."

The number of black rhinos has declined from 60,000 in 1970 to about 2,000 today as a result of poaching. Poachers sell the valuable horn to Chinese medicine shops, where it is prized as a fever reducer (although it works less well than aspirin) and to North Yemen, where it is turned into expensive dagger handles. It is an ironic fact of modern life that hunting may be the only thing that can save the black rhino from extinction.

Zimbabwe, Botswana, Burundi, and South Africa have long managed their elephant herds for profit. Hunters pay about $15,000 to participate in annual elephant shoots, which maintain elephant herds at sizes that keep the animals healthy and prevents them from doing so much damage to local farms that farmers are tempted to kill them. The money is used to pay guides and game wardens who guard the herds from poachers. Unlike the rest of Africa, these countries have kept their elephant populations stable throughout the 1980s.

### Biosphere Reserves

The United Nations' Man and the Biosphere Program has set up 266 preserves since 1976. These are **biosphere reserves**, unusual in that they are people preserves as well as wildlife preserves. Biosphere reserves are designed to conserve typical ecosystems containing human populations, rather than the unusual ecosystems without human residents that are more common in national parks. Biosphere reserves have three missions:

1. **Conservation.** The reserve should contain an ecosystem that is largely undisturbed and big enough to sustain species and genetic diversity. It is not enough merely to maintain a small population of an endangered species. Small populations have small gene pools and become inbred. More than one population is usually needed to maintain healthy genetic diversity.

2. **Logistic.** The logistic role involves research, education, and training. For instance, a reserve can be used to monitor problems such as acid rain and climate changes and to try out experimental management techniques. An important requirement is that research and education should tie in with a network that distributes information internationally.

3. **Development.** The reserve is used to study and promote sustainable uses of the ecosystem by its human population.

Every reserve contains at least one **core zone**, a natural ecosystem that cannot be interfered with. Core zones in many of the reserves overlap national parks or nature preserves. Next to the core is a **buffer zone**, where various activities such as research, tourism, education, and traditional farming can take place as long as they do not interfere with the core zone. Beyond the buffer zone is a **transition zone**, which has no definite boundaries. Ideally it will expand with time. Here the work of the biosphere reserve serves the needs of local communities. The goal is to work with local people toward sustainable rural development.

An example of a biosphere reserve that has gotten off to a good start is Mexico's Sian Ka'an on the Caribbean coast of the Yucatán peninsula. Sian Ka'an contains half a million hectares of several ecosystems, including tropical forest, marsh, mangrove swamp, shallow ocean, and a barrier reef. The reserve protects dozens of endangered species, including tapirs, manatees, jaguars, pumas, and three species of sea turtle. A key to Sian Ka'an's success is its small human population. About 800 Mayans live within the reserve, living much as their ancestors did 1,000 years ago. Some are subsistence farmers and plant gatherers, and some run small tourist operations, but the majority are lobster fishermen. The lobster fishermen have long recognized the importance of sustainable yields of this natural resource and are organized into cooperatives that regulate the number of fishermen and the size of the season's catch.

Besides the research projects that one would expect to find in a wildlife preserve, Sian Ka'an hosts research projects on sustainable use of its ecosystems. One project investigates the effect of artificial reefs on the lobster population. Another is an experimental farm where plants are grown very close together in an attempt to reduce the amount of land needed to grow food. Sian Ka'an residents are involved in these research projects and in the reserve's management. The main hindrance to further work in the park is lack of money. But its managers think that developing tourist facilities for visitors interested in the natural history of the reserve will eventually solve this problem.

The United Nations' Man and the Biosphere Program is based on the belief that conservation is more likely to succeed if it involves an area's human inhabitants.

# SUMMARY

Thousands of species have become extinct as a result of human actions in the twentieth century. Considerable efforts are being made to save what is left. People often think that saving species from extinction requires the sacrifice of money and jobs. In fact, this is often not the case.

## What Are Species and How Do They Form?

A species is a group of organisms sharing a common gene pool. New species form from existing species, usually when a population is cut off from the rest of the species and evolves adaptations to its particular habitat in isolation. Many new plant species form by polyploidy without isolation from parent populations. New species may form very rapidly or extremely slowly and may then survive for brief periods or millions of years.

## Extinction

Nearly all the species that have ever existed are extinct. Large numbers of extinctions sometimes occur in short time periods. None is shorter than the twentieth century, during which time perhaps 1 million species will become extinct, most of them as a result of the destruction of tropical forest.

## Classifying Endangered Species

The IUCN lists endangered species worldwide, giving the status of endangered species and plans to establish sustainable populations. A species is described as endangered, vulnerable, rare, out of danger, or indeterminate.

## Why Are Some Species Endangered?

Some species have characteristics or habitats that make them particularly vulnerable to extinction. The most vulnerable are species with few individuals, species unused to competition, species that need a lot of space or a limited habitat, and species hunted by humans.

## Genetic Resources

As species become extinct, they carry with them gene pools of potential value. Species with useful gene pools are of two main kinds: species for which new uses are discovered and wild species to be used for breeding with domesticated populations. Many plants contain chemicals of medical value, many of which have not been studied by Western scientists. In an attempt to preserve disappearing gene pools, scientists all over the world are establishing germ plasm banks containing genetic samples of every variety of crop plant and domesticated animal they can find. Seed banks and similar institutions collect and save seeds or plants of endangered species.

## Legislation

Most countries have laws designed to protect endangered species. The United States has probably the strongest of these laws, but it could be enforced more effectively. The Endangered Species Act requires that endangered species be listed and a plan to save each listed species be developed and put into action. Only about one third of U.S. species that are actually endangered have even been listed. In practice, state protection for an endangered species may be more effective than federal protection. CITES is an international convention that has had considerable success in controlling international trade in endangered species.

## Zoos and Captive Breeding

Zoos are both good and bad for endangered species. Collecting for zoos has reduced the numbers of many species. Now zoos are playing an increasing role in educating the public and in attempting to breed animals that are almost certain to become extinct if they cannot be bred in captivity. Some people argue that these experiments in preservation seldom succeed in reestablishing a species in the wild and that the money would be better spent on other types of conservation. Conservation groups that include home gardeners often propagate plants to protect a species that is becoming rare in the wild.

## Harvesting Wildlife

As human populations have grown, it has proved difficult to prevent hunters and gatherers from depleting the resources they harvest. Hunting is difficult to control until it becomes unprofitable.

## The New Wildlife Preserves

Habitat destruction is the usual reason species become extinct. Therefore, the most effective way to preserve a species is to preserve some of its habitat and guarantee that the organism will be left alone to reproduce. The world's wildlife preserves and refuges are damaged by increasing numbers of people who encroach on them in various ways. Therefore, the most successful efforts integrate people into the preserve. African park managers have found that the most successful way to preserve large game animals is to give them an economic value to people who live near the park and can prevent poaching. Selling the rights to participate in controlled hunts is one way to do this. Biosphere reserves have multiple purposes, including conservation, research, education, training, and sustainable development for the local population.

## DISCUSSION QUESTIONS

1. Consider the panda loan program. At first sight it appears to be bad for the pandas, removing them to zoos where they hardly ever breed. But consider this: before Westerners showed interest in pandas, the Chinese gave low priority to preserving them. The World Wildlife Fund has contributed millions of dollars to panda conservation efforts. This money came from people who have

seen pandas as a result of the loan program and hope to help save them. Is the loan program a good thing or not? Perhaps it could be better managed. How?

2. European laws to protect endangered species are weaker on paper than similar laws in the United States. But a European collecting a wild plant from the roadside would be scolded by passing motorists. In the United States, millions of endangered plants are collected from the wild, not just by individuals but by companies that sell the plants illegally. A 1988 article in the American magazine *Organic Gardening* praised a garden created by a man who had collected most of his plants from American and Mexican deserts. A note at the end of the article warned that some plant species were protected, but such an article would never be tolerated in a European publication. Why the difference between American and European attitudes?

3. This chapter suggests that education may be part of the solution not just to protecting endangered species but to many environmental problems. How many cases can you think of in which this might be true? There are cases in which education is not enough to solve the problem. What kinds of cases are these?

4. If you were asked to save as much information and as many genetic resources as possible from a tropical forest that was steadily being destroyed, how would you approach the problem?

5. There are various arguments for trying to save endangered species. Which ones do you find most persuasive?

# WILDERNESS AND FOREST

# Chapter 17

## Key Concepts

- Since the beginning of agriculture, people have destroyed about half the forest on Earth and have altered nearly all the grassland from its natural state.

- There is no hope of saving more than a tiny remnant of the world's tropical forests. They are being cut down rapidly to provide subsistence farms, grazing land, fuelwood, and cash crops.

- When tropical forest is cut down, local rainfall decreases, forest dwellers are displaced, useful organisms become extinct, and ecosystems that produce oxygen and absorb carbon dioxide disappear.

- There is an urgent need to protect wilderness areas from human interference.

- Most forests are managed to produce sustainable yields, but the management is often ineffective, leading to environmental problems like soil degradation and floods.

- In forest management, economic considerations are generally given priority over preservation.

- Imaginative reforestation programs are responsible for planting billions of trees a year, although we are still not planting trees as fast as they are being cut down.

*National forest in the Cascade Mountains of the Pacific Northwest.* (Richard Feeny)

*"When you compare the forests of the Pacific Northwest with those of the Brazilian Amazon, the Northwest is in much worse shape, torn up and fragmented by thousands of clearcut areas."*

Compton Tucker, NASA scientist engaged in mapping Earth's forests from satellite pictures, 1992.

This chapter discusses two different natural resources. The first is **wilderness**, habitat that is preserved as nature left it, before human interference. It includes desert, wetland, and prairie as well as **virgin forest**, forest untouched by human hands. The second is forest, on which we depend for much of the oxygen in the air and for our wood supply. Most of the world's forest is not wilderness, it is managed so as to produce harvests of wood and other forest products.

When Europeans first settled in North America, the continent was covered with forest, desert, and prairie. Most of it had existed for millions of years, undisturbed except by fires, lumbering, plant collecting, and hunting by small populations of Native Americans. Although North America is not densely populated, less than 400 years later even the remote Canadian forests have been logged, and farmland has replaced large areas of forest and nearly all the prairie.

## 17-A  U.S. PUBLIC LAND MANAGEMENT

*"Is this any way to manage our national forests? To sell off our natural heritage piece by piece and lose money in the process?"*

Association of Forest Service Employees
for Environmental Ethics

In 1992, the Montana National Forest Management Act was being pushed through Congress. The act aimed to open up Montana's Yaak Valley National Forest to logging. The National Forest Service would move in and use taxpayer dollars to destroy the corridor by which grizzly and black bears, deer, elk, porcupines, wolves, badgers, and wolverines travel between Canada and wilderness areas in the Northern Rockies. Local residents were to vote on whether or not they wanted to retain even part of the valley as wilderness, which would preserve the wildlife corridor. Local newspapers were violently opposed. At an antiwilderness meeting, Dennis Winter, for the Western Environmental Trade Association, said that preserving the land would cause 30% unemployment in the community. "And along with that comes wife battering, child molestation and all the rest of it. Now

do you think that the environmentalists give a damn about the fact that kids are going to be molested as a result of this?" This may make little sense to those not familiar with the boom-and-bust economies of lumber communities, but this argument worked. Only 2% of local residents voted to preserve any of the valley as wilderness.

Why had the forest not already been cut for timber, like most of the American West? Because of a little-known act passed under President Carter's administration, called the Roadless Area Review and Evaluation Program. As its name suggests, this program is designed to make the federal government stop and study the situation before opening up new public land to development. With development, wildlife corridors are destroyed, soil erosion starts, and the ecosystem becomes degraded.

Whether in Montana or the Pacific Northwest, the battles between conservationists and those who want to open public land to development seem never-ending. Some U.S. public land has been preserved as wilderness since 1964 (Close-Up 17-1). Most of this land had already been cut over for timber. But if it is now completely preserved, within a few centuries it will again revert to mature climax vegetation (Section 5-F).

### Multiple Use and Sustained Yield

About one third of the land area of the United States is public land, much of it forested. President Theodore Roosevelt established the U.S. Forest Service in 1905 to manage the national forests. Today the Forest Service manages about 75 million hectares. Commercial timber cutting is permitted on about half of this land. Much of the rest is used for camping and other forms of recreation and for grazing, mining, and hunting. Such uses are beginning to decrease with the realization that they have severely degraded national land (Figure 17-1). The nation is not exactly getting rich on the money earned by using public land in this way. In 1990, the revenues from selling timber and leasing grazing in national forests amounted to less than $1 billion, and less than $100 million of that was turned back into managing public lands.

By 1910, two conflicting philosophies were already apparent among those interested in U.S. public lands, and the conflict causes problems to this day. On one side were the managers, including Roosevelt and Gifford Pinchot, first head of the U.S. Forest Service. This group saw national forests and wildlife refuges as economic resources, to be used for timber-cutting, mining, recreation, and other activities for

(a)

Number of cattle, horses, sheep, goats

(b)

**Figure 17-1** Using Forest Service lands. (a) Cattle grazing in national forest in Idaho. In the background lie dead tree branches left by loggers. (b) The slight decrease in the amount of timber cut and the number of livestock grazed in national forest during the 1980s.

the economic benefit of the country, regulated so that they did not deplete the natural resources. The other group, led by John Muir and later by forester Aldo Leopold, pointed out that a managed forest does not contain the biodiversity of a natural forest. They believed in preserving wilderness with as little human interference as possible for future generations to study and enjoy.

In 1962, preservationists succeeded in passing the Wilderness Act, which (in theory) preserved designated wilderness areas from almost all interference. This was followed in 1972 by the Coastal Zone Management Act, which calls on states to develop plans to protect coastal wetlands. But wilderness areas make up less than 15% of U.S. public land. The managers won the fight for the rest. National forests are now managed under the policy of **sustainable yield and multiple use**, according to guidelines set out in the National Forest Multiple Use and Sustained Yield

Act of 1960. This Act decrees that national forests must be managed so as to achieve the greatest good for the greatest number of people in the long term.

Multiple-use management of public lands was a fine democratic idea, but it is crumbling under the demands of a growing population. For one thing, multiple uses of land and sustained yields of oil, timber, or minerals are often not compatible. The language of the Act leads to numerous conflicts. These often involve people who want the forest left as wilderness on the one hand, and mining, timber, and development advocates, who want to exploit the forest on the other hand. The Forest Service must decide whether a particular area should be left alone for backpackers, hunters, and wildlife; lumbered; or leased to be used as a ski resort or for oil exploration. Sooner or later, we shall have to decide which uses of our national forests take priority. The sooner these decisions are made, the greater the chance that there will be something left to use. Until then, our national forests are not being managed sustainably but are being severely degraded.

The Forest Service does not have enough money or people. Huge areas of national forest are completely deforested and have not been replanted. The Forest Service also manages to lose money for taxpayers by selling timber rights at cut-rate prices. To add insult to injury, the Forest Service then uses taxpayer money to build dirt roads into inaccessible forest so timber companies can reach the trees to cut them down (Figure 17-2). During the 1970s, the Forest Service built about 15,000 miles of roads in national forests. During the 1980s, exploitation of the forest increased: 17,600 miles of roads were built at a cost of more than $500 million. This policy has caused public outcry, but it continues for political reasons: every time a new area is opened up for timber, it provides jobs in areas with high unemployment. The trouble is that these jobs are temporary. As soon as an area of forest has been cut, the jobs are gone. Cutting down our national forest until it is all gone is a poor solution to providing jobs in country areas.

Political and economic pressures have led to our public lands becoming seriously degraded under the policy of sustained yield and multiple use.

### The New Zealand Model

New Zealand seems to have come up with an idea for its public lands that the United States might consider copying. New Zealand started a national parks system

# CLOSE-UP 17-1

## Public Lands in the United States

About 40% of all land in the United States is owned by the public and managed by federal, state, and local governments. More than one third of the country's land is managed by the federal government (Figure 17-A). Nearly all of this land is in the West, particularly in Alaska. Federal land consists of six major types:

1. **National Wilderness Preservation System** (about 120,000 square kilometers). In 1964, the Wilderness Act authorized the government to protect undeveloped public land from development as part of a National Wilderness System. The Act defines **wilderness** as land that has not been seriously disturbed and where people are only occasional visitors. In the late 1970s, President Carter added millions of acres of public land, mainly in Alaska, tripling the area of the National Wilderness System. The system now includes 474 areas where there are no roads. Wilderness is open to hiking, fishing, boating (without motors), and camping. Road building, timber harvesting, grazing, mining, and commercial activities are forbidden. No structures are allowed except those that were there before the area was designated as wilderness. Obviously wilderness areas are havens for wildlife. The forested areas will, after several centuries, turn once more into the climax forest of the biome, where trees are allowed to age and die naturally.

2. **National Parks and Monuments** (about 325,000 square kilometers). The National Parks Service (established in 1912) manages 49 major National Parks, most of the them in the West. In addition, the parks system includes 292 national recreation areas, such as monuments, battlefields, trails, rivers, seashores, and historic sites (Figure 17-B).

   The goal of the Parks Service is to preserve interesting and scenic natural landscapes, preserve and interpret the country's cultural heritage, provide recreation, and protect wildlife habitats, including the Wilderness Areas that lie within many of the parks. National Parks can be used only for camping, hiking, boating, and fishing. Motor vehicles are permitted only on roads.

3. **National Wildlife Refuges** (about 365,000 square kilometers). The Fish and Wildlife Service manages 437 refuges. About one quarter of this land is also protected as wilderness. Most refuges are designed to protect breeding habitat for waterfowl and big-game animals to provide sustainable populations for hunting. A few refuges are specifically designed to protect one or more endangered species. Hunting, trapping, fishing, oil and gas development, and even livestock grazing and farming are permitted on some of the refuges as long as the Secretary of the Interior finds such uses compatible with the purpose of the refuge.

4. **National Resource Lands** (about 1.2 million square kilometers). These lands are mostly scrubland, grassland, and desert in the West and in Alaska. They are managed by the Bureau of Land Management for multiple use. Much of the land is used for livestock grazing, and some of it was brought into the system because it was found to contain strategic minerals and fossil fuels.

5. **National Forests** (about 775,000 square kilometers). National Forest includes 20 areas of grassland and 156 forests, managed by the Forest Service. Apart from the 15% of this land protected as wilderness, the land is managed for sustained yield and multiple use. The lands are used for timber, grazing, agriculture, hunting, fishing, mining, and oil and gas leasing; the land is even leased for such purposes as ski resorts.

6. **Military installations** (about 100,000 square kilometers). Many military bases are little-known treasure troves of virgin habitat and endangered species. They have often been used as firing ranges, undeveloped land for troop maneuvers, and dumping grounds for old trucks and tanks, but they have not usually been deforested or otherwise disturbed. For instance, Wright-Patterson Air Force Base in Ohio contains the largest tallgrass prairie in the state; Fort Stewart, Georgia, contains at least a dozen species of rare amphibians and reptiles and Georgia's largest population of an endangered woodpecker. In 1990, Congress lent support to the work of local sci-

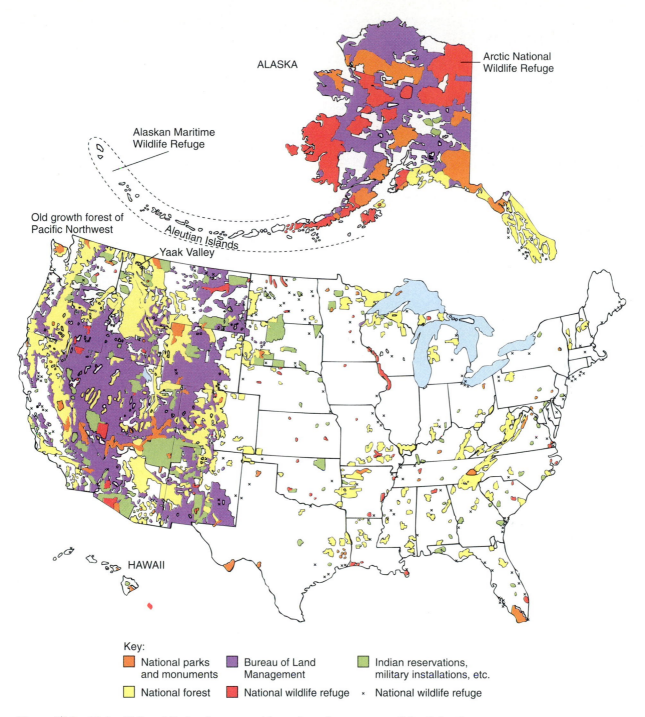

ALASKA

Arctic National
Wildlife Refuge

Alaskan Maritime
Wildlife Refuge

Old growth forest of
Pacific Northwest

Aleutian Islands

Yaak Valley

HAWAII

Key:

■ National parks
and monuments

■ Bureau of Land
Management

■ Indian reservations,
military installations, etc.

■ National forest

■ National wildlife refuge

× National wildlife refuge

**Figure 17-A**  Major U.S. public lands managed by various departments of the federal government. The position of Yaak Valley and of the old-growth forests of the Pacific Northwest are shown. *(U.S. Geological Survey)*

entists and the Nature Conservancy by allotting money for the Legacy Resource Management Program to promote the identification and management of biological resources on Department of Defense Lands. Now Fort Bragg, North Carolina, has an Endangered Species Officer, and 1,200 rare plant populations have been identified on this base alone.

**Figure 17-B**   Tourists exploring an old Spanish fort maintained by the Parks Service at St. Augustine, Florida. The Parks Service employs thousands of people and spends more than $1 billion each year to preserve natural and historic sites throughout the United States. (It also spends more than $15 million each year picking up the litter that people leave in the parks.)

**Figure 17-2** Logging national forests: redwood logs waiting to be cut up at a lumber mill in California. The small logs at the top of the pile are from young trees. The large logs at the bottom show that the Forest Service has recently built more dirt roads and opened up a new area of old-growth forest to logging. *(Richard Feeny)*

at about the same time as the United States and has managed its public lands in similar ways. As a result, it has almost no native forest left. However, the country began to replant deforested public land with softwood plantations (mostly pines) early in this century, and it is now harvesting a lot of timber from these planted forests.

The problems New Zealand faced were much the same as those of the United States. First, the remaining native forest was threatened, which angered conservationists. Second, most of the trees cut down were sold cheaply as logs, not as finished wood products—which employ more people and generate more money. The same is true in the Pacific Northwest, where the United States's remaining old-growth trees are being sold as logs to places like Japan for a price that is much less than the cost of replacing the trees.

New Zealand wanted to preserve native forest. It also wanted to create jobs by turning the country's wood into finished products such as paper, planks, siding, furniture, and similar products, which are more valuable than raw timber. Lumber companies will not invest in paper mills and lumber yards unless they control the supply of wood, which they do not do when the wood comes from public land. This is why most of the furniture-making factories and paper mills in the United States are in the East, where most of the land is privately owned and managed as tree farms and other types of forest (Section 17-C).

New Zealand decided to sell some of its public plantations to encourage job creation and to raise money to protect the rest of its forest. Between 1989 and 1992, it sold nearly all its public plantations to private owners. Two main results have followed. First, private money has flowed into wood processing and replanting. Although only 13,000 hectares of defor-

ested land were replanted in 1991, 30,000 hectares were replanted in 1992. Second, the remaining native forests have finally received the protection environmentalists had been demanding. Under a new Department of Conservation, all logging is banned and the forest is used for wildlife protection and recreation—like U.S. wilderness areas. Arguments between loggers and conservationists, which were once as bitter as anywhere in the world, have virtually ended.

It would certainly make political waves to suggest selling off U.S. National Forest to private owners, but most people believe the land would be better managed by individuals than by the government. Then the Department of the Interior would be able to devote itself to protecting some of the national resources that are irreplaceable, including old-growth forest. They might also do a better job of protecting existing wilderness areas, among the most neglected of the Forest Service's areas (Facts and Figures 17-1).

## Managing Rangeland

**Rangeland** is any land that is used for livestock grazing rather than for arable farming. It is usually land that receives too little rainfall for farming but supports grasses and shrubs that livestock can eat. In the United States, most rangeland is in the West and is owned by the federal government or Indian tribes.

The Navajo live on a 6-million-hectare reservation in Utah, New Mexico, and Arizona. Traditionally they have raised small herds of sheep on the arid land. With growing populations, the Navajo have increased the size of their herds far beyond the carrying capacity of the land, and the overgrazed pasture is becoming desert. Nearly all U.S. rangeland has been severely degraded by overgrazing, and millions

## FACTS AND FIGURES 17-1

### The Wilderness Scandal

A 1989 report from Congress's General Accounting Office (GAO) concluded, "Many Wilderness Areas are deteriorating under Forest Service management. This deterioration includes erosion, damaged vegetation, polluted lakes and streams, invasion of non-native plants, trail gullying, and a proliferation of manmade structures and facilities."

The GAO found that the Forest Service had used large amounts of the money appropriated by Congress for Wilderness Areas for other purposes. For instance, in 1988 Congress set aside $44.9 million for wilderness management. Less than 40% of this money was actually used for day-to-day management of wilderness. The GAO concluded, "The funds spent on wilderness management actually decreased by 4% between 1988 and 1990, despite a 20% increase in the public funds set aside for wilderness management." Where did the money go? To tasks such as building logging roads in national forests, cleaning up litter left by tourists in Yellowstone Park, and similar nonwilderness uses.

Jay Watson of the Wilderness Society said, "Wilderness programs have never been considered a full partner by the Forest Service. While the Service's minerals, rangeland, and forest programs have full-time directors, the wilderness program is merely part of the Service's recreational program. The funds have fallen prey to the demands of competing programs."

(a)

(b)

**Figure 17-3** Degraded rangeland. (a) This rangeland in an Indian reservation in northern Nevada has been degraded by overgrazing and soil erosion. Instead of grass, it is largely covered with sagebrush. One cause of soil erosion is obvious on the hill: trucks and off-road vehicles have been driven over the range, destroying the vegetation. (b) Ranchers at a Labor Day picnic ponder the latest edict from the Department of the Interior. These are my aunts and uncles. They look depressed because they have just been informed that they can put fewer cattle on the national forest where they lease grazing land than they did last year.

of hectares have been converted into desert with little plant cover or soil by processes like those that have desertified much of the Sahel (Section 14-D). A recent study of U.S. rangeland showed nearly half the land to be in poor condition, with privately owned lands even more degraded than public lands (Figure 17-3).

In 1978, the Public Rangelands Improvement Act was passed to improve the management of public range. First, improving rangeland requires **grazing management**, limiting animal herds to sizes the land can support. Second, ranchers can embark on **range improvement**, including practices such as weeding out sagebrush and other weedy plants that invade overgrazed land, planting vegetation where the soil is bare, fencing areas to let them recover from overgrazing, and digging enough small water holes so that livestock do not destroy the vegetation around a single watering area. These measures are not popular with some ranchers who lease grazing on public land and who may not see the benefits from improving rangeland or who dispute claims that they are overgrazing the land.

## 17-B WILDERNESS

Wilderness that has never been significantly altered by human hands is vitally important and fast disappearing. It is our only chance to study ecosystems in their original complexity. Such studies often yield important new products and biological methods of controlling pests (Section 21-F). If we need natural enemies of an agricultural pest, the best place to look for them is in the agricultural crop's original habitat. In addition, many species cannot survive when their habitat is disturbed, and preserving wilderness is our only chance of saving many species from extinction. Let us consider the state of wilderness areas in various parts of the world.

> Natural forests provide us with numerous resources, such as new species and varieties of organisms, absorption of carbon dioxide produced by burning fossil fuel, food, lumber, fuelwood, and beauty.

## U.S. Forest Wilderness

When we look at the vast forests of North America, it is hard to believe that they have nearly all been logged and that nearly all the trees are less than 300 years old. But so it is. Even before settlers began to destroy American forests for lumber or to create farmland, loggers were at work in the forests. In the seventeenth century, Europeans had so degraded their own forests that they turned to America. Expeditions scoured the forests for trees to build the ships that made up the huge navies of Britain, Spain, Portugal, France, and the Netherlands. Particularly valuable were single trees tall enough to be used as masts—the oldest trees in the forest. The trees were cut and removed from the forest in rafts floated down the many waterways. For instance, of the 7 million hectares of forest that survive in Maine today, only 2,400 hectares have never been lumbered. The Maine Chapter of the Nature Conservancy (Close-Up 17-2) recently acquired nearly 1,600 hectares of this virgin forest at Big Reed Pond. Such preserves constitute our only wilderness, our only chance to see what the American forest was once like. Most U.S. wilderness areas are not virgin forest. They have been protected only since 1964, and nearly all of them were logged before that date (Facts and Figures 17-1).

## Desert

The early European settlers understood the value of forests, but they did not understand prairie and desert, biomes not found in Western Europe. The prairie looked like pasture for livestock, and the desert looked like useless rocks, sand, and scrub. As a result, early conservationists did little to preserve such areas. Only in the twentieth century have we realized that these ecosystems have much to offer us beside natural beauty. In a world increasingly short of fresh water, the adaptations of desert plants and animals to drought are now the subject of intensive study. The salt lakes found in some desert areas are home to organisms adapted to life in salty soils like those often created by irrigation.

The desert has become prime land for development. Many Americans choose to retire to hot, dry areas in California, Texas, Arizona, Utah, and New Mexico, and the growing human population threatens to decimate desert wildlife before we have had a chance to study it. The preservation of prairie and desert wilderness should be high on the American agenda, or there will be little left to preserve in our twenty-first century.

## Prairie

What most impressed early European visitors to the United States was the prairie and the wildlife it supported (Figure 17-4). Of all the biomes on this continent, the prairie has probably suffered greatest damage because of its value as agricultural land.

Ten million hectares of **tall-grass prairie** once stretched in a wide band from Alberta through the Dakotas to Texas and included most of Nebraska, Iowa, Kansas, Oklahoma, Illinois, and Indiana. Two hundred years later, less than 1% of this rich grassland remains. This was the American land with deepest soil, formed where there was not enough rainfall to support forest, but decomposition was so slow that organic matter accumulated. In spring, flowers such as the towering compass plant dominated the landscape, to be replaced later by big bluestem, indian grass, sideoats grama, and other grasses, which towered to 3 meters or more by fall. Much of the northern tall-grass prairie lies in the prairie pothole region, where ponds and bogs punctuate the grassland. These are vital staging grounds for millions of waterfowl and shorebirds on their way to Arctic nesting grounds. California's Central Valley was once grassland that was probably tall-grass prairie, home to endangered species such as the San Joaquin kit fox (shown in the frontispiece to this book). But this area has been converted to farmland so completely that we do not know what it was originally like.

In the rain shadow of the Rocky Mountains lies **short-grass prairie**, adapted to dry conditions. The plants are shorter and farther apart than on the Great Plains, enabling their roots to gather just enough moisture to survive (Figure 17-5). The destruction of short-grass prairie is so complete that ecologists are not even sure what plants and animals it originally contained. Dozens of species of prairie animals, such as the Columbian sharptail grouse, burrowing owls, and black-tailed ferrets are endangered or extinct. Small state and private reserves are beginning to save what is left of the prairie and to restore areas of damaged grassland.

## The Nature Conservancy

The Nature Conservancy is a remarkable private organization that uses capitalist tools to preserve land of particular ecological interest or beauty—and spends less than 9% of donations for fund-raising. By 1990, an area larger than Michigan, in the United States and abroad, was protected as wilderness as a result of the Conservancy's efforts.

The Conservancy collects money and land—any land. The money it can always use, but the land is often useless for conservation, in which case, the Conservancy trades the land for land that it wants or sells it for cash. It has put together huge preserves from a few big donations, an exchange or two, and a program of purchasing all the land it cannot get in any other way. The Conservancy does not manage its land if it can help it. Better to turn the preserve over to local management, the state, the Forest Service, or a foundation so the Conservancy can get on with the next urgent project.

Save taxes by giving land and money to the Conservancy in trust? You live tax-free on the land with money from the trust, and when you die, the Conservancy gets the land and the money. A vacation tour of the Amazon jungle in Peru or Costa Rica's parks and beaches (Figure 17-C)? It will be the trip of a lifetime for photographers and nature lovers.

Of course, the trip will cost a little extra because the Conservancy uses some of your money to save yet more rain forest from the chain saw. Lawyers and accountants pore over the tax laws, looking for every provision that might make donations to the Conservancy attractive to corporations and individuals.

You don't have to be wealthy or a landowner to help. With programs in 46 states and several foreign countries, the Conservancy always has work for its one half million members. Members and other volunteers in Illinois are restoring disturbed prairie—carting away garbage and raking the seeds of prairie grass and wildflower species into the ground. Workers are needed to prepare management plans for Natural Area Reserves administered by the Hawaii Department of Land and Resources. More than 1,000 preserves are managed by a nationwide army of volunteers—the largest private system of nature sanctuaries in the world. And then there is the Natural Area Registry program, through which the Conservancy identifies outstanding ecological areas in each state and notifies their owners. In 1993, this program induced private, corporate, and public landowners to protect thousands of hectares from development.

**Figure 17-C** A Nature Conservancy group stops for lunch on a visit to Palo Verde National Park in Costa Rica.

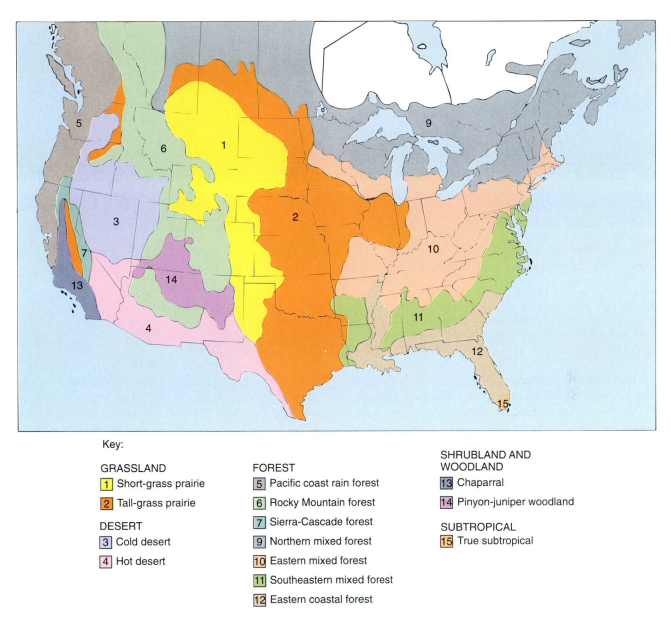

Key:

GRASSLAND
1 Short-grass prairie
2 Tall-grass prairie

DESERT
3 Cold desert
4 Hot desert

FOREST
5 Pacific coast rain forest
6 Rocky Mountain forest
7 Sierra-Cascade forest
9 Northern mixed forest
10 Eastern mixed forest
11 Southeastern mixed forest
12 Eastern coastal forest

SHRUBLAND AND WOODLAND
13 Chaparral
14 Pinyon-juniper woodland

SUBTROPICAL
15 True subtropical

**Figure 17-4** Prairie and other types of natural vegetation in the United States and southern Canada. Prairie once covered a vast area in the middle of North America that is now largely farmland.

Even in the United States, true wilderness is rare. Private conservation groups, rather than government, have saved most of the untouched forest, prairie, and desert that is now protected.

### Tropical Forest

Tropical rain forests are Earth's richest and most diverse ecosystems. They contain about 50% of all plant and animal species and 80% of the biomass of land vegetation, although they cover less than 10% of the land surface (Figure 17–6). In fact, ecologists say that we could preserve much of the world's biodiversity by protecting just 5% of Earth's land area—most of it tropical forest. This finding stemmed from studies by scientists at the International Council for Bird Preservation, who spent three years mapping the breeding distributions of all the world's land birds. They found, to their surprise, that nearly 30% of all birds are confined to 5% of Earth's land surface and never leave that area.

*(a)*                                                                                    *(b)*

**Figure 17-5**   Short-grass prairie. (a) South Dakota. The trees in the distance grow along the bed of a stream that usually dries up in summer. This prairie is somewhat degraded by grazing, but the stand of little bluestem grass in the foreground is quite healthy. (b) Flowers of short-grass prairie include fireweed (purple), sulfur flower (bright yellow), erigeron (white), and rabbitbrush (orange-yellow). The natural prairie vegetation is about 80% grasses and 20% other species of flowering plants. *(Richard Feeny)*

The country with greatest biodiversity is Indonesia, a chain of 17,000 islands lying between Australia and the mainland of southeast Asia and containing more species of mammals, birds, and reptiles than any other country. The nearby Philippines and Malaysia once contained almost as many species and as much forest (Table 17-1). Indonesia ranks as the fourth most populous nation in the world and includes the island of Java, which is more densely populated than Bangladesh. Indonesia is trying to save some of its forest, but with a population growth rate of 1.6% a year, this is not easy. On the island of Sulawesi, half of the original forest remains, and new species are discovered there every year. The government, together with conservation organizations, is working hard to preserve some of the island's forest

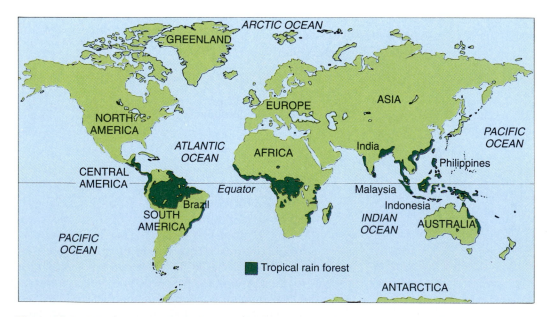

**Figure 17-6**   Where the tropical rain forest is. Although the largest area is in the basin of the Amazon River in South America, the greatest biodiversity is on the islands of the Malay archipelago between Australia and the Asian mainland.

| TABLE 17-1 | Loss of Tropical Forest in Southeast Asia During the 1980s* | | |
|---|---|---|---|
| Country | Forest Area in 1990 (ha x 1,000) | Average Area Deforested Each Year, 1981–1990 (ha x 1,000) | Percent of Area Deforested Each Year |
| Indonesia | 108,600 | 1,315 | 1.21% |
| Malaysia | 18,400 | 255 | 1.39% |
| Philippines | 6,500 | 110 | 1.69% |

*Source: R. Schmidt, *Sustainable Management of Tropical Moist Forests*, 1990.

while providing jobs that involve protecting the forest instead of cutting it down.

## Disappearing Tropical Forest

We know only roughly how fast the world's tropical forest is disappearing or what is happening to the land. Most of the countries with large areas of forest are poor and seldom perform censuses or land surveys. The best estimates of the World Resources Institute on forest loss in eight important countries are shown in Figure 17-7. The loss in 1990, more than 120,000 square kilometers, is an area larger than Pennsylvania or the whole country of Cuba. Three years of destruction at this rate would clear an area about the size of Oklahoma. The only reason there is so much tropical forest left today is that most of it lies in countries with small populations that did not start to expand agriculture into all areas of the country until the population explosion of the last 50 years. More than half of Earth's rain forest has been cut down since 1945, whereas most virgin temperate forest was felled before 1850.

> It is essential to preserve more areas of tropical forest immediately, before most of it is destroyed.

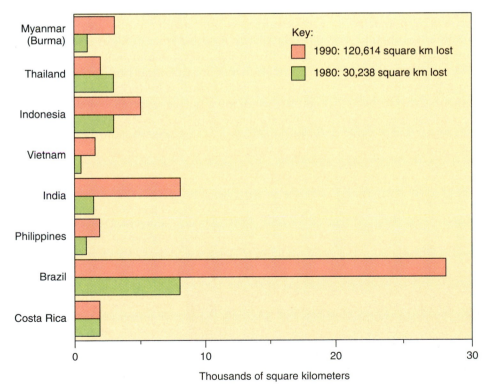

**Figure 17-7**   Tropical deforestation accelerates. Nearly four times as much tropical forest was cut down in 1990 than in 1980. (Figures for 1980 from the U.N. Food and Agriculture Organization and for 1990 from the World Resources Institute.)

Tropical forest has long been home to small, stable human populations. Traditionally the people live in tribes and forage in the forest, which provides them with food, clothing, firewood, and shelter. Some forest tribes plant small plots of crops that provide part of their food. In this **slash-and-burn** agriculture, an area of forest is cut and burned and the land used for agriculture for a year or two. After this, the thin soil has lost its fertility and the plot is abandoned. Trees and shrubs invade the clearing, and forest grows back.

At the end of the nineteenth century, some forest dwellers started to export plant foods such as bananas, coconuts, and cocoa from the forest, and lumbering became important. The world's most beautiful hardwoods, such as teak and mahogany, come from tropical forest. Cutting hardwoods by traditional methods has little permanent effect on the forest because the commercially valuable trees grow scattered through the forest. This makes it hard to remove them but means that cutting them has little effect on the forest.

This relatively stable relationship between forest and people has been altered completely in the last 50 years because the human population of the tropical forest biome has more than quadrupled. The forest does not produce enough food for these large populations. Only intensive agriculture can do so. Sustainable agriculture is undermined as population pressures force farmers to recultivate land before it has recovered from the last crop. In addition, nearly all tropical countries are poor. They need cash crops that they can export to pay their bills for imports of food and oil. As a result, governments provide incentives for their citizens to convert forest into arable land, tree plantations, or ranchland and to log hardwoods that can be sold to other countries.

Poverty-stricken developing countries are not the only culprits, however. Hawaii's last remaining lowland rain forest is being cleared at the rate of several hectares a day to supply wood chips that power an electric generating station. The area will be turned into pasture under tax incentives provided by the Hawaiian and U.S. governments. A further irony is that the University of Hawaii is a center for research on the use of trees and other biomass to produce energy. But all the research is directed toward developing countries, while biomass burning is destroying the forest near the university. When one of the university's scientists wrote to a newspaper to protest the forest's destruction, he was chastised by a state forester for being a traitor to his profession.

> Tropical forest is being destroyed mainly to provide fuelwood, subsistence farms to support rapidly growing populations, and cash crops for export.

**Farming**   Most tropical land today is deforested to produce agricultural land. Plants grow so fast in a tropical rain forest that one might expect agricultural crops would do the same. But they do not. As ecologists have pointed out, most of the poverty-stricken people in the world are poor because they have inherited poor-quality land. High agricultural productivity is possible only in areas with adequate sunshine, rainfall, and soil. Sunshine and rainfall exist in abundance in a tropical rain forest, but most of the soil is very thin. The high temperature and moisture are ideal for decomposer organisms, so organic matter decomposes rapidly. The plants absorb nutrients as fast as they are released by decomposition, so essentially no minerals are left in the soil,

**Figure 17-8**   Rain forest into farmland. This coffee is growing on what was once rain forest on Costa Rica's central plateau. Unlike most tropical forest soil, which is thin and infertile, this is volcanic soil so fertile that three crops of potatoes a year can be grown without fertilizer.

and almost all the nutrients of the forest are locked up in living organisms (Section 4-I). The best tropical soil is found in the limited areas where volcanoes sometimes spew mineral-rich ash and lava over the countryside (Figure 17-8). This is why farmers rapidly return to farmland on the slopes of volcanoes, even after an eruption that has killed many people and threatens to be repeated.

Brazil commissioned a nationwide land survey that showed that about one third of its land area is suitable for intensive agriculture. Another one third has soil such that it can be cropped lightly or used for light grazing, but the soil erodes rapidly if the vegetation cover is removed. The other one third of Brazil is completely unsuited to agriculture of any kind. Most of it contains lateritic soils, which turn to a substance like concrete when the tree cover is removed (Section 14-F). As a result of this study, Brazil set aside a remarkable 100,000 square kilometers of land to be preserved as tropical forest. The economic value of the land would be recovered by using it to generate tourist revenue and by using it for research on rain forest and the resources it contains.

The trouble with this sensible plan is that the Brazilian government finds it impossible to prevent landless, unemployed members of the country's rapidly growing population from moving into the preserved area and cultivating the land, no matter what official policy may be. The situation has been made worse by the World Bank, which financed construction of a road that cuts through the heart of the Amazon rain forest. This makes it easy for illegal settlers to reach this previously inaccessible area. Illegal settlers are a major cause of forest destruction in almost every country with rapid population growth.

The secret to preserving wilderness is to prevent road building. As soon as roads enter a wilderness area, the migratory paths of wildlife are disrupted, and, in developing countries, illegal settlers move in and start destroying natural resources.

> The secret to preserving wilderness is to prevent road building.

Officials often actively encourage deforestation. Many Latin American governments provide economic incentives to people who turn forest into farmland. Between 1971 and 1977, international agencies, funded largely by the United States, provided $3.5 billion in loans and assistance to be used to cut down tropical forest.

> National and international subsidies encourage deforestation.

**Ranching** When you bite into a fast-food hamburger, you may be contributing to the demise of tropical rain forests. Some 20% of the beef we eat comes not from Texas, Kansas, or Saskatchewan but from a pasture that was once rain forest in Central or South America (Figure 17-9). We usually think of locals cutting down the rain forest for firewood or to grow food for themselves, but most Latin Americans cannot afford to eat beef and grow it only to export— mainly to the United States. In Brazil alone, an area of forest the size of South Carolina was converted into cattle ranches between 1966 and 1978.

**Logging** After agriculture, most tropical forest is cut down by commercial logging operations. Gone are the days when a single mahogany tree was cut and dragged out of the forest. Now huge areas of forest are mowed down to provide access to the few commercially valuable species, which often amount to only two or three in every hundred trees. Some of the remaining trees will be converted into charcoal or sold as fuelwood, but most will be burned or rot where they were felled.

**Fuelwood** More than half the world's people depend on wood for fuel, and half of all the wood used in the world is burned as fuel. Someone who depends on wood for heating and cooking burns about a ton of fuelwood a year, so it is not surprising that wood is now being cut for fuel faster than the trees can regrow. The burden on local forests is worst in the tropics, where four fifths of all wood harvested is used directly as fuel or for making charcoal. (Charcoal is lighter and cheaper to transport than raw wood.) The increase in the cost of oil in the 1970s made the fuelwood shortage worse because it raised the cost of kerosene—a petroleum product that is the main alternative to wood as a fuel in most countries.

The extent of the problem is hard for us to imagine. In many African countries, the average laborer's family spends one third of its income on wood—as much as the average North American family spends for both energy and the house it lives in. Fuelwood users collect dead wood and fallen limbs when they can find them. But when there is not enough dead wood, people have to cut down living trees. Charcoal, the main fuelwood supply to urban areas, can be made only by cutting down live trees. It is made by burning fresh, green wood slowly.

> One of the most pressing problems in many developing countries is a fuel shortage caused by the unsustainable harvesting of wood.

**Figure 17-9** Tropical forest into hamburger. These Brahma cattle are grazing in what was once tropical dry forest in Central America. Brahma cattle survive tropical conditions better than temperate breeds such as Herefords, but they produce tough beef used mainly as hamburger for the North American market. The cattle are eating Johnson grass (*Sorghum halapense*), an invasive species introduced from the Mediterranean. Spreading Johnson grass has caused the extinction of hundreds of native plant species (in Texas as well as in Central America) and clogged waterways so floodplains no longer flood every year, which has caused the deaths of millions of migratory water birds.

The extent to which demand for fuelwood exceeds supply can be gauged from the fact that, by 2000, native woodlands will be able to supply only about 20% of North Africa's fuelwood. Every tree has been felled for 90 kilometers around Khartoum, capital of Sudan, and forest rangers in national parks confront armed thieves cutting wood to sell in the city. In Burkino Faso, women often have to walk for 18 hours to find enough wood to cook evening meals for a week, and urban families spend more than a quarter of their incomes on fuelwood or charcoal. In India, special mobile courts and guard squads attempt to capture tree thieves.

The high price of scarce fuelwood makes boiling water an unaffordable luxury in many places. This increases the spread of diseases caught by drinking impure water. It also contributes to malnutrition: people are forced to eat quick-cooking cereals such as cornmeal instead of more nutritious foods such as beans that take longer to cook. The gap between supply and demand is also indicated by the United Nations' calculation that to meet their fuelwood needs, developing countries need to plant nearly 3 million hectares of trees a year, compared with the one half million hectares a year they are actually planting.

The high cost of wood raises the hope that landowners will see the advantage of growing trees to develop a sustainable, labor-intensive business, but there are few signs of such enterprises in most places.

Fuelwood shortages are local problems because it is seldom economical to transport fuelwood more than a few hundred kilometers from where it grows. But the effects of fuelwood depletion are national and international in scope. One unfortunate effect is the increasing use of animal dung for fuel. This dung would otherwise be used to fertilize the soil. Without it, soil erosion accelerates, agricultural productivity declines, and the land is eventually added to the world's spreading deserts.

> The depletion of wood supplies is the largest single cause of environmental degradation in many developing countries. It raises the cost of living and causes malnutrition and disease.

### Saving Tropical Forest

Tropical countries understand the importance of preserving at least some of their forest as tourist attractions; for their genetic resources; and as the producers of rainfall, wood, and other natural resources. The main problem is that these countries do not have the money needed to defend large areas of land against settlers displaced from their ancestral villages and jobs by explosive population growth. Developed countries can help in several ways. Aid programs that emphasize population control are essential. So are imaginative programs such as debt-for-conservation swaps (Section 24-C). Involving multinational corporations in the search for valuable plants and chemicals in tropical forests also provides money for conservation (Close-Up 17-3).

Realistically we cannot hope to preserve more than about 10% of tropical virgin forest. It is important, however, that the area surrounding such preserves should be managed in such a way as not to reduce the biodiversity of the protected area. One step toward this is to set up **extractive reserves**, areas of forest where sustainable harvests of wood, nuts, animals, and other resources are extracted from the forest. Extractive reserves provide employment, while leaving the area forested and capable of supporting wildlife and rainfall as well as preventing soil erosion. This is the idea behind the sale of various tropical products such as Rainforest Crunch ice cream, which contains nuts from the forest, and jewelry and buttons made from ivory-like tagua nuts from Ecuador in a project instituted by Conservation International.

### Temperate Forest

Brazilians are liable to get upset when people point out that an area of forest the size of Belgium is cut down in their country each year. "Yes," they say, "it would be nice if Belgium were reforested wouldn't it?" Brazil contains thousands of times as much virgin forest as any developed country and may well be looking after it better, as the quotation at the beginning of this chapter suggests.

Eighty percent of Europe was once covered by forest—which is hard to believe when we look at the bare, sun-baked hills of Greece and Spain or the wide open spaces of Scotland's moors. By 1800, only 14% of the forest was left. Europe has been farmed for so long that natural ecosystems have essentially disappeared except in parts of the taiga that stretches across Scandinavia and northern Russia.

North America is more like a developing tropical country than like Europe in that it contains a lot of forest, even some virgin forest. The question for the next 50 years is whether we shall protect enough old-growth and virgin forest to save from extinction hundreds of plant and animal species that live in this habitat.

For more than two centuries, settlers waged open warfare against the American wilderness. Then in less than a decade, they discovered that it had limits. In 1880, the Boskowitz Hide Company of Chicago shipped more than 34,000 bison (buffalo) hides east from Montana and Wyoming. In 1884, they shipped only 529. In 1890, a note in the national census stated that the frontier no longer existed and would be dropped from official consideration. Between 1880 and 1890, most of the remaining American wilderness had been transformed into cow pasture. It took the near extinction of the bison to shock the conservation movement into action. Easterners, in particular, rose in arms against the destruction of the last wilderness for the benefit of a small number of settlers. The end of the nineteenth century saw the protection of millions of hectares of forest. What eastern conservationists did not realize was that much of the West was not forest but desert and grassland. Although forests have been protected to some extent, desert and grassland have been steadily degraded in a confusion of public management and private exploitation.

> Nineteenth-century conservationists saved North American forests from the almost total destruction that had occurred in Europe, but our national forests are still being destroyed.

## 17-C MANAGING FOREST

Worldwide, more than 90% of all commercial log production comes from unmanaged forests. The cost

# Costa Rica and the Merck Agreement

Merck is a multinational chemical and drug company that has entered into an agreement with Costa Rica that will probably prove to be a model for many more. Merck paid Costa Rica's National Institute of Biodiversity $1 million for the right to analyze hundreds of extracts from species indigenous to Costa Rica in the hopes of finding useful drugs and other products. If Merck does discover a valuable product and develop it for market, Merck will retain all rights to the product but pay Costa Rica a royalty for each item it sells.

In turn, Costa Rica will spend 10% of the first $1 million, and half of any royalties, on conservation. The country badly needs the money to buy up land and hire park rangers to preserve at least part of the country's rapidly vanishing forests.

Drug companies and developing countries applaud this agreement as an excellent way to exploit and preserve species and habitats. Indonesia and Brazil have expressed interest in similar agreements, and we shall probably see many of them before long—an imaginative way of preserving the environment and profiting from it at the same time.

**Figure 17-D**  Palo Verde National Park in Costa Rica. *(Paul Feeny)*

of this wood does not reflect the cost of replacing this natural resource because most countries are still chopping down irreplaceable virgin forest and not replanting trees. Therefore, the price of wood is bound to rise as natural forests are depleted. In the future, more and more wood will come from plantations or tree farms and managed forests.

Where forests are properly managed, the goal nowadays is sustainable yield. An estimated 5,000 commercial products come from our forests, these include turpentine, lumber, paper, gums, resins, oil, and charcoal. Each year, the United States sells about $25 billion worth of products from its 300 million hectares of forest, making a significant contribution to the national economy.

## Harvesting Methods

Two thirds of the United States's forest is commercial timberland, most of it managed by lumber companies. Although methods of cutting trees have improved, the United States still lags behind parts of the world where trees are harvested by methods that do less long-term damage. There are two main ways of harvesting trees:

**1. Clear-cutting** Clear-cutting means cutting down every tree on an area of land, often using huge machines. Clear-cutting is the method of choice for harvesting conifers (softwood) that are used for many building purposes or converted into pulpwood that is made into paper. After the trees are cut, loggers remove the commercial timber and often burn what is left. Burning returns some nutrients to the soil and clears the land either to reseed itself or to be replanted. Conifers usually grow, or are planted, in monoculture stands of one or a few species (Figure 17-10).

Clear-cutting is the fastest and cheapest method of harvesting trees, and it also increases the populations of some species of deer and wildflowers that thrive in the open spaces of a disturbed forest. The disadvantage of clear-cutting is that it increases the runoff of water, nutrients, and soil from the area, reducing the soil's fertility. It is also very unsightly and destroys the habitat of organisms such as the probably extinct ivory-billed woodpecker that lives in old forest. In tropical forests, clear-cutting may be ruinous, exposing delicate soils to rain and sun and permanently destroying their productivity.

In the United States, public outcry against clear-cutting led to Forest Service regulations restricting clear-cutting to a maximum of 45 contiguous acres. However, the Forest Service does not always enforce

**Figure 17-10** A tree plantation in Oregon with a sign to educate passersby. Foxgloves and ferns have invaded the disturbed habitat at the edge of the plantation. *(Richard Feeny)*

its own regulations, as anyone who flies over national forests in the Pacific Northwest can see for themselves. A 1986 survey showed that on one third of clear-cut national forest land foresters had not replanted even the minimum number of trees required by regulations (Figure 17-11).

American forests could be managed so as to be much more attractive and to provide a better habitat for wild plants and animals, while avoiding much of the soil erosion that is slowly reducing their productivity. Several European countries harvest just as much wood from their forests while avoiding the damage. For instance, in Austria clear-cutting is forbidden on steep land where the soil erodes easily, whether the land is private or public. A private landowner must obtain permission to clear-cut more than half a hectare. Drawing on ecological research such as that at Hubbard Brook in New Hampshire (Close-Up 5-1), Western Europe usually limits clear-cutting to narrow strips so as to minimize the loss of soil, water, and nutrients. The forest regrows much faster than when clear-cut by American methods.

**Tree Farms** About 35 million hectares of the United States are occupied by tree farms. Here fast-growing hybrid pines are planted, usually by machine. The trees are fertilized then sprayed with pesticides. After as little as five years, the trees are

**Figure 17-11**   Clear-cutting is wrecking national forest in the Northwest. Debris is being burned after the area has been clear-cut. You can see that the forest service is not replanting promptly as it should. The patch in the foreground is almost covered with vegetation, meaning that it must have been cut at least a year earlier, but it has still not been replanted. You can also see gullies formed by soil erosion. Clear-cutting should be done in narrow strips along the contours of the hill to prevent soil erosion. These clear-cut areas almost touch each other, which is why from the air this area looks like a checkerboard. *(Richard Feeny)*

harvested by clear-cutting (Figure 17-12). Land managed in this way supports as little wildlife as a field of cabbages. The deer, birds, and wildflowers of lumber company advertisements are not to be found on tree farms.

> Harvesting trees by clear-cutting degrades the soil unless it is carefully managed and destroys wildlife habitat.

**2. Selective Cutting**   Where forest contains trees of different ages or species, it is often cut selectively, removing only the mature trees (Figure 17-13). Selective cutting is the ancient method of harvesting trees. It takes more time and usually costs more than clear-cutting, but it causes much less damage to the

forest. It is the method by which valuable hardwoods, such as maple and beech, must be harvested for sustainable yield. Seedlings of these trees require shade from other trees to survive and cannot be grown on clear-cut land. Owners of small areas of forest usually engage a lumber company to selectively cut the area, harvesting some trees but doing little damage to the soil or to the forest's overall appearance.

## 17-D REFORESTATION

It is unrealistic to hope that the rate at which forest is being felled will slow down before most of the world's natural forest has gone. Our only chance of having enough forest in the twenty-first century to slow climate changes and support rural economies is

(a)

**Figure 17-12** Tree farming in the Southeast. After the area is clear-cut, the debris is gathered into piles and burned. Most forest in the Southeast is privately owned so it is replanted promptly to ensure maximum harvests and to prevent soil erosion.

(b)

**Figure 17-13** Selective cutting. (a) A tree ready to be harvested is cut down by chainsaw and dragged out of the forest by a small bulldozer. (b) Limbs are trimmed from the trunk, which is then loaded onto a truck. Even though selective cutting does much less damage to the forest than clear-cutting, you can see that the use of dirt roads is causing a lot of soil erosion as wind blows clouds of soil away or water washes it down the road.

**reforestation**, planting trees. As people have begun to understand the importance of forest and woodland to our common well-being, massive reforestation has taken place all over the world. An estimated 7 million hectares of land are now planted with trees each year, although this is still less than the area of forest that is felled each year. By 2000, the area reforested during the year is expected to grow to 17 million hectares.

In 1970, the forest area of New England, although smaller than in precolonial times, was 40% greater than in 1890. In the 1950s, the steady decline of forested area in the southeastern United States was reversed, and forested area has slowly increased ever since. China has doubled its forested area in less than 20 years. Similar efforts have been undertaken in other parts of Asia and in Europe.

> By 2000, about as many trees will be planted as are cut down each year.

## Research

Modern research underpins modern reforestation. If the goal is to restore a native forest that is being destroyed, preserves and restoration are the answers (Chapter 16). But if the goal is to grow fuelwood, timber for building, wood to make paper, food for animals or people, or prevent soil erosion, new varieties of trees and new types of forest management can often improve the efficiency of reforestation.

Here we consider some of the more interesting examples.

Frequent dry periods are a problem in many tropical areas. In Sudan, foresters are testing "water grabbers," soil additives that increase the water-holding properties of the soil. Water grabbers are plastic polymers that can absorb up to 600 times their weight of water. They increase the chances that a tree seedling will live after it is planted and before it has developed large enough roots to survive a dry spell on its own. You can now find these water grabbers in American garden centers. If you mix them with the soil when you pot a house plant, the plant will need watering much less often.

Finding the most appropriate trees for the job is an essential part of encouraging rural inhabitants to plant more trees. In North Africa, officials encourage the use of trees instead of grass as animal fodder. The right tree will provide fodder and firewood and retard soil erosion. The thorn tree, *Acacia albida*, is drought-resistant and valuable for fodder. Like the equally useful mesquite relative *Prosopis*, it is a nitrogen-fixing tree that improves the soil. It also has an open canopy so crops can be grown under it with yields similar to those in cultivated fields.

## Private Reforestation

When United Nations workers attempted to assess the state of the world's forests in 1985, they found that less forest was left than had previously been thought. But they also found that reforestation was proceeding faster than anticipated, largely as a result of private activity (Figure 17-14). In Kenya, the number of trees planted by villagers, around farm fields, along roadsides, or as windbreaks, exceeds the number of trees in government plantations. In Rwanda,

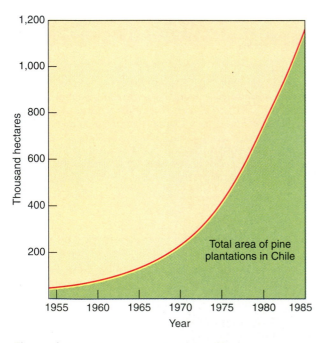

**Figure 17-14** Reforestation in Chile. The increase in area planted with pine trees in Chile between 1955 and 1985. Chile is now self-sufficient in forest products and has surplus to export. The government has subsidized reforestation by private enterprise by about $50 million since 1974, but the subsidies are a wise investment: in 1986 alone, Chile exported more than $400 million worth of forest products.

scattered trees planted by rural people cover more than the total area of the country's remaining natural forest and artificial plantations combined.

In Europe and the United States, millions of trees are purchased from nurseries and planted by homeowners every year, making a significant contribution to the increasing tree cover in towns, suburbs, and the countryside. These plantings are all the more necessary because, even today, developers tend to bulldoze the trees off land before they build on it, thereby reducing the property values of the development before it is even built.

The College of William and Mary in Virginia lacked the space to establish a separate arboretum, or tree collection, so the college first documented all the trees and shrubs that already existed on campus. Then working with grounds staff and landscape architects, the college encouraged groups of donors to pay for particular species of trees that would add to the collection. The result is a campus forested with native trees, redwoods, and many exotic species, with an educational program to match. The University of Idaho has a similar collection of trees and shrubs with labels that identify the species and the donor.

A major thrust in urban tree planting is to increase the number of species used. Planners learned their lesson from the destruction of elms by Dutch elm disease, which left many American campuses and streets denuded of trees. In Georgia, Savannah's Park and Tree Commission, for instance, aims to increase the number of tree species planted by the municipality from about 20 to several hundred by the end of the century. Then a single disease or pest will not leave the city denuded of trees. It is also important to plant new trees all the time so cities contain trees of all ages. Old trees are beautiful, but they require more pruning than middle-aged trees. Also, when an old tree dies or has to be removed, it is helpful if there are young trees nearby that will grow up to take its place.

> Local groups and individuals have planted enormous numbers of trees all over the world.

## Government Programs

Official reforestation attempts can be divided into two main types of program:

**1. Agroforestry**    Agroforestry integrates reforestation with agriculture. One of the barriers to reforestation in the past has often been that it conflicts with agriculture. For instance, the United Kingdom's

Forestry Commission has fenced thousands of hectares of treeless moorland, previously leased to sheep farmers, and planted them with conifers. The farmers were understandably annoyed. Agroforestry aims to enhance local agriculture instead of displacing it. Examples include supplying farmers with trees that can be used for livestock fodder and planting around the edges of fields instead of in large stands.

In the United States, programs encourage farmers to plant trees as windbreaks that will prevent soil erosion. The trees will also supply wood that can be sold as lumber or used as firewood, and the farmer soon begins to experience indirect benefits in the form of increased crop yields. In Costa Rica, farmers are encouraged to plant trees as "living fence posts," along the edges of fields (Figure 17-15).

**2. Community Forestry** Community forestry (also called social or mass forestry) mobilizes communities of people to plant and protect trees, either on public or on private land. Individual care for trees is very important. Without such care, 90% of the trees planted in Chinese government schemes in the 1950s and 1960s died. Half of the trees planted in Los Angeles for the 1984 Olympics died. Small trees cannot stand being walked on or ridden over, and they sometimes need to be watered until they are established. Community care often makes the difference between life and death.

In India, one state instituted a program of planting several rows of windbreak trees on state land beside roads and canals. In the past, such trees were rapidly destroyed by villagers desperate for firewood. Under the new community forestry program, villagers protect the saplings and in return can cut the grass under the trees and take a share of the profits when the trees are harvested. The trees thrive. Also in India, a number of villages were persuaded to establish plantations on communal grazing land. The forestry department supplies seedlings of trees that will produce fuelwood, fodder, and fruit. The villagers nurture the trees and share the proceeds from the harvest. Now the forestry department is contacted by village councils that have seen the benefits their neighbors reap and want to try tree plantations themselves.

The most popular community forestry in dry areas is planting green belts, large windbreaks that protect towns and villages from winds, dust storms, or encroaching sand dunes and that can be harvested for firewood, poles, or fodder. More trees go into the town itself, providing shade and improving air quality. The once hot and dusty desert town of Bouza in Niger is now famous for its cool, tree-lined streets.

In developing countries, villagers form the only labor force large enough to plant trees on the scale that is needed. Villagers have often been reluctant to spend their time planting and caring for trees because they did not understand the benefits. Outsiders could not understand why villagers never seemed to want to plant trees for firewood. One reason is that the villagers may be cutting wood beyond the sustainable yield of the environment but may still not be experiencing an acute shortage. In rural areas

**Figure 17-15** Teak trees act as fence posts in Costa Rica. When they are mature, they will produce a harvest of valuable hardwood. Private reforestation is best encouraged by advising home owners and farmers on trees that have more than one use and provide direct financial benefit.

(a)

(b)

**Figure 17-16**   Community forestry. (a) Volunteers measure the distance at which the hollies in the pots will be planted. (b) Volunteers plant trees in the parking lot of a shopping center while a television cameraman moves back for a better shot.

where firewood is not part of the cash economy, the cost of increasing scarcity is merely greater time spent collecting it. Since women do nearly all the collecting, this cost seldom troubles the village's male decision makers. Not surprisingly, women have been the leaders in attempts to prevent deforestation in many countries, although women seldom have the political power needed to succeed. One of the lessons of forestry programs in developing countries is that people are most likely to plant trees if these provide shade, fruit, fodder, and poles, rather than merely firewood. They know that firewood will be a side product of such plantings.

Traditionally, rural people who have little cash and do not own land have collected their firewood from common land or stolen it from forestry reserves. The government of West Bengal, India, addressed this problem and produced cheap reforestation by letting landless families plant trees on 5,000 hectares of state-owned land that had been illegally deforested. The families were supplied with free seedlings and fertilizer and received small cash incentives if most of the trees survived. The wood was harvested and sold after 5 years, and many of the families used the proceeds to buy plots of land suitable for agriculture. While the trees were maturing, their protectors collected twigs and branches for fuel. In areas where firewood commands a high market price and there is a lot of deforested land, such a strategy can bring unproductive land back into production while providing landless rural families with firewood and income.

Many similarly imaginative programs have increased the tree cover in developed countries, particularly in urban areas. Arbor Day is an American institution, a day when people and organizations are encouraged to plant trees in their communities (Figure 17-16).

> Imaginative agroforestry and community forestry programs have reforested large areas, using the latest research to choose appropriate trees for different purposes.

## SUMMARY

Wilderness and forest have steadily disappeared as the human population has grown. Most has been converted into farmland.

### U.S. Public Land Management

One third of the United States is land owned by the public. Much of it is national forest, managed by the Parks Service for multiple use and sustained yield. Less than 15% of this land is wilderness. Under political and economic pressures, much public land has become seriously degraded. Habitat conservation would probably be improved by selling much of the national land and spending more money on wilderness areas.

### Wilderness

Except in remnants of tropical rain forest, there is hardly any untouched wilderness left in the world. Nearly all the North American forest has been logged at least once. Prairie and desert are even less well preserved. In many cases, we do not even know what organisms these biomes once contained. In the United States, much remaining wilderness is preserved by private groups. Untouched wilderness provides habitat that preserves biodiversity and areas for research and the development of new products. Tropical forests are falling to the demand for fuelwood, timber, and more agricultural land. Government subsidies encourage the destruction of U.S. and tropical wilderness.

### Managing Forest

Forest around the world is managed, often under national regulations, by government agencies and by private individuals and companies. The usual goal of forest management is sustainable yield. In developing countries, lack of resources often means that forest is not effectively managed. Forests are harvested by clear-cutting or selective cutting. In the United States, clear-cutting is often not carefully controlled and does great damage to the land.

### Reforestation

Reforestation has increased rapidly since about 1970, although worldwide more trees are still cut down each year than are planted. In certain areas, such as New England, the southeastern United States, China, and Europe, reforestation has added significantly to the forested area. Research into appropriate tree species for purposes such as fuelwood, lumber, shade, and livestock fodder has contributed largely to encouraging people to plant trees. Agroforestry and community forestry programs have increased tree planting in both developed and developing countries.

## DISCUSSION QUESTIONS

1. Should more of the United States be preserved and allowed to return to wilderness? Why?

2. Suppose you were asked to prepare a plan to double the number of trees in your community at the lowest possible cost. How would you proceed?

3. What incentives and assistance could developed countries offer that would help developing countries preserve some of their tropical forest?

4. Most forest is cut down to make space for agriculture. What are some ways to reduce the conflict between these two uses of land?

5. European settlers destroyed Native Americans and their cultures just as settlers are destroying rain forest natives in South America and Asia. What are some of the things we lost in America as a result of this genocidal destruction?

# POLLUTION

# Part 5

*Navajo Power Plant near Grand Canyon, Arizona.* (Butch Gemin)

# Chapter 18

# TOXIC AND SOLID WASTE

*Remains of cars in a country backyard.*

## Key Concepts

- Most waste produced by pre-industrial societies is biodegradable. In industrial societies, much waste is nondegradable or toxic.

- We need to develop new methods and materials that create fewer toxic and nondegradable wastes.

- The United States produces more waste per person each year than other developed countries.

- Our solid waste problem can be solved by incentives that encourage recovery of resources, conservation, recycling, and composting.

In the next three chapters, we consider an environmental problem that affects everyone: the pollution of our environment with substances that endanger health and kill many organisms. Recent years have seen some strange happenings. In 1992, a train full of New York's garbage took a voyage around the eastern United States in search of a dump. In 1988, a ship full of toxic waste made a similar voyage around the Mediterranean, and several people died of the poisons that oozed from its leaking containers. In many states, returning a can or bottle to a recycling center earns you five cents, and in one Ohio town you may win $1,000 if inspectors find that your garbage contains nothing that could be recycled.

These incidents all reflect the growing problems of waste disposal. Society has always produced solid waste: the coast of the United States is dotted with shell middens—huge piles of the shells of mussels, oysters, and clams eaten by prehistoric people. But since about 1960, the problem of disposing of our waste has reached crisis proportions, particularly in many developed countries.

The problem is partly one of volume. Large populations produce more waste than small ones. Our high standard of living and our throwaway society also mean that each person today generates many times the waste that our grandparents did. One county commissioner calculated that each resident produced one-third more garbage in 1988 than a mere three years earlier. "Our sanitary landfill was supposed to last ten years," he said. "It has been filled up in five."

## Types of Waste

The other part of the problem is the type of waste we produce. Before the industrial revolution, nearly all waste was **biodegradable**, able to be broken down fairly rapidly by decomposer organisms. Today much of our waste is nondegradable or hazardous, or both. **Toxic** (or **hazardous**) **waste** is waste that can injure or kill. It must be disposed of in ways that prevent it from poisoning people and other organisms directly or by polluting food, air, or water.

**Solid waste** is the waste that cannot be disposed of as a liquid by way of a sewage system, and it must be disposed of elsewhere. Some toxic waste is solid. Examples include containers that once held pesticides or cleaning fluids. Toxic waste also comes in the form of gases and liquids, such as the dirty oil left after changing the oil in a vehicle (Lifestyle 18-1).

The problems of disposing of toxic and solid waste are occupying increasing amounts of time and money in every part of the world. There are three main sources of waste:

1. **Mining and Agricultural Wastes** Agricultural and mining wastes are generally disposed of on the land from which they came (Figure 18-1). Three quarters of all waste is produced by mining, much of it rock and minerals from excavation and processing (Chapter 13). On the whole, this waste is left exposed on slag heaps, dumped in the ocean, or disposed of by refilling and landscaping disused mines. When it gets wet, the waste from mines leaches acid water that

**Figure 18-1** Mine and agricultural waste are usually disposed of where they are produced. This photo shows tailings and a pond of toxic runoff from a molybdenum mine near Climax, Colorado. *(Butch Gemin)*

**Figure 18-2**   A manure pile on a farm. The manure will be loaded into a spreader and scattered on fields.

often contains toxic compounds. Disposing of mine waste without polluting the water supply is a continuing problem.

Another 12% of solid waste in the United States comes from agriculture, in the form of things such as crop wastes and manure. This waste contains plant nutrients that traditionally have been returned to the soil (Chapter 14; Figure 18-2). With more livestock being raised in confined areas such as feedlots, however, and with the increased use of concentrated chemical fertilizers, agricultural wastes have come to be seen as a disposal problem rather than as a usable commodity. Some steps are being taken to solve the problem. For instance, some feedlots produce methane for use as fuel from manure, using biogas digesters (Section 10-E). Farmers are beginning to use lower levels of pesticides and chemical fertilizers (Section 21-F). If this trend continues, agricultural waste may increasingly be viewed as a valuable source of soil nutrients rather than as toxic waste.

**2. Industrial Wastes**   Industrial waste requires special treatment when it contains hazardous chemicals. Whether liquid or solid, hazardous industrial wastes are generally treated in the same ways as solid wastes because they must be kept out of the water supply. They must be contained in special disposal areas, either forever or until they are detoxified.

**3. Municipal Wastes**   In this chapter we are concerned mainly with the 13% of solid waste that comes from homes and businesses. The municipal wastes of towns and cities cause pollution of the ocean, litter, water pollution, and air pollution from burning wastes. Most municipal wastes are disposed of in landfill dumps or burned.

> Industrialized societies produce large amounts of toxic and solid waste that are difficult and expensive to dispose of.

## 18-A  SOLID WASTE DISPOSAL

The general public no longer trusts those responsible for waste disposal. The newspapers have carried too many stories of people injured by hazardous waste, of polluted streams, and of clouds of methane from landfills. Nowadays the hunt for sites for new landfills or incinerators is a nightmare for officials, usually involving public meetings with irate citizens and environmental groups, court cases, and huge costs.

In the United States, more than 400,000 tons of municipal waste are now generated each day, more than twice as much as in 1960. Where does it come from, and what does it consist of? Almost every day, I pick up litter in my yard, much of it blown from a litter bin in the softball field across the road. This litter is nearly all paper, plastic, and polystyrene from fast-food restaurants, with a sprinkling of plastic bags, newspapers, and the occasional soft drink bottle or can. The amount of paper littering our yards and cities is not surprising, since paper, and paper products such as cardboard, make up more than one third of municipal waste (Figure 18-3).

The waste gathered on New Jersey beaches during one beach cleanup day was similar to what I find in my yard except that there is little paper amid the garbage on beaches because paper is biodegradable and decomposes fairly rapidly in the ocean (Close-Up 18-1).

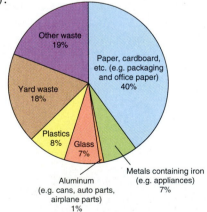

**Figure 18-3**   The composition of municipal solid waste in the United States in 1990. Municipal waste is mainly that collected by garbage trucks from homes and businesses in a community. Paper makes up a large proportion of our waste, partly because almost everything we buy comes in at least one layer of paper or cardboard packaging.

# Beach Cleanup in New Jersey[1]

Around the United States, an army of volunteers turns out once a year to pick up garbage from beaches—much of it washed up from the ocean. And many of these groups record what they find, in the hopes of directing anti-litter education at the right people or merely out of interest. In New Jersey in 1992, 873 volunteers found that 75% of the 6,300 kilograms of garbage they collected was plastic. Among the items collected in one day:

- 6,424 plastic lids and caps
- 7,052 plastic food bags and wrappers
- 9,985 broken pieces of plastic
- 10,705 plastic straws
- 19,726 cigarette butts

Clean Ocean Action and Alliance for a Living Ocean tracked some trends in New Jersey beach garbage over the years:

- Waste associated with sewage, such as plastic tampon applicators, declined from 3.84% of the haul in 1989 to 2.24% in 1992, but this is still much higher than the national average of 1%, reflecting the presence of antiquated sewage treatment systems in the Northeast.
- Broken balloons were found with inscriptions showing that they came from as far away as Illinois, Maryland, and Virginia. New Jersey has been at the forefront of efforts to limit mass balloon releases because of the environmental damage they cause.
- Plastic six-pack holders declined from 2.92% of beach garbage in 1988 to just 0.52% in 1992, following 1982 state legislation requiring beverage holders to be made of photodegradable material (Section 18-B).
- Bottles dropped from 15% of the garbage in 1989 to 8.54% in 1992, undoubtedly as a result of the enactment of bottle bills in the northeastern states (Figure 18-A).

[1]From an article in the *Trenton Times*, August 8, 1992.

**Figure 18-A** Roadside litter, some of it washed up from a nearby river. It is an interesting exercise to gather a group of friends and collect and analyze litter on a beach, riverbank, or roadside. Compare your findings with the New Jersey beach collection. If you can collect for more than one year, do it at the same time of year and see if the collection changes from one year to the next.

There are only three places for wastes to end up: in the ground, in the water, or in the air. Some of our waste ends up in each place. Most nonhazardous waste is burned, dumped at sea, or deposited in **landfills**, areas where mountains of waste are dumped and then covered with dirt.

> Solid waste is usually disposed of by burying it in landfills or by incineration. Both methods have disadvantages.

## Landfills

Before 1970, municipal waste was usually disposed of by taking it to an open dump, where it was bulldozed into position, sometimes burned, and eventually covered with a layer of earth. Dumps were areas where waste decomposed. Wood, paper, food, and lawn clippings, which make up the bulk of our solid waste, eventually decompose if exposed to air. Most metal objects eventually rust away. As the volume of waste increased, however, dumps became inadequate. Piles grew so large that the waste was no longer exposed to the air to decompose rapidly. Nondegradable plastics and substances like aluminum and glass that break down very slowly made up more and more of the waste (Figure 18-4).

Finally, in 1976, Congress passed the Resource Conservation and Recovery Act (RCRA). This did not encourage recycling as its name suggests, but it did prohibit open and burning dumps and require that existing dumps be converted into **sanitary landfills**. A sanitary landfill is designed to reduce the amount of waste that leaks out of it into the environment. The requirements vary with local conditions. For instance, in areas with sandy soil through which water travels rapidly, a landfill may need to be lined with plastic or clay to reduce leaching. It may also need a treatment plant to clean up any water that leaches through the waste before it reaches local surface water or ground water. In some areas, methane produced by decomposition within the dump may need to be vented to prevent possible fire or explosions (Figure 18-5).

Disposing of waste in landfills has many disadvantages:

1. **Wasted resources.**   The waste contains large quantities of paper, glass, metals, plant nutrients, and energy, which could be reused instead of destroyed.

2. **Expense.**   In 1992, state and local governments spent almost $120 for every person in the United States to dispose of solid waste.

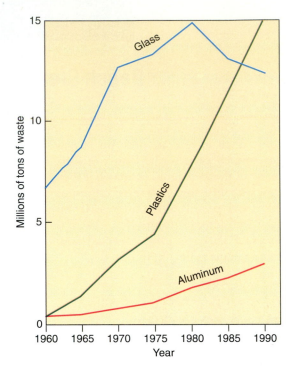

**Figure 18-4**   Plastic, glass, and aluminum in municipal waste between 1960 and 1990. Although plastics and other nondegradable wastes make up a small proportion of municipal waste, plastics are increasing much more rapidly than any other form of waste. The amount of glass has fallen because many glass products such as bottles have been replaced by plastic ones.

3. **Space limitations.**   Many landfills are filled nearly to capacity and are being closed. The U.S. Environmental Protection Agency (EPA) estimates that by 2000, more than 25 states will have no more landfill space. Many materials that would decompose if exposed to air do not do so when buried in a landfill. Landfills contain little air so that most of the decomposers in a landfill are anaerobic decomposers, which break waste down very slowly. They include methanogens, the bacteria that produce methane when they break down organic matter. This is the reason that landfills sometimes contain enough methane gas to make it worthwhile drilling into them for gas that can be used as an energy source.

4. **Leakage and pollution.**   Nearly 7% of municipal waste is hazardous waste that should be specially treated but usually is not. For instance, tires, batteries, and appliances are hazardous in various ways if they are merely added to landfills. They need to be dismantled and recycled. Consider tires, consisting of rubber and steel. The United

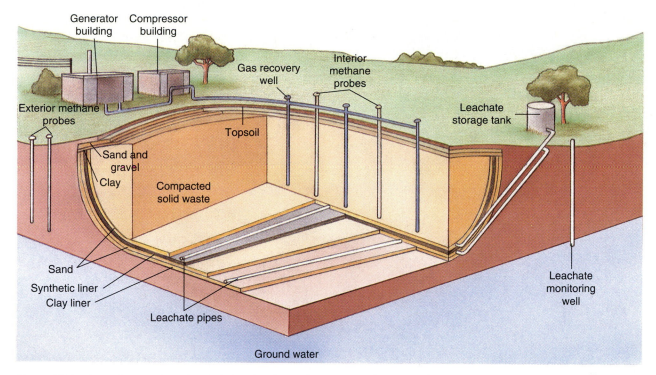

**Figure 18-5**   A modern sanitary landfill that has been closed. This landfill is designed to protect the environment from pollution and to generate electricity. Water dissolves pollutants out of the garbage, forming a solution known as leachate. To prevent leachate from leaking into the ground beneath, the landfill is lined by clay and synthetic materials. Leachate is pumped up from the bottom of the landfill and stored in tanks. Vegetation is planted on top of the landfill to absorb most of the rainfall that forms leachate and to reduce erosion. Anaerobic bacteria in the landfill decompose waste to produce methane. The methane is collected and fuels a turbine to generate electricity.

States discards more than 200 million tires each year. Water accumulates in discarded tires lying near service stations or on a landfill. This water is the preferred breeding place of a mosquito, accidentally introduced from Asia, which transmits the diseases dengue and yellow fever. The mosquito is now found in most parts of the United States, making it more important than ever to dispose of tires properly. Tires are difficult to recycle, but a North Carolina company is now paving roads with asphalt containing chopped up tires.

When we throw away containers that once held household insecticides, cleaning compounds, oil, hair spray, and dozens of other substances, we are putting in the garbage substances that would be treated as hazardous waste if industry produced them in bulk. When these substances leak out of landfills, they pollute soil and water. Even modern landfills often leak. One town in Ohio found muskrats had gnawed through the thick plastic liner of a landfill.

## Incineration

One way to reduce the volume of waste that must go to the landfill is to burn it first, which reduces the volume of the waste by up to 90%. Even modern incinerators cause pollution, adding fly ash, gases, and particulate matter to the air. Burning garbage leaves a residue that must be disposed of. Often, this must be classified as toxic waste because burning has removed biodegradable components, concentrating toxic substances such as heavy metals in the residue (Section 18-B, Close-Up 3-1). In many places, part of the cost of incineration is recovered by using the heat it produces as a source of energy (Figure 18-6).

> The general public no longer trusts those responsible for landfills, incinerators, or toxic waste sites, and the search for a site for waste disposal brings officials into conflict with angry citizens.

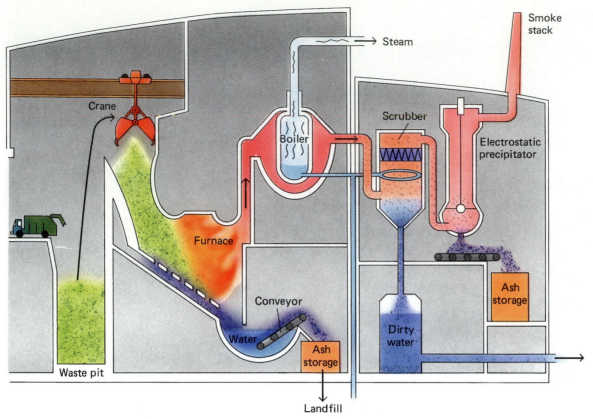

**Figure 18-6**  A garbage incinerator with pollution controls and an energy recovery system. The furnace reduces the garbage to ash, which is removed by conveyor and taken to a landfill. Smoke flows through a scrubber and an electrostatic precipitator (or a filter) that removes toxic and particulate pollutants. (The ash from the pollution control devices must be treated as toxic waste.) Heat from the furnace is used to boil water. The steam is usually used to heat buildings directly, but it may also be used to power a steam turbine that generates electricity.

## 18-B  WHY SOME WASTES ARE TOXIC OR HARD TO GET RID OF

We shall see that the only solution to the problems caused by our ballooning mountains of solid waste is to recover most materials before they reach the waste-disposal system (Section 18-G). However, even if we recycle or reuse almost everything we possibly can, we shall still be left with some categories of waste that are toxic or hard to dispose of for some other reason.

### Nondegradable Waste

Most of the waste produced by preindustrial societies, like those of rural people in developing countries, is biodegradable because it consists of natural products. Cotton, paper, wood, linen, silk, leather, manure, sewage, and even oyster shells are eventually broken down into components that organisms can use, and so they are recycled through an ecosystem. This is not to say that natural products never cause a problem. Biodegradable waste often accumulates faster than it decomposes, and it makes up a large part of the solid waste of modern societies (Section 18-G). The problem that is new today is wastes that never decompose, decompose extremely slowly, or are particularly dangerous while they are decomposing. Most of these are substances such as aluminum, which breaks down very slowly, and plastics, produced industrially from petrochemicals and other fossil fuels.

**Plastics**  Plastics and related compounds are organic polymers synthesized from the long chains of carbon atoms present in fossil fuels (Section 13-G). It is, perhaps, biologically surprising that few manufactured plastics are biodegradable because these polymers resemble molecules produced by living organisms.

**Figure 18-7** Plastic washed up on the bank of a river. This is polystyrene, which breaks up into small pellets that choke fish and other aquatic animals. Polystyrene pellets can be found in almost any body of water in the Northern Hemisphere.

The reason bacteria and fungi cannot break down most plastics is that each decomposer has only a particular array of enzymes. Each enzyme is specific in its action, meaning that it can break only the chemical bonds between certain groups of atoms. Decomposers have evolved the enzymes that permit them to break down the chemical groups they normally encounter, which are those found in living organisms, not those found in artificial plastics.

In recent years, not only have more objects been made entirely of plastics, but plastics also have been incorporated into many goods that were once biodegradable. For instance, much packaging now consists of cardboard with a plastic coating (like the cover of this book). Although the cardboard is biodegradable, the plastic usually is not.

Plastics accumulate in the environment faster than any other form of waste. Scientists estimate that modern polyethylene will not break down for at least 200 years. As a result, surface waters in every part of the world are polluted by plastics (Figure 18-7). One study estimates that 450,000 plastic containers are dumped into fresh water and the ocean every day. Some of these break down into tiny pellets that float near the surface. The pellets injure fish and other organisms that eat them with their usual food. However, the worst damage probably comes from plastic that floats in the water as lengths of fishing net, six-

pack holders, plastic cups, and similar objects. A 1988 treaty, signed by most of the world's major shipping nations, prohibits the disposal of all plastics into the ocean. The United States also prohibits the disposal of plastic into all navigable waterways. The trouble is that a ban of this sort is difficult to enforce. Scientists who have worked on the problem believe that plastic pollution will not be brought under control until governments insist that manufacturers produce only degradable plastics.

Most degradable plastics decompose only under aerobic conditions, and there is little oxygen in most landfills. Others, called **photodegradable plastics**, break down when exposed to sunlight. Photodegradable plastics reduce plastic pollution on the beach or in the park, but they do not help with the problem of plastics in the landfill or deep ocean where there is no light.

Degradable plastics are a vital step toward solving our solid waste problems.

### Toxicity

The usual measure of a compound's toxicity is its **LD50** (lethal dose, 50), the concentration of the toxin that kills 50% of a group of test organisms, often fish or mice. The **LC50** (lethal concentration, 50) is the concentration in the environment that kills 50% of a group of organisms. (Figures for some common pesticides for which we know the toxicities can be found in Table 21-1.)

The other figure that is important in discussing the toxicity of a chemical is its half-life. The **half-life** is the time it takes for half of the chemical to be broken down. For instance, some radioactive compounds have half-lives of only a few months; others have half-lives of thousands of years. Obviously chemicals with long half-lives are persistent in the environment and are the most difficult to dispose of safely.

### Examples of Hazardous Waste

From heavy metals to carcinogenic chemicals, such as radioactive compounds and some pesticides, industrial societies generate millions of tons of toxic waste every year. The manufacture of plastics also produces large amounts of hazardous waste. The EPA lists the six chemicals whose production generates the most hazardous waste. Five of these go into making plastics: propylene, phenol, ethylene, polystyrene, and benzene. Here are the main categories of toxic waste.

**1. Heavy Metals**   Toxic heavy metals are elements, such as mercury, cadmium, zinc, and lead, that are poisonous to organisms when they occur in high enough concentrations. Heavy metals interfere with the actions of enzymes in cells. Cadmium, for instance, used in metal plating and in batteries, can cause kidney and bone marrow diseases and emphysema (lung damage). Heavy metals occur naturally in the environment in small concentrations, combined with other elements and scattered through Earth's crust. The problem is that industrial processes and agriculture concentrate some of these elements to levels at which they are toxic (Section 14-D).

Lead is an example of a metal that is very poisonous and was widely used in many products until recently. Lead binds to thiol (—SH) chemical groups in enzymes. As a result, it reduces the body's ability to synthesize enzymes needed for respiration. Ingesting lead leads to brain damage and may lead to death. Lead is now banned in developed countries in most products in which it used to occur, such as leaded gasoline, paint, and solder, but there is still plenty of it around. We ingest lead in water that has run through lead pipes and even in water run through fancy modern faucets made of bronze and brass. Children often ingest lead in chips of old paint.

About 10,000 tons of mercury are mined each year, and about half of this is lost into the environment. The most notorious cases of mercury poisoning occurred in Minamata Bay, Japan, in the 1950s when mercury in the effluent from a plastics factory was ingested by fish and then by people in communities round the bay, killing many and paralyzing many others. Mercury is used to separate gold from sediment in gold mining. In the 1990s, mercury used for this purpose polluted streams and rivers around mining towns in the Brazilian rain forest.

Since heavy metals are elements, they cannot be broken down, either chemically or by decomposer organisms. The only ways to dispose of them are to dilute them to levels at which they are no longer toxic or to treat them with chemicals that convert them into less toxic compounds.

> Agriculture and industrial processes concentrate some elements to levels at which they are toxic.

**2. Organic Solvents**   Many industrial processes involve compounds that are not salts so they will not dissolve in water but only in organic solvents such as carbon tetrachloride, which is used to clean grease off clothes. These solvents include compounds of carbon and hydrogen, such as benzene, toluene, and xylene, as well as organic compounds containing chlorine, such as carbon tetrachloride and trichloroethylene. These compounds dissolve in the body's membranes and other fat, and many of them cause nerve damage. They also include some of the most powerful known carcinogens. They are formed not only by industrial processes, but also by burning gasoline and garbage.

Waste organic solvents from industry are usually disposed of by storing them in 200-liter steel drums, which eventually corrode, spilling the solvents into the environment. The only ways to dispose of them safely are to convert them chemically into less toxic substances or to purify them so they can be reused.

**3. Organohalogens**   Organohalogens (OHs) are organic compounds of the halogen gases—chlorine, bromine, and fluorine. They are used as solvents, fire-retardants, herbicides, and pesticides. (DDT is an OH.) Not only are they toxic, but they are also exceedingly persistent, remaining in the environment for many years after they have been used. For instance, the OH pesticide kepone contaminated Virginia's James River in the early 1970s. Both the river and parts of Chesapeake Bay had to be closed to fishing from 1975 to 1980 before the kepone had been diluted enough to be safe.

**Polychlorinated biphenyls** (PCBs) are OHs that were first produced commercially in the 1930s as insulating material. They were also used to make plastics and used as solvents, lubricants, and sealants. PCBs enter the body mainly when contaminated water is drunk, and there they remain, dissolved in the body's fat reserves. In 1976, 98% of samples of human fat analyzed around the world contained PCBs. PCBs used in the manufacture of electrical appliances were dumped into the upper Hudson River in New York and the Housatonic River in Massachusetts for many years before 1976. As a result, fishing is still restricted in both rivers.

PCB use was restricted in most Scandinavian countries in the early 1970s and in the United States by the Toxic Substances Control Act of 1976. Nevertheless, Swedish authorities report that levels of PCBs in human tissues in that country did not start to decline until 1980, nine years after use of these chemicals was restricted. PCBs are still used in some lubricants and other products. When PCBs from a polluted river contaminated poultry at a plant in Utah in 1979, the ensuing federal investigation and food recall cost taxpayers nearly $4 million.

> PCBs and other organohalogens are examples of persistent toxic substances that were spread throughout the environment before we knew how toxic they were.

**4. Dioxins and Related Chemicals** Dioxins are unwanted by-products of the manufacture of some organohalogens used in making pesticides, wood preservatives, and other products. They often remain as contaminants in the finished product. This was the case with the dioxin TCDD, found in the herbicide 2,4,5-T, which is often used on lawns and is better known by its code name, Agent Orange. Agent Orange was used as a defoliant in the Vietnam War and caused the deaths of innumerable animals as well as birth defects and genetic disorders among the Vietnamese.

Other dioxin compounds are waste products of paper mills and contaminate large areas of land near some chemical plants. The common plastic, polyvinyl chloride (PVC), produces dioxins when burned. In the 1970s and 1980s, dioxins also contaminated oil sprayed on dirt roads to keep dust down. The town of Times Beach, Missouri, was actually abandoned in 1983 because tests indicated that highway spraying had left high levels of dioxins in the soil.

Although dioxins are extremely toxic to laboratory animals, there is little hard evidence of their effects on people. Pressured by Vietnam veterans groups, the EPA finally began a lengthy review of dioxin toxicity in 1991, but it may be years before the results are known.

**5. Assorted Other Toxins** There are various toxic chemicals that are common in the environment but do not fit into any of the previous categories. They include arsenic, a by-product of extracting zinc, copper, and lead from their ores, often found in coal smoke and waste from printing books and magazines (Close-Up 18-2). Fluoride, while a good thing for teeth in small concentrations, has been known to poison cattle when it leaks into soil after being concentrated during the manufacture of aluminum and inorganic fertilizers. Asbestos is one of the most widely distributed toxic substances in American cities, because buildings built before 1970 often contain asbestos insulation and fireproofing.

Perhaps the most horrifying fact about many toxic chemicals is that so little is known about them. For instance, some 70,000 artificial organic compounds are produced commercially in the United States. Fewer than 1% of them have even been tested for toxicity. We know nothing about the health effects of the others. One of the most extraordinary examples of our ignorance is the case of pesticides that we eat every day in the food we buy. In many cases, we do not know whether or not the pesticides are toxic to humans or what their long-term effects may be (Chapter 21).

> Nothing is known about the toxicity of many chemicals that are common in the environment.

## 18 - C  HOW WE DISPOSE OF TOXIC WASTE

It is infinitely cheaper to dispose of toxic waste safely in the first place than to clean it up later, but toxic waste is still disposed of unsafely for two main reasons. First, the technology of toxic waste disposal is not very advanced, so companies sometimes simply do not know how to dispose of waste safely. This is the case with radioactive waste from nuclear power stations and bomb production. Debates still rage over what to do with this waste. Second, producers and waste disposal companies try to avoid the cost of disposing of their waste safely.

Each year, the United States produces about 275 million tons of hazardous waste that has to be disposed of somehow, and the amount is growing by about 3% each year. Under the 1976 resources act (RCRA), the EPA regulates the handling of substances listed as hazardous. Their movement must be documented from their production through their ultimate disposal. Disposal sites must be designed to ensure they do not contaminate the environment and must be monitored for at least 20 years after they are closed.

This is not an adequate solution to the problem. First, thousands of toxic compounds are not officially listed as hazardous materials. The 1976 Toxic Substances Control Act requires manufacturers to inform the EPA of the estimated volume, uses, and any health studies for every new chemical produced commercially. The EPA can request, but not demand, that the industry supply some toxicity data, and it can regulate or ban dangerous compounds. However, too many new compounds are produced every year for the EPA to investigate them all. The EPA concentrates on substances such as pesticides that are bound to reach the environment and our food and water. It pays much less attention to substances that are not intended to enter the environment, even though most of them will eventually end up in landfills or waterways when they are thrown away.

# A Printer Recycles

by Ned Thomson[2]

Twenty years ago, printers generally just flushed the liquid waste from the camera, plate, and press departments down the drain. Where I used to work, the company had a septic field and a lagoon and everything went in there. We also had well water to drink, and, if you drank too much of it, you'd glow in the dark.

As city water treatment plants became more sophisticated, they realized they couldn't cope with the chromium, silver, lead, arsenic, and other less pronounceable chemicals that printers use and have to dispose of. At this point, printers were told to bury storage tanks and have a disposal service extract the stuff a couple of times a year. The disposal service was supposed to dispose of the chemicals scientifically . . . as well as chemicals from virtually every other manufacturing operation in town. You've probably read newspaper accounts about the job these outfits did. It comes under the heading of toxic dump sites.

In the early 1980s, photographic chemical companies began working on biodegradable chemicals. Today the new products for processing film are both nontoxic and biodegradable. Silver recovery units reclaim 99% of the silver that is removed when film is developed. The silver in the unexposed area of the film is reclaimed when the scrap film goes to a commercial silver recovery firm. They reclaim the silver from the film and recycle the remaining plastic.

The chemicals for plate developing and finishing have been made safe enough environmentally that they can be washed down the drain. Seventy-five percent of the alcohol used in the printing presses' water solutions has been eliminated, and nontoxic, nonacid-based printing inks, made from soybeans, have been developed.

At Thomson-Shore, we recycle virtually every waste material we create. For instance, we recycle all our aluminum printing plates, recovering about 10% of the original price of the plate. All of our press spoilage is recycled. We have special waste bins around the plant that are tagged for the different types of paper we use. Different grades are kept separated because they bring different prices when they are recycled. Recycling has even hit the office. An employee suggested recycling the stuff that goes into office waste baskets. Now virtually anything you put in your office waste basket is bailed and recycled. At the moment, our income from all this recycled material has reached close to $1,500 per week.

For years, we used to have a rubbish hauler pick up the material we could not recycle. This was office waste, trimmings with glue in them, cardboard, and so forth. We filled up six dumpsters a week and paid roughly $2,700 per month to have it hauled away. Now, recycling about two thirds of the office waste and all our cardboard, we have switched from using dumpsters to renting two trash compacters, which we have emptied once every six weeks. This has cut our waste disposal costs down to about $850 a month.

Three out of four of our most commonly used papers contain more than 50% recycled fiber, and about half the books we print are printed on recycled paper. The EPA has produced figures on the savings from using one ton of paper made from recycled fiber—enough to print about 2,000 copies of a 256-page book. To begin with, you save approximately 17 mature trees. In addition, it takes 4,100 fewer kilowatt hours of energy to produce that ton of recycled paper than a ton of virgin paper, and that energy saving is enough to heat and air condition the average American home for six months. The water savings in producing our ton of recycled paper is 7,000 gallons. That's a lot of glasses of water, showers, flushings, and car washings. One ton of recycled paper also produces 60 pounds less of air pollutant effluent than the ton of regular paper does.

The final environmental saving is a hot one right now. It is the volume savings that comes about when you recycle that ton of paper instead of sending it to your local landfill. That savings is 3 cubic yards of landfill space.

I find this whole recycling issue very positive, lucrative, and interesting. To begin with, I had no idea the economics of recycling could be so dramatic for a company our size. Nor did I realize the extent of the "environmental impact" that recycling has. It looks like a prime win-win situation.

[2]Reprinted by permission from articles by Ned Thomson in *Printer's Ink*, Summer 1991. *Printer's Ink* is the newsletter of Thomson-Shore, a printer and book manufacturer with 240 employees in Dexter, Michigan.

Most toxic waste is merely sealed in drums and stored in sites that are monitored to make sure that they are not spilling toxic substances into the environment. The theory is that before the containers leak, which they eventually will, the waste is transferred to new containers that will last for another 20 years or so. But this does not always happen. In a number of cases, the containers have actually been lost—by careless or unscrupulous operators—and are leaking in unknown locations. Consider two cases:

## Love Canal, New York

In 1978, Love Canal in Niagara Falls, New York, became the first waste disposal site ever declared a federal emergency disaster area. For several years, until 1952, the Hooker Chemical Company disposed of some of its hazardous wastes by sealing them in 55-gallon metal drums and burying them in the ditch known as Love Canal. When the ditch was filled, it was covered with clay and houses and schools were built on the site.

Twenty-five years later, wastes were leaking from the site. They leached out into water that penetrated basements in the area and dispersed into the air. More than 300 chemicals, many of them known carcinogens, were identified leaking from the site. High rates of birth defects and miscarriages were reported among the residents as well as many cases of liver cancer and nervous disorders. About 1,000 families had to be relocated. By 1990, portions of the site had been cleaned up sufficiently for some of the houses to be put up for sale, and Love Canal is once more inhabited.

## Oak Ridge, Tennessee

The toxic waste at Oak Ridge is a more complex case than Love Canal because this huge federal site, controlled by the Department of Energy (DOE), manufactured nuclear weapons during the Cold War, so the site contains radioactive as well as other forms of toxic waste. It is one of the most contaminated sites in the United States. Hundreds of cases of diseases associated with weakened immune systems have been found among the residents, including leukemia, nerve damage, and unusual brain tumors.

Some of the health damage probably comes from mercury poisoning. A lot of mercury is used during weapons manufacture, and two buildings on the site have been permanently sealed because they are so contaminated with mercury. In 1993, the DOE admitted that it could not account for 1 million kilograms of mercury used at the site, most of which un-

doubtedly leaked into a nearby stream and is now in the lake that the community uses for drinking water.

> Most of the toxic waste we produce is not disposed of properly.

## Disposal Methods

There are a number of different methods of disposing of toxic waste, depending on the waste.

**Detoxification** Some toxic wastes can be treated so as to convert them into substances that are not hazardous. This can be done chemically, for instance, by adding lime to acids to convert them into harmless salts or by combining cyanides with oxygen to form carbon dioxide and nitrogen.

**Incineration** Some toxic substances can be made harmless by burning. Surplus stores of the herbicide Agent Orange were burned on ships owned by a waste management company. Incinerators for hazardous wastes also exist on land. Such incinerators need pollution control devices and careful monitoring to ensure that they are not releasing toxic by-products into the environment.

**Land Disposal** Hazardous waste landfills are similar to regular landfills but with greater precautions against leaching, periodic monitoring, and insurance against the potential cost of cleanup if the landfill leaks.

Another method of disposing of toxic waste is deep underground. The waste may be injected into deep wells or buried in natural or constructed caverns. No matter how carefully these are built or reinforced, the danger is that these underground dumps will eventually leak or be damaged by earthquakes and release toxic waste into the ground water or into an underground thermal energy source, an oil well, or a natural gas well.

**Bioremediation** Decomposer organisms can detoxify some toxic wastes. Sludge from petroleum refineries is sometimes spread on the soil and left to decay into harmless compounds. We have seen that letting decomposers do the work is often the least harmful method of cleaning up an oil spill (Section 9-C).

In 1990, EPA workers cleaning up hazardous waste sites in Indiana and Ohio used lime (calcium carbonate) to solidify sludge containing PCBs. When they returned to the sites several months later, they could find no trace of the PCBs. Experiments suggested that the lime permitted decomposers to break down the PCBs. Finally, in 1993, researchers isolated freshwater bacteria that break down PCBs into carbon dioxide and water. They found that adding nutrients to a river in the form of fertilizer and adding oxygen to the water by mixing it with air speed up removal of PCBs. Adding the purified bacteria has little effect, probably because there are plenty of the bacteria in the river already. Only about 60% of the PCBs in the river decomposed, probably because the rest is adsorbed on particles of sediment. In laboratory experiments without sediment, more than 90% decomposed. Scientists believe that PCBs on sediment particles will be slowly released and then decompose.

> Microorganisms that detoxify hazardous substances will become an increasingly important part of waste disposal.

### Dumping Waste on Your Neighbors

Until recently, only local laws regulated waste disposal in the United States. Companies would often get rid of toxic waste by dumping it in the next state, which had different laws. During the 1960s and 1970s, industrial states in the northern and midwestern United States got rid of most of their hazardous waste by trucking it to landfills in southern states. The southern states were poor, less densely populated, and had lax environmental laws. They were happy to earn money by storing other people's waste. Then southern populations started to grow, and the states realized that they could not spare the land for toxic waste sites. In the 1980s, attempts to open new landfills in North Carolina, Georgia, and Florida failed because of public opposition. In 1989, both South Carolina and Alabama announced that they would not accept any more hazardous waste from other states.

Now industrial countries are doing to developing countries what the North once did to the South, shipping them hazardous waste. In the midst of food aid to starving Somalis in 1992, a shipload of hazardous waste from Europe arrived in the country. Unscrupulous leaders of poor nations will take the money for accepting the waste, dump it anywhere there is room, and not worry about the health hazard to their citizens.

Since one country's waste often pollutes another country, international attempts to control toxic substances are now being made. Agreements between countries control the release of pollutants into the Rhine River and the Mediterranean Sea. The United Nations has established an International Registry of Potentially Toxic Substances to provide computer data on the environmental hazards of many substances.

## 18-D  TOXIC WASTE LEGISLATION

Over the years, legislation has been enacted that attempts to improve the disposal of toxic waste and to spread the burden more equitably. We have mentioned the 1976 Toxic Substances Control Act, which makes the EPA responsible for overall control of toxic waste.

In 1980, the Comprehensive Environmental Response, Compensation, and Liability Act was passed. It is generally known as the Superfund Act. Realizing that the days of dumping it on your neighbor were almost gone, it contained a provision that, by 1989, every state had to convince the EPA that it could handle all the hazardous waste it would produce in the next 20 years or lose federal funds.

### Accountability

The Superfund Act also attempted to make producers responsible for cleaning up some of their toxic waste, reducing the chance that they would just dump it somewhere illegally and leave, as happened to much toxic waste in the past. In 1988, a court interpreted this to mean that Occidental Chemical Corporation (the owner of Hooker Chemical) is liable for the cost of cleaning up Love Canal. This law will probably help somewhat, but the law is a slow tool, and many companies are long gone into bankruptcy before a court can hold them liable for toxic waste dumps.

Several new laws are designed to educate the public about toxic chemicals and make it easier for local authorities to regulate their use and disposal. A 1986 amendment to the Toxic Substances Control Act requires local committees to monitor the whereabouts of more than 300 ''extremely hazardous substances.'' More than a million manufacturers and farms that produce, store, or use these substances must disclose their presence. The idea of the law is to encourage safer use and disposal of toxic substances by making it more likely that those who misuse hazardous chemicals will be caught. EPA records of toxic chemical release into the environment by industry are placed on a national computer network. Owners of home computers have access to them. As a result, citizens groups now find it easier to take legal action against polluters.

States are also attempting to educate the public. Under California's controversial Proposition 65, California businesses from gas stations to grocery stores were required to post warnings wherever people faced exposure to significant levels of any of various toxic substances (Equal Time 18-1).

# EQUAL TIME 18-1

### Is Prop 65 A Good Idea?

In California, legislation written by citizen groups can be put on the ballot as "propositions" and be voted into law. In 1986, California voters passed Proposition 65, which became the Safe Drinking Water and Toxic Enforcement Act. "Prop 65," as it is known, orders the governor to list chemicals "known to the state to cause cancer or reproductive toxicity." Businesses must then provide "clear and reasonable" warnings to those who may be exposed to any of the chemicals. None of the listed chemicals can be discharged into the drinking water supply unless a chemical poses "no significant risk." Dozens of chemicals, from alcohol and tobacco to dry-cleaning fluids and numerous components of consumer goods, have now been listed.

Proponents of the law, viewing California as the most environmentally progressive state, say:

- Federal laws, such as the Safe Water Drinking Act, the Toxic Substances Control Act, and part of the Clean Air Act, regulate substances one by one, and the burden is on the government to show that a chemical is hazardous. This process is so slow that only a small number of chemicals have been banned or regulated under these acts. Prop 65 does away with this delay by putting the burden on industry to show that it is not exposing people to toxic substances.

- The Environmental Defense Fund and the Sierra Club say that the law was not really intended to cover every conceivable toxic chemical but to put real warnings on "the few things that need them."

- National laws for controlling toxic materials are not providing the margin of safety that people want and it is up to the states to step in.

- The state authorities have used common sense in enforcing the law, focusing first on known health hazards such as alcohol and tobacco.

- The most useful effect of the law will be to pressure industry to find substitutes for potentially toxic chemicals in their products.

Opponents of the law, viewing Prop 65 as another example of Californian bean-dip-on-the-brain, say:

- The law uses resources that would be better spent elsewhere.

- California has lost many jobs during the 1990s. The expense of complying with Prop 65 and expensive lawsuits on the subject are among the reasons industries are leaving the state.

- Warning signs all over the state do not improve the environment; they merely make it more ugly.

- Most substances that cause cancer or birth defects are substances such as tobacco that are produced by living organisms; very few carcinogens are actually produced by industry.

- The law is unscientific and unspecific. How does an administrator decide when a risk is "significant" and when it is not?

- There is no evidence that the law has prevented a single case of cancer or prevented a single birth defect. If the law's goal is education, this is a very expensive way to educate people.

- Businesses often have no way of knowing if they are using substances that might be covered under the law, so many tried to cover themselves against lawsuits by newspaper advertisements advising the public that "[t]he following facilities contain chemicals known to the State of California to cause cancer, birth defects, or other reproductive harm: . . ." This does not even have educational value.

- Experts on consumer behavior say that warning labels don't work. "When people see the same words over and over, they just blank them out," says the University of Texas's Susan Hadden. What do you think?

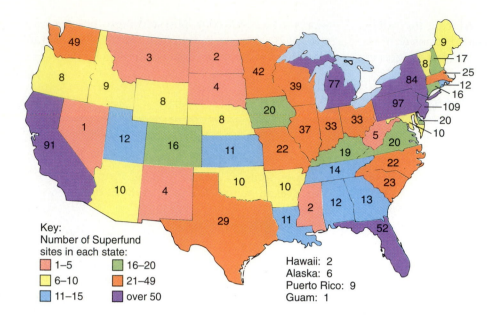

**Figure 18-8** The official Superfund sites that existed in 1992. Since 1980, the EPA has identified more than 1,200 of these sites that need to be cleaned up because they are leaking toxic waste, endangering the environment and human health.

Key:
Number of Superfund sites in each state:
- 1–5
- 6–10
- 11–15
- 16–20
- 21–49
- over 50

Hawaii: 2
Alaska: 6
Puerto Rico: 9
Guam: 1

## Superfund

The Superfund Act was designed to take care of the many abandoned hazardous waste sites in the United States (Figure 18-8). More than 1,200 sites have been proposed or added to the Superfund list of sites that should be cleaned up. What has happened to them?

- Cleanup has officially been completed at 80 sites.
- Cleanup work has started at 357 sites.
- Assessment of the cleanup problem has started at 438 sites.
- No long-term action has started at 103 sites.

About three quarters of Superfund money has been spent on legal fees and administrative costs, which was not what its framers intended. What went wrong?

There are two main faults with the Superfund Act. First, in an effort to make polluters pay for cleanup, the act says that if the EPA can prove that a company dumped even a tiny amount of toxic waste at a site, it can force the company to pay for cleanup of the whole site. Not surprisingly, every company asked to pay such an unfair share of the cleanup cost immediately files a lawsuit, which is why so much money gets spent on lawyers. Second, the act says that each toxic waste site must be made so clean that you could dig a drinking water well in the middle of it. So instead of cleaning up hundreds of sites so they are no longer a danger to their neighbors, workers spend their time cleaning up a single site until it is so clean you could eat your dinner off the ground. What needs to be done?

First, scientists, instead of administrators and lawyers, need to decide which sites most need cleaning and to decide when a site is clean enough. Second, less Superfund money should go to litigation, and more should be devoted to research into methods for disposing of toxic waste and ways of reducing toxic waste in the future. The estimated cost of cleaning up toxic waste skyrocketed during the 1980s as more and more sites were discovered, and new

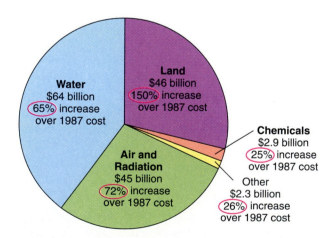

**Figure 18-9** Estimated costs of controlling pollution in the United States in 2000, with the increase over the cost in 1987 (all in 1986 dollars). Land pollution involves toxic products spilled into the soil. Why do you imagine that the cost of controlling land pollution is expected to rise so much faster than for other forms of pollution?

methods are badly needed to reduce the cost (Figure 18-9).

> The Superfund Act would work better if it did not make small-time polluters liable for huge cleanup costs, and if it provided more money for research.

## 18-E PRODUCING LESS HAZARDOUS WASTE

The best way to handle the hazardous waste problem is to produce less of it. Producing less waste can be amazingly simple, using techniques such as evaporating surplus water to concentrate the waste. The chemical industry has started a project known as "33–50." The project is designed to reduce production of toxic waste from the level in 1988 by 33% in 1992 and 50% in 1995. The idea is to get manufacturers thinking about the toxic waste that will be produced before they start to manufacture a chemical. Slight changes in the manufacturing method often result in less waste at the end.

The U.S. Office of Technology Assessment estimates that the volume of hazardous waste produced in the United States could easily be halved. To help with this, a waste-reduction center has been opened in Raleigh, North Carolina, with the collaboration of the EPA, the Tennessee Valley Authority, and eight southern states. It studies and proposes methods to reduce the amount of toxic waste produced. Already some enterprising businesses are making a dent in the problem themselves.

### Waste Brokers

Planar Systems, of Beaverton, Oregon, was paying $300 a drum to get rid of the organic solvent isopropyl alcohol left after producing computer screens. Then they discovered that Western Foundry, of Tigard, Oregon, needed the solvent to clean the casts it uses for making steel products. The two found each other through Pacific Materials Exchange in Spokane, Washington, one of 20 programs in the United States that seeks to match companies who need to dispose of waste materials with those that can use them. The Materials Exchange recently listed 400 waste items for disposal, including 10,000 pounds of mink fat, 100 pounds of gold glitter, and 800 gallons of cyanide solution. Among the "wanted" ads were those for fish waste, 100 gallons of used antifreeze, and sour milk. In 1992, waste producers saved an estimated $27 million in disposal fees by these exchanges.

Some hazardous waste gets reused by such exchanges. In 1992, the Department of Energy sent a first shipment of radioactive cesium-137 from Hanford, Washington, to laboratories and hospitals around the United States. This was part of a $2 billion sale of radioactive waste to a Canadian company that makes capsules for irradiating blood in bone marrow transplants. Hanford is a huge toxic waste site because it was once a major site for nuclear weapons production.

### European Integrated Management

In parts of Europe, land is expensive, and waste disposal technology is consequently more advanced than in the United States. First, there are many things you cannot throw out as municipal garbage, and each household is limited to a small can of garbage each week. Anything else must be taken to a municipal center, where everything that can be recycled is sorted and toxic waste is separated to be sent to a hazardous waste plant (Figure 18-10).

At a typical hazardous waste plant, industrial hazardous waste, sorted by type, is delivered, along with household hazardous wastes. These include bottles that once contained pesticides or dry-cleaning fluid, which Americans drop into their garbage. They also include appliances that contain hazardous compounds. For instance, older refrigerators contain ozone-destroying chlorofluorocarbons, which leak into the atmosphere from the average American landfill. The plant contains facilities for chemical detoxification, and much of the waste is incinerated, generating steam that is used to heat buildings. The residues from incineration, which are relatively nontoxic and have been chemically stabilized, are disposed of in a secure landfill. Industries are required to take all their hazardous waste to collection stations, unless they have special permission to treat it themselves. This is an improvement over the American system, in which each industry disposes of its own waste, with little chance of being called to account by a regulatory agency.

> In the United States, each industry usually disposes of its own waste, with little chance of being called to account by a regulatory agency.

## 18-F HAZMAT TRANSPORT

When it travels around the country, hazardous material is known as "hazmat," as you may have noticed

**Figure 18-10**   A site in Britain where householders deposit waste that the garbage truck will not pick up. Waste is sorted for recycling, and toxic waste is shipped elsewhere for disposal.

| TABLE 18-1   Chemicals Causing Death or More Than 50 Injuries, 1982–1991* | | | | |
|---|---|---|---|---|
| **Chemical** | **Incidents** | **Deaths** | **Injuries** | **People Evacuated** |
| Gasoline | 3,544 | 52 | 82 | 1,648 |
| Compressed gas | 965 | 13 | 89 | 361 |
| Flammable liquids | 5,098 | 6 | 187 | 378 |
| Crude oil | 1,055 | 6 | 14 | 300 |
| Fuel oil | 2,330 | 5 | 10 | 12 |
| Naphtha | 849 | 4 | 13 | 673 |
| Aviation turbine fuel | 238 | 3 | 4 | 15 |
| Hydrochloric acid | 2,807 | 2 | 146 | 1,934 |
| Anhydrous ammonia | 584 | 2 | 81 | 3,125 |
| Paint drier fluid | 1,807 | 2 | 15 | 44 |
| Hexane | 97 | 2 | 1 | 2,518 |
| Corrosive liquids | 4,944 | 1 | 140 | 1,389 |
| Sulfur dioxide | 42 | 1 | 77 | 3 |
| Hydrogen peroxide | 476 | 1 | 49 | 102 |
| Aluminum phosphide | 17 | 1 | 18 | 0 |
| Alcohol | 1,587 | 1 | 7 | 10 |
| Hydrogen gas | 51 | 1 | 4 | 425 |
| Morpholine | 34 | 1 | 4 | 70 |
| Butadiene | 45 | 1 | 3 | 10 |
| Cement | 894 | 1 | 3 | 0 |
| Propellant | 1 | 1 | 0 | 0 |
| Sulfuric acid | 2,779 | 0 | 238 | 57 |
| Sodium hydroxide | 1,688 | 0 | 84 | 208 |
| Chlorine | 67 | 0 | 55 | 6 |
| Pyridine | 141 | 0 | 51 | 37 |
| Ethyl mercaptan | 26 | 0 | 51 | 0 |

*Data from the U.S. Department of Transportation.

on trucks and road signs. The volume of hazmat has increased by 50% in the last five years. A *Los Angeles Times* study estimates the volume at 4 to 8 million tons per day in half a million different shipments, but nobody really knows for sure. Not surprisingly, the amount of death and destruction caused by "incidents" involving hazmat is increasing rapidly (Table 18-1 and Figure 18-11). "Incidents" are usually crashes, but they may also be leaks.

Nearly all the deaths are caused by crashes of tankers, especially those containing gasoline. This is because more gasoline travels than any other form of hazmat. Once, rail inspectors used to worry most about simple explosions, like one that killed several people when a tanker carrying gasoline failed to stop at a railroad crossing in Florida in 1993. Now they worry more about the unknown, unlabeled cargoes hidden in containers.

In 1991, part of a Southern Pacific freight train derailed near Dunsmuir, California, and an unstrengthened freight car labeled "weed killer" slipped into the Sacramento River. Not until 6 hours later did Southern Pacific discover precisely what was leaking from the car into the river. They had to fax the owners to discover that the chemical was metam

sodium, a chemical that produces toxic fumes in the presence of water and sunlight. Luckily, the chemical spilled at night, or the residents of Dunsmuir might not have survived. As it was, several pregnant residents had abortions because the chemical is thought to cause birth defects. The people were luckier than the wildlife along the river. Every tree, weed, fish, and snail along 40 miles of the Sacramento River died. The ecosystem will take 50 years or more to recover.

The chemical in the Dunsmuir accident, metam sodium, is not listed by the EPA as a hazardous material. Representative Barbara Boxer asked a witness at the hearing she held afterward, "Does it strike you as being amazing, as it does me, that a chemical initially developed for chemical warfare wasn't on the hazardous materials list?" EPA assistant Linda Fisher told Boxer that metam sodium had proved useless as a chemical weapon during World War II because it broke down rapidly in the environment—making it useful as a herbicide at low concentrations. How many other materials like this are traveling around, not even labeled as hazardous? Nobody knows.

> The regulation of transportation of hazardous materials is inadequate.

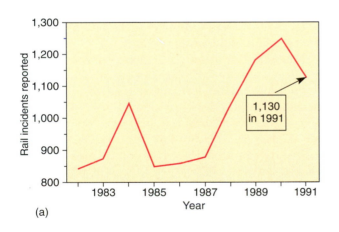

(a)

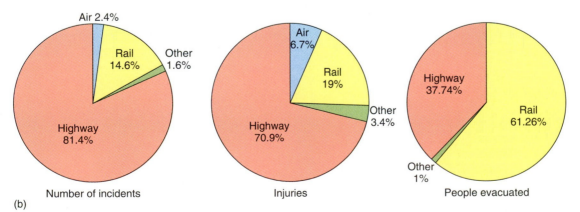

(b)

**Figure 18-11**   "Incidents" involving hazmat transport. (a) Railroad incidents reported. These are nearly all crashes involving large quantities of material. (b) Sources of incidents, injuries, and evacuations. Highway accidents are most numerous and injure the greatest number of people (Table 18-1). Most are crashes of gasoline tankers. But railroad accidents cause a higher percentage of evacuations because they usually involve larger quantities of hazardous material than road accidents.

## 18-G  RESOURCE RECOVERY

In our garbage, we throw away valuable resources such as energy, oil (plastics), metal (appliances), trees (paper), and humus (yard waste). Experts calculate that the energy potential in our solid waste amounts to about one third of the energy potential in all the oil that will ever flow through the Alaska pipeline. A waste-burning plant in Nashville, Tennessee, has reduced the city's landfill requirements by 90% and produces steam used to heat and cool buildings. There are now about 70 energy-recovery plants in the United States.

> Our solid waste contains large amounts of energy and other resources.

The United States has been so rich in space and resources that it has been able to succeed economically, even using its resources wastefully. This is less and less true as the cost of energy, land, and minerals increases. One interesting example shows that change is possible. During the recession of 1974 to 1975, the volume of solid waste generated in the United States actually fell for the first time ever (Figure 18-12). Analysis of the garbage showed that it contained less packaging, newspapers, books, and magazines. The decrease was probably caused by a combination of people purchasing less, reusing what they had, and selling more paper to recycling centers.

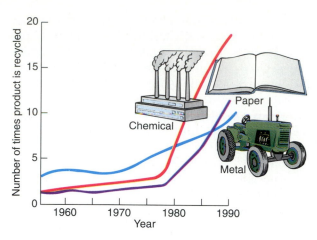

**Figure 18-13**   Increases in the recycling of paper and metal in the United States since 1955. Note that industry is also reusing water more efficiently, collecting it at the end of a process and reusing it, instead of releasing it into the local water supply.

Many substances are recycled industrially—notably metals, such as steel, copper, and aluminum, and various types of paper (Figure 18-13). The proportion of waste recycled by industry will continue to increase as the costs of waste disposal and raw materials rise. We can expect even faster increases in the recycling of domestic waste.

New York City produces 28,000 tons of trash a day and is running out of places to put it. The city has devised an ambitious recycling plan to cope. Residents are required to sort newspapers, magazines, and corrugated cardboard into one bundle, and metal, glass, and plastics into another. By 1995, residents are required to dispose of 25 materials, including batteries and textiles, in recycling bins. Even food scraps are to be collected and recycled.

### Economics of Recycling

In 1986, in one of the first U.S. projects, Salt Lake City, Utah, started collecting newspaper from schools, businesses, and other institutions in an attempt to reduce the cost of using the city landfill. Since the project was started, the city has saved $40,000 a year in reduced landfill costs and earned more than $12,000 a year from selling the papers to the recycling industry.

Recycling is often very cost-effective, using much less energy, water, and other resources to produce material from recycled products than from scratch. For instance, it takes 95% less energy to produce a ton of aluminum from recycled aluminum than from

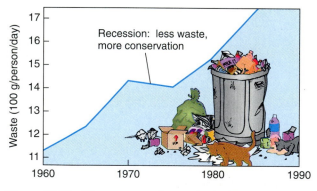

**Figure 18-12**   The amount of waste each American produces each day has more than doubled since the 1960s. This graph shows the volume of municipal waste collected divided by the population. Each American today throws away about five times as much as a European with an equal standard of living. The rise in volume of garbage is not inevitable. Garbage per person dropped in the 1970s as less paper and cardboard reached the garbage truck during a recession. We should not really have to wait for recessions to reduce the volume of garbage!

ore, about 75% less energy to make steel from scrap than from iron ore, and 70% less energy to make paper from recycled paper than from trees. Collecting and separating material to be recycled may be expensive, but communities are undertaking this effort anyway in an attempt to save landfill space.

> Recycling is a vital step in reducing the volume of solid and toxic waste.

## Supply and Demand

Recycling increased so rapidly during the 1980s that the supply of paper, aluminum, glass, and plastic for recycling became greater than the demand. Newspapers pile up in warehouses, and towns that once sold newspaper for $35 a ton are now paying as much to get rid of it.

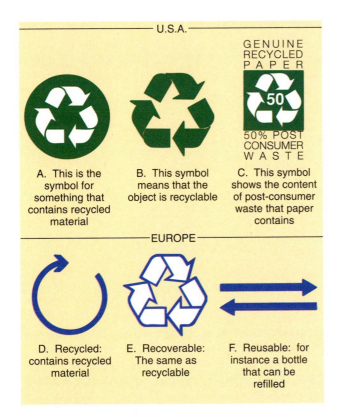

**Figure 18-14**    Recycling symbols in the United States and Europe. (They can be any color.) The U.S. symbols are still not satisfactory. For instance, something can carry symbol A even if it contains only a tiny percentage of recycled material. Symbol B is often misleading; it means that the object can be recycled, which is true of almost everything. Post-consumer waste, referred to in symbol C, is waste generated after something has been sold to the consumer as opposed to waste generated during manufacture. Do you think the European symbols are better or worse?

The demand for finished goods containing recycled products is increasing, partly because companies want to be environmentally friendly—or at least appear to be. That is why this book is printed on recycled paper. But even if the demand is high, it takes time and investment to build the new factories required to turn used plastic or paper into recycled products. The American Paper Institute estimates that supply and demand will reach equilibrium within five years or so. The number of U.S. and Canadian paper mills that can handle used paper is expected to double to about 20 between 1990 and 1997. About one third of all paper consumed in the United States is now collected for recycling, and that figure is expected to reach almost 50% by 1996.

A vital part of recycling is labeling everything so that consumers and recycling industries know what is in it and how it should be recycled. In the 1990s, these symbols have begun to be standardized, at least partly. The symbols used in the the United States and Europe are shown in Figure 18-14.

## Recycling Plastic

Plastics are made from a variety of different chemical compounds called resins. The different resins have different properties and have to be separated for recycling. Figure 18-15 shows the symbols used to distinguish the main resins. Many states now require that manufacturers put the appropriate symbols on plastic products, and the number of goods made from recycled plastic is rapidly increasing (Figure 18-16).

## Bottle Bills

Recycling generally involves breaking down a substance and making it into something else. Even more efficient is **reuse**, simply taking something and using it again for the same purpose: glass bottles are very easy to reuse.

Since Oregon started the trend in 1971, many states have enacted bottle bills, whereby states require a deposit on bottles and cans that the consumer reclaims when the container is returned to the store or a recycling center. New York passed a bottle bill in 1982, and the state's Department of Environmental Conservation believes it has been a remarkable success. Eighty percent of drink containers sold in New York are returned, bottles and cans no longer litter the state's highways, and the state has produced 8% less solid waste since the bill went into effect. Container manufacturers object that the bill is a hardship to grocery stores, which have to store the

| THE TRIANGLES IN THE REFRIGERATOR | | | |
|---|---|---|---|
| Symbol | Name | Typical use and characteristics | Some recycled products |
| 1 PET | Polyethylene terephthalate | Soft drink bottles, peanut butter jars; 25% of all bottles. The most expensive resin because it keeps oxygen out and carbon dioxide bubbles in. | Carpets, soda bottles, fiberfill for parkas, skis, tennis ball fuzz |
| 2 HDPE | High density polyethylene | Milk and water jugs, liquid detergents; 55% of all bottles. Cheap, strong, forms good handles, can be dyed many colors. | Base cups for soft drink bottles, stadium seats, trash cans |
| 3 V | Vinyl or polyvinyl chloride | Blister packs, food wrap, bottles for cooking oils and shampoo; 8% of all bottles. Very clear, not degraded by oils. | Floor mats, pipes, hose, mud flaps |
| 4 LDPE | Low-density polyethylene | Lids, squeeze bottles, bread bags; 7% of all bottles. Flexible. | Garbage can liners, grocery bags |
| 5 PP | Polypropylene | Containers for syrup, ketchup, yogurt, and margarine, bottle caps; 7% of all bottles. Moisture-resistant, flexible, does not deform when filled with hot liquid. | Paint buckets, manhole steps, etc. |
| 6 PS | Polystyrene | Coffee cups, meat trays, packing "peanuts," plastic utensils, videocassette boxes. Light but brittle; can be rigid or made into a foam. | Trays, flower pots, trash cans, pipe |
| 7 | Other | Various resins. Metals, glues, or other contaminants may be mixed with the resin. | Plastic lumber |

**Figure 18-15** Symbols describing the chemical composition of plastics, so different types can be separated for recycling. The right-hand column lists some products made from the recycled resins.

empty containers, and an added expense and inconvenience to consumers. (One clear cost to consumers is the money that manufacturers spend lobbying to defeat bottle bills: $6.2 million in 1986 in California.)

In 1986, container manufacturers spent $6.2 million in California in an unsuccessful attempt to defeat a bottle recycling bill. Consumers paid for this expenditure.

## Composting

To people who understand environmental science or gardening, the money wasted on maintaining the av-

erage suburban lot is extraordinary. Most homeowners spend hundreds of dollars a year on fertilizer and mulch for the garden and then laboriously stuff leaves and grass clippings into expensive plastic bags for the garbage truck to collect. Some bags do not reach the garbage truck: knowledgeable gardeners collect the bags, empty them onto their own compost heaps, spend nothing on fertilizer or mulch, and have the loveliest yards in the neighborhood.

In Europe and Asia, and increasingly in the United States, yard waste is composted, placed in a pile until decomposer organisms have converted it to compost. **Compost** is partially decomposed vegetation, often described as having the appearance and consistency of chocolate cake crumbs. It can be used

**Figure 18-16** Products made from recycled plastics are becoming quite common. These are plastic garbage bags. The white ones in the foreground are made from recycled plastic milk bottles. I have used them, and they seem just as satisfactory as the unrecycled black ones.

to pot houseplants, dug into the soil to add nutrients and improve drainage, or laid on the soil surface as a mulch that controls weeds and helps keep the soil moist (Figure 18-17).

Twigs and branches that have fallen from trees or accumulate when trees and shrubs are pruned are more difficult to dispose of because they do not decompose as rapidly as weeds, leaves, and grass clippings. Europeans usually rescue all the wood that is large enough to use as firewood or stakes. Then they

break the remaining brush into small bits and toss it either on the main compost heap or onto a separate heap for waste that decomposes slowly. An attractive modern alternative is a chipper, a machine that chops branches into wood chips that can be used as mulch, used as surfacing for paths, or merely thrown on the compost heap. Big chippers are expensive, but people often buy them jointly with their neighbors, or the town may supply a chipper for communal use. Small chippers that cost less than most lawn mowers are beginning to appear in garden supply shops.

> Composting yard waste saves millions of dollars in garbage disposal costs and returns nutrients to the soil.

## Municipal Composting Programs

Many American cities, including Burlington, Vermont, and Seattle, Washington, have started composting programs. Seattle aims to recycle 75% of its yard waste by 2000. A corps of city employees teaches people how to compost their waste and maintains a compost demonstration in the city.

In 1993, there were 19 sites in the United States where municipal garbage was sorted and the biodegradable waste composted for use on farms. These plants range in size from about 10 tons a day to a 660-ton-a-day plant in Broward County, Florida. Figure 18-18 shows how a composting plant works. Com-

**Figure 18-17** Backyard composting. This is a rather luxurious composter made by my husband. At any one time, one of the bins contains compost that is being used on the garden, a second contains a completed pile of waste that is decomposing, and the third (at the far end) is the growing pile on which we throw vegetable trimmings, weeds, grass clippings, twigs, newspaper, and just about anything else that is biodegradable. The plant in this bin is an avocado that has germinated from a discarded seed.

## HOW COMPOSTING WORKS

Reuter Recycling in Pembroke Pines is the largest of 19 municipal garbage composting plants in the United States. Four Broward cities send about 500 tons of trash a day to the Reuter plant. About 65% of the trash is composted in six weeks. Processing pulls out cardboard, aluminum, and other metals for recycling. The rest is either dumped in a landfill or incinerated.

This is how large-scale composting works at Reuter:

**1.** Trucks dump trash into a cavernous indoor sorting area, where three people pull out bulky or potentially hazardous wastes. A front-end loader shoves garbage that can be processed onto conveyers that move the trash through six additional steps.

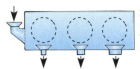

**2.** Trash goes for a roll in screening drums. These tumbling drums, as big as buses, separate wastes into fine, medium and coarse materials.

**3.** The separated wastes move past a dozen people in four lines who pull out aluminum and cardboard for recycling. Fine materials skip the hand-sorting. Powerful magnets grab metals and small batteries.

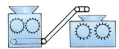

**4.** The sorted trash moves through two sets of shredder blades that churn in into "fluff."

**5.** The fluff is sent into another hangar for composting. The wastes are stored in six 700-foot rows that are turned bottom-to-top by a machine for six weeks.

**6.** Composted trash moves through a huge grinder and a final screening drum that attempts to remove chunks of glass and plastic.

**7.** The cleaned compost is then stored in a separate hangar from which it is distributed to farms.

**Figure 18-18**    How one municipal composting plant in Florida works.

post is applied to Florida tomato fields in large amounts every three years. It reduces the need for chemical fertilizer by 90% and lets the farmer irrigate every five days instead of every three days because compost holds so much water. In Dade County, the use of compost has boosted tomato production by 30%. Of course, old-timers have trouble adapting to these modern methods. "The big problem is cultural," says Cornell University's James Gilbert. "Culturally, we just can't handle putting garbage and hu-

man waste on crops. The Chinese have been using this stuff for 40 generations, and it obviously hasn't wiped them out."

There are still real problems, however. First is the smell. Decomposer organisms produce gases such as methane and hydrogen sulfide that cause neighbors to complain. Second, composting plants need excellent quality control because municipal garbage still contains much toxic waste that can contaminate the compost. In 1993, a farmer in Florida spread municipal compost on 50 acres of strawberries and killed the whole lot. Reuter Recycling in Florida is experimenting with quality control and reducing the smell. Let us hope they succeed because the waste reduction and soil improvement possible from municipal composting are impressive.

## 18 - H   THE FUTURE

To solve our solid and toxic waste problems, we shall have to change the types of waste we produce and the ways we dispose of waste. The main techniques will be manufacturing processes that produce less waste, recycling, and biological decomposition. Degradable polymers will replace persistent plastics. Biological controls will replace many toxic pesticides. New species of decomposers, some probably engineered by biotechnology, will join the battle against waste. The goal will be to minimize the volume of waste that must be stored in landfills or the ocean because there is no other possible way to dispose of it.

We shall see major changes in the way we dispose of our domestic waste. These changes are already clear from experiments that are going on now in progressive towns and cities around the world.

### Producing Less Waste

The first principle of waste management is to generate less waste in the first place. This is because the cost of collecting waste and transporting it to processing centers is so high. In the 1980s, more than two thirds of the cost of disposing of America's municipal garbage was spent collecting the waste from homes and businesses.

In theory, it is not difficult to reduce the volume of American solid waste to about one fifth of what is usually dumped nowadays. This is because up to 70% of what we dump is biodegradable, and another 10% consists of substances that are economical to recycle. Some 80% of our garbage need never reach the garbage truck in the first place.

> In theory, it is easy to solve much of our solid waste problem, since about 80% of our garbage could be composted or recycled.

In one Ohio town, garbage cans containing waste for the landfill are spot-checked every week. If you are one of the lucky people whose garbage is checked, and if your can does not contain anything that could be composted or recycled, you win $1,000. Despite the money it spends on this attractive incentive, the town is saving thousands of dollars every year because its garbage-sorting project reduces landfill costs. Mean-minded towns might encourage people to sort their garbage by fines instead of rewards.

Pennsylvania started a system now used in many places. Garbage bags are picked up by the garbage truck only if they have a stamp on them. Householders have to buy the stamps, just as we buy stamps to mail letters. The less garbage you have, the less you need to spend on garbage stamps. In many places, towns provide different-colored garbage cans for different types of waste. Or residents may have to provide their own cans and write "*recycle*" or "*compost*" on the sides. People can be encouraged to compost their own waste and use the compost in their own yards. Or the city may run a central composting facility where people can drop off their yard waste and pick up wood chips or compost for their own use.

A country's way of life has a major effect on how much solid waste it generates. We tend to blame our affluent society and the invention of plastics for most of our waste problem, but this is not the whole story. The United States generates more than five times as much domestic, industrial, and agricultural waste per person each year as do other developed countries. These countries use less packaging, recycle more waste, compost more vegetable matter, and use more of their waste to generate energy.

In Germany, a 1992 law aims to reduce the amount of packaging that reaches landfills by 80% by 1995. No packaging is allowed in household garbage. Manufacturers and distributors have to take back for recycling any packaging they produce, or they can pay dues to a manufacturers' association, which then lets them put a green dot on packaging. Consumers know that any pack with a dot on it can be left in a green-dot bin at any shop.

All these changes to the garbage rules may sound like a big nuisance, but they are better than having to pay $100 a month to get rid of your garbage, which is where we are headed if we don't take control of the situation soon.

# LIFESTYLE 18-1

### Changing the Oil

Auto mechanics agree that to keep your vehicle in good condition, you should change the oil every 3,000 miles. According to Mobil Oil Corp., 60% of car owners save money by changing the oil themselves. According to Grease Monkey International Corporation, too many of us do not dispose of our used oil properly. American backyard mechanics throw away 450 million liters of used oil every year. The 1992 oil tanker spill in the Shetland Islands polluted the environment with less than one quarter of that amount.

Where does the oil we throw away go? Down the sewer into rivers and streams, into the landfill, and into the ground, where it kills soil organisms and can seep into ground water or waterways. Just 2 liters of old oil (less than half the amount in a car) can make 1 million liters of fresh water, which is a year's supply for a dozen families, undrinkable.

Not only is oil toxic to humans and other organisms, but it is also wasteful to throw it away when it should be recycled. It takes one barrel of crude oil (42 gallons) to make just 2.5 quarts of motor oil. Yet just 1 gallon of used oil can be recycled to make the same 2.5 quarts of oil. Since a large part of the U.S. trade deficit is caused by oil imports, federal inspectors require auto repair shops to account for virtually every drop of used oil and other hazardous fluids, but they have no way to enforce the law about recycling on the average home mechanic.

So what should we do? Where I live, the solution is to take used oil to our local service station for recycling. They also dispose of the container, which would pollute the sewer if rinsed out at home or the garbage if dropped in the trash. In some places, there are municipal oil collection points. Some enlightened cities even have pickup services for used oil, so you might call city hall and inquire.

## SUMMARY

The problem of waste disposal is becoming acute in developed countries as the volume of toxic and solid wastes increases and landfill space becomes harder to find. Agriculture and mining wastes make up most of the waste produced in the United States and are mainly disposed of where they are produced. Industrial and municipal waste is usually transported from the site of production for disposal.

### Solid Waste Disposal

Municipal solid waste is deposited in landfills, sometimes after being incinerated. Modern landfills are expensive because they are designed to prevent pollutants leaking into the ground and water. Nowadays it is hard to find land for a new landfill. Incinerators need excellent pollution-control devices because municipal waste contains many toxic components.

### Why Some Wastes Are Toxic or Hard to Get Rid Of

Most waste of preindustrial societies is biodegradable because it is made of natural substances. Industry produces many nondegradable substances (which persist in the environment) and toxic substances (which may pollute and kill). Plastics are accumulating in the environment faster than any other form of waste. Toxic waste includes inorganic substances, such as heavy metals, arsenic, and cyanide, and thousands of organic compounds, including solvents and organohalogens such as PCBs. We know little about the toxicity of thousands of organic compounds produced by industry.

### How We Dispose of Toxic Waste

Our methods of disposing of toxic waste are still primitive, usually consisting of storing the waste in metal drums or dumping it in a neighboring state or country that has less stringent environmental laws. Storage drums often permit toxic wastes to leak into the air, soil, surface water, and ground water. Modern disposal methods include concentrating the waste to reduce its volume and detoxifying the waste by bioremediation, burning, or chemical means and burying it in secure hazardous waste landfills.

### Toxic Waste Legislation

Legislation in 1976 made the EPA responsible for hazardous substances. The 1980 Superfund Act forced each state to dispose of its own hazardous waste and provided money to clean up hazardous waste sites. Most of the money has been spent on paperwork and lawsuits, partly because the law demands excessive cleanup and tries to hold small polluters responsible for large cleanup costs.

### Producing Less Hazardous Waste

Methods of reducing the volume of toxic waste include altering manufacturing practices, concentrating the waste, and recycling by transferring waste from one industry to another that can use it.

### Hazmat Transport

Unknown quantities of hazardous waste are transported by road, rail, air, and water, causing injury and environmental damage every year. Most accidents involve gasoline tankers because gasoline is the most common hazmat. The transportation of hazmat is so poorly regulated that we do not know how much material is moved or even what is in many shipments.

### Resource Recovery

Our solid waste contains enormous amounts of energy and valuable materials. Industry recycles many substances from its waste, but domestic recycling is in its infancy, especially in the United States. Recycling glass, aluminum, plastic, and paper and composting yard waste would reduce American household and commercial waste to as little as 20% of its present volume. Recycling is usually cost-effective. It requires that plastics be labeled with the resins they contain.

### The Future

We shall see large increases in the fees charged for garbage collection, encouraging individuals to recycle and compost. Regulations will ensure that packaging and most other substances are recycled and that municipal waste does not contain toxic substances. New methods of disposing of toxic and persistent wastes will be developed, and the regulation of toxic wastes will increase.

## DISCUSSION QUESTIONS

1. What can you do to reduce the amount of waste you generate? Have you got room outdoors for a small compost pile or a bucket with holes in it into which you could put vegetable trimmings and dead houseplants for recycling? Does your grocery store accept plastic for recycling as well as returnable bottles and cans? Could you change your shopping habits to reduce waste production? Will your teachers accept papers written on reused paper—paper that has already been used on one side?

2. How would you design a program to reduce the volume of waste your community produces? Remember that most people will not bother to recycle paper and bottles voluntarily and that they do not know much about compost.

3. At the moment, much hazardous and solid waste is disposed of by private ''waste management'' companies. The municipality pays the company to handle its waste. What are the advantages and disadvantages of this system?

4. When we purchase toxic substances such as motor oil, dry-cleaning fluid, bleach, ammonia, and household pesticides, what happens to the container when it is empty? What happens to its toxic contents?

5. How is the waste in your community handled? From industry? From households? How would you go about convincing local authorities to change the system?

6. Were the PCB-munching bacteria described in Section 18-C aerobic or anaerobic? Why?

# Chapter 19

# POLLUTION AND WATER

*A landing place on Lake Ontario.*

## Key Concepts

- The water quality in developed countries has improved rapidly since about 1970.

- Water can be polluted by pathogens, excess nutrients, physical changes, and toxic chemicals.

- Excess nutrients cause eutrophication and lack of oxygen in freshwater systems.

- Nutrients can be removed from waste water by passing it through wetlands.

- Pathogens and nutrients can be removed from waste water by sewage treatment plants before the water is returned to a waterway.

- Most toxic chemicals cannot be removed from water by sewage treatment. They have to be removed before the water enters the sewage system.

- The main source of water pollution in developed countries today is runoff from nonpoint sources such as roads and fields.

*"The public needs to be aware that every time you walk your dog, replace your oil or fertilize your lawn, you create the potential for municipal water pollution."*

Linda Eichmiller, Association of State and Interstate
Water Pollution Control Administrators

In the last 20 years, developed countries have made enormous strides in cleaning up water supplies that had become heavily polluted by the end of the 1960s. Despite this progress, we still have some way to go. In the United States, some of the surface and ground water is still dangerously polluted in 40 states, and the lack of clean water restricts economic development in many areas.

Water pollution in less-developed countries remains a huge and growing problem. The former Soviet-bloc countries are probably the most polluted area on Earth because pollution from industry and agriculture was never controlled. The Polish Academy of Sciences reports that more than half of Poland's water is too polluted for even industrial use. By 2000, it estimates, Poland will contain essentially no water that is fit to drink. The very poor countries are not generally troubled by industrial pollution. Here the problem is that the population has outgrown the water supply so that water used for drinking is often polluted with sewage and agricultural runoff, leading to a terrible death toll from waterborne diseases (Section 8-F). In this chapter, we deal mainly with improvements in the control of water pollution in developed countries.

**Water pollution** is the introduction into fresh or ocean water of chemical, physical, or biological material that degrades the quality of the water and affects the organisms living in it. Water pollution has two underlying causes: industrialization and the human population explosion. Both produce waste products that we cannot dispose of or cannot dispose of as fast as we produce them.

> The two basic causes of water pollution are industrialization and the human population explosion.

### Point and Nonpoint Sources

**Pollutants**, substances that cause pollution, get into the water supply at thousands of different points, which can be divided into point and nonpoint sources. **Point sources** are individual locations such as a factory, a sewage treatment plant, or an oil tanker. Pollution from point sources is relatively easy to control because the source can be identified and regulated. Much of our water pollution comes from **nonpoint sources**, sources that occupy large areas and have many routes into the water supply. Nonpoint sources include streets, highways, construction sites, farmland, forests, and residential areas. Fertilizers, pesticides, oil, animal feces, salts, and other pollutants from nonpoint sources trickle into our water supply by way of the soil, storm drains, and runoff into marshes and streams. Only efforts by every individual can control pollution from such widespread sources. Facts and Figures 19-1 lists some point and nonpoint sources of pollution.

Twenty years of attempts to improve water quality have done much to clean up the water that flows from point sources such as industry and municipal sewers. As a result, nonpoint sources are now believed to account for more than 55% of water pollution. It will be much more difficult to reduce pollution from these sources.

> Much water pollution comes from nonpoint sources such as streets, farmland, and lawns.

## FACTS AND FIGURES 19-1

### Where Does Water Pollution Come From?

Ground and surface water in the United States is exposed to pollution from point and nonpoint sources. Point sources include:

- 23 million septic tank systems
- 9,000 municipal landfills
- 190,000 storage lagoons for polluted waste
- About 2 million underground storage tanks containing pollutants such as gasoline
- Thousands of public and industrial wastewater treatment plants

Nonpoint sources and major pollutants include:

- Highway construction and maintenance: eroding soil and toxic chemicals
- Stormwater runoff from city and suburban streets: oil, gasoline, dog feces, litter
- Pesticides from the 112 million hectares of cropland treated with these substances each year
- 50 million tons of fertilizer applied to crops and lawns
- 10 million tons of dry salt applied to highways for snow and ice control

## Ground Water

About half the nation's drinking water comes from ground water. Ground water usually becomes contaminated when polluted water percolates down through the soil into the ground water. Toxic chemicals from the surface may pollute a huge volume of ground water. In Massachusetts in 1977, an underground storage tank leaked 3,000 gallons of gasoline into ground water used as municipal drinking water. All the wells into the aquifer had to be shut down, leaving two towns with no drinking water supply. Cleaning up the aquifer has cost millions of dollars and is not yet complete. Even the so-called high-technology industries, billed as nonpolluting because they do not belch clouds of smoke into the air, produce many toxic chemicals that have reached the ground water of Silicon Valley in northern California.

More than one quarter of America's ground water is dangerously polluted.

## Surface Waters

Most polluted water runs off the land into surface water: streams, rivers, lakes, marshes, and estuaries and thence into the ocean. Surface water is most easily polluted when it is isolated or moves slowly as it does in many lakes, estuaries, and seas with narrow openings. Pollutants discharged into even a deep lake, such as Lake Tahoe or Lake Superior, tend to accumulate because the water stays in one place for many years. It takes more pollution to damage rivers or streams because water flows through them continually. Similarly, Chesapeake Bay, Long Island Sound, and the Mediterranean Sea have only narrow open-

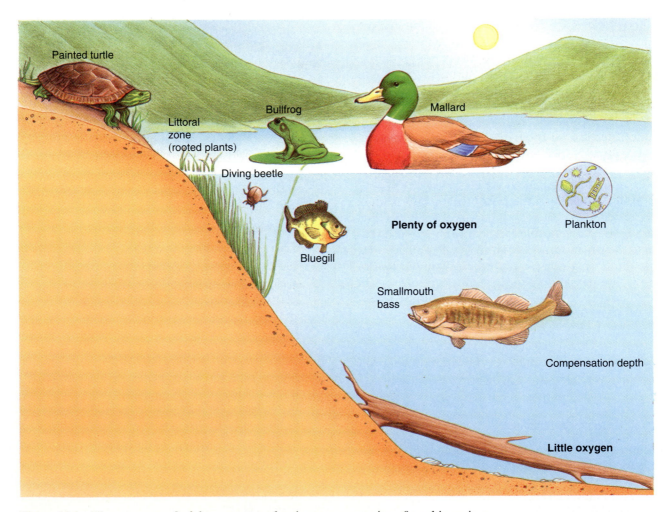

**Figure 19-1**   The structure of a lake ecosystem showing some organisms found in various parts of the lake.

ings to the open ocean, so pollutants tend to accumulate (Close-Up 19-1).

> Most polluted water runs off into surface water and pollutants tend to accumulate in enclosed lakes, estuaries, and seas.

## 19-A FRESHWATER ECOSYSTEMS

Let us examine the natural ecological processes in a body of surface water such as a lake and how various forms of pollution affect such an ecosystem (Figure 19-1).

Sunlight is the lake's source of energy. As light passes down through the water, some of it is absorbed and used in photosynthesis by phytoplankton, and some is absorbed by the water itself. So, as light passes deeper into the water, it becomes dimmer. In deep lakes, there is a depth where the available light is just bright enough for green plants to live: their production of food and oxygen (photosynthesis) exactly offsets their use of food and oxygen (respiration). The clearer the lake, the lower this point. Above this depth, plants produce more oxygen than they use, so extra oxygen is available for the respiration of other organisms; below it, there is not enough photosynthesis to offset respiration, and the water contains little oxygen.

Littoral zones, the shallow edges of the lake, contain **emergent plants**, those with their roots in water and their tops above water. In many lakes and in marshes, photosynthesis in the littoral zone supplies most of the productivity. Fish, tadpoles, insects and other arthropods, snails, and worms live and feed among these plants. Floating plants, which need light, and animals that need abundant oxygen, such as fish and small arthropods, live in the open surface waters of the lake.

### Oxygen

Oxygen affects nearly every aspect of life in the lake, including what animals and plants can live where and the solubility of plant nutrients. Oxygen enters the water from the photosynthesis of aquatic plants and by dissolving into the water from the air. Oxygen leaves the water when it is used in respiration by plants and animals or is used up by aerobic decomposers.

Wherever there is enough oxygen, dead plants and animals falling to the bottom are broken down by aerobic decomposers, which add nutrients to the lake. Deep in the lake, where there is not enough

light for photosynthesis, the chief input of energy is detritus falling from above. Decomposers, fish, and invertebrates consume this detritus or one another. In lakes with highly productive surface waters, so much dead matter falls to the bottom ooze that little or no dissolved oxygen remains in the water near the bottom. Under these conditions, the dominant decomposers are anaerobic bacteria.

### Temperature

Temperature affects the seasonal activities in a lake, particularly where there is considerable temperature difference from one season to another. Water has the peculiar property of being most dense at 4 °C. As a result, water at 4 °C sinks beneath water that is either warmer or colder. In winter, water cooled below 4 °C rises above water at 4 °C, and at 0 °C the surface water freezes (Figure 19-2). Under the ice, the water re-

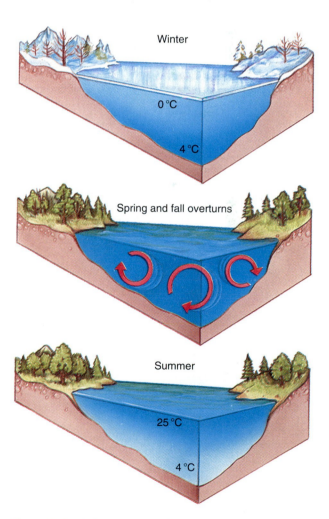

**Figure 19-2** Lake overturn. The warmer water is shown in darker blue.

## Cleaning Up Long Island Sound

In 1991, for the first time in 30 years, people collected shellfish legally from beaches near Greenwich on Long Island Sound. Attempts to clean up the sound are beginning to bear fruit—or, perhaps, seafood.

Molluscs absorb bacteria from their environment, so it is not safe to eat quahogs, oysters, littleneck clams, mussels, or other shellfish from water polluted with sewage. This is the reason that collecting shellfish near the heavily populated western end of Long Island Sound has been illegal since 1960. By that date, the sewage discharged by swelling human populations had made the water look exceedingly unattractive to swim in and the shellfish unsafe to eat.

A host of organizations has been involved in efforts to clean up the Sound. Almost every harbor has a group concerned about water quality. It was the Shellfish Commission of Greenwich that interested officials in testing the water for coliform bacteria and eventually secured state permission to reopen some shellfish areas.

Long Island Sound has nearly 600 miles of shore line and 250 miles of beaches (Figure 19-A). The cleanliness of the water is important to the large recreation and fishing industries. The most important step in cleaning up the sound was the installation of more and improved sewage plants in the 1970s and 1980s. Even small towns now treat their sewage,

**Figure 19-A** Long Island Sound. The sound is trapped between Long Island and the Connecticut shore, so pollutants take a long time to wash out of it.

which is why the bacteria count in the sound has fallen.

Not that every problem is solved. The recessions of the 1980s slowed the flow of money for expanding and upgrading sewage plants and a federal study found low levels of oxygen in much of the sound, a result of nutrient pollution by nitrogen. Low oxygen levels kill aerobic organisms and encourage the growth of less desirable anaerobic species. Phosphate detergents, which also cause nutrient pollution, are banned in New England, but phosphates also reach the water when fertilizer washes off millions of lawns in this densely populated area.

mains at 0 to 4 °C, and the lake's organisms survive here. In spring, the ice melts and the sun warms the surface waters, which sink as they approach 4 °C, forcing colder water below to rise to the surface. A similar overturn occurs as the water cools in fall. In the arctic and in parts of the tropics, lake water reaches 4 °C only once a year, in midsummer or midwinter, and there is thus only one overturn. Overturns bring nutrients from the bottom to the surface and carry oxygen down to deeper waters. Trout fishermen know that spring and fall are the times when trout, which require plenty of oxygen, can be found in surface waters. Warm water holds less oxygen than cold water, so trout spend the summer in the deeper, colder waters.

Overturns contribute to cleaning the lake. Suspended particles and any pollutants attached to them may be carried down to the bottom and deposited in the bottom sediment. Debris from the bottom may be carried to the surface where there is oxygen and where it may be decomposed by aerobic organisms.

### Eutrophication

We can divide lakes into categories based on their productivity. **Eutrophic** ("good food") lakes are relatively shallow and rich in organic matter and nutrients. These lakes are very productive but may contain little oxygen because decomposer organisms rapidly use it up. In contrast, **oligotrophic** ("few food") lakes are usually deeper and have steeper sides and a narrow littoral zone. These lakes are poor in nutrients, so there is little productivity, few organisms, and little organic material. The water is usually very clear, and the deep waters always contain oxygen. In the normal course of events, a lake ages as it is steadily filled in with sediment and organic matter. It usually becomes more eutrophic as it ages and may eventually turn into a bog or marsh and finally into dry land. For a deep oligotrophic lake, such an aging process may take millions of years. A clear, oligotrophic lake can be converted into a smelly, eutrophic sewer in much less than a million years by pollution.

### Some Effects of Pollutants

Toxic chemicals may kill some or all of the organisms in a freshwater ecosystem, but many less obvious forms of pollution can also cause damage. For instance, if eroding soil or other particles are washed into the lake, they block sunlight from deep water, making it impossible for plants to live except very near the surface. At the opposite extreme, extra

plant nutrients or heat increases the rate at which plants grow and may cause the water to become a cloudy green soup full of algae. Heat and nutrients also affect the types of organisms that live in a lake because they alter the oxygen content of the water. Warm water contains less oxygen than cold, and water full of nutrients contains little oxygen because the oxygen is used up by aerobic decomposers. Organisms that need a lot of oxygen cannot live in warm or nutrient-rich waters.

## 19-B POLLUTANTS

Water pollutants can be divided into four main categories:

1. **Pathogens.** These are organisms that cause diseases. Many live only in water. Pollution with pathogens is usually caused by human sewage or animal manure getting into the water.
2. **Nutrients and biodegradable organic matter.** Biodegradable organic matter includes the remains of animals and plants, including feces, leaves, wood waste, fat, and debris from food-processing plants. These substances are broken down by decomposers into mineral nutrients that plants take up.
3. **Physical agents.** Physical agents are things like heat and suspended solids, such as soil, that can destroy the usefulness of water if they are present in large quantities.
4. **Toxic chemicals.** These are many chemicals that are poisonous to organisms. They include metals such as lead and mercury, organic compounds such as some pesticides and waste products of the petrochemical industry, and radioactive waste.

### Nutrients

In a balanced ecosystem, one organism's wastes are another's food and drink. Pollution results when biological wastes and mineral nutrients are not decomposed and used up as fast as they are produced. Some mineral nutrients get into the water supply as fertilizer that washes off agricultural land. Nutrients are also produced by the breakdown of sewage and are added to water by households and many industrial processes.

> Pollution results when biological wastes and mineral nutrients are not used up as fast as they are produced.

Biodegradable organic wastes, the sources of nutrients, can be divided into dry solids and sewage. **Dry solids** include lawn clippings, leftover food, waste wood, and the remains of crops and food processing. **Sewage** is waste that enters the environment in liquid form. The most obvious example is waste from household toilets and sinks that flows into the local **sanitary sewage system**. This is the system of pipes, pumps, and sewage treatment plants that collects liquid waste from buildings and houses and processes it. Many industrialized countries continue to dump untreated human sewage into waterways, and human sewage is still not treated in most parts of poor countries.

## Biochemical Oxygen Demand

Aerobic decomposers use up oxygen when they break down organic matter, so the amount of biodegradable material in water can be measured by determining how rapidly oxygen in the water is used up. The quantity of oxygen-consuming waste in water is usually determined by measuring biochemical oxygen demand. **Biochemical oxygen demand** (BOD) is defined as the amount of oxygen consumed when a biodegradable substance is oxidized in water. BOD is measured by saturating a sample of polluted water with oxygen and keeping it in a closed bottle for several days. The decomposers in the water use up the oxygen as they break down the organic material. The more organic material present, the faster the decomposers use up the oxygen. At the end of the period, the amount of oxygen left in the bottle is measured, and the decrease in the oxygen level is expressed as the biochemical oxygen demand. The higher the BOD, the more organic matter the water contains. Raw sewage has a BOD of 40 to 150 milligrams per liter. Water clean enough for human use has a BOD of less than 0.5 milligram per liter.

The ability of water to support fish and other aerobic organisms is indicated by how much dissolved oxygen it contains. Good-quality water contains 5 to 7 milligrams of dissolved oxygen per liter. It is considered gravely polluted if the level falls below 4 milligrams per liter. Water containing less than 2 milligrams of oxygen per liter kills fish and other aerobic organisms.

## Pathogens

Pathogens and nutrients are natural pollutants because all organisms expel wastes into the environment around them. Some biological pollutants contain pathogens that have been excreted by sick people and animals. Cholera, hepatitis, and typhoid are among the many diseases that people can catch by drinking water polluted with pathogens. Even healthy people contain microorganisms, which can cause disease if they get into the drinking water and are consumed.

## Coliform Bacteria

The most common test used by health departments to see if water is fit to drink involves discovering whether the water contains coliform bacteria. **Coliform bacteria** are bacteria such as *Escherichia coli* that are found in large numbers in the intestines, and therefore the feces, of humans and other animals. Most coliform bacteria are not pathogens, but if coliform bacteria are present in water, it means that the water contains sewage. Water for a coliform test is filtered to collect the bacteria. These are then placed on a medium that provides the substances the bacteria need to grow and divide to form colonies large enough to be visible. The medium is incubated for 24 hours, and any bacterial colonies that form are counted.

According to U.S. Environmental Protection Agency (EPA) standards, water is fit to drink if it produces no more than one coliform bacterial colony per 100 milliliters of water. If this level is exceeded in water from a domestic well or municipal source, the homeowner or municipality must either use a different source for drinking water or add more chlorine to the water to kill the bacteria. The EPA-recommended coliform count for swimming water is 200 colonies per 100 milliliters, but some states and cities permit a higher level. In 1985, the EPA reported that one quarter of swimming areas tested around the United States were not safe for swimming by EPA standards.

## 19-C Nutrient Pollution and Eutrophication

When decomposers break down organic matter in water, they use the energy released for their own metabolism and convert the elements into salts and gases. The chemistry of decomposition is different when it occurs in water containing oxygen and in water containing no oxygen because different decomposers live in aerobic and anaerobic environments. Table 19-1 shows the compounds that decomposers produce from the elements that are most common in organic matter. Ammonia and hydrogen sulfide are gases with a foul smell that are poisonous in high concentrations. Most of the gases produced during decomposition (such as carbon dioxide and

**TABLE 19-1   Fate of the Main Elements in Organic Matter after Decomposition**

| Element in Organic Compounds | Where Each Element Ends Up | |
|---|---|---|
| | After Aerobic Decomposition | After Anaerobic Decomposition |
| Carbon | Carbon dioxide ($CO_2$) | Methane ($CH_4$) |
| Hydrogen | Water ($H_2O$) | Water and methane |
| Oxygen | Water | Water |
| Sulfur | Sulfate salts ($SO_4^-$) | Hydrogen sulfide ($H_2S$) |
| Nitrogen | Nitrate salts ($NO_3^-$) | Ammonia ($NH_3$) |

methane) leave the water and pass into the atmosphere. Sulfates and nitrates are plant nutrients.

If a body of water contains more nitrates than its plant population can use, nitrates may get into the drinking water and then into the human body. In the digestive tract, nitrates are converted into toxic nitrites, which reduce the capacity of the blood to carry oxygen. Hundreds of infants have died of nitrite poisoning since World War II, usually in rural areas where a drinking-water well is polluted by manure from a farm.

Detergents and fertilizers containing phosphate are important sources of nutrient pollution (Lifestyle 19-1). Phosphorus is one of the elements most likely to limit the productivity of an ecosystem. If phosphorus is added to water just once, aquatic plants grow rapidly, but then they die when they have used up the phosphorus. Decomposers eventually break down most of the dead organisms in the water and then die themselves. At this point, the oxygen con-

tent of the water starts to rise, and the lake becomes more oligotrophic again.

Long-term problems arise when large quantities of nutrients continue to drain into a lake or river. The nutrients come from sources such as sewage, phosphate-containing detergents, feedlot waste, fertilizer runoff, and industries such as paper mills, meat-packing plants and industries. These sources often add nutrients to the water continuously, in which case the productivity of the water remains high and eutrophication proceeds rapidly. Such changes have fundamentally altered Lake Erie, which is relatively shallow. Even deep oligotrophic lakes, such as Lake Tahoe, have become noticeably more eutrophic in the last 20 years as a result of nutrient pollution.

Oligotrophic lakes with clear water are much more appealing and useful than eutrophic lakes covered with algae, clogged with weeds, and stinking from the by-products of anaerobic bacteria (Figure

**Figure 19-3** An oligotrophic lake in the Northwest. *(Richard Feeny)*

19-3). For this reason, there has been strong pressure to ban the use of detergents containing phosphorus and to speed the installation of tertiary sewage treatment plants, which remove organic and inorganic molecules from the water. In some cases, these measures have brought lakes back to a more oligotrophic condition. Phosphate pollution from detergents is unnecessary because there are plenty of detergents with adequate cleaning power without phosphates.

The biggest source of the mineral nutrients that pollute our lakes and rivers is fertilizers from farms, lawns, and gardens. As much as 50% of some fertilizer applications ends up in a stream or lake. This is a terrible waste of the money spent on the fertilizer in the first place. Although farms are the largest source of fertilizer runoff, homeowners do proportionally more damage because they are less likely to measure out fertilizers properly. Up to 80% of the pesticides and inorganic fertilizers applied to home lawns and gardens are wasted and end up running into our water supplies.

> The biggest source of the mineral nutrients that pollute our lakes and rivers is fertilizers from farms, lawns, and gardens.

### Hormone Pollution

One odd example of pollution by organic matter was discovered recently in Britain. Researchers have found that male trout downstream from sewage plants produce egg-yolk proteins normally found only in female fish. Production of the proteins is controlled by the female steroid hormone, estradiol. It appears that the fish are being exposed to female steroids. The source of the hormone is thought to be the urine of women taking contraceptive pills and domestic detergents, both of which contain compounds that imitate the effects of estradiol. The researchers do not yet know if wild fish populations are suffering any harm from their inadvertent steroid abuse. But freshwater hormone pollution is not a peculiarly British problem: French researchers studying eels in the River Seine are discovering similar results.

### 19-D   CLEANING UP POLLUTION

The Water Pollution Control Act of 1972 (Clean Water Act) set a national goal of making surface waters fit for swimming and fishing by 1983 (Facts and Figures 19-2). The act charged the EPA with implementing standards that would steadily improve the quality of waste water discharged into waterways. In

**FACTS AND FIGURES 19-2**

#### Water Quality Legislation in the United States

Here is a list of the main federal legislation designed to improve water quality in the United States. Many of the deadlines set by legislation in the 1970s, such as those for sewage treatment, have been extended because states and towns have simply not met them. Sometimes the deadlines were unrealistic, and sometimes money to implement changes has been lacking. The EPA oversees most water quality legislation.

- **1972 Clean Water Act** (CWA). This is technically the **Water Pollution Control Act.** The act set a national goal of making all natural surface waters fit for fishing and swimming by 1983 and of no pollutant discharge into these waters by 1985. The Act required that metals be removed from waste water beginning in the early 1980s. This is also the federal wetlands protection act, which has done such a poor job of saving our wetlands.

- **1972 Marine Protection, Research, and Sanctuaries Act,** amended 1988. This act empowered the EPA to control the dumping of sewage wastes and toxic chemicals in the ocean.

- **1974 Safe Water Drinking Act:** This act introduced programs to protect both ground water and surface water from pollution.

- **1980 Comprehensive Environmental Response Compensation and Liability Act** (CERCLA). This is the Superfund Act, which makes owners, operators, and customers of hazardous waste sites responsible for their cleanup. It has reduced the pollution of water by toxic substances leached from hazardous waste dumps.

the same year, the Marine Protection, Research, and Sanctuaries Act empowered the EPA to control the dumping of sewage wastes and toxic chemicals in the ocean.

Efforts to clean up polluted surface waters generally involve preventing polluted water from reaching the waterway and leaving nature to do the rest. Biological pollutants decompose, nutrient levels eventually fall, and the oxygen content of the water increases. Toxic substances may fall to the bottom. All will be diluted, and some of them may even decompose. We have seen that bacteria decompose toxic PCBs (Section 18-C).

> Once a body of water has become polluted, the main way to clean it up is to stop polluting it and then leave it alone for biological processes to cleanse it.

There are some situations in which leaving it to nature will not work. For instance, an aquifer may contain few decomposers and refill very slowly so that nature may take thousands of years to clean up polluted ground water. Then there are pollutants that disappear very slowly. The Savannah River between South Carolina and Georgia is polluted by radioactive material that has leaked from the Savannah River Plant, a Department of Energy site used to make material for nuclear weapons. Radioactive substances may take millions of years to decompose. These cases, however, are the exceptions. In general, the solution to water pollution is to purify waste water before it reaches the waterway. Waste water comes from several main sources:

1. **Industrial and agricultural waste water.** Most water polluted by industry and agriculture is treated where it is produced and reused or returned to the water system. Government regulations now govern the quality of water that an industry can discharge into waterways, and meeting the regulations requires different treatment for the water, depending on what it contains. Waste water from agricultural sites such as feedlots, chicken farms, and fish farms is usually polluted with biodegradable nutrients and pesticides. It is often stored in lagoons while the pollutants decompose (Figure 19-4). Industrial treatment is much more varied. Here is one example.

   The Clean Water Act required that metals be removed from waste water beginning in the early 1980s. Fees for disposing of waste metals rose. By 1985, the Robbins Company in Attleboro, Massachusetts, was in trouble. This small company makes electroplated objects, using large amounts of water and leaving a host of toxic acids and heavy metals behind. The company had been heavily fined for polluting the local Speedway Brook and Ten Mile River, and it turned to an environmental engineering company for help. L & T Technologies suggested generating fewer pollutants, together with a system that would clean the polluted water and then reuse it, eliminating waste water altogether. The system, finished in 1988, uses 82% fewer chemicals, generates 89% less toxic material for disposal, and reduces the cost of chemicals by 87%. The company is saving more than $70,000 a year in chemicals, water charges, and disposal fees, and the new equipment paid for itself in less than four years. Perhaps the biggest surprise was that Robbins, once hated by the community as a polluter, has improved its public image and finds that local orders are increasing.

2. **Municipal sewage.** Most of the waste water from cities and residential areas is collected in pipes and taken to a sewage treatment plant, where it is purified before being returned to a river or lake. Domestic sewage is biodegradable and easy to purify. The trouble comes when sewage also contains industrial waste water and storm runoff from streets and fields, which contain toxic substances that interfere with sewage treatment.

**Figure 19-4** Industrial wastewater treatment: lagoons for a paper mill. Paper mills use a lot of water so they are usually built beside rivers or lakes. The waste water they release is heavily polluted and cannot be released back into the waterway. Instead, it is usually treated to remove toxic waste, allowed to stand in lagoons while organic matter decomposes, and then reused. *(Steve Bisson)*

**3. Storm runoff.** Periods of heavy rainfall cause special problems with pollution from nonpoint sources. More pollutants than usual run off agricultural land, city streets, and suburban lawns, and sewage systems may overflow, discharging raw sewage. In 1990, the EPA issued rules under the Clean Water Act requiring towns with populations of 100,000 or more to obtain permits for their stormwater discharges.

Naples, Florida, a city of 20,000, is exempt from the rules, but it is nevertheless pursuing a stormwater management program. Naples is concerned about protecting the marine ecosystems of the Gordon River and Naples and Moorings bays from polluted stormwater runoff and preventing the floods that sometimes fill city streets. Part of the solution is to create ponds and wetlands where storm water is stored and purified before being pumped into Naples Bay. This system is designed as an attractive park, with a walkway extending into the bay. Naples plans to fund this program mainly through a stormwater tax. Each parcel of land in the city will be charged a fee based on how much water runs off it during storms. This will encourage industries and home owners to build ponds, plant vegetation, and undertake other steps to reduce runoff.

## 19-E  SEWAGE TREATMENT

The Clean Water Act mandated that all cities that release waste into waterways pass the water through primary and secondary sewage treatment plants before they release it. There are three kinds of sewage treatment: primary, secondary, and tertiary. If you are lucky, your local sewage plant is a tertiary plant, but there are not many of these around because they are expensive to build and operate.

A sewage treatment plant is designed to encourage the same processes that occur when wastes decompose in a lake or river.

> Sewage treatment imitates nature, treating the sewage by encouraging decomposition.

### Primary Treatment

Primary treatment starts when sewage enters the plant through a series of screens that remove debris, such as toothbrushes and jewelry, that we should not have flushed down the toilet in the first place (Figure 19-5). The sewage proceeds to settling tanks, where 80% of the solid matter settles to the bottom. The sewage here is exposed to bacteria that cause particles of organic matter to stick together until they are large enough to settle to the bottom as **sludge**. The liquid from the settling tanks may then flow into a chlorinator where chlorine is added to kill any bacteria. The liquid still contains 40% of its solids and 70% of its organic material. The sludge from the bottom of the settling tank is treated with anaerobic organisms in a sludge digester. These organisms break down some of the organic material, releasing gases such as methane.

### Advanced Primary Treatment

Some cities in Southern California have unusual advanced primary treatment plants, which they claim produce waste water of higher quality than secondary treatment does. In the 1970s, engineers tried adding various chemicals to primary-treated water to remove more of the suspended particles. Nowadays the chemicals of choice are hydrocarbon polymers mixed with ferric (iron) chloride. The polymers are electrically charged so they bind to charged particles and the whole mass precipitates. The system produces less sludge than secondary treatment, but the waste water may contain more toxic chemicals, which precipitate into the sludge. Southern California engineers say that the solution is to keep toxic chemicals out of the sewage in the first place, not to insist that every sewage works should include secondary treatment.

### Secondary Treatment

Secondary treatment starts with the products of primary treatment and exposes both the sludge and the fluid part of sewage to aerobic decomposers. The liquid from primary treatment is run into an **aeration tank**, a tank in which air is bubbled through the liquid to permit aerobic decomposers to work. It is then piped into a settling tank, and the activated sludge that settles out of it is recycled into the aeration tank. Aerobic decomposition reduces the volume of the sludge.

Another method is to spray the fluid from primary treatment onto trickling filters by way of rotating nozzles, which add oxygen from the air. **Trickling filters** are beds of gravel where dozens of aerobic decomposers live and work, including bacteria, fungi, protists, fly larvae, and worms. After passing through trickling filters or an aeration tank, the fluid again passes into a settling tank where any solid matter settles out. The remaining fluid contains little organic

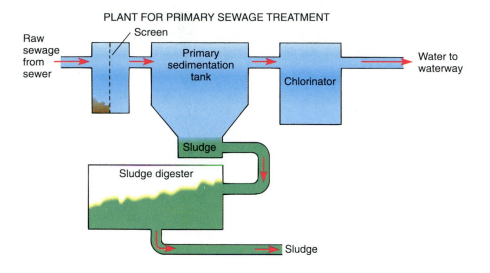

PLANT FOR PRIMARY SEWAGE TREATMENT

PLANT FOR SECONDARY SEWAGE TREATMENT

**Figure 19-5**  Primary and secondary sewage treatment plants. Note that in this plant with secondary treatment, the sewage is aerated in a tank. In other similar plants, it is aerated by trickling filters.

matter and has also lost about half of its inorganic nutrients. It is chlorinated and released.

### Chlorination

Chlorine is used to kill bacteria and algae in water. The trouble with chlorine is that it is very toxic to fish, and it reacts with various organic molecules to form compounds that cause cancer and birth defects in humans. The amount of chlorine in most water at the moment does not seem to be a major threat to human health, but we cannot afford to let the concentration of chlorine compounds in drinking water rise much farther.

Chlorine reacts with various organic molecules to form compounds that cause cancer and birth defects in humans.

### Disposing of Sludge

Sewage sludge often contains dangerous concentrations of toxic chemicals, so it must usually be disposed of as hazardous waste. It is usually incinerated and then buried in a secure landfill. The volume of sludge that has to be disposed of every year is enormous and an expensive burden to municipalities. Many towns, therefore, are attempting to reduce the toxicity of their sludge so it can be disposed of in other ways. Sludge contains plant nutrients, so if it is free of toxic chemicals it can be sold as fertilizer, which is diluted and sprayed on agricultural fields. In another interesting process, the sludge is combined with clay to make bricks that can be used for building. We can expect more industries that consume sludge to develop in the future as part of the solution to our solid waste problem (Chapter 18).

## 19-F  TERTIARY WATER TREATMENT

There are many ways of removing the pollutants that remain after secondary sewage treatment. Some of them involve expensive chemical treatments and are rarely used unless the water will subsequently be used as drinking water. For instance, the water may be treated with aluminum sulfate, which causes phosphates to precipitate, or the water may pass through a filter of activated charcoal, which absorbs many impurities.

One growing method of cleaning up sewage effluent is spreading it on land or passing it through wetlands. This has several advantages over pouring the effluent into a river, where it runs wastefully off into the sea. First, the effluent is purified as it trickles down through soil ecosystems, where it eventually reaches the ground water and helps to refill a depleted aquifer. Second, the effluent may be used for irrigation, thereby saving precious water. Third, any nutrients in the effluent act as fertilizer. The yield of crops growing on treated land increases markedly.

It is not safe to assume that natural wetlands can be used as treatment plants for waste water. Scientists studied an experimental peat bog in Michigan that had absorbed sewage effluent for ten years and found that the vegetation changed considerably in that time. It seems safer to build artificial wetlands or to restore severely degraded wetlands when these are needed for wastewater treatment and to save natural wetlands that are in good condition as wildlife habitat.

### Artificial Wetlands

There is now considerable enthusiasm for building artificial wetlands to purify waste water. This process was pioneered in California, where waste water is piped into marshes, which are kept in place by artificial dikes. The water eventually trickles out of the marsh into a river or reservoir, purified and ready to be used again. Some California marshes have been made into recreational areas, with bike paths along the dikes from which people can watch the birds and other creatures that rapidly make their homes in this artificial ecosystem.

Let us look at a few examples.

**Mesa, Arizona: Water in the Desert**  It is a great benefit to towns in areas with little water if they can make their waste water clean enough so that it does not have to be discharged into the local river but can be reused. Mesa built a new treatment plant that achieves just this. The plant produces water that is clean enough to use for recharging the depleted aquifer under the city and for irrigating parks and lawns, freeing other water to be used for drinking water. The waste water from secondary sewage treatment passes through a series of ponds designed to remove nitrogen. First the water travels through ponds containing lots of air, where aerobic bacteria convert nitrogen compounds into nitrates. Then the water enters an anaerobic system, where nitrates are converted to ammonia gas, which leaves the water (see Table 19-1). After this treatment, the water contains less than the EPA's drinking water standard of 10 milligrams of nitrogen per liter. Finally, the water reaches a series of percolation basins, built from abandoned gravel pits, where it percolates down into the aquifer.

**Incline Village, Nevada: Creating Wetlands**  Incline Village's system for disposing of secondary sewage effluent disposes of the water, expands existing wetland habitat for wildlife, and provides an educational facility for visitors. The primary disposal area consists of newly created wetland—ditches and lagoons surrounded by dikes. Waste water leaves the system through evaporation, transpiration, and percolation and never reaches a waterway.

The new lagoons are designed as habitat for emergent vegetation, whose transpiration disperses much of the water into the air. The center of each lagoon is deeper and free of vegetation to provide a "landing strip" for water birds. Islands, planted with trees and food plants, provide habitat where the birds nest.

This part of the world has distinct wet and dry seasons, so there are wetlands with several depths of water so that the system can absorb storm runoff while still providing wildlife habitat during hot dry summers. One backup area can be emptied by spraying the water it contains onto nearby land during very wet weather. Because of the different types of wetland, the system contains an enormous range of aquatic plants, invertebrates, mammals, and birds, including some endangered migratory species.

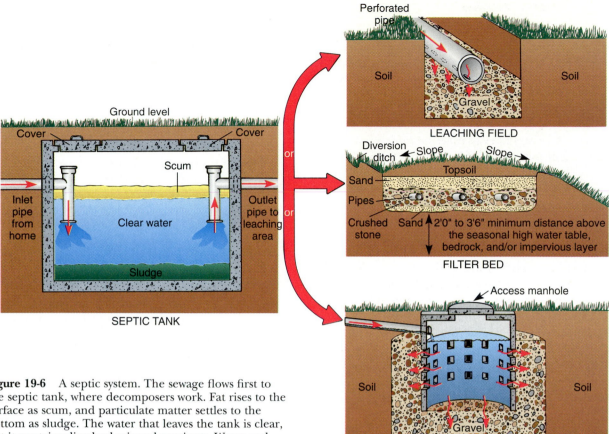

**Figure 19-6** A septic system. The sewage flows first to the septic tank, where decomposers work. Fat rises to the surface as scum, and particulate matter settles to the bottom as sludge. The water that leaves the tank is clear, but it contains dissolved mineral nutrients. Water and nutrients are disposed of by permitting them to leach slowly into the soil by one of the three methods shown on the right, all of which work on the same principle. A pipe releases the water slowly into a bed of gravel, which acts as a trickling filter. The waste and minerals are dispersed into the surrounding soil. Some older systems consist of nothing but a cesspool, which looks like the drainage pit shown here. There is no septic tank, and the sewage decomposes and disperses into the soil from a single tank. If either the septic tank or the drainage field is not large enough for the volume of sewage it receives, a septic system pollutes surface and ground water.

> Artificial wetlands are used to clean nutrients from waste water and also serve as wildlife habitat.

## 19-G RURAL SEWAGE SYSTEMS

Rural areas have fallen behind in the race to produce clean water, largely because it costs much more per person to provide sewage treatment for a town of 500 than of 50,000. Novel solutions are helping with the problem in some areas.

### Ohio: Helping with the Landfill Crisis

By working with the managers of a local landfill, the 1,349 people of Lowellville, Ohio, will manage to meet EPA wastewater standards without even having to raise their sewer rates. Ohio passed strict new landfill laws in 1989, which caused one third of the state's landfills to close. The remaining Lowellville landfill was under pressure to clean up the water that leached out of the landfill. Its owners decided they could save the expense of building a water treatment plant at the landfill by treating the leachate to remove toxic chemicals and then piping it to the Lowellville sewage works. The landfill's owners paid part of the cost of modernizing the sewage plant and now pay part of its operating cost.

### Septic Systems

Millions of Americans in rural areas do not discharge sewage to the municipal sewage system but to their own septic systems (Figure 19-6). **Septic** means that the system contains microorganisms that decompose the sewage. Septic systems are a major source of water pollution because they are not well regulated. Some states have no regulations at all, and most states require that septic systems be tested for leaks and overflows only when a house is sold.

In a septic system, sewage flows into a septic tank, where decomposers break down organic matter. Inorganic materials and solids sink to the bottom as sludge. The water that remains contains mineral nutrients. It flows through an outlet pipe to a **leach field**, a drainage area where it is gradually dispersed in the soil and absorbed by plants. The sludge is pumped out of the septic tank every few years so that the tank does not fill up.

The trouble with septic systems is that they are underground, so owners seldom know if a pipe breaks or the system stops working until sewage starts to backup into the house or flow over the surface of the lawn, by which time the polluted water may have done considerable damage. There are several steps that can be taken to keep septic systems working well, most of them designed to keep the decomposers working.

1. Have the tank pumped and the leach field checked every two years.

2. Don't use colored toilet paper because the dyes it contains are often toxic to bacteria, and most of the water in a septic tank comes from toilets (Figure 19-7).

3. Never put nondegradable things, such as plastic tampon applicators, down the toilet.

4. Check products used in the home for labels saying "Harmful or fatal if swallowed; avoid contact with skin; do not get in open cuts or sores; if comes into contact with eyes, call a physician immedi-

ately." Such labels indicate substances that are bad for humans and worse for bacteria. They may be found on detergents, disinfectants, toilet cleaners, bleach, acids, cleaning compounds, polishes, sink and tub cleaners, and caustic drain openers.

## 19-H PHYSICAL POLLUTANTS

**Thermal pollution** is heat pollution. It results from adding excessive amounts of heat to a body of water. Enormous volumes of water from lakes and rivers pass through the cooling systems of power plants and other industries. This water is usually returned to the waterway unchanged except that it is warm. Adding heat to an ecosystem alters it in several ways. The production of organic matter in a lake depends, among other things, on the water temperature during the summer and on the length of the warm period. If extra heat is put into the water, productivity increases. Greater production of organisms means more work for the aerobic decomposers. The lake's oxygen may all be used up before the fall overturn replenishes the supply, leading to the same kind of anaerobic conditions that result from nutrient enrichment. For these reasons, citizens often fight the construction of power plants on the shores of oligotrophic lakes or insist on cooling ponds or cooling towers that reduce the temperature of the water before it is returned to the waterway.

One of the most widespread water pollutants is soil that erodes from areas such as agricultural land, deforested hillsides, and construction and road-building operations. This soil usually contains both organic and inorganic components, fertilizers, and pesticides. The soil pollutes the water directly, and it also silts up reservoirs, blocks turbines at hydroelectric plants, has to be filtered out of drinking water, and is generally an expensive and dangerous nuisance.

## 19-I REGULATION AND CLEANUP PROGRAMS

In the 1960s and 1970s, public outcry over pollution led to action by governments and industries in many countries. Environmental protection policies and agencies were established. Millions of homes were hooked up to sewage systems for the first time (Figure 19-8). Taxation, fines, and subsidies for pollution control were introduced. Pollution control became big business, accounting for as much as 2% of the gross national product in some industrial nations by the end of the 1970s. The United States alone spends

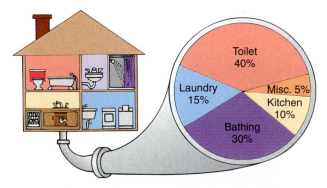

**Figure 19-7** Where the effluent that leaves the average house comes from. Although flushing the toilet produces most of the effluent, toilet waste is biodegradable. Wastewater from kitchen or laundry is much more likely to contain toxic or nondegradable pollutants.

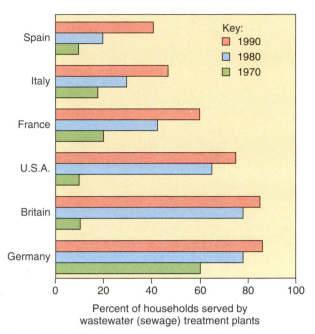

**Figure 19-8** The increase in sewage hookups in various countries from 1970 to 1990. In developed countries, nearly all households not served by sewage treatment plants have septic tanks.

pesticide. The EPA proposes to get around such attitudes by completely banning particular pesticides from areas where the local authorities refuse to regulate their use.

The 1986 amendments to the Safe Drinking Water Act were held up in Congress for many months because they would have required states to prevent industries that might pollute water from being built over ground water that might be used for drinking. The states complained that this infringed on a state's right to use its land as it pleases because it would indirectly require land-use planning, which most states do not have. Many people think that the political problems are so difficult to overcome that we shall never have effective federal control over the pollution of ground water.

The surface waters for which pollution control is most needed are those where industry and people are most abundant. People and industry congregate around ports, which provide jobs and transportation, and around lakes and rivers, which supply fresh water. As a result, populations have grown more rapidly along the shores of bodies of water than anywhere else in the world. Consider the campaign to try to clean up the North American Great Lakes.

about $40 billion a year on controlling water pollution.

## Ground Water

Our ground water is probably in worse shape than our surface water because regulations to protect it hardly existed before the 1980s. In 1986, Congress amended the Safe Drinking Water Act to limit hazardous substances in ground water. There are currently limits on only about 60 of the more than 200 hazardous chemicals that have been detected in ground water. Not until 1988 did the EPA propose a plan to ban or regulate the use of leachable pesticides in areas with porous soils. Leachable substances are those that pass through the soil most rapidly. In porous soils, they run off into ground water and surface water more rapidly than they do elsewhere. Enforcement of the EPA's regulations is left largely to local agencies. This approach will not work without large carrots or sticks to induce the states to comply. Many state governments care more about the financial interests of pesticide makers and farmers than they do about safe drinking water. For instance, a 1988 Georgia law states that farmers are not liable for damage caused by their use of pesticides, even if they are grossly negligent in the way they use the

## Lake Ontario

The Great Lakes basin is one of North America's most important natural resources and home to some 50 million people. The Great Lakes contain one fifth of the world's lake water, and this abundant fresh water is the basis for the prosperity of the northeastern United States and southeastern Canada. The Great Lakes drain from west to east, running eventually into the St. Lawrence River and so to the Atlantic Ocean (Figure 19-9).

Lake Ontario is the easternmost Great Lake. Once, it was surrounded by dense forests and fed by the waters of Lake Erie, which flow over Niagara Falls to join the lake. Sturgeon, pike, bass, and trout lived in the clear water and migrated up the Niagara River below the falls and up smaller streams to spawn. Then the human population started to grow. Forests were cleared, and soil that eroded from farms around the lake brought sediment that blocked waterways and silted up spawning grounds. As the last in the chain of the Great Lakes, Ontario received not only its own waste, but also that from all the other lakes, including notoriously polluted Lake Erie. By the 1960s, the waters of Ontario were polluted with organic waste and with lead, zinc, and other toxic wastes. Stinking blooms of blue-green bacteria were common in the shallows, and by the early 1970s the

**Figure 19-9**   The Great Lakes. Water flows eastward, eventually draining into the St. Lawrence River.

fish contained so much mercury they were unsafe to eat.

In 1972, the United States and Canada produced the Great Lakes Water Quality Control Agreement, under which both governments undertook to stop the discharge of pollutants into the lakes. At first, the agreement appeared to be a dramatic success. Governments, industries, and municipalities cooperated, and the water quality in all the lakes improved rapidly. Algal and bacterial blooms disappeared, raw sewage was no longer to be seen, and fish populations rebounded. A 1992 study painted a less rosy picture. At Toronto and Kingston in Canada and Oswego and Rochester in New York, toxic substances still reach Lake Ontario from industries and cities. Commercial fishing is still not permitted, and people are advised to eat only one fish a week from the lake and less if they are pregnant (Figure 19-10).

An important conclusion from the 1992 study is that cleaning up the water that flows into the lakes is not enough because much of the pollution that reaches the lakes now comes from the sky. Lake Ontario gets the worst of this because the prevailing winds come from the west and it is the easternmost of the Great Lakes. Polluted air brings up to 80% of the toxic chemicals, nitrates, and phosphates that reach the lake each year. This reinforces the conclusion that we need to clean up all forms of pollution.

> It is not enough to prevent polluted water from reaching a lake or river because much of the pollution in a waterway reaches it from the air.

## Economics of Pollution Control

In the early 1970s, governments and industry worried that the cost of controlling pollution would depress growth, competitiveness, job formation, and trade, while driving up prices and inflation. We certainly spend a lot of money controlling pollution. According to a recent EPA report, the United States spent $115 billion to clean up pollution of air, land, and water in 1990. That sum amounts to just over 2% of the gross national product. The EPA projects that by 2000, the cost will climb to about $185 billion, or about 2.7% of GNP.

This sum is the EPA's attempt to put a price tag on complying with all environmental regulations in effect in the United States at the present time. It includes compliance with some state and local as well as federal regulations but does not include the cost of wildlife conservation and other environmental programs. This is not money spent by governments. It is nearly all spent by individuals and companies to comply with regulations. The money has generated a huge pollution-control industry, providing millions of jobs and many new engineering techniques. Longer-term benefits include the prevention of damage to health and property.

Is this money well spent? Those who disapprove of environmental regulation declare that American industry cannot remain competitive in the world market if it has to spend huge sums on pollution control and clean up. But our newfound knowledge of industry in Eastern Europe makes it clear that pollution and industrial competitiveness do not necessarily go together.

432

**Figure 19-10**   A corner of Lake Ontario. The Great Lakes were scoured out by glaciers, which left tumbled granite rocks in many areas like those shown here. Although Lake Ontario is only about 30 meters deep, it was an oligotrophic lake with a rocky bottom and oxygenated water before it was polluted by runoff from farms, municipalities, and industry.

The Iron Curtain's disintegration has revealed environmental horrors in Eastern Europe that are beyond the wildest dreams of most Westerners. We see cities that look like American and European cities in the worst days of the Industrial Revolution, where everything is covered with soot and people die of air pollution in middle age. We see lakes so polluted with toxic chemicals that the water contains no living organism and, if used for irrigation, kills plants. The industry of Eastern Europe certainly did not waste its money on pollution control. Yet the reunification of Germany revealed East German industry as so inefficient that it produced essentially nothing at a cost that could compete with Western industry.

It is much cheaper to prevent pollution at its source than to clean it up later. For instance, the staggering sums allotted to clean up hazardous waste dumps in the United States are almost certainly not enough to do the job and would have been much better spent on prevention in the first place. Our goal today, while we clean up some of the messes created in the past, is to see that industrial development is sustainable and nonpolluting in the future.

> Most experts agree that spending on environmental protection has produced benefits that exceed its cost.

## 19-J  WHAT CAN WE DO?

You may have seen a backhoe digging up your local gas station. In the 1980s, regulations came into effect designed to reduce the amount of gasoline that leaks into the soil and then into the ground water. In 1986, the Steel Tank Institute estimated that across the nation 350,000 tanks filled with gasoline would leak during the next five years. New leakproof gasoline storage tanks are part of the solution. Each of these small-scale efforts helps. We can vote for politicians who will ensure local industries do not dump their wastes into our water. We can encourage farmers, cities, parks, and golf courses to replace chemical pesticides with biological pest controls. As consumers, we can get used to apples with a few flaws in them but free of toxic chemicals. We can stop buying detergents that contain phosphates. As householders, we can stop using pesticides that are not biodegradable.

We can also educate ourselves and others. As the quotation at the beginning of this chapter suggests, the main reason for the pollution of municipal water is that people do not seem to realize exactly what municipal pollution is. We need to learn how many of our actions contribute to pollution, from letting the dog defecate in the street to changing the oil in our cars and fertilizing the lawn. Around home and city, we can remember that any action that reduces the volume of water that runs off the land reduces water pollution and helps to recharge our groundwater supplies. Measures that reduce runoff include sidewalk paving stones that absorb water, gravel areas instead of asphalt, parks containing lakes, and maintaining as much vegetation as possible.

Rural communities are also going to need a lot of help in controlling pollution by actions such as soil conservation, reducing runoff, and building sewage treatment plants. The EPA estimates that 80% of the municipalities in violation of sewage treatment requirements are rural.

## All Washed Up?*

Environmentally concerned dining halls have been replacing disposable polystyrene (styrofoam) coffee cups with plastic or china mugs (Figure 19-B). But there has been endless argument about whether this really is better for the environment. Now the Dutch environment ministry has settled the question (perhaps) with 123 pages of charts, tables, and equations.

The report follows the life history of a china coffee mug from extraction and processing of its raw materials, through production and use, to its final resting place in a landfill. It takes account of the consumption of raw materials; landfill space; energy used for processing, transport, and cleaning; and of air and water pollution.

Reusable mugs have one main disadvantage: they have to be washed. "Washing a reusable mug once in a dishwasher," says the report, "has a greater impact on the water supply than the entire life cycle of a disposable cup." The main reason is the surfactants in detergents, the substances that cause grease to float off surfaces. Surfactants release various toxic gases into the air and are a major cause of water pollution. As a result, whether it is greener to drink from a reusable mug than from a disposable cup depends on two things: how many times the mug is used and how often it is washed with detergent.

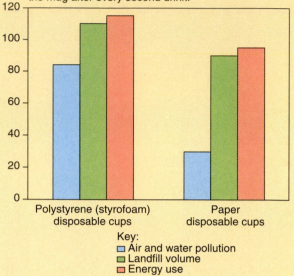

Number of times you must use each china or plastic mug to be a greener citizen than your neighbor who uses a styrofoam or paper cups, assuming you wash the mug after every second drink.

Key:
■ Air and water pollution
■ Landfill volume
■ Energy use

**Figure 19-C** The number of times you must reuse a mug to have less environmental impact than someone drinking the same amount from disposable cups. The graph assumes that you wash the mug with detergent after every second filling. *Source: Netherlands Environmental Ministry.*

The report concludes that if you wash it every time you use it, you need to use a mug 1,800 times before it has less environmental impact than a disposable cup. In theory, this gives the mug the edge because Dutch caterers report that they use each china mug 3,000 times. But if you refill the mug at least once between washings, you need to use it only 114 times before it does less damage. That would mean using your college mug, with a refill, about once a day for six months, which you are quite likely to do. Although you might not expect it, disposable polystyrene cups are greener than those made of paper (Figure 19-C).

Incidentally, if you remember the arguments of the 1980s over disposable diapers, you may be interested to know that washing and detergents are also the reason that reusable diapers are not necessarily better for the environment than disposable diapers that take up landfill space.

**Figure 19-B** Reusable mugs on campus. My daughter holds a mug produced by the Brown University Environmental Action Network and distributed by campus dining services. The mug is designed to replace styrofoam cups used in the dining halls.

*The Economist, August 1, 1992.

## SUMMARY

Water pollution from point and nonpoint sources affects surface and ground water. Pollution from nonpoint sources is particularly difficult to control. Water pollution in poor countries is a massive health problem. In developed countries, the quality of surface water has improved rapidly during the last 20 years, but ground water in many areas is badly polluted.

### Freshwater Ecosystems

In a body of fresh water such as a lake, emergent and floating plants produce oxygen near the surface. Detritus falling from the surface is the main energy source deep in the lake where light does not penetrate. Temperature changes in temperate lakes cause seasonal turnovers, which carry oxygen to the depths of the lake. Eutrophic lakes contain organic matter and nutrients but little oxygen because aerobic decomposers use it up rapidly. Oligotrophic lakes contain few organisms and more oxygen. If organic matter, heat, or nutrients reach the water continuously, the water becomes more eutrophic.

### Pollutants

The main types of pollutants are pathogens, biodegradable organic matter and nutrients, physical agents, and toxic chemicals. Biochemical oxygen demand reveals how much organic matter, including pathogens, a sample of water contains. Coliform bacteria indicate the presence of feces in water.

### Nutrient Pollution and Eutrophication

Water with a high nutrient content is not fit to drink. Nutrient pollution eventually eutrophies surface water, reducing its oxygen content, encouraging the growth of anaerobic decomposers, and killing oxygen-requiring organisms such as trout. Excess nutrients come mainly from sewage and fertilizer.

### Cleaning up Pollution

Under the Clean Water Act, the EPA is charged with making surface water fit for drinking and swimming. This involves preventing polluted water from reaching waterways and leaving them to cleanse themselves naturally by the action of decomposers. Waste water from industrial and agricultural sources is usually treated where it is produced. Many industries have redesigned manufacturing processes and reuse water many times to prevent polluted water from reaching waterways. Waste water from municipalities includes sewage and storm runoff, which may cause a sewage plant to overflow as well as washing toxic chemicals from streets and lawns into a waterway.

### Sewage Treatment

A sewage plant treats sewage with settling and decomposition, adding chlorine to the waste water effluent to kill pathogens. Sewage treatment produces sludge, which is usually polluted with toxic chemicals and is usually disposed of in a landfill. Ideally toxic chemicals should be kept out of sewage systems.

### Tertiary Water Treatment

Tertiary treatment usually involves removing more nutrients from effluent that remains after secondary treatment. It may also involve chemical techniques for removing toxic chemicals. A tertiary system produces unpolluted water. Artificial wetlands are increasingly used as tertiary treatment plants and may serve additional functions such as flood control and providing wildlife habitat. Water from wetlands purification may be used for irrigation or to recharge an aquifer.

### Rural Sewage Systems

Providing sewage plants for small communities is expensive, and thousands of rural areas have not yet met clean water standards under the Clean Water Act. Millions of rural homes dispose of their sewage by septic systems, which are often poorly monitored and cause considerable pollution.

### Physical Pollutants

Heat and silt are physical agents that degrade water quality. Heat hastens eutrophication. Silt, usually eroding soil from agricultural land, fills up reservoirs and blocks filters.

### Regulation and Cleanup Programs

New regulations in the 1970s and 1980s have provided sewage service to more houses, led to the construction of new and improved sewage plants, and increased waste water cleanup by industry. Ground water has been largely neglected and is badly polluted in much of the United States. Regulations to control groundwater pollution are beginning to be enforced. Regional cleanup programs, such as the one to clean up the Great Lakes, have shown that many pollutants reach water from the air, even after wastewater pollution has been greatly reduced. Pollution control has become big business in developed countries, creating economic benefits such as cost savings, millions of jobs, and new techniques.

**What Can We Do?**

Water pollution cannot be solved without individual action because many of the pollutants that reach waterways are washed off streets and yards. Individual actions that help to clean up our water supply include political agitation for better enforcement of antipollution regulations, decreasing water runoff from surfaces in the neighborhood, and disposing of hazardous materials properly.

## DISCUSSION QUESTIONS

1. Where does the water that runs off streets into the storm drains in your neighborhood go? How many substances can you think of that wash into the storm drains after a heavy rain?

2. Some hazardous waste is disposed of by injecting it into caverns and wells far underground. What are the possible disadvantages of this system?

3. Are there any lakes and rivers in your state that suffer from industrial pollution? What have the effects been, and is anything being done about it?

4. What is the source of drinking water in your area? Where does runoff into this water source come from? Is the runoff polluted?

# AIR POLLUTION AND NOISE

## Key Concepts

- Most air pollution in developed countries is produced by industry and motor vehicles.

- Air pollution causes destruction of the ozone layer and increases the greenhouse effect, which warms Earth.

- Destruction of the ozone layer was discovered in the 1980s, and the international community has acted rapidly to halt the process. By the end of the 1980s, emissions of ozone-destroying chemicals were falling fast.

- The number of species in an ecosystem often declines when it becomes polluted.

- The air inside a building is nearly always more polluted than the air outside.

- Progressive hearing loss of people in industrialized nations is the result of a lifetime's exposure to noise.

*Pollution from a smokestack at Kennecott Copper Company near Great Salt Lake, Utah.* (Paul Feeny)

ir pollution often seems like a personal or local matter: a smoke-filled room or the smell from a local factory. Today, however, we realize that the problem is global. From the heart of a crowded city to the top of Mount Everest, the air is polluted by thousands of substances. Air pollution threatens us with long-term changes caused by depletion of the ozone layer and global warming as well as causing ill health, water pollution, and damaged ecosystems. There is no practical way to clean up air before we breathe it. The only solution to air pollution is to prevent it.

Even preindustrial societies pollute their air, mainly by burning organic matter. Acid rain in the mountains of China, far from any industry, was traced to fires of yak dung used locally for cooking and heating. Until about 1950, air pollution in industrial countries was usually caused by burning coal. London's "pea-souper" fogs were a mixture of fog with the sulfur and particulate matter produced by burning coal. They caused thousands of deaths. Today most air pollution is caused by burning gasoline in vehicles and by industrial use of fossil fuels. As a result, developed countries cause nearly all of today's air pollution. As poorer countries develop, however, they add their share to this growing problem. As in the case of water pollution, our modern problem is the amount, as much as the type, of substances that huge modern populations, using enormous amounts of energy, add to the air.

Pollutants in the air have many effects. They injure human health in polluted cities and indoors where we live and work. When they dissolve in water, compounds of sulfur and nitrogen make rain acidic. Once we stop polluting the air, snowfall and rain will remove pollutants from the atmosphere. However, our problems are not over when rain and snow dissolve pollutants and shower them back on Earth. Polluted air drops nutrients, toxic metals, and acids into our water supply; kills fish and forests; and dissolves the stone of which buildings are made.

The main pollutant in an area may vary, depending on the local economy and on what industries lie upwind. The compounds that cause pollution may also change with time. Different human activities produce different pollution problems, and many of these must be solved individually. In this chapter, we discuss some of the most widespread problems as well as a few special cases.

## 20-A POLLUTANTS AND SOURCES

**Air pollution** can be defined, rather imprecisely, as the presence in air of substances that adversely affect its chemical composition. In industrialized countries, most air pollutants are produced by industry and vehicles. Air pollutants are usually divided into a few classes of substances, including oxides, volatile organic compounds, small particles and droplets, and secondary pollutants. Examples of the main groups of air pollutants are shown in Table 20-1.

**Primary air pollutants** are those formed elsewhere and discharged into the air. An example is soot, which is formed in fires. **Secondary pollutants** form in the air when primary pollutants react with each other. For instance, sulfur dioxide is a primary pollutant that forms when fossil fuel is burned. In the air, it may react with oxygen gas to form the secondary pollutant sulfur trioxide ($2 SO_2 + O_2 = 2 SO_3$). This in turn may react with water vapor in the air to form another secondary pollutant, sulfuric acid ($SO_3 + H_2O = H_2SO_4$), which is one of the substances that can make rain acidic.

Pollutants produced by natural processes only occasionally reach high enough concentrations to cause harm. Volcanic eruptions sometimes generate clouds of sulfur dioxide and particulate matter that

| TABLE 20-1   Major Types of Air Pollutants | |
|---|---|
| **Class of Pollutants** | **Examples** |
| Oxides | Carbon monoxide (CO), carbon dioxide ($CO_2$), sulfur dioxide ($SO_2$), sulfur trioxide ($SO_3$), nitric oxide (NO), nitrogen dioxide ($NO_2$), nitrous oxide ($N_2O$) |
| Volatile organic compounds | Hydrocarbons such as methane, ethylene, benzene, benzopyrene; other organic compounds such as formaldehyde, carbon tetrachloride, chloroform, methylene chloride, ethylene dichloride, trichloroethylene, vinyl chloride, ethylene oxide |
| Suspended particles and droplets | Dust, soot, lead, cadmium, asbestos, chromium, arsenic, nitrate salts, sulfate salts, sulfuric acid, nitric acid, oil, pesticides |
| Secondary pollutants | Ozone ($O_3$), acetaldehyde, hydrogen peroxide, hydroxyl radical (HO), sulfuric acid, PANs (peroxyacyl nitrates) |

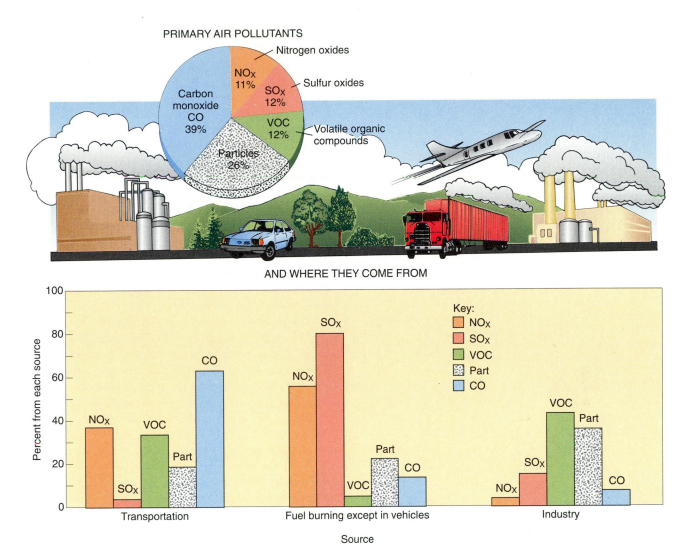

**Figure 20-1**  The primary air pollutants and their sources. In the bottom chart, note that different sources produce different pollutants. Transportation produces most carbon monoxide, other fuel burning produces most sulfur oxides, and industrial plants produce most of the volatile organic compounds.

are dense enough to suffocate animals. But nearly all the dangerous concentrations of pollutants in our air are there because we put them there.

Figure 20-1 shows the main primary pollutants and where they come from. For instance, vehicles produce mainly nitrous oxides and carbon monoxide. Burning fossil fuels in the home and in power stations produces mainly nitrous oxides and sulfur oxides, whereas industry produces mainly volatile organic compounds and particulate matter. You will notice that carbon dioxide ($CO_2$) is not included in the list of pollutants in Figure 20-1. This is because $CO_2$ is a natural component of the atmosphere and is not poisonous. However, most processes that produce air pollutants produce even larger volumes of $CO_2$, and we shall see that controlling the volume of

$CO_2$ emitted into the air is increasingly important as nations try to slow global warming (Section 20-B).

The Clean Air Act, revised in 1992, sets federal Ambient Air Quality Standards, listing the concentrations of various air pollutants that are acceptable in "clean" air. The standards permit areas and industries to decide when they should take measures to reduce air pollution. For instance, Table 20-2 shows the metropolitan areas in the United States that failed to meet the Air Quality Standard for carbon monoxide on one or more days in 1992. Since carbon monoxide is produced mainly by transportation (see Figure 20-1), cities such as Las Vegas and Los Angeles, which have little public transport, know that they have to reduce pollution from automobiles to clean up the air in the city.

**TABLE 20-2    Metropolitan Areas Failing to Meet U.S. Federal Ambient Air Quality Standards for Carbon Monoxide in 1992: Number of Days Exceeding Standards**

| Area | Days |
| --- | --- |
| Albuquerque, NM | 3 |
| Anchorage, AK | 12 |
| Baltimore, MD | 1 |
| Chico, CA | 1 |
| Denver-Boulder, CO | 3 |
| El Paso, TX | 4 |
| Lake Tahoe, CA | 5 |
| Las Vegas, NV | 17 |
| Los Angeles, CA | 47 |
| Memphis, TN | 1 |
| Minneapolis-St. Paul, MN | 1 |
| Modesto, CA | 2 |
| New York, NY | 4 |
| Ogden, UT | 3 |
| Phoenix, AZ | 4 |
| Portland, OR | 2 |
| Provo-Orem, UT | 11 |
| Raleigh-Durham, NC | 2 |
| Reno, NV | 7 |
| Sacramento, CA | 11 |
| San Francisco, CA | 2 |
| Seattle-Tacoma, WA | 2 |
| Spokane, WA | 6 |
| Stockton, CA | 2 |
| Syracuse, NY | 1 |

The main air pollutants are oxides, volatile organic compounds, small particles and droplets, and secondary pollutants. Most air pollution in developed countries is produced by industry and motor vehicles.

## Industrial Emissions

Industry is a major source of air pollution in China, Eastern Europe, and other areas that are industrialized but not wealthy enough to have invested in pollution control. Industrial pollutants sometimes form a gray haze, particularly on cold, wet winter mornings when there is little wind. The haze is mainly particulate matter, sulfur dioxide, and organic compounds.

Industries in North America and Western Europe have reduced their contributions to air pollution rapidly since about 1970 and now produce less than 20% of air pollution. Pockets of industrial air pollution remain, however. One of these is the lower Missis-sippi River valley. This "petrochemical corridor," produces one fifth of the nation's petrochemicals and stretches for 130 kilometers north of New Orleans. Hundreds of tons of toxic materials leak into the ground, are pumped into the river, and spew from smokestacks. The mixture makes the area, in the words of one health specialist, "a massive human experiment." In the 1980s, the residents of Revelletown, Louisiana, started worrying when they woke up gasping for air. Tests showed that their blood contained vinyl chloride—a powerful carcinogen and the main product of the neighboring chemical plant. Plant officials would not listen to their complaints, so the residents filed suit. Faced with the prospect of continuing to live in the polluted air through years of litigation, most of them settled out of court. Revelletown is now deserted, but dozens of neighboring communities are just as dangerous to live in.

Why is the petrochemical corridor located where it is? First, the industrial plants are near oil-producing areas and ports where imported oil is unloaded. Second, the Mississippi River provides the water that industry needs, as well as cheap transport. Third, the plants are located mainly in economically depressed communities, where pollution control is less stringent and less well enforced than in more prosperous areas. These are the conditions that produce centers of pollution anywhere in the world.

The pollutants emitted by industry can be drastically reduced by pollution control techniques. One common method is to change a manufacturing process so it produces fewer pollutants, or the air emitted by a plant may be treated to clean it before it is released. We have mentioned the use of electrostatic precipitators to remove fly ash from industrial emissions (Section 3-B). Another common device is a scrubber, which passes gases through a spray of water that dissolves many pollutants (Figure 20-2). As with so many environmental problems, the absence of pollution control devices is more a political and an economic problem than a scientific one. In one common scenario, industries do not want to spend the money to install the latest pollution control devices or even the controls required by law. When the local community demands steps to clean up the filthy air, the industry threatens that the cost will force it to close down, with a consequent loss of local jobs.

## Motor Vehicle Emissions

Automobile use is a convenience to the individual, but society pays a high price—especially in the United States, where there are more motor vehicles per person than anywhere else in the world.

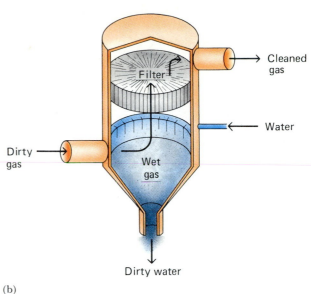

(a)

(b)

**Figure 20-2**  Pollution control. (a) The structures on the outside of this industrial plant are scrubbers that remove sulfur compounds from the gas discharged by the plant. (b) A wet scrubber. *(Science Visuals Unlimited)*

Cities where motor vehicles are the main source of pollution tend to suffer from yellowish **photochemical smog**. Sunlight plays an important part in forming this type of air pollution (Figure 20-3). Cars and trucks give off nitric oxide, which reacts with oxygen to form nitrogen dioxide. Nitrogen dioxide is a yellow-brown gas with a strong, choking smell that produces a colored haze. Ultraviolet rays from the sun release a reactive atom of oxygen from the nitrogen dioxide. Some of these oxygens react with oxygen gas in the air to produce poisonous ozone and highly reactive chemicals such as hydrogen peroxide. (Although ozone in the ozone layer serves a useful function, ozone is nevertheless poisonous to organisms.) In addition, organic compounds, such as industrial solvents and spilled or unburned gasoline, are added to the air as the temperature rises during the day.

The cities that suffer most from photochemical smog have dense traffic and are situated in dry, sunny areas. In the hot, dry summer of 1990, 55 American cities had ozone levels higher than those permitted by federal air pollution standards, although the number was lower in the damper summers of later years. Los Angeles, Denver, and Mexico City are well-known examples of smoggy cities, but there are many others. Cities such as Tokyo and Chicago that suffer from industrial smog in the winter often suffer from photochemical smog in the summer, and the two types of smog sometimes occur together.

Cities with heavy traffic and dry, sunny weather suffer from photochemical smog produced by motor vehicles.

**Emissions control devices** are gadgets that reduce the pollution from motor vehicles. However, the most effective control mechanism is high gasoline prices because when gasoline is expensive, people drive less and buy cars with small engines that produce less pollution. They also use public transportation more. The high price of gasoline in Western Europe (as a result of gasoline taxes) is the main reason this area suffers less from automobile pollution than does the United States. Figure 20-4 shows that carbon monoxide, which comes mainly from motor vehicles, is more concentrated in several American cities than in European cities such as Glasgow and Athens, which are notoriously crowded with traffic but have a lower density of cars, most with smaller engines, than cities in the United States.

An effective way to reduce air pollution from automobile emissions is to increase the price of gasoline.

## Wood Smoke Pollution

As the cost of oil and electricity rose during the 1970s, many people in developed countries saved

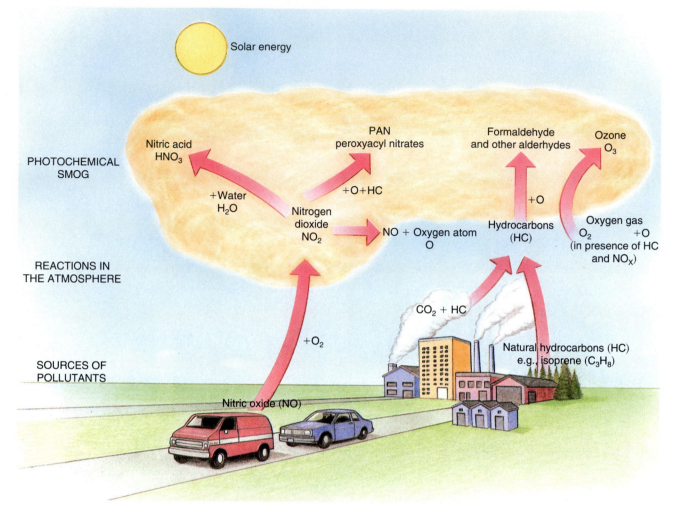

**Figure 20-3**    The formation of photochemical smog. Nitrogen dioxide is a brown gas that gives smog its characteristic color and odor. Most of the reactions shown here take place only in the presence of sunlight. Note that even hydrocarbons given off by trees contribute to photochemical smog. (You can smell some of these compounds if you sniff a pine tree on a hot day.)

money by installing stoves to burn coal or wood for heat. Many of these stoves were poorly designed and belched clouds of pollutants into the air. As a result, in the winter of 1981, the air in parts of rural Vermont and Montana was more polluted than in urban areas such as Los Angeles and Pittsburgh! Rural towns started broadcasting smog alerts just like those issued for large polluted cities. In the United States, it is now illegal to use a stove that does not conform to emission standards, and most heating stoves are fitted with pollution control devices (Section 10-E). However, there are no pollution control standards in most parts of the United States for open fireplaces that burn wood or coal and that pollute many suburbs during the winter months. Many parts of the United States have no rules forbidding the burning of yard waste. In the suburb where I live, the air is

thick with smoke from bonfires of twigs and leaves for several months every fall.

> The air in rural and suburban areas may be as polluted from wood smoke or bonfires as the air of large cities is from automobiles and industry.

## 20-B  OZONE AND GLOBAL WARMING

In Chapter 3, we saw that Earth's atmosphere contains a layer of ozone that reduces the amount of ultraviolet radiation reaching Earth from the sun (Figure 3-12). Ultraviolet radiation damages DNA so it can cause mutation or death of cells near the surface of the body. It is a major cause of skin cancer in

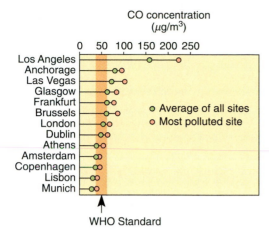

**Figure 20-4** Concentrations of carbon monoxide (CO) in various cities in 1992. CO comes mainly from automobile exhausts. The World Health Organization (WHO) standard for CO concentrations too low to damage health is shown in orange. *(Data from WHO)*

humans. Single-celled organisms, such as phytoplankton in the ocean, may be killed outright by ultraviolet radiation (Figure 20-5). In Chapter 3, we also saw that the only reason Earth is warm enough for life is that it is surrounded by a "greenhouse" of gases in the atmosphere that traps heat near Earth and prevents it from being lost into space. In recent decades, we have learned that air pollution is destroy-ing the ozone layer and adding to greenhouse gases, actions that have worldwide effects.

## Ozone Destruction

During the early 1980s, researchers discovered a hole in the ozone layer over Antarctica. Nowadays information on the ozone layer comes from weather balloons and NASA's Upper Atmosphere Research Satellite. Samples of air in various parts of the atmosphere are also collected by the high-flying ER-2, which is a converted U-2 spy plane.

Damage to the ozone layer comes mainly from **chlorofluorocarbons (CFCs)**, pollutants found everywhere in the atmosphere. When CFCs land on particles of ice, sulfuric acid, or other substances in the atmosphere, their chlorine atoms are converted to chlorine monoxide. When sunlight shines on ozone and chlorine monoxide, the ozone is destroyed. Holes in the ozone layer form wherever the right kinds of particles are present in large numbers, and the holes form fastest in sunny weather. In 1992, researchers found that particles from the explosion of Mount Pinatubo in the Philippines had accelerated the formation of chlorine monoxide. The highest concentrations of chlorine monoxide ever recorded were discovered over eastern Canada and northern New England.

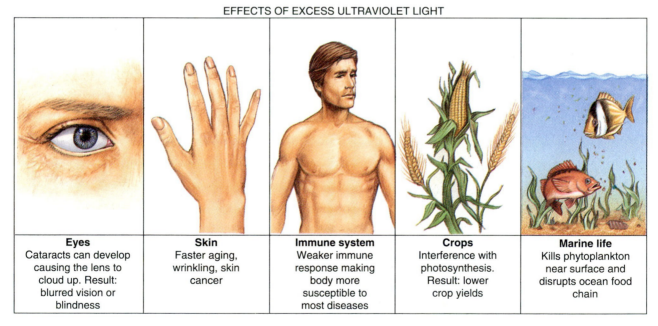

EFFECTS OF EXCESS ULTRAVIOLET LIGHT

| **Eyes** | **Skin** | **Immune system** | **Crops** | **Marine life** |
|---|---|---|---|---|
| Cataracts can develop causing the lens to cloud up. Result: blurred vision or blindness | Faster aging, wrinkling, skin cancer | Weaker immune response making body more susceptible to most diseases | Interference with photosynthesis. Result: lower crop yields | Kills phytoplankton near surface and disrupts ocean food chain |

**Figure 20-5** The main adverse effects of increasing amounts of ultraviolet light reaching the ground because of depletion of the ozone layer. Note that even sunscreens do not block out all the rays that cause the most dangerous form of skin cancer (melanoma). The best protection is to wear zinc ointment on an exposed nose, a hat, and long-sleeved clothes.

CFCs are widely used for producing foam rubber, for cleaning electrical components, and for refrigeration. They are also used in the production of aerosol sprays. CFCs are ideal aerosol propellants because they do not react chemically with whatever is being sprayed. But this same lack of chemical reactivity means that they are not broken down in the air but make their way to the stratosphere. The use of CFCs in aerosols has been banned in the United States since the early 1980s, but other countries still produce aerosols containing CFCs.

## International Ozone Agreements

In the mid-1980s, governments realized they had to act quickly to stop further ozone destruction. In 1987, after a year of negotiations, many countries endorsed the Montreal Protocol, a treaty to halve the use of CFCs by 1998. By 1992, however, scientists reported new holes in the ozone layer that were growing much more rapidly than had been predicted. A 1992 conference in Copenhagen was attended by 93 countries and made three major decisions:

1. Rich countries agreed to eliminate CFCs completely by 1995. Under the Clean Air Act, the United States is committed to banning all substances that pose significant danger to the ozone layer by 2000.

2. Developed countries agreed to set up a fund to help poor countries replace CFCs.

3. Two new substances were added to the list of ozone-destroying chemicals: methyl bromide and hydrofluorocarbons (HCFCs).

**Methyl Bromide** Methyl bromide is a gas widely used as a pesticide (see Close-Up 21-1). It also turns out to be 40 times as effective as CFC at destroying ozone. Bromine and chlorine are chemically related, so it is not surprising that bromine destroys ozone as chlorine does. Susan Solomon of the National Oceanic and Atmospheric Administration estimates that methyl bromide pesticide is responsible for 10% of the annual depletion of the ozone layer over the Northern Hemisphere.

**Hydrochlorofluorocarbons** HCFCs were developed to replace CFCs as coolants in refrigerators and air conditioners. Now we find that they also destroy ozone at an unacceptable rate. The U.S. Clean Air Act of 1990 required people repairing air conditioners to stop releasing HCFCs into the air in 1992. Instead HCFCs from air conditioners must now be collected and detoxified.

Chemical companies have developed substances that can be used in place of CFCs and HCFCs for many purposes, but we are still faced with many older appliances, particularly air conditioners, whose coolants need to be replaced. About a dozen molecules have been identified as likely substitutes for CFCs. Some of these are old-fashioned substances such as ammonia, butane, and propane, which were used as coolants before CFCs were invented. All these are either poisonous or capable of causing pollution, so chemical companies are working to develop other substitutes.

Unlike most environmental problems, which have been understood at least since the 1960s, ozone depletion was not even discovered until the 1980s. It appears that we shall have to solve the problem within 20 years of its discovery if we are to prevent enormous environmental damage. The ozone story is an example of how rapidly countries can act when they are convinced that it is necessary. In 1993, the concentration of CFCs in the atmosphere was still increasing, but much more slowly than in the 1980s. CFC emissions are now down by more than 30% from their peak in 1988, meaning that industry is actually ahead of the Copenhagen schedule for phasing out ozone-destroying chemicals.

> Governments and industry have acted with unprecedented speed to halt the release of chemicals that destroy the ozone layer.

## Global Warming

Imagine that by 2200 our summers are 10 degrees hotter than they are today. Imagine that the homes of millions of people on the East Coast of the United States are under water. Imagine four months without rain every summer in the farm belt. These projections come from the finding that human activities are heating up Earth's atmosphere, with consequences that may prove catastrophic unless we can slow the process to a pace that will give us time to adjust.

During the 1980s, the American media sometimes gave the impression that global warming might be a myth. This is simply not true. The vast majority of scientists have long been convinced that human activities are causing global warming, and new evidence confirms that conclusion every year.

During the last hundred years, the temperature on Earth has risen. This does not mean that this summer will necessarily be warmer than last summer. What it means is that overall, on average, the temperature has increased slightly, as shown in Figure

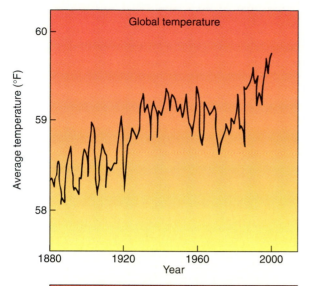

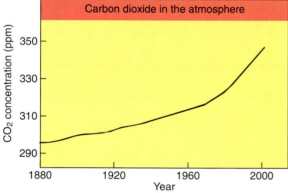

**Figure 20-6** Global warming: the increases in average temperature worldwide and average concentrations of carbon dioxide in the atmosphere since 1880. *(Source: National Oceanic and Atmospheric Administration)*

20-6. The temperature on Earth has changed enormously in the past. Thousands of years ago, the Midwest was covered with a sheet of ice even in midsummer during the last of the Ice Ages. In the Middle Ages, Earth passed through a ''mini–ice age'' when the temperature was lower than usual. Dozens of factors affect the temperature on Earth and cause it to change from time to time. For this reason, it is difficult to predict precisely how rapidly the world is warming up. For instance, aerosols in the troposphere deflect sunlight and tend to keep Earth cooler than it otherwise would be. **Aerosols** are droplets and particles so small that they hang suspended in the air. Over industrial areas, aerosols of smoke and sulfur compounds are common. Volcanic eruptions and burning biomass also produce aerosols that cool Earth.

Table 20-3 lists the main air pollutants that add to the greenhouse gases every year and where they come from. It also lists their relative importance by comparing them with $CO_2$. You will notice that CFCs contribute to global warming as well as destroying ozone. CFCs are powerful greenhouse gases. The table shows that much less CFCs than methane reach the atmosphere every year, yet CFCs contribute the equivalent of more than 1,500 millions of tons of $CO_2$, while methane contributes little more than 1,000.

### Methane

After remaining stable for thousands of years, the concentration of methane ($CH_4$) in the troposphere has doubled in the past 250 years. Most of the methane is produced by methanogens, bacteria that produce methane as a by-product of the reaction they use to obtain energy, which can be represented:

$$CO_2 \quad + \quad 4\,H_2 \quad \rightarrow \quad CH_4 \quad + 2\,H_2O$$
carbon dioxide    hydrogen    methane    water

Methanogens can live only in environments that contain no oxygen. They live in a variety of anaerobic (oxygen-free) habitats, such as sewage-treatment ponds, garbage dumps, the digestive tracts of livestock, and the bottoms of rice paddies and wetlands.

It seems strange that the amount of methane in the air is increasing when the area of wetlands in the world is decreasing. New sewage-treatment plants alone cannot account for the increase. The rise seems to be due to the expansion of agriculture. Methane production from the world's herds of ruminants (mammals such as cattle, sheep, goats, and water buffalo) is believed to have doubled since 1890. Methane production from rice paddies has also doubled since about 1950 as the area devoted to growing rice has increased.

Methane is removed from the atmosphere in several ways. One of these is useful: methane can react with a hydroxyl radical (OH) in the troposphere to produce ozone. It also leaves the air when it is broken down by various soil bacteria. Overall, however, the methane content of the air is growing by more than 1% each year (faster than the atmosphere's $CO_2$ content is growing). Although the methane content of the atmosphere is still small, methane makes a relatively large contribution to the greenhouse effect because each molecule of methane absorbs infrared radiation 20 times more effectively than does a molecule of $CO_2$.

**TABLE 20-3**   **Estimate of Current Greenhouse Emissions from Human Activity***

| Source | Annual Emissions (in millions of tons per year) | | CO₂-Equivalent† (in millions of tons per year) | |
|---|---|---|---|---|
| | World | United States | World | United States |
| **Carbon dioxide (CO₂)** | | | | |
| Commercial energy, etc. | 19,400 | | 19,400 | |
| Tropical deforestation | 2,600 | | 2,600 | |
| *Total CO₂* | *22,000* | *4,800* | *22,000* | *4,800* |
| **Methane (CH₄)** | | | | |
| Rice paddies | 120 | | 2,300 | |
| Livestock | 80 | | 1,500 | |
| Fuel production | 70 | | 1,300 | |
| Landfills | 30 | | 600 | |
| Tropical deforestation | 20 | | 20 | |
| *Total CH₄* | *320* | *50* | *6,700* | *1,050* |
| **CFCs** | | | | |
| *Total CFCs* | *0.6* | *0.3* | *3,200* | *1,640* |
| **Nitrous oxide (N₂O)** | | | | |
| Agriculture (fertilizer, waste, plowing) | 2.3 | | 680 | |
| Coal burning | 1.0 | | 290 | |
| Tropical deforestation | 0.5 | | 150 | |
| Biomass (fuel and industrial) | 0.2 | | 0.2 | |
| *Total N₂O* | *4.0* | *1.4* | *1,180* | *410* |
| **Overall Total** | | | **32,880** | **7,900** |

*Data from R. S. Rubin, et al., "Realistic Mitigation Options for Global Warming," *Science*, 257:148. 1992.

†CO₂-Equivalent is the amount of CO₂ it would take to produce the same amount of global warming.

## Carbon Dioxide

The biggest villain of Earth's increasing temperature is CO₂. The amount of CO₂ humans release into the air is usually reported as "carbon emissions," and each of us makes our own contribution (Lifestyle 20-1). Thanks to population increases and our use of fossil fuel, the amount of CO₂ in the atmosphere is increasing by about 0.5% each year (Figure 20-7).

CO₂ leaves the atmosphere mainly when it dissolves in the oceans or is used up by plants during photosynthesis. Reactions that release CO₂ into the air include the respiration of organisms (we release CO₂ when we breathe) and the burning of organic molecules. Forest fires, grass fires, and burning fossil fuels all produce CO₂. While our burning of organic

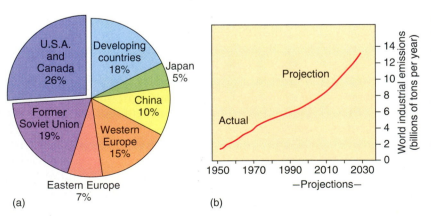

(a)
(b)

**Figure 20-7** Carbon dioxide emissions. (a) North America produces a higher proportion of all CO₂ emissions than any other continent. (b) The increase in worldwide CO₂ emissons since 1950, with projections to 2030. Present emissions are expected to double in the next 30 years, which means that global warming will accelerate.

**TABLE 20-4** Carbon Dioxide Emissions from Burning Fossil Fuels

| Source | Total $CO_2$ Emissions (Millions of Tons per Year) | $CO_2$ per Capita (Tons) | $CO_2$ per Dollar of GNP |
|---|---|---|---|
| United States | 4,800 | 18.37 | 1,010 |
| Russia | 3,711 | 13.07 | 1,563 |
| Western Europe | 2,899 | 7.61 | 651 |
| China | 2,031 | 1.9 | 6,925 |
| Japan | 908 | 7.43 | 564 |
| India | 549 | 0.7 | 2,386 |
| Canada | 388 | 14.93 | 875 |
| **World** | **22,000** | **3.88** | **1,138** |

matter releases $CO_2$ into the air faster than at any other time in history, human activities have also slowed the rate at which $CO_2$ is removed from the air by photosynthesis. Much of the vegetation that absorbs $CO_2$ is found in the forest in wet parts of the tropics. Most of the world's tropical forests have already been destroyed. To make matters worse, when tropical forest is destroyed to make way for agriculture, the trees are usually burned to get them out of the way. This releases large amounts of $CO_2$ into the air. However, burning fossil fuel releases much more $CO_2$ into the air each year than does burning tropical forest.

We have used fossil fuel as our main energy source only since the industrial revolution, about 200 years ago. How have these 200 years affected the $CO_2$ concentration in the air? It is hard to give a precise figure because we have no records of the atmosphere's $CO_2$ content before 1958. Swiss workers decided to look for some old air to analyze. They found air 1,000 years old, trapped in bubbles in the ice of Antarctica and Greenland. With this and other information, scientists today agree that atmospheric $CO_2$ has increased by about 25% since 1860.

Table 20-4 shows the production of $CO_2$ from burning fossil fuels by various countries. Note that the United States is the biggest single producer of carbon emissions. The United States and Canada also use more fossil fuel per person than any other country, mainly because we are so dependent on automobiles for transport. China is typical of countries with rapidly growing but fairly inefficient industry. It uses large amounts of fossil fuel to produce a dollar's worth of gross national product (GNP) but uses little fossil fuel per person, mainly because it contains few automobiles and many families use biomass for cooking and heating. Western Europe and Japan are by far the most efficient users of fossil fuel, thanks to energy-efficient industries and widespread mass transit systems. Table 20-5 shows that the United States could save money and reduce carbon emissions by increasing fuel efficiency and by simple steps to re-

**TABLE 20-5** Reducing Emissions and Saving Money by More Efficient Energy Use*

| Option (Use of Most Efficient Technology in This Area) | $CO_2$ Reduction (Millions of Tons per Year) | Savings from Use ($ per Ton $CO_2$) |
|---|---|---|
| **More Efficient Electricity Use in Home** | | |
| White roofs and trees | 32 | 84 |
| Lighting | 39 | 79 |
| Water heating | 27 | 74 |
| Appliances | 72 | 44 |
| Space heating | 74 | 39 |
| **More Efficient Fuel Use in Transportation** | | |
| Cars and light trucks | 251 | 40 |
| Heavy trucks | 39 | 59 |

*This table shows how much less $CO_2$ would be added to the atmosphere each year and how much money could be saved if everyone in the United States adopted the best technology available in 1993 in these areas.

duce air conditioning bills, such as planting trees and installing light-colored roofs that reflect summer heat.

### Effects of Global Warming

Global warming produces many effects beyond the need to use more sunscreen and winters with less snow. Biologists are most alarmed by the effect of warming on agriculture and wildlife. Rapid climate change probably caused the extinction of the dinosaurs and other mass extinctions of species. Rapid change is always the enemy of living things. Plants and animals can evolve adaptations to new conditions only if they have enough time to do so.

Most of the world's food is grown in a band of land with a temperate climate that stretches across North America, Europe, and Asia. South of this band, summer temperatures are too high to grow important food crops such as wheat. As temperatures increase, the area where these crops grow best will move northward. More land in New England and Canada will have a climate suitable for intensive agriculture, and we shall see more and more years in which crops in the southern part of the American farm belt are devastated by heat and drought. These changes will cause severe social and financial problems if we do not plan for them.

One dramatic effect of global warming will probably be a partial melting of the world's great ice caps, which will cause the sea level to rise. In 1993, the National Research Council reported that sea levels have been rising worldwide for at least the last century. The level has risen about 30 centimeters along the Atlantic Coast of the United States. The sea level is rising partly because coastal land is subsiding and partly from heat, which not only melts ice, but also causes sea water to expand. As an example of what this might mean, a rise of 5 meters in the sea level would affect 40% of Florida's population, putting Miami, St. Petersburg, and Jacksonville under water. We should be planning now to permit several major cities to expand only away from the sea.

Another effect of rising sea level will be the flooding and disappearance of coastal wetlands, such as marshes and mangrove swamps. Without them, we should lose much of our seafood, many species of plants and animals, and much of our ability to protect the ocean from pollution.

### Buying Time to Prepare?

We cannot prevent global warming. Even if we replaced every tree that has been cut down in the last century and stopped using fossil fuels, the tempera-

ture would continue to rise until all the trees were full grown. We are going to have to learn to live with warmer weather. What policy makers hope is that we can slow the rate of change to provide time to adapt. For instance, more people will starve unless we can develop varieties of wheat, maize, and other crops that can survive longer, drier growing seasons. Whole cities such as New Orleans and Jacksonville may have to be moved inland. The areas between the new cities and the coastline should be set aside as preserves where new wetlands can form. All these and many other changes are perfectly possible, but they take planning and they take time. The way to buy that time is a two-pronged attack: reduction in the use of fossil fuel and reforestation.

The European Community is urging industrialized countries to join it in taking tougher measures to tackle global warming than those laid down in the 1992 United Nations treaty signed at the Rio conference. The effort seems likely to embarrass the United States, the biggest producer of greenhouse gases.

In 1988, the American Forestry Association unveiled a program it calls "Global Releaf" to encourage the planting of 100 million new trees. These trees would remove about 18 million tons of $CO_2$ from the atmosphere every year of the 6 billion tons of $CO_2$ released each year from fossil fuels. Those planted in urban areas would also shade buildings, leading to savings on air conditioning estimated at $4 billion a year. Proposed legislation would deny American loans to countries that do not include reforestation in their plans for development and would monitor the extent of tropical forests.

The move to counteract the greenhouse effect by planting trees has caught the imagination of the public. A spokesman for the Environmental Defense Fund suggests that large producers of $CO_2$ should be required to offset their emissions of $CO_2$ by planting trees. One company with plans to do precisely that is Applied Energy Services of Arlington, Virginia. It has contracted with the World Resources Institute to counter $CO_2$ emissions from one of its coal-fired power plants in Connecticut with a reforestation project in Guatemala.

## 20-C  REDUCING AIR POLLUTION

In spring, you may read in a newspaper that a local industry reduced its toxic emissions last year from 35,000 tons of sulfur to 30,000. This annual report is the result of the Environmental Protection Agency's (EPA) Toxics Release Inventory. This is a system by which manufacturers report each year the quantities of particular toxins they release into the air, water,

## FACTS AND FIGURES 20-1

### Russian Pollution

The biggest industrial polluters near my home are paper mills and chemical factories. Last year they spewed 75,000 tons of toxic emissions into the air and water. That may sound like a lot, but it is peanuts by international standards. São Paolo in Brazil subjected its inhabitants to 350,000 tons of pollutants that year. The figures beginning to emerge from the former Soviet Union and East Germany, however, make everyone else look like amateur polluters. In 1993, students gathering the onion harvest in Sverdlovsk, Russia, began to feel dizzy and some of them collapsed. One of the students had worked in a chemical warfare unit of the army and recognized the symptoms of gas poisoning. So much herbicide and pesticide had been sprayed on the onions that the students were being poisoned by the fumes. The residents of Norilsk in Russia were exposed to 2.4 million tons of air pollutants in 1992. Their rates of blood diseases, bronchitis, and endocrine (hormone) disorders are three times the Russian average. Mental retardation among children is more than twice that in cleaner areas nearby. It is going to take enormous amounts of money and international assistance to clean up air pollution in the former Soviet-bloc countries.

(a)

(b)

**Figure 20-8** Fighting smog in California. (a) A service station that gives instant checks of exhaust emissions to see if the vehicle complies with emission control regulations. (b) A gasoline nozzle with a rubber concertina that reduces the amount of pollutants from gasoline that evaporate into the air when a gas tank is filled. *(a, b Richard Feeny)*

and soil. Unfortunately, the system still does not cover mining operations. In 1987, Kennecott Copper, operator of Utah's huge Bingham Canyon copper mine, reported toxic releases to the EPA by mistake. Out of 18,000 places reporting, the mine was in first place for toxic metals emitted and in fourth place for total toxic releases. The company realized its mistake and stopped reporting the following year, but the incident spurred attempts to include the mining industry in the inventory in the future. Compared with industries in the Soviet-bloc countries, industries and mining operations in North America and Western Europe produce few toxic emissions, although they are the world's largest producers of $CO_2$ (Facts and Figures 20-1).

### The Los Angeles Plan

Every summer, warm still air settles over the Los Angeles valley, and the exhaust smoke from millions of vehicles combines with other air pollutants to produce murky brown air that hides the distant mountains. People with lung or heart disorders are told to stay indoors, athletes exercising outdoors suffer chest pains, and teachers are advised to keep schoolchildren indoors during recess. Until the weather changes and blows away the smog, the soup of pol-

lutants in the air becomes thicker and thicker every day. California has the strictest emission standards for automobiles in the world (Figure 20-8).

In 1989, Los Angeles decided that it would have to take stricter measures than merely controlling automobile emissions if it wanted to beat the smog. Emissions controls were proposed for just about every source of pollution, including oil refineries, utility companies, lawn mowers, barbecue starter fluid, dry cleaners, and even underarm deodorants. The plan is designed to cut air pollution by 80%, enabling Los Angeles to meet federal air quality standards for the first time.

The first step is to use energy sources that produce less air pollution. More automobiles sold in the area will have to run on electricity or other clean fuels.

The first stage of a mass transit system of underground electric trains has opened, in an attempt to reduce the use of automobiles by commuters. Household products that produce air pollution, including paints, solvents, and hair sprays, will have to be manufactured so that they do not give off pollutants or they will be banned. Restaurants will have to install more efficient filters on charcoal broilers, and bakers and dry cleaners will have to refit or replace equipment to reduce emissions. Some of these changes will involve the use of chemicals and technology that are not yet available. By publishing the plan in the 1980s for completion in the 2000s, the city aims to give chemical companies and equipment manufacturers time for the necessary research and development.

Industrial leaders estimate that the first part of the plan will cost about $3 billion. Some of them are lobbying to dilute or cancel some of the plan's requirements. Others, like General Motors representative Ronald Sykes, think that the plan will induce research and development into technology that will be needed in many parts of the world in the twenty-first century. Those who develop it first will have the chance to sell it to many others in the years to come.

## Variable Fuel Vehicles

In 1992, rental car companies operating in California began to offer their customers variable fuel vehicles (VFVs) that will run either on ordinary gasoline or on M85, a fuel containing 85% methanol (methyl alcohol) and 15% gasoline. A car using M85 produces about 40% fewer smog-forming emissions than a gasoline-only car, and the price and performance are similar. Auto companies do not make many VFVs at the moment, but they plan to increase their output if the rental car experiment shows that the public accepts these new vehicles.

## Electric Vehicles

Rechargeable batteries are used all over the world as cheap power for vehicles that move slowly and stop frequently, such as golf carts, delivery trucks, and the little trucks used by meter readers. Electric vehicles (EVs) do not belch pollutants out of tailpipes, but they are not truly pollution free. Their batteries must be discarded and cause a solid waste problem (although the lead they contain may be recycled), and producing electricity causes the emission of pollutants by the power station. Nevertheless, electric cars are 97% less polluting than gasoline-powered cars, according to the California Air Resources Board.

Most people think that electric cars will be part of our clean air future. Modern electric cars go much faster than older models and travel up to about 100 miles without recharging their batteries, but they are still slow and still need to be recharged too frequently. Nevertheless, Los Angeles expects to have more than a million electric cars in the city by 2010. Southern California Edison says that it can supply electricity for these cars without the need for new power plants. It is already ensuring that once the cars arrive there will be electric outlets where they can be

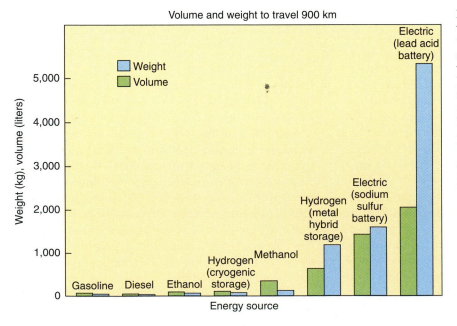

**Figure 20-9**   The volume and weight needed for a VW Golf to travel 900 kilometers on various sources of energy. Note the huge weight of standard lead acid batteries compared with the necessary volume of gasoline. Sodium sulfur batteries are much lighter but are still largely experimental. Two forms of hydrogen are listed: one is generated by metal electrolytes as it is used; the other (cryogenic) is hydrogen gas stored under pressure at low temperature. *(Source: Volkswagen)*

recharged in household garages, parking lots, shopping centers, and airports. An electric car uses about $650 worth of electricity a year, about the same as gasoline for a small car. But utilities expect to offer half-price discounts for nighttime recharging since utilities have surplus electricity at night.

Electric cars have several major problems. For instance, their batteries are very heavy (Figure 20-9). The 400 kilograms of lead-acid batteries in the handsome General Motors electric car called the Impact store less energy than 6 liters of gasoline. The batteries are expensive because they need to be replaced every two years or so at a cost of about $2,000. Electric cars also have very low net energy efficiencies. Most experts think that the best alternative vehicle is not an electric car but a hybrid that uses battery power for slow movement in town and an engine powered by gasoline or diesel on the highway. Highway travel would also recharge the batteries as it does in a standard gasoline-powered vehicle.

## 20-D  EFFECTS OF POLLUTION ON ECOSYSTEMS

Air pollution affects the health of an enormous number of organisms, many of them of economic importance. However, some organisms turn out to be surprisingly resistant to pollution and even capable of absorbing and detoxifying large quantities of acid rain or toxic metals. As air pollution worsens, research into these vital aspects of ecosystems increases.

### Acid Rain

Pure water has a pH of 7. Acid rain has a pH of less than 5, showing that it contains high concentrations of acid substances. The most important of these are sulfur compounds, which form sulfuric acid when they dissolve in water (Figure 20-10). Acid rain corrodes various substances and dissolves calcium carbonate, which makes up a large part of cement, concrete, and the limestone from which many buildings are made (Figure 20-11). It also kills fish and other aquatic organisms when it lowers the pH of water. Acid rain also affects minerals in the soil, for instance by releasing aluminum ions, which encourages the growth of some species and kills others.

One tree species particularly susceptible to acid rain is red spruce. Researchers at Dartmouth College found that acid rain makes the trees more likely to be damaged by cold weather. In winter, needles of the damaged red spruce trees turn brown and fall off. Eventually the tree usually dies.

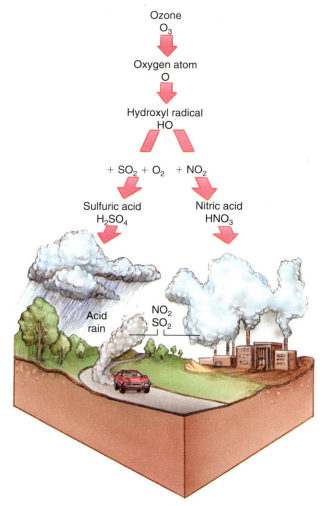

**Figure 20-10**   Acid rain: the formation of nitric acid and sulfuric acid, which will dissolve in water and fall as acid rain. The important compound is the hydroxyl radical, formed when a molecule of ozone breaks apart, releasing an oxygen atom that reacts with water. The hydroxyl radical is very reactive and starts the conversion of sulfur dioxide and nitrogen dioxide into sulfuric acid and nitric acid.

### Lichens as Monitors of Pollution

Prevailing winds in North America are from the west and most U.S. air pollution—including 90% of the sulfur dioxide—is emitted east of the Mississippi. As a result, people assumed that acid rain would spare the western part of the United States. Now, however, there has been growing evidence of acidification in western national forests, parks, and wilderness areas. The evidence comes from lichens that carpet rocks and tree trunks in many areas (Figure 20-12).

A lichen is not a single organism. Each individual consists of both an alga and a fungus. The alga contains chlorophyll and provides the lichen with car-

**Figure 20-11**  Acid rain: limestone carvings on a church in Paris partly dissolved by decades of acid rain. *(Thom Smith)*

bohydrates by photosynthesis, and the fungus absorbs water and nutrients from the surroundings. As a result, lichens are very self-sufficient and can grow on rocks, roofs, tree trunks, and similar surfaces. Because they have no roots, lichens absorb their water and nutrients from the air around them instead of from the soil as plants do.

In the 1950s, European biologists discovered that the presence or absence of particular species of lichen indicated whether or not the air contained pollutants from burning coal. Since then, dozens of lichens have been identified as indicating particular pollutants. Some lichens die at low levels of air pollution, so they provide early warning that air quality is declining. In others, nutrients from the air are metabolized, while pollutants are bound to cell walls in concentrations that mimic concentrations in the atmosphere. Analyzing heavy metals in the lichen indicates the type and concentration of heavy metals in the air. This is more accurate than analyzing samples of the air because air quality changes from day to day depending on factors such as the wind direction.

> Because they absorb their nutrients from the air around them, lichens are excellent biological monitors of air pollution.

### Nutrient Enrichment

Some of the most expensive damage from air pollution is to vegetation—on land and in the water. It is becoming obvious that polluted rain can have many effects.

Pollutants may block the pores through which plants exchange gases with the air, thereby slowing photosynthesis and plant growth. In contrast, acid rain may supply plants with extra nutrients, particularly nitrogen. Occasionally this is a good thing. Crops may be more productive in areas of acid rain. Extra nutrients are almost invariably bad for aquatic ecosystems because they increase plant growth, speeding up eutrophication. The effect of pollution on plants that live for many years (such as trees) is usually detrimental. Forests affected by air pollution may show no signs of damage for years, and then many trees die in one year.

If a tree dies, animals and plants that depend on it may also die. An ecosystem is also disrupted when lichens, algae, or aquatic plants are killed by pollution. Animals that feed on these organisms may starve to death. Several studies in different areas have shown that ecosystems affected by pollution contain fewer species of organisms than they did previously. This loss of species diversity has long-term effects that we do not yet understand properly.

> Ecosystems damaged by air pollution contain fewer species of organisms than they did previously.

### Planting Trees for Cleaner Air

The trees we plant can make a significant difference to the quality of the air we breathe. Many plants have

**Figure 20-12**  The pale patches on these rocks are lichens, sensitive indicators of air pollution.

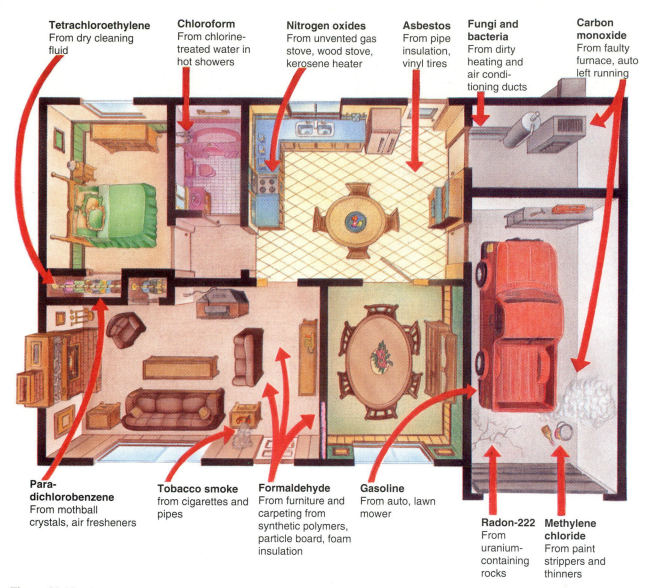

**Tetrachloroethylene**
From dry cleaning fluid

**Chloroform**
From chlorine-treated water in hot showers

**Nitrogen oxides**
From unvented gas stove, wood stove, kerosene heater

**Asbestos**
From pipe insulation, vinyl tires

**Fungi and bacteria**
From dirty heating and air conditioning ducts

**Carbon monoxide**
From faulty furnace, auto left running

**Para-dichlorobenzene**
From mothball crystals, air fresheners

**Tobacco smoke**
from cigarettes and pipes

**Formaldehyde**
From furniture and carpeting from synthetic polymers, particle board, foam insulation

**Gasoline**
From auto, lawn mower

**Radon-222**
From uranium-containing rocks

**Methylene chloride**
From paint strippers and thinners

**Figure 20-13**    Some indoor air pollutants and their sources.

remarkable abilities to absorb pollutants from the air without damage to themselves. For instance, a red maple (*Acer rubrum*) can absorb 20 kilograms of pollutants from the air each year, about twice as much as most other shade trees. A hectare of red maples absorbs about 1 ton of sulfur dioxide each year. When we plant trees in our gardens or city streets, it is worth finding out which are most efficient at this kind of cleanup operation. There is also little point in planting trees that will die rapidly of the local air pollution. *Gingko biloba* is one species favored for street planting because it is highly resistant to air pollution.

Planting trees that can absorb pollutants can improve air quality significantly.

## 20-E INDOOR AIR POLLUTION

For most people, walking in the front door provides no escape from pollution. The air inside a building is nearly always of worse quality than the air outside. This is partly because air in a building moves around less than air outside and partly because buildings contain many sources of pollutants (Figure 20-13).

The air inside a building is nearly always of worse quality than the air outside.

### Sick Building Syndrome

Polk County, Florida, spent $37 million to build a new courthouse in the 1980s. When workers moved in, many of them became sick. Now they have moved out, and the county faces a bill of many millions more to cure what has come to be known as "sick building syndrome." About half the people who worked in the courthouse developed symptoms of allergic reactions. Some developed more serious respiratory problems, with coughing, wheezing, shortness of breath, and fever (Close-Up 20-1).

Sick building syndrome is indoor air pollution on a grand scale. It is caused by air pollutants being trapped inside a building and continuously recirculated by the air-conditioning system. It is particularly likely to occur in buildings that are tightly sealed to reduce energy costs. The allergic reactions in the Polk County courthouse were caused by fungi, which reproduce in the air-conditioning ducts, ceiling tiles, carpets, or furniture. When high concentrations of molds, mildews, and other fungi are breathed in, they often cause allergies and respiratory problems.

Sick building syndrome is particularly common in places like Florida, Southern California, and the desert Southwest where there are many new buildings that are kept tightly sealed against the heat for most of the year. The usual culprits are fungi, but in some buildings formaldehyde insulation installed during the 1970s adds to the problem.

The solution is to reduce the number of places fungi can survive by steps such as cleaning heating and air-conditioning ducts, replacing carpet with tile, and using dehumidifiers to reduce the humidity of the air. Filters in the air-conditioning system may also help. The best solutions, however, involve ventilating the building to get rid of the pollutants, perhaps by turning off the air-conditioning and opening windows at night. This is a sensible practice in homes as well as commercial buildings.

Plastics and other industrial compounds in furniture, drapes, and carpeting are major sources of pollution. It is a good idea to refurnish at a time of year when windows can be left open because many of these pollutants disappear slowly as carpets and furniture age.

We are in a dilemma. We try to seal cracks in windows and walls to keep down the cost of heating and air-conditioning, and yet we want a stream of fresh air to replace the polluted air in our houses and offices. Air filters do not solve the problem because many pollutants are molecules too small to be removed by filters. The only real answer is adequate ventilation combined with low-pollution furniture and appliances. The solution to ventilating a building without losing too much hot or cold air is an efficient heat pump (see Figure 10-6). A heat pump is a heat exchanger. It removes the heat (or cold) from dirty air that leaves the house and transfers it to fresh, incoming air.

## Formaldehyde

Many Americans first became aware of indoor air pollution in the formaldehyde case, an unpleasant example of the fact that we cannot always rely on the regulatory agencies we pay to protect us from pollutants. Formaldehyde is used in the manufacture of particleboard and plywood, which are widely used as building materials. It was also used in foam that was pumped into walls as insulation. Some of the formaldehyde dispersed into the air, irritating skin and throats and producing nausea and headaches. A 1980 study by the Chemical Institute of Toxicology showed that formaldehyde exposure was correlated with nasal cancer in rats. Companies selling formaldehyde insulation went out of business, and manufacturers reduced the amount of formaldehyde used to make plywood and particleboard. Despite these private actions, in 1982 the EPA announced that the risk to human health was trivial and did not even justify a government review to determine if formaldehyde was dangerous. An environmental group sued the agency, and in 1984 the EPA promised to review formaldehyde as required by the Toxic Substances Control Act. Finally in 1987, the EPA announced that studies on humans and other animals showed formaldehyde to be a carcinogen.

## Carbon Monoxide

Anything burning indoors can cause air pollution. Tobacco smoke is an obvious example. But gas stoves, indoor grills, furnaces, and a car running in the garage all use up oxygen and generate toxic gases. The most deadly of these is the poisonous gas, carbon monoxide (CO), which kills an estimated 4,000 people each year. Carbon monoxide is colorless and odorless and produces flu-like symptoms of fatigue and headaches. Properly used, kerosene heaters and modern stoves that burn coal or wood do not produce much pollution. But old-fashioned kerosene heaters and older stoves release enormous amounts of pollution into the house.

## Radon

Thousands of Americans live in homes containing invisible, tasteless, odorless, radioactive radon gas. **Radon** is one of the elements produced by the radioactive decay of uranium or radium. Each radon atom decays a few days after it forms, but its decay products may attach to dust particles in the air and be inhaled into the lungs. Here they decay by emitting radiation products called alpha particles, which can destroy the genetic material in cells lining the air passages.

# How Air Pollution Reaches the Lungs

The respiratory system is a marvelously efficient method of permitting gas exchange between the body and the environment (Figure 20-A). Air enters the body through the nose or mouth. It passes through the pharynx, a common passageway for both air and food, and enters the trachea (windpipe) by way of the larynx, also known as the voice box or Adam's apple. The walls of the trachea contain cartilaginous rings, which hold the tube open. On the inner surface of the trachea, hair-like cilia keep the air passages clear by moving foreign particles up into the pharynx, where they can be swallowed.

The lower end of the trachea divides into two bronchi, which divide into finer and finer tubes, the bronchioles. The smallest bronchioles end in a myriad of tiny sacs, the alveoli, whose thin walls are the actual respiratory surfaces. Here blood picks up oxygen for transport to the rest of the body and gives up carbon dioxide.

The diaphragm extends across the bottom of the chest cavity, beneath the lungs, so the chest cavity is closed off from the abdominal cavity. During inhalation, the muscles between the ribs contract and lift the ribs outward. At the same time, the diaphragm contracts and so moves lower. Air then rushes into the nose, down the trachea, and into the lungs.

During exhalation, relaxation of the rib muscles and diaphragm decreases the volume of the chest cavity, increasing the pressure inside it and forcing air back out of the lungs. (You can get an idea of how negative-pressure breathing works by closing your mouth and holding your nose while you expand your rib cage and lower your diaphragm. You will feel the partial vacuum created. Then remove your fingers from your nose, and you can hear and feel the air rushing in as the pressure equalizes.)

Negative-pressure breathing allows an animal to eat and breathe at the same time. If it were necessary to push air from the mouth into the lungs, any food in the mouth might be pushed into the trachea and cause an obstruction. Negative-pressure breathing creates a more gentle stream of air, which is less apt to pull food along into the air passages.

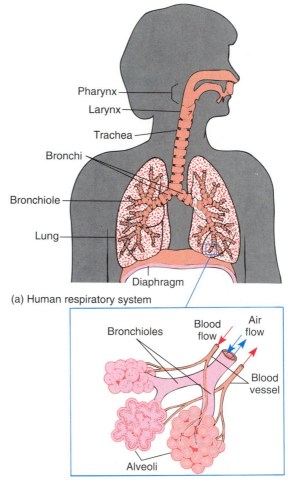

(a) Human respiratory system

(b) Respiratory surfaces and their blood supply

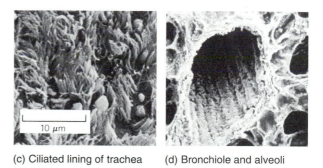

(c) Ciliated lining of trachea   (d) Bronchiole and alveoli

**Figure 20-A**   The anatomy of the human respiratory system. Air travels from the nose or mouth to the lungs by way of the air passages: the trachea, bronchi, and bronchioles (lavender). The actual respiratory surfaces are the walls of the numerous alveoli in the lungs (pink).

This genetic damage can contribute to causing cancer.

Radon is a natural product that reaches the air wherever uranium or radium are decaying in rocks in Earth's surface. Radon is most concentrated in porous soils overlying rocks that contain uranium. It is considered an indoor pollutant because radon concentrations are higher in houses than they are outdoors. Radon outdoors quickly disperses and blows away from Earth's surface.

A 1991 study by the National Research Council found that radon gas in homes does not cause as many cases of lung cancer as was once thought. In previous studies, researchers assumed that radon exposure in the home was equivalent to exposure in uranium mines, where the rate of lung cancer caused by radon was known. The new study took another fact into consideration: that mines are dusty places, full of particles that absorb radon decay products and carry them into the lungs. The new calculations suggest that if you are exposed to the average amount of radon found in an American home, you have about one chance in 1,000 of dying of lung cancer caused by radon. This is a much greater risk than the risk from typical exposures to such environmental health hazards as asbestos, vinyl chloride, pesticides, chloroform in drinking water, or benzene in the air. It is much less than the chance of dying of lung cancer caused by smoking.

### Smoking

Studies on uranium miners who have been exposed to radon for many years and develop lung cancer suggest that about half of these people would not have gotten lung cancer if they had not also smoked. Interestingly, this means that the most effective way to reduce the environmental risk from radon is to not smoke. If you have been exposed, both the risk from smoking and the risk from radon decline steadily after you stop smoking and get rid of the radon. Smoking, by itself, is the single biggest source of indoor air pollution and the type of air pollution that causes most deaths—several million each year worldwide.

> Smoking is the single biggest source of indoor air pollution and the type of air pollution that causes most deaths.

### Asbestos

The relationship between radon and smoking is similar to that between asbestos and smoking. **Asbestos** is the name given to several fibrous minerals containing silica, which are valued for their strength and resistance to heat. Asbestos has been widely used to reinforce cement and to make vinyl floor tiles, brake linings, and garments to protect fire fighters. Exposure to asbestos dust is very dangerous. The fibers scar the lungs, causing the disease asbestosis. Victims find it harder and harder to breathe and may eventually die of heart failure. If asbestos workers smoke, however, they are more likely to develop and die of lung cancer than of asbestosis. If they do not smoke, asbestos workers are no more likely to get lung cancer than any other nonsmoker.

## 20-F NOISE

Most pollutants are chemicals. Noise is different. It consists of pressure waves that travel through the air and that we perceive as sound. (Close-Up 20-2 describes how the human ear works.) Noise is measured in decibels, which compares the intensity of sounds. Table 20-6 shows the decibel level of some common sounds. Industrialized society surrounds us with sounds. Even alone in a "silent" house, we can usually hear an air conditioner, furnace, fan, refrigerator, and the sound of traffic. Appliances, stereo systems, televisions, and lawn mowers often make the average house as noisy as a busy factory.

Noise pollution is one of the prices we pay for modern living. It is a day-to-day irritation, and it damages our hearing. As we age, our hearing becomes less and less acute, and many people become almost totally deaf in old age. In most people, this progressive hearing loss is not the result of aging. It is the result of a lifetime of noise. Studies of hunter-gatherers in Africa and other people living in environments with little noise show that most 80-year-olds in these societies have better hearing than 15-year-old Americans. Many rock musicians have lost most of their hearing by the age of 40. Besides its physiological effects, noise affects us psychologically. Someone accustomed to sleeping in a 50-decibel city apartment where sirens, traffic noises, and music continue throughout the night may be unable to sleep in a country bedroom where the only noise is the 10-decibel sound of leaves rustling outside the window. More commonly, honking horns and other people's radios produce stress reactions, such as irritation or anger, raised blood pressure, and a low attention span.

> Progressive hearing loss in industrial societies is the result of a lifetime of noise, not the result of aging.

More than 1,000 years ago, Julius Caesar banned iron-wheeled chariots from the Roman cobblestones during the night. Mufflers were added to early au-

# How We Hear

The ear is really two organs in one. It controls both our sense of balance and our hearing. The semicircular canals are the parts of the ear concerned with balance (Figure 20-B). They detect movements of the body. The rest of the ear is a sense organ that permits us to hear.

Sound waves reach the body in the form of pressure waves (vibrations) in the air. They are detected (as are all stimuli) by receptor cells, cells that convert the energy of a stimulus into electrical energy, the only form of energy that can be transmitted by the nervous system. The receptor cells that permit us to hear are located in the cochlea of the inner ear. Before it reaches the cochlea, sound passes through several other structures whose role is to transform the stimulus of sound waves in air into pressure waves in the fluid in the cochlea.

Vibrations in the air first strike the tympanic membrane, or eardrum. Vibration of the tympanic membrane moves three small bones that span the

cavity of the middle ear. The third bone presses against the oval window of the cochlea. Vibration of the oval window in turn moves the fluid inside the cochlea.

The entire cochlea is lined by receptor cells and is shaped like a snail shell, a coiled tube of increasing diameter. The gradual change in diameter permits the nervous system to tell the pitch of a sound— how high or low it is. Each pitch produces a maximum vibration at one point in the cochlea, with a particular width. The brain "knows" the pitch of the sound because it knows the location within the cochlea of the receptor cells that are stimulated. The volume (loudness) of the sound is signaled by the frequency of electrical activity in the nervous system.

The hair cells in the cochlea are delicate and easily damaged by loud noises. As we age, we progressively lose hair cells and therefore hearing.

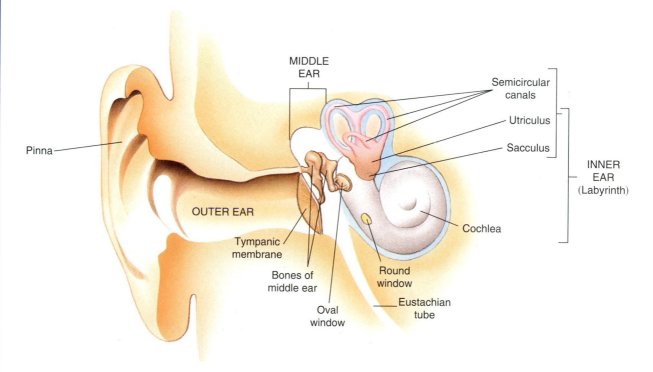

**Figure 20-B**   The human ear. Sound waves enter the outer ear, causing vibrations of the tympanic membrane, which transmits them to the three bones in the air-filled middle ear. This, in turn, causes vibrations of the fluid in the cochlea in the inner ear. The inner ear also contains the semicircular canals, utriculus, and sacculus, where acceleration and gravity are detected.

tomobiles to prevent them from frightening horses. Regulations designed to control noise have proliferated throughout the world as the noise problem has increased. In the United States, the Noise Control Act of 1972 directed the EPA to regulate noise. The EPA's job is to establish standards for manufacturers of all sorts of motor vehicles and to establish guidelines for the workplace. People who work with noisy equipment must wear ear protectors, and machinery is designed to be as quiet as possible.

Prolonged exposure to 90-decibel noise damages the hearing. But noise well below this level causes psychological damage. Ninety decibels was the guideline level for the workplace for many years. It clearly needed to be reduced. But how far? Studies showed that reducing the noise level from 90 to 85 decibels would cost American industry billions of dollars. Environmentalists claimed it would save money—by lowering the rate of absenteeism and reducing ill health. What dollar value should we place on the effects of noise: a mistake made by an irritable employee, someone not hearing an instruction? And can we be sure that the mistake was actually caused by noise in the first place? As with so many environmental issues, the cost-benefit analysis is extremely difficult.

**TABLE 20-6**   **Comparison of the Average Intensities of Various Sounds**

| Source | Intensity (in decibels) |
|---|---|
| Jet aircraft at takeoff | 145 |
| *Pain occurs* | *140* |
| Air hammer | 130 |
| Jet aircraft 200 meters overhead | 120 |
| Unmuffled motorcycle | 110 |
| Subway train | 100 |
| Maglev at 500 km/hour at a distance of 25 meters | 100 |
| Farm tractor | 98 |
| Gasoline-powered lawn mower | 96 |
| Food blender | 93 |
| Heavy truck 15 meters away | 90 |
| Heavy city traffic | 90 |
| Vacuum cleaner | 85 |
| *Hearing loss occurs with long exposure* | *85* |
| Garbage disposal unit | 80 |
| Maglev at less than 300 km/hour | 80 |
| Dishwasher | 65 |
| Window air conditioner | 60 |
| Normal speech | 60 |

# LIFESTYLE 20-1

## How Many Tons of Carbon?

If you want to calculate your own contribution of carbon to global warming, here are some of the (approximate) figures you need. Americans generate 18.4 tons of $CO_2$ per capita each year, but that includes government, industrial, corporate, and personal production, so your total should be a lot less. You can see from Tables 20-3 and 20-4 that the United States produces nearly one quarter of all greenhouse emissions and nearly one quarter of $CO_2$, even though we make up less than 5% of the world population. Go plant some trees!

- Burning 1 (U.S.) gallon (3.8 liters) of gasoline produces 9 kilograms of $CO_2$.

- Using 1 kilowatt-hour of electricity (see your utility bill) from a coal-fired power plant produces 1 kilogram of $CO_2$. Hydropower and nuclear electricity produce no $CO_2$ directly. Your utility company can tell you which you have.

- Natural gas: burning 100 cubic feet (= 1 therm = 100,000 BTU) produces 5.5 kilograms of $CO_2$.

- Flying 7 kilometers in an airplane generates about 1 kilogram of $CO_2$ per passenger.

When you have calculated your direct production of $CO_2$, double it to account for the production of $CO_2$ you use indirectly by purchasing goods and services.

## SUMMARY

Air pollution has steadily increased as larger and larger human populations burn increasing amounts of fossil fuel and organic matter and as industry has grown.

### Pollutants and Sources

The main classes of air pollutants are various oxides, volatile organic compounds, suspended particles and droplets, and secondary pollutants that are formed in the air from primary pollutants. Most air pollution in industrialized countries is produced by industry and motor vehicles. The wealthy industrial countries have reduced industrial emissions rapidly since 1970 by changing manufacturing processes so they emit fewer toxic compounds and by adding pollution control devices that remove pollutants from the air that leaves a factory. Photochemical smog is common in cities with dense traffic in dry, sunny areas and has increased rapidly as the number of vehicles has grown. It can be reduced by encouraging the use of mass transit instead of automobiles and by emission control devices.

### Ozone and Global Warming

Air pollution is changing the climate by destroying the ozone layer and accelerating global warming caused by the greenhouse effect. Ozone depletion is happening rapidly, permitting increasing levels of ultraviolet radiation to reach Earth's surface. International agreements are rapidly reducing the emission of ozone-destroying chemicals such as CFCs, HCFCs, and methyl bromide. The most important greenhouse gas is $CO_2$, produced by burning fossil fuels. We shall have to learn to live with a warmer climate and a rising sea level, but we can slow down the rate of global warming by reducing $CO_2$ emissions and by reforestation.

### Reducing Air Pollution

We can reduce air pollution most effectively by pollution control of industry, more efficient energy use, and replacing gasoline-fueled vehicles by those that run on fuels that do not generate $CO_2$. Los Angeles, one of the world's most polluted cities, has an ambitious plan to reduce air pollution, which includes encouraging mass transit and the use of electric vehicles and variable fuel vehicles as well as controlling the pollutants emitted from thousands of sources, from hair sprays and deodorants to home barbecues, lawn mowers, and bakeries.

### Effects of Pollution on Ecosystems

One common sign of air pollution is acid rain, which kills organisms and dissolves structures containing calcium carbonate. The presence or absence of particular species of lichens is widely used to monitor the presence of particular pollutants in an ecosystem over periods of time. Pollution leads to reduced species diversity in many polluted ecosystems and speeds eutrophication of freshwater ecosystems. Many organisms absorb pollutants from air and water. Planting trees that absorb large quantities of pollutants is a practical way to reduce air pollution.

### Indoor Air Pollution

Air indoors is usually more polluted than that outdoors because there is less air movement. Sources of indoor pollution include tobacco smoke, fungi and allergens from air conditioners, volatile organic compounds from new furniture and carpets, formaldehyde from building materials, carbon monoxide from burning, radon from underground uranium, and asbestos from old floor tiles and stove installations. The main solution is to ventilate the building.

### Noise

Noise consists of pressure waves that travel through air and are perceived as sound. Noise causes psychological problems such as inattention and stress and causes progressive hearing loss, which occurs quite rapidly in modern societies. Noise control regulations are becoming common. Most are designed to protect workers in noisy occupations.

## DISCUSSION QUESTIONS

1. Litmus paper or a pH meter will permit you to search for acid rain. If you try it, sample rain and soil water in as many places as you can. Can you account for the differences? Is any of the water you test neutral pH? Is most of it acid or alkaline? (The pH scale is shown in Figure 3-7.)

2. As our oil supply runs out, we are generating more and more electricity with coal. Coal can be one of the most polluting of fuels. Do you think that air pollution will get worse in the future for this reason?

3. What would it take to clean up our air? Is it an impossible task either physically or politically?

4. Is air pollution getting worse or better in your community? Why?

5. Should smoking be banned in all indoor public places? Why?

6. What measures would you take to try to improve indoor air quality in a local school and at home?

# Chapter 21

# PESTS AND PESTICIDES

*A tent caterpillar web on a pear tree. The caterpillars produce the web, which protects them from birds. They leave the nest to feed on the pear leaves.*

**Key Concepts**

- Controlling pests is essential to the production of food and other crops.

- Since World War II, developed countries have relied almost exclusively on chemical pesticides to control pest populations. The most widely used pesticides are herbicides.

- Pesticides are poisonous, causing diseases, deaths, and pollution.

- The effects of most pesticides on human health have never been tested.

- Chemical pesticides harm many organisms besides those they are designed to kill.

- The use of chemical pesticides in developed countries is not increasing, and growers are increasingly using biological and mechanical pest controls.

- The use of chemical pesticides is still growing rapidly in many developing countries.

*"Trying to control pest outbreaks with insecticides is like pouring kerosene on a house fire."*

Peter Kenmore, Manager of United Nations Food
and Agriculture Organization's Inter-Country
Program for Integrated Pest Control in Rice.

In 1985, the gypsy moth, tussock moth, southern pine beetle, and spruce budworm destroyed enough trees to build nearly a million houses in the United States alone. In North America, insects destroy approximately 10% of all crops grown. Crops in tropical climates suffer even greater insect damage because insects grow and reproduce faster in tropical heat. In Kenya, officials estimate that insects destroy 75% of the nation's crops. Insects are among the pests that compete with humans for resources and reduce the harvest of natural resources and farm crops.

A **pest** is any organism that occurs where you don't want it or in populations that are large enough to cause damage. For instance, a poppy is a garden flower when we plant seeds in the garden but a weed when it pops up in a wheat field. One raccoon in the garden is wildlife; a family of raccoons in the vegetable garden is a pest problem. The most important pests fall into several groups, including weeds, rodents such as rats and mice, insects, nematode worms in the soil, and fungi. **Pesticides** are things that kill pests. They include **herbicides** that kill plants, **nematocides** that kill nematodes, **insecticides** that kill insects, and **fungicides** that kill fungi (Figure 21-1). Although viruses and bacteria cause many diseases of plants and animals, they are not usually regarded as pests because they are controlled mainly by breeding organisms that are resistant to disease, not by treatment with chemicals.

Many crops are descendants of wild plants that defend themselves by growing scattered among other plants so that few of their predators find them. These predators do more damage to cultivated plants than they do in the wild because crops are planted in **monocultures**, stands containing only one kind of plant. Pests are all around us all the time. They cause problems only when their populations become large enough to cause significant damage. The goal of pest control is to keep pest populations small enough so that they do little harm.

## 21-A  A BRIEF HISTORY OF PEST CONTROL

There is nothing new about pesticides. Ancient Romans sprinkled diatomaceous earth in stored grain to kill insects that eat the grain. Diatomaceous earth is the ground-up shells of fossil diatoms (protists). The dust scratches the waterproof cuticle that covers an insect, and the insect dies of water loss.

I have a 1939 book on vegetable crops from the Cornell College of Agriculture that sums up the pest control methods in common use before World War II, most of which are still used today. These are the methods it lists:

1. **Crop rotation.**  Planting different crops in a field from year to year prevents the pests that live on any one crop from building up large populations.

2. **Destroying plant debris.**  Many pests pass the winter in piles of leaves and similar debris. You can sometimes reduce their populations by cleaning up debris in a field or garden. In areas where it freezes in winter, turning over the soil in autumn to expose pests to the air may permit some of them to freeze to death.

4. **Mechanical protection.**  Covering beds of vegetables with cheesecloth permits rain to reach the plants while preventing many insects from reaching the vegetables. Today farmers and gardeners often use "floating row covers" for the same purpose. These are fabrics, originally from France, that are made of artificial polymers, which are lighter than cheesecloth.

5. **Soil sterilization.**  Pests in seed beds can be killed by sterilizing the soil with steam or formalin. Nowadays we "solarize" soil to kill nematode worms by covering it with plastic that traps the sun's heat and warms the soil.

6. **Growing resistant varieties.**  The easiest way to preserve a crop from a particular pest is to grow a variety that is not damaged by the pest. For instance, many varieties of tomato are seldom affected by nematodes and various fungi. A tomato called "Sweet Million VFNT" is resistant to verticillium and fusarium wilts (fungi), nematodes, and tobacco mosaic virus. Breeding pest-resistant varieties of plants is big business today (Figure 21-2).

7. **Poisoned baits.**  A mixture of bran, arsenic, molasses, and lemon juice is recommended as a bait to attract and kill cutworms, which eat the stems of young tomato and pepper plants. Today gardeners fight cutworms more safely by surrounding young plants with paper collars or sticking toothpicks in the soil touching the plant stem, so the cutworm cannot get right around the stem.

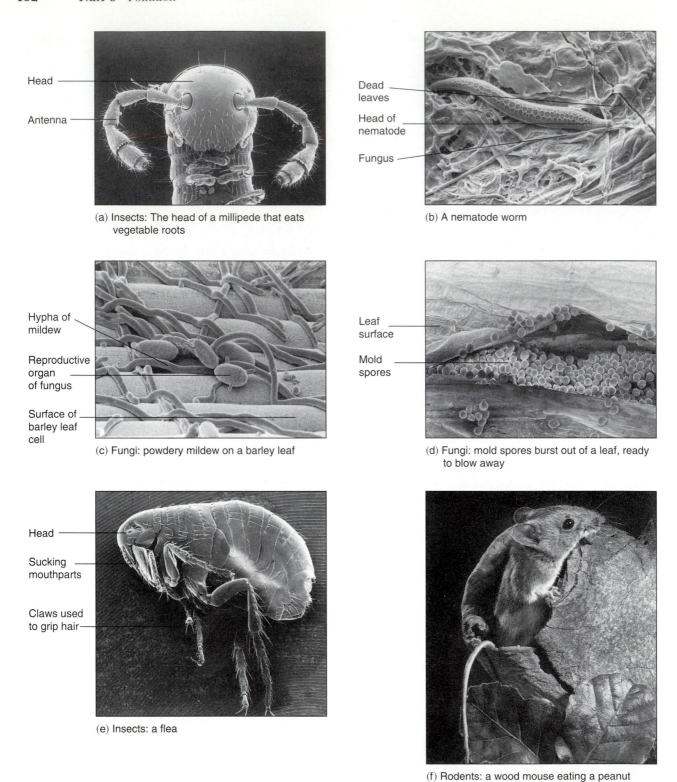

(a) Insects: The head of a millipede that eats vegetable roots

Head
Antenna

(b) A nematode worm

Dead leaves
Head of nematode
Fungus

(c) Fungi: powdery mildew on a barley leaf

Hypha of mildew
Reproductive organ of fungus
Surface of barley leaf cell

(d) Fungi: mold spores burst out of a leaf, ready to blow away

Leaf surface
Mold spores

(e) Insects: a flea

Head
Sucking mouthparts
Claws used to grip hair

(f) Rodents: a wood mouse eating a peanut

**Figure 21-1** Pest portrait gallery. These are photographs taken with an electron microscope (except for the photo of the mouse). An electron microscope uses electrons instead of light so it takes only black and white photographs, but electrons have very short wavelengths, so they can be used to photograph extremely small objects. *(Biophoto Associates)*

**Figure 21-2** A cucumber plant infested with fungus—a mildew that looks like gray powder on the leaves—which will eventually die. This is an old variety that I grew one year. Most modern varieties of cucumber have been bred to be much more resistant to fungus infections.

8. **Use of trap crops.**   The example the book gives is planting kale or mustard before cabbages to attract harlequin cabbage bugs and then spraying the plants with kerosene to kill the bugs. Then when the cabbages grow up, the bugs that eat them will be gone.

9. **Chemical pesticides.**   The kerosene used with trap crops is a chemical pesticide. Formaldehyde, copper oxide, and mercury chloride are recommended as fungicides. Nicotine, sulfur, and cyanide gas are suggested for killing greenhouse pests. The book also lists arsenic, lead arsenate, and copper sulfate as pesticides.

If you think today's pesticides are dangerous, this 1939 list of chemicals makes you realize that we have made some progress at developing less toxic pesticides. This is a very poisonous collection. Interestingly, this pre-war book also mentions diatomaceous earth as well as the "new pesticides," pyrethrins and rotenone. These two are plant **secondary chemicals**, chemicals produced by plants that protect them against diseases and pests. Today pyrethrins and rotenone are the most commonly used household insecticides, forming the active ingredients in flea shampoos and sprays for reducing insect populations in homes. They are **biological pest controls**, natural pesticides produced by living organisms (Section 21-F).

Nowadays a farmer or home gardener would use many of the pest control methods in this list except that most of the chemicals would be different. But we passed through a period after World War II when many of these pest control methods were almost forgotten. The invention of DDT ushered in an era of almost complete dependence on chemicals. At first, the new pesticides seemed wonderful. They killed the mosquitoes that carried malaria and increased yields of the crops needed to feed the exploding human population. But by the 1960s, people like Rachel Carson were sounding the alarm. Farmers were using larger quantities of pesticides each year as pests became resistant to them. Reports of human deaths from pesticides began to accumulate. This chapter tells the story of the uses and abuses of pesticides and how we can use them more responsibly.

Farm chemicals are big business. In the United States, nearly $5 billion worth of synthetic pesticides were sold in 1992 (Equal Time 21-1). There are about 1,000 chemical pesticides in common use worldwide, and the market is growing rapidly. In 1993, developed countries still used 80% of pesticides produced, and the amount of pesticides used each year was steady or even declining (Figure 21-3). In developing countries, however, pesticide use is growing by 7 to 9% a year, and in some countries it is expected to triple by 2000.

> Common older methods of pest control include crop rotation, mechanical protection, soil sterilization, growing resistant varieties, poisoned baits, and chemical pesticides.

> During the last 50 years, farmers have become almost completely dependent on chemical pesticides, using increasing amounts every year.

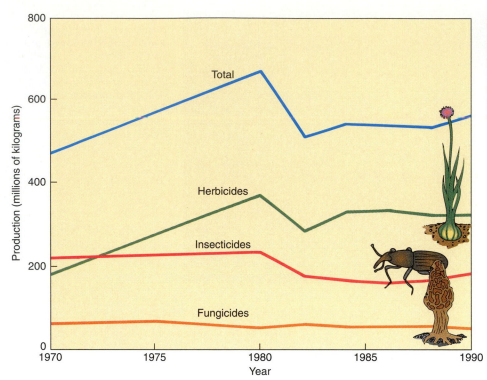

**Figure 21-3** Pesticide production in the United States from 1970 to 1990. Note that since the early 1970s, herbicides have been the most widely used pesticides. This is the case all over the world.

## 21-B PESTICIDE TOXICITY

Pesticides are chemicals designed to kill organisms. Not surprisingly, many of them are also toxic to humans and organisms they are not designed to kill. The usual measure of a compound's toxicity is its LD50, the concentration of the toxin that kills 50% of a group of test organisms (Section 18-B). Figures for some common pesticides for which we know the toxicities are shown in Table 21-1. Many pesticides are neurotoxic, meaning they poison the nervous system. They may be safe for human use because it takes a much larger dose of a nerve poison to kill a human than an insect. But many small animals will be killed by most insecticides. This is why you are advised to spray many insecticides on plants only in cool weather when beneficial insects such as bees are not active.

| | LD50 | LC50† at 11°C | |
|---|---|---|---|
| **Insecticide** | **(mg/kg in White Rats)** | **(mg/l in Fish)** | **Toxicity Class** |
| **Chlorinated hydrocarbons** | | | |
| Aldrin | 40.0 | 0.0082 | Extremely toxic |
| DDT | 250.0 | 0.005 | Very toxic |
| Lindane | 125.0 | >0.03 | Very toxic |
| **Organophosphates** | | | |
| Malathion | 1,500.0 | 0.55 | Moderately toxic |
| Parathion | 8.0 | 0.065 | Extremely toxic |
| **Carbamates** | | | |
| Carbaryl | 540.0 | >1.0 | Moderately toxic |
| Zectran | 30.0 | >1.0 | Extremely toxic |

**TABLE 21-1** Toxicities of Some Insecticides*

*The higher the LD50, the less toxic the pesticide is.

†LC50 is the concentration that kills 50% of a group of test organisms.

The effects of many pesticides are increased by **synergistic interaction** with other chemicals, meaning that the effect of two things together is more than the combined separate effects of the components. For instance, the pesticide malathion is usually detoxified by enzymes in the liver and so is not toxic to mammals. However, if some other chemical, such as a drug often used to treat ulcers, inhibits the action of the liver's enzymes, malathion is much more toxic.

## Health of Farmers

To date, we have seldom had the scientific evidence to prove that particular pesticides cause particular human health problems. The evidence usually consists of a finding that a particular disease is much more common in an area where a pesticide is used than where it is not or that the incidence of a disease increases when more of the pesticide is used. For instance, a doctor in an Armenian village, where DDT was the only pesticide used from 1960 until the 1980s, found that the incidence of cardiovascular disease, diabetes, cancer, and allergies increased enormously during that period. This only suggests, but does not prove, that the DDT caused the increase in disease.

Some areas of the world have been particularly plagued by illness probably caused by pesticides. The San Joaquin Valley in California is a center of fruit and vegetable growing and has very high levels of pesticide use. Cancer rates among children in the valley are eight times the national average. In the 1980s, farm workers claimed that the insecticide parathion was responsible for poisoning 650 workers in California's fruit and vegetable fields, 100 of them fatally.

Farmers are generally healthier and live longer than other workers, but farmers do have unusually high rates of several cancers, such as leukemia, melanoma, brain cancer, and multiple myeloma (a cancer of the immune system). In 1993, the U.S. government announced a study on farm chemicals and health. Epidemiologists will examine the work habits and lifestyles of 100,000 farmers and their families for five years, looking at exposure to such things as pesticides, fertilizers, dust, and viruses. They will also compare "safe" farmers, those who wear recommended protective clothing and respirators to apply chemicals and also follow directions on chemical labels with those who do not. By 2000, we should have a much clearer idea than we do now of the health hazards from using farm chemicals and how much protection current safety standards provide.

We know very little about the toxicity of most pesticides to humans and other organisms.

## Pesticide Manufacture: Bophal

Pesticide manufacture may involve substances that are much more poisonous than the pesticide itself. The most disastrous single instance of air pollution we know of occurred when methyl isocyanate gas was accidentally released in Bophal, India, in 1984, killing 2,500 people, permanently disabling 17,000, and injuring 200,000. The gas was used to produce the relatively safe pesticide Sevin, which breaks down rapidly in the environment. The gas was released from a chemical plant belonging to American Union Carbide Corporation when water was pumped into a storage tank in which the cooling and backup systems were not working.

## Persistence and Pollution

Many artificial chemical pesticides do not break down rapidly in the environment. Such **persistent pesticides** may be with us for a long time and cause most of the environmental damage attributed to pesticides (Close-Up 21-1). About 60 different pesticides have so far been detected in the ground water of 30 states. One class of pesticides often found in ground water is nematocides, which are designed to be mobile, persistent, and toxic. They include EDB (ethylene dibromide), DBCP (1,2-dibromide-3-chloropane), and aldicarb, all of which have caused wells to be shut down in various parts of the United States. EDB causes sterility in males and cancer. New York banned the use of the nerve poison aldicarb in 1982 after it was discovered that the only aquifer on Long Island was contaminated by this pesticide. Now aldicarb is showing up in the ground water of many states.

Pesticides that pollute waterways and soil come from many sources other than farm fields. For instance, the waste from a factory that produces applesauce contains traces of the pesticides used on the apples. The Environmental Protection Agency (EPA) lists 110 pesticides that can be used on apples. In one study, one third of apples sampled contained pesticides, and 43 different pesticides were detected. Even if you wash a commercial apple in the sink at home, you are likely to be washing low concentrations of pesticide into the sewage system or septic tank. This does not matter much with a pesticide that is rapidly degraded, but persistent pesticides accumulate in the soil, soil water, or waterway.

Persistence is a particular problem with chlorinated hydrocarbon pesticides such as DDT (Table 21-2). DDT was used in the 1940s to kill the mosquitoes that cause malaria and the lice that spread

**TABLE 21-2  Persistence of Some Pesticides**

| Pesticide | Persistence (half-life)* |
|---|---|
| **Chlorinated hydrocarbons** | Persistent |
| DDT, aldrin, chlordane, dieldrin, lindane, mirex, kepone, heptachlor | 2–5 years |
| **Organophosphates** | Degradable |
| Malathion, Azodrin, miazinon, Phosdrin, Diazinon, parathion | 1–10 weeks |
| **Carbamates** | Degradable |
| Carbaryl (Sevin), Zineb, maneb, Zectran, Temik | 1 week |
| **2,4-D, 2,4,5-T herbicides** | Moderately degradable 2–4 months |

*The half-life is the time it takes for half of the applied quantity to be broken down.

typhus. Millions of lives were undoubtedly saved by this program. The trouble is that DDT is broken down only extremely slowly by decomposers. It and its breakdown products DDD and DDE remain in the environment indefinitely. As a result, each use of this pesticide increases the amount in the biosphere. DDT can now be detected everywhere in the world: from the Antarctic ice to our own bodies.

After World War II, DDT was used in massive amounts. In the United States, more than 30 kilograms of DDT per hectare were applied in unsuccessful attempts to control populations of the beetle that carries the spores of Dutch elm disease. The Soviet Union banned DDT in 1970, two years before the United States, but actually continued to use it secretly until 1988. In Siberia, DDT was sprayed from airplanes to control forest ticks that carried a virus that causes encephalitis.

Persistent pesticides eventually reach waterways. Birds that eat fish were the worst sufferers from DDT because DDT interferes with calcium metabolism. Birs laid thin-shelled eggs that broke when they sat on them. Penguins, pelicans, peregrines, and eagles were all endangered or eliminated from some areas as a result. DDT was eventually banned in many places. When people stop using a persistent pesticide, the pesticide remains in the environment, but it disperses, becoming less concentrated in local areas but more widespread. Now that DDT has dispersed so that pelicans and other birds do not eat concentrated doses of it, these birds again lay eggs they can sit on, and the populations of many have increased.

# CLOSE-UP 21-1

## Methyl Bromide and the Ozone Layer

Every time he fumigates the walnuts in his storage shed with methyl bromide gas to kill worms, California farmer Mark Gibson worries that he is also destroying the ozone layer. His fumigation shed is across the road from Claveras Elementary School, and he also worries that a release of the toxic gas could endanger schoolchildren.

Methyl bromide is a persistent, odorless, very poisonous gas that is one of the most widely used pesticides. It is used on crops from alfalfa and strawberries to nuts and wheat. In urban areas, it is the pesticide of choice for destroying termites, wood-eating insects that may infest and destroy wooden houses. Methyl bromide is a small molecule, so it penetrates some hard surfaces such as walnuts and wooden walls. It kills just about every animal exposed to it, including 16 people who walked into

houses treated with methyl bromide in the late 1980s. It is also suspected of causing birth defects. Most pesticides damage the environment when they sink into the soil or drain into waterways, but methyl bromide rises into the atmosphere, where it destroys the ozone layer (Section 20-B).

Despite a $3 million campaign by the makers of methyl bromide, in 1992 the gas was added to the list of chemicals to be phased out worldwide by 2000 to protect the ozone layer. Farmers and agriculture officials would be happy to see the dangerous gas banned but only when an equally effective pesticide has been introduced to replace it, which has not yet happened. It appears that Gibson and other users will have to find some other way to protect walnuts from insect damage within a few years.

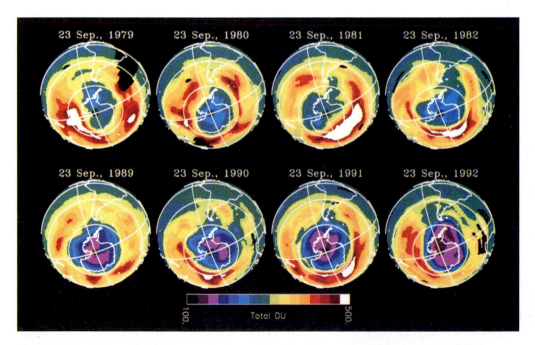

**Figure 21-A** Ozone depletion: computer models of ozone concentration near the South Pole (where the white lines cross toward the bottom of each model). Between 1979 and 1992, the blue area of low ozone levels increased to 8.9 million square miles, more than twice the area of the U.S.A. and 15% greater than in 1991. In 1989, the first area of extreme depletion (purple) appeared. By 1992, this area had approximately doubled in size. Ozone concentration is expressed in Dobson Units (DU). The U.S. Nimbus satellite has measured ozone with a spectrometer since 1978. In 1991, the task was taken over by a similar instrument on the Russian Meteor-3 satellite. (*National Aeronautic and Space Administration*)

Persistent pesticides are those most likely to damage ecosystems. They have caused widespread pollution of surface and ground water.

### Bioaccumulation

The DDT story is sometimes cited as an example of **bioaccumulation**, the concentration of a substance as it passes through a food chain, but it is not. DDT does not become more concentrated as it passes from one organism to the next, although it is not broken down either. Indeed, there is no good example of poisons being concentrated to a harmful level by passage through a land food chain. As far as we know, bioaccumulation occurs only in aquatic food chains, when toxins such as heavy metals that dissolve in fat sometimes do become concentrated as they pass through the food chain (Lifestyle 5-1).

## 21-C  RESISTANCE AND ITS EVOLUTION

One of the problems with pesticides is that they often have to be sprayed at the right time in the pest's lifespan if they are to work. For instance, many pests spend part of their lives inside plants where pesticides cannot reach them (Figure 21-4). As a result, farmers and gardeners who do not take the time to study the pests they find on their plants often waste pesticides by spraying them when they will do no good.

The more a pesticide is used, the worse becomes the problem of **resistance**, the evolution of pests that are not killed by the pesticide. Consider the example

**Figure 21-4**  Insect galls on a leaf. The insect injects its eggs into the leaf, which stimulates the leaf to form a gall of plant tissue that the insect larva eats before it matures into an adult.

of cotton. Cotton has long been sprayed with more pesticides than any other major crop (Lifestyle 21-1). In 1988, the U.S. Department of Agriculture joined with cotton farmers in the most expensive program to date, an attempt to eradicate the boll weevil. Cotton crops, from North Carolina south to Alabama and Florida, were sprayed with the insecticide Guthion (azinphosmethyl) every five to ten days for several weeks. Guthion is sprayed from crop-duster planes flying about 15 feet above the ground, from which height the insecticide also drifts into neighboring suburbs, fields, and streams.

Will this dangerous and expensive program significantly reduce the boll weevil problem? Any biologist would predict that it will not. Boll weevils will become resistant to Guthion as they have to many other pesticides sprayed on them over the years.

When you expose a population to a poison for long enough, either all its members die or the population evolves resistance to the poison. For example, a scale insect feeds on citrus trees in California. In the 1880s, growers sprayed the trees with cyanide, and this killed the scale. But then farmers noted that some of the insects were surviving the spraying. Research showed that cyanide did not kill them because they possessed a single gene that permitted them to break cyanide down into harmless substances. As spraying continued, more insects with the gene than without it survived to reproduce, and they passed the gene on to their offspring. The frequency of the new gene in the population increased until the whole population was resistant to the spray. Like many insects, scale insects breed rapidly, producing more than one generation a year. As a result, genes for pesticide resistance can spread through an insect population in a single growing season.

Pesticide resistance is now a widespread and extremely expensive problem. Chemical companies estimate that a new fungicide has a useful life of about four years in commercial agriculture before most of the fungi it kills become resistant. (After that it may still be useful to home gardeners.) Insects generally evolve faster than do pesticides. It takes years to develop a new pesticide, whereas it may take only weeks for an insect population to evolve resistance, so the pests tend to win this battle. We have noted (Section 8-F) that many mosquito species have evolved resistance to many pesticides, making it much more difficult to control malaria.

Pests invariably evolve resistance to any pesticide that is used on them frequently in large quantities. Resistance often evolves very rapidly, making a pesticide useless.

## 21-D REGULATION

The need for strict regulation of hazardous substances in the environment is obvious, and we are moving in that direction—but rather slowly. In the United States, this is still an area controlled by a patchwork of inadequate federal and local regulations.

Pesticides are controlled under the 1947 Federal Insecticide, Fungicide, and Rodenticide Act (FIFRA), now administered by the EPA. Let us briefly consider the steps a new pesticide goes through before it becomes a "registered" pesticide. This process may take six to seven years and costs millions of dollars.

### Pesticide Registration

Chemicals that might become useful pesticides are first tested in the laboratory to see if they kill the pest in question. Promising chemicals are then tested on a small scale on laboratory animals to see if they are toxic or cause genetic mutations. The new chemical is then tested on a small scale in the field or in a greenhouse to see how it performs under actual growing conditions. The chemical's toxicity is determined to find out what kind of protective clothing and other precautions should be used by people who apply the pesticide. In addition, the pesticide is tested for persistence in the environment by examining what happens to the chemical when it is broken down by decomposer organisms. The most useful pesticides are broken down rapidly into harmless compounds. Environmental tests must then be performed to see, for instance, if the pesticide has damaging effects on aquatic ecosystems if it leaches into ground water or surface water.

Eventually, if the pesticide passes all these tests, a label must be written bearing instructions for the safe use of the pesticide. These include requirements for protective clothing, any special training needed to use the pesticide, instructions and warning statements, how long workers must wait before entering a treated field, warnings of possible environmental damage, and how to dispose of the pesticide container properly.

**Tolerance** If a pesticide is to be used on food crops, the maker must determine how much of the chemical remains on the food after it has been harvested and processed for the consumer. Under the Federal Food, Drug, and Cosmetic Act, the EPA establishes **tolerance levels**, the maximum levels of pesticide residues permitted to be on or in a food crop at the time the crop is harvested.

Pesticide tolerances are determined to produce an **acceptable daily intake** level, the amount of the chemical that can be ingested by an average person every day for a lifetime without causing ill effects. For pesticides to be registered, residue levels in foods must fall below the acceptable daily intake.

> In the United States, new pesticides for use on major crops must be registered with the EPA after tests for safety and efficacy.

### Minor-Use Pesticides

A registered pesticide that is used according to the instructions poses relatively little danger. One of the biggest holes in pesticide regulation is that registration applies only for pesticides for pests of crops, such as soybeans, wheat, and maize, that are grown on huge areas of land. Maize, for instance, covers nearly 30 million hectares in the United States. Many crops are grown on such small areas that it is not cost-effective for the manufacturer to register a pesticide for use on these crops—including our fruit and vegetables, which occupy only about 2 million hectares of U.S. farmland. Some of the same pesticides are used on minor and major crops, but this still means that many pesticides used on minor fruit and vegetables have not been properly tested for safety. The EPA and the pesticide industry are working on methods to register pesticides for minor uses, but so far the only testing for pesticides on foods is limited tests for residues in and on the food.

> Pesticides used on crops such as fruit and vegetables that are grown on small areas are usually not registered and are not properly tested for safety.

## 21-E INADEQUATE REGULATION

Scientists have argued for years that the EPA is not doing its job of keeping us safe from pesticide poisoning. The simple fact is that very little is known about the health or environmental effects of thousands of chemicals that we encounter as pesticides, food additives, and cosmetics.

For instance, a 1992 report by the National Research Council summarized the evidence that neurotoxic pesticides, while they may not kill people, may contribute to the growing number of cases of Alzheimer's disease, Parkinson's disease, amyotrophic lateral sclerosis (Lou Gehrig's disease), and other nervous disorders. At the moment, only one fifth of pesticides are rated for neurotoxicity: those produced in large volume and used on crops likely to reach consumers. Neurotoxicity is determined only by comparing the chemical structure of the molecule with that of a known neurotoxin, not by actually testing it. Some members of Congress have called for better testing.

The only pesticides that you can be sure are tested and registered are new pesticides used on major crops. Many older pesticides and those used on minor crops are not effectively regulated. In 1972, Congress directed the EPA to reregister all pesticides to ensure that they met modern standards. But the task has hardly begun. There are hundreds of pesticides, and the law has made the EPA's task very difficult. Consider the following example.

## Carcinogenic Pesticides and the Delaney Clause

In 1993, Carol Browner, Administrator of the EPA, identified 35 pesticides that cause cancer in laboratory animals and that are commonly found in processed foods. Their presence in food is illegal under the Delaney Clause of the Food and Drug Act, which prohibits the addition to processed foods of any compound that causes cancer in test animals. But the Delaney Clause applies only to processed food not to fresh food, so the EPA could permit carcinogenic pesticides on fresh tomatoes but ban them from tomatoes processed into tomato sauce. The National Academy of Science estimates that nearly all the cancer risk from pesticides in the diet comes from older pesticides used on fresh produce. Nearly 80% of the cancer risk from pesticides comes from the residues of only ten pesticides (Table 21-3).

| TABLE 21-3   The Cancer Risk from Various Fresh Foods as a Result of the Pesticide Residues They Contain* | |
| --- | --- |
| **Food** | **Lifetime Cancer Risk†** |
| Tomatoes | $8.75 \times 10^{-4}$ |
| Beef | $6.49 \times 10^{-4}$ |
| Potatoes | $5.21 \times 10^{-4}$ |
| Oranges | $3.76 \times 10^{-4}$ |
| Lettuce | $3.44 \times 10^{-4}$ |
| Pork | $2.67 \times 10^{-4}$ |
| Wheat | $1.92 \times 10^{-4}$ |

*The result of a study by the National Research Council.

†These figures mean that eating tomatoes, for instance, causes an estimated 8.75 cases of cancer for every 10,000 people exposed throughout a 70-year lifetime.

In an attempt to do something about this ridiculous double standard for fresh and processed food, the EPA announced that it would permit the use of new, safer, weakly carcinogenic pesticides banned under the Delaney Clause but ban older, more dangerous pesticides. The EPA's new policy would eventually lower the overall cancer risk from pesticide residues in food, but in 1993 a court ruled the policy illegal because it violates the Delaney clause against carcinogens in food. In its efforts to improve the safety of our food, the EPA had achieved little except to embroil itself in expensive and time-consuming lawsuits.

The end result of this regulatory muddle will probably be that the Delaney Clause will have been revised by the time you read this book to permit the EPA to implement its new policy. However, the EPA cannot merely list and ban all the old dangerous pesticides. Lobbyists for pesticide producers insisted that the law be written so that the EPA has to examine each existing pesticide on a case-by-case basis. This requires a complex legal process and will take years. We shall be eating foods containing fairly carcinogenic pesticides for many years to come.

## Human Error

It seems odd that pesticides that end up in food are not as closely regulated as food additives and drugs, which are regulated by the Food and Drug Administration (FDA). Part of the reason is that there are regulations about the *way* pesticides are used. A pesticide is poisonous by definition and has to be used properly if it is not to endanger human health or the environment. If you look at the label of many pesticides in a garden center, you will find a booklet describing how to use the product safely, when and on what, and stating that using the pesticide in any other way is against the law. This works fine on a prosperous, well-run farm. Here workers who apply pesticides wear masks, respirators, and protective clothing and dispose of containers safely.

The trouble is that vast amounts of pesticides are used every day by people other than well-educated farmers. If you stroll down any suburban street on a Saturday afternoon, you will probably see householders in shorts and sandals spraying pesticides and herbicides on their yards with reckless disregard for their own health and that of the environment. Only the least toxic pesticides are approved for use by home owners because regulators know that people don't read the labels and therefore use the products improperly. Nevertheless, in 1992 a man died after spraying his lawn with a popular herbicide. Research-

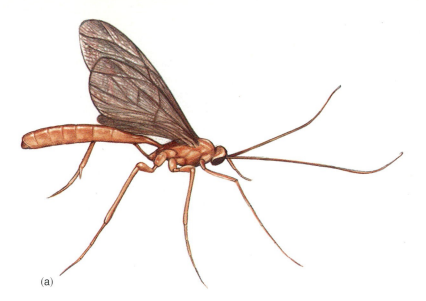

(a)

(b)

**Figure 21-5**  Parasitic wasps. Many species of wasps inject their eggs in other insects. When the larvae hatch, they feed on the host insect, eventually killing it. (a) Many parasitic wasps look like this ichneumid, with long legs and mouthparts. (b) The white things on the outside of this caterpillar are the pupae of a wasp. The wasp larvae, which look like tiny maggots, have fed on the caterpillar and have then moved to the skin, where they form pupae. Inside each pupa, an adult wasp is maturing and will eventually emerge as a free-living adult like the one in part (a). The caterpillar will then die. *(b, Rod Planck/ M. L. Dembinsky Jr., Photography Associates)*

ers believe that the herbicide interacted with the widely used drug Tagamet that the man was taking for an ulcer. The same herbicide has been implicated in the deaths of dogs from cancer. Pets and small children are more vulnerable than adults because they are smaller and more likely to roll on a lawn covered with pesticides.

If things are bad in the health-conscious American suburbs, you can be sure they are worse in most other places. Many U.S. farm workers are migrant laborers who work on a farm for brief periods. When handed a chemical and told to spray it, they are in no position to demand to see the safety instructions before they start, and many cannot read or cannot read English. In developing countries, the supervisor and the workers often cannot read, and the warning instructions are of no use at all. The Soviet Union encountered problems because central authorities ordered workers to apply more and more fertilizer and pesticide each year in the hope of increasing farm production. For example, in 1988, 87% of Russia's cultivated land was treated with pesticides. In contrast, 61% of farmland in North America was treated and with lower levels of pesticides. U.S. farmers reduced their pesticide usage between 1976 and 1983, while Soviet use almost doubled. In one amazing incident in Uzbekistan, tons of fertilizer and pesticides for cotton arrived at a collective farm after the cotton had been harvested. The workers were ordered to spread it all on the fields anyway.

> Much of the damage done by pesticides results from people using them improperly, either because they cannot or will not follow safety instructions.

### Circle of Poison

The U.S. government eventually regulated the sale and use of pesticides such as DDT that prove to be particularly dangerous, but this does not stop chemical companies from selling these products abroad. However, we live in a global economy and many foods produced abroad, laced with these banned pesticides, turn up on our grocery shelves.

Pesticides produced in North America and applied in Central and South America also poison North American birds and butterflies that migrate south for the winter. Texans attempted to save an endangered songbird by making a wildlife preserve of its breeding area. Their efforts were frustrated by deforestation and unregistered U.S. pesticides in Guatemala, which decimated the population in its winter feeding grounds.

## 21-F  NEW METHODS OF PEST CONTROL

Pest control is changing. In Nigeria, crop-duster planes spread parasitic wasps instead of chemicals to control a mealybug that eats cassava (Figure 21-5).

In 1990, for the first time, the U.S. Department of Agriculture registered more biological pest control products than chemical ones. America's largest lawn care companies now offer customers biological pest control as an alternative to chemicals. Their trials have shown that some nematode worms are as effective as chemicals at controlling Japanese beetles, armyworm, sod webworm, and black cutworm. Home owners spread compost on lawns to fertilize it and suppress fungus infections of lawn grasses at the same time. We are moving slowly toward methods of controlling pests that are less dependent on chemicals. Some of the new ways of controlling pests are ancient methods, such as crop rotation. Others are modern inventions.

Many modern methods of controlling pests are described as **biological pest control**. Their main difference from chemical control is that they involve living organisms or chemicals that are naturally produced. Their use involves some understanding of the pest's biology. This usually includes knowing how it reproduces and how it finds the plants it eats.

### Chemical Versus Biological Pest Control

Chemical pesticides and biological controls have different effects on prey populations. Figure 21-6 compares biological and chemical control of the spider mites attacking a crop of cucumbers in a greenhouse. Note that after a chemical pesticide is used, the prey population crashes and then explodes. The biological control, a predatory mite that eats spider mites, reduces the population in small oscillations to a lower level. In this experiment, the predatory mite was much more effective and 25% cheaper to use than the pesticide (Dicofol), even though the cost of the pesticide was minimized by applying it only when it was needed—whenever the spider mite population grew to the size at which it caused visible damage to the plants.

On the island of Jersey, off the coast of France, large greenhouses produce early tomatoes and cucumbers for the British market. For years, the owners spent large sums on chemical pesticides and on pollinating the plants by hand, transferring pollen from male anther to female stigma with paintbrushes. Hand pollination was needed because the pesticides killed insect pollinators. Then a few years ago, the owner read up on biological pest control and decided to try it. Now when pest damage appears on the plants, the pest is identified and the appropriate species of predator is released into the greenhouse. The plants are no longer pollinated by hand but by bees who buzz around the greenhouse. Pest damage is half what it was in the days of chemical pesticides, and the yield of tomatoes and cucumbers has almost doubled because the bees are much more efficient pollinators than people with paintbrushes.

> The use of predators to control pests stabilizes pest populations at lower levels than the use of chemical pesticides.

### Predators and Parasites

Many specialized predators and parasites keep their prey species at low density and can be used to keep pest populations under control. This may involve the use of herbivorous insects for weed control, bacteria and viruses that cause disease to control insect pests, and parasites. The ideal biological control agent is illustrated by the vedalia beetle used to control scale insects (Section 6-G). The beetle rapidly reduces the population of the pest organism and then remains at a steady, low density ready to prevent future population explosions of the pest.

The first recorded case of pest control by a predator was weed-control in India in the mid-1800s. The American prickly pear cactus had been introduced to India to feed cochineal insects, which produce a valuable red dye. With no natural enemies in their

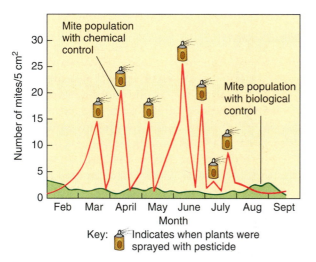

**Figure 21-6**   Biological and chemical control of pests: the number of cucumber-eating mites found on cucumber plants during the growing season. The green line shows the mite population with biological control by a parasite of the mite: the mite population remained small and stable. The red line shows the mite population with chemical control. The mites were sprayed with a chemical pesticide whenever they became so numerous that they could be seen to be damaging the plants. After each spraying, the population crashed but was soon exploding again as the few survivors reproduced.

**Figure 21-7** A prickly pear cactus. You can see why this plant is a terrible pest when it spreads uncontrollably: the spines pierce even thick leather gloves, and they are barbed so they are very hard to get out of your fingers.

new home, the prickly pears escaped and spread (Figure 21-7). They were finally controlled by the introduction of a *Dactylopius* beetle that eats the cactus.

Introducing a predator from a foreign country, however, is not always a good idea. In 1991, students at the University of Wyoming pointed this out to the U.S. Department of Agriculture (USDA). Every seven years or so, rangeland grasshoppers undergo population explosions in western states. Since the grasshoppers feed on the same grasses as cattle, this makes ranchers unhappy, and the grasshoppers are usually

controlled with chemical pesticides. This year the USDA decided to try biological control by introducing from Australia a wasp that fed on grasshoppers. Jeffrey Lockwood's students wrote a paper pointing out that the USDA should examine the effects of introducing exotic species before going ahead with the plan. Perhaps the Australian wasps would also kill beneficial insects such as hoppers that eat weeds. A USDA memo stated that the Wyoming work was so compelling that the USDA must reconsider the idea of importing exotic parasites.

## Pathogens

Pathogens are organisms that cause disease, and they can be used to control pests biologically. One of the most common pest-control pathogens is the bacterium *Bacillus thuringiensis*, usually abbreviated BT or Bt. BT causes disease that kills the caterpillars of moths and butterflies and other insect larvae (Figure 21-8). More than 1,000 strains of BT are known, each attacking specific larvae. Several strains of BT attack the caterpillars of gypsy moths and are the main spray used to control this pest.

**Genetic Engineering** A fascinating extension of the use of pathogens and predators is to build the working parts of the pathogen into the prey by genetic engineering. In 1987, the first field trials of a plant containing a transplanted pesticide gene took place. The gene came from the BT strain used to kill gypsy moth caterpillars. The bacterium produces a toxin that kills the caterpillar but is not toxic to most other animals. The problem with the BT spray is that it rapidly washes off plants, and it also breaks down in sunlight. So the gene for the toxin was transplanted into tobacco so that the toxin is produced within the tobacco leaf, and this has neither of the disadvantages of the spray. This technology is still in its infancy, but we can look forward to the day when many plants will be genetically engineered to make them more resistant to attacks by pests.

> Biological pest control is safer than chemical pest control because the organisms or chemicals are usually very specific and biodegradable.

## Biopesticides: Plant Secondary Chemicals

Many plants produce secondary chemicals that protect them from pests. Several of these chemicals have been isolated and marketed as **biopesticides**, pesticides produced by living organisms. Because they are produced by organisms, these pesticides are biode-

**Figure 21-8** Controllable by BT: a tomato hornworm (named for the "horn" at its tail) eating one of my tomato plants. Tomato hornworms are easy to control by just picking them off the plant and squashing them because the caterpillars are quite large and easy to see. But hand-picking is not practical if you have a lot of tomatoes.

gradable and usually less hazardous to humans, pets, and larger forms of wildlife than are artificial pesticides. Examples include rotenone and the pyrethrins.

For about 70 years, Indian researchers have studied the secondary chemicals produced by the neem tree, revered for centuries for its ability to ward off insects. Neem seeds produce chemicals that repel more than 200 species of insects and many bacteria. One of these chemicals is now marketed as Margosan-O. It kills various species of insects by interfering with their molting, reproduction, and digestion.

## Hormones and Pheromones

One effect of chemicals from the neem tree is to mimic insect hormones and pheromones. **Hormones** are chemical messengers within the body that have specific effects on other cells in the body. **Pheromones** are chemical messengers produced by one animal that affect the behavior of another.

Insects produce juvenile hormone, which normally prevents the larva from turning into an adult before it is mature. An accidental discovery in the 1970s showed that some plants contain substances similar to insect juvenile hormone. A Czech scientist visited Harvard University, bringing with him some insects that he had been raising for several generations in his laboratory in Europe. When he attempted to raise them in the United States, the insects grew into bigger larvae than usual but never turned into adults. After much searching, the scientists discovered that the paper towels placed at the bottom of the insect cages were the cause. The paper

manufacturer traced the paper to balsam fir trees, which, when tested, also prevented the insects from maturing. The active substance was extracted from hundreds of paper towels. It is a hormone-like substance that prevents maturation of members of a particular family of insects.

Several plants are now known to produce compounds that act as insecticides because they interfere with the hormones that control insect growth. Some of these are now produced on a large scale and sprayed as insecticides. They have the advantage of not being toxic to organisms except the particular insect species they affect.

Female moths find mates by releasing pheromones into the air that attract males from miles away. Some pheromones are now used for controlling pests on crops. In many cases, pheromones control insects much more effectively than chemical insecticides. At a 1990 conference, Manuel Jimenez of the University of California reported on cherry tomato crops treated with a pheromone that disrupts mating between tomato pinworm moths. The treated fields had only one tenth as many moths as control fields sprayed with insecticides.

The wine industry will accept grapes to make wine only if fewer than 2% of the grapes show insect damage. This low level of damage is difficult to achieve with insecticides, but vineyards treated with pheromones usually have even lower levels of damage. Pheromones are delicate chemicals that have to be manufactured and stored with care if they are to work. As a result, they are not yet widely used, but they clearly have great potential in reducing pest damage and will become more common in the future.

### Sterilization

Many female insects mate only once. They then lay their eggs and die. Knowing this, workers breed large numbers of insects and then treat the males with x-rays to make them sterile. The males are then released. In the field, they mate with normal females but do not fertilize the eggs. The females lay eggs that will not develop. Sterile males are being used in the attempt to control medflies in California (Close-Up 6-2).

### Traps

When you cook cabbage, broccoli, or mustard greens, they give off the odor of mustard oils, secondary chemicals found in this family of plants. Scientists discovered that flea beetles are attracted to the smell of mustard oils. They use them to pick out the cabbages and broccoli plants they feed on from a field containing many different plants. Traps containing solutions of pure mustard oils plus a sticky glue to catch the beetles will attract just as many flea beetles as will a real plot of cabbages, whereas beetles are not attracted to a trap containing glue alone. Mustard oils also attract cabbage white butterflies, which lay their eggs on plants that contain them. The caterpillars that hatch from the eggs eat the plants. Traps containing chemicals that attract insects are now used to control several insect species.

> Biological control of pests may involve the use of predators, parasites, hormones, sterilization, or compounds that attract the pest.

### Integrated Pest Management

Since the 1960s, county extension agents, who are government employees who advise farmers and home owners, have been promoting new methods of pest control. **Integrated pest management** (IPM) is a collection of methods for minimizing pest populations while minimizing cost, damage to the environment, the evolution of pesticide resistance, and damage to beneficial insects. Every year, extension agents tell growers what pests to look for and when particular control methods should be used. Their recommendations are based on computer records of pests, their whereabouts, population sizes, and breeding cycles every year. Biological control is used whenever possible. Growers are encouraged to spray chemical pesticides only when large populations of the pest are actually seen to be damaging a crop and to use different pesticides in different months and years.

> Integrated pest management minimizes the use of pesticides while keeping pest populations under control. It is much cheaper than chemical pest control.

In Pangyingkiran, Indonesia, residents gather to watch a play about rice paddies. Villagers act the roles of paddy workers, pesticide distributors, brown planthoppers, and spiders that prey on the planthoppers, teaching their audience about the rice paddy ecosystem. This unusual drama is the result of work by the United Nations to bring integrated pest management to Asian rice growing. In 1986, the brown planthopper had become a devastating pest in Indonesia, consuming four times as much rice as in the previous year, enough to feed 3 million people. In 1984, Indonesia for the first time grew enough rice to feed its population without imports, and the planthopper infestation threatened this progress. Research showed that chemical pesticides were making the problem worse by killing predators of the planthopper and causing the evolution of pesticide resistance. So in 1986, President Suharto banned 57 of the 63 pesticides used on rice and eliminated government subsidies for pesticide distribution. The money was used instead to educate farmers about integrated pest management.

Farmers are taught to recognize insects, such as the spiders that prey on planthoppers, as well as diseases and pests that do not reduce the rice yield. They learn how to calculate the right amount of fertilizer and when to apply it and to forecast how much pest damage the crop will sustain. If the damage will be unacceptably high, they learn to apply the minimum necessary amount of the right pesticide at the right time. When the white rice stemborer began to infest fields in Java in 1990, the government ignored cries from the villagers for massive distributions of pesticides and rallied 300,000 people to pick the stemborer's white eggs off the rice plants. The following year, the stemborer had disappeared from all but a few hectares of rice.

Education in IPM is an attempt to wean farmers in developing countries from the dependence on pesticides that has been thrust on them and their governments by chemical company advertising since World War II. The message, says Peter Kenmore of the United Nations, is that "trying to control pest outbreaks with insecticides is like pouring kerosene on a house fire." Where the message has spread, it seems to be working as well as in Indonesia. In 1992, farmers in Bangladesh who received IPM training spent 75% less on pesticides than untrained farmers and produced 14% more rice.

# LIFESTYLE 21-1

## Give Up Wearing Cotton?

Cotton is often the fabric of choice for environmentalists who see it as a natural product, not a synthetic fabric made from petroleum. But cotton is one of the most destructive crops as far as farmland and the health of workers is concerned. It requires more fertilizer and pesticides than any other major crop (Table 21-A) and usually ends up destroying the soil it is grown on. Some retailers offer organically grown cotton, but this is twice as expensive as cotton grown with pesticides.

In Uzbekistan, cotton is sprayed with defoliants to remove the leaves and make harvesting easier. Schoolchildren are brought in to harvest the cotton by hand and are sometimes working in the fields when the defoliants are sprayed. In the cotton-growing area, life expectancy is less than 60 years, and the infant mortality rate is 66 per thousand, more than twice the world average. Tens of thousands of hectares have been abandoned be-

cause plants will not grow in soil so polluted by fertilizer and pesticides.

Conditions are better in other parts of the world. Nevertheless, even in the United States, more pesticides are used on cotton than on any other crop, and pesticides are applied 15 to 20 times each season. As recently as 1988, the federal government attempted to exterminate the boll weevil by spraying cotton with huge amounts of Guthion. Nobody knows how dangerous Guthion is. The EPA is still reviewing studies designed to show whether fears that it causes cancer, birth defects, mutations, and sterility are justified. Guthion is certainly very toxic. Significant concentrations of the insecticide reached the water supply, and Guthion was blamed for killing large numbers of fish in more than 50 incidents in Georgia and Alabama.

**TABLE 21-A   Global Use of Pesticides on Some Major Crops**

| Crop | Herbicides ($ spent per million tons of crop) | Insecticides ($ spent per million tons of crop) | Fungicides ($ spent per million tons of crop) |
|---|---|---|---|
| Wheat | 1.6 | <1 | <1 |
| Maize (corn) | 2.8 | 1 | <1 |
| Rice (paddy) | 1 | 1.2 | <1 |
| Sugar beets | 1 | <1 | <1 |
| Soybeans | 13 | 1.3 | <1 |
| Sorghum | 1.3 | <1 | <1 |
| Cotton | 17 | 57 | 1.2 |

# SUMMARY

Enormous amounts of crops and livestock are destroyed each year by pests. Growers attempt to control pest populations by mechanical, chemical, biological, and other means. Since World War II, widespread use of chemical pesticides has done massive damage to human health and the environment.

## A Brief History of Pest Control

Before 1940, pests were controlled by methods including crop rotation, covering crops so pests cannot reach them, sterilizing the soil with heat, and poison-

ous chemicals, including plant secondary chemicals. All these methods are still used, but the chemicals have changed. Since World War II, nearly all pesticides have been organic compounds produced from petrochemicals. Since the 1960s, scientists have warned that chemical pesticides cause enormous damage to people and the environment.

## Pesticide Toxicity

Many pesticides are known to be toxic to humans and other organisms. Areas with heavy pesticide use have greater incidence of many human diseases than other areas. The environmental danger posed by a pesticide depends on its persistence as well as its toxicity. DDT caused the deaths of millions of birds even after it was banned. Persistent pesticides have polluted water supplies, rendering them unusable.

### Resistance and its Evolution

Pests evolve resistance to most types of pesticide. Generally the more pesticide is used, the faster resistance evolves. Resistance usually evolves faster than new pesticides can be developed.

### Regulation

In the United States, new pesticides for major crops are registered by the EPA for particular uses after extensive testing for safety and efficacy. Older pesticides and pesticides for minor crops such as fruit and vegetables are not registered and are less thoroughly regulated.

### Inadequate Regulation

Regulation fails in many ways. For instance, pesticides used in small amounts are not tested to see if they are nerve poisons or cause other illnesses. Most pesticides have not been tested for neurotoxicity. A law banning carcinogens in food actually frustrated the EPA's efforts to ban the most carcinogenic pesticides from food. Although most pesticides are reasonably safe if used correctly, they are often not used correctly. U.S. companies are allowed to export pesticides that are banned as too dangerous for use in the United States. American consumers are exposed to these pesticides when food treated with them is imported.

### New Methods of Pest Control

In developed countries, the use of chemical pesticides is beginning to decline. More and more biological controls are being developed and found to control pests more cheaply and effectively than chemical controls. Biological controls include pesticides made of secondary plant chemicals; predators and parasites that attack pests; and pathogens, hormones, and pheromones that kill the pest or prevent it from reproducing. Sterilized males may be released to control insect pests, and many insects can be trapped to reduce their populations. Across the world, extension agents are encouraging farmers to practice integrated pest management instead of widespread spraying of chemicals.

## DISCUSSION QUESTIONS

1. Are there any pesticides around your home? What type are they? Are they used according to the label directions?

2. Why don't we just give up using pesticides, which are expensive and time-consuming to use?

3. Can you think of anything you can do to reduce pesticide use?

4. Most people do not realize how much damage they do to themselves. They protest when they read that there are trace amounts of pesticides in the food they buy, and then they spray their houseplants with high concentrations of toxic pesticides when they find a few aphids on a leaf. How would you design a campaign to encourage people to use fewer pesticides around the house and to use them more carefully?

5. Less than 10% of U.S. food is organically grown, and it is usually more expensive than nonorganic produce. Why is it more expensive? Is this just a ripoff by organic growers, or are there economic reasons? Are you prepared to pay more for food or cotton grown organically?

6. If pests evolve resistance to pesticides, why do we continue to use pesticides?

7. The work of the students at the University of Wyoming (Section 21-F) recalls our earlier discussion of the problems of introducing exotic species. What are some other examples of introductions and their consequences?

# SOCIETY AND ENVIRONMENT

*An East Coast beach.*

# Chapter 22

# SOLVING THE PROBLEMS

## Key Concepts

- Many environmental problems are caused by a conflict between the short-term welfare of individuals and the long-term welfare of society.

- People are most inclined to work for the general welfare when their actions are effective and do not cost them very much.

- There is no obvious correlation between the political system of a country and its ability to solve environmental problems.

- A cost-benefit analysis is performed before committing human and financial resources to solving environmental problems.

- Our assessment of environmental risks is often flawed by poor understanding of the chances of events occurring; it gives extra weight to control, certainty, and avoiding loss.

- Regulation is a costly and sometimes ineffective method of solving environmental problems.

- Many governments are beginning to expand financial rewards and deterrents that encourage people and businesses to take steps that benefit the environment.

*A sign warning motorists to watch out for endangered diamondback terrapins, turtles that cross this road through a salt marsh during the spring.*

*EPA's funding priorities are more closely aligned with public opinion than with scientific assessments. Many environmental problems of relatively low risk, such as hazardous waste sites, receive public attention and federal resources, while problems the EPA's Scientific Advisory Board consider a greater risk, such as indoor air pollution and pesticides, receive far less attention and resources.*

Report to Congress of the
U.S. General Accounting Office

Why have we not done a better job of solving environmental problems as they arose? If population control had been more widespread during the nineteenth century, none of today's environmental problems would be nearly as pressing. If governments and businesses had always insisted on installing the best available industrial pollution control, we should not now be looking for expensive solutions to polluted air and water. By and large, we have not been good at predicting problems before they reach crisis proportions or at solving them efficiently when they become obvious. In this chapter, we investigate facets of human nature that affect how we solve, or fail to solve, environmental problems.

## 22-A THE TRAGEDY OF THE COMMONS

In 1968, ecologist Garrett Hardin published an influential parable entitled *The Tragedy of the Commons*, which became the theoretical backbone of the environmental movement. In it, Hardin argued that the main difficulty in our attempts to solve problems such as overpopulation and pollution is the conflict between the short-term welfare of individuals and the long-term welfare of society.

Hardin illustrated this conflict with the example of commons, such as those of medieval Europe or Colonial America. A common was grazing land that belonged to the whole village. Any villager could graze cows and sheep there. It was in the interest of each individual to put as many animals on the common as possible, to take advantage of the free animal feed. However, if too many animals grazed on the common, they destroyed the grass. Then everyone suffered because no one could raise animals on it. For this reason, common land was eventually replaced by individually owned, enclosed fields. Before the era of subsidized agriculture, owners were careful not to put too many cows on one patch of grass because overgrazing one year would mean that fewer cows could be supported the next year.

The tragedy of the commons is not some moral failing. It is a result of natural selection. During the course of evolution, people who gave priority to increasing the health and wealth of their own families raised more children than those who put their village or nation first. They also raised more children than those who reduced their consumption to conserve resources for the future. The parable of the commons throws fresh light on a broad range of public problems. Congress as a whole deplores the budget deficit, but each member goes on increasing it with expenditures that benefit his or her own district. Whalers and loggers often contribute to the future destruction of their own jobs.

Hardin was sure that the commons (which we can now think of as all natural resources) are limited and that people will pursue their own self-interests, destroying the commons to the point of society's collapse. He thought that the only possible solution was the emergence of a ruling elite with the power to control individual freedom in the collective interest.

> Conflict between the short-term welfare of individuals and the long-term welfare of society makes it difficult to solve environmental problems.

Hardin was certainly right in thinking that individuals acting separately cannot be expected to solve all environmental problems. If we, as individuals or as corporations, knew we should pay directly for the overpopulation and the pollution each of us causes, we would each have fewer children and contribute less to pollution. An example comes from laws that made employers liable for on-the-job injuries. When a firm knows that injuries will cost it money, it takes steps to ensure greater safety so fewer injuries occur. In the case of ecological problems, however, it is often not possible to assign responsibility directly to the people who cause the problem. Future generations will pay most of the price for the fact that we have too many children and cause soil erosion now, and our individual contributions to water pollution cannot be easily distinguished. Thus, many environmental problems can be solved only by actions by groups such as governments, agencies that, at least in theory, can look beyond the immediate interests of individuals and plan for the long-term welfare of society.

### Lifeboat and Foreign Aid

Garrett Hardin pursued his argument to its logical conclusion in a 1974 essay entitled *The Ethics of a Lifeboat*, which explores the way the wealthy countries of the world treat their impoverished neighbors. An allegorical lifeboat filled with the world's rich is surrounded by the struggling poor, who attempt to clamber aboard, like present-day environmental refugees (Section 8-C). How should the rich behave? They cannot let everyone into the lifeboat because it would sink. If they allow some into the boat, they remove the lifeboat's safety margin and are presented with the further ethical dilemma of whom to save and whom to condemn. To Hardin, the only sensible course of action is to ignore the pleas for help and maintain the boat's safety margin.

The lifeboat argument proposes that supplying economic aid to poor countries makes things worse. The argument is particularly strong in the case of aid to countries, such as Tunisia and Sudan, that are not effectively curbing population growth. If a poor country can call on aid in times of need so its people do not starve to death, its population and its need continue to grow indefinitely, reducing still further the resources that are left for future generations. Lifeboat philosophers argue that the most humane course, in the long run, is to withhold aid so that starving people die. The smaller population that remains will have a greater chance of developing sustainable agriculture that does not degrade the environment and can feed the population.

> One argument against foreign aid is that if a poor country can call on aid in times of need, its population and its need will continue to grow indefinitely.

Happily for those who do not like the lifeboat solution to environmental problems, there are flaws in Hardin's analysis. He applied the idea to the population explosion, arguing that relying on voluntary birth control would not "rescue us from the misery of overpopulation." But his analysis assumes that children are an economic benefit to their parents, so people want to have as many children as possible. On subsistence farms in developing countries, children are a financial asset. However, in developed countries, and in many urban areas in all countries, children have become expensive luxuries. As a result, reproductive rates have fallen rapidly in the second half of the twentieth century. The lifeboat argument also assumes that countries are content to remain on the verge of starvation and dependent on other countries. This is just not so. National pride is an outstanding feature of the twentieth century, and it is hard to think of any government that does not want a higher standard of living for its people, leading to increased international respect.

## 22-B  SOLUTIONS TO THE TRAGEDY OF THE COMMONS

Even though having children is not always in the immediate self-interest of an individual, if Garrett Hardin's analysis of human nature were the whole story, it would make the task of solving environmental problems almost impossible. Acting in our own self-interest, we should continue to destroy Earth's resources until we became extinct. However, Hardin ignored important additional facts about the nature of human beings: we are intelligent, deeply social animals, and we are capable of altruistic behavior.

### Social Behavior and Altruism

Altruistic behavior is behavior that benefits others at the expense of the actor (Figure 22-1). Altruistic behavior is part of our makeup because we are social animals. The biological basis of social behavior is that we depend on cooperation with other people for our very survival. This is ultimately why we can prevent ourselves and each other from damaging the environment. The megalomaniac chieftain who burns

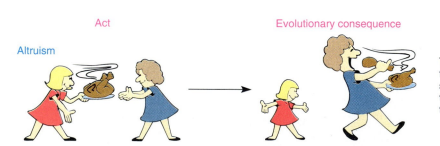

Act    Evolutionary consequence

Altruism

The altruist reduces her own fitness but increases that of her sister so much that the genes they share are more common in the next generation.

**Figure 22-1**   The evolution of altruistic behavior. Altruism evolves only in animals that live in social groups with their relatives. The group is represented here by an actor (*red*) and her sister. In such groups, an individual contains many genes that other members of the group also carry. These common genes may become more common in the next generation and spread through the population if individuals in the group help each other at slight sacrifice to themselves.

down the local woodlot to enlarge his wheat field runs the risk that he will be thrown out of the village, and if he is ejected from human society, he will die. The many heads of nations who have abused their citizens and natural resources and then been deposed (and often killed) are examples of this fact. So if our urge to work first for ourselves and our own families endangers the environment, our need to be accepted by society can save it. When enough people, in a village or a nation, want a particular reform, the rest must go along with it. You can make almost any change in society if you can convince enough people that it is the right thing to do.

As we would also predict from this analysis, people are more inclined to work for the general welfare if their actions are enjoyable or are effective and do not cost them very much. Volunteers will band together to plant trees or pick up roadside litter because such social occasions are effective and fun. Public opinion polls show that, even during a recession, the vast majority of people will accept tax increases to pay for pollution control and the conservation of resources. Politicians are often surprised at the results of such polls, but they should not be. People are behaving logically in voting for effective environmental action that will not cost each of them very much.

> Because humans cannot survive except in social groups, you can make almost any change in society if you can convince enough people that it is the right thing to do.

## Political Systems and Environmentalism

Environmental scientists have long asked whether particular political ideologies make it more or less likely that a country will develop in a sustainable manner. The answer seems to be that there is no simple connection between ecologically responsible economics and political systems. Consider the world's two major economic systems in relation to the tragedy of the commons and human altruism. Capitalism works on the assumption that individuals work best when they pursue their own interests. Communism assumes that people work best when they work for the good of society. In fact, neither of these assumptions reflects both sides of human nature, and capitalist and communist governments tend to fail unless they take this into account.

For instance, communist governments find it is extremely difficult to galvanize an individual to work for the good of the community when that work con-

flicts directly with the individual's interest. Communist governments have to regulate the amount of time individuals can spend on their own gardens instead of on the state farm, or little would get done on the farm. Frequently little gets done on the state farm anyway because people may put in the hours, but it is difficult to make them put in the work. The Chinese government found that its economy improved rapidly when it permitted people to work for their own individual benefit as well as for the state.

Capitalist governments find that if individuals are given the freedom to do so, they will reenact the tragedy of the commons, cutting down forests, building houses on wetlands, and generally squandering natural resources. Capitalist governments have had to take over many functions that benefit society but not the individual. Laws are needed to prevent individuals from killing turtles and whales, marketing dangerous drugs, and polluting rivers.

Political philosophy is not a good predictor of whether a government is progressive or backward in protecting the environment. The Soviet Union had one of the world's worst environmental records, with dangerous nuclear power plants, disastrous agricultural policies, and appalling air and water pollution. China, in many ways a more rigidly communist state, leads the world in experiments with population policies, reclaiming sand dunes, reforestation, and organic agriculture. In the capitalist democratic world, the United States and Costa Rica provide world leadership in research and conservation, whereas Brazil seems unable to prevent wholesale destruction of the Amazon rain forest.

> Governments that are progressive in environmental affairs are to be found among capitalist and communist nations, democracies, and dictatorships.

## Actions of Governments

The solution to our environmental problems lies in modifications of individual and collective behavior sponsored by a government that can enforce compliance of everyone concerned. In a democratic society, this can be done only with the support and consent of a well-informed public.

A government can act in four main ways to solve environmental problems:

1. **Research.** Before any problem can be solved, it has to be researched. For example, water must be analyzed to find out if it is polluted. Then the source of the pollution must be found. Experi-

**Figure 22-2** Educating the public. A sign on a California bridge in 1991 during a drought reminds motorists that every drop of water should be conserved. *(Richard Feeny)*

ments are needed to find out how rapidly the water supply will purify itself if the pollution is stopped. Although much environmental research applies to many different situations, some of it has to be done at the local level. For instance, some construction projects cannot be approved without a study of how the project will affect a particular environment (Close-Up 22-1). Not surprisingly, those countries and localities that devote the most resources to research tend to have the best environmental records.

2. **Education.**   Educated people can understand the tragedy of the commons and do not really want to make Earth an expensive, unpleasant place for future generations or for their own old age. So programs to educate the public about environmental problems are vital. For instance, no town can afford enough workers to keep the streets free from litter all the time. Yet many urban areas are spotlessly clean, although they are seldom swept—because the people who live there have been raised not to drop litter themselves and to pick up garbage in the streets when they see it. The photograph at the beginning of this chapter shows a road sign, put up every spring by a conservation group, that alerts motorists to avoid diamondback terrapins that cross this road during the breeding season. Every year, the local media cover the group's efforts, educating the public about the endangered status and biology of the

terrapins. During the drought of the 1980s, California resorted to water rationing, but it also educated residents and visitors about the need to conserve water (Figure 22-2).

3. **Incentives.**   Probably the most effective way for a government to reduce pollution, for example, is to apply incentives (often in the form of tax reductions) to those who pollute less and to levy a "pollution tax" against individual and corporate polluters (Section 22-D).

4. **Regulation.**   When all else fails, governments must resort to regulation, banning particularly dangerous practices altogether. Regulation becomes necessary when a practice is so damaging that no financial value can be placed on it. For instance, it is not enough to permit a firm to market thalidomide (a drug that caused severe birth defects) or to leave poisonous pesticides in our food and then to tax the firm for it afterward. Second, regulation is the only solution when it is impossible to assign responsibility with any degree of justice. How should the price for sulfur discharged into the air by a coal-fired power station be distributed among the utility that owns the station and the consumers who use its electricity?

> Governments can solve environmental problems mainly by research, education, incentives to private action, and regulation.

# Environmental Impact Statements

In the United States, the National Environmental Policy Act of 1969 led to one of the best-known environmental regulations: the requirement that nearly all federal agencies file an environmental impact statement (EIS) for any proposed legislation or project that would have a significant effect on the quality of the environment. The U.S. government has seldom enforced this regulation, but the idea of producing such statements has been very influential. Most states now require EISs for state projects, and the legislation has been imitated in about a dozen other countries.

In general, a draft EIS must be made available to the public and the relevant regulatory agencies 90 days before a proposed action. A final statement, taking into account comments and objections, must be made public 30 days before the action. Every EIS must include the justification for the proposed action, the resources that will be committed to the project, the effects of the proposed action on environmental quality in the long run, any unavoidable adverse effects on the environment, and any possible alternatives to the action that might have different environmental consequences.

EISs usually do not work as their designers intended. Too often they are prepared only to justify a decision that has already been made. Because an EIS may be required for a trivial project, a lot of time and money is wasted generating heaps of paper that no one ever reads. In theory, an unfavorable EIS should lead to a project's being canceled or a less damaging alternative used in its place. Often this does not happen. The shortcomings of EIS regulations should be corrected. But this does not mean that producing an EIS is a useless exercise. On the contrary, the need to produce an EIS has spurred enormous amounts of environmental research. It has made planners consider environmental questions in ways that would never have occurred to them before 1969.

As an example of the uses of an EIS, the Army Corps of Engineers was asked to dredge part of the Savannah River so as to extend the Port of Savannah. The EIS showed that the dredging would flood a local wildlife sanctuary and drain wetlands vital to the local shrimping industry (Figure 22-A). Planners are now searching for ways to enlarge the port without doing this kind of damage. In the old days, we should probably have dredged first and discovered the damage when it was too late to correct it. Hundreds of plans for dams, highways, and airports have been canceled or modified as a result of public and government reaction to EISs.

**Figure 22-A** A plan to enlarge the Port of Savannah would have destroyed this marsh, home to oysters, crabs, shrimp, and flounder. This egret would suffer, and so would the local fishing industry.

## 22-C COSTS, BENEFITS, AND RISKS

Before deciding how best to solve environmental problems, governments and individuals take into account the cost of a project, the benefits to be gained from it, and the risks of not undertaking it. Economists point out that it is often worth spending large sums to prevent environmental damage. Suppose a billion dollars worth of trees are killed by air pollution each year. In estimating how much to spend controlling air pollution, a government should add the value of the trees killed, the cost of purifying fresh water polluted from the air, the cost of cleaning and renovating buildings damaged by air pollution, the cost to human health, and all the other indirect costs of air pollution. As we discover more and more indirect costs of pollution, we find that pollution costs us enormous sums. The United States spends more than $100 billion a year on pollution control, but economists are convinced that much greater investments would pay large dividends.

### Cost-Benefit Analysis

When we decide how much to spend controlling air pollution, we perform a **cost-benefit analysis**: balancing the cost of the action against the benefits we expect from it. The chance of unfortunate outcomes

### FACTS AND FIGURES 22-1

#### How Do You Assess Risk?

Here is one study on how people assess risk. Answer the questions yourself and then see how the doctors who took part in the study responded (at the end of the Discussion Questions at the end of this chapter).

This will be most interesting if you can get a whole class to answer the questions and find out how most of them respond.

1. Which of the following programs would you support to deal with a disease that would kill 600 people if left untreated?
   a. Program A will save 200 people.
   b. Program B has a one-third chance of saving everyone and a two-thirds chance of no one being saved.
2. For the same disease, choose between
   a. Program C, which would result in the death of 400 people.
   b. Program D, which gives a two-thirds chance that all 600 people will die but a one-third chance that nobody will die.

of the action is called the **risk**. The cost-benefit analysis of solving an environmental problem tends to depend on who is doing the analysis. To an industry, the cost of pollution control or a safety measure may seem all-important. Industry officials often argue that the cost will be passed on to the consumer or taxpayer, who will refuse to pay it. The public tends to view the cost of any single measure as trivial and the public purse as bottomless.

### Risk Assessment

Cost-benefit analysis involves assessing risks because risk of an undesirable outcome is one of the costs of any action. Risk assessment helps us to create cost-effective laws to protect human health and the environment. It is important that the risk a hazard actually poses should equal the risk the public perceives. If the public perceives the risk as greater than it really is, legislators may enact overprotective laws that cost more than they need to achieve their goal. How good are we, the general public, at risk assessment? Recent studies and the quotation at the beginning of this chapter suggest that we are erratic in our judgments and overinfluenced by media coverage of risks (Table 22-1). Before reading further, you might like to find out how you assess one variety of risk by taking the quiz in Facts and Figures 22-1.

**TABLE 22-1  Rankings of Some Environmental and Health Risks by General Public and Scientists**

| Scientists | General Public |
|---|---|
| **Higher Risk** | |
| 1. Global warming | 1. Chemical waste disposal |
| 2. Indoor air pollution | 2. Water pollution |
| 3. Exposure to chemicals in consumer products (e.g., pesticides, food additives, plasticizers) | 3. Chemical plant accidents |
| 4. Surface water pollution (risk to the environment) | |
| 5. Pesticides | |
| **Lower Risk** | |
| 1. Hazardous waste sites | 1. Indoor air pollution |
| 2. Underground storage tanks (mainly of petroleum) | 2. Exposure to chemicals in consumer products |
| | 3. Global warming |
| | 4. Pesticides |

## Perception of Risk and Gain

The manner in which a risk is presented affects how we assess it. Losses strike us as more important than gains, so we are more willing to gamble to avoid losses than to achieve gains. In addition, few people understand the rules of probability (chance), which is what determines risk. Because we don't understand probability, certainty is appealing. In one study, people playing a simple game of chance would pay more money to increase their chance of success from 90 to 100% than they would to increase it from 60 to 70%, although the 10% increase in performance gives the same result in both cases. Our desire for absolute certainty tends to make low probabilities seem higher and high probabilities lower than they actually are. A high probability of success, such as 85%, may seem insufficient, and a low probability of failure, such as 5%, may seem unacceptable.

The most important risk we usually have to consider is the risk of death. When asked about causes of death, however, most people overestimate the death rate from sensational causes such as homicide, tornadoes, botulism, and drowning (Figure 22-3).

**Figure 22-3** Where is the risk in this photograph of a sailboat race? Drowning? No. The worst risk, from this and many other outdoor sports, is sunburn. Sailors are much more likely to die from melanoma (a form of skin cancer) than from drowning.

This is because we get most of our information from the media, which tend to emphasize the spectacular.

Because we are often ignorant of the risk of death, a technology may be replaced by one that is actually riskier. For instance, experts estimate that coal-generated electricity costs some 10,000 lives in the United States each year through mining, transportation, and pollution. Nuclear reactors have caused fewer than 500 deaths in 30 years. Why then has public pressure brought the nuclear power industry almost to a halt? Is it because we have an inordinate fear of radiation? On the contrary, in one study in which people were asked to evaluate the risk of death caused by various technologies, experts considered medical x-rays far more dangerous than did members of the public. Yet the public voted nuclear power the riskiest technology on the list, whereas experts ranked it twentieth as less risky than railroads and riding a bicycle.

Are people really as ignorant about the risk of death from nuclear power stations as this study implies? Probably not. Millions of people live within a few miles of a nuclear power station without worrying about it at all. In describing nuclear power as risky, the general public is obviously giving much more weight than the experts to several factors. One of these is the risk of a cataclysmic nuclear accident killing large numbers of people. Another is the possibility of radiation-caused cancers in the future and of genetic damage to future generations (although this applies to medical x-rays as well). The public is also probably expressing distrust of governments, who tell them that nuclear power plants are safe. Many people have learned not to trust their governments. In the end, the public dislike of nuclear power is an example of the fact that we will take bigger risks to avoid losses than to achieve gains. We risk the lives of the many people whose job is to produce coal-generated electricity to avoid the small risk of massive death and disaster from nuclear plants.

## Voluntary Risk and Inadequate Data

We often take very high risks when they are voluntary but avoid low risks that are involuntary. For instance, many people smoke cigarettes but are frightened of flying.

We are more prepared to take a risk when we think we understand or can control the situation than when we do not. In one study, respondents said that the risks of nuclear power were involuntary, uncontrollable, unknown, inequitably distributed, likely to be fatal, potentially catastrophic, and dreaded.

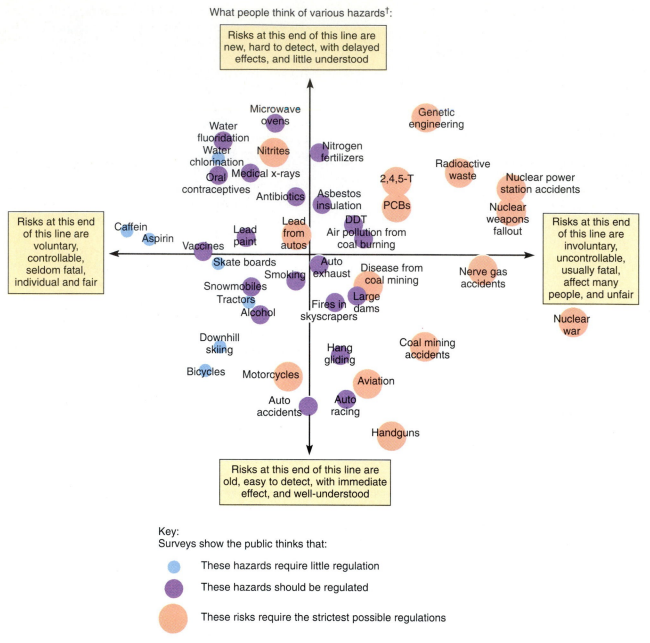

**Figure 22-4** The results of a study to find out what people think of the risk from various activities and whether those activities should be controlled by government regulations.

Automobiles evoked few of these concerns, although, objectively, they have many of the same characteristics (Figure 22-4). The driver often has no control over a road accident; there is little justice in the distribution of road deaths; and crashes actually, not just potentially, kill many people. It turns out that part of the judgment in this case involves the riskiness of assessing the risk. We consider technologies such as nuclear power, toxic wastes, and genetic engineering highly risky, partly because we know that there is no proven track record by which anyone can assess the risk accurately. So we do not trust the judgment of the "experts" who tell us what they think the risks are. This is part of our yearning for certainty—we prefer knowing that coal-generated electricity on average kills a certain number of people each year to not knowing how many people nuclear power might kill.

> When people assess risk, they tend to give extra weight to factors such as control over the risk, certainty of the risk, and avoiding loss.

## Tacoma's Copper Smelter and the Price of Life

Consider the cost-benefit analysis of one environmental decision. The benefit of a copper-smelting plant owned by Asarco, Inc., in Tacoma, Washington, was that it provided about 600 residents with jobs and contributed some $30 million each year to the local economy. The cost was that the smelter poured out smoke containing poisonous arsenic.

In 1987, Asarco had reduced its arsenic emissions in accordance with the Clean Air Act and now argued that the cost of any further reduction would force it to close the smelter. Regulatory officials appeared to have a choice: put up with the arsenic emissions or lose 600 jobs. William Ruckelshaus, then head of the Environmental Protection Agency (EPA), suggested that Tacoma residents hold town meetings to discuss the question. This proposal angered some, who argued that residents were being asked to rate the cost-effectiveness of their lives. (The argument became moot shortly thereafter when Asarco went bankrupt and closed the smelter.)

In fact, putting a price on life is something that people and governments do all the time. When a jury awards damages in an injury or wrongful death suit, it decides how much money a particular life is worth. When a government writes environmental regulations, it decides, at least in theory, how much money it is worth spending to prevent a death. Table 22-2 shows approximately how much it costs to save one

life by complying with various U.S. government regulations. Notice that, as in the case of altruism, the most cost-effective measures are those like seat belts that cost little and are very effective. Treating wood-preserving chemicals as hazardous waste is an expensive way to save lives because the chemicals kill few or no people anyway.

## 22-D REGULATION VERSUS MARKET FORCES

Table 22-2 suggests that many government regulations are amazingly expensive and save few lives. This does not mean that the regulations do no good because they may have goals other than saving lives. Banning hazardous waste from standard landfills, for instance, has goals such as preventing pollutants leaking into ground water, ensuring that the land over a closed landfill is safe to build houses on, and saving money by recycling waste (Lifestyle 18-1).

Once a regulation is in place, it is hard to get rid of because there are many people whose livelihoods depend on regulations. In 1993, the federal government employed 125,000 people just to write up the details of more than 4,000 new regulations demanded by Congress. Hundreds of thousands more people are employed to enforce regulations. Nevertheless, there is growing revolt against environmental regulations viewed as costly and ineffective. The EPA is charged by Congress with issuing environmental regulations (Close-Up 22-2). State and local governments must comply with these regulations, although Congress often provides no money to pay for the program. For instance, California calculates that it will cost the state nearly $4 billion and the nation about $20 billion to meet federal standards for radon in drinking water. Even after this vast sum was spent,

| TABLE 22-2 Estimates of the Cost-Effectiveness of Selected U.S. Regulations* | | |
|---|---|---|
| **Regulation** | **Year Issued** | **Cost per Premature Death Prevented** |
| Car seat belt standards | 1984 | $100,000 |
| Aircraft floor emergency lighting standard | 1984 | $600,000 |
| Banning flammable children's night clothes | 1973 | $800,000 |
| Side-impact standards for trucks and buses | 1989 | $2.2 million |
| Electrical equipment standards for coal mines | 1970 | $9.2 million |
| Limits on exposure to coke (coal) ovens in workplace | 1976 | $63.5 million |
| Limits on exposure to arsenic in workplace | 1978 | $106.9 million |
| Asbestos ban | 1989 | $110.7 million |
| Ban on disposal of hazardous waste in standard landfills | 1988 | $4.19 trillion |
| Hazardous waste listing for wood-preserving chemicals | 1990 | $5.7 quadrillion |

*Source: Center for the Study of American Business, *Regulatory Programme of the U.S. Government, 1991–1992.*

# The Environmental Protection Agency

The EPA is charged with administering a large and growing number of environmental laws. Table 1 summarizes the main legislation that the EPA administers. The Agency is very overworked, which is one reason that environmental laws are not enforced as well as many people would like. Figure 22-B shows that the agency's operating budget has remained essentially constant even as Congress has presented it with new environmental laws to enforce.

| TABLE 1    Major Laws Administered by the EPA | |
| --- | --- |
| **Law** | **Purpose** |
| Clean Air Act | Protect and enhance air quality |
| Federal Water Pollution Control Act Amendments of 1972* | Restore and maintain the chemical, physical, and biological integrity of the nation's waters |
| Safe Drinking Water Act | Protect the quality of all sources of drinking water |
| Federal Insecticide, Fungicide, and Rodenticide Act (FIFRA) | Regulate the distribution, sale, and use of pesticides |
| Resource Conservation Recovery Act of 1976 (RCRA) | Regulate the generation, transportation, treatment, storage, and disposal of hazardous wastes; correct leaking underground storage tanks |
| Toxic Substances Control Act of 1976 | Require testing and, if necessary, restrictions on chemicals |
| Marine Protection, Research, and Sanctuaries Act of 1972 | Regulate the dumping of all materials into the ocean and prevent or limit the dumping of materials that adversely affect the environment or human health |
| National Environmental Policy Act of 1970 | Require EPA to review environmental impact statements |
| Environmental Research, Development, and Demonstration Authorization Acts | Authorize appropriations for various EPA research programs |

*The date given is the date the act was first passed. All these acts have subsequently been amended.

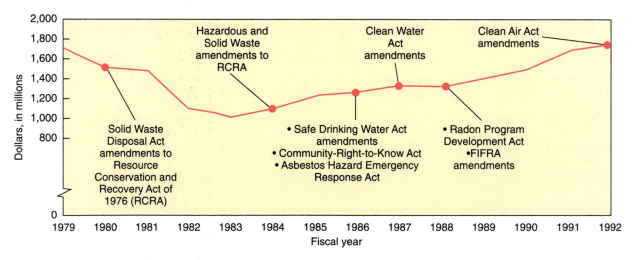

**Figure 22-B** The operating budget of the EPA from 1979 to 1992 (in 1982 dollars) with new responsibilities during that period. (The budget figures do not cover extra funds for Superfund, construction, and leaking underground storage tanks.)

public exposure to radon would be reduced by only 1%.

A 1993 study found that complying with EPA regulations costs many cities more than one fifth of their budgets. As examples of regulations that are expensive to comply with and produce little benefit, an Ohio study questioned a requirement under the Safe Water Drinking Act that local governments continually monitor drinking water for 133 specific pollutants, many of which have never been detected in Ohio water. Even if the pollutants are present, the study questions the safety standards. The drinking water of Columbus, Ohio, contains an average of less than 3 parts per billion of the herbicide Atrazine. But the level in some water sources occasionally rises above 3 parts per billion. Therefore, the city is supposed to install "best available technology" to remove Atrazine from the water at a cost of $80 million—which is not expected to save a single life or prevent an illness.

> Regulation involves a cost-benefit analysis of benefits to be gained compared with the cost of regulation.

## Setting Standards

Who decides on the health risk from Atrazine, radon, or any other pollutant in air or water, and how do they set safety standards? Regulators start with tests of the effect of the pollutant on bacteria and on animals such as rats and mice. Then they use these data to estimate health effects in humans and add in a safety factor. These estimates of risk are often highly uncertain. Consider the 1989 case of Alar, a chemical used to make apples red. Scientists had showed that Alar caused mutations in the genetic material of bacteria (the "Ames test"), suggesting that Alar might be carcinogenic (cancer-causing). Environmentalists clamored for the EPA to ban Alar for fear it was causing cancer, particularly in children, who eat more apples than adults. How does the EPA proceed?

To determine the risk of a child developing cancer from Alar, we need to know how carcinogenic Alar is: how many cases of cancer are caused by exposure to a given amount of Alar? Then we need to know how much Alar a child ingests when eating an apple and how many apples a child eats in a year. Carcinogenicity is tested by feeding the chemical to animals to see how many of them develop cancer. The rats and mice used in these tests are bred to develop cancer very easily. If they did not, the chemical would have to be fed to millions of animals for

many years to obtain just a few cases of cancer. Nevertheless, the tests require many animals and several years, and the tests on Alar were not complete in 1989.

How many apples a year does a child eat? There were several conflicting studies on this, and regulators had to decide which one to trust. The EPA used a 1977 study of 30,000 people. The environmentalists preferred a 1985 study of only 2,000 people. The EPA discounted the 1985 study because so few people were questioned, and only 65% of them responded to the survey. The environmentalists criticized the EPA for using an outdated study when people apparently eat more apples now than they did in 1977. A new, larger study of apple eating was commissioned to settle the question.

The end result was that growers gave up using Alar on apples so that they could advertise "Alar-free" apples. Then the EPA banned the use of Alar on apples. Then the health studies were finished and showed essentially no risk of cancer from eating Alar-treated apples.

Paul Portney, director of the Center for Risk Management, sums up the argument. "No one was right or wrong. There were plausible arguments on both sides. I suspect the EPA risk estimate is usually high because of the conservatism built into their procedures. But it might be low. Once a risk assessment is done, people tend to forget all the assumptions they made along the way and attach too much certainty to the final number. But there are few alternatives. If we don't try to make quantitative risk assessments, how do we decide what to regulate—by rolling dice?" Perhaps not by rolling dice, but we certainly should recognize that regulation is an expensive and often inaccurate way to solve environmental problems and should be used only when we cannot think of any other solution.

> A great deal of uncertainty is involved in setting standards for permissible levels of pollutants in air, water, and food.

## Solutions Other than Regulation

The main alternative to regulations for solving environmental problems is the use of financial incentives. This is usually fairer and more cost-effective than regulation. For instance, when European countries wanted to change from leaded to unleaded gasoline, several of them increased the tax on leaded gasoline and lowered it on unleaded gasoline. This provided a financial incentive for motorists to have their cars converted so they would run on unleaded gas or to turn in the old jalopy for a new car because these use unleaded gas. After a time, cars that would run only on leaded gasoline disappeared from the roads. Compare this with a regulation proclaiming that all motorists must convert their cars to run on unleaded gas. This would cost the owner of a new car nothing but would be very expensive for the owner of an older car. Incentive schemes are gaining favor in many countries. They include charging factories for the volume of waste water they discharge, charging extra-noisy aircraft higher landing fees in Switzerland, and instituting returnable deposits on car bodies (to stop them being dumped) in Sweden and Norway.

Even where regulations on pollution are justified, they are effective in dealing with point pollution but give little control over nonpoint pollution. Consider the regulation that used motor oil must be recycled and not spilled on the ground. It takes relatively few inspectors to force service stations to account for all the oil they use, but there is no practical way to force millions of car owners to comply with the regulation. A much more effective solution would be to provide an incentive for home owners to recycle their oil, for instance, by repaying a deposit when a home owner deposited used oil at a collection center. This is the principle behind bottle bills: charge the consumer a deposit on each can or bottle and repay it when the container is returned. The best hope of getting small polluters to clean up their act is ensuring that the price is right. Here are two examples:

1. Germany is considering car registration fees based on the dirtiness of the exhaust. In every country, 10% of vehicles produce 90% of exhaust emissions. So when you go to pay your annual car registration, the clerk would stick an emissions meter on your tailpipe, and it would ring up the registration fee—very high for polluting vehicles. You can repair the car so as to reduce emissions or pay the high fee. This is a much cheaper solution to emissions control than California's system. In California, emissions are checked regularly, and those whose cars fail the test are fined. Nobody much likes this system, which generates bureaucracy, paperwork, and cases of bribery.

2. One snag with regulations is that they provide no incentive to reduce the level of pollution below that required by the regulations, even if that would be easy to do. Part of the U.S. Clean Air Act of 1990 tackles this problem, encouraging industries to clean up their emissions as much as

possible. The Act in essence says to industries, "We want no more than 10 million tons a year of industrial emissions in the air by 2000. That is 20,000 tons for each company. You can either get your own emissions down to that level or you can buy spare emissions from a cleaner company." Thus, a company that generates 35,000 tons of emissions each year can spend the money to reduce its own emissions to less than 20,000, or it can pay a company that produces only 5,000 tons each a year to buy their spare 15,000 tons. A company engaged in reducing its emissions has a strong incentive to reduce them as far as possible, not merely to 20,000 tons a year. If it can reduce its emissions to 2,000 tons a year, it will have 18,000 tons left over to sell to other industries and can make a lot of money doing so.

There are a lot of possibilities in this approach. For instance, a state might abolish the sales tax on food and clothing (necessities) and double it on all containers that are nonreturnable (polluting). The state would raise about the same amount of revenue, and the volume of solid waste would fall, at least slightly. A state could charge towns $1 for each liter of waste water they dumped into the local waterway. The town could either pay up, providing revenue for the state, or it could invest the money in a water treatment plant.

> Governments are experimenting with controlling environmental hazards by inducing private action instead of by regulation.

## 22-E  PLANNING FOR THE FUTURE

This review of the ways people and societies analyze and solve environmental problems gives us some insight into the future. We understand the scope of the major environmental problems, and the developed nations started to solve these problems at the beginning of the twentieth century. We have never been in a better position to solve our problems, and the need has never been more urgent. A noteworthy development is the beginning of international cooperation in this area. For instance, the North American Free Trade Agreement (NAFTA) depends on Mexico, Canada, and the United States to enforce much the same pollution control laws. Otherwise, a company could move to the country where it did not have to spend money on pollution controls, giving it an unfair advantage over companies in the other countries.

One of the most important features of the twentieth century is communications. Telephone, radio, and television (even for those who cannot read) and publishing (for those who can) reach essentially everyone in the world. Staggering amounts of information are gathered and disseminated to a vast audience. We are so used to worldwide broadcasts by satellite that it is hard to remember how recent they are. The first global satellite hookup was made in 1967. It broadcast throughout the world a live performance by the Beatles: "All You Need Is Love."

An important effect of communication is to broaden the political power base. Even under totalitarian regimes without a free press, there are few corners of the globe in which people do not hear of developments elsewhere and where they cannot bring some influence to bear on their leaders. This trend can only continue as time goes by. Its main advantage is massive informal education. Never before have so many people understood why overpopulation, pollution, and soil erosion imperil their future or been in such a good position to do something about them.

> The increase in worldwide communications is increasing the number and proportion of people who understand and care about environmental issues.

## SUMMARY

Societies have failed to solve most environmental problems as they arise. This is partly because there are aspects of human nature that complicate our efforts to solve such problems.

### The Tragedy of the Commons

Garrett Hardin's parable of the tragedy of the commons argued that the main difficulty in our attempt to solve environmental problems is the conflict between the short-term welfare of the individual and the long-term welfare of society. Hardin later argued that food aid to poor nations makes their problems worse in the long run.

### Solutions to the Tragedy of the Commons

Part of Hardin's argument seems flawed by inaccurate assumptions. Humans need social acceptance to survive, and they are capable of altruism. These two factors make environmental solutions possible. None of the world's political or economic systems seems better than any other at solving environmental problems. This may be because all systems emphasize certain aspects of human nature but ignore others. Governments can act in four main ways to solve environmental problems: by using research, education, incentives, and regulation.

### Costs, Benefits, and Risks

To make sensible decisions about how much money and effort to spend on particular environmental problems, we balance the cost of the effort against the benefits we can expect from it. The risk of death or injury is one of the costs of an environmental problem or its solution. People's assessments of risk are often illogical, sometimes causing illogical use of resources. In addition, we will put the greatest effort into avoiding risks that we consider are involuntary, uncontrollable, unknown, inequitably distributed, potentially catastrophic, and dreaded.

### Regulation Versus Market Forces

Many environmental regulations are very expensive and relatively ineffective. The burden of paying for regulations tends to fall on local and state governments, industry, and individuals, not on the federal government that promulgates the regulations. The tests and calculations involved in setting environmental standards involve many uncertainties and inaccuracies. Many governments are experimenting with the use of market forces to improve the environment by making it financially desirable for individuals to act in the desired manner. This approach is better than regulation at things such as controlling pollution from nonpoint sources, spreading the burden fairly, and encouraging the achievement of the lowest possible levels of pollution.

### Planning for the Future

For the first time in history, communications ensure that knowledge is essentially universal. Environmental problems are understood by many people all over the world, and everyone shares in the solutions. A noteworthy development is the beginning of international cooperation to improve the environment.

## DISCUSSION QUESTIONS

1. How does the "tragedy of the commons" apply to efforts by individuals or governments to limit human family size?

2. What obligations, if any, do you feel that we have to future generations?

3. Julian L. Simon, a professor of business, is a "cornucopian," the name given to people who believe that environmental problems have been exaggerated and pose no immediate threat to human life or economic growth. These people believe that history teaches that "the nature of the world's physical conditions and the resilience in a well-functioning economic and social system" enable us to overcome environmental problems as they arise. What do you think of this point of view? What do you think is likely to be the educational background of people with these ideas?

4. List five changes in your lifestyle that you would be prepared to make as a contribution to solving environmental problems. Does your list confirm or refute the idea that individuals are most likely to help solve environmental problems by actions that are effective and cost them little?

5. The tragedy of the commons throws light on the difficulty of solving many national and international problems. What examples can you think of?

6. Table 22-1 shows the results of a study by the EPA. Members of the public and government scientists were asked to rank various hazards to human health and the environment as posing either a high or a low risk. The EPA was trying to answer this question: does the agency spend its resources and money on the problems that pose the greatest risk to public health and the environment? What do you think the answer is? Why does the EPA spend its money and efforts as it does?

7. Suppose we decide to solve the world's population problem by encouraging people to have no more than two children. Would you recommend regulation (require two or fewer children and fine or imprison those who have more)? Or would a carrot work better? For instance, you might pay a woman $1,000 every year that she does not have a child, or you could pay her for every year she uses a long-lasting contraceptive. Can you come up with a scheme that would be effective but would not be morally offensive to too many people? Would you recommend different schemes for different countries? Why or why not?

### Study results, Facts and Figures 22-1

In question 1, the majority of doctors chose program A (saving 200 out of 600 people), which sounds like less of a gamble than B (one-third chance of saving everyone). For question 2, the majority of doctors chose program D (two-thirds chance of all 600 people dying) over C (400 people dying).

*In fact, all four programs, A, B, C, and D, are the same.* Each produces the same probability that 200 people will live and 400 will die. The doctors apparently chose as they did because A and B were presented in terms of lives *saved* and C and D in terms of lives *lost.*

# POLITICS, ECONOMICS, AND ETHICS

# Chapter 23

## Key Concepts

- Our ethical values influence the decisions we make about environmental affairs.

- We shall probably never be able to solve the world's environmental problems without first greatly reducing poverty.

- The increase in military spending since 1950 has impaired our ability to deal with environmental problems.

- International economics at the moment require poor countries simultaneously to accept growing poverty and to export growing amounts of natural resources.

- Exploitation of developing nations by industrialized nations has led to deep mistrust of the Western world by most inhabitants of the poorer nations.

- We have built our standard of living by depleting our natural resources.

*A shrimp boat puts to sea from the Florida coast. Do we owe it to our children to leave shrimp and fish in the sea and birds in the air for them to enjoy?*

*The world cannot wait before embarking on an immediate action program, including*

1. *A large-scale transfer of resources to developing countries.*
2. *An international energy strategy.*
3. *A global food program.*
4. *A start on some major reforms in the international economic system."*

> Independent Commission on International
> Development Issues (1988)

Environmental science is not pure science. It has a goal. That goal is to develop ways of living that permit the human species and other species to survive and prosper into the indefinite future. The human population is still growing, and many people are still without the necessities of life. Our future prosperity depends on our producing even more energy, food, and other essentials than we do now. This can be done only if we can find ways to preserve and expand our environmental resources.

In this chapter, we consider how the ethical, economic, and political systems within which we live affect our progress toward this goal. Must we change our ethical standards, our economic organizations, or our political systems if we are to realize the dream of a sustainable world? Many experts believe that we must.

## 23-A  HOW DO ETHICS COME INTO IT?

*Human survival could depend on elevating sustainable development to a global ethic.*

> United Nations World Commission
> on Environment and Development

An **ethic** is a principle or value that we use to decide whether an action is good or bad. Consider how ethics come into this definition by economist Paul Samuelson:

> *Economics is the study of how men and society choose to employ scarce productive resources to produce various commodities and distribute them for consumption, now and in the future, among various people and groups in society.*

Our values influence how we choose to employ and distribute scarce resources. For instance, fresh water is a scarce resource. Charleston, South Carolina, recently faced a request from tobacco farmers to double the volume of water they withdraw from a river each year. One effect of this withdrawal would be to lower the water level in a nearby wildlife refuge. Naturalists estimated that this would halve the number of bird species nesting in the refuge. How would you reply to the farmers? Your decision will depend partly on the relative values you place on growing tobacco, preserving bird species, increasing farmers' incomes, and preserving wildlife refuges as recreation for nearby city dwellers.

### Ethics and Economics

Discussing depleted stocks of wild animals, *The Economist* magazine pointed out that the Atlantic cod's problem was that, although scarce, it had few endearing habits. It was therefore unlikely to get as much protection from a 1992 international conference as the much more common minke whale. The fuss in the United States and Europe about Japan and Iceland hunting the whale was caused less by the threat of its extinction than by the fact that the minke is an attractive fellow mammal (Figure 23-1).

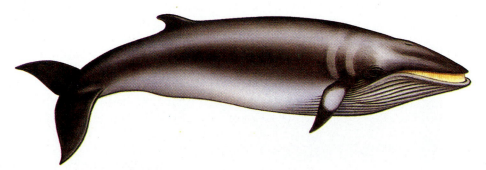

**Figure 23-1**  A minke whale, *Balaenoptera acutorostrata.* Also known as the piked whale or lesser rorqual, the minke is the smallest of the finback whales and was considered too small to bother with when larger whales were plentiful. At about 10 meters long, it is less than one third the length of the blue whale, which is the largest animal that has ever lived. Minkes are found in both the Atlantic and Pacific oceans, usually in coastal waters.

Values change, and they differ from one country to another. In the last century, the United States had a huge whaling fleet. Now Americans dislike the idea of killing and eating whales. Norwegians, strong environmentalists on most issues, cheerfully eat minke steak. Why is it any worse than eating pig? Pigs are pretty intelligent. What if pigs were shown to be as intelligent as whales? Such conflicts of values were at the root of the row between the United States and Mexico over the killing of dolphins, which are not endangered, in the nets of tuna fishermen. *The Economist* suggested that money might solve the whale problem just as it solved the argument about dolphins: Mexico gave up using nets that endanger dolphins in return for financial compensation from the United States to help Mexican fishermen. Perhaps rights to whales and elephants could be put up for sale? Those who value them could purchase the right to protect them. In any case, we have to solve environmental problems while remembering that different people may have different ethical values and that two sets of values may conflict.

### Ethical Compromise

Decisions about the environment often involve us in conflicts between incompatible values. For instance, one ethic common within the environmental movement is that we have a duty to the future: we should leave thriving forests and fertile soil for our descen-

dants. Some people argue that this means we should not feed people who would otherwise die in a famine. If a country contains more people than it can support, the people will cut down the forests, turn the land to desert, and increase the chance that even more people will starve to death in the long run. The opposing argument is that it is unethical to permit people to starve to death if we have the means to save them.

The practical solution to ethical dilemmas of this kind is usually some sort of compromise. In the North African famines of recent years, some of the aid supplied by concerned people all over the world has been used to feed the starving, but some of it is going to projects designed to prevent famine in the future: agricultural advice, reforestation, and family planning programs. In this particular case, most observers agree that too little of the aid has been spent on population control. There is so little fertile soil and fresh water in North Africa that famines and starvation will continue unless the population is substantially reduced.

> The practical solution to ethical dilemmas in environmental affairs is usually some sort of compromise.

### 23-B  TOLERATING POVERTY

Many politicians and scientists have pointed out one ethical shortcoming of the twentieth century that is going to cause problems for future generations: our failure to abolish poverty at home and abroad.

At a time when international cooperation is ever more important, it becomes ever less possible because of the growing gulf between poor and wealthy nations. The developing countries no longer trust the industrial countries to treat them fairly, and the gap between the resources and wealth of the developing and industrial nations actually widened during the 1980s (Figure 23-2). The industrial world has already used much of the planet's environmental capital, and industrial nations have dominated decision making in key international bodies such as the United Nations, the Law of the Sea Conference, and the Antarctic Treaty group.

Already those living in developing nations far outnumber those in developed nations, and this gap too is growing. Most people in the industrialized world consume far more than their fair share of the world's natural resources, while millions of people in other countries live in hideous poverty. Even if we do not view improving the lot of the have-not nations as a

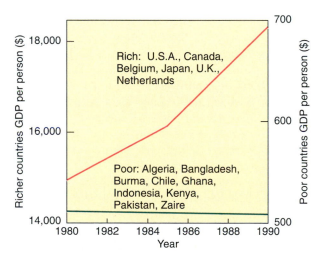

**Figure 23-2**  The wealth gap between those in rich and poor nations widened during the 1980s. The graph shows the increase in gross domestic product per person in some rich countries and its decline in some poor countries. If these figures were adjusted for inflation, economic growth among rich countries would appear much less and the decline in poor countries would be sharper. *U.S. Department of Commerce*

moral imperative, we should view it as a matter of national security. The industrialized nations will have to attempt to bridge the philosophical and economic gap and to alleviate poverty in the Third World if they do not wish to be confronted with dangerous enemies in the next century.

> Even if we do not view improving the lot of the have-not nations as a moral imperative, we should view it as a matter of national security.

We are equally callous toward our neighbors, neglecting substandard housing, inadequate medical care, and malnutrition in our own towns. Most people, if asked, would say that they have no power and no duty to alleviate such suffering. The power and the duty belong to other people—to local or national officials. Even if our morality does not drive us to fight poverty, however, our self-interest should, because poverty is both a cause and an effect of environmental problems.

During the nineteenth century, philanthropists had high hopes of eradicating poverty throughout the world. For most countries, that hope has faded during the twentieth century. Some of the wealthiest nations have not managed to abolish poverty within their boundaries. A few countries have come close to developing economies that guarantee even the poor-

est member of the population adequate food, housing, and medical care. Japan, China, most western European countries, and some developing Asian countries fall into this group. In the poorer nations, the goal of eradicating poverty was overwhelmed long ago by the rate of population increase.

Poverty and overpopulation are a vicious cycle. In every society in the world, the poor have more children than the well-off. In developed nations, this may be because the poor do not have the money to buy contraceptives or have the education to use them. In developing countries, children help gather food and tend the fields and are the only security for one's old age. Without money or social services, children are the only provision for their old age that the poor can make.

A city in which poverty is endemic will always be victimized by high crime rates, inadequate educational systems, and a political rift between the haves and have-nots. Similarly, a world containing massive poverty will always be prone to ecological catastrophe and war.

## 23-C MILITARY EXPENDITURE

Military expenditures provide an interesting example of the choices we make about how to use scarce resources. Many countries have chosen to devote resources to armaments that could have solved environmental problems and improved standards of living.

The military future is always uncertain. No country can be absolutely sure that a foreign power will not try to conquer it. Nevertheless, more than 90% of the wars fought during the last 20 years have been civil wars within countries. There have been only a few wars between countries. When war is imminent, most people will agree to spend large sums on weapons. Sometimes having powerful weapons may deter potential aggressors, but how much should one spend when there is no immediate threat? People have always disagreed on the answer.

After World War II, the military big spenders were headed by the United States and the Soviet Union. These two countries contained about one tenth of the world's population but spent more than half the money spent for arms throughout the world every year. The United States has spent between 5 and 10% of its gross domestic product ($200 to $250 billion a year) on military expenditures in each of the last 50 years (Figure 23-3). Many smaller countries, such as Israel, Iraq, North Korea, and Saudi Arabia, also spend large proportions of their budgets on arms. In contrast, countries such as Brazil, Mexico, and

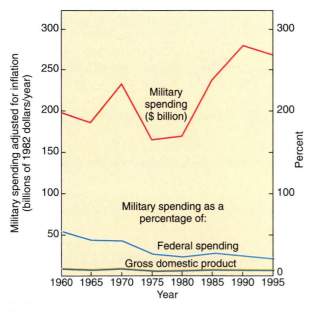

**Figure 23-3** U.S. military spending since 1960. Despite fluctuations in the dollars spent, military spending has remained at between 5 and 10% of GDP. It has declined as a percentage of the federal budget because spending for social security and medical programs has grown much faster. *U.S. Department of Commerce*

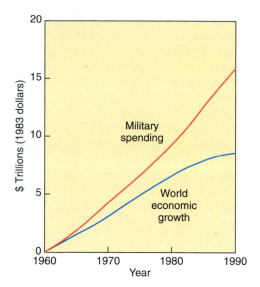

**Figure 23-4**   Worldwide military spending and economic growth between 1960 and 1990. Military spending absorbed more money than total economic growth produced and has put many nations deep in debt.

Switzerland spend less than 2% of their gross domestic products on defense. We cannot be sure what choices lie behind these different decisions on military spending. Probably many different values are involved.

The resources—financial, human, and material—that the big spenders have put into armaments are so vast that they are hard for us even to imagine. Military spending has absorbed more than the total increase in global productivity. Between 1960 and 1990, worldwide military spending grew nearly three times as fast as economic output (Figure 23-4).

Where then did the rest of the money that was spent on weapons come from? It came from cuts in standards of living and from enormous increases in national debts. Choosing to spend money on arms means choosing not to spend it on other things. For instance, both the former Soviet Union and the United States have high infant mortality rates, ranking forty-eighth and seventeenth in the world. This is not surprising since the Soviet Union spent more than twice as much on defense as on education and health combined, and the United States also spends more on defense than on health.

More than 500,000 scientists worldwide are engaged in military research and development, which is part of the reason research on environmental problems is so inadequate. The United States and United Kingdom spend more government money on military research than on all other forms of research put together (Figure 23-5). Not surprisingly, those countries that lead the world in economic growth and in sustainable development spend less on arms.

Poor countries suffer worst if they join the arms race. In 1960, arms imports cost developing countries less than $5 billion, which was less than one third of the economic aid they received. (Aid from foreign countries can be divided into economic and military aid.) In 1991, arms imports by India, North Korea, and Thailand alone cost $30 billion, more than the total of all foreign aid. We tend to blame the debts of developing nations on population growth and the high cost of energy, but military spending is almost equally responsible (as it is for much of the debt of the United States). Arms imports between 1975 and 1990 accounted for half of the increase in foreign debt of developing countries.

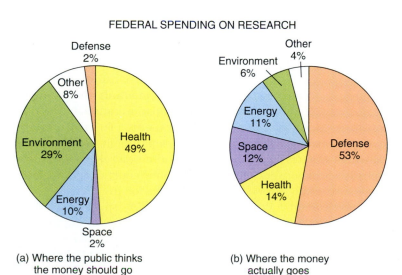

(a) Where the public thinks the money should go

(b) Where the money actually goes

**Figure 23-5**   U.S. federal spending on research: where the public thinks the money should go and where it actually goes. Defense absorbs more federal research dollars than all other research combined. Note that the second largest category is space, money spent by NASA on projects such as the shuttle and the planned space station. The public thinks that environmental research should be in second place. *Data from a 1993 poll by Louis Harris Associates Inc. and the National Science Board.*

> Military spending has absorbed more than the total increase in global productivity in the past 50 years.

Few people realize that environmental destruction may present a country with a greater security threat than military enemies. For instance, the loss of life and of agricultural land in the African Sahel region since 1980 was greater than the damage that any invading army would have caused if it had pursued a policy of complete destruction. Most of the governments facing this kind of environmental problem spend much more to protect their countries from invading armies than from the invading desert (Close-Up 23-1).

In other cases, the invasion of a country by refugees from the environmental crisis is a greater threat to a nation's security than an invading army. During the 1980s, millions of refugees from land devastated by war in Afghanistan poured into Pakistan, and refugees from the Ethiopian famine poured into Sudan. Neither of these countries had the resources to cope. Today there are more people living in refugee camps because they have fled ecological disasters than there were refugees from World War II.

The United States has begun to realize that vast military expenditures cripple economic development. Since 1945, disposable income and standard of living have risen faster in countries such as Norway, Switzerland, Japan, Canada, France, Taiwan, China, and Sweden than they have in the United States. If we compare the international positions of Germany and Japan during World War II and now, we might reasonably conclude that it is cheaper and easier to buy up the rest of the world than to conquer it by military force.

## 23-D POLITICAL SYSTEMS

How can any country spend its resources on weapons while its soil erodes and its people starve? This seems like such illogical behavior that it is hard to understand. The answer lies in the realization that it is not all the inhabitants of a country that decide how to spend the country's money, it is a small number of individuals. And people's actions are often guided more by what they think will be good for themselves than by what will be good for the country as a whole. Thus, the dictator of a developing country may be prepared to spend most of the country's wealth to defend the country against a neighbor or against local insurgents because not to do so might cost the dictator power or money.

The situation is less obvious in an industrialized democracy, but again the choices that are made can be explained as enhancing the wealth or power of individuals. In August 1992, members of Congress met with scientists who told them that the hot, dry summers that plagued American farmers in the 1980s were a reflection of the greenhouse effect. Droughts would continue to get worse and to disrupt farming unless we drastically reduced our emissions from burning fossil fuel. The representatives cut the meeting short. It was an election year, they needed to be out campaigning, and no television cameras were present. It would be much easier and more popular to spend the taxpayers' money bailing out farmers for another year than to plan for a warmer future and reduce air pollution. A future Congress could worry about that.

> A few individuals, rather than the population as a whole, decide how to spend a country's money.

### Political Reorganization Needed

The organization of most governments may have suited the eighteenth century, but it is woefully inadequate to deal with the twenty-first century. For instance, if our goal is sustainable development, environmental issues should be taken into account in nearly all economic decisions. Yet in most governments, those responsible for managing natural resources and protecting the environment are separated from those responsible for managing the economy. There is one department that deals with environmental issues, one with agriculture, another with industry, and yet another with ensuring adequate energy supplies. The department of industry views its goal as maximizing industrial development, leaving the environmental department to worry about the pollution that industry causes. The environmental department too often finds itself trying to clean up the resource degradation caused by the policies of the other departments. To make things worse, every government department in the United States spends much of its time merely dealing with paperwork and trying to obey court orders. A 1992 study showed that employees of the U.S. Environmental Protection Agency (EPA) spent more than half their time writing regulations, working to obey court orders, and responding to memoranda from other agencies. Figure 23-6 shows where tasks undertaken by the EPA come from.

# Operation Desert Disaster

Operation Desert Storm may have been a military triumph for the United States and its allies, but it was an environmental disaster for the Middle East: toxic chemicals from weapons and from bombed Iraqi factories, oil spills in the Gulf, and smoke-filled skies are only some of the problems the war caused.

Large areas of the Arabian Peninsula are uninhabitable because the fine desert sand blows about, covering fields, oases, and roads. Kuwait and surrounding areas are livable because the sand is stabilized by a "desert shield" of pebbles too large to be moved by the wind, which holds the fine sand in place. This crust has taken thousands of years to develop. Now this protective crust has been broken up by the fortifications built by the Iraqis in Kuwait and by the passage of thousands of military vehicles. Farouk El-Baz, head of Boston University's Center for Remote Sensing, predicts dust storms and moving sand dunes as a result.

Moving sand dunes have engulfed airports, agricultural settlements, and even whole cities. Similar effects have been seen before. In 1973, during the Yom Kippur war, Israel bulldozed a 10-meter wall of sand east of the Suez Canal to slow Egyptian troop movements. Sand dunes in the Sinai desert began moving all along the coast. Similarly, Iraq started farming its western desert in the 1970s, in an attempt to feed its growing population. The entire Gulf region experienced an increase in sand storms from the disturbed desert soil.

**Figure 23-A**  These are just a few of the hundreds of oil wells that were set afire or damaged by the retreating Iraqi army in the Gulf War of 1991. Clouds of smoke caused massive atmospheric pollution. All the fires were extinguished by the end of 1991, but leaking wells have left lasting pollution in Kuwait. *(Steve McCurry/Magnum)*

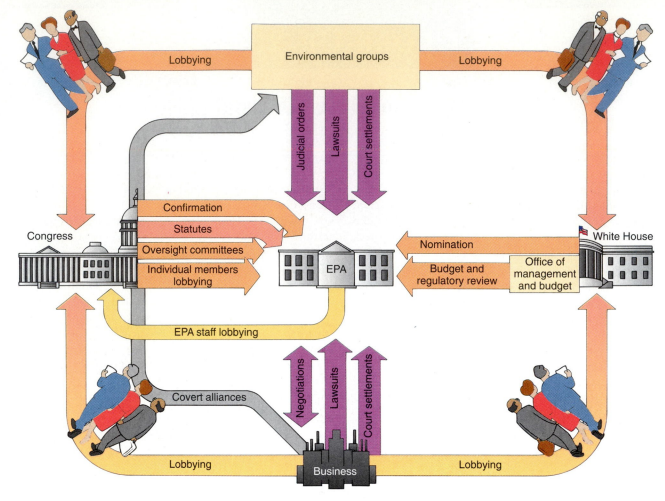

**Figure 23-6**  The U.S. Environmental Protection Agency and outside influences that its employees have to deal with. Lawsuits and dealing with lobbyists takes up large amounts of the agency's time and resources. Lobbyists are people employed to influence the writing and enforcement of laws and regulations so as to benefit the businesses, industries, or other groups that they represent. *(Adapted from an article in* Forbes, *March, 1993)*

In an ideal world, the agriculture department should be working not merely to maximize agricultural yield but to maximize sustainable yield. It should be studying and researching biological controls, ways to avoid runoff and water pollution, and methods of recycling organic matter back into the soil. It should be working with similar agencies in other countries to determine worldwide demand for particular products, instead of leaving foreign relations to other departments of export or foreign affairs. Under the present system, millions starve, while the United States stores tons of unsold grain and western Europeans lament the agricultural surpluses they call their "wine lake" and "butter mountain."

> Government departments need to be reorganized so they can act more efficiently on environmental affairs.

## 23-E ECONOMICS

In the twentieth century, economists have been highly influential in planning for the future. But economic policies cannot guarantee our future prosperity if they do not take environmental affairs into account. In the 1980s, many pointed out that standards of living have begun to fall and will continue to fall unless overpopulation and the depletion of natural resources can be stopped.

People in poor countries have finally stopped looking on environmental protection as something only the rich countries can afford. They have good reason because poor countries live directly off the environment. Although most of the gross domestic product (GDP) of developed countries comes from the sale of services and manufactured goods, most of a poor country's income comes from the sale of its natural resources.

For instance, a developing country borrows money to fund its development. To pay its debts, it must export, and what it exports are its natural resources. Typically, farming, forestry, fishing, and mining account for more than one third of GDP and more than two thirds of employment. Most of these poor nations face enormous economic pressures to overexploit their natural resources. In addition, trade barriers in many countries make it hard for poor nations to sell their goods abroad for reasonable returns, putting yet more pressure on their land and forests. With their forests and agricultural lands destroyed, many developing nations will be even less able in the future to support a billion more people every 12 years. When the environment is degraded, the poor countries have much more to lose than the rich.

One of the solutions to this dilemma that is increasingly popular is to encourage ecotourism. **Ecotourism** is the name given to travel designed for those who want to see the natural beauty and wildlife of an area. If a country can acquire enough income and employment from ecotourism, it has the motive and the means to preserve its natural resources. Flamingo Lodge in Florida offers an ''Eco-Tour,'' advertised as ''Take a Green Vacation to Everglades National Park.'' Guests who like to fish learn to release the fish they catch, others take guided nature tours of Florida Bay. Pro Dive of Fort Lauderdale bans spear guns and teaches divers to photograph fish and to respect ocean life. ''It's a matter of protecting our assets,'' says the company president. Ecotourism is the main source of foreign income for some developing nations such as Botswana and Costa Rica and is growing rapidly all over the world (Figure 23-7).

## Environmental Accounting Systems

Economists are beginning to recommend new methods of accounting that take resource degradation into account. For instance, Russia estimates that its fresh water is so polluted by industrial waste that less than one third of it is fit to drink. Eventually there will not be enough drinking water in the country, and Russia will have to spend enormous sums cleaning up its water. You can look on the act of polluting the water as taking on a debt that will have to be paid off by cleaning up the water in the future.

One way to make sure that we run up no more unpaid debts on our natural resources is to set the present price of a resource, such as water for irrigation or drinking, taking into account its long-term scarcity and the environmental effects of its use. This means pricing the water to include all its hidden

(a)

(b)

**Figure 23-7** Ecotourism. (a) A guide in Costa Rica explains that country's coffee plantations to a group of tourists. (b) A sign in Santa Rosa National Park, Costa Rica, explains the symbiotic relationship between the acacia trees and ants that hikers can see along this trail—and which is typical of the complex relationships between species that are found in tropical forests. The acacias provide food and nesting spaces for the ants, and the ants kill other herbivores that attack the tree.

costs (Figure 23-8). What does this mean in cold hard cash? It probably means water at about $1 a gallon and gasoline at $20 a gallon. When we buy a gallon of gasoline, we pay for the immediate cost of extracting, refining, and transporting the oil. If we also paid for the hidden cost, we should have to pay for drilling new wells when the local water supply is polluted by gasoline from a leaking service station tank. We should have to pay the cost of cleaning up spilled oil and the value of resources destroyed by oil spills (Close-Up 23-2). These hidden costs are not fully

# CLOSE-UP 23-2

## The Alaska Oil Spill and the Choices We Make

In March 1989, one of Exxon's oil tankers, the *Exxon Valdez*, was steered onto a rock and spilled oil into Prince William Sound, Alaska. Salmon fry (juveniles) and herring eggs died of toxins in the oil. Other animals died of cold, asphyxiation, and starvation. The oil was deposited in marshes, bottom sediments, and the gravel of beaches. Here it entered the food chain as it was absorbed by plankton, crustaceans, molluscs, and other organisms. Ten years after the *Amoco Cadiz* spilled oil on the coast of France, the toxic effects of the oil are still evident. Toxic effects are bound to last much longer in the cold Alaskan waters, where decomposer organisms work more slowly.

Perhaps we can learn some lessons in the politics, economics, and ethics of environmental affairs by examining the motives and behavior of some of those involved in the disaster. Although everyone involved deplored the accident and vowed to prevent it from happening again, there were other forces at work.

The state of Alaska wanted to preserve its scenery and the income from its oil, fishing, and tourist industries at the lowest possible cost. So the state decried the oil company's inadequacies and called for better federal regulation while omitting to mention that the Alaska legislature had cut the budget of the state agency that was supposed to inspect the oil terminal and its safety precautions.

The oil companies wanted to ensure that environmentalists did not use the accident to impose expensive restrictions on the oil industry. Accordingly Exxon apologized for the spill and promised swift cleanup and compensation. But the oil companies also wanted to minimize the money they spent. Why not save a lot of money on cleanup crews by just letting the oil disperse and then compensating the fishermen for their lost revenues later? Instead of rapidly cleaning up the oil, the oil companies raised the price of oil to consumers and gave the oil time to disperse before they began cleaning it up (Figure 23-B). Everyone who drove a car or owned a gas station had more than paid the bill for compensation and cleanup within a few weeks after the accident.

The federal government was motivated to protect President Bush's "environmentalist" image, to preserve the oil industry, and to rescue wildlife threatened by the oil. The Reagan administration had cut funding for the Coast Guard, which therefore lacked the radar sets that would have permitted it to see that the *Exxon Valdez* was in danger of running on the rocks. The nearest Coast Guard oil cleanup equipment was in San Francisco, 2,000 miles away.

The motive of environmentalists was to protect the environment. But they must take responsibility for a political atmosphere in which the Alaska oil spill was almost inevitable, an atmosphere that now threatens one of the world's most important wildlife refuges. The success of environmentalists in limiting the production of nuclear power is one of the main reasons that so many oil tankers now travel the coasts of the United States and, inevitably, sometimes run onto rocks. Almost exactly 10 years before the *Exxon Valdez* ran aground, the nuclear accident at Three Mile Island in Pennsylvania generated an outcry that crippled the United States's nuclear power program.

After the collapse of nuclear power, the United States was unable or unwilling to practice sufficient conservation, so it came to rely more and more on coal and oil. The federal government feared that the *Exxon Valdez* accident might slow the flow of domestic oil and make the country even more dependent on oil imports. It hastened plans to permit oil drilling in the Arctic National Wildlife Refuge between Prudhoe Bay and the Canadian border. The Refuge is home to grizzly bears, musk oxen, caribou, polar bears, and wolves and ranks in ecological importance with the world's greatest game preserves. A major oil spill in the reserve would do incalculable damage.

Human error caused the Three Mile Island accident. Human error caused the damage from the *Exxon Valdez*. Human error is unavoidable. In the debate about our energy future, we have to ask the ethical, political, and economic question: what degree of environmental risk should we accept for the sake of increasing supplies of particular forms of energy? What can we do to decrease the risk? And what consequences might follow from one choice rather than another?

**Figure 23-B** Cleaning up the *Exxon Valdez* oil spill by using high-pressure hoses to wash oil off rocks. *Al Grillo/Picture Group*

**Figure 23-8**   Hidden costs. Although we seldom think of it in this way, one of the hidden costs of soft drinks is the labor of the volunteers who pick up the discarded bottles that litter this beach.

paid by the manufacturer or its customers but have to be paid for in some way by society.

Hidden costs are paid for mainly through taxation. Governments tax their citizens to raise the money to provide clean water, subsidize farming, or clean up toxic waste sites. This is rarely an economical way of handling hidden costs, for two reasons. First, governments usually spend money less efficiently than industries or individuals. Second, pollution and resource depletion are cheaper to prevent than to cure. It is more efficient to provide incentives to prevent environmental degradation than it is to clean up the mess later. Because of these factors, many environmental laws today attempt to reduce the hidden costs of products by making manufacturers financially responsible for cleaning up oil spills or adding scrubbers to smokestacks. This raises the direct costs of the manufacturer's products, but it minimizes the cost of products in the long run, as many studies have confirmed.

At the moment, humanity is not just failing to repair the destruction caused by overuse of natural resources, it is actually subsidizing the destruction. In every country, the wasteful use of natural resources is encouraged in some way. Many tropical countries pay settlers to cut down forest and turn it into cattle range. Developed countries subsidize farmers to plant crops that need pesticides, which pollute the water system. Subsidized energy and gasoline encourage the wasteful use of energy. It all amounts to running up debts that will have to be paid in the future although there is nowhere in the national budget to take account of this invisible debt.

> Economic theories are being developed that take into account the fact that the world's resources are limited.

Many world leaders are beginning to understand that environmental considerations constrain economic growth. We can expect the following among future trends:

1. **Research.**   The number of scientists in the world has grown three times as fast as the world population. But the more we learn, the more clearly we realize that today's research efforts are inadequate. For instance, biologist E.O. Wilson points out, "We do not know the true number of species on Earth even to the nearest order of magnitude." Scientists are not doing an adequate job of assessing how humans affect their environment, and research efforts will have to increase.

2. **Conservation.**   The standard of living of people in the prosperous part of the world has risen remarkably in the last hundred years. Some of this economic progress was made by subsidizing the use of resources, and it cannot be repeated. Supplies of oil, coal, fresh water, and minerals that can be extracted from the earth cheaply are just about used up, and what remains will prove more expensive. We can therefore expect that conservation will play an increasing part in our lives as the prices of resources increase to reflect hidden cost. In the future, we shall conserve energy, forests, soil, water, and other resources on an impressive scale.

3. **Recycling.** To make minerals and other resources go further, we shall reuse them.

4. **International cooperation.** The day has gone when any nation can expect a prosperous future unless it plays an increasing role in cooperative ventures with other countries. Multinational corporations, instant worldwide communications, and growing international trade ensure that all the economies of the world are linked. If the New York stock market plunges, the effects are felt around the world. Economically powerful nations such as Japan now have as much say in world affairs as militarily powerful nations.

We can expect the power of the poorer nations in South America, Africa, and Asia to grow because they will soon contain more than twice as many people as the developed nations. The developed nations will increase their efforts to help these countries modernize their environmental strategies and economies, not only for humanitarian reasons but also because these huge populations can become vast new consumer markets.

All of these developments, and many others that you can probably think of, will produce many interesting jobs and opportunities for workers in the twenty-first century.

### The Economics of Improving the Environment

Environmental science has given rise to huge industries. In the United States, pollution control is big business. The industry employs scientists who study environmental problems, government workers who administer environmental laws, and people who produce studies such as environmental impact statements. It includes those who invent, manufacture, and operate pollution control devices for factories, agriculture, and the home and those who build sewage treatment plants, units to dispose of solid and hazardous wastes, and water purification devices.

## 23-H CONCLUSION

In our attempts to guard our own standard of living, we have borrowed environmental capital from the future with no intention of paying our debt. Politicians have not yet realized that voters understand this. A recent poll of U.S. opinion showed that Americans believed they would personally be better off the following year than they were then. Yet they overwhelmingly said that economic disaster loomed in the near future. Because unemployment and interest rates were low and all the economic indicators optimistic, analysts could not understand this response. The voters know that our present prosperity is a debt to the future. A majority of the poll's respondents even said that more money should be spent on environmental preservation than on economic development.

Today's decision makers will be dead before the bankruptcy of our economic systems becomes obvious. We act irresponsibly because we can get away with it. Generations yet unborn have no power to challenge our decisions. But today's young voters should; they will suffer for our greed.

## SUMMARY

We cannot solve our environmental problems merely by studying the scientific principles involved. Economic systems, ethical values, and political structures all influence our chances of building a sustainable future.

### How Do Ethics Come into It?

Our ethical values, shaped by our cultural history, influence the choices we make and how we choose to use scarce resources. Values change with time, and they differ from one group of people to another. When decisions involve conflicts between incompatible values, the outcome is often a compromise.

### Tolerating Poverty

Our ability to tolerate widespread and increasing poverty undermines our efforts to solve the world's environmental problems. The poor are much more numerous than the wealthy, reproduce more rapidly, degrade natural resources in the attempt to survive, and give low priority to improvements such as pollution control. Our pattern of economic development leaves increasing numbers of people in poverty and an increasingly degraded environment.

### Military Expenditure

We choose how to spend limited human and financial resources. The United States, the Soviet Union, and some of the poorest nations in the world have chosen to spend large sums on armaments. Military spending has cost more than the total increase in productivity of the world during the last 50 years. Countries that have chosen to restrict military expenditures have had more money left for other purposes and lead the world in economic growth and sustainable development. Environmental decline may be a worse threat to national security than military engagements.

## Political Systems

It is hard to explain the neglect of environmental problems without realizing that it is individuals, not countries or states, that make decisions. The practical result of this is that governments do little about environmental affairs until their people understand some of the problems and demand action. Governments are organized in ways that make it difficult to take environmental affairs into account when decisions are made.

## Economics

Environmental policies have a major impact on economic affairs. In the twentieth century, the countries with a prosperous future appear to be those that are working toward sustainable agriculture, rural development, pollution control, and conservation. Older accounting systems do not take the degradation of resources into account. Newer methods attempt to build the hidden cost of using resources into economic planning.

## Conclusion

We have built our present standard of living by consuming our environmental capital. Future generations will have to pay for our greed because they cannot protest it now.

## DISCUSSION QUESTIONS

1. What new moral values might spread through the world and improve our chances of building a sustainable future?

2. Many biologists think that if our ethical systems do not change and we carry on operating as we have during the twentieth century, the human species will be extinct within another few centuries. Why do they think this? Do you think the human species could become extinct?

3. Is it realistic to think that we could change the world's moral values to include the ethics that would be necessary to develop a sustainable future?

4. Think back through history. Can you think of occasions when the moral values of large numbers of people changed in a relatively short time?

5. Suppose a nuclear war or agricultural disaster wiped out most, but not all, of the people on Earth. What kinds of ways of life would be most likely to permit people to survive after the disaster?

6. If you had a lot of power in the United States (more power than any individual in fact possesses in a democracy), what would you do to try to reduce the economic gap between the poor and wealthy nations in the hopes of solving international environmental problems? How would you go about convincing your voters that these were the right things to do?

7. Choose any everyday item and list as many hidden costs of using it as you can think of.

# TOWARD A SUSTAINABLE FUTURE

# Chapter 24

**Key Concepts**

- Unless we get on with the job of building a sustainable future, standards of living everywhere will fall until humans again live in an essentially Stone Age society.

- Living standards are declining largely because of resource depletion and overpopulation.

- Overpopulation and environmental degradation are partial causes of the wars in many parts of the world.

- Massive environmental programs are needed if living standards are ever to improve for most people.

***Ecola State Park, Oregon.*** (David Muench)

*Act, act, act. You can't just watch.*

Angeles Serrano, community activist
from Manila's Leveriza slum.

**F**ailure to plan for our future is a major reason for the difficulties that beset us. So in this chapter we plan. We plan for a future with a sustainable way of life, a future in which the human population does not exceed the carrying capacity of Earth, a future in which we are living within our environmental income.

Physical laws dictate that we cannot go on depleting natural resources forever. A sustainable world is bound to be part of our future, even if that future is horrible to contemplate. Perhaps the carrying capacity of Earth for humans will eventually be zero, as many believe, and we shall have caused our own extinction. Perhaps the sustainable future will be a miserable existence eked out by the survivors of a nuclear war. Perhaps it will be an international society with a less wasteful style of living than we now enjoy but with greater security. This might be a society in which we have fewer material possessions than we do today, but no one starves; in which hope for the future involves repairing the damage done by the twentieth century, but no one wonders when the bomb will drop. If we are to reach this last version of a sustainable future, *Homo sapiens* will have to evolve from a pioneer species that is aggressive, fast growing, and greedy into a climax species that is stable in size, recognizes ecological constraints, and accepts cooperation and compromise.

> A sustainable future is bound to come, even if that future is horrible to contemplate.

When we look at our spend-now/pay-later society and at human nature, it is easy to conclude that we cannot get there from here or that even if we can, we shall not. Many people believe that we have so overdrawn our ecological capital that we have nowhere to go but down. They believe the human population will double, or more than double, within the next century and then decline rapidly as billions of people die of starvation, disease, and war, leaving a pitiful remnant to a primitive existence. The most pessimistic believe that the political upheaval resulting from environmental degradation and overpopulation will bring about a nuclear war and that the human species will be extinct within a few centuries.

The pessimists forget how rapidly humans can change. We have seen that not until the 1930s did many people start teaching that we inhabit a finite

**Figure 24-1**   Do we owe it to our children to leave them a sustainable world?

biosphere and had better learn to live within it. Not until the 1950s was this view shared by a significant number of people. We have come a long way in the 50 or so years since that first realization. There are probably now few people anywhere in the world who do not find the struggle for a sustainable world influencing some part of their lives. The idea that we owe our children a decent future is now widely discussed (Figure 24-1). In this chapter, we consider what will happen to the world if we do not change our ways. Then we consider some examples of progress toward a sustainable future and where they may lead us.

> Pessimists forget how rapidly humans can change.

## 24-A  WAR

The most obvious scenario for the end of human society as we know it is nuclear war. The world now contains enough nuclear weapons to kill every person on Earth many times over and to make Earth uninhabitable almost overnight.

## Nuclear Proliferation

Many people believe that the greatest threat to world peace has always been the small nuclear arsenals of many developing nations rather than the huge nuclear stockpiles of superpower nations. Nuclear weapons are owned by a number of nations that either are at war with their neighbors or are threatened by war. Some fear that terrorist groups will acquire nuclear weapons and hold the world to ransom. It seems impossible to prevent the slow spread of nuclear weapons throughout the world.

One reason for the popularity of nuclear weapons is that they are relatively cheap. You can wreak more destruction with a billion dollars' worth of nuclear weapons than with conventional weapons purchased for the same price. After World War II, ever more powerful and sophisticated nuclear weapons were developed by the original "nuclear club" countries— the United States, the Soviet Union, France, China, and the United Kingdom. These countries tried to prevent other countries from developing the bomb, but they were not completely successful. The nuclear nonproliferation treaty of 1968 was eventually signed by 130 nations that agreed to do without nuclear weapons in exchange for help in developing nuclear power for energy.

Among the nations that did not sign the 1968 treaty, and even among those that did, several have since developed their own nuclear weapons and the means to deliver them. Bomb development is usually pursued with great secrecy, but there is circumstantial evidence that a number of countries are developing nuclear weapons. Some new members of the nuclear club are involved in long-standing conflicts. Suppose India launched a nuclear missile at China or North Korea attacked South Korea. Imagine that Israel launched a nuclear missile at Cairo. All these actions are now technically possible, even if not politically probable. The victim nations are all linked by defense treaties to other countries that might be obliged to come to their aid, possibly with other nuclear weapons. Even if a nuclear war was fought on a strictly regional level, the damage could make most of Earth essentially uninhabitable for humans.

## Nuclear Winter

The world contains nuclear warheads with explosive power more than a million times greater than the 0.013 megatons that destroyed Hiroshima in Japan at the end of World War II. (Bomb size is compared with the explosive power of 1 ton of the conventional explosive TNT.) A single American Trident submarine carries weapons with an explosive power of 24 megatons—eight times the total explosive power detonated by all countries during the six years of World War II.

When a nuclear bomb explodes, in the air or on the ground, its blast creates a shock wave that kills people and flattens buildings around the center of the explosion. A variable number of people beyond reach of the shock wave subsequently die of radiation sickness. A nuclear explosion also generates a pulse of electromagnetic radiation that knocks out electrical equipment. The potential power of this electromagnetic pulse has only recently been recognized. Members of the nuclear club are now engaged in an expensive scramble to shield some of their military and government electrical equipment from the pulse. Shielding civilian equipment is prohibitively expensive. Civilization as we know it in the Western world simply would not exist after the loss of our electrical power.

Bombs start fires when they explode near things that will burn, such as oil fields, refineries, cities, or forests. These fires propel a cloud of smoke, soot, and other debris into the atmosphere. In a major nuclear war, experts estimate that the nuclear cloud would be dark enough to blot out most of the sunlight over the Northern Hemisphere, producing a **nuclear winter**. Experts do not agree about all the details of a nuclear winter, which would depend on the number, power, and locations of nuclear explosions. Evidence for the nuclear winter comes from sources such as volcanic explosions, which have lowered global temperatures by spewing huge clouds of sunlight-blocking ash into the air (Figure 24-2). Continuous twilight for many months would halt photosynthesis even in cold-resistant plants, with the result that animals would starve.

If even 10% of the nuclear weapons in the Northern Hemisphere were detonated, the nuclear cloud would not be confined to the Northern Hemisphere. Traveling in the atmosphere, it would reach the Southern Hemisphere. It would probably wipe out a complete growing season for agriculture. This version of World War III might kill less than one quarter of the world's human population immediately, but the survivors would probably wish that they had been at the center of an explosion.

In the nuclear winter that would follow a large-scale nuclear war, loss of sunshine would lower the temperature on Earth, making it almost impossible for humans, other animals, and crops to survive.

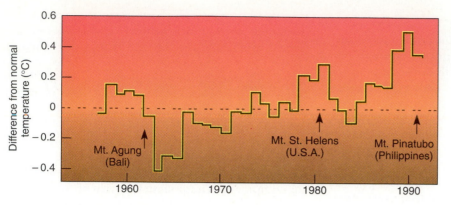

**Figure 24-2**    The nuclear winter effect. This graph shows the effect of dust in the atmosphere from three volcanic eruptions on average global temperature. The graph shows deviation of the average global temperature from normal (0 °C).

## Environment and War

You may sometimes wonder if civil wars and other conflicts are becoming more common. Civil strife in Angola, Somalia, Armenia, Bosnia, Sudan, and Croatia and terrorist bombings in India, Israel, Britain, and even the United States seem to dominate the news. In fact, such conflicts *are* becoming more common and involve more and more people every year. They are partly the result of environmental problems. Scientists have long warned that population growth was causing shortages of natural resources that increased the chance of conflict within and between countries. In 1993, the University of Toronto and the American Academy of Arts and Sciences published the first study confirming that this is happening.

The study concluded, "scarcities of renewable resources are already contributing to violent conflict in many parts of the developing world. These conflicts . . . foreshadow . . . violence in coming decades, particularly in poor countries where shortages of water, forests, and especially fertile land, coupled with rapidly expanding populations, already cause great hardship." The researchers conclude that the single biggest contributor to these conflicts is the growing shortage of fertile agricultural soil on which to grow food (Figure 24-3).

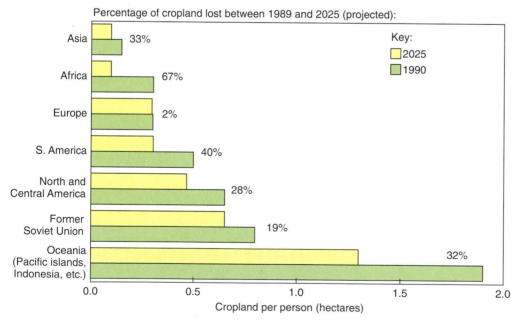

**Figure 24-3**    The loss of cropland per person in different parts of the world between 1990 and 2025. Although much cropland will be lost to soil destruction, most of the change is due to projected population growth. Thus, North and Central America, with growing populations, will lose much more cropland per person than Europe, where the population has almost stopped growing. (*Source: World Resources Institute*)

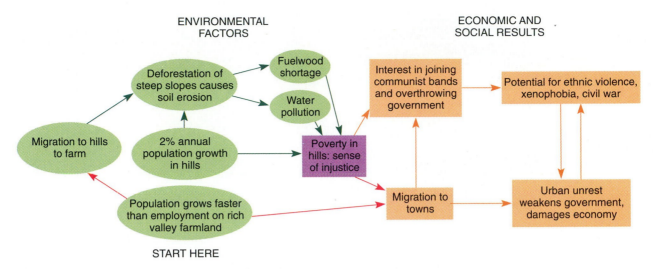

**Figure 24-4**   The environmental cycle in the Philippines: how population growth depletes natural resources, leading to urbanization; further depletion of resources; armed conflict; and sometimes terrorism, ethnic violence, and civil war.

Consider just one example: the Philippines, where the government is not in control of the whole country because communist bands control most of the hill country and have been launching attacks on military stations and villages for years. Spanish and American occupation of the Philippines left most of the fertile land in the valleys in the hands of a few families. The green revolution increased food production from this land, but the population grew faster. Millions of those who could not find work on the farms migrated to shantytowns around the capital, Manila, and to less productive land in the hills. Here deforestation from logging caused massive soil erosion, leading to starvation and poverty. During the 1970s and 1980s, the communist bands found upland peasants receptive to revolutionary ideas, especially where greedy landlords left them little choice but rebellion or starvation (Figure 24-4). Similar scenarios are being played out in other countries such as El Salvador, Nicaragua, Angola, Senegal, and Mauritania. Especially where the affected country is part of a continent, the unemployed may emigrate to neighboring countries, as Bangladeshis emigrate to India, often causing ethnic violence when they compete for land and jobs in the second country.

## 24-B  IF WE FAIL TO CHANGE

Even ignoring the possibility of nuclear war, much of the world is doomed to violent conflict in the next 50 years by shortages of fertile soil and water, caused by the population explosion (Figure 24-5). Even countries not directly involved will be affected by influxes of environmental refugees and worldwide political turmoil. If we fail to make ecological progress, we are all doomed to slide into poverty. Without sustainable development, our way of life will revert to one more like that of our Stone Age ancestors than of industrialized nations in the twentieth century.

Already the standard of living enjoyed by most people in the world is lower than it was 20 years ago. This means that most people have fewer of the necessities of life than they used to or that they now spend more of their incomes or time to obtain those necessities. For instance, in 1993 American workers spent more hours at work each week than ever before. As water, food, housing, clothing, garbage disposal, energy, and medical care use up more and more of what we earn, we are left with less of the disposable income and spare time that determine how well we live.

Standard of living is measured by determining what proportion of income is spent on a list of necessities, including food, shelter, medical care, and clothing. A recent report gives an arbitrary value of 100 to the United States's standard of living in 1972. By 1988, the standard of living had fallen to 78.9, more than 20% lower than it had been 15 years before. The United States's decline is due partly to the large sums spent on armaments, but it is also partly environmental. Supplies of natural resources such as fresh water have dwindled since 1972, and the cost of these necessities (and of everything derived from them) has risen in consequence.

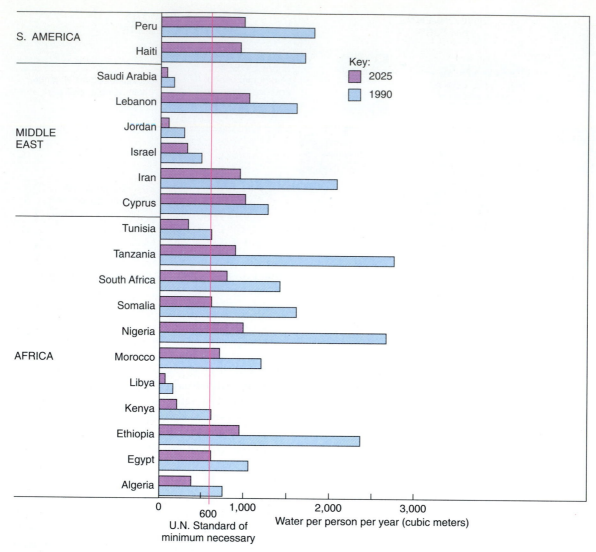

**Figure 24-5** Water shortages. The expected decrease in fresh water per person between 1990 and 2025. The U.N. minimum of 600 cubic meters per person per year is shown in red. Fresh water does not disappear. Why do you think the decrease is so sharp, especially for countries like Haiti, Lebanon, Iran, and essentially all the African countries? *(Source: Pacific Institute, Oakland, CA)*

> The decline in the United States's standard of living is due partly to spending so much of our money on armaments and partly to environmental degradation.

Standards of living will rise in the future, at least for a while, in a few countries. Some of these have huge reserves of natural resources, such as fossil fuels in Russia and Saudi Arabia. Others are already launched on paths to sustainable development. These countries may include China, Singapore, Taiwan, and Thailand. But in the vast majority of developed and developing nations, standards of living will decline until these countries adopt policies of sus-

tainable development and put them into effect, at which point standards of living may again begin to rise.

## 24-C SUSTAINABLE DEVELOPMENT

Whether it knows it or not, every country has the goal of sustainable development. **Development** describes changes aimed at improving standards of living and the country's economic position. **Sustainable** describes changes that do not deplete resources and therefore can continue indefinitely. Various paths to sustainable development are possible from our present unsustainable economies. Many of the necessary steps have been discussed throughout this book. They include conservation, education, and population control. In this section, we suggest a few routes to the future that seem promising and that have been pioneered by one country or another.

## Planning and Priorities

Even when we plan carefully, things do not turn out as we expect, but if we fail to plan at all, the surprises are worse. When drought disrupted American farming in the early summer of 1988, farmers and others asked what plans the U.S. government had made for coping with the results of the greenhouse effect. There were no plans. Most politicians had never even heard of the greenhouse effect.

Democratic governments tend to be short on planning. They seem to have a naive faith that market forces will take care of the future. Communist countries often produce five- and ten-year development plans. The problem here is that once the plans are written, they are treated as if cast in stone. If their economic prophecies fail to come true, the plans are quietly buried. Neither of these approaches will help us toward sustainable development. We need to decide on specific goals and plan how to achieve them. Then as circumstances change and uncertainties are resolved, we must reexamine our plans and modify our methods. Many people find it depressing to plan. They tend to believe that anything that is planned costs money that is not available. But as Costa Rican parks director Alvaro Ugalde says, "There are always resources for concrete ideas. If I say I want $10 million for reforestation, the money will be found. If I say I want it for general development, it will not."

Several national and international commissions have suggested that individual governments and international bodies such as the United Nations should, at the very least, have plans that include the following:

1. Research to determine what the world's natural resources actually are.

2. An international energy strategy.

3. A global food program.

4. Plans to reduce military spending so as to free the money to fund sustainable development.

5. A large-scale transfer of resources from developed to developing countries that will decrease poverty and increase national security.

6. A global reforestation program to slow global warming and replace lost resources.

7. Improved conservation and recycling of almost everything.

8. Ways to respond to the consequences of global warming, such as rising ocean levels and losses of agricultural land.

> The most important steps toward sustainable development involve conservation, education, and population control.

## Policy Changes: Costa Rica

Many countries, including the United States, will have to change their national policies drastically before the future is secure. Costa Rica provides an interesting example of the kinds of changes that are needed.

Costa Rica is a largely agricultural country about twice the size of Vermont, lying between Nicaragua and Panama. It contains enormous diversity, including mountains, seashore, rain forest, and dry forest. For this reason and because it has a long history as a stable socialist democracy, Costa Rica has been a favorite research area for American biologists and agriculturalists. The country also has a good educational system so that scientists have developed a considerable understanding of many aspects of the country's ecology.

In the early 1970s, the government realized that Costa Rica's peaceful charm was threatened. Under the influence of a national health system, the death rate had fallen rapidly. The population was spiraling out of control, overwhelming social services, threatening economic development, and destroying the environment. The government instituted population control policies, emphasizing education and making family planning available to all. The rate of population growth fell immediately, although it is still uncomfortably high. Then Costa Rica embarked on a series of experiments in conservation and sustainable development designed to support its growing population. When Oscar Arias was elected president in 1986, these efforts intensified.

In the 1970s, a series of national parks was set up that now covers 10% of the country's area and a great diversity of habitats. Ecotourists visiting these parks contribute about $150 million each year to the country's economy. Arias established a new Ministry of Energy, Mining, and Natural Resources in an attempt to put responsibility for sustainable development under one roof. Alvaro Umana, head of the new ministry, faced enormous problems. For instance, more than 90% of the country has been deforested by farmers forced into marginal areas by population pressure and unemployment, as in the Philippines. For decades, residents have torn down forest, inside and outside the parks, for cattle ranching, for crop farming, and simply to claim squatters' rights to the land. The country relies on reservoirs and hydroelectric dams for energy, but deforestation has caused soil erosion that has silted up the reservoirs.

"Sustainable development is important to meet these problems," said Umana. "It used to be that conservation meant preservation without man; that land should be kept in a bubble. . . . Sustainable development represents a new style."

Among the government's aims are increasing the area of the national parks by 50%, massive reforestation in the parks and on private land, and the development of less destructive agriculture. One of the tools is widespread education. If new generations of schoolchildren and older neighbors of the parks learn that they can make a better living by not destroying their environment, resources such as the vegetation and animals in the parks will be preserved. Mario Boza, who founded the park service in 1970, says, "You can argue that conservation is important for our . . . grandchildren. But if you do, you're lost. No one is interested in future generations. People want their reward now."

### Swapping Foreign Debt for Conservation

A major burden of many nations is their debts to other countries. A new financial technique uses this fact to provide land and money for conservation. An American environmental group purchased some of Bolivia's foreign debt. In exchange, Bolivia agreed to protect three parcels of land adjoining that country's Beni Biosphere Reserve. Part of the area will be preserved for research. Another part will be used by the indigenous Chimane people and for experiments in sustainable agriculture and forestry.

Debt swapping is particularly important at a time when economic aid (although not military aid) to developing countries has been falling. Banks sell debt notes from developing countries at a big discount from their original value because they doubt the country can pay off the note. In the case of Costa Rica, foreign conservation groups have bought these notes at a cost of little more than 15 cents on the dollar and donated them to the Costa Rican foundation for conservation, the Fundación Neotrópica. The country's foreign debt has now been reduced. In exchange, the Costa Rican government agrees to exchange the notes for bonds in local currency that

are worth 75 cents on the dollar face value of the notes. This multiplies the value of the notes to the Fundación five times. The Fundación uses the bonds for conservation work.

Banks sometimes get into the act directly. The Fleet Bank of Rhode Island donated $250,000 to Costa Rica for land acquisition and park management by canceling part of the country's debt to the bank. The bank writes off a poor loan and takes a charitable tax deduction. Costa Rica reduces its foreign debt and converts the loan into $1 million in local funds. It is hard to overestimate the value of schemes such as these, which provide for sustainable development while benefiting everyone involved immediately. We need every such bright idea from the world's financial whiz kids in our push for a better future.

> Swapping foreign debt for conservation is a scheme that provides for sustainable development while benefiting everyone involved immediately.

### Slowing the Loss of Biological Diversity

The loss of species and of biological diversity in general has become an urgent problem. Attempting to rescue part of Earth's gene pool is a race against time that has produced some remarkable techniques. Gene samples of vanishing species and varieties are being saved in many kinds of preserves throughout the world.

Some of these preserves are as near as your back garden. Many seed companies now sell "heirloom varieties" of vegetables and flowers. These are old-fashioned varieties, such as our grandparents planted, that have not been produced on a large scale for many years. Some have disappeared, but others have been preserved by gardeners who saved their own seed or exchanged it with friends over the decades (Figure 24-6). Seed companies now want samples of these varieties. Some they will use for breeding, to see if qualities of old varieties will enhance modern varieties. Some they will propagate unchanged because the public is interested in growing old varieties. On a larger scale, agricultural stations all over the world collect and breed varieties of crops collected from vanishing habitats or from farmers who still grow varieties adapted to local conditions.

Attempts to preserve diversity may be more high-tech. Some plant seeds are frozen to preserve them indefinitely. A number of zoos have undertaken breeding programs with endangered species. In the

**Figure 24-6**   Heirloom varieties of zinnias and black-eyed Susans growing in an Idaho garden.

case of mammals, some zoos are experimenting with surrogate mothers, females pregnant with embryos of other species. They also freeze embryos of endangered species produced by fertilization in a test tube. If the species becomes extinct, preserved embryos could be implanted into surrogate mothers and become the ancestors of a new population.

> Attempting to rescue part of Earth's gene pool is a race against time that has produced some interesting new techniques.

### International Cooperation

It has been obvious throughout this book that one of the barriers to progress is competition, instead of cooperation, between nations. International plans for the future are probably essential to a sustainable future. In some ways, barriers between nations separate governments but not people. People tend to

care more about their work than they do about politics. Scientists from different nations often exchange ideas and work together on agriculture or endangered species, even if their governments are not speaking to each other. Most businesspeople care more about profits than about politics and find ways to sell their goods even to countries with which they are at war.

In many ways, this internationalization by individuals and businesses has left governments at the starting gate. For instance, the U.S. government has only the vaguest plans for supplying the country's energy needs in 2050. Multinational oil companies have very detailed plans, but the primary concern of oil companies is profit. Unless international agreements force them in new directions, they will continue to use up fossil fuels in whatever manner produces most profit. If governments agree on sustainable development, the oil companies will fish from the file drawer their plans for less destructive energy production. The type of development we shall see depends to a large extent on whether or not governments can get together and make decisions about our future.

There are many signs of progress. International agreements cover Antarctica, the management of ocean fisheries, the reduction of ozone depletion, and hundreds of other equally important areas. Unfortunately, these agreements often do not involve countries whose contribution is vital to the agreements' success. India and Israel refused to sign the nuclear nonproliferation treaty, and North Korea has withdrawn from it; the United States turned down the Law of the Sea and watered down the United Nations environmental treaties; Japan avoided the ban on whaling. Countries have to move beyond narrow views of their national interests if we are to progress.

> Free enterprise will continue to destroy Earth's resources unless we can develop international plans for the future and environmental regulations that are enforced.

### Education

Most people know very little about the environment. If you have read this book, you know more about environmental issues than most world leaders. More than one quarter of the U.S. population does not even believe that evolution occurs. Local politicians fail to find solutions to the rising cost of landfills, although imaginative solutions are all around them.

**Figure 24-7**    Education. Environmental education takes many forms. Here a marine science staff member teaches environmental biology to beach resort visitors. The group has pulled a seine net through the surf. The teacher will use the organisms they have caught to explain the interactions between some of the plants and animals found on the beach and the effect of local building projects on seaside habitats.

If we are to progress toward sustainable development, this widespread ignorance has to be dispelled.

Education is probably the most important single step on our staircase to a better future (Figure 24-7). We do not have to start teaching environmental issues in elementary school, although this is happening in some places. Education is a more general weapon. Educated people understand why pollution control and conservation save them money. They know that solutions can be found to problems such as water shortages, high utility bills, and ugly landscapes. They understand environmental discussions in the media and educate their elected representatives through telephone calls and letters. Educating women is vital to population control, lower health-care costs, and lower crime rates: the uneducated bear more expensive premature babies, have more abused or neglected children, and look after their health less well than the educated. Education is vital to better international cooperation. Fear and distrust of foreigners are the product of ignorance. It is sometimes difficult to decide how to spend our limited resources in the battle for sustainable development, but no single item produces a better return on our investment than education.

> Education is probably the most important single step on our staircase to a better future.

## Private Actions

The scale of our environmental problems is so vast that we tend to despair. Perhaps none of us alone can save the world, but the actions of individuals and small groups can add up surprisingly rapidly to the kind of progress that makes a difference. Consider a few examples.

America's farmers are using fewer pesticides than they were in 1975. A local council in South Carolina refused to allow the state to cut down 400-year-old oak trees to widen a road. A citizen's group induced a national fast-food chain to stop buying beef from cattle ranches established on clear-cut tropical forest. Billions of trees were planted by individuals on their own land last year. Women in India prevent deforestation by throwing their arms around trees about to be cut down. Private conservation groups preserve millions of hectares from the bulldozer every year and rescue millions of plants from destruction. Citizens' lawsuits have forced companies to clean up the toxic waste they produce.

These actions add up, and some of them have a snowball effect. When they are reported, locally or internationally, many people cheer and pressure local politicians to prevent environmental abuse in the future. Every bit of progress starts as a bright idea. A few people decide to save the local marsh, to recycle newspapers, to spend less on armaments, to swap debt for nature, or to open a family planning clinic, and many of these decisions generate further progress.

> Although none of us alone can save the world, individual actions rapidly add up to the kind of progress that can make a difference.

## 24-D  A VISION OF THE WORLD IN 2040

Suppose that we make reasonable progress toward a sustainable future starting with the world outlined in Close-Up 24-1. What might life in the United States be like toward the middle of the twenty-first century?

You can undoubtedly come up with your own prophesies, but here is my version of what the future may be like. Imagine the State of the Union in 2040.

## The United States

The cost of living in the United States is high. Loss of agricultural land and the cost of irrigation have raised food prices, and the average water bill is high. Those who can afford them have installed water-saving appliances and gray water systems. You pay $30 for every pound of waste the garbage collectors remove from your premises, so most people take paper, metals, and glass to recycling centers and compost all their yard waste. Gardeners and farmers are desperate for organic soil additives, so people and municipalities make money by selling compost.

Farming and gardening have changed because water and fertile land are expensive, pesticides closely regulated, and chemical fertilizers heavily taxed to discourage their use. Irrigation is something most Americans see only on European vacations. At home, they plant only drought-resistant and disease-resistant plant varieties and keep them in good health using compost, manure, and biological pest controls. Lawns have disappeared from the drier parts of the country. They have been replaced by native ground covers or old-fashioned "swept yards," with bushes and trees. Many of these bushes and trees bear edible fruit because anything that helps out the food budget is welcome. Summers are warmer and drier than they were a century ago so that agriculture has been abandoned in huge areas of the south and central United States, which are now covered with forest. Canada and Russia are the world's largest grain producers.

The chemical composition of the atmosphere has stabilized. Global warming appears to have been halted—mainly by massive reforestation. The sea level on the United States's East Coast has risen, and millions of homes and businesses have fallen into the sea. It would have risen further but for a number of volcanic explosions during the past 40 years, which have slowed global warming by spewing huge clouds of sunlight-blocking ash into the air. Seawalls and similar attempts at erosion control have been outlawed, minimizing the losses caused by the rising sea level (Figure 24-8). The coast is a recreation paradise of protected public beaches and wetlands teeming with wildlife.

Energy comes from a multitude of sources: electricity-generating plants driven by hydroelectric, fossil, or nuclear power and from solar batteries, biodigesters, and windmills. Energy prices have risen enough to make conservation worthwhile. Individu-

**Figure 24-8**   Beach erosion. Snow fence has been installed on this beach to trap sand blown in by the wind in the hope of rebuilding sand dunes that once protected the shore from erosion but have since been destroyed by building roads and houses on the dunes. On sandy beaches, the sand moves up and down the coast, depending on the tides and the weather. If the dunes are destroyed, nothing is left to catch the sand and large areas of the coast may be washed away. Destroying dunes also wipes out the nesting habitat of many endangered shore birds and sea turtles. *(Steve Bisson)*

als who can afford it, and all successful businesses, live in passive solar buildings and use public transport much of the time. Planes, buses, maglevs, and trains are fast and energy-efficient. Most motor vehicles run on methanol or on liquefied natural gas that is imported from Russia through a pipeline across the Bering Strait. Cars go more than 100 kilometers on a liter of fuel. However, few households have more than one car, and most have none because car ownership is discouraged by a tax. In many places, it is faster to travel by bus, bicycle, or train than by car. Cars are permitted only in a single lane of each multilane road, leaving the space once oc-

# The State of the World

Figure 24-A grades various countries for environmental achievement on the basis of population growth rate, carbon emission, the access of the population to clean drinking water, and protection of habitat to preserve forest and other ecosystems as well as endangered species. Notice that carbon emissions are caused mainly by burning fossil fuel. For instance, although China is much more densely populated than Europe or North America, it produces much less carbon per person because it uses less fossil fuel. What other trends and generalizations can you detect?

## Questions for Discussion

African countries are not graded on this map, although it would be interesting to compare them with the countries that are graded. Why do you suppose Africa is not represented here?

### Population

- Australia, the United States, and Canada are the only developed countries that don't get top grades in population growth. What makes them different?

- Why do you suppose Latin America gets worse grades for population growth than Asia and Oceania?

- What do the countries with top grades for population have in common?

- Why does Mexico get a worse grade for population than Argentina? European settlers in Mexico came mainly from Spain. Those in Argentina were more likely to be from northern Europe. Could this account for the difference? (Take into account that Spain gets top marks for population.)

### $CO_2$ Emissions Per Capita

- Why are Europe's grades for emissions worse than those of Latin America?

- Former Soviet countries (Poland, Russia, Romania) get poor grades for emissions because their fossil fuel-based industries burn fuel inefficiently and produce large amounts of air pollution. Why do Canada, Australia and the U.S. get worse grades even though they have lower levels of air pollution?

- In 1993, the U.S. Congress raised the federal tax on gasoline by less than 5¢, leaving the price of gasoline, adjusted for inflation, lower than it was in 1980. How will this affect the U.S. grade for emissions?

### Water Quality

- The U.S. and Israel are the only developed countries without perfect grades for drinking water. What population groups in the U.S. do not have access to safe drinking water?

- Argentina and Indonesia get failing grades for water because less than 70% of the population has access to safe drinking water. If you were minister of health in either country, how would you try to improve the situation as cheaply as possible?

- Why do Poland, Russia, and Romania get poor grades for access to safe drinking water?

### Habitat Conservation

- Why does Brazil get a failing grade for conservation although it has announced that more than 5% of the country's area is set aside as wildlife habitat?

- Japan and Britain are island nations with similar levels of development and high population densities. Yet one gets a high and the other a low grade for conservation. Can you think of any possible reasons?

- Do these grades for conservation support the generalization that capitalist countries value their natural resources more highly than do communist countries?

- Protecting habitat often involves protecting lakes and streams that are drinking water sources from development and industry. So we might expect a correlation between a country's grade for conservation and its grade for water quality. Is there such a correlation?

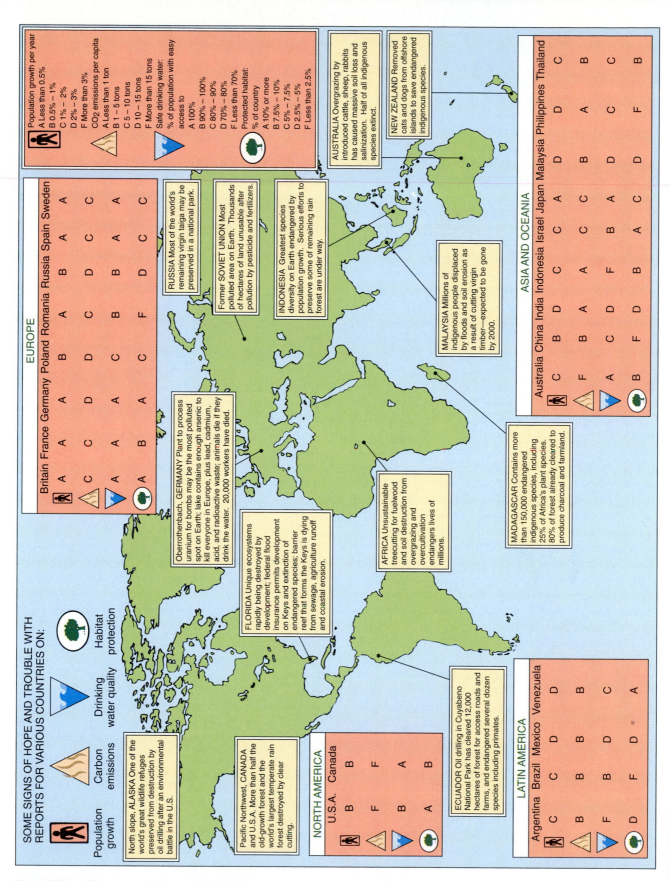

**Figure 24-A** The state of the world, 1993. Environmental reports on selected countries with some of the world's trouble spots and signs of hope. (*Sources:* Time, *U.N. Development Programme, U.N. Population Fund, World Resources Institute, Worldwatch Institute.*)

cupied by other lanes for high-speed buses, electric trains, and maglevs.

Millions of items, from televisions to automobiles, are made of plastics instead of metals. All plastics are recyclable and biodegradable, with varying life spans depending on the purpose for which they are intended. As a result, your personal possessions sometimes fall apart before you want them to, but that is just one of the minor snags of modern life. Most plastics are made using polymers extracted from wood and maize instead of from petroleum. Trees are more valuable than ever before, and the lumber industry is booming, caring for enormous nurseries of seedlings and planting millions of hectares of trees each year. The trees killed by acid rain have been replaced by pollution-resistant species and varieties, and forests are recovering, although they look different from twentieth-century forests.

Social habits and ethical standards have changed. The cost of garbage disposal means that antilittering laws are strictly enforced. In any case, social disapproval would greet anyone seen dropping litter. Antinoise ordinances are also enforced so that a radio is as rare as a beer can on the beach. Sunbathers are rare too. Skin cancer rates soared as the ozone layer thinned. Now broad-brimmed hats and baggy coverups set leisurewear fashions. Sales of tobacco are illegal, and only a few people grow their own. You would never be so impolite as to smoke in public or in anyone else's house. No one would dream of being seen in a coat made from the pelt of a wild animal. Ten years in jail without parole awaits anyone caught with a protected animal or plant. AIDS is still incurable, but the number of new cases each year stopped increasing some time ago because sexual behavior is much changed since the promiscuous old days. Contraceptives and family planning clinics are widely approved, advertised on television, and supported by local taxes. As a result, teenage and illegitimate births have fallen to their lowest rate in history, and public costs for infant medical care and welfare are low.

Society invests heavily in the one or two children that most people have. School is in session year-round, but grade-school education ends at the age of 14. After this, an enormous range of free higher education, paid for by government and industry, is available throughout life. Most people completely retrain themselves at least once in their lives.

Unemployment is low because there is a permanent shortage of skilled labor. The demand for research workers is enormous. They study every imaginable aspect of environmental science. They develop new plant varieties, plastics, energy sources, educational methods, and water systems, not only for the industrial world but also for developing nations. The health-care industry is also crying out for research workers because cancer rates from environmental pollutants are still rising.

Communications are efficient and cheap. Video telephones and computer translators permit you to see and talk to almost anyone in the world at a moment's notice. The fax machine will send a four-color copy of the document at your side to your colleague in Bangkok as you discuss it. You have learned at least one Asian language in school so that you do not have to rely on translators for all business dealings, and you improve your skills in other languages on foreign vacations.

Military spending is much reduced. People finally decided that improving standards of living took priority over deterring foreign aggression. But taxes are no lower than during the twentieth century because the money saved from the military budget is needed for other things. These include medical expenses for millions poisoned by pesticides, toxic wastes, and heavy metals. They also include large sums for water treatment, waste disposal, reforestation, and soil preservation. The large elderly population is another burden on the taxpayer, although this burden might be worse because people do not live as long as they did 50 years ago, and few people can afford to retire. Many remain taxpaying wage earners until they die, a situation resembling that before World War II.

## Developing Countries

There are signs of hope in some developing countries. The rate of population growth has continued to fall since the 1970s and in developed countries has now reached zero, meaning that population growth in those countries is now due solely to immigration (Figure 24-9). The growth rate in most developing countries was still 1% a year in 2030, but populations in some countries in Africa, South America, and the poorer nations of Asia actually fell slightly in the 2030s, largely because AIDS, starvation, and wars raised the death rate. This tragic period may have proved a blessing in the long run. Small-scale nuclear weapons were used in the war in Africa, and the nations of the Northern Hemisphere finally took fright. Exporting arms to developing countries, so popular in the twentieth century, suddenly seemed like a bad idea. Foreign aid to be spent on sustainable development came back into fashion.

The industrial countries now spend large sums on education, population control, agriculture, and conservation in the tropics. These investments are beginning to bear fruit. Drought-resistant plants, discovered and developed in Africa, are now important

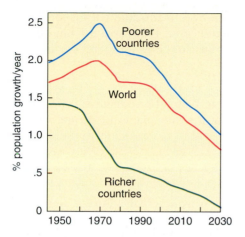

**Figure 24-9** The decline in population growth rate worldwide and in poor and rich countries, projected until 2030. Note that this is not a decline in population. Population will not decline until the growth rate becomes negative. The graph shows that even in 2030, the world population will still be growing by 1% a year.

crop plants for many farms in the Northern Hemisphere and produce license fees for African nations. Cures for liver cancer and mercury poisoning were discovered in the forests of Cameroon, and each

Cameroon citizen shares in the license fees—an effective encouragement to conservation and research in that country. In many developing nations, literacy rates, agricultural productivity, and life expectancies have begun to rise, and reforestation and research have supplied employment for millions of rural residents. It is a long and crooked road to prosperity, but we appear to be traveling in the right direction at last.

## 24-E CONCLUSION

We are fresh out of excuses for our failure to build a sustainable society. We cannot shift the blame onto the shortsighted politician, the power-hungry military planner, the profit-crazed industrialist, or the ignorant educator. We have been, are, or shall be the voter, the teacher, the scientist, the miner, the engineer, the farmer. "We have met the enemy, and they are ours."

If we had started earlier on the task, the world could have had a higher standard of living than it can now ever hope to achieve: the giant panda and the Florida Everglades once gone are gone forever. But it is still within our power to build a society that does not consume our remaining environmental capital and that provides a reasonable standard of living for everybody.

## SUMMARY

Physical laws dictate that a sustainable future is bound to come. It is up to us to decide whether this will occur only when the human species is extinct or whether we can build a human society that does not destroy our environment faster than it is restored.

### War

Nuclear war becomes more likely as more and more countries and groups become capable of building nuclear weapons. Such a war might start by accident. Even if it did not kill many people directly, a nuclear war would kill millions indirectly and drastically lower standards of living by wiping out electrical power over large areas and by changing the climate in such a way as to cripple agriculture. Environmental degradation is already causing armed conflict in many countries.

### If We Fail to Change

If we continue to consume our environmental resources faster than they are replaced, living standards everywhere will decline until the whole world is living in what we would now think of as poverty.

### Sustainable Development

We can move toward sustainable development by way of planning, education, conservation, and population control. Our plans for the future should make international provision for research, energy, food, reduced military spending, abolishing poverty, reforestation, conservation, and recycling. Sustainable development requires policy changes by governments. Costa Rica's government reorganization provides an interesting example. Imaginative ideas, such as swapping foreign debt for conservation, come from all segments of society.

**A Vision of the World in 2040**

If we progress toward a sustainable future, life in the middle of the twenty-first century will be very different from life today but quite pleasant to contemplate.

We shall see differences in education, transportation, employment, health care, energy systems, our ethical standards, our habits, and how we spend our money.

**Conclusion**

We have no one to blame but ourselves if we fail to save what is left of the only planet we've got.

## DISCUSSION QUESTIONS

1. The view of the world in 2040 presented here is, I think, probably overoptimistic. Do you agree? Come up with your own "state of the world" based on what you have learned in this book.

2. Can we do anything to prevent the increase in violent conflicts between members of different tribes, religions, races, and nationalities that have broken out in so many parts of the world during the 1980s and 1990s?

3. Education in environmental affairs is not widespread in the developed world. What do you think people should be taught about this subject and at what stage in their educations?

4. Suppose a nuclear war or agricultural disaster wiped out most, but not all, of the people on Earth. What kinds of ways of life would be most likely to permit people to survive after the disaster?

# APPENDIX A
## Measurements and Units

Environmental science is full of measurements and statistics. Even the international scientific and business communities cannot agree on how things should be measured or how the measurements should be spelled, so it is not surprising that everyone gets thoroughly muddled from time to time. This appendix is a guide to the units used in this book.

## 1. THE METRIC SYSTEM

This book uses the metric system of measurements (as does everyone in the world except some segments of the United States). This is a decimal system, in which all of the units are expressed as powers of ten times some basic unit.

To establish a uniform set of units, the General Conference of Weights and Measurements in 1960 recommended a single base unit to be used for each measured quantity. The resulting system is called the *Système International d'Unités*, abbreviated **SI**. Base units of the SI are listed in Table A-1. Other units are derived from the base units.

Quantities larger and smaller than the base units are expressed by using the appropriate prefix (Table A-2) with the base unit. For instance, road distances are given in kilometers, multiples of meters, which are the base units. A kilometer is 1000 meters (1000 m or $10^3$ m). *Kilo* means 1000 times. Laboratory ob-

### TABLE A-2 Selected Prefixes Used in the Metric System

| Prefix | Symbol | Meaning | Example |
|--------|--------|---------|---------|
| Giga- | G | $10^{12}$ | 1 gigawatt = $10^{12}$ watts |
| Mega- | M | $10^6$ | 1 megaton = $10^6$ tons |
| Kilo- | k | $10^3$ | 1 kilogram (kg) = $10^3$ grams |
| Deci- | d | $10^{-1}$ | 1 decimeter (dm) = 0.1 m |
| Centi- | c | $10^{-2}$ | 1 centimeter (cm) = 0.01 m |
| Milli- | m | $10^{-3}$ | 1 milliliter (mL) = 0.001 liters |
| Micro- | μ | $10^{-6}$ | 1 micrometer (μm) = $1 \times 10^{-6}$ m |
| Nano- | n | $10^{-9}$ | 1 nanometer (nm) = $1 \times 10^{-9}$ m |
| Pico- | p | $10^{-12}$ | 1 picogram (pg) = $1 \times 10^{-12}$ g |

jects are often measured in centimeters. *Centi* means one hundredth.

The only exception to my use of metric units is in American laws in which units such as feet and acres are used. These have not been converted into their metric equivalents. Table A-3 shows conversion into metric units and Table A-4 shows conversion from metric units.

## 2. MONEY AND INFLATION

At various places in this book you will see terms such as "in 1983 dollars." These dollars with dates reflect inflation, which must always be taken into account when talking about money at various dates.

Inflation means that a given sum of money almost invariably has less purchasing power at a later date. For instance, $1 in 1900 would buy what $25 would

### TABLE A-1 SI Base Units

| Physical Quantity | Name of Unit | Symbol |
|-------------------|--------------|--------|
| Mass | kilogram | kg |
| Length | meter | m |
| Time | second | s |
| Temperature | kelvin | K |
| Amount of substance | mole | mol |
| Electric current | ampere | A |
| Luminous intensity | candela | cd |

| TABLE A-3   Conversion into Metric Units | | |
|---|---|---|
| **Multiply Number of** | **by** | **To Obtain Equivalent Number of** |
| **Length** | | |
| inches (in) | 25.4 | millimeters (mm) |
| feet (ft) | 30.5 | centimeters (cm) |
| yards (yd) | 0.9 | meters (m) |
| miles (land 5,280 ft) | 1.6 | kilometers (km) |
| **Area** | | |
| acres (4,840 sq yd) | 0.4 | hectares (ha) |
| square miles | 2.6 | square kilometers (km$^2$) |
| **Volume** | | |
| cubic inches | 16.4 | cubic centimeters (cm$^3$) |
| U.S. gallons | 3.8 | liters (L) |
| **Mass** | | |
| ounces (oz) | 28.4 | grams (g) |
| pounds (lb) | 0.45 | kilograms ("kilos") (kg) |
| short tons (2,000 lb) | 0.9 | metric tons |
| long tons (2,240 lb) | 1.02 | metric tons |
| **Speed** | | |
| miles per hour (mph) | 1.61 | kilometers per hour (km/h) |
| miles per hour | 0.87 | knots (k) |

| TABLE A-4   Conversion from Metric Units | | |
|---|---|---|
| **Multiply Number of** | **by** | **To Obtain Equivalent Number of** |
| **Length** | | |
| millimeters | 0.039 | inches |
| centimeters | 0.033 | feet |
| meters | 1.09 | yards |
| kilometers | 0.62 | miles |
| miles, nautical (nm) | 1.9 | kilometers |
| **Area** | | |
| square meters | 1.2 | square yards |
| hectares | 2.48 | acres |
| **Volume** | | |
| cubic centimeters | 0.06 | cubic inches |
| liters | 0.26 | U.S. gallons |
| **Mass** | | |
| grams | 0.035 | ounces, avoirdupois |
| kilograms | 2.2 | pounds |
| tons, metric | 1.1 | short tons (2,000 lbs) |
| **Speed** | | |
| kilometers/hour | 0.62 | miles/hour |
| knots | 0.54 | miles/hour |

buy in 1980 and about what $37 would buy in 1989. If, in 1989, we mentioned that someone spent $10,000 in 1900, this might not sound like a very large sum. We would get a much better idea of the amount of money involved if we converted it into modern money and called it "$375,000 (in 1989 dollars)."

In the United States, the rate of inflation is measured by the consumer price index (CPI). To determine the CPI on a given date, the government calculates how many dollars it takes on that date to buy a specified collection of goods and services (such as a medium-priced house, a given amount and type of food and clothes, and a particular amount of medical care). This number of dollars can be compared with the cost of the same collection of goods and services a year earlier. If the cost of the goods and services is 6% higher in 1994 than in 1993, we say that the rate of inflation during 1993 was 6%. In 1994, we would say that a $1 donation to charity in 1993 cost "$1.06 in 1994 dollars."

Another term used in this book that refers to the value of money is "hard currency." This refers to the fact that the money of some countries is more desirable than the money of other countries. Each nation uses its own currency (dollars, pesos, liras, francs, or whatever it may be). The rate of exchange between two currencies describes how much money of one currency you have to pay to purchase money in a different currency at any particular time. Some currencies are more sought after than others. These are the hard currencies such as American dollars, Japanese yen, French francs, and German marks. You can buy or sell these hard currencies at almost any bank in the world. But some countries have currencies that few people want because little trade is conducted in them. This means, for example, that if Norway sells oil to Zimbabwe, it will accept payment only in U.S. dollars (or some other hard currency). Zimbabwe cannot obtain the dollars it needs by selling its own currency for dollars because few people want Zimbabwe currency. The only way Zimbabwe can acquire dollars to pay for its oil imports is to export its own products to a country that is prepared to pay for the products in a hard currency. This is why earnings from exports are so important to developing nations.

## 3. ENERGY

Energy values are measured in a bewildering array of different units. Here are some definitions that will help:

**1 joule (J)** is the work done when 1 kilogram is raised 1 meter.

**1 Btu** (sometimes written BTU) = 1,054.8 J.

**Power** (measured in watts) is rate of energy delivery.

**1 watt (W)** = 1 joule per second ($Js^{-1}$).

**1 kilowatt-hour (kWh)** = 1,000 watts for 1 hour.

**1 million (metric) tons of oil equivalent (mtoe)** generates 4 billion kilowatt-hours of electricity.

**1 barrel** (of oil) = 42 U.S. gallons = 0.159 cubic meters.

**1 Calorie** = 1000 calories = **1 kilocalorie**.

**1 calorie** = heat needed to raise one gram of water from 14.5° C to 15.5° C. The energy values of foods are expressed in Calories.

# APPENDIX B
## References, Further Reading, and Sources of Information

## CHAPTER REFERENCES

(Many apply to more than one chapter.)

### Chapter 1    Science and Environment

BROWN, L.R., and others. *State of the World 1994.* New York: W.W. Norton, 1994. A volume of this book is produced each year by the Worldwatch Institute. It contains about a dozen chapters on environmental problems and progress.

CARSON, R. *Silent Spring.* Boston: Houghton Mifflin, 1962. The environmental classic that alerted the world to the dangers of pesticide misuse.

EHRLICH, P.R. *The Population Bomb.* New York: Ballantine Books, 1968. Now a classic of environmental science, this is the sensationalist book that popularized the explosive situation created by our toleration of rapid human population growth.

ELKINGTON, J., J. HAILES, and J. MAKOWER. *The Green Consumer.* New York: Viking Penguin, 1990. Loads of details on consumer issues and how they affect the environment: from buying energy-efficient appliances to using pesticides safely.

GRANT, L. "The cornucopian fallacies: The myth of perpetual growth." *The Futurist,* 17:16, 1983. A rebuttal of the optimistic, "cornucopian" views of Julian Simon and the late Herman Kahn, who have denied that environmental problems are destroying our economic prospects.

GROVE, R.H., "The origins of western environmentalism." *Scientific American,* July 1992. A little history: the destruction of beautiful tropical islands by western settlers in the seventeenth and eighteenth centuries induced scientists to study human effects on the environment and to recommend preservation instead of destruction.

SAMUELSON, P.A., and W.D. NORDHAUS. *Economics.* 13th ed. New York: McGraw-Hill, 1991. The first few chapters of this classic textbook provide an excellent introduction to economics.

SIMON, J. "Life on earth is getting better, not worse." *The Futurist,* 17:7, 1983. Simon is a leading cornucopian who denies that environmental problems lead to economic disaster. This is an optimistic projection of the future based on past events that environmentalists would argue will not continue.

STUDENT ENVIRONMENTAL ACTION COALITION. *The Student Environmental Action Guide.* Berkeley, CA: Harper-Collins Publishers and the Earth Works Group, 1991. A list of environmental actions that college students can take on campus, with suggestions and success stories.

U.S. BUREAU OF THE CENSUS. *Statistical Abstract of the United States.* Washington, D.C.: USBC. This book of tables contains statistics on population, health, production, agriculture, trade, foreign aid, and so on. It is a mine of information on everything that happens in the United States, with a few tables devoted to the rest of the world. A new edition is published every year.

### Chapter 2    How Humans Affect Their Environment

ARMS, K., and P.C. CAMP. *Biology: A Journey into Life.* 3rd ed. Philadelphia: Saunders College Publishing, 1994. An introductory textbook for more detail on evolution, classification, biodiversity, ecology, and other biological topics.

BAR-YOSEF, O., AND B. Vandermeersch. "Modern humans in the Levant." *Scientific American,* April 1993. Evidence from Israel that modern humans preceded tool-using Neanderthals in the Middle East.

BEAUNE, S.A. DE, and R. WHITE. "Ice age lamps." *Scientific American,* March 1993. The story of the prehistoric domestication of fire and the importance of the development of portable lamps.

GHIGLIERI, M.P. "The social ecology of chimpanzees." *Scientific American,* June 1985. A fascinating study of our nearest relatives.

528

HARDIN, G. *Filters Against Folly*. New York: Viking, 1985. A noted environmentalist writes on ecology, science, and economics.

MOOREHEAD, A. *Darwin and the Beagle*. New York: Harper & Row, 1969. A short, beautifully illustrated account of Darwin's travels based on his diaries.

PILBEAM, D. "The descent of hominoids and hominids." *Scientific American*, March 1984. A clear account of how thinking on the subject changed during the preceding five eventful years, summarizing a new emerging consensus and pointing out what we do and do not know about our human "roots."

TOBIAS, M., ed. *Deep Ecology*. San Diego: Avant Book, 1985. "Deep ecology" is the name given to the belief that non-human entities such as animals, plants, and mountains have the right to exist. This is a collection of writings on environmental problems and solutions.

## Chapter 3   Energy, Chemistry, and Climate

HOUGHTON, R.A., AND G.M. WOODWELL. "Global climatic change." *Scientific American*, April 1989. Presents the evidence that the greenhouse effect is already warming Earth and discusses how to control it.

KOTZ, J.C., AND K.F. PURCELL. *Chemistry and Chemical Reactivity*. 2nd ed. Philadelphia: Saunders College Publishing, 1992. A general chemistry text, useful for reference.

POLLACK, H.N., AND D.S. CHAPMAN. "Underground records of changing climate." *Scientific American*, June 1993. Holes drilled into rocks underlying the continents are providing evidence of global temperatures in the past, leading to more accurate descriptions of global warming.

SCHNEIDER, S.H. "Climate modeling." *Scientific American*, May 1987. Discusses how computer models of Earth's climate permit predictions of future climatic possibilities, such as the effect of a nuclear war and global warming.

MOHNEN, V.A. "The challenge of acid rain." *Scientific American*, August 1988. Argues that acid rain must be stopped by controlling emissions, particularly from coal-fired power plants. The technology to reduce emission of acid-forming oxides of nitrogen and sulfur is available.

POPE, C. "An immodest proposal." *Sierra*, 70:43, 1985. Discusses economic incentives that will induce people to institute pollution control.

## Chapter 4   Biodiversity and Habitats

BUCHSBAUM, R. *Animals Without Backbones*, 2nd ed. Chicago: University of Chicago Press, 1976. A classic introductory textbook on invertebrates.

DORIT, R.L., and others. *Zoology*. Philadelphia: Saunders College Publishing, 1991. A general textbook on animal biodiversity.

MARGULIS, L., and K.V. SCHWARTZ. *Five Kingdoms: An Illustrated Guide to the Phyla of Life on Earth*. San Francisco: W.H. Freeman, 1982. An introduction to the diversity of life on Earth.

MAUSETH, J.D. *Botany: An Introduction to Plant Biology*. Philadelphia: Saunders College Publishing, 1991. A general textbook on plant biodiversity.

WILSON, E.O., ed. *Biodiversity*. Washington, D.C.: National Academy Press, 1988. Essays by many authors on the crisis caused by the extinction of species in the twentieth century.

## Chapter 5   Ecosystems and How They Change

BREWER, R. *The Science of Ecology*. 2nd ed. Philadelphia: Saunders College Publishing, 1993. A general textbook on ecology, including ecosystems.

COUSINS, S. "Ecologists build pyramids again." *New Scientist*, 106, 4 July 1985. The value of pyramids of energy and matter in understanding energy flow through ecosystems.

EHRLICH, P.R., AND J. ROUGHGARDEN. *The Science of Ecology*. New York: Macmillan Publishing Company, 1987. A readable introductory text.

MITSCH, W.J., and J.G. GOSSELINK. *Wetlands*. New York: Van Nostrand Reinhold, 1986. An analysis of wetlands ecosystems and the steps necessary to preserve them.

MYERS, N., ed., *The Gaia Atlas of Planet Management*. London and Sydney: Pan Books, 1985. Our environmental problems portrayed in the form of maps. Some excellent illustrations and loads of interesting information.

PIMM, S.L., J.H. LAWTON, and J.E. COHEN. "Food web patterns and their consequences." *Nature* 350: 25, April 1991. Review of ecological understanding of food webs.

## Chapter 6   Natural Populations

BEGON, M., J.L. HARPER, and C.R. TOWNSEND. *Ecology: Individuals, Populations, and Communities*. Sunderland, MA: Sinaeur Associates, 1986. A population-oriented ecology textbook.

COLE, L.C. "The population consequences of life history phenomena." *Quarterly Review of Biology*, 29:103, 1954. The classic paper in which Cole showed how much more a female contributes to the growth of a population if she breeds as young as possible.

COLINVAUX, P.A. *Ecology*. New York: Wiley, 1986. An introductory text with good chapters on population biology.

RICKLEFS, R.E. *Ecology*. 3rd ed. New York: W.H. Freeman and Company, 1990. A general ecology text with good chapters on natural populations.

## Chapter 7   Human Populations

CALDWELL, J.C., AND P. CALDWELL. "High fertility in sub-Saharan Africa." *Scientific American*, May 1990. A discussion of why birth rates have declined much more slowly in the southern half of Africa than in most other parts of the world.

COMMISSION ON POPULATION GROWTH AND THE AMERICAN FUTURE. *Population and the American Future*. Washington, D.C.: U.S. Government Printing Office, 1985. Analysis of the need to control U.S. population growth.

CROLL, E., and others. *China's One-Child Family Policy*. New York: St. Martin's Press, 1985. Fascinating description of Chinese population policy.

DJERASSI, C. *The Politics of Contraception*. New York: W.W. Norton, 1980. The answer to the question posed in this chapter about why modern contraceptives are so primitive compared with our other technology: control of their development by (mostly male-run) drug companies and governments.

EHRLICH, P.R., and A.H. EHRLICH. *The Population Explosion*. 1990, New York: Simon and Schuster. The sequel to *The Population Bomb*, 20 years later. I don't find this quite such good reading as the first book, but it updates a lot of the statistics and scenarios.

GRUPTE, P. *The Crowded Earth: People and the Politics of Population*. New York: W.W. Norton, 1984. Excellent summary of population problems and attempts to tackle them in various countries.

HATCHER, R.A. *It's Your Choice*. New York: Irvington Press, 1982. Discussion of birth control methods.

SEN, A. "The economics of life and death." *Scientific American*, May 1993. Argues that mortality data should be incorporated into measures of a country's economic performance. For instance, women have long life expectancies in India and Costa Rica meaning that the quality of life is better than would be predicted from these countries' low GDP's.

### Chapter 8    Population, Environment, and Disease

BRENNEMAN, R.L., AND S.M. BATES, eds. *Land-Saving Action*. Covelo, CA: Island Press, 1984. Collection of articles on how individuals can save land from development.

ESPENSHADE, T.J. "A short history of U.S. policy toward illegal immigration." *Population Today* 18:2, 1990. The increase in illegal immigration into the United States and various attempts to control it.

FABOS, J.G. *Land-Use Planning*. New York: Chapman & Hall, 1985. A general book on planning with interesting examples.

HOBHOUSE, H. *Seeds of Change: Five Plants That Transformed Mankind*. New York: Harper & Row, 1986. Fascinating book; includes the story of quinine.

LAMM, R.D., AND G. IMHOFF. *The Immigration Time Bomb*. New York: E.P. Dutton, 1985. Analysis of the problems, argues the need for greater control of immigration into the United States.

MENKEN, J., ed. *World Population and U.S. Policy: The Choices Ahead*. New York: W.W. Norton, 1986. Papers on U.S. population policy, its strengths and weaknesses.

O'RIORDAN, T. *Environmentalism*. London: Pion Ltd., 1989. A British approach to the philosophy of environmentalism, containing interesting studies on land use planning.

### Chapter 9    Fossil Fuels and Nuclear Energy

ATOMIC INDUSTRIAL FORUM. *Nuclear Power Plant Response to Severe Accidents*. Bethesda, MD: Atomic Industrial Forum, 1985. Analysis of the safety of American nuclear power plants.

BALZHISER, R.E., and K.E. YEAGER. "Coal-fired power plants for the future." *Scientific American*, September 1987. By 2000, 70% of U.S. energy may come from coal-fired plants. To reduce the environmental damage from this source, fluidized bed combustion and coal gasification plants are urgently needed.

BURNETT, W.M., and S.D. BAN. "Changing prospects for natural gas in the United States." *Science*, 244:305, 1989. Discusses prospects for increasing use of natural gas as estimates of reserves have become more optimistic and presents new technologies that will be needed to use natural gas for electricity generation, transportation, and cooling.

EDMONDS, J., and J.M. REILLY. *Global Energy: Assessing the Future*. New York: Oxford University Press, 1985. Energy projections, supplies, alternatives, and policy during the next century.

HIRSCH, R.L. "Impending United States energy crisis." *Science*, 235:1467, 1987. A look at energy shortages that will probably arise during the 1990s.

OFFICE OF TECHNOLOGY ASSESSMENT. *Managing the Nation's Commercial High-Level Radioactive Waste*, 1990. Washington, D.C.: Government Printing Office. An excellent survey of the problem.

### Chapter 10    Energy Conservation and Renewable Energy

GREENBERG, D.A. "Modeling tidal power." *Scientific American*, November 1987. A discussion of the world's largest tides in the Bay of Fundy and how harnessing these tides with tidal power plants can be expected to affect tides in other areas.

HAFELE, W. "Energy from nuclear power." *Scientific American*, September 1990. The current state and future potential of nuclear power.

HEEDE, H.R.L., and others. *The Hidden Costs of Energy*. Washington, D.C.: Center for Renewable Resources, 1985. Discusses overt and hidden federal subsidies and how they distort the cost of different forms of energy.

HUBBARD, H.M. "Photovoltaics today and tomorrow." *Science*, 244:297, 1989. Discusses hard and soft paths to supplying future energy needs and the increased contribution to be expected from photovoltaics as a result of technological improvements.

PENNEY, T.R., and D. BHARATHAN. "Power from the sea." *Scientific American,* January 1987. A discussion of how ocean thermal-energy conversion works and the prospects that it will be used to produce significant amounts of energy.

ROSENFELD, A.H., and D. HAFEMEISTER. "Energy-efficient buildings." *Scientific American,* April 1988. Excellent illustrations of ways to improve energy efficiency in an article arguing that properly designed homes and offices can reduce energy bills, free capital for investment, and avoid the expense of building new power plants.

ROSS, M. "Improving the efficiency of electricity use in manufacturing." *Science,* 244:311, 1989. Conservation by industries as a means of minimizing energy costs in the future and a discussion of the policy changes that would encourage conservation.

TAYLOR, J.J. "Improved and safer nuclear power." *Science,* 244:318, 1989. The author argues that small, safe, long-lived nuclear power plants will be a cost-effective means of supplying part of our future energy needs. Different kinds of nuclear power plants are described.

## Chapter 11    Fresh Water

ANDERSON, T.L., ed. *Water Rights: Scarce Resource Allocation, Bureaucracy, and the Environment.* San Francisco: Pacific Institute for Public Policy, 1986. Excellent discussion of the problems of sharing this limited resource.

DOLAN, R., and H. LINS. "Beaches and barrier islands." *Scientific American,* July 1987. A description of why building seawalls and groins to protect East Coast beaches, wetlands, and barrier islands does not work. Argues that the best way to preserve coastal features is to let nature take its course.

FRANCO, D.A., and R.G. WETZEL. *To Quench Our Thirst: The Present and Future Status of Freshwater Resources of the United States.* Ann Arbor, MI: University of Michigan Press, 1993. Packed with information on U.S. freshwater resources.

GOULDING, M. "Flooded forests of the Amazon." *Scientific American,* March 1993. The amazing ecosystems of the flood plain of the Amazon River.

HUNDLEY, N. *The Great Thirst: Californians and Water, 1770s–1980s.* Berkeley, CA: The University of California Press, 1992. The story of how the battle to control water has shaped the landscape and people of California.

PENNSYLVANIA ACADEMY OF NATURAL SCIENCES. *Ground Water Contamination: Sources, Effect, and Options to Deal with the Problem.* Philadelphia: Pennsylvania Academy of Natural Sciences, 1989. Proceedings of the Third National Water Conference, with contributions by industry, scientists, and state and federal regulators.

RAMADE, F. *Ecology of Natural Resources.* New York: Wiley, 1984. A good general book on natural resources with a chapter devoted to worldwide water resources.

WORSTER, D. *Rivers of Empire: Water, Aridity, and Growth of the American West.* New York: Pantheon, 1992. The history of water development and the politics of water in the American West.

## Chapter 12    The Oceans

BORGESE, E.M. *The Future of the Oceans.* New York: Harvest House, 1986. Discussion of ocean resources and pollution.

BROWN, B.E., and J.C. OGDEN. "Coral bleaching." *Scientific American,* January 1993. Evidence that rising seawater temperature caused by global warming causes bleaching, which often kills coral reefs.

CARSON, R. *The Sea Around Us.* New York: Oxford University Press, 1961. Rachel Carson's delightful description of the sea and its life.

CUSHING, D.H., and J.J. WALSH, eds. *The Ecology of the Seas.* Philadelphia: W.B. Saunders Company, 1987. A number of papers covering various aspects of marine ecology, including food chains, water circulation, and productivity.

KIMBALL, L.A. *Southern Exposure: Deciding Antarctica's Future.* Washington, DC: World Resources Institute and the Tinker Foundation, 1990. The case for more effective international management of Antarctica.

O'HARA, K.J., and S. LUDICELLO. *A Citizen's Guide to Plastics in the Ocean.* Washington, D.C.: Center for Marine Conservation, 1988. Sources and control of plastics pollutions.

ROEMIMICH, D. "Ocean warming and sea level rise along the southwest U.S. coast." *Science,* 257: 373, 1992. The physical and social effects of the rising sea level.

## Chapter 13    Minerals

BORGESE, E.M. *The Mines of Neptune: Minerals and Metals from the Sea.* New York: Abrams, 1985. Covers the deep-sea resources that are the focus of political battles.

ESPENSHADE, E.B., and J.L. MORRISON, eds. *Goode's World Atlas,* 17th ed. Chicago: Rand McNally & Company, 1986. Contains many thematic maps showing the location of mineral reserves; trade routes; and also information on agriculture, population, and similar topics.

HAMRIN, R.D. *A Renewable Resource Economy.* New York: Praeger Scientific, 1983. An overview of the principles of resource supply and conservation.

KAHN, H., and others. *The Next 200 Years: A Scenario for America and the World.* New York: William Morrow, 1984. Cornucopians argue that we do not face a shortage of minerals or other resources.

LEONTIEF, W., and others. *The Future of Nonfuel Minerals in the U.S. and World Economy: 1980–2030.* Lexington, MA: Lexington (Heath), 1983. An overview of the problems of mineral resources.

MAURICE, C., and C.W. SMITH. *The Doomsday Myth.* Stanford, CA: Hoover Institute Press, 1984. Two economists, who believe that market forces solve all problems of supply and demand, present the argument that we do not face a mineral shortage.

## Chapter 14 Agricultural Land

BATIE, S.S. *Soil Erosion: Crisis in America's Croplands?* Washington, D.C.: Conservation Foundation, 1991. An analysis of soil erosion in America and the problems it poses.

GRAINGER, A. *Desertification: How People Make Deserts, How People Can Stop, and Why They Don't.* London and Washington, D.C.: International Institute for Environment and Development, 1985. The title of this little monograph is self-explanatory. Well written and with some horrifying statistics and examples.

PADDOCK, J., and others. *Soil and Survival: Land Stewardship and the Future of American Agriculture.* San Francisco: Sierra Club Books, 1991. An excellent discussion of the soil restoration needed to keep American agriculture productive.

POINCELOT, R.P. *Toward a More Sustainable Agriculture.* Westport, CT: AVI Publishing Co., 1986. A textbook for farmers on how to preserve soil fertility and use less fertilizer.

REGANOLD, J.P., R.I. PAPENDICK, and J.F. PARR. "Sustainable agriculture." *Scientific American,* June 1990. The advantages and present state of low-input agriculture.

## Chapter 15 Food and Hunger

BROWN, J.L. "Hunger in the U.S." *Scientific American,* February 1987. This article makes it clear why growing enough food is not sufficient to stop world hunger. Economic and political influences play a vital role. Hunger in the United States was virtually eliminated in the 1970s, but it returned with federal budget cuts during the Reagan administration. By 1987, some 12 million children and 8 million adults in the United States were hungry by federal definition.

DOBBING, J., ed. *Infant Feeding: Anatomy of a Controversy, 1973–1984.* New York: Springer-Verlag, 1988. The fascinating history of the battle between infant formula producers, such as Nestlé, and those who believe their marketing practices increase infant malnutrition. One aim of the book is to explain the process used to resolve the controversy—as a possible model for other types of conflict resolution.

FUSSELL, B. *The Story of Corn: The Myths and History, the Culture and Agriculture, the Art and Science or America's Quintessential Crop.* New York: Alfred Knopf, 1992. Everything you didn't realize you wanted to know about maize. Despite its intimidating title, this is an engrossing book, chock full of fascinating anecdotes, facts, and even fantasies.

GLANTZ, M.H. "Drought in Africa." *Scientific American,* June 1987. It takes more than climate to make a drought a disaster: policy and agricultural practices play major roles.

HEISER, C.B. *Seed to Civilization: The Story of Man's Food.* San Francisco: W. H. Freeman, 1973. A fascinating little book

on the plants and animals that make up human food: their history, biology, and uses.

HINMAN, C.W. "Potential new crops." *Scientific American,* July 1986. Discussion of efforts to develop new commercial crops.

INDEPENDENT COMMISSION ON INTERNATIONAL HUMANITARIAN ISSUES. *Famine: A Man-Made Disaster.* London and Sydney: Pan Books, 1985. A study of famine and starvation in the world today. Presents the evidence that crop failures and famine are not "acts of God" but disasters created by human mismanagement and overpopulation.

PACEY, A., and P. PAYNE, eds. *Agricultural Development and Nutrition.* London: Hutchinson, 1985. This United Nations FAO and UNICEF book defines malnutrition, discusses its causes and distribution, and discusses the economic and agricultural policy changes needed to overcome it.

SIMPSON, B.B., and M. CONNOR-OGORZALY. *Economic Botany: Plants in Our World.* New York: McGraw-Hill, 1986. My favorite botany book: full of interesting facts about food plants and other plants of economic importance.

## Chapter 16 Species and Extinction

BAILEY, J.A. *Principles of Wildlife Management.* New York: Wiley, 1984. A basic textbook on managing and preserving species.

*Bioscience,* vol. 38, 1988. An entire issue devoted to conservation at home and abroad.

BUSH, M. "The cheetah in genetic peril." *Scientific American,* May 1986. Even quite large populations of animals and plants may be in danger of extinction if they are so inbred that they retain little genetic diversity.

CULLINEY, J.L. *Islands in a Far Sea: Nature and Man in Hawaii.* San Francisco: Sierra Club Books, 1988. A detailed but nontechnical book on the spectacular biology of Hawaii. Includes the impact of species introduced by Polynesian and European colonizers and lambasts developers and the state government for their continued and knowing destruction of endemic species and irreplaceable ecosystems.

HARRIS, L.D. *The Fragmented Forest: Island Biogeography Theory and the Preservation of Biotic Diversity.* Chicago: University of Chicago Press, 1989. Deals with the theory behind attempts to create wildlife preserves with sufficient biodiversity to preserve particular groups of organisms.

PLUCKNETT, D.L., and others. *Gene Banks and the World's Food.* Princeton, NJ: Princeton University Press, 1987. Interesting background on the history and status of germ plasm conservation, plant breeding programs, and the international debate on the ownership and management of germ plasm collections.

SOULÉ M.E., and B.A. WILCOX, eds. *Conservation Biology.* Sunderland, MA: Sinauer Associates, Inc., 1990. A collection of short articles on various aspects of the conservation of biodiversity.

WORLD WILDLIFE FUND. *Buyer Beware.* A 1989 brochure prepared in conjunction with the U.S. Fish and Wildlife Service and the American Society of Travel Agents that warns tourists about the dangers of attempting to import ivory, coral, tropical birds, sea turtles, and other protected species into the United States. Discusses the rationale for the U.S. laws and penalties for breaking them.

## Chapter 17    Wilderness and Forest

ANDERSON, D., and R. FISHWICH. *Fuelwood Consumption and Deforestation in African Countries.* Washington, D.C.: The World Bank, 1985. Documentation of the fuelwood crisis in Africa.

COLINVAUX, P.A. "The past and future Amazon." *Scientific American*, May 1989. An ecologist argues that the species diversity of the Amazon rain forest is partly a result of frequent disturbances of the region's climate and topography. The Amazon's resilience to change gives it a fighting chance of recovering from human depredations—as long as these can be kept within reasonable limits.

FORSYTH, A., and K. MIYATA. *Tropical Nature.* New York: Charles Scribner's Sons, 1984. Delightful introduction to tropical rain forests.

HOLLOWAY, M. "Sustaining the Amazon." *Scientific American*, July 1993. Reconciling the need for economic development with preservation of biodiversity.

MOORS, P.J., I.A.E. ATKINSON, and G.H. SHERLEY. *Prohibited Immigrants: The Rat Threat to Island Conservation.* Washington, D.C.: IUCN-US, 1989. The New Zealand experience in removing introduced species from island preserves.

MYERS, N. *The Primary Sources: Tropical Forests and Our Future.* New York: W.W. Norton, 1984. An excellent summary of the tropical forest crisis.

PERRY, D.R. "The canopy of the tropical rain forest." *Scientific American*, November 1984. Story of how scientists are beginning to develop technology that permits them to work in this hard-to-reach ecosystem.

POSTEL, S., and L. HEISE. *Reforesting the Earth.* Washington, D.C.: Worldwatch Institute, 1988. A booklet on how to increase the rate of reforestation in many different countries.

REPETTO, R. "Deforestation in the tropics." *Scientific American*, April 1990. Argues that the destruction of tropical forest is caused largely by government policies that encourage people to chop down forest.

## Chapter 18    Toxic and Solid Waste

BLOCK, A.A., and F.R. SCARPATTI. *Poisoning for Profit: The Mafia and Toxic Waste in America.* New York: William Morrow, 1984. The cost of disposing of toxic waste is so high that organized crime is making money by removing the problem from the hands of the businesses and municipalities that should be responsible for it.

ENVIRONMENTAL DEFENSE FUND. *To Burn or Not To Burn.* New York: EDF, 1985. The pros and cons of garbage incineration.

ENVIRONMENTAL PROTECTION AGENCY. *Hazardous Waste Generation and Commercial Hazardous Waste Management Capacity: An Assessment.* Washington, D.C.: EPA, 1986. Masses of data on hazardous waste in the United States.

ENVIRONMENTAL PROTECTION AGENCY. *Operating a Recycling Program: Citizens Guide.* Washington, D.C.: EPA, 1985. A useful handbook for those who want to start recycling projects.

MACEACHERN, D. *Save Our Planet: 750 Everyday Ways You Can Help Clean Up the Earth.* New York: Dell Publishing, 1990. Includes many good ideas for recycling, reusing, and reducing waste.

O'LEARY, P.R., and others. "Managing solid waste." *Scientific American*, December 1988. Discussion of the steps needed to reduce the solid waste problem: reduced packaging, recycling, new incinerators, and new types of landfill.

## Chapter 19    Pollution and Water

AGERWAHL, A., and others. *Water, Sanitation, Health—For All? Prospects for the International Drinking Water Supply and Sanitation Decade, 1981–1990.* Washington, D.C.: Earthscan, 1986. The struggle to provide clean drinking water and sanitation for all.

ASHWORTH, W. *The Late, Great Lakes: An Environmental History.* New York: Alfred A. Knopf, 1986. The story of the Great Lakes' environmental problems.

ENVIRONMENTAL PROTECTION AGENCY. *A Ground-Water Protection Strategy.* Washington, D.C.: Government Printing Office. Outline of the actions the EPA proposes to minimize groundwater pollution.

LOER, R.C. *Pollution Control for Agriculture*, 3rd ed. New York: Academic Press, 1991. The massive problem of pollution from agriculture and how to prevent it.

ORGANIZATION FOR ECONOMIC COOPERATION AND DEVELOPMENT. *Water Pollution by Fertilizers and Pesticides.* Washington, D.C.: OECD, 1992. Discussion of the most important sources of nonpoint water pollution.

## Chapter 20    Air Pollution and Noise

BARDEN, R.G. *Sound Pollution.* St. Lucia: University of Queensland Press, 1976. A technical monograph on the nature of sound, mechanism of hearing, and noise pollution standards and legislation.

BARTH, M.C., and J.G. TITUS. *Greenhouse Effect and Sea Level Rise.* New York: Van Nostrand Reinhold, 1984. The effect of rising temperatures on rising sea levels and coastal areas.

NERO, A.V. "Controlling indoor air pollution." *Scientific American*, May 1988. A rather horrifying description of the pollutants in air in buildings and a discussion of whether and how indoor pollution might be regulated.

PAWLICK, T. *A Killing Rain: The Global Threat of Acid Precipitation.* San Francisco: Sierra Books, 1986. Easy-to-read survey of the problem.

SCHIEFELBEIN, S., and others. *The Incredible Machine.* Washington, D.C.: National Geographic Society, 1986. Fascinating account of human biology, disease, and life.

TOON, O., and R. TURCO. "Polar stratospheric clouds and ozone depletion." *Scientific American,* June 1991. A clear account of the chemistry of ozone depletion.

WHITE, R. "The great climate debate." *Scientific American,* July 1990. The prospect for global warming. Do we need to act now?

## Chapter 21   Pests and Pesticides

HUSSEY, N.W., and N. SCOPES. *Biological Pest Control.* Ithaca, NY: Cornell University Press, 1986. An overview of biological pest control.

IKER, S. "The promise and peril of pesticides." *International Wildlife,* July–August 1982. Problems of pesticide use in developing nations.

LIFTON, B. *Bug Busters: Poison-Free Pest Controls for Your House and Garden.* Berkeley, CA: Avery Publishing Group, 1991. A readable guide to doing without chemical pesticides in house and garden.

MARKKULA, M., and others. *Annales Agriculture Fenniae,* 11:74, 1972. Predatory mites as biological controls in cucumber production.

METCALF, R.L., and A. KELMAN. "Integrated pest management in China." *Environment,* 23:6, 1981. One of the articles that alerted the West to the progressive experiments taking place in China.

OLKOWSKI, W., S. DAAR, and H. OLKOWSKI. *Common-Sense Pest Control: Less Toxic Solutions for Your Home, Garden, Pets, and Community.* Berkeley: Taunton Press, 1991. A guide for homeowners.

STROBEL, G.A. "Biological control of weeds." *Scientific American,* July 1991. Assesses the prospects for controlling weeds by introducing insects and fungi that attack them.

## Chapter 22   Solving the Problems

HARDIN, G. *Exploring New Ethics for Survival.* New York: Viking, 1978. Explores ethical aspects of population control policies: the lifeboat model.

HARDIN, G. "The tragedy of the commons." *Science,* 162:1243, 1968. The thought-provoking argument that ecologically ethical individuals are evolutionarily doomed.

MCKAY, B.J., and ACHESON, J.M., eds. *The Culture and Ecology of Communal Resources.* Tucson, AZ: University of Arizona Press, 1987. A collection of papers evaluating Hardin's parable, "The Tragedy of the Commons," in the light of studies of fisheries, grazing lands, and similar commons.

The authors conclude that sustainable management of commons is possible if the users of the commons make up a single group of accountable people and if the limits of the commons are under the group's control.

U.S. GENERAL ACCOUNTING OFFICE REPORT TO CONGRESS. *Environmental Protection: Meeting Public Expectations with Limited Resources.* June 1991. The GAO's report on rising public expectations for the environment and ways for government to respond without spending undue sums of money.

## Chapter 23   Politics, Economics, and Ethics

INDEPENDENT COMMISSION ON INTERNATIONAL DEVELOPMENT ISSUES. *North-South: A Program for Survival.* London: Pan Books, 1980. The report of a group of world leaders, headed by Willy Brandt, on the failure of the world economic systems to reduce the inequality of wealth between the world's haves and have-nots.

MEEKER-LOWRY, S. *Economics as if the Earth Really Mattered.* Santa Cruz, CA: New Society Publishers 1991. Rethinking economics to take environmental affairs into account.

MILLBRATH, L.W. *Environmentalists: Vanguard for a New Society.* Albany, NY: State University of New York Press 1992. Contains several chapters on the social changes that will be necessary to build a sustainable society.

NASH, R.F. *The Rights of Nature: A History of Environmental Ethics.* Madison: University of Wisconsin Press, 1989. A thought-provoking history of the ethical theory that everything on this planet, including soil and water, is morally equivalent to us and shares an equal "right to continued existence." The lunatic fringe of this persuasion views killing a flower as morally equivalent to killing a person. It includes the Animal Liberation Front, which has destroyed laboratories (and killed animals) in the name of animal rights.

SCHUMACHER, E.F. *Small Is Beautiful: Economics as If People Mattered.* New York: Harper & Row, 1973. Something of a classic, this book helped alter attitudes toward Third World development. Schumacher argues that more foreign aid should be devoted to small-scale projects such as the building of better cooking stoves, plows, water pumps, and tools that assist individuals to earn a living. Aid is often wasted when it is poured into massive infrastructure projects.

WORLD COMMISSION ON ENVIRONMENT AND DEVELOPMENT. *Our Common Future.* New York: Oxford University Press, 1987. The report of a United Nations commission asked to propose long-term strategies for achieving sustainable development. The commission did not fulfill its ambitious goal, but it has identified the political and economic obstacles to sustainable development. Emphasizes the conflicts between developed and developing countries that stand in the way of sustainable development and outlines proposed international legal principles for managing such "commons" as Antarctica and the oceans.

## Chapter 24   Toward a Sustainable Future

BERGER, J.J. *Restoring the Earth*. New York: Alfred A. Knopf, 1986. A collection of case studies on the ways small groups have restored degraded habitats.

FESHBACH, M., and A. FRIENDLY. *Ecocide in the USSR: Health and Nature Under Siege*. New York: Basic Books, 1992. The amazing story of the damage to human health and natural resources caused by decades of uncontrolled pollution by industry and agriculture in the Soviet Union.

HOMER-DIXON, T.F., J.H. BOUTWELL, and G.W. RATHJENS. "Environmental change and violent conflict." *Scientific American*, February 1993. The evidence that environmental degradation is the main cause of violent conflict in many parts of the world.

WADE, N. *A World Beyond Healing: The Prologue and Aftermath of Nuclear War*. New York: W.W. Norton, 1987. A readable account by a journalist who has studied the technical aspects of the effects of nuclear war. Also explains the theory of nuclear deterrence that has guided Western military policy since World War II.

# APPENDIX C
## Environmental Organizations and Periodicals

Air Pollution Control Association, P.O. Box 2861, Pittsburgh, PA 15230. Publishes *Journal of the Air Pollution Control Association*.

American Demograpics, Inc., P.O. Box 68, Ithaca, NY 14851. Publishes *American Demographics*.

American Forestry Association, 1319 18th Street, NW, Washington, D.C. 20036. Concerned with soil and forest conservation, reforestation, creation of parks, and the role of trees in combating pollution. Publishes *American Forests*.

American Institute of Biological Sciences, 730 11th Street, NW, Washington, D.C. 20001. AIBS is made up largely of biology teachers. It publishes *Bioscience* as well as various other publications on biology, teaching, conservation, and related subjects.

American Society for Environmental Education, P.O. Box 800, Hanover, NH 03755. Educational materials for teachers and the public. Publishes *Environmental Education Report*.

American Water Resources Association, 5410 Grosvenor Lane, Suite 220, Bethesda, MD 20014. Publishes *Water Resources Bulletin*.

Association of Forest Service Employees for Environmental Ethics, P.O. Box 11615, Eugene, OR 97440.

Bio-Integral Resource Center, P.O. Box 7414, Berkeley, CA 94707. Educates people on how to control household and garden pests without using toxic chemicals.

Center for Action on Endangered Species, 175 West Main Street, Ayer, MD 01432.

Center for Marine Conservation, 1725 DeSales Street, NW, Washington, D.C. 20036. Publishes the quarterly *Marine Conservation News*, which reports progress and problems with marine mammals, fish stocks, offshore drilling, and other matters that affect the environmental health of the ocean.

Citizen's Clearinghouse on Hazardous Waste, P.O. Box 926, Arlington, VA 22216. A source of information.

Clean Water Action, 1320 18th Street, NW, Washington, D.C. 20036.

Conservation Foundation, 1717 Massachusetts Avenue, NW, Washington, D.C. 20036. Publishes *Conservation Foundation Letter*, a newsletter containing good summaries of major issues.

Conservation International, 1015 18th Street, NW, Suite 1000, Washington, D.C. 20036. A private nonprofit scientific organization dedicated to saving biodiversity in endangered rain forest and other ecosystems worldwide. Activities include funding and technical assistance for local conservation efforts in developing countries. CI arranged the first debt-for-nature swap in 1987.

Earth First, P.O. Box 5176, Missoula, MT 59806. Publishes what bills itself as "the radical environmental journal." It's certainly not what you might call mainstream.

Earth Island Institute, 200 Broadway, Suite 28, San Francisco, CA 94133.

Earthscan, 1717 Massachusetts Avenue, NW, Washington, D.C. 20036. A news and information service covering global environmental and development issues. Publishes numerous books and other publications.

*Environment*, Heldref Publications, 4000 Albemarle Street, NW, Washington, D.C. 20016. Good articles analyzing environmental issues in some depth.

Environmental Action, 1525 New Hampshire, NW, Washington, D.C. 20036.

Environmental Defense Fund, 257 Park Avenue South, New York, NY 10010.

Environmental Law Institute, 1616 P Street, NW, Suite 200, Washington, D.C. 20036. Dedicated to advancing environmental protection by improving law, management, and policy. Publishes *Environmental Law Reporter,* a vital resource for all environmental lawyers.

Food and Agriculture Organization (FAO) of the United Nations, UNIPUB, Inc., 650 First Avenue, P.O. Box 433, New York, NY 10016. Publishes *Ceres,* with articles on the problem of feeding a growing population.

Greenpeace USA, 1436 U Street, NW, Washington, DC 20009.

International Union for the Conservation of Nature and Natural Resources, Avenue du Mont-Blanc, CH-1196 Gland, Switzerland. A union of governments, governmental agencies, and nongovernmental agencies promoting conservation all over the world. Produces a quarterly journal with environmental news from countries ranging from Nepal to Vietnam.

League of Conservation Voters, 1707 L Street, NW, Suite 550, Washington, D.C. 20036. For $5 this group will send you an environmental scorecard that lists how all U.S. senators and representatives voted on key environmental bills.

National Audubon Society, 950 Third Avenue, New York, NY 10022. Operates wildlife sanctuaries; supports research on endangered species, education, and lobbying. Publishes an excellent annual review of wildlife conservation, *Audubon Wildlife Report*, and a popular magazine, *Audubon.*

National Environmental Health Association, 1600 Pennsylvania Avenue, Denver, CO 80203. Publishes *Journal of Environmental Health.*

National Wildflower Research Center, 2600 FM 973 North, Austin, TX 78725. Lady Bird Johnson is the leading light of this organization that studies and encourages the use of native wildflowers in public and private landscaping.

National Wildlife Federation, 1400 16th Street, NW, Washington, D.C. 20036. Education and research on preservation of wildlife and biodiversity. Publishes *Conservation News* and the annual *Conservation Directory*, which contains a detailed list of national, state, and local organizations concerned with conservation. The Federation's *Cool It!* group is designed to motivate college students to establish campus and community models of environmentally sound practices by providing resources, organizing tools, and regional coordinators.

Native Americans for a Clean Environment, P.O. Box 1671, Tehlequah OK 74465.

Natural Resources Defense Council, 40 West 20th Street, New York, NY 10112.

Nature Conservancy, 1815 North Lynn Street, Arlington, VA 22209. One of the most effective conservation organizations, with chapters in every American state and an expanding interest in international conservation, particularly in the Caribbean and Central and South America.

Population Institute, 110 Maryland Avenue, NE, Suite 207, Washington, D.C. 20002. Education on global overpopulation and its solutions.

Population Reference Bureau, Inc., 1875 Connecticut Av-

enue, NW, Suite 520, Washington, D.C. 20009-5728. A nonprofit educational and scientific organization that tracks and analyzes masses of data on human populations and related public policy questions in the United States and throughout the world. Publishes *Population Today,* a monthly newsletter containing readable articles on all aspects of population as well as bulletins, educational charts, and slide shows.

Rainforest Action Network, 301 Broadway, Suite A, San Francisco, CA 94133. A nonprofit activist organization founded in 1985 and working to save the world's rain forests. Methods include letter-writing campaigns, boycotts, consumer action, demonstrations, and education. This was the group that organized the push to induce Burger King to stop buying beef raised in former rain forests. Produces educational materials for use by teachers.

Rural Alliance for Military Accountability, 6205 Franktown Road, Carson City, NV 89704. An information clearinghouse for people concerned about the increasing noise pollution caused by (often illegal) low flying by military aircraft.

Save America's Forests, 1742 18th Street, NW, Washington, D.C. 20009.

Sierra Club, 730 Polk Street, San Francisco, CA 94109. A conservationist organization with considerable political power. The national office will put you in touch with your local chapter, which probably organizes field trips, work parties, social activities, and local political efforts.

Solar Lobby, 1001 Connecticut Avenue, NW, Suite 638, Washington, D.C. 20036. Lobbies for use and development of solar energy and other renewable energy sources.

Student Conservation Association, Inc., P.O. Box 550, Charlestown, NH 03603. Not an association to conserve students as the name suggests, but a provider of high school and college student volunteers for natural resource conservation projects in the United States. They also place adults in 12-week paid positions. An excellent place for students interested in summer jobs in conservation to start looking.

Union of Concerned Scientists, 26 Church Street, Cambridge, MA 02238. Research, education, lobbying; particularly knowledgeable about nuclear power and safety.

United Nations Environment Programme, P.O. Box 30552, Nairobi, Kenya. In charge of the United Nations' work to pass international treaties protecting the environment.

United Nations Publications Sales Section, New York, NY 10017. Produces numerous environmental publications, including Population and Vital Statistics Report.

Wilderness Society, 1400 I Street, NW, Washington, D.C. 20005.

World Bank, Publications Department, 1818 H Street, NW, Washington, D.C. 20433. Produces numerous publications, mainly on economic rather than environmental aspects of developing countries, including the annual World Development Report.

World Resources Institute, 1709 New York Avenue, NW, 7th Floor, Washington, D.C. 20006. Publishes the excellent annual report *World Resources*.

World Wildlife Fund, 1250 24th Street, NW, Washington, D.C. 20037. An influential international organization devoted to conservation, research, and education.

Worldwatch Institute, 1776 Massachusetts Avenue, NW, Washington, D.C. 20036-1904. Founded in 1975, the institute is a nonprofit research organization that develops public policy positions on environmental affairs worldwide. It produces an annual *State of the World* report (see Chapter 1 references); *World Watch*, a magazine on environmental affairs; and various bulletins and pamphlets.

Zero Population Growth, 1400 16th Street, NW, Washington, D.C. 20036. Works for a stable world population via family planning services and education. Produces excellent *Fact Sheets* that describe briefly how population growth affects other problems such as transportation, energy use, and water shortages.

## U.S. GOVERNMENT AGENCIES

Bureau of Mines, Columbian Plaza, 2401 E Street, NW, Washington, D.C. 20506.

Bureau of the Census, U.S. Department of Commerce, Washington, D.C. 22161.

Conservation and Renewable Energy Inquiry and Referral Service, P.O. Box 8900, Silver Spring, MD 20907. Information on conservation and renewable energy sources.

Consumer Information Center-K, P.O. Box 100, Pueblo, CO 81002. Will send you a catalog of pamphlets (many of them free) produced by the government for consumers on subjects ranging from used cars to endangered species and national parks.

Department of Agriculture, Washington, D.C. 20250. Includes soil conservation service.

Environmental Protection Agency, 401 M Street, SW, Washington, D.C. 20460.

Fish and Wildlife Service, Department of the Interior, Interior Building, Room 3256, Washington, D.C. 20240.

Food and Drug Administration, 5600 Fishers Lane, Rockville, MD 20857.

Geological Survey, U.S. Geological Survey National Center, 12201 Sunrise Valley Drive, Reston, VA 22092.

National Academy of Sciences, 2101 Constitution Avenue, NW, Washington, D.C. 20418.

National Energy Information Center, Department of Energy, Forrestal Building, 1000 Independence Avenue, SW, Washington, D.C. 20585. Many publications, statistics, and studies on energy, including historical aspects, reserves, and future plans. Publishes *Annual Energy Review*.

U.S. Government Printing Office, Washington, D.C. 20402. Produces thousands of government publications each year. It will send you a catalog listing publications in particular areas.

# Glossary

**abiotic**  Nonliving; occurring neither within living cells nor under their influence.

**abortion**  Process whereby a mammalian embryo or fetus becomes detached from the uterine wall and expelled from the female's body either naturally ("miscarriage," spontaneous abortion) or by medical means (induced abortion).

*Acacia*  Large genus of trees, most of which are tropical. Common name: thornwood.

**acid**  1. Substance that releases hydrogen ions ($H^+$) in aqueous solution. 2. Substance that accepts electrons.

**acid rain**  Rain containing pollutants that give it a pH of less than 7.0.

**activated sludge**  Substance produced during secondary sewage treatment by mixing wastewater with bacteria-rich sludge and air to encourage aerobic decomposition of organic matter.

**adaptation**  1. Process by which populations evolve to become suited to their environments over the course of generations. 2. Characteristic that increases an organism's evolutionary success.

**adaptive radiation**  Formation of two or more new species, macromolecules, or physiological pathways, adapted to different ways of life, from one ancestral species, molecule, or pathway.

**adhesion**  Sticking together (e.g., of cells, molecules of different substances).

**aerial**  In the air.

**aerobic**  1. Requiring molecular oxygen ($O_2$) to live. 2. In the presence of molecular oxygen.

**Agricultural Revolution**  Change in human society from mainly hunter-gatherer to mainly agricultural means of acquiring food.

**agriculture**  Practice of breeding and caring for animals and plants that are used for food, clothing, housing, and other purposes.

**albedo**  (of Earth)  Shininess of Earth viewed from space, produced by light reflecting off surfaces such as oceans, ice caps, or clouds.

**alcohol**  Organic compound containing an alcohol group (COH).

**alga**  (*pl.*, algae)  Photosynthetic organism with a one-celled or simple multicellular body plan and lacking a multicellular embryo protected by the female reproductive structure.

**alkali**  (=base)  Substance that releases hydroxide ions ($OH^-$) in water, accepts hydrogen ions ($H^+$), or gives up electrons.

**alkaline**  Having a (basic) pH of more than 7.0.

**alloy**  Mixture of a metal with one or more other elements.

**alpine**  Of high mountain regions.

**altitude**  Height above sea level.

**altruistic behavior**  Behavior that favors the reproductive success of some member of the species other than the actor.

**amino acid**  Small organic molecule containing both a carboxyl and an amino group bonded to the same carbon atom. Amino acids are the monomers from which polypeptides and proteins are made.

**ammonia**  $NH_3$.

**ammonium**  $NH_4^+$.

**amphibian**  Member of the vertebrate class Amphibia; e.g., frogs, toads, salamanders, newts, and their relatives.

**anaerobic**  (*n.*, anaerobe)  1. Without oxygen. 2. Not requiring molecular oxygen for extraction of energy from food (respiration).

**angiosperms**  Flowering plants, including trees, grasses.

**anion**  Negatively charged ion.

**annelid**  Segmented worm; e.g., earthworm, polychaete, leech.

**annual**  Flowering plant that grows from seed, flowers, sets seed, and dies within one year.

**Anthophyta**  Plant division containing the flowering plants (= angiosperms).

**anthropogenic**  Produced by human action.

**anthropoids**  Monkeys, apes, and humans, which make up the suborder Anthropoidea of the order Primates.

**anthropology**  Study of the origin and of the physical, social, and cultural evolution of humans.

**Anura**  Order of amphibians containing frogs and toads.

**aquatic**  Of water (fresh or salt).

**aqueous**  Containing or composed largely of water.

**arboreal**  Living in trees.

**arid**  Dry; of areas where more water leaves the ecosystem (by evaporation and transpiration) than enters it (as precipitation).

**arthropod**  Member of the phylum Arthropoda: segmented animals with jointed appendages and stiff chitin-containing external skeletons; e.g., crabs, lobsters, barnacles, insects.

**atmosphere**  (of Earth)  Layer of gases that surrounds Earth, including the troposphere (next to Earth's surface) and the stratosphere.

**autotroph**  (*adj.*, autotrophic)  Organism that can make its own (organic) food molecules from inorganic constituents.

**bacterium**  (*pl.*, bacteria)  Any member of the kingdom Monera, containing organisms without nuclear envelopes.

**base**  (*adj.*, basic)  See Alkali.

**bicarbonate ion**  $HCO_3^-$.

**bioaccumulation**  Concentration of a chemical in particular parts of the body.

**biodegradable**  Capable of being broken down by living organisms into inorganic compounds.

**biodiversity**  Number of different species of organisms in an area.

**biogeochemical cycle**  Description of the geological and biological processes that affect an element in an ecosystem.

**biological pest control**  Use of naturally occurring chemicals or organisms to reduce the populations of pest organisms.

**biomass**  Amount of material that is part of the bodies of living organisms.

**biome** Major type of terrestrial community of organisms; e.g., tropical rain forest, desert.

**bioremediation** Use of organisms to degrade pollutants.

**biosphere** Total of all areas on Earth where organisms are found; includes deep ocean and part of the atmosphere.

**biotic** Having to do with living organisms.

**biotic (reproductive) potential** Rate at which a population of a species can increase in size under ideal conditions.

**bipedalism** Habit of walking on two legs.

**bloom** Generally means flower; ecologists also use it to mean a rapid increase in population of microorganisms.

**broad-leaved evergreen** A tree or shrub that is evergreen but not a conifer; e.g., live oak, rhododendron, southern magnolia.

**bromeliad** Member of a large family of epiphytic plants.

**bryophyte** Member of the Bryophyta: mosses and other small plants without transport tissues.

**buffer** Something that lessens or absorbs the shock of an impact. In chemistry, a substance that, within limits, prevents the pH of a solution changing when acid or alkali is added to it.

**bug** Member of the insect order Hemiptera.

**calcareous** Composed of, or containing, calcium carbonate ($CaCO_3$).

**calorie** (*adj.*, caloric) 1 Calorie = 1,000 calories = 1 kilocalorie or kcal. A calorie is the amount of heat needed to raise 1 gram of water from 14.5° C to 15.5° C. The energy values of foods are often expressed in Calories.

**carbon dioxide** Gas ($CO_2$) produced by respiration of organisms, burning of organic matter or other carbon-containing substances.

**carcinogen** Something that causes cancer.

**carnivore** 1. Animal that eats other animals. 2. Member of the mammalian order Carnivora; e.g., dogs, cats, weasels, bears, raccoons, skunks.

**carrying capacity** (of an area for a species) Number of individuals of a species that the environment can support.

**cation** Positively charged ion.

**cellular respiration** Stepwise release of

energy from food molecules, accompanied by storage of the energy in short-lived energy intermediates such as adenosine triphosphate (ATP).

**census** Count of a human population.

**cereals** Those grass species of the plant family Gramineae that make up the bulk of human food.

**Cetacea** Order of mammals whose members are most highly adapted to aquatic life, including whales and porpoises.

**chaparral** Dry shrub lands of temperate coastal regions such as California and the Mediterranean coast.

**chlorophyll** Green pigment that traps light energy during photosynthesis.

**chondrichthyes** Cartilaginous fish with jaws; elasmobranchs: the sharks, skates, and rays.

**chordate** Member of the animal phylum Chordata, with a notochord and pharyngeal gill slits at some stage of its life.

**chromosome** (*adj.*, chromosomal) Thread-like structure in the nucleus of a eukaryotic cell, consisting of DNA and proteins and carrying genetic information.

**cilium** (*pl.*, cilia; *adj.*, ciliary) Thread-like organelle containing microtubules present on the surfaces of many eukaryotic cells.

**clay** Mineral particles with a diameter of less than 0.002 millimeters consisting of aluminum and silica.

**climate** Those aspects of the weather such as temperature, rainfall, light, humidity, and air movement that influence the life of organisms.

**clone** Population of cells or individuals descended from one original cell or individual by asexual propagation and hence genetically identical.

**Cnidaria** Phylum of simple animals with only two well-developed layers of cells, only one opening into the gastrovascular cavity, tentacles, and stinging nematocysts (e.g., jellyfish, hydra, corals, sea anemones).

**coevolution** Evolution together of two or more species whose members exert selective pressures on one another.

**cohesion** Sticking together of molecules of the same substance.

**community** All the organisms living in a particular habitat.

**concentration** Proportion of one substance found in the total of a mixture of

several substances; may be given in terms of weight, of proportion of molecules, and so on. Concentration is symbolized: [sugar] = concentration of sugar.

**conifer** Member of the plant division Coniferophyta: cone-bearing gymnosperms, including pines and spruces (as well as junipers and yews, whose reproductive structures do not resemble cones).

**conservation** Careful use and management of resources, so as to maximize the benefit from them now and in the future. Methods include preservation, reducing waste, recycling, reuse, and decreased use.

**consumer** In ecology, an organism that eats other organisms. In economics, someone who buys something.

**contraceptive** Something that prevents fertilization of an egg.

**corn** Common name for the cereal maize, *Zea mays*.

**cost-benefit analysis** Estimate of the expected costs or losses resulting from a particular project with the benefits and gains to be expected from it.

**crude oil** Unprocessed petroleum.

**crustacean** Member of the arthropod order Crustacea; e.g., lobsters, shrimp, crabs, barnacles, pill bugs, copepods, ostracods.

**cubic centimeter** (*abbr.*, cc) Equal to 1 milliliter.

**cultural evolution** Change in an animal society from one generation to the next as a result of learning.

**Cyanobacteria** Blue-green bacteria; prokaryotic organisms that use chlorophyll in their photosynthesis and produce $O_2$ as a by-product.

**Cyanophyta** Former name for cyanobacteria.

**cyst** 1. Dormant organism within a resistant covering; stage in which some organisms pass through adverse conditions. 2. Sac-like nonmalignant tumor.

**deciduous** Of plants that lose their leaves during one season of the year; not evergreen.

**decomposer** In ecology, an organism that feeds on the dead bodies, body parts, or wastes of other organisms, thereby breaking down and recycling the nutrients they contain.

**defoliant** A substance that causes a plant to lose its leaves.

**defoliate** Of a plant, to remove all the leaves.

**deforestation** Removal of trees from an area without replacing them.

**degradable** Capable of being broken down by natural processes (usually by decomposer organisms).

**delta** Land formed where a river deposits silt as it enters a lake or sea.

**demographic transition** Change from a population with a high birth rate and death rate to one with low rates of birth and death that has occurred in many countries as they became more industrialized.

**demographics** Study of human populations.

**dependency ratio** Ratio of the number of people under 15 and over 65 years of age to the rest of the population.

**depletion** Using up of a significant portion of a natural resource.

**desiccation** Drying out.

**detritus** Molecules and larger particles of dead organic matter.

**development** (economic) Change from a society that is typically agricultural, rural, poor, and illiterate to one that is largely industrial, urban, wealthy, and educated.

**dinoflagellate** One-celled organism with two flagella; most are photosynthetic.

**diploid** Containing twice the number of chromosomes found in a gamete; having paired homologous chromosomes.

**division** Taxon of plants or fungi equivalent to a phylum in the animal kingdom.

**drought** Period with less than average precipitation.

**Eastern-bloc countries** Countries that were formerly part of the Soviet Union, such as Lithuania, Russia, Ukraine, together with countries controlled by the former Soviet Union, such as Bulgaria, Romania, Poland.

**echinoderm** ("spiny-skinned") Member of the invertebrate phylum Echinodermata; e.g., sea stars, sea urchins, sand dollars, sea cucumbers.

**ecology** Study of the relationships of organisms with other organisms and with their physical environment.

**economics** Study of the production, distribution, and consumption of commodities.

**ecosystem** All of the organisms present in a particular area, together with their physical environment.

**effluent** Flowing out of: usually used to mean liquid waste; e.g., from a factory.

**elasmobranchs** Cartilaginous fish with jaws; sharks, skates, and rays. Another name for Chondrichthyes.

**electrical potential** Difference in concentration of electrically charged particles on two sides of a membrane.

**electrolysis** Separation of a compound or mixture into its components by passing an electric current through it, which causes negatively and positively charged components to separate.

**electromagnetic radiation** Radiant energy (including solar energy) that can move through a vacuum or through space as waves of oscillating electrical and magnetic fields.

**electron** Fundamental negatively charged particle found around the nucleus of an atom.

**electrostatic precipitator** Device for removing pollutants from smoke by imparting an electrical charge to particles, which are then attracted to an oppositely charged plate, where they are precipitated.

**embryophyte** Plant lacking vascular (transport) tissue: mosses and liverworts.

**emigration** Leaving an area of residence for some other place.

**emission** Releasing something; the thing that is released. Usually applied to waste gases released into the atmosphere from a vehicle or factory.

**endemic** (*adj.*, sometimes used as a noun) Peculiar to a particular population or locality where it originated.

**energy** Capacity to do work and transfer heat.

**energy efficiency** Percentage of the total energy input to a system or process that does useful work.

**entomology** Study of insects.

**environment** Organism's physical and biological surroundings.

**enzyme** Protein that catalyzes a particular biochemical reaction between specific substrate (reactant) molecules.

**epiphyte** Plant rooted not in the soil but on the surface of another plant.

**Equator** Imaginary circle around Earth's surface formed by the intersection of a plane passing through Earth's center perpendicular to its axis of rotation.

**essential amino acid** Amino acid that an animal must obtain in its diet because its body cannot make enough of the molecule to survive.

**eutrophic** Of a body of fresh water rich in nutrients and hence in living organisms.

**eutrophication** Process in which organic matter accumulates in a body of water, making it richer in nutrients and hence in organisms, until eventually it fills in and becomes dry land.

**evaporative cooling** Reduction of temperature as a result of the escape of the fastest moving (that is, warmest) water molecules as water vapor.

**evolution** 1. Descent of modern species of organisms from related, but different, species that lived in previous times. 2. Change in the gene pool of a population from generation to generation.

**exajoule** One billion joules, approximately equal to the amount of heat released by the combustion of 22 million tons of oil.

**externality** In economics, costs of a transaction that are borne by people other than those engaged in the transaction.

**extinction** (of a species) Disappearance of a species from Earth caused by the death of its last surviving member.

**family planning services** Agencies that provide information and birth control methods that permit people to choose the number and spacing of their children.

**famine** Food shortage sufficiently widespread to cause ill health in many members of a human population.

**federal deficit** Amount of money the government has to borrow to make up the difference between the sum it spends and the sum it receives (mainly in the form of taxes).

**fermentation** Anaerobic breakdown of food molecules to release energy, in which the final electron acceptors (and end products) are organic molecules.

**fertility** 1. Of a female, the ability to reproduce. 2. Of soil, the ability to supply plants with the nutrients they need. 3. Of a population, the number of offspring produced per female per unit time.

**fertilization** 1. Union of an egg with a sperm. 2. Supplying nutrients to crop plants.

**fertilizer** Substance containing plant

nutrients that is added to soil or sprayed on leaves.

**filtrate** Substance that has passed through a filter.

**fix** In chemistry, to incorporate into a less volatile compound.

**flagellum** (*pl.*, flagella) Long, thin projection from a cell that moves in spiral undulations and propels the cell.

**flatworm** Member of the invertebrate phylum Platyhelminthes; e.g., planarians, flukes, tapeworms.

**flora** Plant life of an area (often includes bacteria and fungi).

**fluidized bed combustion** Method of burning coal efficiently and cleanly by suspending a mixture of powdered coal and limestone in a stream of hot air during combustion.

**fly ash** Small particles of ash and soot generated when coal, oil, or waste materials are burned.

**forest** Region where trees grow as a result of adequate temperature and annual precipitation of 75 centimeters or more.

**fossil fuel** Fuel created by decomposition and geological processes from the remains of dead organisms, including peat, oil, coal, and natural gas.

**fossils** Remains of organisms, or other evidence of once-living organisms, preserved in rocks.

**freons** Chlorofluorocarbon compounds composed of carbon, chlorine, and fluorine.

**fruit** Structure that develops from the ovary of a flower, surrounding one or more seeds.

**fugitive species** Species that occurs in an area for only a short time.

**fumigate** To kill pests by releasing poisonous gas into an enclosed space.

**Fungi** Kingdom of organisms containing eukaryotic, multicellular saprobes (heterotrophs feeding by absorbing nutrients across their cell walls).

**gasoline** Liquid petrochemical fraction used as vehicle fuel.

**gastropod** Member of the molluscan class Gastropoda, including snails, slugs, nudibranchs, and limpet.

**GDP** *See* gross domestic product.

**gene** Length of DNA that functions as a unit.

**gene bank** Institution where plant material is stored in a viable condition. Usually seeds are dried and frozen in a sealed container. Some plants have seeds that will not survive this treatment, and they must be maintained as growing plants or in tissue culture.

**gene pool** All of the genes present in a population of organisms.

**genera** Plural of genus.

**generation time** Time elapsed from production (birth) of a new individual to production of its first offspring; usually estimated at 20 years for humans.

**genetic engineering** Isolation of useful genes from a donor organism or tissue and their incorporation into an organism that does not normally possess them.

**genome** All the genetic material contained by an individual (or by a representative member of a population).

**genus** (*pl.*, genera; *adj.*, generic) Taxon above species in the hierarchical classification of organisms. The genus name is the first word of the Latin binomial for a species; e.g., *Ursus* is the generic name of the grizzly bear, *Ursus horribilis.*

**germ cells** Cells that give rise to eggs, sperm, or spores (reproductive cells).

**germ plasm** Term used by botanists for genetic material, especially that contained within the reproductive (germ) cells.

**Glaciation** ( = Ice Age) One of four cold periods during the Pleistocene era when ice and glaciers extended farther south from the North Pole than they do now.

**GNP** *See* gross national product.

**grain** Fruit of members of the monocotyledonous plant family Gramineae from which most human food comes.

**green algae** Members of the division Chlorophyta: plants with unicellular or multicellular body plans, using chlorophylls *a* and *b* for photosynthesis and storing food as starch.

**green plants** Photosynthetic organisms that give off oxygen during their photosynthesis, including prokaryotic cyanobacteria and all photosynthetic eukaryotes (protists and plants).

**greenhouse effect** Heating of Earth caused by gases in the atmosphere that trap infrared radiation from Earth and prevent it from escaping into space.

**gross domestic product** (GDP) Value of all the goods and services produced in a nation during a specified period. *Compare* gross national product.

**gross national product** (GNP) Total market value of all the goods and service produced by a nation during a specified period. Includes profits earned by the nation's companies in other countries, which gross domestic product does not.

**gymnosperm** Nonflowering plant that produces seeds, e.g., pines, redwoods, cycads.

**habitat** Physical area where an organism lives.

**haploid** Containing the number of (unpaired) chromosomes found in a gamete; equal to half the number of chromosomes found in a body cell of most higher plants and animals.

**hectare** 10,000 square meters, or about 2.5 acres.

**hemoglobin** Respiratory pigment that carries oxygen in the blood of vertebrates and various invertebrates.

**herbaceous** Of plants with nonwoody stems.

**herbicide** Chemical used to kill plants.

**herbivore** Animal that eats plants or parts of plants.

**heterotroph** (*adj.*, heterotrophic) Organism dependent on other organisms to produce its organic (food) molecules.

**homeostasis** (*adj.*, homeostatic) Maintenance of conditions within specified limits.

**hominids** Humans and their direct ancestors: members of the primate family Hominidae, with large brains, small teeth, and bipedal locomotion.

**hominoids** Humans and apes; large tailless primates.

**hormone** Chemical messengers produced in one part of the body and specifically influencing certain activities of cells in another part of the body.

**humus** The dark organic matter, resistant to further decomposition by microorganisms, that stays in the soil after plant and animal remains have decomposed.

**hunter-gatherers** Members of human society who obtain their food by collecting plants and hunting wild animals.

**hybrid** Offspring of a mating between genetically different individuals.

**hydraulic pressure** Force exerted by a fluid.

**hydrogen bond** Weak link between two molecules, or two parts of the same molecule, due to the attraction of a hydrogen with a partial positive charge to an oxygen or nitrogen with a partial negative charge.

**hydrolysis** Breaking apart of a molecule into its monomer subunits by addition of the components of a water molecule into each of the covalent bonds linking them.

**hydrophilic** ''Water-loving''; able to dissolve in water.

**hydrophobic** ''Water-hating''; unable to dissolve in water; nonpolar.

**hydrostatic pressure** Pressure exerted by confined fluid.

**Hymenoptera** Order of insects that includes bees, wasps, and ants.

**hypothesis** (*pl.*, hypotheses) Proposed answer to a question.

**immigration** Movement of new individuals into an area.

**inbreeding** Mating of related individuals that share many of the same genes.

**Industrial Revolution** Social and economic changes brought about when mechanization of production results in a shift from home manufacturing to large-scale factory production.

**infant mortality rate** Death rate for humans in the first year of life.

**infanticide** Killing of a child less than one year old.

**ingest** To take into the body through the mouth.

**inorganic** 1. Not produced by living organisms. 2. In chemistry, all compounds that do not contain carbon atoms plus very small carbon-containing atoms.

**insect** Arthropod with three distinct body areas (head, thorax, abdomen); adult has three pairs of legs and usually two pairs of wings attached to the thorax.

**insecticide** Substance that kills one or more species of insects.

**intertidal** Between tidemarks; covered by water at high tide and exposed to the air at low tide.

**invertebrate** Animal that lacks a backbone; e.g., earthworm, snail.

**ion** (*adj.*, ionic) Particle carrying one or more positive or negative electrical charges.

**irrigation** Human methods of supplying water to plants.

**isotopes** Two or more forms of an element, differing by the numbers of neutrons in the nucleus.

**K** 1. Chemical symbol for potassium, which usually occurs as an ion ($K^+$). 2. Symbol used for carrying capacity.

**kelps** Large marine brown algae with considerable differentiation of tissues.

**kinetic energy** Energy of movement.

**krill** *Euphasia superba*, small marine planktonic arthropods common in the Southern Ocean, up to several centimeters in size.

**labeled** Prepared in such a way that it can be traced; e.g., containing a high proportion of a particular isotope, which is usually rare, allowing the fate of the atoms to be traced by methods that can distinguish between the isotopes of an element.

**landfill** Land area used as a dumping ground for solid waste.

**larva** Immature stage of an animal with different appearance and way of life from the adult.

**laterite** Hard crust that may develop when vegetation is removed from the surface of soil containing metals such as aluminum and iron in tropical regions with wet and dry seasons. In the dry season, soil solution rises to the surface by capillarity, and aluminum and iron oxides accumulate and combine to form the crust. Laterization results in an infertile soil called latosol.

**latitude** Distance north or south of the Equator.

**leaching** Washing out of soluble substances by water passing down through soil. Leaching occurs when more water falls on the soil than is lost by evaporation from the surface. Leached soil has lost mineral nutrients that have dissolved in rainwater running through the soil and away into streams.

**leeward** (*pronounced* loo-ard) (side) The side of an object from which the wind is not coming.

**legumes** Members of the plant family Leguminosae, including beans, peas, peanuts, alfalfa, clover, and acacia.

**life expectancy** Number of years a particular person can expect to live, calculated from actuarial statistics.

**liquefied petroleum gas** (LPG) Mixture of propane and butane derived from natural gas and compressed to form a liquid.

**litter** Plant remains (mainly leaves) that have fallen on the surface of the soil.

**macromolecule** Large molecule; usually used to describe a polymer with a molecular weight of many thousands made up of many (identical or different) monomers.

**macronutrient** Nutrient needed in relatively large amounts.

**macroscopic** Visible to the unaided eye.

**magma** In geology, the molten matter under Earth's crust from which igneous rock is formed by cooling.

**maize** *Zea mays*, corn, important cereal crop.

**mammal** Warm-blooded vertebrate with lower jaw consisting of only one bone (the mandible) on each side, with fur or hair, with young nourished by milk from the mammary glands of the female parent, e.g., humans, rabbits, cattle.

**marine** Of the sea.

**metabolism** All the chemical reactions taking place within an organism (often measured as the amount of energy or oxygen the organism uses in a given time).

**metal** An element characterized by a tendency to give up electrons and by good thermal and electrical conductivity.

**methane** Gas ($CH_4$) produced by the metabolism of anaerobic methanogen bacteria.

**micrometer** (*abbr.*, μm) $10^{-3}$ millimeters = $10^{-6}$ meters.

**micronutrient** Nutrient needed in relatively small amounts in the diet.

**microorganisms** Small unicellular or simple many-celled organisms: bacteria, fungi, protists, or small algae.

**milbroke** Paper pulp residues that are produced and recycled during papermaking.

**milligram** (*abbr.*, mg) One thousandth of a gram.

**mineral** Any naturally occurring homogeneous inorganic substance having a definite chemical composition and a particular crystalline structure, color, and hardness.

**mites** Small arthropods in the class (Arachnida) that includes spiders. Mites have eight legs, and the body is not divided into two parts as it is in spiders.

**mitigation** Reducing the adverse effect of something.

**molecular biology** Study of the molecular basis of inheritance.

**molecular weight** Sum of the atomic weights of the atoms in one molecule of a compound.

**Mollusca** Phylum of soft-bodied invertebrate animals with a muscular head-foot and a mantle, which usually secretes a shell; e.g., snails, clams, squids.

**Monera** Kingdom containing all prokaryotic organisms (bacteria).

**monogamy** Mating of one male with one female, usually for life or at least for the duration of one breeding season.

**monomers** Small molecules that may become joined together to form large (macro) molecules; e.g., amino acids are the monomers that make up polypeptides.

**mortality** Death.

**mulch** Anything placed on the surface of the soil between plants, e.g., straw, gravel, black plastic, grass clippings. Mulch holds water, reduces soil erosion, and (if it is organic) adds nutrients to the soil.

**multicellular** Composed of more than one cell.

**mutagen** Agent (e.g., chemicals, certain kinds of radiation) that causes mutation.

**mutation** Inheritable change in the genetic material (DNA).

**mycelium** (*pl.*, mycelia) Body of a fungus.

**mycology** Study of fungi.

**mycorrhiza** (*pl.*, mycorrhizae) Mutualistic association between a fungus and the roots of a higher plant; the fungus takes up mineral nutrients from the soil and passes them to the plant, receiving some organic (food) molecules made by the plant in return.

**nanometer** (*abbr.*, nm) $= 10^{-9}$ meters.

**natural gas** Gaseous fossil fuel containing 50% to 90% methane and lesser amounts of other organic gases such as propane and butane.

**natural resource** Anything that is produced naturally that is needed by a group of organisms.

**natural selection** Differential reproduction among the variety of genotypes in a population. Natural selection enhances reproductive success.

**negative feedback** Mechanism whereby the change detected in some condition stimulates compensating activity that brings the condition back toward its average value.

**nematode** Roundworm, member of the animal phylum Nematoda.

**niche** Way of life of a species; includes the habitat, food, nest sites, and so on that it needs to survive.

**nitrogen fixation** Conversion of gaseous nitrogen ($N_2$) to ammonia ($NH_3$). Carried out in ecosystems mainly by bacteria of the genus *Rhizobium.*

**nomadic** Moving from place to place, with no single fixed home.

**nondegradable** Not broken down by natural processes (or broken down only over a time span of hundreds of years).

**nonrenewable resources** Natural resources that can be used up completely or else used up to such a degree that it is economically impractical to obtain any more of them.

**nuclear power** Power produced from the energy generated when the nuclei of atoms split (fission) or fuse (fusion).

**nutrient** Any chemical that an organism must take in from its environment because it cannot produce it (or cannot produce it as fast as it needs it).

**oligotrophic** Of a body of fresh water that contains few nutrients and few organisms.

**omnivore** Animal that eats both plants and animals.

**organic** 1. Produced by living organisms. 2. In chemistry, of a compound containing carbon except for very small molecules such as carbon dioxide ($CO_2$) that contain only a few atoms.

**Osteichthyes** Class of vertebrates containing the bony fishes.

**ozone** $O_3$. Poisonous gas, a common pollutant in smog; also formed by the action of sunlight on oxygen in the ozone layer of the atmosphere.

**ozone layer** Layer in the upper stratosphere where solar radiation converts some oxygen atoms into ozone atoms. Ozone absorbs much ultraviolet radiation and prevents it from reaching Earth.

**parasite** Organism that feeds on another living organism without killing it.

**pathogenic** Disease-causing.

**per capita** Per person; Latin for "of each head."

**perennial** Of a plant, living for many years and surviving normal seasonal changes.

**permafrost** Permanently frozen layer in the soil, found in arctic and antarctic regions.

**pest** Any organism that is undesirable at the time and in the place where it ex-

ists. Includes plant-eating insects, molluscs, and nematodes on crop plants; parasites on animals; and weeds in fields of crops.

**pesticide** Substance used to kill undesirable organisms; includes insecticides, herbicides, and nematocides.

**petrochemical** Any chemical derived from petroleum or natural gas.

**petroleum** (=crude oil) Natural mixture of hydrocarbons found beneath Earth's crust.

**pH** Measure of how acidic or basic a solution is, on a scale of 0 to 14 (0 = very acidic, 14 = very basic, 7 = neutral).

**photochemical reaction** Chemical reaction powered by light energy.

**photosynthesis** Process whereby plants capture solar energy and store it as chemical bonds in carbohydrate molecules, using $CO_2$ to build the carbohydrate.

**physiology** Processes by which an organism carries out its various biological functions; how an organism works.

**phytoplankton** Plants and photosynthetic protists floating in the upper layers of a body of water.

**plankton** Collective noun for organisms that drift around in water because they are not capable of swimming against currents in the water.

**Platyhelminthes** Phylum of animals containing the flatworms; e.g., planarians, tapeworms, flukes.

**polar** Electrically asymmetrical; polar molecules dissolve in water.

**pollutant** Substance that causes pollution.

**pollution** Change in the physical, chemical, or biological properties of air, water, or soil that can adversely affect the health, survival, or activities of humans and other living organisms.

**polymer** Large molecule made up of many subunits that are smaller molecules similar or identical to one another.

**population** All members of a species living in a particular area and making up one breeding group.

**postconsumer materials** Materials removed from the waste stream (garbage) after being used by consumers. In the case of paper, postconsumer waste nearly always has to be de-inked before it can be reused.

**potential energy** Energy due to position.

**preconsumer materials** Things that are recycled before reaching the consumer;

e.g., paper trimmings from printing may be incorporated in recycled paper.

**predator**  Animal that feeds on other organisms (usually animals).

**primary productivity**  Rate at which food is made from inorganic substances by photosynthetic and chemosynthetic organisms.

**Primates**  Order of mammals that contains monkeys, apes, humans, and their relatives.

**primitive**  Showing features believed to have arisen early in evolution.

**producers**  Photosynthetic and chemosynthetic organisms.

**productivity**  Amount of organic matter produced by members of a given trophic level during a given period of time.

**progeny**  Offspring.

**Protista**  Kingdom (under the five-kingdom system of classification) of unicellular or colonial eukaryotes (e.g., Ameba, Paramecium, Euglena).

**protozoa**  Heterotrophic protists, unicellular eukaryotes.

**recycled paper**  Any paper pulp that is recycled, including both preconsumer and postconsumer waste.

**reduction**  Addition of electrons to; often takes the form of adding entire hydrogen atoms.

**renewable resource**  Natural resource whose supply can essentially never be exhausted, usually because it is continuously produced.

**replacement level fertility**  Of a population, the fertility level (about 2.1) at which each generation replaces itself. If replacement level fertility continues for several generations, a population becomes stable, neither growing nor declining if immigration and emigration are ignored.

**reproductive potential**  See biotic (reproductive) potential.

**reptile**  Vertebrate with dry, scaly skin and eggs laid on land; e.g., snakes, lizards, alligators, turtles.

**respiration**  Series of processes by which organisms break the chemical bonds in food molecules to release energy.

**rice**  Cereal (usually *Oryza sativa*) widely grown for human food.

**rodent**  Member of the order of mammals with gnawing teeth; e.g., mice, rats.

**ruminants**  Mammalian herbivores in which the stomach contains fluids of an alkaline (basic) pH and is divided into fermentation chambers housing microorganisms that digest the food.

**runoff**  The water that runs off the surface of land into a waterway; as opposed to water that percolates down into ground water.

**salinity**  Saltiness.

**salt**  Substance that yields neither hydrogen nor hydroxyl ions when it dissociates in water.

**saprobe**  Organism that absorbs nonliving organic matter for food. Fungi are saprobes.

**scavenger**  Animal that eats dead organisms or organic matter.

**seaweed**  Multicellular alga found in marine habitats.

**seed**  Dispersal unit of gymnosperms and angiosperms, consisting of a seed coat, an embryonic plant, and a food supply.

**seed coat**  Outer covering of a seed, developed from the outer layers of the ovule.

**semiconductor**  A substance having an electrical resistance somewhere between that of electrical insulators, which do not conduct electricity, and conductors, which readily conduct electricity.

**sewage**  Liquid waste.

**siliceous**  Containing silica ($SiO_2$).

**species**  Group of organisms whose members share a common gene pool, usually because they can breed with one another.

**spore**  Single reproductive cell that reaches a new habitat by traveling through the air. Common method of reproduction in fungi and bacteria.

**Squamata**  Order of reptiles including lizards and snakes.

**standing crop**  Mass of organisms actually present at any one time.

**sterilization**  1. Any more or less permanent change that prevents an animal from reproducing sexually. 2. Cleaning of an object by destroying all the organisms on it.

**stormwater**  Water that runs off the land as the result of high levels of precipitation in a short period of time. In urban areas, usually carried away by storm drains.

**stratosphere**  Layer of the atmosphere that extends from the troposphere outward (with increasing temperature) to the point (about 70 kilometers from Earth's surface) at which the temperature starts to fall again.

**subsidy**  Monetary assistance granted by a government to a private commercial enterprise. A subsidy may be money actually paid to the private enterprise, or it may consist of forgiving the payment of sums that must be paid by other members of the society, as in the case of tax credits. Subsidies are sometimes spoken of as "federal payment" or "government money," but in fact governments do not have any money except what they borrow or collect in taxes, so subsidies are actually paid by all members of society except the enterprise to which they are paid. *See also* externality.

**subtropical** (or semitropical)  Lying near to, but not within, the tropics.

**succession**  In ecology, process by which the inhabitants of an area that has been disturbed change with time in a regular sequence; succession finishes when the organisms of the climax community of the area have become established.

**sustainable**  Capable of continuing indefinitely in approximately its present form.

**synfuels**  Synthetic liquid and gaseous fuels, usually made from coal.

**taxon** (*pl., taxa*)  Any one of the hierarchical categories into which organisms are classified; e.g., species, order, class.

**taxonomy**  Study of the classification and identification of living organisms.

**technology**  Use of tools and machines.

**temperate region**  Region of Earth that is neither tropical nor arctic.

**terrestrial**  Of land (as opposed to water or air).

**territory**  Area defended by one or more animals against intruders.

**thermochemical reaction**  Chemical reaction whose rate varies with the temperature.

**Tragedy of the Commons**  Conflict in the use of a resource ("common") between the short-term welfare of an individual and the long-term welfare of society.

**transpiration**  Loss of water by evaporation through pores (stomata) in the shoot system of a plant.

**trophic level**  Level in the food chain at which an organism functions; e.g., herbivores, members of the second trophic level, eat autotrophs, members of the first trophic level.

**tropics** That part of Earth lying between the Tropic of Cancer (at latitude 23 degrees, 27 minutes north of the equator and the Tropic of Capricorn (at the same latitude south of the equator). These latitudes mark the limits of the sun's apparent movement north and south during the year.

**troposphere** Layer of the atmosphere that contains about 95% of Earth's air and extends about 12 kilometers up from Earth. The troposphere ends at the tropopause, the point at which atmospheric temperature starts to increase instead of decrease as one moves farther from Earth.

**tumor** Abnormal (cancerous) growth of tissue; sometimes malignant but may be benign.

**ultraviolet radiation** (UV) That part of the electromagnetic spectrum with wavelengths shorter than those of visible light (below 400 nanometers) but higher than those of x-rays (about 100 nanometers), with sufficient energy to break hydrogen bonds and disrupt the structure of many biological molecules.

**ungulates** Hoofed mammals.

**unicellular** With a body consisting of only one cell.

**variety** Subdivision of the species in the hierarchy of taxonomic classification; the variety (sometimes called subspecies) name is written after the species name.

**vascular** Use of tissues that transport fluids around the body; e.g., veins, arteries, xylem, phloem.

**vegetative reproduction** Reproduction by growth of an individual's body or fragments of its body; reproduction without production of gametes or spores.

**venereal disease** (VD) Traditional name for a disease, such as syphilis or AIDS, transmitted by sexual activity. These diseases are now generally known as sexually transmitted diseases (STDs).

**vertebrate** Animal with a backbone; e.g., fish, human.

**virus** Particle composed of nucleic acid and protein that can reproduce only in a living cell.

**vitamin** Organic micronutrient.

**waste paper** A term used by the EPA to refer to a mixture of preconsumer and postconsumer waste that is generated once the papermaking process is complete.

**watershed** The land area from which runoff drains into a particular river or stream.

**wavelength** Light and sound may be considered as traveling in wavy lines. The distance between adjacent peaks of the line is the wavelength, symbolized by $\lambda$.

**wheat** Type of cereal, the most widely grown type of human food plant, usually *Triticum aestivum* or *T. durum*.

**windward** (side) The side of an object from which the wind is coming.

**zooplankton** Animals and heterotrophic protists floating in the surface layers of a body of water.

# Index

Part of North America Enlarged

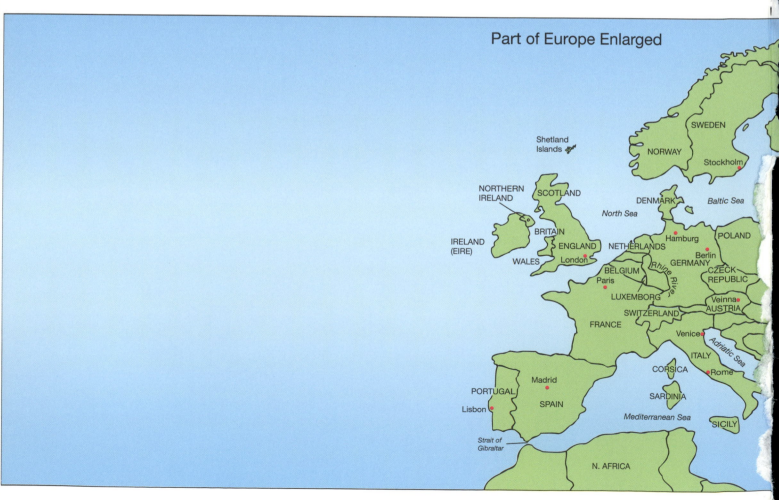

Part of Europe Enlarged

**Gulf of Mexico**

Yucatan
Peninsula

Sian Ka'an
Biosphere Reserve

**MEXICO**

**BELIZE**

**GUATEMALA**

**HONDURAS**

**EL SALVADOR**

**NICARAGUA**

Managua

Guanacaste
National Park

**COSTA RICA**

San Jose

Panama
Canal

**PANAMA**

**CUBA**

HAITI

**DOMINICAN
REPUBLIC**

San Juan

**JAMAICA**

Santo
Domingo

**PUERTO
RICO**

**U.S. VIRGIN ISLANDS**

**BR. VIRGIN
ISLANDS**

**MARTINIQUE**

**BARBADOS**

*Caribbean Sea*

**VENEZUELA**

**COLUMBIA**

**Central America Enlarged**

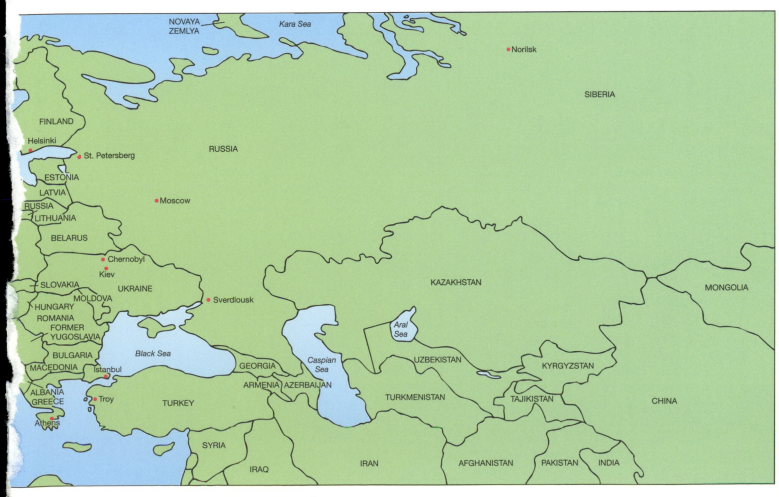

**NOVAYA
ZEMLYA**

*Kara Sea*

Norilsk

**SIBERIA**

**FINLAND**

Helsinki

St. Petersberg

**RUSSIA**

**ESTONIA**

**LATVIA**

**RUSSIA**

**LITHUANIA**

Moscow

**BELARUS**

Chernobyl

Kiev

**SLOVAKIA**

**MOLDOVA**

**UKRAINE**

Sverdlousk

**KAZAKHSTAN**

**MONGOLIA**

**HUNGARY**

**ROMANIA**

**FORMER
YUGOSLAVIA**

*Aral
Sea*

**BULGARIA**

*Black Sea*

**MACEDONIA**

Istanbul

**GEORGIA**

*Caspian
Sea*

**UZBEKISTAN**

**KYRGYZSTAN**

**ALBANIA**

**GREECE**

Troy

**ARMENIA** **AZERBAIJAN**

**TURKMENISTAN**

**TAJIKISTAN**

**CHINA**

Athens

**TURKEY**

**SYRIA**

**IRAQ**

**IRAN**

**AFGHANISTAN**

**PAKISTAN**

**INDIA**